Volcano Deformation
Geodetic Monitoring Techniques

Eruption column from Guagua Pichincha Volcano as seen from Quito, the capital city of Ecuador, on 7 October 1999. The column, which rose to a height of ∼ 12 km (∼ 7.5 miles), exhibits many of the features described by Pliny the Younger in his eyewitness account of the 79 CE (Common Era; substitute for AD, Anno Domini) eruption of Vesuvius (Preface), including a narrow 'trunk', spreading 'branches', white areas caused by condensation of water vapor as a result of rapid expansion of humid air, and dark areas laden with tephra. Such columns are characteristic of *plinian* eruptions, so called in recognition of Pliny the Younger's early contribution to descriptive volcanology. Photograph by Daniel Andrade Varela, Departamento de Geofísica de la Escuela Politécnica Nacional, Quito, Ecuador.

Daniel Dzurisin

Volcano Deformation

Geodetic Monitoring Techniques

 Springer

Published in association with
Praxis Publishing
Chichester, UK

Dr Daniel Dzurisin
United States Geological Survey
David A. Johnston Cascades Volcano Observatory
Vancouver
Washington
USA

SPRINGER–PRAXIS BOOKS IN GEOPHYSICAL SCIENCES

SUBJECT *ADVISORY EDITOR*: Dr. Philippe Blondel, C.Geol., F.G.S., Ph.D., M.Sc., Senior Scientist, Department of Physics, University of Bath, Bath, UK

ISBN 3-540-42642-6 Springer-Verlag Berlin Heidelberg New York

Springer is part of Springer-Science + Business Media (springeronline.com)

Bibliographic information published by Die Deutsche Bibliothek

Die Deutsche Bibliothek lists this publication in the Deutsche Nationalbibliografie; detailed bibliographic data are available from the Internet at http://dnb.ddb.de

Library of Congress Control Number: 200593703

Cover design: Jim Wilkie
Project management: Originator Publishing Services, Gt Yarmouth, Norfolk, UK

Printed on acid-free paper

USGS scientist Jack Kleinman, typically attired in shorts and a smile, sets up EDM reflectors on Novarupta lava dome in the Valley of Ten Thousand Smokes, Alaska, with Baked Mountain in the distance. The rhyolite dome was emplaced during the waning phase of the largest volcanic eruption of the 20th century, which produced about 20 km³ of air-fall tephra and 11–15 km³ of ash-flow tuff within about 60 hours in June 1912. Jack helped to establish a geodetic network near Novarupta in 1989 and served as crew chief for follow-up surveys in 1990 and 1993. His energy and zest for life inspired all those who knew him. Jack died in a kayaking accident on the White Salmon River, Washington, in 1994. Photograph by John Eichelberger, University of Alaska Fairbanks.

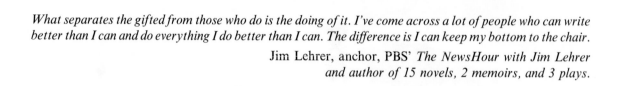

What separates the gifted from those who do is the doing of it. I've come across a lot of people who can write better than I can and do everything I do better than I can. The difference is I can keep my bottom to the chair.

Jim Lehrer, anchor, PBS' *The NewsHour with Jim Lehrer* and author of 15 novels, 2 memoirs, and 3 plays.

Contents

Contributors

Daniel Dzurisin, Michael Lisowski, Evelyn A. Roeloffs, and Steve P. Schilling
US Geological Survey
David A. Johnston Cascades Volcano Observatory
1300 S.E. Cardinal Court
Vancouver, WA 98683-9589, USA

Robert Fournier
US Geological Survey (retired)
108 Paloma Road
Portola Valley, CA 94028, USA

Alan T. Linde
Carnegie Institution of Washington
Department of Terrestrial Magnetism
5241 Broad Branch Road, NW
Washington, D.C., 20015, USA

Zhong Lu
US Geological Survey
EROS Data Center
47914 252nd Street
Sioux Falls, South Dakota 57198, USA

Robert I. Tilling
US Geological Survey (retired)
27749 Altamont Circle
Los Altos Hills, CA 94022, USA

Ren A. Thompson
US Geological Survey
Box 25046
Denver Federal Center, MS 913
Denver, CO 80225-0046, USA

Foreword

Volcanoes and eruptions are dramatic surface manifestations of dynamic processes within the Earth, mostly but not exclusively localized along the boundaries of Earth's relentlessly shifting tectonic plates. Anyone who has witnessed volcanic activity has to be impressed by the variety and complexity of visible eruptive phenomena. Equally complex, however, if not even more so, are the geophysical, geochemical, and hydrothermal processes that occur underground – commonly undetectable by the human senses – before, during, and after eruptions. Experience at volcanoes worldwide has shown that, at volcanoes with adequate instrumental monitoring, nearly all eruptions are preceded and accompanied by measurable changes in the physical and (or) chemical state of the volcanic system. While geochemical methodologies of volcano monitoring have shown increasing sophistication and promise in recent decades, seismic and geodetic (ground-deformation) techniques remain the most widely used tools in volcanic surveillance. These two geophysical workhorses have proven to be robust and, generally, the most diagnostic and reliable techniques for volcanologists.

In the early 20th century, systematic measurements of ground deformation were initiated at a few active volcanoes in Japan and the USA. Since then, as this book illustrates, *volcano geodesy* – a specialized field of the still-young science of volcanology – has come a long way, not only in terms of greatly increased diversity and precision of measurements, but also in the number of volcanic systems being monitored worldwide. Having had some hands-on experience myself with 'classical' techniques of geodetic monitoring (e.g., tilt measurements, leveling, and laser electronic distance measurements), while a staff member at the USGS Hawaiian Volcano Observatory in the mid-1970s, I am overwhelmed by the tremendous advances in instrumentation, new techniques, data-acquisition telemetry and processing, and volcano-deformation source models over the past three decades. There has been a virtual explosion of volcano-geodesy studies and in the modeling and interpretation of ground-deformation data. Nonetheless, other than selective, brief summaries in journal articles and general works on volcano-monitoring and hazards mitigation (e.g., UNESCO, 1972; Agnew, 1986; Scarpa and Tilling, 1996), a modern, comprehensive treatment of volcano geodesy and its applications was non-existent, until now.

In the mid-1990s, when Daniel Dzurisin (DZ to friends and colleagues) was serving as the Scientist-in-Charge of the USGS Cascades Volcano Observatory (CVO), I first learned of his dream to write a book on volcano geodesy. DZ asked me whether he should undertake this book project in addition to his CVO managerial duties and research commitments. As his immediate supervisor at the time, my answer to him was a no-brainer: Absolutely! With the advent of 'space geodesy' (e.g., GPS and InSAR), such a book would be timely and fill a long-existing need. Most importantly, however, DZ, with his expertise and experience in both classical and emerging geodetic techniques, was clearly the right guy for the job. With the passing years, his dream gradually assumed tangible form, with decisions on content and format, identification of possible collaborators, obtaining a suitable and interested publisher, and the actual writing and technical reviews of each individual chapter. DZ was able to assemble a stellar team of knowledgeable experts to author several of the chapters (6, 8, 9, and 10) covering related topics beyond his own areas of specialization. With the reawakening of Mount St. Helens Volcano in October 2004 and its ongoing eruption, the book required some last-minute additions and updating. The rest, as they say, is history. However, completion of the book took somewhat longer than DZ and I had anticipated; Praxis and its editors and production staff have been patient and cooperative throughout a sporadic and lengthy process.

In the *Preface*, DZ states that the book's intended audience is primarily undergraduate and graduate students interested in volcanology. Doubtless this book will admirably fulfill the academic needs of professors and students in the geosciences, but *Volcano Deformation* has much more to offer. Not only is it a content-rich, clearly written and profusely illustrated textbook, but it also provides useful contextual information for the general volcanologist, like myself, who is not involved full time with geodetic studies. Moreover, because volcano geodesy is a multidisciplinary pursuit, this book should be valuable to volcano-deformation specialists wanting to know more about technique(s) apart from those they themselves employ. This volume is not a how-to compendium for making and interpreting ground-deformation measurements, but it contains pertinent references to classical surveying manuals and other works containing the theoretical basis, instrumentation and specifications, equations, and data-processing procedures related to measurement of the Earth. The book is organized and written in an easily approachable manner, even for the non-specialist reader, in that any discussion heavily laden with technical detail or mathematics can be skimmed, or even skipped, with minimal loss of the chapter's principal theme and message.

For me, the major strength of the book is its focus on how volcano geodesy complements other volcano-monitoring approaches and tools. As emphasized throughout, optimum monitoring of a restless or an erupting volcano is achieved by a combination of techniques, rather than by uncritical reliance on any single one. An even greater strength of *Volcano Deformation* is its extensive treatment (Chapter 7) of the key lessons volcanologists have learned at well-monitored deforming volcanoes. These lessons amply demonstrate that geodetic

and other volcano-monitoring data not only contribute to advancing our scientific understanding of how volcanoes work, but that they also can benefit society in reducing the risks from hazardous eruptions. The results of volcano geodesy provide basic research data on our dynamic planet and, at the same time, have practical, at times life-saving, applications. Yet, as the book cautions, despite the impressive strides made in volcano geodesy and other monitoring techniques in recent decades, volcanologists still lack the capability to reliably forecast the outcome of sustained or escalating volcanic unrest. Lamentably, at present we cannot provide definitive answers to vital questions invariably asked of us by emergency management officials and populations at potential risk: Will the volcanic unrest culminate in eruption? If so, how large will the eruption be? Will the eruption be explosive or non-explosive? How long will it last? Chapter 11 addresses how we might better answer these and related questions in the 21st century.

Finally, the readers of this book cannot help but notice a sense of adventure and excitement – above and beyond scientific curiosity – that pervades the pages of the book, especially the chapters solely authored or co-authored by DZ. For me at least, this excitement is palpable. Clearly, the author has been fascinated by how and why volcanoes deform throughout his career, and in this book DZ and his co-authors convey that fascination and wonderment to the readers. *Volcano Deformation* is hardly a dry, information-packed scientific treatise; instead, it provides vibrant testimony that doing good, important science can also be a lot of fun. It is sure to inspire some readers to become volcanologists, if not volcano geodesists, and to pick up where this book leaves off.

Robert I. Tilling
Menlo Park, California
October 2005

Preface

He [Pliny the Elder] was at Misenum in his capacity as commander of the fleet on the 24th of August, when between 2 and 3 in the afternoon my mother drew his attention to a cloud of unusual size and appearance. He had had a sunbath, then a cold bath, and was reclining after dinner with his books. He called for his shoes and climbed up to where he could get the best view of the phenomenon. The cloud was rising from a mountain – at such a distance we couldn't tell which, but afterwards learned that it was Vesuvius. I can best describe its shape by likening it to a pine tree. It rose into the sky on a very long 'trunk' from which spread some 'branches'. I imagine it had been raised by a sudden blast, which then weakened, leaving the cloud unsupported so that its own weight caused it to spread sideways. Some of the cloud was white, in other parts there were dark patches of dirt and ash. The sight of it made the scientist in my uncle determined to see it from closer at hand.

From Pliny the Younger's eyewitness account of the 79 CE (Common Era; substitute for AD, Anno Domini) eruption of Vesuvius, translated from his Letter 6.16 to the historian Tacitus by Prof. Cynthia Damon (2000), Amherst College, Massachusetts.

Almost two millennia after Pliny the Younger's insightful description of the Vesuvius eruption that resulted in the death of his uncle and entombed the residents of Pompeii and Herculaneum, modern volcanology is still in its infancy. Like a child with boundless enthusiasm and seemingly endless opportunities, volcano science has yet to develop fully an identity of its own. Instead, it freely draws upon such diverse disciplines as geology, seismology, hydrology, geochemistry, and geophysics. Fewer than half of the scientists who study volcanoes would call themselves *volcanologists*, and therein lies a great strength. Volcanology is inherently a multidisciplinary pursuit with ample room for specialists and generalists alike – physicists, chemists, biologists, hydrologists, limnologists, dendrochronologists, ecologists, remote-sensing specialists, and many others. Scientists who study volcanoes often find themselves looking beyond their own expertise for answers, surrounded by others doing the same. For novices and experts alike, volcanoes are humbling but very exciting subjects to explore.

In spite of the best efforts of countless volcano watchers dating back at least to Pliny the Younger, no unifying theory yet exists to explain the diversity and complexity of magmatic and volcanic systems. This may be due in part to the tremendous breadth of volcanic phenomena (e.g., bubbling mudpots and lava lakes; steaming hot springs and jetting lava fountains; lavas that flow freely, others seemingly too stiff to move, and others still that explode into towering ash columns and swift pyroclastic flows; geysers that entertain onlookers by erupting on schedule and rare, caldera-forming cataclysms that threaten entire civilizations). To make matters worse, all of this diversity is served up with a heavy dose of complexity. Earth's deep interior is not only hidden from direct view, but it is also heterogeneous in virtually every way imaginable. Before it explodes or oozes onto the surface, magma forms deep within the Earth, rises through extreme conditions of pressure and temperature, interacts physically and chemically with diverse crustal rocks, cools, partially crystallizes and degases, sometimes mixes with other magmas of similar complexity, responds to regional tectonic influences, and interacts to varying degrees with groundwater. Each of these processes leaves an imprint on the eruptive or intrusive end products, so the complete sequence of events can be exceedingly difficult to decipher.

All of this says nothing about the additional complexity that arises when magma finally reaches the surface, where it interacts with the atmosphere, hydrosphere, and biosphere. During the past two centuries, more people have died as a result of tsunamis, lahars (volcanic mudflows), and post-eruption starvation and disease associated with eruptions than as a direct result of eruptive processes. On the island of Hawai'i, many residents' complaints of unusual respiratory problems have been attributed to 'vog' (volcanic fog) or 'laze' (lava haze), a product of previously undocumented interactions between seawater and basaltic lava flows as they pour into the Pacific Ocean from a long-lived eruption of Kīlauea Volcano (Monastersky, 1995).[1] Atmospheric scientists still debate the relative contributions of anthropogenic and volcanic sources of several gases that play an important role in sustaining life on Earth. Because they reflect our planet's inherent complexity from mantle to stratosphere, volcanoes present an immense challenge to those who strive to understand them.

Responding to crises that threaten the lives and livelihoods of thousands of people, volcano scientists at the beginning of the 21st century still rely to a great extent on collective experience and intuition to anticipate the outcome of volcanic unrest. In the words of two scientists who played key roles in the mitigation effort surrounding the historic 1991 eruption of Mount Pinatubo, Philippines, '*Successful mitigation of volcanic risks requires correct forecasts, effective warnings, and a willingness of public officials and citizens to take necessary precautions. Volcanologists should be prepared for obstacles and delays at each step and should be aware that, even when progress is being made toward such forecasts, warnings, and precautionary actions, the margin of safety can be alarmingly narrow.*' (Newhall and Punongbayan, 1996, p. 807). Happily, this situation is improving rapidly. The past 25 years have witnessed several important advances in our understanding of volcanic processes, products, and

---

[1] For the most part, spellings of Hawaiian words throughout this book are in accordance with guidelines established by Heliker *et al.* (2003) in their Preface to USGS Professional Paper 1676, *The Pu'u 'Ō'ō – Kupaianaha eruption of Kīlauea Volcano, Hawai'i: The first 20 years.* The modern spellings were adopted '... both to satisfy the new standards of the [US] Board on Geographic Names and to honor the Hawaiian language after more than a century of neglect' (Heliker *et al.*, 2003, p. iii). Exceptions here include formal names (e.g., Hawaiian Volcano Observatory), publication citations, and reproduced figures, where the original spellings have been retained.

attendant hazards. Progress has been spurred both by technological improvements and by inexorable encroachment of cities onto the flanks of dangerous volcanoes. Tragically, a major impetus for this change was the highest death toll from a single eruption since the beginning of the 20th century. In 1985, more than 23,000 people died as a result of lahars triggered by an eruption of Nevado del Ruiz Volcano in Colombia – the deadliest eruption since 1902, when pyroclastic flows from Mont Pelée claimed nearly 30,000 lives in the town of St. Pierre on the island of Martinique, West Indies (Fisher *et al.*, 1980; Fisher and Heiken, 1982).

In response, volcanologists focused their efforts on eruption prediction and volcano-hazards mitigation, with encouraging results. Volcano seismology has evolved from a descriptive science based on cataloging events toward a deterministic theory of the mechanisms of several distinctive types of earthquakes that commonly occur beneath volcanoes. Volcano geochemistry has taken its place with seismology and geodesy as a mainstay of modern volcanology, in part by providing some of the earliest and most definitive indicators that fresh magma was involved in several recent episodes of volcanic unrest that culminated in eruptions. Remote-sensing techniques have been applied to volcanoes with increasing frequency and success, promising a future of global volcano surveillance by spaceborne sensors.

The global telecommunications revolution, especially the Internet and World Wide Web, has profoundly changed the way in which scientists monitor volcanoes and share information. Although the presence of trained scientists at volcano crises will continue to be essential for effective hazards mitigation, in most places it is now possible to establish a virtual volcano observatory to share information globally in near-real time. Scientists monitoring a remote volcano can draw upon the expertise of colleagues around the world, even as hazardous events are unfolding. At the same time, advances in volcano geodesy have brought powerful new tools to bear on problems of volcano monitoring and hazards assessment. At the start of a new millennium, volcanology seems to be passing from infancy to adolescence, anxious to establish a unique identity among its more mature siblings in the Earth sciences.

This book focuses on one aspect of volcanology's recent advances that I have been fortunate to experience firsthand – a revolution in volcano geodesy. During my career, not one but two breakthrough techniques for measuring surface deformation from space have burst on the scene: the Global Positioning System (GPS) and interferometric synthetic-aperture radar (InSAR). Earth scientists have coined the phrase 'space geodesy' and put the concept to good use studying geodynamic processes, including volcanism. Other technological advances have made it possible to measure crustal strain *in situ* with amazing precision over a very wide range in frequency, blurring the distinction between seismology and geodesy. Although traditional techniques are still the best choice for many tasks, a growing collection of new tools has greatly extended the reach and capability of classical geodesy.

In one sense, volcano geodesy is a very specialized field with relatively few practitioners. On the other hand, the subject is sufficiently intuitive to have widespread appeal among many non-specialists who are fascinated by volcanoes and curious about how they work. This book is by no means a comprehensive treatment of modern geodetic techniques applied to volcanoes, nor is it a historical account of the development of volcano geodesy. For a more systematic treatment of geodetic principles and practices, the reader might want to consult a reference book on geodesy or surveying such as Bomford (1980), Davis *et al.* (1981), or Krumm *et al.* (2002). For an overview of geodetic techniques applied to volcanoes, the article on ground deformation methods and results by Van der Laat (1996) is a good choice. Readers interested in a more general treatment of volcanology or volcano hazards might want to consult excellent texts on those subjects by Williams and McBirney (1979), Latter (1989), Francis (1993), Scarpa and Tilling (1996), Fisher *et al.* (1997), or Decker and Decker (1998).

This is neither a how-to book nor a reference manual. Instead, it describes some widely used techniques for measuring ground movements at volcanoes and attempts to place volcano geodesy in the broader context of volcano monitoring and hazards assessment. It tries to make the point that useful geodetic information is where you find it, and sometimes an old-fashioned tape measure or leveling rod is a better tool than a state-of-the-art GPS receiver or remote-sensing satellite. Most of all, I have tried to convey a sense of the excitement that comes from knowing that the ground beneath your feet is moving, for reasons you can only guess – slowly at first, but inexorably, in response to forces that defy comprehension. This is the essence of volcano geodesy – the thrill of exploring unstable ground – that I have tried to capture on these pages.

Many of the examples cited in the book are from my personal experience, because these are the ones I know best. At the same time, I have tried to acknowledge others' work wherever possible, as a guide to readers in search of other perspectives or more detailed information. Toward that end, I have included an extensive list of References in the hope that some readers will use the book as an entrée to the volcanological research literature. Also included is a Glossary of technical terms that should be helpful for non-specialists seeking to get beyond jargon to an understanding of volcanic processes and phenomena. The book's intended audience is primarily undergraduate and graduate students interested in volcanology, but hopefully it will also be of interest to anyone else who wonders, as I do, how and why volcanoes deform.

Acknowledgements

I am indebted to my co-authors Bob Fournier, Alan Linde, Mike Lisowski, Zhong Lu, Evelyn Roeloffs, Steve Schilling, and Ren Thompson, who contributed important chapters to the book. Their work helps to bridge the shrinking gaps between geodesy, hydrology, seismology, geochemistry, and remote sensing, and to demystify the potentially intimidating subject of numerical modeling. Christine G. Janda created or improved most of the book's illustrations – a monumental task that required equal measures of creativity and persistence. David E. Wieprecht tracked down several photographs for the book and worked digital magic on others, turning marginal images into good ones. Every author should have such talented partners; I am deeply grateful for my good fortune and your hard work. Thanks also to reviewers Elliot T. Endo, John W. Ewert, John O. Langbein, Kristine M. Larson, Michael Lisowski, Jacob B. Lowenstern, Asta Miklius, Christopher G. Newhall, Gary C. Perasso, Seth C. Moran, James C. Savage, and Wayne R. Thatcher for catching mistakes, sharpening fuzzy prose, and recommending numerous improvements. Michael P. Poland and Steve P. Schilling read a draft of the manuscript, and suggested many changes that enhanced the book's readability and interest. My thesis adviser, friend, and mentor Robert P. Sharp taught me the importance of careful observation and clear prose in the Earth sciences. Bob's many grateful students are his enduring legacy. His honesty, integrity, and good humor will be missed but never forgotten. Robert I. Tilling was a steadfast and caring supporter throughout this lengthy project. He critiqued the entire manuscript, wrote the Foreword, and helped me to believe that I would finish the book someday. Clive Horwood of Praxis and Philippe Blondel of the University of Bath (UK) waited patiently until I did, for which I am genuinely grateful. During my 25-year tenure at the USGS David A. Johnston Cascades Volcano Observatory, dozens of students and other volunteers have contributed their exuberance, fellowship, and good humor to volcano-deformation surveys throughout the western USA, thereby turning hard work into good fun. This book was written, first and foremost, for you and others like you. Carry on. Family members Linda, Jason, and Jill gave me many reasons to smile throughout the lengthy writing process, put up with my curmudgeonly demeanor, and repeatedly forgave my churlish behavior. To each of you, heartfelt thanks for helping to make this book possible.

DVD

A DVD supplied with this book includes all of the figures in jpeg and pdf format. Photographs without annotation are supplied in tiff format. The reader is free to use the figures, with proper attribution, for presentations or reproduction. Also included on the DVD is a *Mathematica* notebook with an expanded version of Chapter 8, Analytical volcano deformation source models, by Michael Lisowski. *Mathematica*, by Wolfram Research Inc., is one of several mathematics software packages capable of symbolic mathematics. A free reader to access the notebook is included on the DVD. Those with a licensed version of *Mathematica* will have greater functionality, including the ability to specify parameter values, calculate surface displacement and its derivatives for several analytical models, and modify plots that appear in Chapter 8. Readers should direct questions or requests concerning the figures, including the availability of other file formats, to Daniel Dzurisin. Questions concerning Chapter 8 or the *Mathematica* notebook should be directed to Michael Lisowski.

Figures

Tables

Symbols

Common symbols used in the text are listed alphabetically, first in the Latin, then in the Greek alphabets. In a few cases, the same symbol is used differently in different chapters. All symbols are defined on first use in each chapter, and the meaning should be clear from context. Units are shown in square brackets, mostly in the International System of Units (SI). See also Abbreviations and Acronyms.

a = semi-major axis (ellipsoid) [m], radius [m], or acceleration [$m\,s^{-2}$]

A = area [m^2]

b = semi-minor axis (ellipsoid) [m] or perpendicular component of the baseline between two image-acquisition points for SAR images (Chapter 5)

B = Skempton's coefficient [dimensionless] or radar-pulse frequency bandwidth (Chapter 5) [s^{-1}]

c = speed of light in a vacuum [299,792,458 $m\,s^{-1}$]

d = distance or depth [m]

f = ellipsoid flattening (Chapter 2) [dimensionless], frequency [s^{-1}], or focal length (Chapter 6) [m]

g = local gravitational acceleration at Earth's surface [$m\,s^{-2}$]

G, μ = shear modulus (one of two Lamé constants; also called rigidity, modulus of rigidity, or torsional modulus) [Pascals, $kg\,m^{-1}\,s^{-2}$]. In Chapter 2, G = universal gravitation constant [$6.6742\pm0.0010\times10^{-11}\,m^3\,kg^{-1}\,s^{-2}$] and μ = shear modulus. Elsewhere, G = shear modulus.

h = height [m], ellipsoidal height (Chapter 2) [m], or thickness [m]

h_a = altitude of ambiguity (Chapter 5) [m]

H = orthometric height (Chapter 2) [m] or altitude (Chapter 5) [m]

K = bulk modulus (Chapters 8 and 9) [Pascals, $kg\,m^{-1}\,s^{-2}$]

K_{magma} = effective bulk modulus of magma (Chapter 8) [Pascals, $kg\,m^{-1}\,s^{-2}$]

K_u = undrained bulk modulus (Chapter 9) [Pascals, $kg\,m^{-1}\,s^{-2}$]

K_s = bulk modulus of solid grains in porous medium (Chapter 9) [Pascals, $kg\,m^{-1}\,s^{-2}$]

L = length [m]

m = mass [kg]

M = earthquake magnitude, assumed to be local magnitude, M_L, unless specified otherwise [dimensionless]

M_E = mass of the Earth [kg]

M_L = local earthquake magnitude [dimensionless]

M_b = earthquake body magnitude [dimensionless]

M_s = earthquake surface-wave magnitude [dimensionless]

M_w = earthquake moment magnitude [dimensionless]

M_g = Mach number [dimensionless] (Chapter 3)

M_0 = seismic moment [joules, $kg\,m^2\,s^{-2}$]

$M_0^{(g)}$ = geodetic moment [joules, $kg\,m^2\,s^{-2}$]

N = Newton, Standard International (SI) unit of force: $1\,N = 1\,kg\,m\,s^{-2}$ (Chapter 2)

N = ellipsoid–geoid separation (Chapter 2) [m] or initial ambiguity at first observation, cycle ambiguity, phase ambiguity, or integer ambiguity (Chapter 4) [dimensionless]

p, P = pressure [Pascals, $\mathrm{kg\,m^{-1}\,s^{-2}}$]

P_f = pore-fluid pressure (Chapter 10) [Pascals, $\mathrm{kg\,m^{-1}\,s^{-2}}$]

R, r = pseudorange (Chapter 4), slant range (Chapter 5), or radial distance [m]

R = gas constant (Chapter 9) [8.314472 $\mathrm{J\,mol^{-1}\,^{\circ}K^{-1}}$ (joules per mole per degree Kelvin)]

R_E = mean radius of the Earth (Chapter 2) [6.371×10^6 m]

S = scale (Chapter 6) [dimensionless]

S_V = lithostatic load or vertical stress (Chapter 10) [Pascals, $\mathrm{kg\,m^{-1}\,s^{-2}}$]

t = time [s]

T = period [s], torque (Chapter 3) [$\mathrm{kg\,m^2\,s^{-2}}$], or absolute temperature (Chapter 10) [degrees Kelvin]

u, v, w = displacements along the x-, y-, and z-axes, respectively (Chapter 8) [m]

v = velocity [$\mathrm{m\,s^{-1}}$]

V = volume [$\mathrm{m^3}$]

$\Delta V_{surface}$ = integral of surficial vertical displacement (Chapter 8) [$\mathrm{m^3}$]

$\Delta V_{chamber}$ = source-chamber volume change (cavity volume change) (Chapter 8) [$\mathrm{m^3}$]

ΔV_{magma} = volume of intruded magma (Chapter 8) [$\mathrm{m^3}$]

$\Delta V_{compression}$ = net volume change of stored magma due to pressure change in chamber ($\Delta V_{compression} = \Delta V_{magma} - \Delta V_{chamber}$) (Chapter 8) [$\mathrm{m^3}$]

V_p = seismic compressional velocity [$\mathrm{m\,s^{-1}}$]

V_s = seismic shear-wave velocity [$\mathrm{m\,s^{-1}}$]

W = weight [$\mathrm{kg\,m\,s^{-2}}$]

W_a, W_r = width of the radar antenna footprint in the azimuth and range directions, respectively (Chapter 5) [m]

α = radius (Chapter 8) [m]

β_a, β_r = angular beam width in azimuth and range directions, respectively (Chapter 5) [degrees or radians]

γ = gravitation constant (Chapter 8) [$6.6742 \pm 0.0010 \times 10^{-11} \mathrm{N\,m^2\,kg^{-2}}$ or $\mathrm{m^3\,kg^{-1}\,s^{-2}}$]

Δ = change or difference (e.g., Δg = change in local gravitational acceleration, Δh = height difference between bench marks

ε_{ij} = strain component [dimensionless]

ε_{rr} = radial strain (Chapter 8) [dimensionless]

$\varepsilon_{\theta\theta}$ = tangential strain (Chapter 8) [dimensionless]

$\varepsilon_{\Delta V}$ = volumetric strain (Chapter 8) [dimensionless]

$\dot{\varepsilon}$ = strain rate (Chapter 10) [$\mathrm{s^{-1}}$]

Θ = bearing (degrees) (Chapter 2)

λ = wavelength [m], one of two Lamé constants [Pascals, $\mathrm{kg\,m^{-1}\,s^{-2}}$], or coefficient of friction (Chapter 10) [dimensionless]

μ, G = shear modulus (one of two Lamé constants; also called rigidity, modulus of rigidity, or torsional modulus) [Pascals, $\mathrm{kg\,m^{-1}\,s^{-2}}$]. In Chapter 2, G = universal gravitation constant [$6.6742 \pm 0.0010 \times 10^{-11} \mathrm{N\,m^2\,kg^{-2}}$ or $\mathrm{m^3\,kg^{-1}\,s^{-2}}$] and μ = shear modulus. Elsewhere, G = shear modulus.

ν = Poisson's ratio [dimensionless]

π = pi [dimensionless]

ρ = density [$\mathrm{kg\,m^{-3}}$]

ρ_c = density of Earth's crust (Chapter 8) [$\mathrm{kg\,m^{-3}}$]

σ = standard deviation

π = tilt (Chapter 2) [microradians] or radar pulse duration (Chapter 5) [s]

σ_{ij} = stress component [Pascals, $\mathrm{kg\,m^{-1}\,s^{-2}}$]

σ_1 = maximum principal stress [Pascals, $\mathrm{kg\,m^{-1}\,s^{-2}}$]

σ_3 = least principal stress [Pascals, $\mathrm{kg\,m^{-1}\,s^{-2}}$]

φ = longitude [degrees]

ω = tilt (Chapter 8) [microradians]

Abbreviations and acronyms

a	annum or year (e.g., mm a^{-1}, millimeters per year)
A/D	analog-to-digital (e.g., A/D converter)
ADGGS	State of Alaska Division of Geological and Geophysical Surveys
AKDA	Alaska Deformation Array (continuous GPS network)
ALOS	Advanced Land Observing Satellite (Japan)
ANSS	Advanced National Seismic System (USGS)
AS	anti-spoofing (GPS)
ASAR	Advanced Synthetic-aperture Radar (Envisat, European Space Agency)
ASI	Agenzia Spaziale Italiana (Italian Space Agency)
AKST	Alaskan Standard Time. AKST = GMT − 9 hours
Auto GIPSY	Automated online GPS data-processing service provided by the Jet Propulsion Laboratory: *http://milhouse.jpl.nasa.gov/ag/*
AVO	Alaska Volcano Observatory (USA)
BARD	Bay Area Regional Deformation Network (continuous GPS network)
BARGEN	Basin and Range Geodetic Network (continuous GPS network)
BCE	Before the Common Era (substitute for BC, Before Christ)
BM	bench mark
BS	backsight (leveling or triangulation)
°C	degree(s) Celsius *or* degree(s) Centigrade
C&GS	US Coast and Geodetic Survey
CAVW	Catalog of Active Volcanoes of the World
C/A-code	coarse/acquisition code (binary sequence used to modulate GPS carrier signals)
CCD	charge coupled device
CDMA	Code Division Multiple Access
CE	Common Era (substitute for AD, Anno Domini)
CGPS	continuous GPS
cm	centimeter(s)
CNES	Centre National d'Etudes Spatiales (France)
COSPEC	Correlation spectrometer (trade name) used to measure SO_2 concentration
CORS	Continuously Operating Reference Station (GPS)
CP-FTIR	closed-path Fourier transform infrared (spectrometer)
CRT	cathode ray tube
CSA	Canadian Space Agency
CVO	David A. Johnston Cascades Volcano Observatory
CWAAS	Canadian Wide Area Augmentation System for GPS, analogous to WAAS (USA), EGNOS (Europe), MSAS (Japan), GAGAN (India), and SNAS (China)
d	day (e.g., mm d^{-1}, millimeters per day)
DARA	Deutsche Agentur für Raumfahrtangelegenheiten (former German Space Agency, reorganized as DLR in 1997)
dB	decibel(s)
DC	direct current
DEM	digital elevation model
DLP	deep long-period (earthquake)
DLR	Deutsches Zentrum für Luft- und Raumfahrt (German Aerospace Center, established 1997)
DRAO	continuous GPS station located at the Dominion Radio Astrophysical Observatory south of Pentictin, B.C., Canada; part of the Western Canada Deformation Array
DT	differential transformer

DTED	digital terrain elevation data
DTM	digital terrain model
EBRY	Eastern Basin-Range (Wasatch Front) and Yellowstone Hotspot (Yellowstone–Snake River Plain) Network (continuous GPS network)
EDM	electro-optical distance meter, electronic distance meter, or electronic distance measurement
EGNOS	European Geostationary Navigation Overlay Service for GPS (Europe), a regional augmentation service analogous to WAAS (USA), CWAAS (Canada), MSAS (Japan), GAGAN (India), and SNAS (China)
EEPROM	Electrically Erasable Programmable Read Only Memory (computing)
EPROM	Erasable Programmable Read Only Memory (computing)
ERS	European Remote-Sensing Satellite (ERS-1 and ERS-2)
ESA	European Space Agency
ETS	episodic tremor and slip
FAA	Federal Aviation Administration (USA)
FBN	Federal Base Network
FDMA	Frequency Division Multiple Access
FFT	Fast Fourier Transform
FLIR	forward looking infrared radiometer
FM	frequency modulated
FOC	full operational capability (GPS and GLONASS)
FS	foresight (leveling or triangulation)
FTIR	Fourier transform infrared spectrometer
FTP	File Transfer Protocol
g	gram(s) *or* gravitational acceleration at Earth's surface ($g \sim 9.81 \, \mathrm{m \, s^{-2}}$)
GAGAN	GPS and Geo Augmented Navigation system for GPS (India), analogous to WAAS (USA), CWAAS (Canada), EGNOS (Europe), MSAS (Japan), and SNAS (China)
Galileo	Global Navigation Satellite System being developed by the European Space Agency (ESA)
GEONET	GPS Earth Observation Network (continuous GPS network, Japan)
GHz	gigaHertz (frequency unit, 10^9 Hertz)
GIS	geographic information system
GLONASS	Global Navigation Satellite System (Russia)
GLORIA	Geologic Long-Range Inclined Asdic (side-scanning sonar system)
GMT	Greenwich Mean (or Meridian) Time, defined as the mean solar time at the Royal Greenwich Observatory in Greenwich near London, England, which by convention is at 0 degrees geographic longitude
GNSS	Global Navigation Satellite System (see Glossary)
GPS	Global Positioning System, specifically the US NAVSTAR GPS Global Navigation Satellite System
GPa	gigaPascal(s)
GRS 80	Geodetic Reference System 1980
GST	Galileo System Time, a timescale generated by atomic clocks for the European Space Agency's Galileo global navigation system. GST is steered toward International Atomic Time (TAI), and is specified to be within 50 nanoseconds of TAI for 95% of the time over any yearly time interval
GVN	Global Volcanism Network
HF	high-frequency (earthquake) *or* hydrogen flouride
hPa	hectopascal (1 hPa = 1 millibar (mbar) = 9.86923×10^{-4} atmospheres)
HST	Hawaiian Standard Time, also known as Hawaiian/Aleutian Standard Time. HST = GMT − 10 hours
HVO	Hawaiian Volcano Observatory (USA)
Hz	Hertz, frequency unit equal to one cycle per second ($\mathrm{s^{-1}}$)
IAU	International Astronomical Union
IAVCEI	International Association of Volcanology and Chemistry of the Earth's Interior
IERS	The IERS was established as the International Earth Rotation Service in 1987 by the International Astronomical Union (IAU) and the International Union of Geodesy and Geophysics (IUGG). In 2003 it was renamed to International Earth Rotation and Reference Systems Service. One of its primary objectives is to provide the International Terrestrial Reference System (ITRS) and its realization, the International Terrestrial Reference Frame (ITRF).
IGS	International GPS Service

InSAR	Interferometric synthetic-aperture radar
IOC	initial operational capability (GPS and GLONASS)
ITRF	International Terrestrial Reference Frame (currently ITRF2000, updated regularly)
IUGG	International Union of Geodesy and Geophysics
IUSS	Integrated Undersea Surveillance System (US Navy)
JAMSTEC	Japan Marine Science and Technology Center
JAXA	Japan Aerospace Exploration Agency (formerly NASDA, National Space Development Agency of Japan)
JERS	Japanese Earth Resources Satellite
JPL	Jet Propulsion Laboratory
JRO	Johnston Ridge Observatory (US Forest Service visitor center near Mount St. Helens)
JRO1	Continuous GPS station at the Johnston Ridge Observatory near Mount St. Helens
$°K$	degree(s) Kelvin
ka	thousands of years before present
kbar	kilobar(s)
kg	kilogram(s)
km	kilometer(s)
KPa	kilopascal(s)
L1, L2	Primary carrier frequencies for NAVSTAR and GLONASS satellite signals. See Chapter 4 and Table 4.1 for details
L3	Military signal broadcast discontinuously at 1381.05 MHz by NAVSTAR satellites. Also, a particular linear combination of the L1 and L2 carrier frequencies. See Chapter 4 for details
L5	A third civilian carrier frequency to be broadcast in addition to L1 and L2 by Block IIF NAVSTAR satellites starting in 2006 and by Block III satellites starting in 2012. See Chapter 4 for details.
LAN	local area network
LBT	long-base tiltmeter
Lidar	light detection and ranging
LF	low-frequency (earthquake)
LP	long-period (earthquake)
LVO	Long Valley Observatory (USA)
m	meter(s)

M	magnitude (earthquake) on the Richter scale, assumed to be local magnitude, M_L, unless specified otherwise
M_L	local magnitude (earthquake) as originally defined by C.F. Richter and B. Gutenberg in 1935; the scale is based on the maximum amplitude of a seismogram recorded on a standard Wood–Anderson torsion seismograph
M_b	body magnitude (earthquake), based on the amplitude of P body-waves; this scale is most appropriate for deep-focus earthquakes
M_s	surface-wave magnitude (earthquake), based on the amplitude of Rayleigh surface waves measured at a period near 20 s; appropriate for distant earthquakes
M_w	moment magnitude (earthquake), based on the moment of the earthquake, which is equal to the rigidity of the Earth times the average amount of slip on the fault times the area of the fault that slipped
M_g	Mach number of the gas in a separated gas–liquid flow through a nozzle under choked conditions (Section 3.1.4)
M-code	Military code modulation structure, analogous to C/A-code and P-code, implemented on Block IIR-M and subsequent NAVSTAR satellite series starting in September 2005
Ma	millions of years before the present
Mbyte	megabyte (10^6 bytes)
mg	milligram(s)
mGal	milliGal (gravitational acceleration unit, 10^{-3} Gal)
MHz	megaHertz (frequency unit, 10^6 Hertz)
MITI	Ministry of International Trade and Industry (Japan)
mm	millimeter(s)
MPa	megaPascal (pressure unit, $1\,\mathrm{MPa} = 10^6$ Pascals $= 10$ bar)
ms	millisecond(s)
μGal	microGal(s) (gravitational acceleration unit, 10^{-6} Gal)
μrad	microradian(s) (angular tilt unit, 10^{-6} radian)
MSAS	MTSAT Satellite-based Augmentation System for GPS (Japan), MTSAT being Multi-functional Transport Satellite, analogous to WAAS (USA),

	EGNOS (Europe), CWAAS (Canada), GAGAN (India), and SNAS (China)
N	newton (force unit, equivalent to $\mathrm{kg\,m\,s^{-2}}$)
NAD 27	North American Datum 1927
NAD 83	North American Datum 1983
NASA	National Aeronautics and Space Administration
NASDA	National Space Development Agency of Japan (reorganized as JAXA in 2003)
NAVD 88	North American Vertical Datum of 1988
NAVSTAR	Navigation Satellite Time and Ranging (US GPS satellite)
NBAR	Northern Basin and Range (continuous GPS network)
NDIR	Non-dispersive infrared
NEXRAD	Next Generation Weather Radar (National Weather Service, National Oceanic and Atmospheric Administration, USA)
nGal	nanoGal(s) (gravitational acceleration unit, 10^{-9} Gal)
NGS	National Geodetic Survey (USA)
NGVD 29	National Geodetic Vertical Datum of 1929 (USA)
nm	nanometer $= 10^{-9}$ m
NOAA	National Oceanic and Atmospheric Administration (USA)
ns	nanosecond(s) $= 10^{-9}$ s
NSF	National Science Foundation (USA)
nT	nanoTesla(s) (magnetic field unit, equivalent to 1 gamma or 10^{-5} gauss)
OFDA	Office of Foreign Disaster Assistance (US State Department)
OPUS	On-line Positioning User Service, a GPS data-processing service provided by the National Geodetic Survey: *http://www.ngs.noaa.gov/OPUS/*
OTF	on-the-fly (GPS)
P-code	Precise code (binary sequence used to modulate GPS carrier signals)
PALSAR	Phased Array L-band Synthetic-aperture Radar (aboard the Japanese ALOS)
PANGA	Pacific Northwest Geodetic Array (continuous GPS network)
PBO	Plate Boundary Observatory (EarthScope component)
PDT	Pacific Daylight Time. PDT = GMT − 7 hours
ppb	part(s) per billion

ppm	part(s) per million
PRI	pulse repetition interval (radar) = 1/PRF
PRF	pulse-repetition frequency (radar) = 1/PRI
PRN	pseudorandom noise
PROM	Programmable Read Only Memory (computing)
psi	pounds per square inch
PST	Pacific Standard Time. PST = GMT − 8 hours
radar	radio detection and ranging
RAR	real-aperture radar
RINEX	receiver independent exchange format (GPS)
ROM	Read Only Memory (computing)
RTK	real time kinematic (GPS)
s	second(s)
S-code	Standard code (Russian GLONASS, analogous to C/A-code for American GPS)
SA	selective availability (GPS)
SAFOD	San Andreas Fault Observatory at Depth (EarthScope component)
SAR	synthetic-aperture radar
SBAS	Satellite Based Augmentation System (regional civilian GNSS augmentation system; see Glossary)
SCIGN	Southern California Integrated GPS Network (continuous GPS network)
Scintrex	Scientific Instruments, Research and Exploration. LaCoste & Romberg–Scintrex, Inc. produces over 90% of the world's gravimeters
SEAN	Scientific Event Alert Network
SI	Système International (International System of Units)
SLAR	side-looking airborne radar
SLC	single-look complex (a type of SAR image, distinct from a multi-look image)
SLR	side-looking radar, or single-lens reflex (camera)
SNAS	Satellite Navigation Augmentation System for GPS (China), analogous to WAAS (USA), CWAAS (Canada), EGNOS (Europe), MSAS (Japan), and GAGAN (India)
SOPAC	Scripps Orbit and Permanent Array Center
SOSUS	Sound Surveillance System (US Navy)
SP	self-potential, streaming potential, or spontaneous potential

SPOT — 'Système Pour l'Observation de la Terre' satellite (France)

SRTM — Shuttle Radar Topography Mission

STS — Space Transportation System *or* Shuttle Transport System (Space Shuttle missions are numbered sequentially starting with STS-1 in April 1981)

SV — satellite vehicle *or* space vehicle (GPS)

SVN — Satellite Vehicle Number *or* Space Vehicle Number (GPS)

t — metric ton(s) or tonnes, unit of mass equivalent to 1,000 kg

TAI — Temps Atomique International (International Atomic Time), based on a continuous counting of the SI second by a large number of atomic clocks. TAI is currently (1 January 2006) ahead of UTC by 33 s $(TAI - UTC = +33 \, s)$

TIN — triangulated irregular network (GIS)

TOMS — total ozone mapping spectrometer

TopSAR — topographic synthetic-aperture radar

UAFGI — University of Alaska Fairbanks Geophysical Institute

UNAVCO — University NAVSTAR Consortium formed in 1984 under auspices of the Cooperative Institute for Research in Environmental Sciences (CIRES) at the University of Colorado and funded by the National Science Foundation (NSF). UNAVCO, Inc. was created as an independent, non-profit, membership-governed corporation in April 2001

USAID — United States Agency for International Development (US State Department)

USArray — United States Seismic Array (EarthScope component)

USGS — United States Geological Survey

USNO — US Naval Observatory

UTC — Coordinated Universal Time, the modern implementation of Greenwich Mean Time (GMT)

UTC (Russia) — Coordinated Universal Time, Russia

UTC (SU) — Coordinated Universal Time, Soviet Union

UTC (USNO) — Coordinated Universal Time, US Naval Observatory

UTM — Universal Transverse Mercator

VDAP — Volcano Disaster Assistance Program (USA)

VEI — Volcanic Explosivity Index

VHP — Volcano Hazards Program (USGS)

VLP — very long period (earthquake)

VT — volcano–tectonic (earthquake)

W-key — Secret encryption used to form the Y-code from the unclassified P-code (GPS)

WAAS — Wide-Area Augmentation System for GPS (USA), analogous to CWAAS (Canada), EGNOS (Europe), MSAS (Japan), GAGAN (India), and SNAS (China)

WCDA — Western Canada Deformation Array (continuous GPS network)

WGS 84 — World Geodetic System 1984

WOVO — World Organization of Volcano Observatories

wt% — weight percent (i.e., percent by weight, as in magma containing 2 wt% H_2O)

Y-code — Binary code modulation scheme for GPS signals formed by encrypting the P-code using a secret W-key. Y-code is the basis for the anti-spoofing (AS) feature of GPS.

YVO — Yellowstone Volcano Observatory (USA)

CHAPTER 1

The modern volcanologist's tool kit

Daniel Dzurisin

It's an exciting time to be a volcanologist, particularly if your specialty is volcano geodesy. Volcanology is in the midst of a revolution, and geodesy is helping to lead the way. Faster than ever before, new technologies and techniques are changing our understanding of how volcanoes work by revealing how they behave – before, during, and after eruption – in unprecedented detail. Capabilities that would have seemed far-fetched just a few decades ago – such as watching volcanoes deform from space, tracking the rise of a magma-filled dike in real time, or predicting eruptions accurately and early enough for people to move out of harm's way without unduly disrupting their everyday lives – are fast becoming realities. I wrote this book because I wanted to share my excitement over these advances, not only with serious students of volcanology but also with anyone who has admired the beauty of a snow-capped volcano or marveled at the raw power of a volcanic eruption.

This first chapter provides an overview of the rest of the book and a scientific rationale for studying volcanoes, both to better understand how they work and to mitigate volcano hazards. There are ample opportunities for basic curiosity-driven research in volcanology, and I believe strongly that the pursuit of knowledge for its own sake would be justified even if there were no obvious short-term payoffs. Of course, such is not the case in volcanology. Volcanic disasters claim lives and property somewhere in the world every year, and progressively more people put themselves at risk as cities increasingly encroach on dangerous volcanoes. The results of basic volcano research, however esoteric they might seem to a non-specialist, provide insights into volcanic processes that eventually lead to more effective mitigation of volcano hazards.

In this chapter, I first describe some remarkable

examples of extreme volcano deformation, hoping to whet the reader's appetite for what is to come. Next I discuss some of the challenges faced by scientists and other groups during volcano crises, the terrible cost of failure in one particularly tragic case, and a generalized model for effective volcano crisis response. Then I move on to a survey of volcano-monitoring techniques from such diverse fields as seismology, geochemistry, and hydrology, before ending with a brief overview of volcano geodesy. Treating the other techniques first helps to put geodesy's contribution in context (i.e., to show where and how geodesy fits into the modern volcanologist's tool kit).

Chapters 2–6 deal with various techniques and instruments for studying volcano deformation, including classical surveying techniques (Chapter 2), continuous sensors such as tiltmeters and strainmeters (Chapter 3), the Global Positioning System (GPS) (Chapter 4), interferometric synthetic-aperture radar (Chapter 5), and photogrammetry (Chapter 6). Chapter 7 discusses four examples of well-studied volcanoes (Mount St. Helens, Kīlauea, Yellowstone Caldera, and Long Valley Caldera) to illustrate the importance of a comprehensive monitoring strategy that includes a healthy dose of geodesy. Chapter 8 addresses numerical modeling of deformation sources and serves as a blueprint for the design and analysis of geodetic networks and sensor arrays at volcanoes. Chapter 9 deals with borehole observations of strain and fluid pressure, which can be extremely sensitive indicators of volcanic conditions and volcano–tectonic interactions. Hydrothermal systems and volcano geochemistry – topics that are sometimes overlooked by geodesists, but which in some cases are inextricably linked to volcanic unrest – are covered in Chapter 10. A recurring theme throughout the book is the

Figure 1.1. Remains of the Roman market Serapeo near the center of the Phlegraean Fields Caldera, Italy. Here, marine mollusk borings (visible on the lower third of large columns in the lower left of the photograph) record more than two millennia of relatively steady subsidence punctuated by episodic uplift. Modern-day sea level is indicated by the Tyrrhenian (Mediterranean) Sea, which is visible behind trees in the background. Dramatic uplift associated with eruptions in 1198 and 1538, and with an episode of caldera unrest from 1982 to 1984, raised the formerly submerged columns several meters above sea level. Photograph by Michael P. Poland, June 2000.

multidisciplinary nature of modern volcanology, as evidenced by the emerging links among geodesy, seismology, geochemistry, hydrology, and the study of hydrothermal systems. The final chapter poses some challenges and opportunities for volcanology in the early part of the 21st century, which promises to build on the excitement of recent decades.

1.1 VOLCANOES IN MOTION – WHEN DEFORMATION GETS EXTREME

For me, the idea that volcanoes stretch, bulge, shrink, sink, or rise wholesale out of the sea is emotionally stirring and intellectually captivating. Knowing that the ground we live on is constantly being shoved around by subterranean processes, including the buoyant ascent of molten rock from deep within our planet's interior, is wondrous beyond words. The next four sections attempt to capture a sense of the visceral excitement that fast-moving volcanoes elicit in those who study them. From bradyseisms[1] at a Roman marketplace to the famous bulge at Mount St. Helens – such are the 'true legends' of volcano geodesy.

[1] The term bradyseism refers to slow uplift or subsidence of the ground surface in response to inflation or deflation of a magma reservoir, or to pressurization or depressurization of a hydrothermal system. The best known example is the remarkable motion of the Phlegraean Fields Caldera near Naples, Italy, which has persisted since Roman time.

1.1.1 The ups and downs of a Roman market – Phlegraean Fields Caldera, Italy

Consider the case of the coastal town of Pozzuoli, Italy, which sits near the center of a partly submerged caldera called Campi Flegrei (Phlegraean Fields). Serapeo, an ancient Pozzuoli marketplace constructed by Roman artisans in the first century BCE,[2] has two generations of floors and columns: the second was built after the first sank below sea level (Figure 1.1). Borings in the columns by *Lithodomus lithophagus*, a marine mollusk that burrows into rock for shelter, indicate as much as 11 m of subsidence by about 1000 CE[3] (Parascandola, 1947). The area then rose about 2 m from 1000 to 1198, when an eruption occurred at nearby Solfatara Crater (Parascandola, 1947; Caputo, 1979). Uplift continued unabated after the eruption and accelerated in about 1500 CE. By 1503, the amount of uplift was so great that a portion of newly emerged shoreline was deeded to the local university! Net uplift during the period 1000–1538 CE was about 12 m.

The uplift rate accelerated dramatically during several years before an eruption that began on 29 September 1538. The most rapid uplift occurred on 26–27 September, when the ground level at Pozzuoli rose at least 4–5 m in 48 hours, causing the shoreline

[2] Before the Common Era (substitute for BC, Before Christ).
[3] Common Era (substitute for AD, Anno Domini).

to recede 400 m. On 28 September, hot and cold water poured from fissures in the uplifted block, and on the morning of 29 September the uplifted ground subsided by as much as 4 m. Land near what was soon to become the eruptive vent stopped subsiding about noon on 29 September and began to rise again, so that '*... by eight o'clock it was as high as Monte Ruosi – i.e., it was as high as that hill where the little tower stands*' (del Nero, ca. 1538; English translation in Horner (1846, p. 19)). For a fascinating eyewitness account of the eruption and related phenomena, see Newhall and Dzurisin (1988, p. 97).

1.1.2 Remarkable uplifts in the Galápagos Islands – Fernandina and Alcedo Volcanoes

There are other notable examples of volcanoes deforming at a breakneck pace, including two from the Galápagos Islands, Ecuador. In 1927, three fishing boats anchored for the night at Punta Espinoza near the northeast point of Fernandina, a frequently active shield volcano that forms one of the islands. During the night, the crews were awakened by violent bubbling of the water. Two of the boats put to sea immediately, but the third was unable to leave because, by the time the crew recognized the danger, the seafloor had risen and their keel was grounded! In the morning, their boat was stranded by an uplift of 'several feet' (Cullen *et al.* 1987), presumably caused by a shallow magmatic intrusion. The rusted

remains of the hapless sailors' boat can still be seen on the beach (Figure 1.2).

At nearby Alcedo Volcano, a 6-km length of shoreline including Urvina Bay was uplifted by as much as 4.6 m, probably in early 1954 (Couffer, 1956). More than 1 km^2 of coral reef became exposed above sea level. The uplift was so rapid that fish were stranded in pools before they could escape to the sea.

1.1.3 Rabaul Caldera, Papua New Guinea, 1994

Another striking, more recent example of a volcano rising noticeably out the sea occurred in September 1994 at Rabaul Caldera, Papua New Guinea, prior to simultaneous eruptions from two vents on opposite sides of the caldera rim. The following composite account is excerpted from Lauer (1995, pp. 7–14), with my explanatory notes in [brackets]. '*There was a massive jerking of the house as if it had been slammed by a truck. I leaped up, wide awake. It was the biggest guria [earthquake] we'd had for months and months. Nick (husband) promptly went back to sleep. I looked at the clock – 2.50 a.m. [Sunday, 18 September 1994] ... The gurias continued all day Sunday ... Around 5 p.m. ... the ground was in almost constant motion ... The seismographs [at Rabaul Volcano Observatory, RVO] were a mass of almost unreadable black scribble and the recording pens were having difficulty holding up with the constant movement ... About 7.45 p.m. I got into my car to drive around to a friend's house for a game of cards. The whole car rocked. I noticed a*

Figure 1.2. Remains of a fishing boat grounded in 1927 at Punta Espinoza near the northeast point of Fernandina, Galápagos Islands, when a magmatic intrusion caused the seafloor to rise 'several feet' overnight. Wooden parts of the boat have rotted away, leaving just the rusted motor shown here. The lower flanks of Volcano Darwin are visible in the distance beneath the clouds. (*inset*) Closer view of the ill-fated boat motor at higher tide, with Volcano Ecuador on the skyline in the distance. Photographs by Michael P. Poland, January 2001.

number of people walking and carrying bed-rolls and bags. Interesting – they obviously knew something I didn't . . . Next thing there was a loud thump-thumping noise. It was 5.50 a.m. [the following morning, 19 September] and a helicopter was landing next to the bedroom . . . They were going to check around the harbour. Even from where we were standing the uplift was visible . . .' Incredibly, a tide gauge on the eastern side of Vulcan had been lifted nearly out of the water overnight, indicating uplift of the seafloor by about 5.5 m (Blong and McKee, 1995; McKee *et al.*, 1995) (Figures 1.3 and 1.4).

'. . . After the helicopter took off I went to ring a couple of friends who were still in their homes to warn them. Seconds later Lynden [daughter] came running in, yelling that there was black smoke coming out of Tavurvur . . . It had been just 27 hours since the first guria to the eruption of Tavurvur. Not long after [71 minutes] there followed a mighty BOOM! Vulcan, on the opposite side of the harbour to Tavurvur, had decided to join the eruption. Explosion followed explosion. Rocks and ash were hurled with amazing force skywards. Day seemed to become night as the heavy black cloud oozed its way across town. The eeriness was broken by fork lightning and claps of thunder.'

A massive eruption column quickly rose into the stratosphere (Figure 1.5). During the peak of the eruption, the column towered 20–30 km above the residents of Rabaul town and nearby communities, who self-evacuated in the face of obvious danger. About 90,000 people were displaced, but fewer than 10 died as a result of the eruption or its immediate effects (Blong and McKee, 1995). The number would surely have been much higher had it not been for the vigilance of RVO and for the Rabaul Disaster Plan, completed in 1983, which spelled out four alert stages, established evacuation routes, and called for practice evacuations – the last of which was held in April 1994, just five months before the eruption.

1.1.4 The bulge at Mount St. Helens, 1980

Among the better known examples of runaway volcano deformation is the famous bulge on the north flank of Mount St. Helens, which developed during the two months preceding the catastrophic debris avalanche and eruption of 18 May 1980 (Figure 1.6). Electronic distance meter (EDM) and theodolite measurements tracked the surface of the

Figure 1.3. Vulcan, a ring-fracture vent at Rabaul Caldera, Papua New Guinea, as it appeared from a helicopter at approximately 6:05 a.m. on 19 September 1994. The brown color at the base of the cone is uplifted seafloor that emerged overnight following 27 hours of intense seismicity and ground deformation, including 5–10 meters of uplift in the vicinity of Vulcan. The circle marks a tide gauge that was lifted out of the water overnight (Figure 1.4). Tavurvur, on the opposite side of the caldera, began erupting about 10 minutes after this photograph was taken, and Vulcan joined in 71 minutes later. Photo was taken from an Islands helicopter and published by Sue Lauer (1995, p. 13).

Figure 1.4. Tide gauge on the eastern side of Vulcan, a vent located along the western ring fracture of Rabaul Caldera, Papua New Guinea. The gauge was lifted nearly out of the water during the night preceding the start of simultaneous eruptions at Vulcan and Tavurvur, which can be seen venting in the background. Before the 1994 eruptions, mean sea level was approximately at the third step of the ladder, as marked by the top of the light-colored accumulation of mussel shells. USGS photograph by Elliot T. Endo, November 1999.

bulge marching steadily and nearly horizontally northward at rates of 1.5–$2.5\,\mathrm{m\,day}^{-1}$ from 23 April through the early morning of 18 May (Chapter 7). By 12 May, the high point of the bulge stood $150\,\mathrm{m}$ above pre-existing topography, and by 18 May the net volume increase was a staggering $0.11\,\mathrm{km}^3$ (Moore and Albee, 1981). Working on the volcano during the build-up to the 18 May eruption was an exhilarating experience. Titanic forces were at work, preparing to disassemble the volcano, while we struggled to understand what was happening beneath our feet. Such are the memories that inspired this book.

If these stories don't quicken your pulse, there's no reason to read further. Honestly, it doesn't get any better than this.

1.2 VOLCANOLOGY IN THE INFORMATION AGE

1.2.1 Volcano hazards mitigation – a complicated business

Faced with the daunting task of trying to unravel the tremendous variety and complexity of volcanic processes, especially during a rapidly evolving volcano crisis with lives and property at risk, today's volcanologist can ill afford to rely on any one specialty to furnish the necessary tools and information. Instead, diverse techniques from geology, hydrology, seismology, geodesy, geochemistry, and related fields must be brought to bear and, if necessary, tailored to the situation at hand.

Experience worldwide has shown that no single technique can provide unambiguous information about the cause and likely outcome of volcanic unrest. For example, earthquakes near a volcano might indicate fracturing of brittle rock around a pressurizing magma body, intrusion of magma or other fluid into brittle host rock, or release of tectonic strain not directly related to magmatic processes. Even the occurrence of long-period (LP) earthquakes that are commonly associated with magmatic or hydrothermal processes is inherently ambiguous. Such earthquakes are generally thought to indicate the presence of a two-phase fluid in a resonating crack (Chouet, 1988, 1996a, 2003). However, this fluid could be gas exsolved from fresh magma that is rising and decompressing, hypersaline brine derived from stagnant magma that is cooling and crystallizing, or gas-charged magma itself. Without complementary data from other monitoring techniques to help resolve this ambiguity, the precise significance of LP earthquakes for volcano hazards is difficult to discern, especially during a crisis. Furthermore, some models for LP earthquakes require a relatively narrow range of conditions for resonance to occur, suggesting that some (most?) fluid transport occurs without generating LP earthquakes. Even when they do occur, swarms of shallow LP earthquakes could be caused by pressurization of a magma body as a result of intrusion, boiling within a hydrothermal system in response to magmatic or tectonic changes, or migration of fluids released from a stagnant, cooling magma body. The hazards implications vary widely and can be explored further only by combining seismic information with that from other disciplines.

By itself, any single volcano-monitoring technique provides information that is ambiguous or

Figure 1.5. Eruption plume from Vulcan, Rabaul Caldera, Papua New Guinea on 20 September 1994, as witnessed by astronauts aboard Space Shuttle Discovery. In this photograph, the plume is seen spreading in the stratosphere approximately 20 km above Earth's surface. The cloud-covered island in the foreground is New Ireland. National Aeronautics and Space Administration, STS064-116-064.

subject to alternative explanations. For example, an increase in the flux of sulfur dioxide gas might indicate the recent arrival at shallow depth of fresh, gas-charged magma, but the absence of such an increase does not preclude the same interpretation. This is because SO_2 is very soluble in water and can be effectively 'scrubbed' by groundwater (Doukas and Gerlach, 1995). Carbon dioxide is less soluble in water than SO_2, so an increase in CO_2 followed by an increase in SO_2 might be attributed to newly emplaced magma that released both CO_2 and SO_2 and eventually dried out a conduit to the surface, allowing SO_2 to escape. For confirmation, though, the volcanologist is likely to reach for additional tools from seismology and geodesy.

No amount of monitoring is likely to provide definitive information about the explosivity of newly emplaced magma, and thus about the extent or degree of hazards posed by an impending eruption. For this, the volcanologist relies on the tools of field geology and stratigraphy, under the assumption that the volcano will likely behave in a manner similar to the ways it has behaved in the past. Past performances are chronicled in the stratigraphic record of past eruptive products, but the record is always incomplete and impossible to read unequivocally. In some cases, past eruptive behavior can be highly variable and therefore is a poor guide to the next event. Moreover, there is always the possibility of an unusual or unprecedented event, such as the catastrophic landslide and lateral blast at Mount St. Helens in 1980. In hindsight, similar events had occurred before but the evidence lay buried until the landslide and subsequent erosion cut deeply into the volcano's north flank (Hausback and Swanson, 1990). Only by combining information from a variety of sources can scientists hope to decipher the cause of volcanic unrest, and thus provide useful information to public officials charged with the difficult task of reducing risk while minimizing social and economic disruption.

1.2.2 Lessons from Armero, Colombia

It should come as no surprise that efficient information flow is essential for effective hazards assessment and risk mitigation during a volcano crisis. That fact was made terribly clear on the night of 13 November

Figure 1.6. West-southwest looking view of Mount St. Helens on 25 April 1980 (Krimmel and Post, 1981). The now famous bulge is a prominent feature below and to the right of the summit. Fresh snow covers higher portions of the volcano; lower slopes are covered with ash from a series of phreatic eruptions that began on 27 March. Catastrophic failure of the volcano's north flank on 18 May 1980 produced the largest debris avalanche in recorded history. USGS photo 80S2-122 by Austin Post.

1985, when huge lahars triggered by a small eruption of ice-clad Nevado del Ruiz Volcano in Colombia killed more than 23,000 people, most of them asleep, and virtually erased the hapless town of Armero from the map (Figure 1.7).

The first warnings had come almost a year earlier. Climbers on the 5,200-m peak noticed increased fumarolic activity and felt earthquakes starting in November 1984 and, on 22 December, three widely felt earthquakes were accompanied by small explosions from Arenas Crater near the volcano's summit. Felt earthquakes and abnormal fumarolic activity persisted into 1985. Seismometers installed at the volcano in July confirmed a high level of earthquake activity that intensified in early September before a strong phreatic eruption on 11 September. A preliminary volcano hazards map published in October warned that a magmatic eruption would almost surely trigger damaging lahars. The final version of the map, issued publicly on 8 November 1985, placed the town of Armero squarely in a zone of high lahar hazard. Five days later, and more than 8 hours after the beginning of the paroxysmal eruption, Armero and most of its residents disappeared

under a river of mud (Herd and Comité de Estudios Vulcanológicos, 1986).

In hindsight, this terrible tragedy could have been avoided. The volcano provided ample warning, an accurate hazards zonation map was available, the volcano was being continuously monitored, an evacuation order was issued hours before the lahar reached Armero, and the safety of higher ground was within walking distance for most residents. What went so terribly wrong? In spite of hard work by many and heroic efforts by some, essential information about how to mitigate the impending disaster was not available where and when it was critically needed. The advance warning was not understood by many of those at risk, and they received no short-term instructions to move to higher ground as the giant lahar bore down on them. In the words of one scientist who experienced the Armero tragedy firsthand: '... *the catastrophe was not caused by technological ineffectiveness or defectiveness, nor by an overwhelming eruption or an improbable run of bad luck, but rather by cumulative human error*' (Voight, 1996, p. 719). Shortly after the disaster, the US Geological Survey (USGS)

Figure 1.7. Remains of the city of Armero, Colombia, following the 13 November 1985, eruption at Nevado del Ruiz Volcano. In large parts of the town, a boulder-laden lahar sheared off the upper part of all structures at foundation level, even though some had steel-reinforced connections between concrete walls and foundations (Voight, 1996). More than 23,000 people perished as a result of the eruption. USGS photograph by Richard J. Janda, November 1988.

and the Office of Foreign Disaster Assistance of the US Agency for International Development (USAID OFDA) established the Volcano Disaster Assistance Program (VDAP) to mitigate losses from future eruptions worldwide by applying the lessons learned so tragically at Armero.

1.2.3 Communication – a key to effective hazards mitigation

Experience during various types of natural disasters, including wildfires, floods, storms, large earthquakes, and eruptions, has shown that communication is improved when a single crisis management center forms the hub for information retrieval and dissemination by a trained crisis response team. Ideally, the team includes scientists, land managers, representatives of involved government agencies, affected members of the private sector, civic officials, and emergency service providers such as fire, law enforcement, and medical personnel. During quiet times, the crisis management center usually exists in concept only and information flow among these groups includes consideration of long-term issues such as land use policies, emergency response plans, and public education concerning natural hazards. During a crisis, representatives of each group gather at the crisis management center to facilitate intensive exchange of information and ideas. The focus of discussions necessarily becomes more immediate and the information flow more structured. In particular, the crisis management center is re-

sponsible for issuing all official information about the emergency and measures being taken to mitigate its effects.

This generalized model has been applied successfully to volcano crises, both conceptually and in practice (Peterson, 1996; Figure 1.8). A team of volcanologists, preferably encompassing broad expertise and with prior experience during volcano emergencies, assembles to monitor the unrest, assess the hazards, and communicate with other groups involved in the response. To save time and avoid confusion, most groups appoint representatives who meet regularly at the crisis management center to exchange information and take necessary action. Appointing a spokesperson and staging the scientific response from a separate location has the added advantage of shielding the team from the inevitable hubbub and stress that accompany a crisis. Close coordination among scientists and the other groups is essential to avoid confusion or the appearance of mixed messages to the news media and public. This does not mean that honest differences of opinion among scientists are stifled; rather, that a serious attempt is made to reach consensus within the scientific team and any remaining differences are presented in a balanced way, within the context of the consensus view, at the crisis management center.

Information that flows to the scientific team from many sources forms the basis for analysis and interpretation of current activity and its possible outcomes. Information sources typically include monitoring instruments located on or near the volcano,

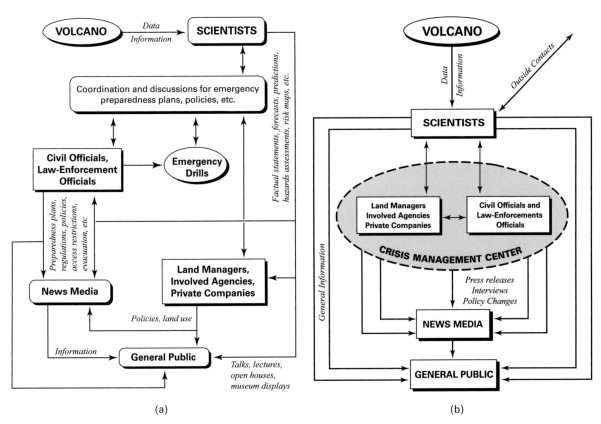

Figure 1.8. Idealized flow of volcano hazards information during periods of relative quiescence: (a) modified slightly from Peterson (1996) and during a volcano crisis (b). During quiescence, emphasis is on multichannel two-way communication, public education, and long-term preparedness. During a crisis, information flow necessarily becomes more structured and emphasis shifts toward short-term preparedness and hazards mitigation. The science team, usually through a designated spokesperson, acts as a conduit of interpreted information from the volcano to the crisis management center and, to a lesser extent, directly to the news media and general public.

field observers, remote-sensing instruments (e.g., infrared sensors for monitoring thermal areas or multispectral instruments capable of tracking ash clouds), pilot reports and sources of weather information pertinent to volcanic ash dispersal, and contributions from scientific colleagues elsewhere.

To be useful during a crisis, information about volcanic unrest must be both accurate and timely. All types of information, whether quantitative or qualitative in nature, have associated uncertainties that must be factored into the crisis team's analysis of the situation. Scientists are accustomed to such uncertainties and, in some cases, have developed formal methods for dealing with them. On the other hand, most scientists are unaccustomed to working under the severe time constraints and high-stress conditions that typically prevail during a volcano crisis. In a rapidly evolving situation with lives or livelihoods at stake, scientists often are forced to make interpretations or recommendations

in the absence of complete supporting information. Such occurrences underscore the importance of timeliness in the acquisition and initial analysis of monitoring information with a minimum of scientific equivocation. Effective communication of the consensus view and associated uncertainties in language understandable to nonscientists at the crisis management center is a challenging task that usually falls to the science team leader.

Generally, adequate information *now* is much more valuable than better information *later*. 'Now' in this context might mean that information is delayed a few minutes or hours, but seldom more than a few days. Most valuable of all is information available in 'real time' (i.e., information acquired by automated sensors and telemetered to the science center for rapid analysis). Modern sensors, telemetry systems, and computer based analysis systems make it possible to monitor key parameters either continuously (e.g., earthquake activity using seismometers) or discretely every few seconds or minutes (e.g., ground deformation using tiltmeters or GPS stations). Additional examples of real time geodetic monitoring are discussed in following chapters.

Information that is both accurate and timely is necessary but not sufficient for successful hazards mitigation during a crisis. Another essential ingredient is effective communication of scientific concepts, interpretations, and associated uncertainties from scientists to those directly responsible for reducing risk (e.g., public officials, land-use managers, emergency service providers). Recent history includes a few widely publicized examples of scientists' failure to reach consensus and provide useful guidance to those at risk, but also some striking successes in this regard (Peterson and Tilling, 1993). Among the latter can be counted the 1980–1986 eruptions at Mount St. Helens, Washington; the 1986 eruption of Oshima volcano, Japan; the 1991 eruption at Mount Pinatubo in the Philippines; and several recent eruptions in the Aleutian volcanic arc where serious threats to aircraft from far-traveled volcanic ash clouds were averted.

Clearly, a balance must be maintained between the public's need to know and scientists' responsibility to deliver a clear and consistent message. Within the crisis management center, full disclosure and discussion of differing viewpoints is both appropriate and constructive. This process can foster mutual understanding, respect, and trust among scientists, public officials, and emergency service providers that will be critically important if the situation worsens. When a wider audience is involved, discussion is understandably more circumspect but still honest and forthcoming. In my experience, citizens who are well informed about volcanic activity that could affect their livelihood and about the uncertainty inherent in any forecast are more likely to deal with the situation constructively than are those with less access to information. That said, the need to keep multiple audiences informed during a rapidly evolving crisis could severely tax the resources of scientists and public officials. Representatives of the news media should be willing to forego personal attention from individual scientists or officials when more efficient means of communication are in the public interest (e.g., press conferences, live broadcasts of official statements, or sharing of information provided to a few designated reporters). When lives are at stake, the needs of local residents who are at risk from the volcano should supersede the curiosity of the public at large.

1.3 A BRIEF SURVEY OF VOLCANO-MONITORING TECHNIQUES

Most volcano-monitoring techniques derive from the fields of seismology, geodesy, or geochemistry. These three, plus the multidisciplinary fields of remote sensing, non-seismic geophysics, volcano hazards assessment, and volcanic risk mitigation, provide the most commonly used tools in the modern volcanologist's tool kit. In addition, a wide variety of techniques from other fields can sometimes be used to good advantage. An excellent example is water-well monitoring, a staple of groundwater hydrology, which can provide important information about hydrologic responses to strain in active volcanic areas (Chapter 9). The following brief discussion of volcano-monitoring techniques (Sections 1.3 and 1.4) is intended to give the reader an introduction to the broader subject before subsequent chapters address various geodetic techniques in more detail.

1.3.1 Seismology – cornerstone of volcano monitoring

Seismology enjoys several clear advantages over other approaches to volcano monitoring, and therefore it is the technique of first choice when time or other resources are in short supply. Standard seismometers (geophones) are widely available and relatively affordable. A network of 6–8 short-period seismometers within 10 km of a volcano, including one very close to the epicentral area to better constrain the focal depths of earthquakes, is adequate in most cases to track seismic activity leading up to possible eruption (Figures 1.9 and 1.10). Conventional radio, microwave, or telephone links can be used to send continuous seismic signals to the science center for routine analysis, including automated calculations of earthquake magnitude, location, and frequency content. From such information, inferences can be drawn in near-real time concerning the amount of seismic energy being released, possible migration of earthquake sources, and the mechanism(s) causing the earthquakes (McNutt, 1996).

Easily overlooked, but nonetheless extremely important for volcano monitoring, is the fact that seismic waves from even small (Richter magnitude

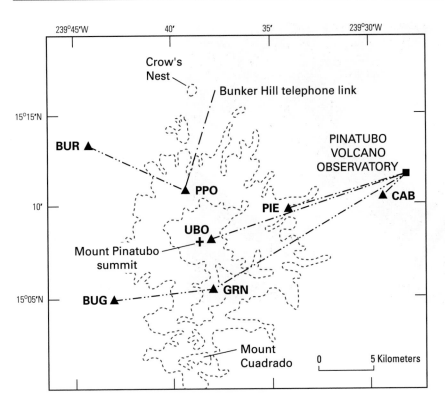

Figure 1.9. Location of seismic stations and radiotelemetry links at Mount Pinatubo, Philippines, before the paroxysmal eruption of 15 June 1991 (Lockhart *et al.*, 1996). Signals from most stations were telemetered to the Pinatubo Volcano Observatory by radio, either directly or by way of a repeater. The signals from stations BUR and PPO were sent to the observatory via the Bunker Hill telephone link. Ideally, a local network like this one is embedded in a regional network that provides better capability for locating deep earthquakes and functions as a back-up in case local stations are destroyed in an eruption.

$M < 1$) earthquakes can easily be recorded several kilometers from their source within seconds of their occurrence.[4] This is not usually the case for small geodetic or geochemical signals, which typically attenuate more quickly with distance from source and, in the latter case, can be delayed or prevented from reaching the surface by interactions with groundwater or other factors. Because seismic waves travel quickly and efficiently through the Earth, an adequate network of geophones usually can be established in just a few days without undue risk to field personnel – a tremendous advantage over techniques that require sensors much closer to the center of unrest.

For example, intense deformation of the dome and adjacent crater floor accompanied a series of mostly extrusive eruptions that built a dacite lava dome at Mount St. Helens from 1980 to 1986. Because the source was shallow, ground deforma-

tion did not extend more than about 1 km from the dome.[5] At times when parts of the dome were moving several meters per day and the crater floor near the base of the dome was tilting hundreds to thousands of microradians per day (μrad day^{-1}), any movement of points outside the 1980 crater (i.e., more than about 2 km away) was too small to measure with an EDM or with short-base watertube tiltmeters. In fact, observed line-length and ground-tilt changes generally were less than 1 part per million (ppm) (Swanson *et al.*, 1983; Chadwick *et al.*, 1983; Dzurisin *et al.*, 1983). But at the same time, thousands of shallow earthquakes were recorded by seismometers on the outer flanks of the volcano and beyond (Malone *et al.*, 1983).

The precursors to every eruption are to some extent unique, but the most consistently reported precursor is earthquake activity. Except possibly for some phreatic eruptions or eruptions triggered by gravitational collapse, virtually all eruptions are preceded by earthquakes large enough to be detected by seismometers within a few kilometers of the volcano. A changing pattern of seismic activity in space or time can provide important clues to the cause of the earthquakes and to the timing of possible eruption. Sharp increases in the rate of seismic energy release, the occurrence of distinctive long-period events and volcanic tremor, and shallowing

---

[4] There are several measures of earthquake magnitude, including local magnitude M_L, body-wave magnitude M_b, surface-wave magnitude M_s, and moment magnitude M_w. Unless specified otherwise, all earthquake magnitudes in this book are local magnitudes $M = M_L$, as originally defined by Richter (1935).

[5] For the elastic case, the scale of surface deformation is proportional to the depth of the source. Most deformation occurs within one source depth, although the details depend on the shape and orientation of the source (Chapter 8).

Figure 1.10. Kinemetrics Inc. PS-2 portable drum recorder showing the onset of a swarm of earthquakes at Soufriere Hills Volcano, Montserrat, West Indies. A small network of seismographs, telemetry links, recorders, and an automated data-analysis system can be installed quickly during a volcano crisis to provide essential seismic information in real time. USGS photograph by C. Daniel Miller, August 1995.

of earthquake hypocenters are common patterns observed in the final days before magma reaches the surface (Mori, 1995; Chouet, 1996a). It is important to note, however, that some volcano deformation is known to occur aseismically, especially when the source is relatively deep or the strain rate is relatively low. In such cases, geodesy and gas geochemistry offer the most promising techniques for detecting volcanic unrest (Chapter 11).

Modern equipment and analysis techniques make it possible to extract a tremendous amount of information from seismic signals emanating from volcanoes. Most seismometers at volcanoes are short-period, single-component instruments capable of sensing the vertical component of ground vibrations at frequencies between $\sim 0.7\,\mathrm{s}^{-1}$ (cycles per second) and $30\,\mathrm{s}^{-1}$, with optimal performance near $1\,\mathrm{s}^{-1}$ (1 cycle per second is equivalent to 1 Hertz (Hz)). Usually coupled with analog radiotelemetry systems with about 40 dB of dynamic range (comparable to an 8-bit digital system), short-period seismometers have been a mainstay of volcano-monitoring efforts for decades and still are widely used around the world. Their relatively low cost, proven reliability and durability, wide availability, and simplicity make them a good choice for most volcanic applications, especially in severe environmental conditions such as those prevalent in Alaska and the Pacific Northwest, USA, and in the Kamchatka region, Russia.

Modern three-component, digitally telemetered, high-dynamic-range, broadband seismometers make it possible to record signals over a much broader range of frequencies, from about 0.02 Hz to 50 Hz, with up to 145 dB of dynamic range (McNutt, 1996). These instruments are opening up the exploration of 'long-period' processes such as unsteady fluid transport in pressurized conduits (Chouet, 1996a,b). This approach has provided important new insights into fluidal processes beneath volcanoes and promises to become an increasingly important monitoring and prediction tool where favorable logistics permit the operation of broadband seismic networks (e.g., Kawakatsu *et al.*, 1992; Chouet, 1996a,b). Where year-round operation is impractical but seismic activity is persistent, temporary installations are useful to study such topics as the mechanisms of volcanic tremor and long-period earthquakes (e.g., Chouet *et al.*, 1997, 1998; Ohminato *et al.*, 1998).

Broadband seismometers are helping to bring the fields of volcano seismology and volcano geodesy closer together. Modern seismometers are capable of recording signals with frequencies as low as 0.02 Hz (period = 50 s), and borehole strainmeters can record signals with frequencies as high as 10 Hz (period = 0.1 s). With data from both types of instruments, volcanologists for the first time can study ground movements over the continuous range of timescales from about 10 ms to 100 days (Chouet, 1996b). As a result, geodesists are starting to think about earthquakes as very fast deformation events, while seismologists think about ground deformations as very slow earthquakes.

Figure 1.11. USGS scientists collect a gas sample from a fumarole in the crater at Mount St. Helens in 1981. Repeated sampling and analysis of fumarolic gases provide a useful means to track any long-term changes in the proportion or composition of magmatic gases. USGS photograph by Katherine V. Cashman.

1.3.2 Volcano geochemistry

The emission of volcanic gases is a common manifestation of volcanic activity and therefore geochemical techniques figure prominently in most volcano-monitoring programs. These techniques mostly fall into four broad categories (Sutton *et al.*, 1992): (1) field sampling of fumaroles or thermal springs with subsequent laboratory analyses; (2) measurements of the emission rates of magmatic gases in volcanic plumes; (3) continuous on-site monitoring of the concentrations of selected volcanic gases; and (4) measurements of the concentrations and fluxes of volcanic gases in soils.

Repeated field sampling and laboratory analyses provide detailed chemical information about fumarolic gases at specific sites and times (Figure 1.11). By way of models that track the chemical interactions among magmatic gases, groundwater, and the atmosphere, such information can be used to infer the original gas content of the parental magma body and thus to monitor subsequent slowly developing changes in a magmatic system. Because the technique requires repeated visits to a field site and time for laboratory analyses,

it is not well suited to monitoring chemical changes during a volcanic crisis. This limitation can be circumvented by using a field-portable gas chromatograph to quickly analyze samples as they are collected (LeGuern *et al.*, 1982), but logistical considerations and hazards to field crews still preclude this approach during most crises. On the other hand, the technique is appropriate and useful for routine monitoring because changes in fumarolic gases are sometimes clear, early indicators of newly emplaced magma.

Because emission rates of volcanic gases can be measured from an aircraft flying below or through a volcanic plume, or sometimes from a safe vantage point on the ground, this approach is often used to monitor geochemical changes during periods of volcanic unrest (Figure 1.12). The relative safety of an airborne gas-sampling platform is particularly advantageous during volcano crises, when safety concerns often preclude ground access for other types of geochemical monitoring. Historically, the most commonly measured plume constituent is sulfur dioxide (SO_2), because its concentration can be measured by flying under a plume and using a correlation spectrometer (COSPEC). The

Figure 1.12. A COSPEC can be used to measure the concentration of sulfur dioxide (SO_2) gas in a volcanic plume. The instrument can be mounted on a tripod, on a wheeled vehicle, or on an aircraft (*inset*). COSPEC measurements can be combined with wind-speed data to calculate the emission rate of SO_2, which sometimes is a good indicator of shallow magma intrusion. USGS photographs by Lyn Topinka and Kenneth A. McGee (*inset*).

instrument measures the amount of solar radiation absorbed by SO_2 molecules in a specific wavelength band in the ultraviolet, using clear sky as a reference (Figure 1.13A). The COSPEC has been used widely to measure SO_2 emission rates from volcanoes since the 1970s (e.g., Stoiber and Jepsen, 1973; Casadevall *et al.*, 1981, 1983, 1987; Gerlach and McGee, 1994; Gerlach *et al.*, 1999, 2002). The amount of absorption is proportional to the concentration of SO_2 in the atmospheric column. The product of SO_2 concentration and wind velocity, determined from meteorological reports or by simultaneously measuring air speed and ground speed while flying through the plume, gives the SO_2 flux as a function of distance across the plume. Integrating this value yields the total SO_2 flux, typically expressed in units of kilograms per second ($kg\,s^{-1}$) or metric tons per day ($t\,day^{-1}$) (Stoiber *et al.*, 1983).

Less commonly measured until recently, but equally useful, is the emission rate of carbon dioxide (CO_2). These measurements are made by flying *through* the plume with a spectrometer system that collects a small sample of the plume, passes a light beam through the sample, and measures the amount of attenuation within a CO_2 absorption band in the infrared part of the spectrum (Figure 1.13B). The CO_2 concentration in the sample is proportional to the amount of absorption. By repeating this process many times at different altitudes while the aircraft flies back and forth through the plume, a 2-D profile of CO_2 concentration for a cross section of the plume is determined. Multiply-

ing by the wind speed yields the total CO_2 flux that, like SO_2 flux, is typically expressed in units of $kg\,s^{-1}$ or $t\,day^{-1}$. A recent improvement makes use of GPS navigation to better determine the ground speed of the aircraft and thus the velocity of the plume (Gerlach *et al.*, 1997, 1999).

The emission rate of SO_2 has been used to infer the volume of degassing magma (Casadevall *et al.*, 1983) and magma supply rates (Casadevall *et al.*, 1987) during both eruptive and non-eruptive periods. Recently, however, Gerlach and McGee (1994) and Gerlach *et al.* (1996) have questioned the validity of this so-called 'petrologic' method, which relies on the concentration of sulfur in melt inclusions to relate SO_2 flux to the volume of degassed magma. They suggest instead that the primary source of SO_2 in plumes at Mount St. Helens and Mount Pinatubo was a vapor phase that developed above vapor saturated magma. If so, melt inclusions significantly underestimate SO_2 emissions or, stated differently, calculations based on melt inclusion and SO_2 emission data may significantly overestimate the volume of magma degassed to produce the gas plume.

Measurements of hydrogen sulfide (H_2S) emission rate from volcanoes are important in their own right as an indicator of magma degassing, and they also can be used to improve the accuracy of CO_2 emission-rate measurements. The relatively high background concentration of atmospheric CO_2 makes it difficult to separate volcanic sources of CO_2 from other sources in airborne surveys. On the other hand, background concentrations of H_2S are typic-

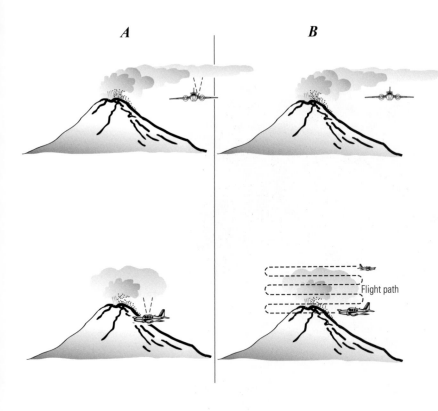

Figure 1.13. Flight patterns for making airborne measurements of volcanic gas emissions, adapted from Sutton *et al.* (1992). To measure sulfur dioxide (A), an aircraft flies under the plume, perpendicular to the prevailing wind, with a vertically aimed correlation spectrometer. The COSPEC measures the amount of sunlight in a specific ultraviolet wavelength band absorbed by SO_2 in the plume. To measure carbon dioxide and other gases (B), an aircraft flies under, through, and above the plume to obtain a concentration profile of the plume in cross section. In both cases A and B, top shows a view looking crosswind and bottom shows a view looking upwind.

ally very low, so a plume of magmatic H_2S generally will be well defined and easily identified. By using the H_2S signal as a marker for the entire plume, the CO_2 signal can be distinguished from the ambient CO_2 background more easily and accurately. McGee *et al.* (2001) used this approach at Mount Baker, Washington, and concluded (p. 4479): '*This technique is sensitive enough for monitoring weakly degassing volcanoes in a pre-eruptive condition when scrubbing by hydrothermal fluid or aquifers might mask the presence of more acid magmatic gases such as SO_2.*'

For volcano-monitoring and hazards assessment purposes, more important than absolute gas fluxes are any *changes* in the emission rates of SO_2, H_2S, and CO_2. These can be related to the ascent of fresh, volatile-rich magma or to changes in the groundwater system beneath a volcano. SO_2 is relatively soluble in water, so a groundwater system can effectively 'scrub' most SO_2 before it enters a volcanic plume. CO_2 and H_2S, on the other hand, are less soluble in water and mostly pass undiluted through the groundwater system and into the plume. At very 'wet' volcanoes, such as Mount Spurr, Alaska, groundwater scrubbing of SO_2 can be so effective that no precursory increase in SO_2 flux was observed before a series of eruptions in 1992 (Doukas and

Gerlach, 1995). In other cases, when the groundwater envelope is boiled off by magmatic heat to create dry passageways to the surface for SO_2, an increase in CO_2 flux is followed eventually by an increase in SO_2 flux. This pattern can be diagnostic of the arrival of fresh magma within a few kilometers of the surface.

A new tool that shows great promise for extending studies of gas emissions at volcanoes is called a Fourier transform infrared (FTIR) spectrometer (Figure 1.14). Unlike the COSPEC, which measures only SO_2, an FTIR spectrometer can determine the concentrations of many common magmatic gases, including SO_2, CO_2, CO, COS, HCl, and HF. Open-path FTIR spectrometers analyze infrared light from a distant source, either natural or artificial, that has passed through a plume of volcanic gas. Closed-path FTIR spectrometers use a system of mirrors to 'fold' an internal infrared beam inside a cell that can be filled repeatedly with sample gas from a plume. Open-path instruments measure across an entire plume. Because the path length is poorly known, absolute gas concentrations are seldom calculated. Instead, the results from open-path systems are typically expressed as a ratio of one gas concentration to another. Closed-path instruments repeatedly measure gas concentrations in discrete

Figure 1.14. Scientists use an open-path FTIR spectrometer to measure the concentrations of volcanic gases from vents along the East Rift Zone of Kīlauea Volcano, Hawaii. This instrument can determine the concentrations of many common magmatic gases, including SO_2, CO_2, CO, COS, HCl, and HF. USGS photograph by Michael P. Doukas.

samples of the plume while traversing and sampling it. A closed-path system thus can produce gas concentration profiles across the plume, similar to the method described above for CO_2, which can be combined with wind speed data to calculate emission rates.

Mori *et al.* (1993, 1995) used an open-path FTIR spectrometer to remotely monitor HCl:SO_2 ratios at Vulcano, Italy, and during a dome-building eruption at Unzen Volcano, Japan. McGee and Gerlach (1998a) adapted the technique for airborne use with a closed-path FTIR spectrometer and demonstrated its utility at Kīlauea Volcano, Hawai'i. By simultaneously measuring SO_2 flux using a COSPEC and the relative concentrations of SO_2 and other gases using an FTIR spectrometer, volcanologists can obtain accurate flux estimates for a large suite of volcanic gases without exposure to hazardous conditions on the ground. For example, Gerlach *et al.* (2002) significantly improved earlier estimates of the CO_2 emission rate from the summit of Kīlauea by combining synchronous SO_2 emission rates measured with a COSPEC with CO_2:SO_2 concentration ratios measured with a non-dispersive infrared (NDIR) CO_2 analyzer and a closed-path Fourier

transform infrared (CP-FTIR) spectrometer. Their result (8,500 metric t day^{-1}) is several times larger than previous estimates. Gerlach *et al.* (2002) also showed that the summit CO_2 emission rate is an effective proxy for the primary magma supply rate, and, therefore, might provide advance warning of intrusions from a rapidly filling summit magma reservoir.

Some volcanic gas-emission events have durations as short as a few minutes, and, therefore, can escape detection by repeated sampling or measurement techniques such as COSPEC and FTIR. Fortunately, commercial chemical sensors are available for many common volcanic gases, including H_2, SO_2, CO_2, H_2S, CO, COS, HCl, and HF. With some modification for volcanic environments, these can be used to monitor fumarolic gas concentrations continuously (Figure 1.15). Continuous gas measurements can be made in fumaroles, in air near active fumaroles, or in soils (Sutton *et al.*, 1992). A widespread gas-emission event was detected by sensors installed in low-temperature fumaroles about 10 days before the start of a long-lived eruption at Kīlauea Volcano, Hawai'i, in January 1983 (McGee *et al.*, 1987).

Figure 1.15. Gas and temperature monitoring station in the summit area of Kīlauea Volcano, Hawai'i. Commercially available chemical sensors have been modified to operate in harsh volcanic conditions, thus providing a continuous record of the concentrations of several fumarolic gases. USGS photograph by James D. Griggs.

Soil-gas measurements also can be made in survey mode (Figure 1.16). This approach is especially useful for mapping areas of anomalously high CO_2 flux, such as a tree-kill zone near the base of Mammoth Mountain, California (Farrar *et al.*, 1995; McGee and Gerlach, 1998b). McGee *et al.* (2000) combined soil-gas flux surveys with continuous monitoring of soil CO_2 concentration at 7 sites within 5 km of Mammoth Mountain and concluded that high soil CO_2 flux in the tree-kill area was the result of a magma-degassing event in September–December, 1997.

1.3.3   Volcano geophysics

Several non-seismic geophysical techniques have been adapted for use at volcanoes, including microgravity, differential magnetics, and electrical self-potential. Here I describe these techniques very briefly; the interested reader is referred to Chapter 2 and references therein for additional information.

Microgravity measurements can be made in either continuous or survey mode. In continuous mode, a gravimeter (gravity meter) is operated at a fixed site and its output is corrected for the effects of Earth tides and, if known, fluctuations in the level of the local groundwater table. The remaining signal is sensitive to changes in both station height (i.e., to uplift or subsidence of the ground surface) and subsurface density near the station. The latter might occur, for example, if magma were intruded into or

withdrawn from host rocks in the vicinity of the gravimeter. The same type of measurement can be made in survey mode by visiting an array of bench marks with a portable gravimeter and correcting for instrumental tares and drift during the survey (Figure 1.17).[6] In either case, microgravity measurements are most useful when made in conjunction with leveling, GPS, or total station measurements to determine land-surface height changes, and with water-well measurements to determine changes in the height of the groundwater table. When these effects are removed, the remaining gravity signal can be interpreted in terms of subsurface density changes caused by fluid migration (including magma), dilatation caused by fracturing, or other processes. Otherwise, microgravity measurements reflect the combined gravitational effects of changes in subsurface density, land surface height, and groundwater table height.

Magnetic field measurements, like gravity measurements, can be made in either continuous or survey mode (Figure 1.18) and are sensitive to various processes that commonly occur at volcanoes. The

---

[6] Portable gravimeters are subject to sudden reading shifts ('tares') and gradual instrument drift. Tares can be caused by a physical shock, vibration, or a change in instrument temperature, for example. Instrument drift results from slow mechanical aging of components. By repeatedly looping back to a reference station during the course of a survey, preferably using two or more gravity meters, tares and drift can be identified and usually removed from the observations. The effect of Earth tides also can be removed, based on theoretical calculations.

Figure 1.16. Portable set-up for measuring soil CO_2 flux in use near Mammoth Mountain, California. This is an efficient means to map CO_2 distribution in volcanic areas by contouring data from a grid of accessible measurement points. USGS photograph by Richard Kessler.

can produce charge separation, called the streaming potential phenomenon, occurs when groundwater moves through permeable rock. Rock surfaces tend to capture excess negative charge from ions in solution, which is balanced by excess positive charge that develops in a layer of electrolytic groundwater near the rock–water interface. If a vertical pressure gradient forces groundwater to move relative to the rock, some of the excess positive charge is swept along with the water to produce a convection current. The resulting electrical charge at the ground surface is positive if the water rises and negative if it descends.

At an active volcano, subsurface magma movements can affect groundwater flow by heating the water or creating new fracture permeability during intrusive events. Rising water produces a positive SP at the surface, so we would expect to find positive SP anomalies within thermal areas and negative anomalies adjacent to them (i.e., above the descending limbs of the groundwater convection cell). If magma intrusion or some other process enhances

Figure 1.17. Geophysicist Daniel J. Johnson makes a precise gravity reading with a LaCoste & Romberg portable gravimeter at a station near Thurston Lava Tube, Kīlauea Volcano, Hawai'i. Photograph by Eileen Llona.

most important magnetic effects are caused by changes in the state of stress (i.e., the piezomagnetic effect) or in the temperature of volcanic rocks. Both processes affect the magnetic field, which provides an opportunity to infer changes in subsurface stress or temperature by monitoring the surface magnetic field. For example, Johnston *et al.* (1981) reported that the magnetic field at Blue Lake, about 5 km west of the summit of Mount St. Helens, increased by 9 ± 2 nanoteslas at the time of the 18 May 1980 eruption. They concluded that the change was caused by regional elastic stress release in response to the large landslide and eruption.

Electrical self-potential (SP) is another geophysical parameter that is sensitive to subsurface changes at volcanoes. A pair of connected electrodes in contact with the ground at different places sometimes will measure a small electrical potential difference (typically hundreds of millivolts) caused by separation of ionic charges at depth. One mechanism that

Figure 1.18. Repeated magnetic field surveys can track changes in the state of stress or temperature of volcanic rocks. This technique has been used successfully in the crater at Mount St. Helens, Washington (a) and along the East Rift Zone of Kīlauea Volcano, Hawai'i (b). USGS photographs by Lyn Topinka (a) and Kaye Shickman, 2 November 1989 (b).

the upward flow of water, the electrical SP is expected to increase in areas where rising water approaches the surface. This effect can be measured either in survey mode, by repeating traverses across the active area with a pair of electrodes, or continuously by leaving the electrodes in contact with the ground and recording the potential difference between them. Both approaches have been used successfully at Kīlauea Volcano, Hawai'i, to map zones of upwelling groundwater and thus to monitor changes caused by shallow magma intrusions (Dzurisin *et al.*, 1980).

1.3.4 Hydrologic responses to stress and strain

Various hydrologic phenomena have been reported in association with volcanic activity, including dramatic changes in groundwater levels, stream flow, and spring discharges (see, e.g., Newhall and Dzurisin, 1988). Such occurrences are not surprising, because deformation is likely to affect subvolcanic aquifers in several ways. For example, the water storage capacity of an aquifer will increase or decrease in proportion to the amount of dilatational or compressional strain, causing groundwater to flow into or out of the aquifer. In addition to such poroelastic effects, changes in permeability or other hydrologic properties caused by deformation can be expected to produce responses in the groundwater system that might be inelastic or nonlinear.

Fortunately, there exists a relatively straightforward means to probe the groundwater system at restless volcanoes. Whether for some other purpose or specifically for this one, wells drilled into a volcano can be instrumented to record the groundwater level, temperature, and various geochemical parameters. The unique response of each well to known forcing functions, such as precipitation and the solid Earth tides, can be analyzed to determine various properties of the aquifer, including its permeability and degree of confinement. This information in turn can be used to remove such effects from the data, which can be used to study ground deformation and other processes that affect the groundwater system (Chapter 9).

1.3.5 Remote-sensing techniques

Observations made from a distance can enhance even an intensive on-site volcano-monitoring program. Spaceborne sensors have the advantage of frequent albeit distant access to hundreds of volcanoes that are not adequately monitored by ground-based techniques. Though never an adequate substitute for a well-equipped and trained volcano response team, remote sensing has become a valuable tool for detecting unrest and eruptions at remote volcanoes, monitoring thermal phenomena, and sometimes observing both the gaseous and particulate components of volcanic plumes and flows. For a thorough treatment of remote-sensing techniques used in volcanology, including future capabilities that are not discussed here, the reader is referred to Francis et al. (1996) and to Mouginis-Mark and Domergue-Schmidt (2000).

Remote sensors can be ground-based, airborne, or spaceborne. Ground-based applications include acoustic monitoring of pyroclastic flows (Manley, 1993), radar observations of pyroclastic flows (Pendick, 1993) and ash clouds (Rose et al., 1995), time-lapse photography and video monitoring of eruptive products (Thornber, 1997), measurements of gas concentrations in volcanic plumes (Section 1.3.2), and thermal-infrared monitoring of volcanic activity (Harris et al., 2003).

Harris et al. (2003) noted that ground-based thermal sensors have the following advantages over satellite and airborne sensors: (1) they can be located

beneath clouds that obscure aerial views; (2) they allow small thermal targets to be resolved; (3) they observe targets with a constant viewing geometry for long periods of time; and (4) they can provide data at high sample rates. A prototype thermal-infrared monitoring system, including two 1° field-of-view sensors and one 60° field-of-view sensor, has been in continuous operation since March 2001 near the crater rim of Pu'u 'Ō'ō, the active vent for the long-lived eruption along Kīlauea's East Rift Zone that began in January 1983 (see Heliker et al., 2003). Success of the prototype system led to the installation of a second permanent system at Stromboli Volcano, Italy, and to the development of a portable system with an interchangeable 1°, 15°, or 60° field-of-view sensor. By October 2003, the portable system had been deployed for short-duration (10–21 days) experiments at Kīlauea, Masaya (Nicaragua), Erta Ale (Ethiopia), Soufrière Hills (Montserrat), Villarrica (Chile), Santiaguito, Fuego, and Pacaya (all in Guatemala), and Stromboli, Etna, and Vulcano (all in Italy). The resulting data comprise a growing library of thermal waveforms for volcanic events, including Strombolian eruptions, lava flows, gas jetting, crater floor collapse, persistent degassing, lava flow in tubes, rockfalls, pyroclastic flow events, and lava lake activity (Harris et al., 2003).

Ground-based radars similar to those used for tracking rainstorms are capable of 'seeing' through weather clouds and darkness to detect ash plumes. The presence or absence of airborne ash (i.e., whether or not the volcano is erupting) is sometimes hard to determine otherwise, especially at night or during bad weather. Even with continuous seismic, geodetic, and geochemical monitoring in place, the question of whether a volcano is actually erupting cannot always be answered unambiguously without direct detection of an ash plume by radar. This technique has been used to track ash clouds at Mount St. Helens, Washington (Harris et al., 1981); Mount Pinatubo, Philippines (Oswalt et al., 1996); and Popocatépetl, Mexico (Centro Nacional De Prevención De Desastres, Mexico, 1997, unpublished data). Given the high stakes involved during encounters between jet aircraft and ash clouds (Casadevall, 1994), the use of radar for this purpose is likely to become more widespread in the future.

Sensors aboard satellites orbiting Earth can be used to detect eruptions, monitor thermal phenomena, track ash clouds, measure the flux of certain volcanic gases, and image ground deformation by radar interferometry. Examples of high spatial resolution sensors that operate in visible, near infrared, and short-wavelength infrared parts of the spectrum include the United States' Landsat Thematic Mapper and the French Système Pour l'Observation de la Terre (SPOT) satellite. These provide spatial resolution of 10–30 m and look cycles of 1–3 weeks, which limit their utility for tracking rapidly developing volcanic unrest or eruptive activity.

Among the most useful satellite sensors for volcano monitoring are those designed for routine meteorological observations. These work in the visible and thermal infrared parts of the spectrum, have relatively low spatial resolution (1 to several km), but provide frequent, broad spatial coverage with repeat times of a few hours or less. Thus, they are well suited to tracking volcanic plumes. Also important in this regard is the total ozone mapping spectrometer (TOMS) on the Nimbus 7 satellite, which is capable of measuring SO_2 load in stratospheric eruption plumes. Many of the June 1991 eruptions of Mount Pinatubo, Philippines, were observed by U.S.A. and Japanese operational meteorological satellites (NOAA-10, -11, -12, and GMS), which provided information on plume heights and dispersal patterns (Lynch and Stephens, 1996).

Of special interest to volcano geodesy are synthetic-aperture radar (SAR) instruments, which illuminate Earth's surface with microwave radiation and capture the returned signals to produce an image of the terrain, even when darkness or clouds obscure the surface (Chapter 5). Two SAR images of the same scene taken from a similar vantage point (on successive satellite orbits, for example) can be used to construct a digital elevation model (DEM) with horizontal resolution of 10–30 m and vertical resolution of 1–2 m. Especially at remote volcanoes where detailed topographic information is lacking, this product is useful for assessments of slope related hazards and support of field operations. With a third SAR image of the same scene, or a DEM from some other source, a radar interferogram can be constructed that, under ideal conditions, represents movements of the ground surface in the direction of the satellite (i.e., the interferogram is essentially a snapshot of the surface deformation field as seen from orbit (Massonnet et al., 1993, 1995; Massonnet and Feigl, 1998)). Surface factors other than ground movement, including erosion, deposition, or changes in vegetation, soil moisture, or snow cover can destroy the coherence of the interferogram, limiting the technique's usefulness in some environments. On the other hand, surface change detection based on coherence loss can in some cases be used to map active flows or new tephra deposits.

1.3.6 Volcano hazards and risk assessment techniques

A thorough discussion of this important topic is beyond the scope of this book; interested readers should consult the excellent treatments by Crandell et al. (1984), Latter (1989), and Scarpa and Tilling (1996). Two especially useful hazards assessment techniques are worth mentioning here, however. The first is careful geologic mapping and dating of young volcanic deposits to determine the frequency and character of past eruptions. At Mount Pinatubo, Philippines, for example, geologic reconnaissance in the weeks before the climactic eruption on 15 June 1991, revealed that extensive pyroclastic flow deposits surrounding the volcano were remarkably young and voluminous (Newhall et al., 1996). The picture that quickly emerged was of a volcano that erupted not often, but usually with a vengeance. Against this backdrop, which was assembled from intensive but abbreviated fieldwork and a few radiocarbon age determinations for key samples, the escalating unrest that followed took on an ominous tone. By correctly recognizing Mount Pinatubo's penchant for highly explosive and potentially devastating eruptions, scientists were able to quickly prepare an accurate hazards zonation map that later helped in saving thousands of lives (Punongbayan et al., 1996).

In situations where little is known about the eruptive history of a volcano and even reconnaissance fieldwork is not feasible, a useful approximation to the extent of hazards from debris flows can be made on the basis of topography and an empirical relationship between the volume of debris flows and the distance they travel. Data from many volcanoes in a wide variety of climatic settings show that the runout distance for debris flows is roughly proportional to the cube root of debris flow volume: $d = k \times V^{1/3}$, where k is a constant. With this relationship and a DEM of the volcano, hypothetical debris flows of various sizes can be routed down valleys using a geographic information system (GIS) (Iverson et al., 1997, 1998). The resulting hazards zonation maps (Figure 1.19), though not as accurate or reliable as those based on detailed fieldwork, are readily available and helpful, especially during a crisis.

1.3.7 A mobile volcano-monitoring system

All of the techniques discussed above have both strengths and weaknesses, so an integrated approach can significantly increase the effectiveness of any volcano-monitoring program. Toward that end, the USGS has developed a mobile volcano-monitoring system for the Volcano Disaster Assistance Program (VDAP). This integrated system includes instrumentation to monitor seismicity, ground deformation, certain geochemical parameters, and surface flow phenomena such as debris flows and lahars (Murray *et al.*, 1996a). Such a system is an essential part of the modern volcanologist's tool kit, because it provides the means to quickly and efficiently acquire, telemeter, analyze, compare, and store monitoring information during a volcano crisis.

Most monitoring data are transmitted in real or near-real time via a combination of analog and digital telemetry systems (Murray, 1992; Murray *et al.*, 1996a) (Figure 1.20). Data types that do not lend themselves to telemetry, such as leveling, EDM, and sulfur dioxide flux measurements, can be entered into the system manually. A network of personal computers running specially designed software is used to acquire, analyze, and display the data. Software design emphasizes ease-of-use and flexibility so the entire monitoring database will be accessible to all members of the crisis response team. The software not only allows volcanologists to compare and analyze data rapidly: it also facilitates presentation of critical data in easily understandable form to civil authorities managing the crisis. The VDAP 'mobile volcano observatory' concept, with its integrated monitoring and analysis system, has been effective during many recent volcano crises, most notably the 1991 eruption of Mount Pinatubo (Murray *et al.*, 1996b; Wolfe and Hoblitt, 1996).

1.4 AN INTRODUCTION TO GEODETIC SENSORS AND TECHNIQUES

As discussed above, no single technique can address adequately the full spectrum of information that volcanoes offer for study. Interesting things happen beneath volcanoes that do not always produce earthquakes, gas emissions, hydrologic changes, or geophysical anomalies. The immediate seismic precursors to an eruption can be brief and compelling (e.g., 27 hours of remarkably strong seismicity at Rabaul in 1994; Blong and McKee, 1995; McKee *et al.*, 1995). In other situations, deformation occurs

aseismically and might be detectable long before the strain or strain rate is high enough to produce earthquakes. A good example is the broad, virtually aseismic uplift in the Three Sisters area, Oregon, which was discovered by radar interferometry in spring 2001 (Wicks *et al.*, 2002a,b). In such cases, geodesy can make a unique contribution, but more often it provides information that complements other sources and thus adds to the overall understanding of volcanic processes. What follows is a brief introduction to volcano geodesy, with emphasis on tools and techniques that are discussed more thoroughly in subsequent chapters.

1.4.1 The emergence of volcano geodesy

Geodesy is the science concerned with determining the shape and size of the Earth, and the precise location of points on its surface. Of course, the first issue was decided to most people's satisfaction long ago. Pythagoras (born 582 BCE) first declared Earth to be a globe, and Aristotle (383–322 BCE) also concluded that Earth must be spherical. Eratosthenes (276–195 BCE) often is credited with the first measurement of Earth's size. His estimate for the planetary circumference (250,000 stadia) sparked a long and still unresolved debate over its accuracy, which depends upon the length of the 'stadium,' an ancient, now long-forgotten unit of measure. Estimates for a stadium's length range from 157.2 m to 166.7 m, which correspond to a planetary circumference by Eratosthenes' reckoning of 39,300 km to 41,675 km. This compares remarkably well to the modern value for Earth's equatorial girth (40,075 km) – an impressive feat for its time. With a modern GPS receiver, it is now possible to measure the distance from Earth's center of mass to any point on its surface to within ~1 cm. That is a precision of about 1.6 parts per billion. Not bad for someone like myself who struggled through solid geometry!

I've since moved on to more practical pursuits, such as studying the *changing sizes and shapes of volcanoes* – the subject matter of volcano geodesy. The origins of the trade are obscure, perhaps because its early practitioners were more interested in volcanoes than in documentation (some things never change). Nonetheless, it is crystal clear that a milestone was achieved in 1958, when Japanese seismologist Professor Kiyoo Mogi published his classic paper entitled *Relations between the eruptions of various volcanoes and the deformation of the ground surfaces around them.* Mogi (1958) presented a

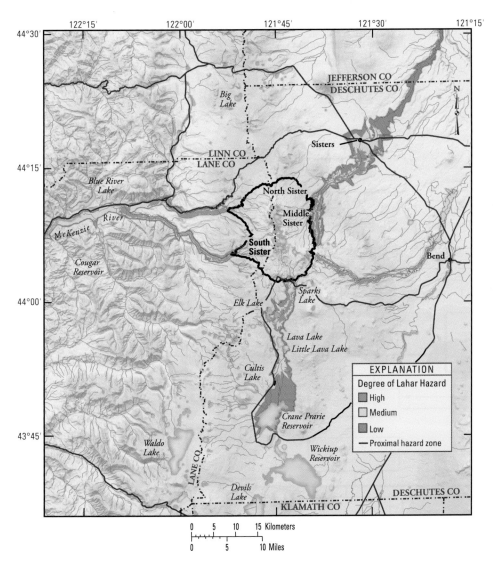

Figure 1.19. Areas near Three Sisters volcanic center, central Oregon, at risk from inundation by debris flows, estimated from an empirical relationship between debris flow volume and runout distance, $d = k \times V^{1/3}$, using a GIS and a DEM of the terrain (Iverson et al., 1998). During a volcano crisis, such maps can be made quickly, without risk to field observers, if an assessment of debris flow hazards based on geologic studies is unavailable or impractical.

mathematical model that now bears his name (the famous 'Mogi model' in the vernacular of volcanology), even though Mogi attributed its derivation to Yamakawa (1955). An interesting historical note is that the equations governing the model had been independently derived at least twice before – by Anderson (1936) who applied them to the formation of cone sheets, ring dikes, and cauldron subsidences, and by McCann and Wilts (1951) who used them to analyze ground subsidence caused by oil extraction in the Long Beach–San Pedro area of California. Nonetheless, Mogi's contribution was revolutionary. He was first to relate surface displacements

to pressure changes in a spherical magma chamber, assuming that: (1) the chamber is very small compared to its depth, and (2) Earth's crust is a uniform elastic medium of semi-infinite extent (Chapter 8). This concept, extended and refined over several decades to include more realistic magma chamber shapes and crustal rheologies, lies near the heart of modern volcano geodesy. Mogi applied his model to two classic volcanic eruptions – the great 1914 eruption of Sakurajima Volcano, Japan, and the 1924 phreatomagmatic eruption of Kīlauea Volcano, Hawai'i – and thus earned his place among the pioneers of volcano geodesy.

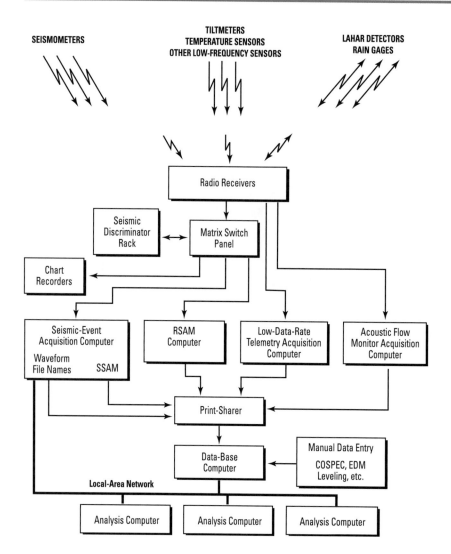

SEISMOMETERS

TILTMETERS
TEMPERATURE SENSORS
OTHER LOW-FREQUENCY SENSORS

LAHAR DETECTORS
RAIN GAGES

Radio Receivers

Seismic
Discriminator
Rack

Matrix Switch
Panel

Chart
Recorders

Seismic-Event
Acquisition Computer

Waveform
File Names · SSAM

RSAM
Computer

Low-Data-Rate
Telemetry Acquisition
Computer

Acoustic Flow
Monitor Acquisition
Computer

Print-Sharer

Data-Base
Computer

Manual Data Entry
COSPEC, EDM
Leveling, etc.

Local-Area Network

Analysis Computer · Analysis Computer · Analysis Computer

Figure 1.20. Block diagram of a mobile volcano-monitoring system (Murray et al., 1996a). A low-power, frequency modulated (FM) analog-telemetry system is used to collect data from various sensors in the field. Received signals are routed via acquisition computers to a database computer, where analysis computers can access the complete dataset. A local area network (LAN) facilitates user access to data in near-real time.

The world's first volcano observatory had been in existence for more than a century when, in 1958, Mogi's classic paper altered the course of volcano geodesy. In 1847, the Osservatorio Vesuviano (Vesuvius Observatory) was established on the western slope of Vesuvius near Naples, Italy, marking an important milestone in the history of volcanology. Vesuvius is notorious for its 79 CE eruption that destroyed the Roman cities of Pompeii and Herculaneum – the eruption described by Pliny the Younger, whose account is excerpted in the Preface to this book. Among the observatory's illustrious past directors is Giuseppe Mercalli (director from 1911–1914), who devised the Mercalli Intensity Scale for earthquakes.[7] Today, the Vesuvius Observatory is among the world's leading volcanological institutions. In 1970, the original building was converted to a museum, exhibit hall, and library, and a new building was constructed that houses the current observatory.

Starting in the early part of the twentieth century, volcanologists in Japan and the U.S.A. adapted techniques from surveying and seismology to the emerging field of volcano geodesy. They used transits and levels to repeatedly measure the changing shapes of geodetic networks on volcanoes, and pendulum seismographs to track ground movements more closely as a function of time. For example, the US Coast and Geodetic Survey (C&GS), precursor to the modern National Geodetic Survey

---

[7] The Mercalli Intensity Scale is used to classify the intensity of an earthquake by examining its effects on people and structures. It was conceived by Italian volcanologist Giuseppe Mercalli in 1902, and was in general use before the Richter scale was developed by Charles Francis Richter and Beno Gutenberg in 1935. The form currently used in the U.S.A. is the Modified Mercalli (MM) Intensity Scale, which was developed in 1931 by American seismologists Harry Wood and Frank Neumann.

(NGS), did the first leveling survey on the Big Island of Hawaii in 1912. The traverse extended from a tide gage in the coastal city of Hilo to Volcano House near the summit of Kīlauea. Subsequent surveys confirmed the suspicions of Dr. Thomas A. Jaggar, who founded the world's second volcano observatory – now the USGS Hawaiian Volcano Observatory (HVO) – at Kīlauea in 1912, that the summit areas of active volcanoes rise and fall in response to changes in subsurface magma reservoirs. In 1926–1927, the C&GS completed the first leveling and gravity surveys from Hilo to the summit of 4,170 m (13,680 ft) Mauna Loa Volcano – an arduous task under difficult conditions (Figure 1.21). The Big Island leveling network has since grown to nearly 400 km and leveling surveys continue to play an important role in HVO's geodetic monitoring program.

Jaggar recognized the importance of measuring ground tilt at volcanoes and immediately established a tilt-monitoring program at HVO (Jaggar and Finch, 1929). Seismologist Harry O. Wood first noticed the effects of ground tilt on the scribing pens of HVO's Bosch-Omori pendulum seismographs. Tilt computations based on the seismic records were made almost daily at HVO from 1913 to 1963. In 1932, three clinoscopes were installed in Kīlauea's summit area (Apple, 1987). The clinoscope was essentially a large plumb bob suspended from a tripod and dipped into an oil bath for damping. These relatively crude precursors to modern tiltmeters amplified tilt signals by a factor of ~50, but frequent earthquakes and temperature fluctuations reduced their effectiveness. In 1956, the watertube tiltmeter (Chapter 3) was developed at HVO using calibrated water pots built from spent artillery shells. The first electronic tiltmeter was introduced to HVO in 1965. It detects changes in capacitance caused by changes in the airspace gap between a plate and the surface of a pool of mercury. The Ideal–Aerosmith mercury capacitance tiltmeter has a sensitivity of ~0.1 μrad and still is in use at HVO today.

The first line-length measurements in Hawai'i were made at Kīlauea between October 1964 and March 1965 by Dr. Robert W. Decker, then at Dartmouth University, using a microwave Tellurometer and a tungsten bulb Geodimeter on loan from the US Army's Cold Regions Laboratory in Hanover, New Hampshire (Decker et al., 1966). The instruments were used to measure strain on the Arctic ice pack during the summer and were available to Decker for use in Hawaii during the winter.

The main line across Kīlauea Caldera from HVO to Keanakāko'i Crater was only about 3 km long, but it had to be measured at night with the Geodimeter because of its weak light source! The first Geodimeter measurements across Mauna Loa's summit caldera, Moku'āweoweo, also were made at night in 1965 (Decker and Wright, 1968). The HVO trilateration network has since grown to more than 750 lines, which are being measured increasingly with GPS instead of EDM. Trilateration data have been key to unraveling complex surface motions at Kīlauea and Mauna Loa, including rapid seaward movement of Kīlauea's south flank (Swanson et al., 1976; Owen et al., 1995).

1.4.2 Continuous sensors and repeat surveys

Volcano geodesy is a craftsman's trade. Its tools range from a simple tape measure to the sublime mechanical precision of a modern gravimeter, strainmeter, or tiltmeter, and to the exotic apparatus of remote sensing. Equally important are laboratory devices for determining the deformational behavior of rocks, and sophisticated modeling techniques for relating surface observations to the inner workings of volcanoes.

This section provides a brief overview of the geodetic sensors and techniques commonly used to monitor ground deformation at volcanoes. Subsequent chapters deal with these topics in more detail. Geodetic techniques can be broadly classified as either continuous or discontinuous. For our purposes, continuous measurements include those that are literally made continuously (e.g., the analog output of an electronic tiltmeter), made continuously but sampled at short, discrete intervals (GPS phase measurements typically made at 15–30 s intervals), or made often enough to provide useful information during a crisis (hourly or daily EDM measurements).

Measurements made discontinuously every few days, months, or years fall into the category of repeat surveys. Leveling, GPS, and EDM surveys are good examples. This type of measurement can be very useful during periods of relative quiescence to document the magnitude and extent of changes that are monitored continuously at a few sites. At Long Valley Caldera, California, for example, where ground deformation is monitored continuously with tiltmeters, borehole strainmeters, and GPS stations, leveling and GPS surveys are repeated every few years to determine the extent and shape of the deformation field. These surveys place continuous

Figure 1.21. (*top*) US C&GS leveling crew beginning a spur line into Moku'āweoweo Crater near the summit of Mauna Loa Volcano, Hawai'i, in 1927. (*bottom*) The last set-up on the summit of Mauna Loa (elevation 4,169.7 m). The survey began two months earlier at a tide gauge in Hilo. Skip Theberge, NOAA Central Library, Silver Spring, Maryland kindly provided these images from Historic C&GS Catalog of Images, C&GS Season's Report Simmons 1927.

measurements in a broader context and provide important constraints for deformation source models. In recent years, radar interferograms spanning one or more years (from summer to summer, to avoid problems caused by snow and ice) have served the same purpose with much better spatial resolution (Chapter 5).

1.4.3   Tiltmeters, strainmeters, and continuous GPS

The most commonly used instruments for continuous monitoring of volcano deformation are tiltmeters, strainmeters, and GPS receivers. Various designs are available, each with advantages and disadvantages in terms of cost, complexity, reliability, power requirements, sensitivity, and ease of use. Simple bubble tiltmeters installed within a few meters of the surface are capable of recording ground tilts caused by diurnal Earth tides, on the order of one part in ten million (10^{-7} radian or 0.1 μrad). Sophisticated borehole and long-base tiltmeters can measure tilt changes as small as one part per billion (10^{-9} radian or 10^{-3} μrad), although greatly improved precision comes at the price of more elaborate and expensive installations (Bilham *et al.*, 1981).

Often the choice of monitoring instruments is dictated by practical considerations. In situations where volcanic unrest can be expected to continue for years or decades and the stakes are especially high (i.e., people or property at risk), expensive instruments that offer superior precision are usually a wise choice. In such cases, permanent GPS stations and borehole strainmeters can provide very precise information on subsurface strain and surface displacement in near-real time. On the other hand, during a crisis response at a long-dormant but restless volcano, or in cases where monitoring instruments are likely to be destroyed by volcanic activity, less expensive instruments with lower but still adequate precision are sometimes a better alternative. Another factor that influences the choice of monitoring instruments is the availability of scientific, economic, and infrastructure resources. These vary considerably, especially between developed and developing countries. All of these issues must be considered when selecting instrumentation and developing a monitoring strategy tailored to a particular situation.

The greatest advantage of tiltmeters in a crisis is that their output is continuous and directly proportional to ground tilt. This means that tiltmeter data can be telemetered and analyzed easily in near-real time, without specialized processing like that required for GPS or interferometric SAR (InSAR) data. Additional advantages of bubble tiltmeters are low cost (a few thousand dollars) and ease of installation (typically 1–2 days). Disadvantages of most designs include long-term drift, susceptibility to environmental effects such as temperature and precipitation, and uncertainty over how representative the tiltmeter site may be of the surrounding area. The latter two problems are shared by most types of continuous sensors, but they can be overcome by: (1) careful site selection and station design, (2) redundancy, and (3) comparing results from continuous sensors with those from repeated geodetic surveys that cover a wider area encompassing the continuous-sensor sites. For a more complete discussion of the advantages and disadvantages of short-base tiltmeters, see Dzurisin (1992a).

Strainmeters are becoming more common near volcanoes and they, too, come in a wide variety of designs. Perhaps the simplest approach is to attach a displacement transducer to a wire stretched across a crack or a zone of localized deformation (Figure 1.22). Such an arrangement can be very useful where intense deformation is occurring, but without substantial effort to protect and insulate the instrument from the environment its effective precision is limited to about ±1 mm (Iwatsubo et al.,

1992b). Another simple but much more precise instrument is a volumetric strainmeter. These devices are fluid-filled cylinders with a very thin tube (manometer) that allows fluid to rise or fall in response to squeezing or stretching of the cylinder (i.e., compressive or dilatational strain) (Chapters 3 and 9). Installed in a borehole 30 m or more deep and rigidly coupled to the borehole wall, a volumetric strainmeter can measure strain changes as small as 10^{-12} over an extremely wide range of frequencies from less than 1 Hz (i.e., periods of seconds to minutes) to about 100 Hz (Linde et al., 1993, 1996). Even over periods of a few months, these hypersensitive instruments are capable of resolving strain changes on the order of 5×10^{-8} (Chapter 9). Disadvantages include relatively high cost of the sensor (>$10,000) and of drilling operations, difficulty drilling in remote areas, and the time required for station installation and equilibration (several months or more).

During the 1990s, continuous GPS (CGPS) stations came into common use in the Earth sciences, especially for measuring strain related to tectonic processes. Networks consisting of hundreds of CGPS stations have been established in Japan and southern California, and smaller networks are being used to measure strain accumulation across ever-shifting plate boundaries or within active tectonic provinces. At the same time, CGPS is being used to monitor ground movements at numerous volcanic systems, including Kīlauea, Hawai'i (Owen et al., 2000a,b; Larson et al., 2001); Augustine, Alaska (Pauk et al., 2001); Long Valley, California (Dixon et al., 1997; Endo and Iwatsubo, 2000); Mount St. Helens, Washington (Dzurisin, 2003); Yellowstone, Wyoming (Meertens et al., 2000); Mount Etna, Italy (Ferrucci et al., 1994); Popocatépetl, Mexico (Cabral-Cano et al., 1996); Rabaul, Papua New Guinea (Endo, 2005); and Taal, Philippines (Lowry et al., 2001). Starting in 2003, 875 CGPS stations are being installed in western North America as part of the Plate Boundary Observatory (PBO), including clusters of 5–20+ stations at selected volcanoes in the Aleutian Arc and Cascade Range, and at the Long Valley and Yellowstone Calderas (Section 4.10.3).

GPS is a very flexible tool that can be used in several ways to study deforming volcanoes. Generally, an independent solution for the 3-D position of a GPS station is calculated once per day or, at most, once per hour. The precision of such positions is typically a few mm horizontally and ~10 mm vertically. Such solutions fit our definition of

Figure 1.22. The author inspects a crack meter on the crater floor at Mount St. Helens. A wire is stretched across the crack inside a pipe and attached to anchor points on either side. Changes in the length of wire played out between the anchor points are sensed by a linear displacement transducer and converted to a digital signal inside the box. A radiotelemetry system (not shown) transmits the signal to the USGS David A. Johnston Cascades Volcano Observatory in Vancouver, Washington. A dacite lava dome that grew in the crater from 1980 to 1986 steams in the background. Photograph reprinted with permission from photographer Steve Raymer and the National Geographic Society.

'continuous' measurements and can be very useful during a volcano crisis. By employing predictive filtering techniques in firmware[8] within the receiver, it is also possible to calculate a 3-D solution for each measurement of GPS carrier phase made by the receiver (typically every 15–30 s), and to do so in the field without human intervention. This approach, sometimes called real time kinematic (RTK) processing, promises to extend the usefulness of GPS during rapidly evolving crises, when human resources are usually at a premium. In situations where the risk to field personnel is acceptable, another approach called rapid static surveying can provide data from many more sites than is possible with CGPS stations. For a rapid static survey, a GPS receiver can be set up at each bench mark in an array for only a few minutes to obtain cm-scale accuracy (Chapter 4). Which mode of GPS monitoring is best

suited to a particular situation depends on many factors, including the size of the deformation signal, the deformation rate, available resources (equipment and personnel), and logistics (e.g., accessibility of field stations, electrical power, and telemetry options).

1.4.4   Repeated surveys – leveling, EDM, and GPS

The most common types of geodetic surveys made at volcanoes are leveling, EDM or total station, and GPS. Leveling and GPS surveys usually involve at least 4–6 people, while EDM or total station surveys typically involve about half as many. These relatively high demands on human resources and the unavoidable exposure of field crews to hazardous conditions near volcanoes limit the utility of such surveys during a crisis. On the other hand, they are invaluable for mapping the pattern of deformation before and following a crisis, and for providing reliable measurements of ground movements over periods of several months to years, during which instrumental drift poses a problem for most contin-

---

[8] The term 'firmware' refers to programming instructions that are stored in read-only memory chips within a computing device, in this case a GPS receiver, for the purpose of controlling the device or adding functionality. Read-only memory chips, including ROM, PROM, EPROM and EEPROM technologies, hold their content without electrical power. Firmware can be thought of as 'hard software' that is retained until it is modified by the user.

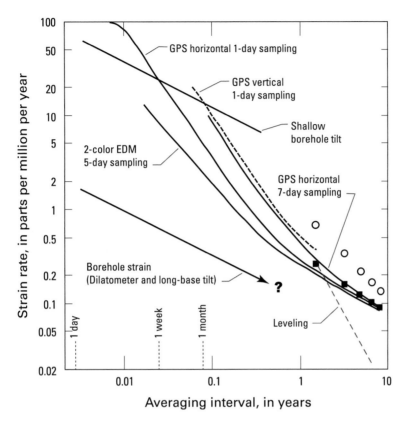

Figure 1.23. Strain rate detection threshold for various deformation-monitoring techniques as a function of the interval over which re-peated measurements are averaged (J. Langbein, USGS, written commun., 1998). An 8-km baseline was assumed for GPS and EDM. Open cir-cles and solid squares represent annual GPS measurements and annual two-color EDM measurements, respectively. Most geodetic measurements are subject to both random-walk noise (e.g., monument drift) and white noise (e.g., atmospheric effects, instrument noise). Using a high sampling rate and averaging a large number of samples can sometimes mitigate the effect of white noise; in such cases, random-walk noise becomes a limiting factor. Borehole strainmeters and long-base tiltmeters provide the most accurate information over timescales of minutes to days, while repeated leveling or GPS surveys are best over periods of months to years.

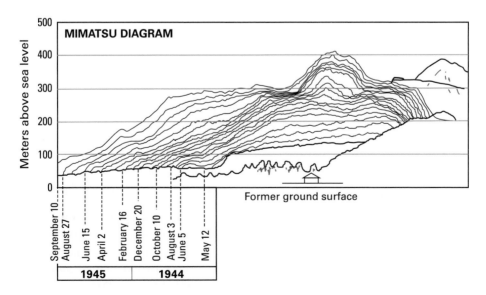

Figure 1.24. Detailed sketch showing the growing profile of the Showa-Shinzan lava dome as viewed from the same vantage point 2.5 km east of the vent. After Mimatsu (1962).

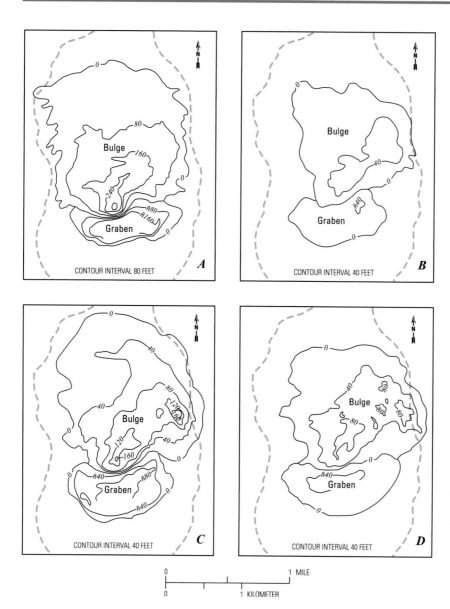

Figure 1.25. Elevation change near the summit of Mount St. Helens (in feet) between sequential topographic maps (Moore and Albee, 1981). (A) 15 August 1979 to 7 April 1980. (B) 7 April to 12 April. (C) 12 April to 1 May. (D) 1 May to 12 May. Contour interval is 80 ft (24.4 m) in A and 40 ft (12.2 m) in B, C, and D. Dashed line shows edge of crater that formed when bulging area collapsed and slid northward on 18 May 1980.

uous sensors (Figure 1.23). Surveying techniques are covered in detail in Chapters 2 and 4. Chapter 7 gives some examples of how repeated surveys can be combined with continuous monitoring to study patterns of volcano deformation in space and time.

1.4.5   Photography, photogrammetry, and water-level gauging

Two additional techniques for measuring ground movements at volcanoes deserve mention: repeated photography or photogrammetry (Chapter 6) and differential or absolute water-level gauging (Chapter 9). Where deformation is extreme, it can be measured by comparing repeated photographs or

sketches made from the same vantage point or by using photogrammetry to produce two or more topographic maps of the deforming area. Sequential sketches of the Showa-Shinzan dome that grew at Usu Volcano, Japan, during 1944–1945 are a classic of the first genre (Mimatsu, 1962) (Figure 1.24). Perhaps the best-known example of using photogrammetry to measure volcano deformation occurred at Mount St. Helens in 1980, where a growing cryptodome produced a bulge on the volcano's north flank with displacements of more than 100 m over a period of two months (Moore and Albee, 1981; Figure 1.25).

For volcanoes bordering the sea, tide gauges can be a very effective monitoring tool for vertical ground movements, as evidenced by the preceding accounts from the Phlegraean Fields and Rabaul Calderas. Differential water-level monitoring near volcanoes also can be useful, even in cases where ground movements are much less dramatic. For example, if a suitable body of water is available, it can be used as a natural tiltmeter by measuring lake stage at two or more points and correcting any changes in the difference for the effects of wave and wind action, seiche, and differential water temperature and atmospheric pressure. By placing pressure transducers in a lake that are sensitive to the height (weight) of the water column above them, this technique can be automated to eliminate the need for human observers in the field and to produce data that are easily telemetered and analyzed. This approach has been put to good use at Lake Taupo in New Zealand (Otway, 1989; Kleinman and Otway, 1992). For the case of summit crater lakes, absolute lake gauging is sometimes useful to monitor the displacement of lake water caused by deformation of the lake bottom or by extrusion of magma onto the lake floor. This approach has been useful at Mount Ruapehu in New Zealand and Kelut Volcano in Indonesia. Of course, water-level gauging is only possible where a suitable body of water is available, which limits the technique's applicability to a fraction of the world's volcanoes.

CHAPTER 2

Classical surveying techniques
Daniel Dzurisin

Modern technological advances notwithstanding, volcano geodesy rests on a solid foundation of time-tested and proven techniques inherited from the exacting and relatively staid discipline of classical geodesy. From time immemorial, wanderers on land and sea have needed to know where they were and where they were going. The early tools they developed to locate and navigate themselves evolved into the surveyor's chain, transit, and various derivatives that still have a place in the volcanologist's tool kit. A noteworthy difference is that, whereas traditional geodesists generally concern themselves with fixing the locations of stable points on the ground, volcano geodesists are energized by where and how fast their points are moving!

2.1 EARLY GEODETIC SURVEYS

The first notable geodetic survey occurred during the latter part of the 17th and early 18th centuries, and stirred a decades-long international controversy (Dracup, 1995). Frenchman Jean Picard began a triangulation survey near Paris in 1669 and worked southward until his death about 1683. Beginning in 1700 and continuing for more than two decades, Jean-Dominique Cassini[1] extended Picard's survey southward to the Pyrenees on the Spanish border and northward to Dunkirk on the English Channel. The results seemed to indicate that the Earth was prolate (i.e., that its polar radius is greater than its equatorial radius). In 1687, Britain's Sir Isaac Newton, on theoretical grounds, had reached the opposite conclusion (i.e., that the spinning Earth is oblate (flattened at the poles)).

To resolve the ensuing controversy, the French Academy of Sciences staged two expeditions, one to Peru (now Ecuador) at the equator, and the other to Lapland on the Swedish–Finnish border in the Arctic region. The objective was to compare the north–south curvature of the Earth at each location's latitude, and thus to determine the planet's shape. The observations, which took nearly a decade to complete (1736–1738 in Lapland and 1735–1744 in Peru), came down firmly on the side of Newton (i.e., Earth is an oblate ellipsoid). In its debut on the international stage, geodetic surveying had both created and resolved an international controversy – an auspicious start for a discipline that seldom grabs today's headlines.

At the start of the 19th century, most countries in Europe were either planning, or on their way to establishing, national triangulation networks. By 1842, network coverage extended from the Mediterranean Sea on the south to the Arctic on the north, and from Ireland, England, and the Atlantic Ocean on the west to the interior of Russia on the east. The first triangulation survey in the U.S.A. was carried out near New York City in 1816–1817 by the fledgling Survey of the Coast, currently known as the National Geodetic Survey (NGS).

Triangulation was the mainstay of geodetic surveying in the U.S.A. and elsewhere throughout the 19th and early 20th centuries. In the U.S.A., it was replaced during the period 1917–1927 by first-order traverse[2] (Dracup, 1995). When the

---

[1] Jean-Dominique Cassini's given name was Giovanni Domenico, but he also used Gian Domenico and, later, Jean-Dominique. He was the first of the famous Cassini family of astronomers, and as such he is also known as Cassini I.

[2] Traverse is a method of surveying in which lengths and directions of lines between points on the Earth are obtained by or from field measurements, and used in determining positions of the points. A first-order traverse is one that extends between (*cont.*)

electronic distance meter (EDM) was introduced after World War II, trilateration became the preferred technique for establishing horizontal control. Today, most geodetic surveys make use of the satellite-based Global Positioning System (GPS) (Chapter 4).

The first level datum for precise vertical control was established in the Netherlands in 1682, little more than a decade after Jean Picard began his renowned triangulation survey near Paris. Great Britain began a national geodetic leveling campaign in 1841 and completed it about 20 years later. In the U.S.A., adoption of the National Geodetic Vertical Datum of 1929 (see Section 2.2) was a major milestone, and first-order triangulation and leveling were basically complete by the start of World War II (Dracup, 1995).

Development and widespread application of triangulation, leveling, and trilateration, in that order, took nearly three centuries from the time of Jean Picard. In three short decades near the end of the 20th century, GPS and interferometric synthetic-apperture radar (InSAR) would revolutionize geodesy in ways that Picard and his comtemporaries could never have imagined. Those stories are told in Chapters 4 and 5, respectively. First, though, we need to discuss the concepts of reference systems, datums, and geodetic networks, which apply generally to all geodetic measurements. Then we'll discuss several classical surveying techniques and their applications to volcano monitoring, before moving on to modern techniques in subsequent chapters.

2.2 REFERENCE SYSTEMS AND DATUMS

Frankly, this isn't the exciting part. Discussions of reference systems and datums are better left to *real* geodesists, who can be downright passionate about such things as deflection of the vertical and the ellipsoid–geoid separation. For me, such issues pale in comparison to the mystery of a deforming volcano, which is spellbinding in *any* reference system! Nonetheless, a few words on this topic are necessary to set the stage for what follows. Additional information on the significance of the refer-

ence ellipsoid and geoid in volcano geodesy is provided in Chapter 4.

Bomford (1980, p. 93) introduces the topic of reference systems with the following levelheaded observation: '*To compute and record the positions of points on or above the Earth's surface, some coordinate system is necessary. Such a reference system may take many forms . . .*' An obvious extension of the same idea is that, in order to measure any *change* in the relative positions of points on the ground (i.e., deformation), we need to specify a reference system (also called a reference frame). Knowing that point A moved 3 cm is not very useful unless we also know the direction of movement and the reference frame. For example, knowing that point A moved 3 cm up with respect to point B, which moved 5 cm up with respect to a local tide gauge, is useful information about the motion of point A. By the way, you probably have never wondered what 'up' means exactly, but real geodesists worry about such things all the time (see below).

What should we choose for the origin of our geodetic reference frame, and what coordinate system should we use to specify the locations of points on the Earth's surface? Bomford (1980) lists several possibilities, including Earth's center of mass and the coordinate set (latitude, longitude, height), respectively. The first choice was problematic before the arrival of satellites and especially GPS, because it was difficult to measure the distance to Earth's center of mass from points on the surface. Instead, geodesists took advantage of the fact that most of our planet is covered with a fluid, which assumes the shape of an equipotential surface under the influence of gravity (ignoring the ephemeral effects of tides, storms, etc.). So '100 m above mean sea level' means the same thing in Greenwich as it does in Greenland, and we can use mean sea level as a global reference for height measurements. Mean sea level closely approximates a widely used reference surface called the geoid, which is everywhere perpendicular to the direction of the local gravity vector. In general, the geoid differs from the mean shape of the Earth (i.e., the reference ellipsoid, see below) as a result of density variations that affect the local gravity vector (Figure 2.1).

In the U.S.A., the NGS is responsible for establishing and updating national control grids and reference frames. NGS has produced two official versions of the geoid: the National Geodetic Vertical Datum of 1929 (NGVD 29) and the current North American Vertical Datum of 1988 (NAVD 88). As a

---

(*cont.*) adjusted positions of other first-order control surveys and conforms to the current specifications of first-order traverse. For surveys in the U.S.A., see Federal Geodetic Control Committee (1984); similar documents exist to guide geodetic surveys in other countries (e.g., New Zealand, Bevin (2003); Australia, ICSM (2000); Canada, Surveys and Mapping Branch (1978)).

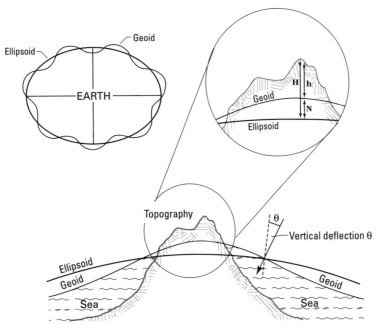

Figure 2.1. Relationship between the ellipsoid and geoid. The geoid is an equipotential surface that is everywhere perpendicular to the direction of the local gravity vector. It closely approximates mean sea level. The ellipsoid is a mathematical figure that is the best-fit surface to the mean shape of the solid Earth. Unlike the geoid, the ellipsoid is not affected by the inhomogeneous mass distribution inside the Earth. The ellipsoid is widely used as the reference surface for horizontal coordinates (e.g., latitude and longitude) and for the GPS. Leveling observations, on the other hand, by their nature are referenced to the geoid. The difference between the ellipsoid and geoid, which is called the geoid undulation or ellipsoid–geoid separation, accounts for the difference between heights determined by GPS and leveling (Seeber, 1993). The angular difference between normals to the geoid and ellipsoid at any point is called the vertical deflection.

result of improved control for NAVD 88, these two surfaces differ from one another by as much as −40 cm to +150 cm in the conterminous U.S.A. and +94 cm to +240 cm in Alaska. In most cases, the relative difference between adjacent bench marks is small (generally <1 cm), so for small networks a single bias factor describing the local mean difference between NGVD 29 and NAVD 88 can be used to convert elevations from one reference surface to the other (Zilkoski *et al.*, 1992).

But the solid Earth isn't a fluid, so its shape might differ from that of the geoid. It does, in fact, and not only because of topography. If we were to smooth out such roughness elements as the Himalaya (including Mount Everest, Earth's highest peak), the Marianas Trench (deepest part of any ocean), and all of Earth's volcanoes (perish the thought!), the resulting figure would not match that of a hypothetical global ocean surface. Instead, density inhomogeneities within the solid Earth and the flattening effect of Earth's rotation would cause the two surfaces to differ by an amount that varies with location. A commonly used best-fit approximation to the shape of a smooth Earth is an ellipsoid of revolution called the reference ellipsoid.

There is a very practical reason for adopting the

reference ellipsoid rather than the geoid as a reference surface for most mapping and geodetic applications. Geodetic instruments that rely on gravity to establish the vertical, such as spirit levels and theodolites, give results that are implicitly referenced to the geoid. This can pose a problem, because the irregular shape of the geoid is reflected in any reference system that might be derived from it. For example, meridians or parallels based on astronomical observations, which rely on the geoid, would not be equally spaced. Two such 'astronomical parallels' that were 10.0 km apart at one longitude might be 9.9 km apart at another (Bomford, 1980). To avoid such complexities, cartographers and geodesists long ago adopted the concept of a reference ellipsoid, which is a best-fit surface to the mean shape of solid Earth. The reference ellipsoid approximates the shape of the geoid, but is not affected by the inhomogeneous mass distribution inside the Earth. Eight independent constants are necessary to define the reference ellipsoid: three to specify the location of the origin of the coordinate system, three to specify the orientation of the coordinate system, and two to specify the dimensions of the ellipsoid (Figure 2.2). The most commonly used set of constants includes the coordinates of the center of the ellipsoid, which is defined to be the center of the Earth $(X, Y, Z) \equiv (0, 0, 0)$; the lengths of the semimajor and semiminor axes $a \approx 6378\,\text{km}$ and $b \approx 6357\,\text{km}$ (or equivalently the length of the semimajor axis and the flattening $f = \dfrac{(a - b)}{a} \approx \dfrac{1}{298.3}$); the minor axis direction, which is taken to be parallel to Earth's mean polar axis and is specified by two constants (e.g., the coordinates (x, y) where the minor axis intersects the ellipsoid near the pole); and the zero of longitude ($\varphi = 0$ at the Greenwich meridian, by convention).

Until recently, reference ellipsoids were fit to the Earth's shape only over a particular country or continent. As a result, they differed from one another by an amount that reflected different solutions for the Earth's center location and flattening. Such differences meant that coordinates from different regions were not directly comparable. In recent decades, satellite determined coordinate systems have resulted in internationally accepted geocentric ellipsoids. The center of a geocentric ellipsoid is the same as the geometric center of the Earth, so there is no offset between coordinates referenced to the same geocentric ellipsoid. Alternatively, coordinates in different reference frames can be compared by specifying the datum for each coordinate set. A datum is a set of constants used to specify the coordinate system used for geodetic control, including the reference ellipsoid. In the U.S.A., commonly used datums include the North American Datum 1927 (NAD 27) and the North American Datum 1983 (NAD 83). Standard reference ellipsoids, which are used for almost all GPS applications, include the Geodetic Reference System 1980 (GRS 80) and the World Geodetic System 1984 (WGS 84). Tabulations of the relationships among these and other reference ellipsoids and datums, and utilities to convert from one to another, are available from the NGS and other sources.

The difference between the geoid and reference ellipsoid is called the ellipsoid–geoid separation. An elevation measured with respect to the geoid (e.g., by leveling) is called an orthometric height H, and an elevation determined with respect to the reference ellipsoid (e.g., by GPS) is called an ellipsoidal height h (Figure 2.1). The difference is the ellipsoid–geoid separation N:

$$H = h - N \qquad (2.1)$$

The value of N ranges from $-100\,\text{m}$ in Sri Lanka to $+70\,\text{m}$ in the Marianas Trench.

By the way, in case you've been wondering what 'up' really means, consider Figure 2.1. Intuitively, 'down' refers to the direction of the local gravity vector, which we can ascertain with a pendulum, and 'up' is simply the opposite of 'down' (even for geodesists). But isn't 'up' equally well defined as being perpendicular to the surface (i.e., the reference ellipsoid) at any point? The answer is no, because the geoid is lumpy with respect to the reference ellipsoid, so the normal vectors to those two surfaces generally do not coincide. The angle between the vertical at any point on the Earth's surface (i.e., local 'up') and the normal to the reference ellipsoid is called the vertical deflection or deviation of the vertical. Vertical deflections of over 70 seconds of arc have been measured near Mount Everest (Bomford, 1980, p. 466). Now that we're clear about what's 'up', let's move on to less esoteric topics enroute to a discussion of deforming volcanoes (the exciting part).

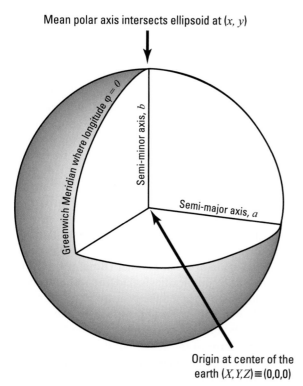

Mean polar axis intersects ellipsoid at (x, y)

Greenwich Meridian where longitude $\varphi = 0$

Semi-minor axis, b

Semi-major axis, a

Origin at center of the earth $(X, Y, Z) \equiv (0, 0, 0)$

Figure 2.2. The reference ellipsoid is specified by eight constants: its origin at the center of the Earth, $(X, Y, Z) \equiv (0, 0, 0)$; the lengths of the semimajor and semiminor axes a and b (or the length of the semimajor axis and the flattening $f = ((a - b)/a) \approx (1/298.3)$; the minor axis direction, taken to be parallel to the Earth's mean polar axis, which intersects the ellipsoid at (x, y); and the zero of longitude $\varphi = 0$ at the Greenwich meridian.

2.3 GEODETIC NETWORKS

Deformation can be broadly defined as a change in the shape or dimensions of a body resulting from stress. It follows that, to monitor deformation, we must determine and keep track of a body's changing size and shape as a function of time. The concept is clear if we consider a laboratory rock sample of known geometry subjected to compressional, dilatational, or shear stress, or layered rocks in the field that have been folded or faulted in obvious ways. In the first case, we have the luxury of creating a sample with known size and shape and then measuring changes as a function of applied stress. In the field, we can infer the original geometry of an outcrop (e.g., flat-lying sediments) and devise various schemes to determine the amount of shortening, offset, and so forth.

The situation is considerably more complicated if the deforming body is a restless volcano. How can we determine the size and shape of an entire volcano and then repeat the measurement with sufficient accuracy to monitor changes as small as a few millimeters? Even the most detailed topographic maps seldom are accurate to 1 m, and the logistics of producing a series of such maps during a crisis are prohibitive. Synthetic-aperture radar (SAR) interferometry can in some cases produce 'snapshots' of mm-scale surface displacements (Chapter 5), but for now the repeat times for satellite-based SAR observations are measured in months – too long to be useful during most volcano crises. In some cases, extreme deformation causes part of a volcano to move many meters and we are faced with the challenge of tracking such large movements without compromising our ability to detect subtle deformation elsewhere.

To make the task of monitoring volcano deformation more manageable, we can approximate the complex shape of a volcano with a network of recoverable points on its surface, and then monitor the geometry of the network rather than the entire volcano. The concept is akin to draping a flexible, irregular grid over the volcano and measuring the locations of the grid points as a function of time. We assume that the network represents the volcano to the extent that any change in the volcano's size or shape is reflected in a corresponding change in the spatial relationship among points in our network. Clearly, this assumption is more likely to be valid for dense networks than for sparse ones. In practice, it is usually possible to derive useful information from a manageable number of survey points, although more is generally better.

The number of points in a network usually is determined by practical considerations, including accessibility and the time required to survey the network relative to the desired survey frequency. The points must be recoverable because the spatial relationships among them define our simplified volcano geometry, and thus our ability to measure deformation. The effect of poorly defined points is to blur spatial relationships and thereby reduce the accuracy of the survey. Henceforth, I'll refer to recoverable points as bench marks (sometimes written 'benchmarks'), marks, or stations, regardless of their specific design. The NGS defines the term 'bench mark' as a marked point whose height above or below an adopted datum is known (National Geodetic Survey, 1986, p. 24), but the term often is applied more widely to any reference point used for surveying purposes (American Geological Institute, 1987, p. 65). Among the more elegant markers used to establish bench marks are metal discs stamped with an agency name and an unmistakable '+', '×', or '.' to precisely define the survey point (Figure 2.3A). Hardened nails driven into rock or concrete, short stainless steel rods secured in a drill hole with mortar or epoxy, or longer rods driven to refusal into soil can serve the same purpose. If necessary, a center punch can be used to stamp a specific, recoverable point on the mark. Where safety considerations or difficult access prevent a leisurely installation, even a pointed rock or a spot of paint can be useful as a mark where deformation is extreme, so long as it is recoverable and stable with respect to its immediate surroundings.

Bedrock is generally the preferred choice for mark setting, but a partly buried boulder, an engineered structure such as a bridge headwall or building foundation, a specially cast concrete monument, or a rod driven to refusal in well-drained soil can be equally effective (Figure 2.3B). The main goal is long-term stability, which requires an installation that is relatively free from near-surface environmental effects (e.g., freeze–thaw deterioration, soil creep), and also from disturbance or burial (e.g., by road construction or burrowing animals). A thorough site description, including coordinates, access directions, and recovery notes, is essential so the mark can be located for future surveys. Datasheets with such information for all survey control stations maintained by the NGS are available online at the NGS website (*http://www.ngs.noaa.gov/*).

Figure 2.3. Examples of survey marks used to establish geodetic control points (*top photograph*). The most common type is a circular metal disk ~10 cm in diameter with a stem that can be secured in a drill hole with mortar or epoxy, or attached to the top of a metal rod driven to refusal into soil (hollow stem). Most survey marks include an 'x', '+', or '.', usually within a triangle or square near the center of the mark, which defines the specific point to be surveyed. For leveling, by convention, the survey point is the highest point on the mark. To avoid confusion, some marks used for leveling include a central nub to accommodate a leveling rod. Other types of survey marks include hardened nails and short stainless steel rods that can be driven into rock or concrete or secured in a drill hole. Yet another type, not shown here, is a long metal rod that can be driven to refusal into soil. Sections of rod ~2 m long are driven with a sledge or reciprocating hammer and coupled together sequentially until the coupled rods can be driven no farther. The top of the rod serves as the survey point for leveling, or is marked with a punch for horizontal surveys. USGS photograph by David E. Wieprecht. (*bottom*) Survey marks can be secured to bedrock (a), a partly buried boulder, a specially cast concrete monument (b), an engineered structure such as a bridge wing wall (c), or a rod driven to refusal. Most marks are stamped with the agency name, a unique identifier that often includes the year the point was established, and in some cases the elevation of the mark when first surveyed (d). USGS photographs a–d by Daniel Dzurisin and Michael P. Poland.

Once a network is established, the next task is to determine the spatial relationship among the marks as a basis for recognizing any future changes. Ideally, we would like to measure the 3-D position of each mark with respect to some invariant reference frame. This can be accomplished by making GPS observations or a combination of slope-distance and vertical-angle measurements. Where 3-D control is impractical or unnecessary (e.g., monitoring crack widening or lateral spreading of a lava flow), slope-distance measurements alone can be very useful for tracking ground movements.

2.4 TRILATERATION AND TRIANGULATION

Among the most time honored of geodetic techniques are trilateration and triangulation. Early geodetic control networks were established by triangulation starting in the latter part of the 17th century by hardy souls who climbed countless peaks and crossed vast expanses with theodolites, tripods, and other surveying equipment in tow. In the U.S.A., the adventure began in 1807 with the creation by Congress of the Survey of the Coast (predecessor to the current NGS, following several name changes) during the presidency of Thomas Jefferson. For a fascinating account of early geodetic surveys in the U.S.A. and elsewhere, I heartily recommend the summary by Dracup (1995), which includes such engaging anecdotes as the first great geodetic controversy (a British–French dispute over the shape of the Earth, which was eventually resolved in favor of Sir Isaac Newton), the use of the telegraph as a geodetic instrument (for determining time differences and thus astronomic longitudes), and the first (and last) spirit-leveling survey to the summit of Mount Whitney.

Trilateration is a surveying technique used to establish or extend horizontal control by measuring the lengths of the three sides of a series of touching or overlapping triangles. Some of the angles formed by the sides may also be measured; others are computed from the measured lengths. Distances are usually measured electronically with an instrument called an electro-optical (or electronic) distance meter, or EDM (Section 2.4.1). The coordinates of each point in a trilateration network are computed from the known position of a reference station in the network, thus fixing the positions of all points in the network. Any deformation of the network can be measured by repeating the survey at a later date.

One limitation of trilateration surveys for measuring volcano deformation is the requirement for line-of-sight visibility among stations. This is often a problem at volcanoes, which tend to obstruct shots that are anything other than nearly radial from the summit. This turns out to be an especially frustrating problem on gently sloping shield volcanoes such as Mauna Loa, Hawai'i, where on a clear day it seems you can see forever but the station you're looking for is always just beyond the horizon! Most EDM networks at volcanoes involve compromises to accommodate the volcano's shape and thus are not strictly trilateration networks (i.e., measured lines generally do not form overlapping triangles). Nonetheless, the technique is very useful for tracking line-length changes and horizontal displacements.

Triangulation, which is based on measuring angles rather than lengths, is another surveying technique that has been used widely for volcano monitoring. Traditionally, the points whose positions are to be determined, together with at least two points whose positions are known, are connected in such a way as to form the vertices of a network of triangles. The angles in the network are measured with a theodolite or similar instrument and the lengths of the sides are either measured or calculated from known angles and lengths. Combined measurements of horizontal angles, vertical angles, and line lengths can be used to determine the locations and 3-D motions of a network of points from repeated surveys. Required line-of-sight visibility among stations often constrains the layout of such networks, but the method can provide valuable information in cases where repeat visits are feasible and the risk is acceptable.

Following the great 1906 San Francisco earthquake, initial triangulation observations made in the 1850s at the Presidio and elsewhere in the San Francisco Bay area were repeated to determine the amount of crustal motion – the first time in the U.S.A. that triangulation was repeated for this purpose. It was discovered that surface displacements in the earthquake were largest at the fault and decreased with distance from it, which provided the basis for the landmark elastic rebound theory introduced by Professor H.F. Reid of Johns Hopkins University (Hayford and Baldwin, 1908; Reid, 1910a,b). Even today, historical triangulation data continue to be an important source of information on crustal motions associated with large earthquakes, including the 1868 Hayward Fault earthquake and the 1906 San Francisco earthquake (Thatcher, 1974, 1975; Segall and Lisowski, 1990;

Yu and Segall, 1996; Thatcher *et al.*, 1997; Kenner and Segall, 2000). In addition, volcano geodesists have borrowed freely from historical geodesy to study crustal motions of a different sort, as discussed below.

2.4.1   EDM and theodolite surveys, with examples from Mount St. Helens and Long Valley Caldera

Until recently, the most widely used instrument for making slope-distance measurements at volcanoes was the electro-optical (or electronic) distance meter. The term refers to a class of photoelectric or laser devices commonly called EDMs. By the 1990s, their role had been partly taken over by GPS receivers, which provide comparable horizontal accuracy plus full 3-D positioning. On the other hand, modern laser EDMs are portable, inexpensive, and still preferable where deployment of GPS receivers is not a viable or prudent option (Figure 2.4). In such cases, one-time installation of EDM reflectors at dangerous or difficult-to-access sites provides an opportunity to make repeated slope-distance measurements from more comfortable sites for as long as the reflectors survive – ideally until the immediate crisis is over.

An EDM directs a narrow beam of amplitude-modulated, monochromatic light (modern EDMs use lasers) onto a distant reflector and measures the EDM-to-reflector distance by comparing the phases of outgoing and reflected beams. The result depends on the speed of light in air, which in turn depends on atmospheric density (a function of temperature and pressure) and on water vapor concentration (humidity). These parameters should be determined during the line-length measurement by an aircraft flying along the line-of-sight (Figure 2.5), or at least at the end points of the line. Standard corrections then can be applied to compensate for differing atmospheric conditions between surveys. The standard deviation of a corrected line length can be expressed as:

$$\sigma = \sqrt{\alpha^2 + \beta^2 \times L^2} \tag{2.2}$$

where α is a constant that includes setup error and length independent instrument error, β is a measure

Figure 2.4.   Line-length monitoring in the 1980 crater at Mount St. Helens. A Lietz Red 1A EDM is set up over a survey point on the crater floor to measure distances to fixed reflectors on the lava dome (steaming, *upper right*). During the period of dome growth from 1980 to 1986, inexpensive highway reflectors (*inset*) or single corner cube reflectors were used for this purpose; repeated measurements could be made without visiting the reflectors each time. The practice was also a money saver, because reflectors were destroyed frequently by rockfalls or eruptive activity and had to be replaced (Iwatsubo and Swanson, 1992). EDMs like this one can measure line lengths up to several kilometers with an accuracy of the order of $[(5\,mm)^2 + (1\,mm/km \times L(km))^2]^{1/2}$ (i.e., ± 11 mm for a 10 km line). USGS photograph by Steven R. Brantley, 9 June 1982.

Figure 2.5. EDM survey at South Sister Volcano in the Three Sisters region, central Oregon. A Hewlett-Packard 3808A infrared EDM is used to measure line length while a helicopter flies along the line-of-sight making refractivity measurements. A fixed-wing aircraft can also be used for this purpose. USGS photograph by Lyn Topinka, 1985.

of systematic length dependent error, and L is the line length in kilometers (Savage, 1975). Typical values of α and β are 1–7 mm and 0.2–5 mm/km (parts per million, ppm), respectively.

EDM network configurations that provide desired redundancy include interlocking doubly braced quadrilaterals (i.e., four-sided figures with observations of the sides and diagonals, with a common side between adjacent quadrilaterals) and centered figures (e.g., quadrilaterals, pentagons, or hexagons with observations of the sides and also from a central mark to each of the vertices, with a common side); chains of simple triangles are less desirable but sometimes difficult to avoid (Bomford, 1980, pp. 1–19). At volcanoes, an ideal configuration is seldom achieved in practice because topography usually precludes the required intervisibility between marks. Instead, the conical shape of a typical stratovolcano lends itself to a star-shaped network consisting of several distal EDM stations (points of the star) and a set of reflector stations higher on the volcano, each visible from two or more of the distal EDM stations.

The layout of a typical slope-distance monitoring

network at Mount St. Helens, which was used during the 1980s, is shown in Figure 2.6.[3] From such an arrangement, both relative line-length changes and 2-D displacement vectors can be determined by repeated surveys, if we assume that the positions of two or more points in the network do not change. This assumption is not always justified and can obscure the true nature of the deformation field, especially if the source is relatively deep. For this reason, it is important to include several distant stations in the network (e.g., Harrys Ridge, Figure 2.6), or preferably to test the stability of the reference stations by observing them with GPS or by including them in a larger network to be measured less often.

At Mount St. Helens and other high volcanoes, thick accumulations of winter snow and ice must be factored into the design of geodetic networks (Figure 2.7). Clearing the line-of-sight between two marks buried under several meters of snow can be a daunting task, even for the most ambitious volcanologist (Figure 2.8). Where time and resources permit, this problem can be addressed by constructing a network of observation platforms and reflector stations high enough to stay above the winter snow pack. This approach was used at Mount St. Helens during the 1980s with good results (Figures 2.9 and 2.10). The stability of this type of station relative to a bench mark embedded in the ground is debatable, but harsh winter conditions necessitated a pragmatic approach. For example, the effects of loading by rain and snow and thermal

---

[3] Network configurations comprising a set of closed geometric figures with common sides (e.g., a chain of triangles, braced quadrilaterals, centered figures, double-centered figures, or some combination thereof) are favored, because they provide the redundancy required for geodetic adjustment (Bomford, 1980, pp. 3–6). Such configurations are seldom realized at volcanoes, however, for reasons of difficult access or lack of station intervisibility. The latter difficulty can be avoided by observing the network with GPS (Chapter 4).

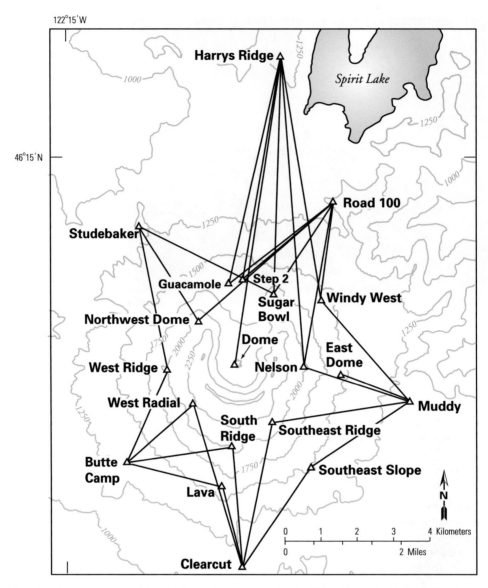

Figure 2.6. Line-length monitoring network at Mount St. Helens established in 1980 and expanded through 1990 (Swanson *et al.*, 1981; Iwatsubo *et al.*, 1992a). To survey the network, an EDM was set up in turn at each of 6 low-elevation stations ringing the volcano. Line-length measurements were made from each station to several reflectors set at higher elevations. The line from Harrys Ridge to Dome represents several lines to closely spaced targets on the lava dome that grew from 1980 to 1986. The shape of the volcano precluded an optimal, doubly braced quadrilateral network geometry. However, most reflectors were visible from at least 2 EDM stations, and displacements of the reflectors could be calculated by assuming that the EDM stations did not move. This assumption was tested occasionally by surveying a larger aperture network that included the stations shown here.

expansion–contraction of the observation platforms were ignored, given the compelling need to monitor the volcano year-round. A good rule of thumb is to do the best job possible within the constraints imposed by time, resources, and environment. Even so, success is never assured (Figure 2.11).

In some cases, deformation is sufficiently localized and obvious that it isn't necessary to monitor the entire volcano. Instead, a local EDM network span-

ning just the deforming area can be established and measured more frequently than would be possible with a larger network. Such was the case at Mount St. Helens from 1980 to 1986, when a dacite dome grew episodically on the floor of the 18 May 1980 crater. Frequent dome-building eruptions meant that surface displacements on or near the dome were likely to be large, and the lifespan of reflectors was likely to be short. For these reasons, the invest-

Figure 2.7. Flat Top radio repeater station near Mount St. Helens in January 1982. Harsh winter conditions can hamper volcano-monitoring efforts, but such difficulties can be overcome through experience, ingenuity, good planning, and hard work. USGS photograph by Daniel Dzurisin.

Figure 2.8. Wintertime EDM survey at station Sauna in the crater at Mount St. Helens. USGS photograph by Lyn Topinka, 22 March 1983.

ment in reflectors was kept to a minimum and normal concerns about station stability were held in check. In some cases, corner-cube reflectors were attached to solid rock on the dome with expansion bolts, or wedged into cracks and wired in place – a nonstandard, pragmatic approach that suited the situation well. Even inexpensive plastic highway reflectors from the local hardware store, which produced enough signal return out to distances of about 200 m, proved to be adequate (Figure 2.4).

To increase the number of reflector sites without leaving more reflectors on the dome, some stations were marked with a rebar or a nail driven into a crack. These were surveyed by holding a reflector at the marker in a consistent way while the line length was measured from an EDM station on the crater floor. The 'holder' then walked to the next reflector station and the procedure was repeated. After all of the reflector stations that were visible from a given EDM station had been measured, the EDM was moved and as many reflector stations as possible were measured again. By measuring most reflector stations from at least two EDM stations, displacement vectors could be calculated by assuming that the EDM stations were stable and any line-length changes could be attributed to deformation of the

Figure 2.9. (A) Welded-steel instrument station STUD at Studebaker Ridge on the lower northwest flank of Mount St. Helens, with Rangemaster III EDM mounted on top. Temperature sensor is attached to top of pole projecting above tower. Walkway provides access to all sides of tower, so shots can be made easily to reflectors at various azimuths. (B) Reflector tower GUAC in the crater at Mount St. Helens, with two corner-cube reflectors mounted near the top. Each reflector points to a different EDM tower, as in (A). Below the two reflectors is a light attached to a battery, which is part of an unrelated experiment. Tower heights range from 3 to 4 m. Each leg is cemented into the ground for optimum stability. USGS photographs by Lyn Topinka, reprinted from Iwatsubo et al. (1992a).

dome. This assumption was verified periodically by measuring lines to fixed reflectors elsewhere in the crater or beyond. Invariably, any changes involving distant stations were small compared with changes on the dome (Iwatsubo and Swanson, 1992).

One disadvantage of this procedure is that it exposes the holder and, to a lesser degree, the EDM operator, to hazardous conditions and events on or near the lava dome. At Mount St. Helens in the 1980s, these included rough terrain, changeable weather, hot ground, deep snow, infrequent ballistic showers from small explosions, and rockfalls. Today, the risk could be minimized by substituting helicopter-slingable GPS stations (Section 11.3.4) for observers on the ground. That said, the 'boots on the ground' approach employed at Mount St. Helens from 1981 to 1986 resulted in an unprecedented series of successful eruption predictions (Swanson et al., 1983), with no serious injuries to field personnel.

EDM results for one of several dome-building

episodes at Mount St. Helens are shown in Figure 2.12. The radial pattern of displacement vectors was caused by intrusion of magma beneath and into the dome that culminated with an extrusion onto the dome's east flank in September 1981. Combined with other deformation-monitoring techniques adapted for use in the crater, repeated EDM surveys facilitated a series of successful predictions of dome-building episodes (Swanson et al., 1983). Among the other techniques that proved to be useful were triangulation surveys, leveling on the crater floor, and steel-tape measurement of displacements on radial cracks and thrust faults (Iwatsubo and Swanson, 1992) (Figures 2.13–2.16).

Measuring line lengths with both red and blue light can greatly reduce the uncertainty associated with changes in atmospheric water vapor content between EDM surveys. The difference in travel time accurately reflects the density of the dry atmosphere (Slater and Huggett, 1976), which is used to adjust the travel time for one of the wavelengths to obtain

Figure 2.10. (A) EDM station Clearcut at Mount St. Helens in summer. (B) Same station in winter with ~2 m snowpack. USGS photographs by Lyn Topinka, reprinted from Iwatsubo *et al.* (1992a).

distance (Langbein *et al.*, 1993). So-called 'two-color' EDMs designed to take advantage of this fact can be very accurate, but the market for such devices is small and very few have been manufactured. One such instrument has been used since 1983 to monitor inflation at the Long Valley Caldera, California, with excellent results. The two-color EDM is not easily portable, however, so the network was designed around a few central instrument stations each surrounded by several permanent reflector stations. Most of the lines span the caldera's resurgent dome and south moat, where rapid deformation and repeated earthquake swarms have been centered since 1980 (Hill *et al.*, 2002a) (Figures 2.17–2.19).

The two-color EDM results from Long Valley are noteworthy not only for their precision, which is represented by:

$$\sigma(mm) = \sqrt{[0.48]^2 + [0.18 \times L(km)]^2}$$

(i.e., 1.9 mm for a line length of 10 km) (Langbein *et al.*, 1987a,b), and for their measurement frequency (a few times each week), but also for the detailed insight they provide into subsurface processes. For example, modeling of the two-color data reveals four separate deformation sources that have been active episodically since 1980: (1) a Mogi-type or

Figure 2.11. EDM reflector stations at Mount St. Helens covered with rime ice. Climate and topography often play important roles in determining what types of geodetic measurements are possible at volcanoes. USGS photographs, David A. Johnston Cascades Volcano Observatory.

ellipsoidal inflation source 7–10 km beneath the center of the resurgent dome; (2) a shallow, right-lateral fault beneath the south moat; (3) a second Mogi or ellipsoidal source 10–20 km beneath the south moat near Casa Diablo Hot Springs; and (4) a dike intruded beneath Mammoth Mountain to within 1–3 km of the surface during 1989–1990 (Langbein, 1989; Langbein *et al.*, 1993, 1995; Hill *et al.*, 2002a) (Figure 2.20).

In support of the dike intrusion hypothesis, Hill (1996) cited the following attributes of the 1989–1990 seismic swarm beneath Mammoth Mountain:

(1) a dike-like distribution of hypocenters at depths between 6 and 9 km with a north–northeast strike essentially perpendicular to the T-axes of the local earthquake focal mechanisms (Hill *et al.*, 1990); (2) deformation in the vicinity of Mammoth Mountain consistent with a north–northeast-striking dike extending to within 2 km of the surface beneath Mammoth Mountain (Langbein *et al.*, 1993); (3) frequent spasmodic bursts of seismic energy suggesting rapid-fire brittle failure driven by transient surges in local fluid pressure (Hill *et al.*, 1990); and (4) an increase in ^{3}He/^{4}He ratios from fumaroles

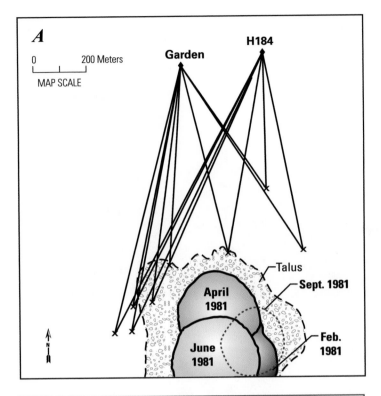

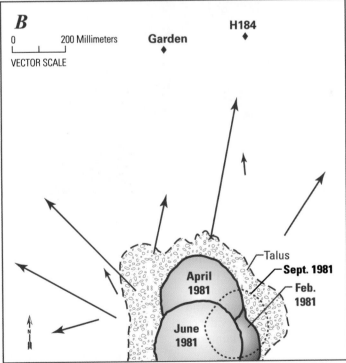

Figure 2.12. Line-length monitoring network and data spanning the September 1981 dome-building episode at Mount St. Helens (Iwatsubo and Swanson, 1992). (A) Locations of EDM stations (diamonds) and reflector stations (×'s). (B) Displacement vectors calculated from repeated slope-distance measurements, assuming Garden and H184 did not move. Also shown are the outlines of lava lobes extruded onto the dome in February, April, June, and September (dotted) 1981. Displacement vectors represent cumulative displacements from 25 June 1981 to 18 September 1981. The September 1981 dome growth event began on 6 September 1981, lasted ∼5 days, and increased the volume of the dome by 3.9×10^6 m^3 to 18.5×10^6 m^3.

Figure 2.13. Dome-building episodes at Mount St. Helens produced extreme localized deformation of the dome and adjacent crater floor, including large radial cracks and thrust faults that developed in a matter of days to weeks. Here, a steel tape is used to measure the distance between rebar stakes driven into the ground on either side of a fresh crack. Repeated measurements of this type revealed accelerated movement prior to the onset of extrusive activity. USGS photograph by Lyn Topinka, 12 May 1984.

Figure 2.14. Monitoring the development of a thrust fault on the crater floor at Mount St. Helens. A steel tape is used to measure the distance between one rebar stake on the upper plate of the thrust and another on the lower plate. The upper plate over-rode the lower one as an extrusive episode neared, causing the distance to shorten at an increasing rate. A remarkable set of thrust faults like this one developed on the southwest crater floor adjacent to the dome in 1981 (Figure 2.15). Photograph by Terry A. Leighley, Sandia National Laboratories, 1981.

on Mammoth Mountain detected in late 1989 (Sorey *et al.*, 1993). In addition, Farrar *et al.* (1995) linked the anomalous forest kill at Mammoth Mountain to diffuse CO_2 emission from a magmatic source, and McGee *et al.* (2000) reported geochemical evidence for a magmatic CO_2 degassing event at Mammoth Mountain in 1997.

The two-color EDM results (Figure 2.18) clearly show that the extension rate across the resurgent dome started to increase in early October 1989, about 5 months after the start of the Mammoth Mountain earthquake swarm on 4 May and 2 months before a marked increase in seismic activity

beneath the caldera starting in December 1989. The distribution of earthquake hypocenters combined with the distinctive swarm characteristics listed above make a compelling case for an intrusion beneath Mammoth Mountain in the latter half of 1989, followed by re-inflation of a magma body beneath the resurgent dome. The two-color data confirm a basic tenet of rock mechanics (i.e., that rocks always strain before they break). This implies that surface deformation caused by intrusions should be detectable before the onset of unusual earthquake activity, if geodetic measurements are sufficiently frequent and precise, and if the intrusion is shallow enough

Figure 2.15. Aerial view of the southwest crater floor at Mount St. Helens in June 1981, showing thrust faults, bounding tear faults, and radial cracks. The width of the view is about 200 m. Scarps define toes of the upper plates of thrust faults (Figure 2.14) and are mostly directed away from the base of the lava dome, which is located in shadow at the top of the photograph. Circle marks the location of the Christina 2 thrust-monitoring station, which yielded the data shown in Figure 2.16. USGS photograph by William W. Chadwick, Jr.

to produce measurable deformation. This is especially true if the brittle–ductile transition in the crust is relatively shallow (i.e., the host-rock temperature is high enough and the strain rate produced by intrusion is low enough that the host rock deforms in non-brittle fashion).

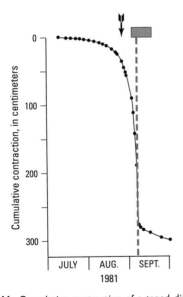

Figure 2.16. Cumulative contraction of a taped distance across the toe of Christina 2 thrust fault (Figure 2.15) prior to an extrusive eruption at Mount St. Helens in September 1981 (Swanson *et al.*, 1983). Arrow indicates the date (August 26) on which the USGS David A. Johnston Cascades Volcano Observatory issued an eruption prediction, and box indicates the time window in which the eruption was predicted to occur (one to three weeks after the prediction was issued). Dashed line represents the start of the eruption on 6 September.

Another important finding at Long Valley is that the geodetic moment represented by deformation within the caldera has far exceeded the cumulative seismic moment release throughout the current episode of unrest, which began in 1978 (Langbein *et al.*, 1993; Hill *et al.*, 2003). The cumulative seismic moment release for a sequence of earthquakes is:

$$\Sigma M_0 = \mu A d \qquad (2.3)$$

where μ is the shear modulus of the surrounding rock, d is the average slip across a fault, and A is the area of the fault surface. The geodetic moment for a volume source producing a given deformation pattern is:

$$M_0^{(g)} = a\kappa\Delta V \qquad (2.4)$$

where a is a constant depending on source geometry, κ is an elastic modulus, and ΔV is the change in the source volume required to produce the observed deformation (Aki and Richards, 1980; Hill *et al.*, 2003). For a spherical or Mogi source, for example, $M_0^{(g)} = (\lambda + 2\mu)\Delta V$, where ΔV is the volume change of a spherical magma body and λ is one of two Lamé constants that appear in stress–strain relationships.[4] For a shear dislocation (slip on a

---

[4] Lamé constants λ and μ are elastic parameters that express the relationships between the components of stress and strain for linear elastic behavior of an isotropic solid; μ is identical with rigidity (also called modulus of rigidity, shear modulus, or torsional modulus), and λ is equivalent to the bulk modulus or stiffness K minus $2\mu/3$.

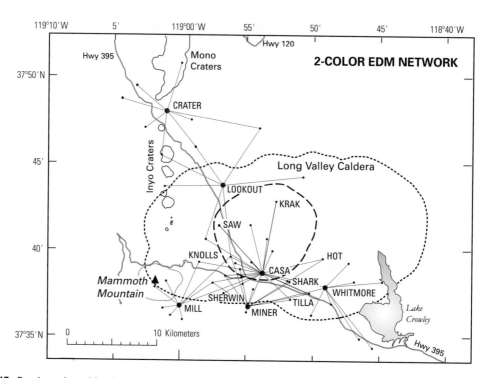

Figure 2.17. Bench marks and baselines in the two-color EDM network in the Long Valley region (Hill *et al.*, 2002a). Large dots are instrument sites; small dots are reflector sites. Baselines radiating from the instrument site, CASA, to the labeled reflector sites are measured several times per week, weather permitting. Other baselines are measured monthly or annually (Langbein *et al.*, 1993).

fault), $M_0^{(g)} = M_0$. At Long Valley, where the geodetic moment is much larger than the cumulative seismic moment, deformation is driven primarily by aseismic processes associated with magma movement (Hill *et al.*, 2003). Hill (1992) showed that the brittle–ductile transition occurs at depths of less than 5 km beneath the resurgent dome at Long Valley, so an inflation source at 7–10 km depth might well produce far fewer earthquakes than would be expected to occur in cold, brittle crust elsewhere. The upshot is that deformation monitoring, including the use of classical geodetic techniques described in this chapter, might in some cases provide earlier warning and a better estimate of the energy involved in volcanic unrest than seismic monitoring. In any case, both techniques are essential to understanding the spectrum of processes operative at restless volcanoes.

2.4.2 Triangulation and total-station surveys

Triangulation is another useful tool for monitoring volcano deformation. In cases where difficult access or safety concerns preclude setting up conventional triangulation targets, features such as distinctively shaped rocks or fractures can be used instead. During a prolonged period of dome growth at Mount St. Helens in 1983, paint bombs dropped from the open door of a helicopter served to mark targets on inaccessible parts of the dome (Iwatsubo and Swanson, 1992). A theodolite was used to measure horizontal and vertical angles to each target from two or more instrument stations (Figure 2.21). By assuming that the instrument stations were stable during the observation period, displacement vectors could be derived from repeated measurements (Figure 2.22). This approach is insensitive to movements toward or away from an instrument station, so large angular separation between stations is desirable. As is the case for trilateration surveys, the assumed stability of instrument stations for triangulation surveys should be checked periodically by measuring a larger network of targets that surround the instrument stations or by using GPS.

For better control on 3-D displacements, angular and slope-distance measurements can be combined by mounting an EDM and theodolite together on a

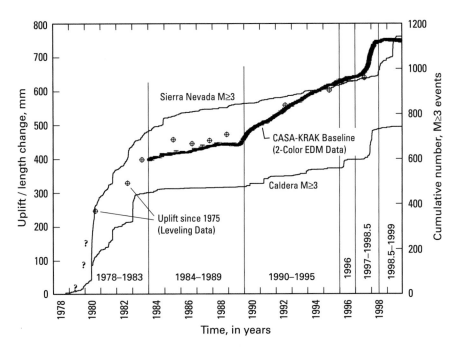

Figure 2.18. History of earthquake activity and swelling of Long Valley's resurgent dome from 1978 through 1999, including: (1) the cumulative number of $M \geq 3$ earthquakes within both Long Valley Caldera (Caldera $M \geq 3$) and the Sierra Nevada block to the south (Sierra Nevada $M \geq 3$); and (2) deformation of the resurgent dome reconstructed from the uplift history of a bench mark near the center of the resurgent dome from leveling surveys from 1980 through mid-1997 and the extension (thick line) of the 8 km long baseline spanning the dome between CASA and KRAK (Figure 2.17) since 1983 based on frequent measurements with a two-color EDM (Hill *et al.*, 2002a).

tripod (Figure 2.23), or by using a total station surveying instrument for the same purpose. Modern total stations such as the Leica T2002 combine an electronic theodolite and EDM in a compact unit with point-and-shoot features that greatly reduce the amount of training required for efficient use (Figure 2.30).

2.5 LEVELING AND TILT-LEVELING SURVEYS

Leveling is a tried and true technique used to measure height differences within a network of bench marks and, by repeated surveys, to measure height changes as a function of time. Height differences can be determined by summing incremental vertical displacements of a graduated rod (differential leveling), or by measuring vertical angles (trigonometric leveling). Kozlowski (1998) provides an excellent introduction to the principles and capabilities of trigonometric leveling using a total station. Henceforth, I use the term leveling to mean differential leveling unless specified otherwise. Two common forms of differential leveling are spirit leveling

and compensator leveling, which are distinguished by the type of level used. These and several related terms are defined in the Glossary at the back of this book. The discussion that follows is mostly general in nature and applies to any type of differential leveling. Compared with modern space-based techniques such as GPS and SAR interferometry (Chapters 4 and 5), leveling is a decidedly low-tech approach to geodetic monitoring. Nonetheless, it remains among the most precise means available to measure vertical height changes over the range of distances of interest for volcano monitoring (i.e., from tens of meters to tens of kilometers (Dzurisin, 1992b)).

In this section, I discuss the advantages and disadvantages of leveling at volcanoes and illustrate the utility of leveling surveys through two examples. Unfortunately, staffing constraints plus the availability of GPS and other modern geodetic techniques have combined to reduce dramatically the amount of leveling being done in the U.S.A. during the past few decades. Nonetheless, as a long-time practitioner, I strongly advocate leveling studies at volcanoes. Like all geodetic tools available to volcanologists, leveling should be used judiciously.

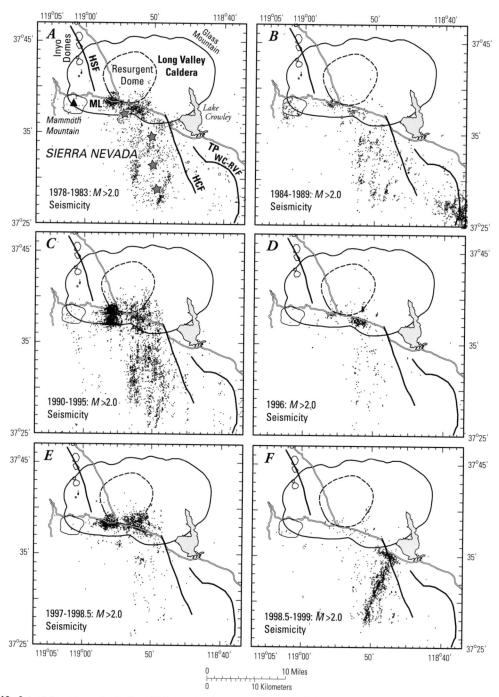

Figure 2.19. Seismicity patterns in the Long Valley region for the six time intervals bracketed in Figure 2.18. ML, Mammoth Lakes; TP, Tom's Place; HCF, Hilton Creek Fault; HSF, Hartley Springs Fault; WC-RVF, Wheeler Crest-Round Valley Fault. Stars in (A) represent epicenters of the four $M = 6$ earthquakes of 25–27 May 1980 (Hill et al., 2002a).

In most cases, other monitoring techniques are preferable or more feasible. But in some specialized circumstances, repeated leveling surveys are the best means to accurately – and most precisely – measure vertical surface displacements and thus help to understand subsurface processes. For this reason, leveling is a tool that should continue to be exploited whenever possible. Near the end of this section, I offer some suggestions for making the best use of leveling at volcanoes.

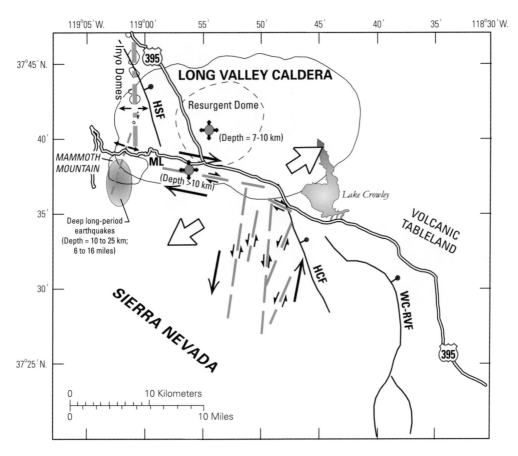

Figure 2.20. Map showing sources contributing to unrest at Long Valley Caldera and vicinity starting in 1978 (Hill *et al.*, 2002a). Orange circles with radial arrows indicate inferred pressure centers, most likely inflating magma bodies, beneath the resurgent dome and south caldera moat. Long-dashed orange lines in center portion indicate near-vertical fault zones. Small, single-barbed arrows show relative sense of slip across individual fault zones. Larger double-barbed arrows indicate average sense of slip across all fault systems. Large open arrows indicate direction of regional, east–northeast extension. Short-dashed orange line in upper left portion (near orange ellipse representing deep long-period earthquakes) represents the dike intruded beneath Mammoth Mountain in 1989, with small black arrows indicating opening direction. Heavy green dot–dashed line near upper left corner represents the dike that fed the Inyo Dome eruptions 500 to 600 years ago. Major range-front normal faults indicated by solid black lines with the ball on the down-dropped block: HSF, Hartley Springs Fault; HCF, Hilton Creek Fault; and WC-RVF, Wheeler Crest-Round Valley Fault. ML marks the town of Mammoth Lakes.

2.5.1 Field procedures and accuracy

During a leveling survey, a leveling instrument such as a spirit level, compensator level, or digital level, and a pair of leveling rods are used to measure the height difference between permanent bench marks by accumulating the height differences between a series of temporary turning points (Figures 2.24–2.26). Leveling rods are typically 3 m long, graduated every 0.5 or 1.0 cm for use with an optical level or with a barcode design for use with a digital level. The forward turning point (relative to the direction of the level line) is called the foresight and the backward turning point is called the backsight. Typically, bench marks are spaced 1–3 km apart

and turning points are 10–100 m apart (depending on the desired accuracy of the survey and on the steepness of the terrain). Adjacent turning points bound a setup and adjacent bench marks bound a section; measurement of a series of sections constitutes a level line. A level line or series of lines that forms a loop back to the starting point is called a circuit. Circuits provide the advantage of a built-in check on accuracy. The accumulated height difference around any circuit should, of course, be zero. The height difference actually measured is called the circuit misclosure or closure error. For logistical reasons, level lines usually follow roads, railroads, or other land corridors wherever available; however,

Figure 2.21. Using a theodolite to measure horizontal and vertical angles in the 1980 crater at Mount St. Helens. Under ideal conditions, the precision of such measurements is ±1 arc second, which corresponds to a displacement of ±5 mm at a distance of 1 km. Displacement vectors for a set of targets can be derived from repeated measurements from two or more theodolite stations by assuming that the theodolite stations did not move. USGS photo by Daniel Dzurisin, 20 May 1982.

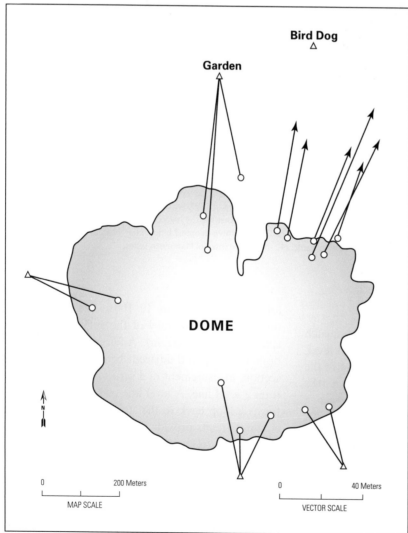

Figure 2.22. Horizontal displacements from repeated theodolite and EDM measurements to targets on the lava dome at Mount St. Helens from mid-July to late September 1983 (Iwatsubo and Swanson, 1992). During that period, dome growth was concentrated in the northeast sector, where paint bombs dropped from a helicopter served as targets for theodolite measurements. Triangles and circles represent instrument stations and targets, respectively. Where these are connected by lines, both theodolite and EDM measurements were made to targets like the one shown in Figure 2.4. Elsewhere (northeast sector), only theodolite measurements were made to paint-bomb targets because continuous endogenous dome growth in that sector made traveling on foot to install EDM reflectors inadvisable. Arrows represent horizontal displacements at 6 targets measured from both Garden and Bird Dog. Displacements at all other targets were within measurement error. Dome shape is schematic.

Figure 2.23. Line-length and angular measurements can be made in tandem by mounting an EDM and conventional (i.e., non-electronic) theodolite together on a tripod. Here, such a set-up is being used on the lava dome at Mount St. Helens, with the snow covered wall of the 1980 crater in the distance. USGS photograph by Steven R. Brantley, March 1986.

if necessary, leveling can be done cross country where terrain permits (Figures 2.24 and 2.25).

Spirit level leveling instruments and newer compensator-leveling instruments include an optical telescope with a line-of-sight that can be adjusted precisely to horizontal either manually (spirit level type) or automatically (compensator type). In both cases, the operator uses the telescope to make readings along the horizontal line-of-sight to each of two graduated rods, thus determining the height difference between the rods. The Federal Geodetic Control Committee (1984) specifies that for first-order leveling surveys (see next paragraph for a description of various orders and classes of leveling surveys) the rods shall use a double-scale

strip (i.e., two sets of graduations) made of Invar, a metallic alloy with a very small thermal coefficient of expansion (10^{-6} per °C)[5] (Figure 2.24). Reading two scales on each rod provides a measure of the reading precision, and using Invar reduces the magnitude of temperature effects that must be removed from the observations based on temperature measurements made at each setup. Next the leveling instrument and one of the rods are moved ahead along the level line and the procedure is repeated. The rods are leapfrogged in this manner through a series of temporary turning points to the next mark, thus establishing its height with respect to the first (Figure 2.26). A series of bench marks is connected in this way to form a level line and, in some cases, several lines are linked to form a leveling network. Any height changes within the network can be measured by repeating the survey at a later date. For a given survey, sections or lines that are measured once are referred to as single-run and those measured twice as double-run. Double running helps to identify any blunders, cancel or identify systematic errors, and reduce the magnitude of random errors (see below).

The accuracy of leveling surveys is determined by the combination of equipment, field procedures, allowable misclosure for double-run sections or circuits, and the types of corrections applied to field observations. Vanicek *et al.* (1980) provide a readable and useful discussion of these factors, including typical standard deviations and maximum allowable misclosures for various orders and classes of leveling surveys. Another useful document is a report by the Federal Geodetic Control Committee (1984) that discusses standards and specifications for geodetic control networks, including leveling networks and surveys. The FGCC report divides leveling into five types based on the procedures and equipment used and on the expected accuracy. The most stringent specifications apply to first-order, class I networks and surveys, which provide the greatest precision. Next in order of decreasing precision are first-order, class II; second-order, class I and II; and third-order. The report specifies, for each order and class of leveling, the requirements for network geometry, instrumentation, calibration procedures, field procedures, and office procedures. Included are such things as the average and maximum bench mark spacing (1.6 km and 3 km, respectively, for all first-order, class I or II and second-order, class I surveys), the type of leveling rod construction (Invar for all first- and second-order surveys), the combined instrument and rod

---

[5] The thermal coefficient of expansion is the change in length per unit length of a material for a 1 °C change in temperature. For example, a 1 m long Invar rod, with a thermal coefficient of expansion of 10^{-6} per °C, would stretch or contract 10^{-6} m, or 1 micron, for each 1 °C of temperature increase or decrease, respectively.

Figure 2.24. Leveling near the Pu'u 'Ō'ō vent along the East Rift Zone of Kīlauea Volcano, Hawai'i, using an optical level and dual-scale Invar rods (*inset*). Modern levels are capable of measuring the height difference between graduated rods in a single set-up to a precision of 0.1 mm or better. Using such an instrument, an experienced crew can cover 3–5 km day^{-1} in relatively flat terrain, or 1–3 km day^{-1} in steeper terrain. USGS photograph by James D. Griggs, 12 January 1985.

Figure 2.25. Leveling in backcountry terrain can be physically challenging but nonetheless rewarding for a hardy crew. Here, a Leica NA 3003 digital level and barcode rods (*inset*) are used to level along a trail in the Three Sisters Wilderness, Oregon. USGS photograph by Michael P. Poland.

resolution (at least 0.1 mm for first-order surveys), the maximum leveling instrument collimation error (0.05 mm m^{-1} for all first- and second-order surveys), the maximum sight length (50 m for first-order, class I; 60 m for first-order, class II and second-order, class I), minimum ground clearance of the line-of-sight (0.5 m for all orders and classes), and literally dozens of other specifications. Suffice it to say that the FGCC report is an essential reference and 'how to' manual for anyone conducting leveling surveys in the U.S.A., including volcanologists. Similar documents exist to guide geodetic surveys in other countries (e.g., New Zealand, Bevin (2003); Australia, ICSM (2000); Canada, Surveys and Mapping Branch (1978)).

Leveling surveys are subject to both random error and various types of systematic error. Random error, which accumulates with distance along the level line, cannot be removed from the observations but its magnitude can be estimated for a given order and class of survey (see below). Systematic errors arise for several reasons. They can be reduced by following standard field procedures and by applying specified corrections to the observations (Federal Geodetic Control Committee, 1984). The National Geodetic Survey recommends that the following corrections be applied whenever applicable (Balazs and Young, 1982): rod scale, rod temperature, level collimation, refraction, astronomic, and orthometric. The rod scale correction is intended to remove the effect of small errors in the placement of graduations (i.e., any discrepancy between the

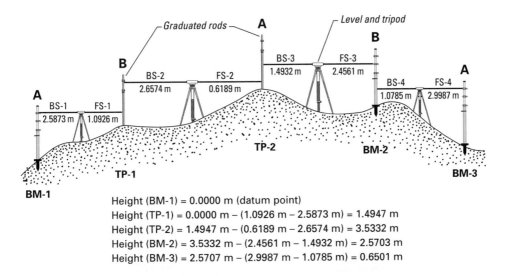

Height (BM-1) = 0.0000 m (datum point)
Height (TP-1) = 0.0000 m – (1.0926 m – 2.5873 m) = 1.4947 m
Height (TP-2) = 1.4947 m – (0.6189 m – 2.6574 m) = 3.5332 m
Height (BM-2) = 3.5332 m – (2.4561 m – 1.4932 m) = 2.5703 m
Height (BM-3) = 2.5707 m – (2.9987 m – 1.0785 m) = 0.6501 m

Figure 2.26. In differential leveling, height differences between a datum point (BM-1) and other bench marks are determined by accumulating the difference between foresight (FS) and backsight (BS) readings on one or more (usually two) graduated rods (A and B), which are leapfrogged through a series of temporary turning points (TP-1 and TP-2) between bench marks.

marked and actual height of graduations above the bottom of the rod). For first-order surveys, annual rod calibrations are recommended for this purpose. The rod temperature correction takes account of scale changes induced by stretching or shrinking of rods as a function of temperature. The use of Invar in modern leveling rods minimizes the effect of temperature induced scale changes. The refraction correction accounts for bending of the light path between the leveling instrument and rod caused by near-surface vertical temperature gradients (see below). Both the rod temperature and refraction corrections are based on temperature measurements made in the field during a leveling survey. Likewise, the level collimation correction is based on a simple calibration procedure, sometimes called a peg test, which should be performed at least once a day during a survey (see the Glossary at the back of this book for a description of how a peg test is done). The astronomic correction accounts for the effect of tidal accelerations due to the Moon and Sun on Earth's equipotential surfaces. The correction is small ($\leq 0.1\,\mathrm{mm\,km^{-1}}$), but it accumulates in the north–south direction. It is not usually necessary to apply the astronomic correction to local surveys at the scale of interest for volcano monitoring. Lastly, the orthometric correction accounts for non-parallelism between the geoid and other equipotential surfaces at higher elevations. Because it depends on latitude and average height of the section, it does not change significantly from one

survey to the next and can be ignored for most volcano-monitoring applications.

The standard deviation of a height difference between bench marks measured by leveling is given by (Pelton and Smith, 1982):

$$\sigma(h) = \frac{\beta}{\sqrt{j}} \times \sqrt{L} \qquad (2.5)$$

where $\sigma(h)$ is in millimeters, β is a constant factor for each order, class, and vintage of survey, $j = 1$ for single-run surveys or $j = 2$ for double-run surveys, and L is the distance in kilometers along the leveling route from a reference bench mark to the bench mark in question. For double-run surveys, β can be calculated from observed section misclosures (Pelton and Smith, 1982):

$$\beta^2 = \frac{1}{2m} \times \sum_{i=1}^{m} \frac{\Delta l_i^2}{L_i} \qquad (2.6)$$

where m is the number of double-run sections. For single-run surveys, Vanicek et al. (1980) assigned a value of β for each order, class, and vintage of survey based on extensive experience by NGS and USGS.

The standard deviation of a height change measured by comparison of two leveling surveys is given by (Pelton and Smith, 1982):

$$\sigma(\delta h) = \sqrt{\sigma_1(h)^2 + \sigma_2(h)^2} \qquad (2.7)$$

where $\sigma_1(h)$ and $\sigma_2(h)$ are the standard deviations of the height differences measured by the first and second surveys, respectively. These statistics describe only random errors that remain after appropriate corrections have been made to field observations. They do not account for uncorrected systematic errors. Following proper and consistent field procedures can reduce the impact of both types of error.

It is important to recognize that the expected standard deviation in height differences measured by leveling differs from the maximum allowable misclosure for double-run sections or circuits. The maximum allowable misclosure is by definition the worst acceptable case, so the expected standard deviation is only a fraction of the maximum allowable misclosure. As shown in Table 2.1, the standard deviation in height difference is approximately one-sixth as large as the maximum allowable misclosure for each order and class of leveling. So for a first-order, class I survey, the maximum allowable misclosure is $3.0 \, \text{mm} \times \sqrt{L}$, and the standard deviation is $0.5 \, \text{mm} \times \sqrt{L}$ (L is the distance in kilometers along the level line).

Another important distinction to keep in mind is that between the standard deviation of height differences measured by a single survey and the standard deviation of height *changes* measured by repeated surveys (i.e., between $\sigma(h)$ and $\sigma(\delta h)$). In the first case, the standard deviation of height differences among bench marks measured by any one survey of a given order and class is approximated by the values given in the last row of Table 2.1. For example, the standard deviation from random error of a height difference between two bench marks separated by 9 km, measured by a first-order, class I leveling survey, is approximately $0.5 \, \text{mm} \times \sqrt{L}$, or 1.5 mm. I say approximately because this value is representative of most first-order, class I surveys that adhere to commonly accepted standards and procedures, but the value for an individual survey can vary. A survey-specific value can be calculated from observed misclosures for double-run sections, using the equations given above. If we repeat our hypothetical first-order, class I survey, the standard deviation from random error of the measured height change is, from Equation 2.7:

$$\sqrt{(1.5)^2 + (1.5)^2} = 2.1 \, \text{mm}$$

A plot of $\sigma(\delta h)$ as a function of distance for various pairs of survey types is shown in Figure 2.27. For a line length of 20 km, $\sigma(\delta h)$ is 3.2 mm for two first-order, class I surveys; 5.0 mm for a first-order, class I survey paired with a second-order, class I survey; and 9.2 mm for a first-order, class I survey paired with a third-order survey. The attainable accuracy of

vertical surface displacements measured by repeated leveling surveys is controlled to a large degree by the accuracy of the higher order (i.e., less accurate) survey. Because the accuracy of past surveys is beyond our control and older surveys tend to be of higher order, there is a limit on the accuracy of vertical displacements attainable with even the most accurate repeat survey. Thus, the strength of modern, first-order surveys is compromised to some extent when their results are compared with those from less accurate historical surveys. On the other hand, for deformation that accumulates slowly over several decades, this shortcoming might be outweighed by the leverage of time. In such cases, there is simply no substitute for a survey of any order done long ago. That said, future generations of geodesists and volcanologists will have to rely on 'historical' data acquired during our lifetimes for comparison with their more-precise survey results, so it is incumbent on those of us working in the field today to adhere to the highest possible standards.

Decisions about the order and class of future surveys should be based on the desired accuracy and the expected deformation rate. In many cases, the increased accuracy of first-order surveys justifies the additional effort. On the other hand, if deformation rates are high or resources are limited, second- or even third-order surveys are sometimes a useful alternative for acquiring important time series data. In my experience at the USGS David A. Johnston Cascades Volcano Observatory (CVO), first-order, class II surveys have proven to be a good compromise between accuracy and effort. In Hawai'i, where deformation rates typically are higher than in Cascadia, the USGS Hawai'ian Volcano Observatory (HVO) generally has opted for second-order, class II surveys.

An important requirement for first-order surveys is that the near-surface vertical temperature gradient be measured at each setup. Temperature differences along the line-of-sight from the leveling instrument to the leveling rods create a vertical gradient in the index of refraction of air, which in turn causes light paths to bend slightly. The result, especially along sloping sections where surface heating is strong and there is little wind to mix the lower atmosphere, is a systematic error that accumulates with distance and elevation change. If the temperature is measured at two levels on the tripod that holds the leveling instrument or, better, on each leveling rod, a correction can be applied after the fact to partly remove this source of systematic error. Other potential sources of systematic error include rod scale error

Table 2.1. Specifications for vertical control surveys in the U.S.A., maximum allowable misclosures, and estimated standard deviations associated with surveys that meet these specifications (Vanicek *et al.*, 1980). L is the distance along a traverse in kilometers.

Specifications		Classification		
	First order, Class I and II	Second order		Third order
		Class I	Class II	
Specifications				
Instrument standards	Automatic or tilting levels with parallel plate micrometers; Invar scale rods	Automatic or tilting levels with optical micrometers or three-wire levels; Invar scale rods	Geodetic levels and Invar scale rods	Geodetic levels and rods
Field procedures	Double-run; forward and backward, each section	Double-run; forward and backward, each section	Double or single run	Double or single run
Section length	1–2 km	1–2 km	1–3 km for double run	1–3 km for double run
Maximum length of sight	50 m class I; 60 m class II	60 m	70 m	90 m
Maximum difference in forward and backward sights per set-up	2 m class I; 5 m class II	5 m	10 m	10 m
... per section	4 m class I; 10 m class II	10 m	10 m	10 m
Maximum allowable misclosure				
Section; forward and backward	$3\,\mathrm{mm} \times \sqrt{L}$ class I $4\,\mathrm{mm} \times \sqrt{L}$ class II	$6\,\mathrm{mm} \times \sqrt{L}$	$8\,\mathrm{mm} \times \sqrt{L}$	$12\,\mathrm{mm} \times \sqrt{L}$
Loop or line	$4\,\mathrm{mm} \times \sqrt{L}$ class I $5\,\mathrm{mm} \times \sqrt{L}$ class II	$6\,\mathrm{mm} \times \sqrt{L}$	$8\,\mathrm{mm} \times \sqrt{L}$	$12\,\mathrm{mm} \times \sqrt{L}$
Standard deviation				
Accuracy of height difference between directly connected bench marks	$0.5\,\mathrm{mm} \times \sqrt{L}$ class I $0.7\,\mathrm{mm} \times \sqrt{L}$ class II	$1.0\,\mathrm{mm} \times \sqrt{L}$	$1.3\,\mathrm{mm} \times \sqrt{L}$	$2.0\,\mathrm{mm} \times \sqrt{L}$

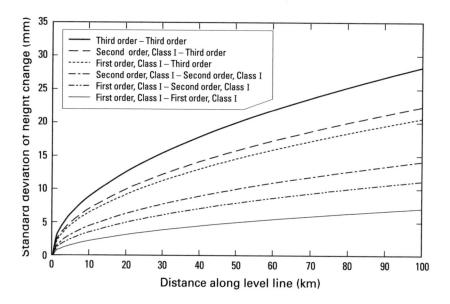

Figure 2.27. Standard deviation of height change measured by pairs of leveling surveys of various orders and classes as a function of distance along the level line. For example, the second curve from the top (long dashes) shows the standard deviation of height changes obtained by differencing results from a second-order, class I survey with those from a third-order survey. Standards for various types of surveys are from Table I of Vanicek *et al.* (1980).

and pin settling (the tendency for a rod to settle slightly between the time it is read as a foresight in one setup and then as a backsight in the next) (Bomford, 1980, pp. 213–216). Having the rods calibrated annually and applying the resulting corrections to field observations can largely remove the effect of rod scale error. Pin settling can be reduced by using well-designed turning plates or turning pins, by ensuring that these are firmly placed at each turning point, and by observing other appropriate field procedures (Federal Geodetic Control Committee, 1984). The cumulative amount of pin setting increases systematically in the direction that a given section is observed, so its effect can be reduced greatly by running adjacent sections in opposite directions. For example, if the level line starts at bench mark A and runs through marks B, C, and D before ending at E, the effect of pin settling will be reduced by running from A to B, then C to B, then C to D, and finally E to D. This is the best approach, but it can be time consuming and logistically difficult for the leveling crew to reverse direction so often. Alternatively, for longer surveys, the running direction can be reversed from morning to afternoon each day or on successive days. The important thing is to divide the entire survey into a large number of adjacent segments to be run in opposite directions, where 'large' refers to a number typically between 10 and 100.

Systematic error in leveling surveys received considerable attention during the 1970s and early 1980s as a result of the so-called Palmdale Bulge, a broad uplift feature discovered by leveling surveys in the Palmdale–Lancaster area about 80 km north of Los Angeles, California. For years, a debate raged over the true amount of uplift, if any, and the relative contributions of systematic error and real surface displacement (Strange, 1981; Stein, R.S., 1981; Holdahl, 1982; Castle *et al.*, 1983; Stein *et al.*, 1986; Castle, 1987). The issue was never completely resolved to everyone's satisfaction, but eventually a consensus emerged that uncorrected systematic errors were at least partly responsible. Much was learned about sources and mitigation strategies for systematic errors, and modern surveys now pay careful attention to field procedures, equipment calibration, and data processing to reduce the magnitude of uncorrected errors. Of course, problems persist in older surveys that may not be correctable in hindsight, but even so it is possible to recognize the existence of various types of systematic error by applying such tests as circuit misclosures, comparison of results for relatively flat and steep terrain, and correlation between slope and tilt (i.e., between the height difference and the height change within a given section measured by two surveys). An unfortunate shortcoming of the slope versus tilt technique applied to volcanoes is that deformation often mimics topography (i.e., uplift or subsidence is often centered near the summit), making it difficult to discriminate between real displacement and slope dependent error. In such cases, an independent technique such as GPS should be used to verify the result (assuming the apparent displacement is large enough to be measured in this way).

The only techniques for measuring vertical displacements that might approach the precision of first-order leveling over distances of tens of kilometers or more are combined GPS + radiometer + barometer surveys (Alber *et al.*, 1997) and GPS-guided aircraft laser altimetry. Neither technique is currently in wide use by volcanologists, and this situation is unlikely to change in the near future. Logistical and technical challenges (e.g., the need for costly specialized equipment, aircraft, and data processing) will continue to limit the utility of these state-of-the-art techniques during volcano crises. Other factors, including steep terrain and difficult access, mitigate against leveling at many volcanoes, but in some cases the effort is justified if vertical accuracy is especially important. Often leveling is best used in conjunction with other geodetic techniques to improve the quality of a geodetic dataset. For example, leveling and GPS surveys can be done concurrently to obtain mm-scale precision for vertical and horizontal displacements, respectively. Alternatively, as at Long Valley Caldera, GPS can be calibrated and bracketed in time by less-frequent leveling surveys to track 3-D displacements without sacrificing vertical accuracy.

2.5.2 Single-setup leveling

Leveling surveys can be used to serve two different but often complementary purposes. Long level lines (typically tens to hundreds of km) measure near-absolute height changes with respect to a distant bench mark or group of bench marks, which is assumed to be stable. Short lines (typically 1 km or less and L- or T-shaped) measure relative displacements among a small group of marks, usually for the purpose of determining local ground tilt. This approach is sometimes referred to as tilt leveling or *dry tilt*.[6] Single-setup leveling is a special type of tilt leveling in which relative displacements within a small array of marks are measured from a single instrument setup to determine local ground tilt. Each form of leveling has advantages and disadvan-

---

[6] The term *dry tilt* was coined at the USGS HVO in the late 1960s to distinguish the technique from the *wet tilt* method, in which water and air hoses were used to connect brass containers (pots) attached to separate concrete piers, thus forming a long-base watertube tiltmeter (Eaton, 1959; Yamashita, 1992). The air hoses assured equal pressure between pots. Micrometers were used to measure the height of water in the pots simultaneously, so elevation differences between piers could be calculated precisely. Wet tilt measurements are no longer made so the term dry tilt is obsolete and its origin is becoming obscure, but it still appears in the volcanological literature.

tages that determine which is most appropriate for a specific application. One effective strategy is to measure short segments of a long level line periodically, then measure the entire line when significant changes are detected.

Single-setup leveling is particularly useful when deformation rates are high or surveying time is very limited (e.g., for safety reasons). If terrain permits, small arrays of 3–6 bench marks with 30–50 m spacing can be installed and height differences measured from the center of each array (Yamashita, 1992) (Figure 2.28). The length of the leveling rods limits the maximum height difference among the marks, typically to about 2.5 m. Therefore, such an array requires an area of relatively flat ground that might not exist on a steep-sided volcano.

A variant on the technique called single-setup trigonometric leveling can be used where flat sites are unavailable. In this case, a total station consisting of an electronic theodolite and EDM is used to measure height differences among a small array of bench marks (Figures 2.29–2.31). Advantages of trigonometric leveling include: (1) height differences between marks are not limited by the length of leveling rods, so site selection is more flexible; (2) reflector targets are more portable than leveling rods; (3) the same equipment can be used for trigonometric leveling, EDM, and triangulation surveys; and (4) bench mark arrays for single-setup trigonometric leveling can be larger than for single-setup differential leveling, which lessens the effect of random-walk bench mark instability (Ewert, 1992). The accuracy of the two techniques is about the same; therefore, the trigonometric variant is preferable, especially in uneven or steep terrain.

The layout of a typical single-setup leveling station is shown in Figure 2.32. The simplest array for measuring two components of the tilt vector (usually north and east) comprises three bench marks arranged in an equilateral triangle. A single tilt component can be measured with a linear array of two or more marks. For differential leveling, this is sometimes necessary where flat areas suitable for triangular arrays are rare, but linear arrays are possible (e.g., along narrow ridges). Where topography permits, it is always a good idea to build redundancy into the array by including additional bench marks. A particularly strong layout in this regard includes two triangles that overlap to form a six-pointed star. In this case, two independent tilt vectors can be compared to estimate the

Figure 2.28. A crew from the USGS HVO measures a single-setup leveling array at Sand Spit near the summit of Kīlauea Volcano, Hawai'i. Three self-stabilizing leveling rods (open arrows) are set on marks secured to bedrock (in this case, a 1919 lava flow with 1924 explosion ejecta from nearby Halemaumau crater scattered on its surface), forming a triangular array with approximately 40 m sides. A compensator-type leveling instrument is set up at the center of the array (solid arrow) and shaded with an umbrella to reduce thermal effects on the instrument and tripod. Height differences between the rods are measured and compared with results from earlier surveys to determine ground tilt. Such measurements are well suited to basaltic volcanoes, where the terrain is relatively flat and the risk to field crews is usually acceptable even during periods of unrest or eruptive activity. USGS photograph by James D. Griggs, January 1981.

measurement precision and reveal any bench mark instability within the array.

The following equations (2.8–2.10), modified slightly from Yamashita (1992) and Ewert (1992), can be used to calculate the north and east components of the tilt vector from relative height changes within a triangular array as measured by repeated surveys:

$$\tau(N) = \left[\frac{\cos\varphi}{XY\sin(\phi - \theta)} \times \Delta(Y - X) \right.$$
$$\left. + \frac{\cos\theta}{XZ\sin(\varphi - \theta)} \times \Delta(X - Z) \right] \times 10^6 \quad (2.8)$$

$$\tau(E) = \left[-\frac{\sin\varphi}{XY\sin(\phi - \theta)} \times \Delta(Y - X) \right.$$
$$\left. - \frac{\sin\theta}{XZ\sin(\varphi - \theta)} \times \Delta(X - Z) \right] \times 10^6 \quad (2.9)$$

where $\tau(N)$ and $\tau(E)$ are the north and east components of the tilt vector, in microradians; XY and XZ are the lengths of two sides of the array, in meters; $\Delta(Y - X)$ and $\Delta(X - Z)$ are the changes in height difference between pairs of bench marks, in meters; and angles φ and θ define the orientation

of the array, as shown in Figure 2.32. The magnitude of the tilt vector is:

$$\tau = \sqrt{(\tau(N)^2 + \tau(E)^2)} \quad (2.10)$$

To calculate the bearing of the tilt vector, we first have to determine whether it lies in the first (NE), second (SE), third (SW), or fourth (NW) quadrant by using the following rules. If $\tau(N)$ is positive, the tilt vector is in the first or fourth quadrants. If $\tau(N)$ is negative, the tilt vector is in the second or third quadrants. If $\tau(E)$ is positive, the tilt vector is in the first or second quadrants. If $\tau(E)$ is negative, the tilt vector is in the third or fourth quadrants. Thus, if $\tau(N)$ equals -10.3 microradians and $\tau(E)$ equals $+5.2$ microradians, the tilt vector lies in the southeast quadrant. Now, we can calculate the vector's bearing Θ, using the equation:

$$\Theta = \tan^{-1}\frac{\tau(E)}{\tau(N)} + n\pi \quad (2.11)$$

where $n = 0$ if the vector lies in the first quadrant, $n = 1$ in the second or third quadrant, and $n = 2$ in the fourth quadrant. In the example above,
$$\Theta = \tan^{-1}\frac{+5.2}{-10.3} + \pi = 153°. \text{ By convention, } \Theta \text{ is}$$
measured clockwise from north and ranges from

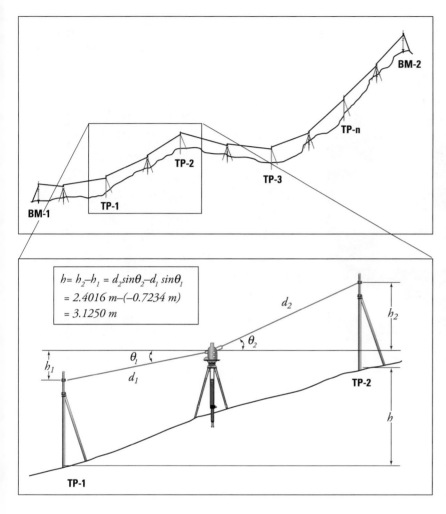

$$h = h_2 - h_1 = d_2 \sin\theta_2 - d_1 \sin\theta_1$$
$$= 2.4016\ m - (-0.7234\ m)$$
$$= 3.1250\ m$$

Figure 2.29. In trigonometric leveling, height differences between a series of adjacent bench marks (BM-1, BM-2) or turning points (TP-1, TP-2, etc.) are determined by measuring vertical angles θ_1 and θ_2 and line-of-sight distances d_1 and d_2, using a total station and specially designed reflector targets placed on each point (Figures 2.30 and 2.31). Modern total stations combine an electronic theodolite for measuring vertical angles and an EDM for measuring line-of-sight distances. This technique is especially useful in steep terrain, because the height difference between points that can be measured at each set-up is not limited by the length of leveling rods, as is the case for differential leveling. Single-setup trigonometric leveling can be used to measure tilt changes within a small array of bench marks where the terrain is too steep for single-setup differential leveling (Ewert, 1992). Adapted from Kozlowski (1998).

$0°$ to $360°$. If $\tau(N)$ is positive (or negative), tilt is *down* to the north (or south). If $\tau(E)$ is positive (or negative), tilt is *down* to the east (or west).

With care, these equations can be applied to single-setup leveling using either a differential leveling instrument or a total station. However, there is considerable potential for errors in sign (i.e., $+/-$) to occur, as a result of several confusing factors related to historical precedent and arbitrary sign conventions. In the equations above, $\Delta(Y - X)$ and $\Delta(X - Z)$ refer to *changes in height difference* between pairs of bench marks. Thus, if during some interval Δt, X moved up with respect to Y, and X moved down with respect to Z, $\Delta(Y - X)$ and $\Delta(X - Z)$ are both negative. Confusion arises from the fact that the same notation has been used to refer to *changes in the difference between rod readings* when a leveling instrument is used to

measure the array (e.g., Yamashita, 1992). In the example above, both $\Delta(Y - X)$ and $\Delta(X - Z)$ would be positive because a lower rod produces a higher reading, and vice versa. If this convention is used, the signs of both terms in (2.8) and (2.9) must be reversed.

Another source of potential confusion and error is the fact that angles θ and φ are traditionally measured counterclockwise from east (with X as the southernmost bench mark in the array), whereas the bearing Θ is measured clockwise from north. In addition, it is somewhat counterintuitive that *positive* tilt is taken to mean *downward* movement (i.e., $\tau(E) = +5.2$ microradians means the ground tilted down to the east). Finally, the $n\pi$ ambiguity in the tangent function (2.11) offers yet another opportunity for error when calculating the tilt vector's bearing. Suffice it to say that meticulous adherence to

Figure 2.30. T-2000 electronic theodolite with a DI-5 EDM mounted on the theodolite telescope. This setup is useful for single-setup trigonometric leveling. From Ewert (1992).

Figure 2.31. Reflector target for single-setup trigonometric leveling set up over an expansion bolt (*lower inset*) in a lava flow (Ewert, 1992). The target-cross and prism are set apart vertically (*upper inset*) the same distance as the theodolite telescope and EDM (Figure 2.30).

conventions and consistent field procedures will produce the correct result, but devising some means to double-check the answer is always a good idea.

2.5.3 Geodetic leveling

The more conventional form of differential leveling, which is often referred to as geodetic leveling or simply leveling, also can be applied to volcanoes in some cases. The field procedures are similar to those used for single-setup differential leveling, but in this case the bench mark array is larger and takes the form of a level line, a circuit, or several interconnected lines or circuits that encompass the area of interest. Height differences between adjacent marks are measured by leapfrogging the leveling instrument and rods through a series of setups in the manner described in Section 2.5.1. All of the sections in the network are measured in this way, not necessarily sequentially, to complete a survey. By repeating the survey at a later date and assuming that the height change at some mark or group of

marks in the network (usually distant from the volcano) is zero, height changes relative to this assumed reference or datum can be determined. For modeling purposes, however, it is important to use the height changes between adjacent marks, rather than the cumulative height change at each mark with respect to the datum, to avoid correlated errors that accumulate along the level line (Arnadottir *et al.*, 1992).

For the reader with experience on a leveling crew, the prospect of surveying on a steep-sided stratovolcano, especially one threatening to erupt, probably conjures up visions of excruciatingly short setups and slow progress at a time when speed or a safer location would be a primary concern. The considerable merits of leveling notwithstanding, leveling on steep, restless volcanoes is to be avoided. In addition to safety concerns, the slow pace dictated by steep topography usually makes such surveys impractical or cost-ineffective. For this application, single-setup leveling or short ($L \sim 1$ km) level lines

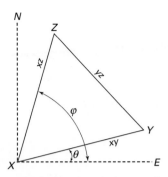

Figure 2.32. Layout of a typical triangular array for single-setup leveling. Bench marks are located at X, Y, and Z. A differential level or total station is set up near the center of the array to measure height differences between each of three bench mark pairs. Lengths xy, xz, and yz are typically 30–40 m if a differential level is used or up to 100 m if a total station is used. Additional bench marks can be added for redundancy.

to measure tilt changes is usually a better alternative. Triangles, L-shaped, or T-shaped arrays are preferable, but in steep terrain linear arrays are often the only practical alternative to trigonometric leveling.

2.5.4 Tilt-leveling results at South Sister Volcano, Oregon

Such is the case at South Sister Volcano, one of five large cones that make up the Three Sisters volcanic center in central Oregon, U.S.A. (also Middle Sister, North Sister, Broken Top, and Mount Bachelor) (Figure 2.33). The most recent eruptions in the area occurred about 1,500 years ago (Scott *et al.*, 2001). In 1985, the USGS Volcano Hazards Program (VHP) established four linear leveling arrays at South Sister to monitor ground tilt (Yamashita and Doukas, 1987). They range in length from 196 m to 317 m and are located along radial ridges on the north, east, south, and west flanks of the volcano. Better precision might have been realized with longer lines, but limited resources required a compromise. Despite numerous short setups (5–10 m) necessitated by the steep terrain, the four arrays can be double-run in a few days with helicopter support. They were measured in 1985, 1986, and 2001 following the discovery by radar interferometry of broad, aseismic uplift in the area (Wicks *et al.*, 2001, 2002a,b) (Chapter 5).

The uplifting area is roughly circular and centered about 5 km west of the summit of South Sister, but it extends eastward across all four of the tilt-leveling stations. Owing to their radial orientation with respect to the summit, the lines on the north and south

flanks of the volcano are mostly insensitive to uplift centered to the west. Tilt changes at both stations were less than 3 microradians from 1986 to 2001. On the other hand, the stations on the east and west flanks are well situated with respect to the uplift center, and the changes there were about 8 and 22 microradians, west up, respectively. The tilt changes observed at all four stations for 1986–2001 match those predicted by a best fit Mogi model based on the radar interferometry observations for 1995–2001 (Figure 2.34), so the uplift probably started sometime after 1995. In fact, Wicks *et al.* (2002b) produced several more interferograms of the area that span 1992–2001 and concluded that the uplift most likely began in 1997 or 1998. Continuous GPS observations from a single station within the deforming area showed that uplift continued at a relatively steady rate of 3–4 cm yr^{-1} at least through the end of 2003. The background level of seismic activity throughout central Oregon has been very low historically, and to date only one swarm of ∼350 small earthquakes ($M < 2$, 23–25 March 2004) has accompanied the Three Sisters uplift.

2.5.5 Repeated leveling surveys at Medicine Lake Volcano, California

At gently sloping shield volcanoes or large silicic calderas, leveling can be both practical and extremely effective for measuring vertical surface displacements over a wide area. Consider our experience at Medicine Lake Volcano, a large Pleistocene–Holocene shield volcano located about 50 km east–northeast of Mount Shasta, between the crest of the Cascade Range to the west and the Basin and Range tectonic province to the east (Figure 2.35). The great majority of Medicine Lake Volcano's lavas are basaltic andesite or andesite, although rhyolite and rare dacite are found on the upper parts of the edifice (Donnelly-Nolan, 1988; Donnely-Nolan *et al.*, 1990, 1991). During the most recent eruptive episode between 3,000 and 900 years ago, eight eruptions produced approximately 2.5 km^3 of lava ranging in composition from basalt to rhyolite (Donnelly-Nolan *et al.*, 1990). The broad, gently sloping edifice is capped by Medicine Lake Caldera, a 7 × 12 km depression interpreted by Donnelly-Nolan (1988) as a collapse feature that formed in response to repeated extrusions of mostly mafic lava. A 640-acre (2.6 km^2 or 260 hectares) lake called Medicine Lake is situated in the southwest part of the caldera (Figure 2.35).

Figure 2.33. Aerial view looking northwest at the Three Sisters volcanic center, including (left to right) South Sister, Middle Sister, North Sister, and Broken Top (foreground). Nearby Mount Bachelor is out of view to the lower left. Light-colored areas on the south flank of South Sister are 2,000-yr-old lava flows. Much of area on the lower flanks of Broken Top is mantled by pumice and ash erupted just prior to emplacement of these lava flows. USGS photograph by William E. Scott.

The relatively gentle slopes of Medicine Lake Volcano are amenable to leveling and in 1954 the US Coast and Geodetic Survey measured a 193-km-long leveling circuit that crosses the volcano, part of a nationwide vertical control network. The leveling route climbs from the small town of Bartle at about 1,200-m elevation, up the volcano's south flank to the south caldera rim near 2,000 m, skirts the shore of Medicine Lake before crossing the west caldera rim at 2,130 m, descends the west flank to 1,400 m, turns south and climbs over Stephens Pass at 1,720 m, then descends to 1,100 m before turning east back to Bartle. The idea of leveling across an entire shield volcano, especially one as large as Medicine Lake Volcano (surface area ~2,000 km², height above surroundings ~800 m) was daunting, but in 1989 the USGS CVO decided to repeat the

1954 survey to determine if there had been measurable height changes. The 1989 survey required a considerable amount of effort – about one person-year (10 people for 4–6 weeks each) – but the results were spectacular.

Comparison of the 1954 and 1989 surveys revealed a broad area of subsidence centered at Medicine Lake Caldera and extending across the entire volcano (Dzurisin et al., 1991, 2002) (Figure 2.36). The maximum subsidence measured was 302 ± 27 mm at bench mark T502 near the center of the caldera. This corresponds to an average rate of -8.6 ± 0.8 mm yr^{-1} during the 35-year interval spanned by the surveys. In addition to broad, relatively symmetric subsidence of the entire edifice, localized subsidence by as much as 20 cm occurred at two sites near the periphery of the volcano

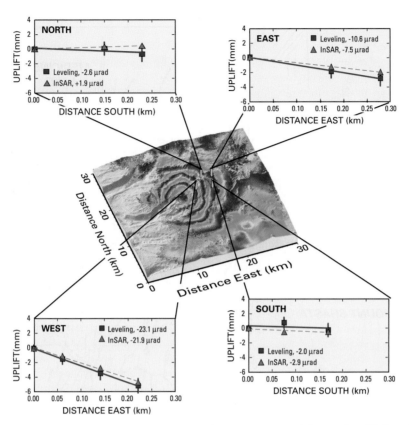

Figure 2.34. Radar interferometry and tilt-leveling results at South Sister volcano, central Oregon, from Wicks and others (2001). The interferogram in the center of the figure is for the period 1995–2001 and shows about 12 cm of surface uplift centered 5 km west of the summit of South Sister. The summit is near the center of four yellow bars, which mark the tilt-leveling stations. Each interferometric fringe (full color cycle) represents 2.83 cm of range change between the ground and the satellite. The data plots show relative height changes and average surface tilts for the period 1986–2001. Red triangles are predicted values based on a best-fit Mogi model of the interferogram. Blue squares are values observed by leveling.

(Figure 2.36): (1) at one bench mark located in the epicentral area of the 1978 Stephens Pass earthquake swarm about 30 km southwest of the summit; and (2) at three adjacent bench marks about 30 km south–southeast of the summit, amidst a group of Quaternary north-trending normal faults. Both cases are attributed to Basin-and-Range style faulting, the former during the 1978 Stephens Pass earthquake swarm (Dzurisin *et al.*, 1991, 2002).

The shape of the subsidence profile in Figure 2.36 mirrors that of the topography along the leveling circuit to a significant degree, suggesting the possibility of a slope dependent systematic error in one of the leveling surveys. Dzurisin *et al.* (1991) took pains to evaluate this possibility and concluded that slope dependent error could account for no more than one-third of the apparent subsidence. A follow-up survey in 1999 of a 20-km segment of the Medicine Lake circuit showed that the subsidence rates for 1954–1989 and 1989–1999 were essentially identical

(Dzurisin *et al.*, 2002). The 1999 survey was centered on the caldera and spanned the area of maximum subsidence, including about 75% of the total subsidence revealed by the 1954 and 1989 surveys. Because the 1989 and 1999 surveys both followed first-order, class II procedures but were done with different equipment, the likelihood that systematic error produced the same result as the 1954 and 1989 surveys is remote. Therefore, we can safely conclude that the summit has been subsiding at an average rate of 8–9 mm yr^{-1} for at least the past 50 years.

The leveling results for Medicine Lake Volcano pose several interesting questions. Why is the summit area sinking at a seemingly steady rate of 8–9 mm yr^{-1}? What mechanism accounts for the fact that subsidence extends across the entire volcanic edifice? What is the relationship between broad, mostly aseismic subsidence of the volcano and abrupt, fault related movements around its periphery? Dzurisin *et al.* (1991, 2002) hypothesized

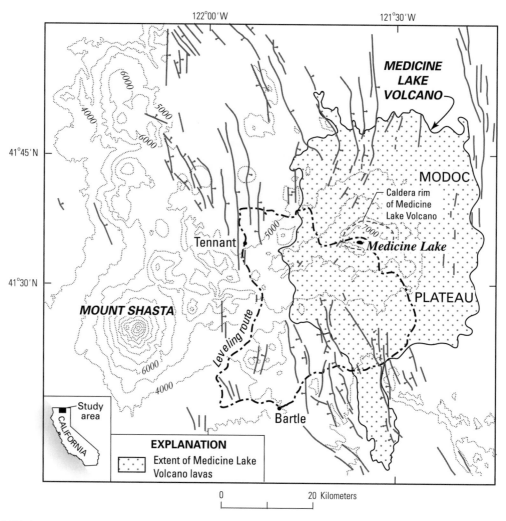

Figure 2.35. Location map showing the Medicine Lake Volcano–Mount Shasta region in the southern Cascade Range, northern California. Stipple pattern indicates extent of lavas of Medicine Lake Volcano, which cover nearly 2,000 km². Dashed line near summit, approximate outline of Medicine Lake Caldera; black dot inside caldera, Medicine Lake (body of water). With a volume of at least 600 km³, Medicine Lake Volcano is the largest volcano by volume in the Cascade Range (Donnelly-Nolan, 1988). Dash–dot line from Bartle counterclockwise past Medicine Lake, through Tennant, and back to Bartle represents a leveling circuit that was first measured in 1954 and remeasured in 1989. Contour interval is 1,000 feet (305 m).

that subsidence is caused by: (1) loading of the crust by the volcanic edifice and dense subvolcanic intrusions, and (2) crustal thinning due to Basin and Range extension. They envisioned a roughly columnar region beneath the volcano that has been intruded extensively by basalt. The addition of mass and heat causes the volcano and underlying crust to soften and subside. Basin and Range extension thins the weakened crust and facilitates subsidence. Earthquakes are rare within the heated crustal column but occur around its cooler, more brittle periphery. The inner zone deforms steadily and mostly aseismically, while the outer zone deforms

episodically by brittle failure during occasional earthquake swarms.

My purpose in describing this model is not to defend it, but rather to illustrate the type of inferences that can be drawn from repeated leveling surveys at volcanoes. This is especially true for volcanoes like Medicine Lake, where a considerable amount is known about the eruptive history, crustal structure, and recent seismicity. Leveling surveys can reveal processes that otherwise might go unnoticed, and provide a stimulus for further study. At Medicine Lake, for example, the leveling results spurred us to establish a regional network of GPS

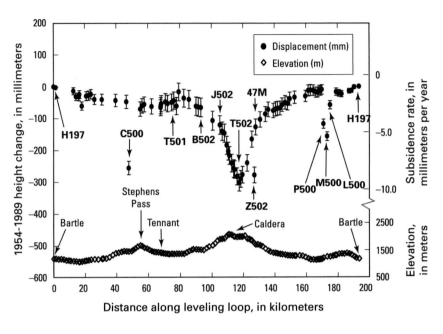

Figure 2.36. Height changes from 1954 to 1989 (*top*) and an elevation profile (*bottom*) along a 197-km leveling loop at Medicine Lake Volcano, northern California (Figure 2.35) (Dzurisin *et al.*, 1991). Bench mark H197 near Bartle was held fixed. Error bars represent one standard deviation from random error (Vanicek *et al.*, 1980).

stations in 1990 and to reoccupy it in 1996 and 1999 (Poland *et al.*, 1997; Dzurisin *et al.*, 2002).

2.6 PHOTOGRAMMETRY

Although photogrammetry is not used often to monitor volcano deformation, it can be a very useful tool in special situations. Conventional photogrammetry makes use of parallax that occurs between sequential photographs taken from an aircraft to construct a topographic map of the ground surface. All common points in a pair of stereo photographs that lie at the same elevation exhibit the same parallax shift, so it is possible to draw contour lines of equal elevation by viewing the photo pair with a specialized stereoscope and moving a cursor along lines of equal parallax. The Kern PG 2 Stereo Plotting Instrument was used for this purpose for decades by the USGS, which is widely known for its topographic quadrangle maps of the U.S.A. Attachments to the instrument produce electronic output in digital format, compatible with modern GISs. The next two sections include a few examples of photogrammetric applications to volcano monitoring. Chapter 6 provides a thorough treatment of the subject, including state-of-the-art photogrammetric equipment and products such as softcopy stereoplotters and high-resolution digital elevation models (DEMs).

2.6.1 Mapping the 1980 north flank 'bulge' at Mount St. Helens

Any difference between topographic maps made at different times represents surface displacements that occurred during the interval spanned by the maps. Most USGS topographic maps at 1:24,000 scale have a contour interval of 20 ft or 40 ft (about 6.1 m or 12.2 m), which is much too coarse for most volcano-monitoring applications. A striking exception was a series of six topographic maps of Mount St. Helens that were made from aerial photographs taken before and during the volcano's reawakening in 1980. These special-purpose maps, at 1:24,000 scale with a contour interval of 80 ft (24.4 m), clearly showed the development of a large bulge on the volcano's north flank and the removal of 2.7 km^3 of material from the summit area by the rockslide–debris avalanche and eruption of 18 May 1980 (Moore and Albee, 1981).

Much higher resolution of topographic and structural changes at a volcano is attainable by comparing detailed topographic maps made from low-altitude, high-resolution aerial photographs taken especially for this purpose. Following the 18 May 1980 eruption at Mount St. Helens, a series of topographic maps was constructed to document the morphologic evolution of the 1980 crater and the growth of a lava dome on the crater floor. From October 1980 to November 1986, four maps of the entire 1980 crater were made at 1:4000 scale with a contour interval of 5 m or 10 m, and eighteen maps of the lava dome and surrounding crater floor were made at 1:2000 scale with a contour interval of 2 m. The 1:4000-scale maps are very useful for measuring surface volume changes caused by dome growth, erosion of the crater walls and floor, and accumulation of rockfall debris and snow on the crater floor. By digitizing these maps and using the ArcINFO GIS, Mills and Keating (1992) and Mills (1992) calculated that, between 1980 and 1988, $30 \times 10^6 \, m^3$ of rock were eroded from the crater walls and deposited on the crater floor as $40 \times 10^6 \, m^3$ of talus. This corresponds to a mean retreat rate for the crater walls of $2.1 \, m \, yr^{-1}$, one of the highest rates of mass wasting in the world. During the same interval, an additional $28 \times 10^6 \, m^3$ of snow accumulated on the south crater floor, intercalated with rockfall debris. In 1988, the thickness of talus plus snow on the crater floor mostly exceeded 60 m on the west, south, and east sides of the dome, and part of the original south crater floor was buried by more than 100 m (Figure 2.37). Numerous large rockfalls from the crater walls had traveled entirely across the floor and impinged on the dome. These results were updated through 2000 and are analyzed further in Section 6.7.

With their 2-m-contour interval, the 1:2000-scale maps were well suited for documenting episodic growth of the lava dome at Mount St. Helens. Holcomb and Colony (1995) produced 18 geologic maps that showed how the dome changed in size and shape as it grew. These maps show that each growth episode was partly exogenous (lava extruded onto the surface of the dome) and partly endogenous (lava intruded within the dome), a fact that was discovered by repeated EDM and theodolite measurements of the dome during growth episodes. Exogenous growth produced a series of stubby lava flows (lobes) on the surface of the dome, while endogenous growth caused sectors of the dome to expand. Both processes triggered rockfalls from the dome that built a talus apron around it. This apron eventually grew large enough to be intercalated with rockfall debris from the crater walls. The 1:2000-scale maps also record changes in eruptive style through time. An early period of growth during 1980–1982 was episodic and dominated by exogenous growth. This was followed by a period of continuous, dominantly endogenous growth from February 1983 to February 1984. Thereafter, dome growth became episodic again while the frequency of growth episodes declined. During this waning phase, endogenous growth was dominant again.

These changes in eruptive style presumably reflected changes in magma supply rate and in the dome's steadily increasing capacity for internal growth.

Comparison of additional 1:4000-scale topographic maps of the crater that were created from aerial photographs acquired in 1988, 1992, and 2000 reveals that accumulation of snow, ice, and talus was continuing at rates that are very high by most geologic standards, especially in the south part of the crater that is mostly shielded from direct sunlight by the steep south crater wall. By 2002, the total accumulation since 1980 was more than 240 m on part of the south crater floor, where a $\sim 1 \, km^2$ rock-and-ice glacier was growing and advancing northward around the dome. The total volume of snow and ice stored within the crater since 1980 was approximately $80 \times 10^6 \, m^3$ (Schilling et al., 2002, 2004) (Chapter 6).

When this was being written in April 2005, a remarkable thing was happening to Mount St. Helens' crater glacier. Following a brief period of intense seismicity and uplift of the south crater floor beginning on 23 September 2004, a new dacite lava dome began to emerge from beneath the thickest part of the glacier, immediately south of the 1980–1986 dome, in October 2004 (Dzurisin et al., 2005; Section 11.3.4). As the whaleback-shaped extrusion grew, it bulldozed its way through the glacier to the south crater wall, splitting the glacier in half but causing surprisingly little melting. The glacier's east and west arms were deformed severely, and the east arm surged northward under the onslaught (Figures 11.7 and 11.8). More than six months after it began, the eruption was continuing and the fate of the glacier was uncertain.

Returning to the subject of tracking geomorphic changes with sequential topographic maps: even if sequential maps are impractical because, for example, resources are scarce or map production cannot be made timely enough to monitor a rapidly developing situation, useful information can be obtained from repeated stereo photographs taken with a calibrated, photogrammetric camera from the air or ground. If the viewing geometry is controlled carefully or several surveyed control points can be identified in the photographs, information on surface displacements can be derived from parallax measurements at an array of targets or recognizable features. For example, a growing lava dome or deforming area can be photographed repeatedly from two vantage points on the ground separated by an appropriate distance. Sequential parallax measurements relative to points that are surveyed

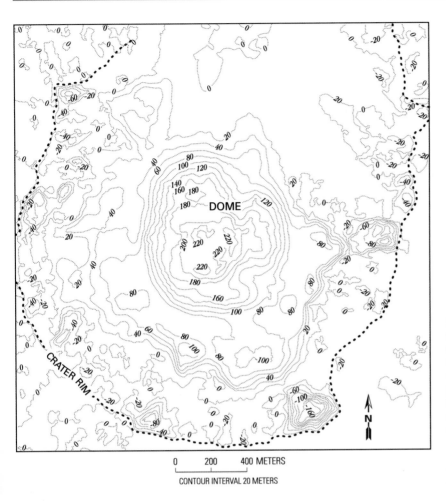

0   200   400 METERS

CONTOUR INTERVAL 20 METERS

Figure 2.37. Elevation changes in the 1980 crater at Mount St. Helens from 1980 to 1988 determined from sequential topographic maps (Mills, 1992). The dome attained a height of more than 220 m above the 1980 crater floor, while retreat of the crater walls reduced spot elevations by more than 100 m in places near the crater rim (dashed line). Accumulation of rockfall debris and snow buried parts of the crater floor by more than 100 m.

by some other means or are assumed to be fixed can be made to determine point displacements. Under favorable conditions an ~10 cm accuracy is achievable with this technique and relative motions of a few tens of cm are resolvable.

2.6.2 Oblique-angle and fixed-camera photogrammetry

In some cases, useful information about ground deformation can be gleaned from photographs even without using a photogrammetric camera. With specialized hardware and software, measurements of surface displacements with a precision of about 1 m can be obtained from oblique photographs taken with a standard, small format (35 mm or 70 mm film size) camera (Dueholm and Pillmore, 1989) or a high-resolution digital camera. Such photographs are much easier to obtain and less expensive than conventional vertical aerial photographs, and therefore the technique might be more widely applicable to volcano monitoring. Small for-

mat, oblique-angle photogrammetry has been tested at Augustine Volcano, Alaska, and at Mount St. Helens, with encouraging results. During the 2004–2005 dome-building eruption at Mount St. Helens, oblique stereo photographs obtained from a helicopter with a single-lens-reflex digital camera were used to supplement conventional vertical air photos for the production of sequential DEMs. Unfortunately, there have been few other applications because the hardware and software required to analyze the photographs are not yet widely available.

A second variation on standard aerial photogrammetry has been used to measure movements of a growing lava dome at Unzen Volcano, Japan. Rather than using two camera stations to produce parallax, Yamashina *et al.* (1999) took repeated photographs of the Unzen dome from a single station and took advantage of the parallax shift produced by relative movements of points on the dome between photos. They used a standard 35-mm camera with a 500-mm focal length lens at a fixed vantage point 2.3 km away from the dome. The technique is called time-differential stereoscopy and had been used to track the motion of an ash plume at Sakurajima Volcano, Japan (Murai, 1979). When pairs of the Unzen photographs are viewed stereoscopically, relative movement of points is perceived as relief. Points that moved up or down (or left or right, depending on the orientation of the photos in the viewer) are seen as standing above or below more stable parts of the dome. Parallax measurements can be made on the photos to obtain quantitative estimates of surface displacements. The accuracy of the technique depends on such factors as the equipment used, viewing conditions, and quality of the photographs. Yamashina *et al.* (1999) were able to measure accelerated movement of a large block before it collapsed to produce a pyroclastic flow, and to track the growth of a lava spine at an average rate of $0.8 \, \text{m} \, \text{day}^{-1}$. As long as a safe vantage point for photography is available, this simple technique would seem to be widely applicable at volcanoes where rapid motions are occurring.

2.7 MICROGRAVITY SURVEYS

2.7.1 Physical principles

Repeated gravity surveys provide another useful tool for studying ground deformation at volcanoes, and they have the added advantage of providing information about subsurface mass changes (Figure 2.38). In the absence of other effects that will be addressed shortly, repeated gravity surveys are capable of measuring surface height changes over distances of a few tens of kilometers with a precision of about 2 cm, which is roughly comparable to GPS and theodolite surveys. More important for most applications, including volcano monitoring, is the capability to detect and quantify changes in subsurface mass or density, if surface height changes are determined independently. Such changes can be caused by a variety of processes that commonly occur at volcanoes, including magma movements and variations in the groundwater table. Simultaneous, independent measurements of gravity and height changes, therefore, can provide an especially strong constraint on models of volcanic processes. Following a brief discussion of the physical principles involved, I cite a favorite example from Hawai'i that illustrates the utility of this approach for helping to decipher complex events at volcanoes.

The local gravitational acceleration at any point

on the Earth's surface g, is the force exerted on a unit mass m, by the mass of the Earth M_E:

$$g = \frac{F}{m} = \frac{GM_E}{R^2} \qquad (2.12)$$

where G is the universal gravitation constant, $6.67 \times 10^{-11} \, \text{N} \, \text{m}^2 \, \text{kg}^{-2}$), R_E is the mean radius of the Earth in meters, m and M_E are expressed in kilograms, and F is expressed in newtons (N).[7] Because the Earth is neither a perfect sphere nor uniform in terms of its mass distribution, the value of g at the surface varies as a function of both elevation and subsurface density. Thus, g has a worldwide average of about 980 Gal ($9.8 \, \text{m} \, \text{s}^{-2}$), with a range from the equator to the poles of about 5 Gal.

More important for volcano-monitoring purposes, the value of g at any point on the surface can vary as a function of time as a result of changes in surface height and (or) subsurface mass distribution – two processes that commonly occur at volcanoes. Temporal gravity changes also occur as a result of solid Earth tides, ocean tides, and variations in atmospheric pressure, but these effects are not of primary interest for volcano monitoring and, therefore, are removed routinely from field observations during data analysis. Repeated gravity surveys corrected for these effects have the potential to provide information on both surface displacements (i.e., uplift or subsidence) and subsurface mass changes (caused, e.g., by accumulation or withdrawal of magma or groundwater). What's more, if surface height changes are measured independently (e.g., by leveling surveys or GPS), the free-air correction can be applied to the gravity data to remove the effect of those changes, and any remaining signal can be interpreted solely in terms of subsurface mass changes. Thus, we can obtain direct information on magma or groundwater movements that otherwise might go undetected. Independent surface-height measurements are particularly important for modeling and unambiguous interpretation of microgravity data.

Now, let us address the same topic a little more quantitatively to see what precision is required to measure the kinds of changes we might expect at volcanoes. The theoretical free-air gravity gradient is $-3.086 \, \mu\text{Gal} \, \text{cm}^{-1}$ (1 Gal = 1 cm s^{-2} = $10^6 \, \mu\text{Gal}$).

---

[7] The Newton (N) is the Standard International (SI) unit of force: $1 \, \text{N} = 1 \, \text{kg} \, \text{m} \, \text{s}^{-2}$. Thus, the universal gravitation constant, $G = 6.67 \times 10^{-11} \, \text{N} \, \text{m}^2 \, \text{kg}^{-2}$, can also be expressed as $6.67 \times 10^{-11} \, \text{m}^3 \, \text{kg}^{-1} \, \text{s}^{-2}$. Dimensional analysis of (2.12) using the latter expression for G shows that the units of g are m s^{-2}, as expected.

Figure 2.38. The author using a gravity meter at East Dome on the east flank of Mount St. Helens during a small phreatic eruption in April 1980. Ash covering snow is from a series of such eruptions that preceded the large debris avalanche and magmatic eruption on 18 May 1980. At each gravity station, measurements were made at two sites separated by no more than a few meters with each of two LaCoste & Romberg gravimeters. This redundancy made it easier to identify measurement errors or instrument tares (Jachens et al., 1981). USGS photograph by Carter W. Roberts.

This means that, in the absence of other effects, for each centimeter of height increase the local gravitational acceleration g decreases by 3.086 μGal. Departures from this value of up to ~40% can be caused by local terrain effects and shallow density anomalies (Rymer, 1996, and references therein), so the actual free-air gradient should be measured at each station if possible. This can be accomplished by measuring gravity first at the surface and then at some known distance above the ground (e.g., ~1 m by setting the gravity meter on a surveying tripod). For repeated surveys, departures from the theoretical free-air gradient can be assumed to be constant between surveys. Modern portable gravity meters such as those produced by LaCoste & Romberg–Scintrex, Inc. are capable of measuring g with a precision of about 1 μGal, which corresponds to a height difference of about 0.3 cm. In practice, though, uncorrected instrument drift or other unmodelled effects typically limit the repeatability (accuracy) of this type of measurement to a few mGal, corresponding to a height uncertainty of ±1–2 cm.

The foregoing discussion applies if our gravity meter rises into thin air, but in the more interesting case that the ground beneath it also rises, we have to account for the Bouguer gravity effect. We can imagine that uplift is caused by the addition of a horizontally infinite flat slab of material with density ρ and thickness h. The slab's gravitational effect is given by:

$$g = 2\pi G \rho h \qquad (2.13)$$

where G is the universal gravitation constant ($6.67 \times 10^{-11}\,\mathrm{N\,m^2\,kg^{-2}}$). In this case, the ratio of the gravity change to the height change at a given station $\Delta g / \Delta h$ is a combination of the free-air effect and the Bouguer effect (Bomford, 1980, pp. 451–453):

$$\frac{\Delta g}{\Delta h} = -3.086\,\mu\mathrm{Gal\,cm^{-1}} + 2\pi G \rho \qquad (2.14)$$

For a typical crustal value of $\rho = 2.7\,\mathrm{g\,cm^{-3}}$, $\Delta g / \Delta h = -1.95\,\mu\mathrm{Gal\,cm^{-1}}$.

For volcanologists, the infinite horizontal slab model corresponds nicely to the case of an intruding sill, but a spherical model might be more appropriate for a shallow magma reservoir. In this case, the Bouguer gradient is given by (Telford *et al.*, 1990; Turcotte and Schubert, 2002):

$$\frac{\Delta g}{\Delta h} = -3.086\,\mu\text{Gal}\,\text{cm}^{-1} + \frac{4\pi G\rho}{3} \qquad (2.15)$$

For $\rho = 2.7\,\text{g}\,\text{cm}^{-3}$, which is appropriate for non-vesiculated molten basalt, this corresponds to $\Delta g/\Delta h = -2.33\,\mu\text{Gal}\,\text{cm}^{-1}$. Similar calculations can be done for other shapes, and any departure from the expected gradient can be interpreted in terms of a departure from the preferred model.

For example, if a spherical reservoir were inflated by magma with a bulk density of $2.7\,\text{g}\,\text{cm}^{-3}$, gravity and height changes measured at the surface above the reservoir would follow a Bouguer gradient of $-2.33\,\mu\text{Gal}\,\text{cm}^{-1}$. If, instead, the intruding magma had a density of $2.3\,\text{g}\,\text{cm}^{-3}$, the Bouguer gradient would be $-2.44\,\mu\text{Gal}\,\text{cm}^{-1}$. The corresponding values for a horizontal infinite slab model are $-1.95\,\mu\text{Gal}\,\text{cm}^{-1}$ ($\rho = 2.7\,\text{g}\,\text{cm}^{-3}$) and $-2.12\,\mu\text{Gal}\,\text{cm}^{-1}$ ($\rho = 2.3\,\text{g}\,\text{cm}^{-3}$). By plotting Δg as a function of Δh for a network of stations and determining the slope of the best-fit line by linear regression, we can distinguish among an array of possible models (Figure 2.39).

To separate the effects of height changes and mass changes on gravity data, it's useful to define the free-air corrected gravity change $\Delta g'$ as the difference between the observed gravity change Δg and the free-air gravity change corresponding to a measured height change Δh (Rymer, 1996):

$$\Delta g' = \Delta g(\mu\text{Gal}) - 3.086\,\mu\text{Gal}\,\text{cm}^{-1} \times \Delta h(\text{cm}) \qquad (2.16)$$

Because the free-air effect has been removed, $\Delta g'$ represents that portion of an observed gravity change associated with a subsurface mass change: $\Delta g' > 0$ means that subsurface mass has increased and $\Delta g' < 0$ means that subsurface mass has decreased. Of course, this assumes that the local free-air gradient is $-3.086\,\mu\text{Gal}\,\text{cm}^{-1}$. Any departure from this value based on measurements of the actual free-air gradient would have to be factored into the analysis. With this in mind, we can plot $\Delta g'$ as a function of Δh and interpret the results directly in terms of subsurface mass changes. For this type of analysis, an accurate measurement of height changes, preferably by leveling or GPS, is essential to adequately separate the free-air and Bouguer gravity effects.

Williams-Jones and Rymer (2002) used a similar approach as the basis for a model that predicts how the observed $\Delta g/\Delta h$ gradient might evolve as the state of a volcano changes from dormancy through unrest to eruptive activity. Their work assumes that the resulting surface uplift conforms to the Mogi model, but a similar analysis could be done for other types of volcano sources (e.g., ellipsoid, cylinder). A key result is that the observed gravity gradient is sensitive to changes in both mass and density within a magma reservoir. Data that plot above the free-air gradient line in a $\Delta g/\Delta h$ diagram (e.g., Figure 2.39) reflect a mass increase, while data that plot below the line reflect a magma decrease. Similarly, data that plot above the Bouguer corrected line for a particular source and magma density reflect a density increase, while data that plot below the line reflect a density decrease (e.g., Rymer *et al.*, 1995; Rymer and Williams-Jones, 2000).

It follows that, for the case of inflation (i.e., $\Delta h > 0$ and $\Delta g < 0$, as in the lower right quadrant of Figure 2.39), data that plot between the two lines are diagnostic of a process in which mass increases but density decreases. This would be the case, for example, if turbulent mixing occurred between magma that was resident in a reservoir and new magma that was intruding the reservoir. The resulting heating, convection, vesiculation, and expansion of the reservoir would be conducive to an eruption, which might be heralded by a diagnostic change in the $\Delta g/\Delta h$ gradient. Intrusion of magma that did not mix turbulently with the reservoir magma would result in an overall mass increase, but a density change that was positive or negligibly small. In this case, the likelihood of an eruption, especially a large or explosive one, would be low (Rymer and Williams-Jones, 2000; Williams-Jones and Rymer, 2002). Testing such a model requires meticulous fieldwork in a variety of volcanic settings during both eruptive and non-eruptive episodes of unrest, but the promise of more accurate eruption forecasting justifies the effort.

2.7.2 Results from Kīlauea Volcano, Hawai'i

Although concurrent gravity and leveling surveys can provide a wealth of information about subsurface processes at volcanoes, often their interpretation is less straightforward than we might hope. This is because volcanoes are complex systems where several processes can occur simultaneously, each of which might produce a net change in surface height or gravity. In a simple scenario, magma enters

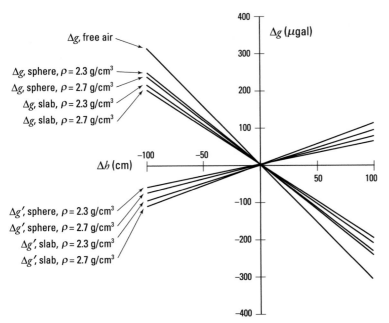

Figure 2.39. Relationship between gravity change Δg and height change Δh for several source models. Lines with negative slopes correspond to the free-air gradient, $-3.086\,\mu\mathrm{Gal\,cm}^{-1}$, which applies in the case of no net mass change, and to a slab or sphere with $\rho = 2.3\,\mathrm{g\,cm}^{-3}$ or $\rho = 2.7\,\mathrm{g\,cm}^{-3}$. Lines with positive slopes represent the free-air corrected gravity change $\Delta g'$, which is the difference between the measured gravity change and the free-air gravity gradient.

or leaves a shallow reservoir and the surrounding rock deforms elastically. The volume of surface uplift or subsidence is related to the volume change of the reservoir through Poisson's ratio,[8] and resulting changes in surface height and gravity follow a linear Bouguer gradient determined by the density of the magma. Often we can use information from other sources to guide our interpretation of such changes. At Kīlauea Volcano, for example, we know from long-term seismic, geodetic, and petrologic studies that the summit magma reservoir is approximated by a sphere[9] centered at about 3 km depth and that the density of magma entering or leaving the reservoir is likely to be $2.7 \pm 0.1\,\mathrm{g\,cm}^{-3}$ (e.g., Eaton and Murata, 1960; Wright, 1984; Decker, 1987; Yang *et al.*, 1992). Therefore, we would expect to measure a

Bouguer gradient during inflation or deflation cycles in the range from -2.30 to $-2.36\,\mu\mathrm{Gal\,cm}^{-1}$.

This would be the case if the spherical model were exactly correct, deformation were perfectly elastic, and no other processes were involved in producing additional changes in surface height or gravity. However, at real volcanoes the situation is seldom so clear-cut. For example, during the interval between surveys the groundwater table might rise or fall in response to some volcanic, tectonic, or climatic change. The mass decrease associated with a falling water table will produce negative Bouguer gravity changes that may or may not be accompanied by ground subsidence. Conversely, a rising water table will produce positive Bouguer gravity changes but not necessarily uplift. At the same time, any subsidence caused by groundwater withdrawal will produce free-air gravity *increases*, and any uplift caused by rising groundwater will produce free-air gravity *decreases*. The disconnect between changes in gravity and height as a result of groundwater fluctuations is especially pronounced where the porosity (i.e., water-bearing capacity) of near-surface materials is high. This is the case at many volcanoes that are composed mainly of fragmental debris or highly fractured lava flows and domes.

We can imagine an analogous process occurring in a shallow magma reservoir. Unlike water, which is a nearly incompressible liquid, magma is a multiphase mixture of liquid (silicate melt with dissolved

---

[8] When a sample of material is stretched in one direction, it tends to get thinner in the other two directions. Poisson's ratio (ν) is a measure of this tendency. It is defined as the ratio of the lateral unit strain to the longitudinal unit strain in a body that has been stressed longitudinally within its elastic limit. For most crustal rocks, Poisson's ratio is approximately 0.25.

[9] Davis (1986) showed that an ellipsoidal source produces a better fit to surface deformation data at Kīlauea than a spherical source. Subsequent studies support the idea that Kīlauea's summit reservoir is elongated vertically, with an eccentricity of about 2. The difference in $\Delta g / \Delta h$ for spherical and ellipsoidal models is small, and the spherical model is mathematically simpler and more generally useful. For these reasons, the spherical model is used here.

volatiles), gases (exsolved volatiles in bubbles), and solids (crystals). As such, it is compressible to a degree that depends mainly on its gas content. In particular, if the gas content is high, magma can be highly fluid and move in or out of a reservoir without appreciably deforming the host rock. The effect is to change the bulk density of the reservoir without significantly changing its volume, and therefore without deforming the ground surface. This alters the relationship between gravity and height changes that we would observe otherwise during periods of inflation or deflation.

Johnson (1992) and Johnson *et al.* (2000) treated this topic quantitatively and showed that magma compressibility can be an important factor affecting both gravity and surface deformation measurements, especially at volcanoes that store a significant quantity of gas-rich magma in a shallow reservoir. Such a reservoir can accommodate a considerable amount of new magma through bulk compression without significantly deforming the host rock. For example, based on repeated gravity and leveling surveys during the 1983–1986 period of the ongoing Pu'u 'Ō'ō eruption at Kīlauea, Johnson *et al.* (2000) showed that 75–80% of the volume of magma transferred into (or out of) the summit reservoir was accommodated by compression (or decompression) of stored reservoir magma. The remaining 20–25% by volume was accommodated by reservoir enlargement (or contraction), and only this small fraction resulted in deformation of the edifice. Such effects can alter substantially the relationship between gravity and height changes, as we shall see shortly.

When all of the possibilities are considered, it might seem that the relationship between gravity and height changes at volcanoes is hopelessly complex and therefore not very useful for understanding volcanic processes. Fortunately, the situation is not as bleak as it might seem at first glance. Like all volcano-monitoring techniques, this one has both strengths and weaknesses. It can provide information about subsurface mass changes that are important and sometimes undetectable with other techniques, but the information is inherently ambiguous and best used in conjunction with other monitoring results. To illustrate how this is done, a case will be described in which gravity surveys played an important role in deciphering what happened beneath Kīlauea during two eventful years in the mid-1970s.

With several colleagues at the USGS HVO, I analyzed results from repeated gravity and leveling surveys at Kīlauea during three periods from November 1975 to October 1977 (Dzurisin *et al.*, 1980). Surveys in November and December 1975 bracketed a magnitude 7.2 earthquake on 29 November 1975, the largest in Hawai'i for more than a century. The quake triggered a brief summit eruption, large summit subsidence, and meter-scale seaward movements of the volcano's south flank (Chapter 7). The second period, from December 1975 to April 1977, included several intrusions of magma from the summit reservoir into the upper East Rift Zone. These were accompanied by shallow earthquake swarms and sharp summit deflations, but no eruptions. This was a clear change from 1969–1974, when a long-lived eruption along Kīlauea's upper East Rift Zone built the prominent Mauna Ulu lava shield and a lava delta along the south coast where flows from Mauna Ulu spilled into the sea. This change suggested that the powerful 1975 earthquake changed Kīlauea's magma plumbing system in some way that favored intrusions over eruptions. The volcano apparently recovered by September 1997, when the first eruption since the 1975 earthquake occurred along the middle East Rift Zone. The September 1977 eruption was bracketed by leveling and gravity surveys in April and October 1977, the third period included in our study.

The combined gravity and leveling results for these three periods are strikingly different (Figure 2.40). During the first period, which included the 1975 earthquake, changes in gravity and height followed the relationship:

$$\Delta g(\mu\text{Gal}) = -1.75 \times \Delta h \text{ (cm)}$$

This is a surprising result, because it suggests the removal of a slab (height changes were negative during this period) of bulk density $3.2\,\text{g cm}^{-3}$, which is implausibly high for basalt. For a spherical model, the required density is even higher, $\rho = 4.8\,\text{g cm}^{-3}$.

How can we make sense of this seemingly nonsensical result? In a companion paper to ours, Jachens and Eaton (1980) proposed that catastrophic draining of Kīlauea's summit magma reservoir triggered by the November 1975 earthquake resulted in an incomplete collapse of the reservoir, thus lowering its bulk density. This is one of the situations described by Johnson *et al.* (2000), in which partial draining of a relatively large, gas-rich reservoir causes mainly decompression of the remaining magma with relatively little

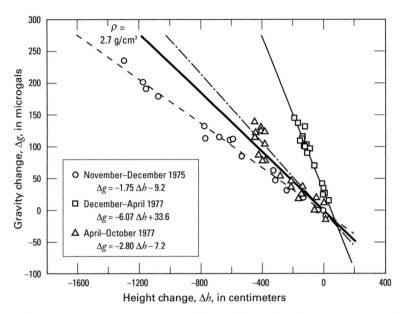

Figure 2.40. Gravity and height changes in the summit region of Kīlauea Volcano, Hawai'i, for three time periods (Dzurisin et al., 1980). Surveys bracket a M 7.2 earthquake in November 1975 (circles), a series of intrusions into the volcano's East Rift Zone (squares), and the September 1977 eruption (triangles). Heavy line labeled $\rho = 2.7\,\mathrm{g\,cm^{-3}}$ represents theoretical changes that occur when magma of that density is added to or withdrawn from a spherical magma reservoir beneath the summit. Departures from this line are explained by: (1) decompression of magma in the summit reservoir during and immediately following the earthquake as magma drained into the East Rift Zone, and (2) recompression of reservoir magma by subsequent intrusions. A density of $2.7\,\mathrm{g\,cm^{-3}}$ is reasonable for Kīlauea magma under normal reservoir conditions.

reservoir contraction.[10] As described above, the effect would be to reduce g without changing h very much, thereby partly counteracting the net increase in g associated with partial subsidence. Thus, $\Delta g/\Delta h$ would be less negative than the expected Bouguer gradient ($-2.33\,\mu\mathrm{Gal\,cm^{-1}}$ for $\rho = 2.7\,\mathrm{g\,cm^{-3}}$, compared with the observed value of $-1.75\,\mu\mathrm{Gal\,cm^{-1}}$). Stated differently, observed values of Δg were less than expected from the free-air effect, which means that values of $\Delta g'$ were negative and there was a net decrease in mass beneath the summit area.

If this interpretation is correct, we might expect the opposite effect to occur when magma returned to the summit reservoir (i.e., magma in the reservoir should have recompressed to produce a more nega-

tive gravity gradient than would be expected otherwise). In fact, we observed just such a gradient during the period from December 1975 to April 1977. Several intrusions from the summit reservoir into the East Rift Zone clearly indicated that magma again was entering the volcano from a deep source. Nonetheless, the summit area experienced only a small net subsidence while gravity values increased dramatically: $\Delta g/\Delta h$ was an amazing $-6.07\,\mu\mathrm{Gal\,cm^{-1}}$. The observed gravity changes were approximately twice as large as the free-air effect, values of $\Delta g'$ were positive, and it was clear that a mass *increase* had occurred beneath the summit area during a period of modest *subsidence*. We hypothesized that the combined effects of slight reservoir contraction, injection of fresh magma from below, and consequent recompression of stored magma had increased the bulk density of the reservoir, thus producing the unusually steep gravity gradient. Stated differently, as discussed by Johnson *et al.* (2000), an influx of fresh magma had been accommodated entirely by compression of reservoir magma rather than by edifice expansion.

If the September 1977 eruption along Kīlauea's Middle East Rift Zone marked the end of a reservoir healing and refilling process following the

[10] Jachens and Eaton (1980) and Dzurisin et al. (1980) envisioned that incomplete collapse occurred when magma drained out of fractures comprising the summit magma reservoir, thus creating voids and lowering the reservoir's bulk density. Subsequently, Johnson et al. (2000) called attention to the importance of magma compressibility, which suggests a more plausible scenario in which reservoir magma decompressed as a result of partial draining, thus producing the same effect. The remainder of this discussion is couched in these terms for brevity, but Johnson et al. (2000) are credited with revising my thinking about the physics of the process.

November 1975 earthquake, then the gravity gradient for a period bracketing the eruption should likewise have returned to near normal. Our result fits this scenario relatively well:

$$\Delta g(\mu\text{Gal}) = -2.80 \times \Delta h \text{ (cm)}$$

This gradient is a little steeper than we would expect for inflation of a spherical reservoir with $\rho = 2.7\,\text{g cm}^{-3}$ magma ($-2.33\,\mu\text{Gal cm}^{-1}$), which suggests some additional reservoir compression prior to the eruption, but for the most part Kīlauea's plumbing system had recovered from the effects of the November 1975 earthquake.

Our interpretation of activity at Kīlauea from November 1975 to October 1977 is not unique, but I think the picture we were able to paint with the help of gravity and leveling data is plausible, consistent with our knowledge base for Kīlauea, and considerably more detailed than would have been possible otherwise. Our experience suggests that this approach might be productive at other volcanoes, and there have been numerous studies of this type in the intervening years. Gravity surveys can be made before and after eruptions or other significant events without exposing scientists to hazardous conditions. GPS can be substituted for leveling where the latter technique presents logistical problems (e.g., steep terrain or difficult foot access). Of course, all necessary care should be taken to ensure the acquisition of reliable data. This includes returning to a local base station at least twice a day during gravity surveys to identify tares and determine instrument drift, and following standard field procedures for leveling or GPS surveys.

Another important consideration is groundwater monitoring. Volcanoes are notoriously porous and often waterlogged. Gravity changes caused by variations in the groundwater table can be large relative to other effects of more direct concern. For example, the gravity change caused by a 10 m fall in the groundwater table in rocks of 10% porosity is about $-42\,\mu\text{Gal}$ (assuming no surface height change), which is about the same as the Bouguer effect associated with 18 cm of uplift ($\rho = 2.7\,\text{g cm}^{-3}$). An array of continuously monitored water wells is ideal for this purpose, but periodic measurements at just a few wells, or even anecdotal evidence of changes in wells or springs can be useful. An example of the utility of concurrent microgravity, leveling, and groundwater measurements for studying magmatic processes at Long Valley Caldera (Battaglia *et al.*, 1999) is discussed in Chapter 7.

2.7.3 Results from Miyakejima Volcano, Japan

A remarkable sequence of volcano–tectonic events occurred near the Izu Peninsula, central Japan, following the *M* 6.9 Izu-Hanto-Oki earthquake in 1974, which was preceded by 40 years of relative quiescence. The 1974 shock was followed by the 1976 *M* 5.4 Kawazu earthquake, the 1978 *M* 7.0 Izu-Oshima-Kinkai earthquake, the 1980 *M* 6.7 Izu-Hanto-Toho-Oki earthquake, and the 1990 *M* 6.5 Izu-Oshima-Kinkai earthquake. Meanwhile, there have been more than a dozen earthquake swarms associated with magmatic intrusions, broad uplift of the eastern part of the peninsula, and two offshore eruptions. The first was a submarine eruption near Ito City, which began on 13 July 1989 (Okada and Yamamoto, 1991). The second was accompanied by large gravity changes and formation of a small collapse caldera atop Miyakejima Volcano (also called Miyake Island or Miyakejima) in July–August 2000 (Kumagai *et al.*, 2001).

Quasi-periodic eruptions had been recorded at Miyakejima since 1085, most recently in 1940, 1962, and 1983. On 26 June 2000, an anomalous tilt and an earthquake swarm beneath the island indicated that magma intruded close to the surface, but an eruption did not ensue immediately (Ukawa *et al.*, 2000). Instead, the swarm migrated toward Kozu Island, 40 km to the northwest, before a small eruption and dramatic subsidence started at Miyakejima on 8 July 2000. The summit depression gradually sank and expanded through mid-August, producing repetitive very long period (VLP) earthquakes and a 1.7-km-diameter caldera (Nakada *et al.*, 2001). A phreatomagmatic eruption on 18 August 2000, was the largest of the sequence, with an eruption column that reached as high as 15 km (Furuya *et al.*, 2003a,b).

Furuya *et al.* (2003a,b) used spatiotemporal gravity changes to constrain a detailed model for the Miyakejima eruption that fits several other datasets as well. They attributed a gravity decrease in the summit area by as much as 145 µGal, which was detected 2 days prior to the start of caldera collapse, to lateral outflow of magma from beneath the volcano. The outflow fed a dike intrusion to the northwest, as suggested by the earthquake swarm that migrated toward Kozu Island starting in late June 2000. Rapid withdrawal of magma from a reservoir beneath the volcano caused the roof of the reservoir to collapse, forming a small caldera. Large gravity changes detected after initiation of collapse, including an extraordinarily large decrease of 1,135 µGal at one station near the caldera rim, can be explained mostly by the collapsed topography. After correcting for topography, Furuya *et al.* (2003a) identified a temporal gravity decrease that occurred from mid-July to late August despite ground subsidence during the same period. They attributed the decrease in gravity to a reduction in subsurface density caused by groundwater inflow to a porous conduit beneath the summit, which promoted magma–water interaction and subsequent explosive eruptions. From September to at least November 2000, gravity values increased to a degree that cannot be explained by groundwater displacement alone. Instead, the increase probably reflected refilling of the conduit with magma.

Kumagai *et al.* (2001) proposed a similar model based mainly on seismic observations. They suggested that a vertical piston of rock intermittently sinks into the magma reservoir, producing VLP earthquakes and intrusions marked by migrating earthquake swarms. Increasing pressure from the weight of dense cumulate magma eventually causes the reservoir to rupture, resulting in an intrusion, large gravity decrease, and caldera collapse. Groundwater inflow to the magma plumbing system fuels explosive eruptions. Subsequent upwelling of magma results in a large gravity increase, emission of volcanic gases, and volcanic glow, which were observed at Miyakejima in that order. Microgravity data played a key role in the development of both models.

2.8 MAGNETIC FIELD MEASUREMENTS

Magnetic field measurements, like gravity measurements, can be made in either continuous or survey mode and can be affected by several processes that commonly occur at volcanoes. The most important magnetic effects are caused by changes in the state of stress or temperature of volcanic rocks. A change in the magnetization of rocks caused by a change in stress is called a piezomagnetic effect. Thermal demagnetization of rocks occurs when subsurface temperatures exceed the Curie point ($\sim$580$°$C), which is common beneath volcanoes. Both processes affect the magnetic field as measured at the surface, which provides another opportunity to infer changes in subsurface stress or temperature by monitoring the surface magnetic field. The interested reader will find an excellent review of the physical mechanisms involved, including a discussion of results from sev-

eral volcanoes, in the article by Johnston (1997). A brief summary follows.

2.8.1 Physical mechanisms

The fractional change in magnetization per unit volume as a function of stress can be expressed as (Johnston, 1997):

$$\frac{\Delta I}{I} = K\sigma_{ij} \qquad (2.17)$$

where ΔI is the change in magnetization in a body with net magnetization I due to a stress σ_{ij}, and K is a proportionality constant called the stress sensitivity. For reasonable values of I and K, models of elastic stress changes caused by magmatic intrusions suggest that magnetic anomalies of a few nanoTeslas (nT) should be expected to accompany volcanic activity as a result of piezomagnetic effects (Johnston, 1997).

Another process that causes magnetic field changes at volcanoes is thermal demagnetization (or remagnetization) of crustal rocks. The solidus temperature of magma exceeds the Curie point (i.e., magma is hot enough to destroy ferromagnetism in most rocks by thermal agitation), so the potential exists for an intruding magma body to demagnetize a volume of rock around its periphery and thus to alter the magnetic field at the surface. In the absence of other effects, this process occurs very slowly because the low thermal diffusivity of rock limits the migration rate of the Curie point isotherm to less than 1 m yr^{-1}. However, fracturing and fluid transport during magmatic intrusions can dramatically increase the Curie point migration rate, and, accordingly, the magnitude and rate of magnetic field changes measured at the surface. The reverse process, remagnetization of rocks during cooling below the Curie point, can provide useful information on cooling rate and subsurface thermal structure (Dzurisin *et al.*, 1989). I have limited the following discussion to stress-induced magnetic changes at Mount St. Helens and Long Valley Caldera, because the inferred mechanism is related to the subject of this book. Readers interested in a general discussion of magnetic and electric effects at volcanoes are encouraged to read the informative review article by Johnston (1997).

2.8.2 Changes associated with eruptions at Mount St. Helens

Stress-induced magnetic changes have been observed on two occasions at Mount St. Helens:

(1) at the time of the catastrophic debris avalanche and eruption on 18 May 1980, and (2) during an extrusive dome-building episode in October–November 1981. Johnston *et al.* (1981) reported that the magnetic field at their Blue Lake station, 5 km west of the 1980 crater, increased by 9 ± 2 nT at the time of the 18 May 1980 eruption. They concluded that the change was caused by regional elastic stress release in response to the events of that day, because the gross redistribution of magnetic material by the debris avalanche would have caused changes of opposite sign. This result suggests the possibility that regional stress-induced changes might have preceded the eruption as well, but the existing data are inadequate to test this idea because the Blue Lake magnetometer and two others were installed just 10 days before the eruption.

The Blue Lake magnetometer survived the 18 May eruption, and in August 1980 another was installed at North Ape Cave, 3 km south of the dome growing in the crater. In October 1981, two more magnetometers were installed on the floor of the 1980 crater within 1 km of the dome, and a third at Nelson's Ridge, on the volcano's east flank about 2 km east of the dome (Davis *et al.*, 1984). When recording began on 23 October 1981, the magnetic field difference between the two crater stations was observed to increase at the rate of 1 nT day^{-1} until 27 October, when the trend reversed. The accumulated 4 nT change then relaxed over a period of 16 hours while a nearby tiltmeter recorded rapid tilting of the crater floor. An extrusive dome-building eruption began 3 days later on 30 October 1981.

Davis *et al.* (1984) interpreted the magnetic-field changes as a piezomagnetic effect (i.e., stress-induced changes in the magnetization of the volcano). A point source Mogi (1958) model (Chapter 8) fit to radial tilts measured by three tiltmeters on the crater floor placed an inflation source about 1 km beneath the dome. Using this as an approximation to the stress source that caused the magnetic field changes, Davis *et al.* (1984) placed an upper limit on the stress change of ∼300 bars.

2.8.3 Results from Long Valley Caldera

A compelling case for piezomagnetic changes at a restless volcano can be made at Long Valley Caldera, California, the site of persistent but variable activity since 1978. The unrest has included several earthquakes of $M > 5$, frequent swarms of smaller earthquakes, ∼80 cm of uplift centered on the caldera's resurgent dome, and increased outgassing of magmatic CO_2 sufficient to kill vegetation in some areas (Hill *et al.*, 1990; Hill, 1996). Detailed analysis of this fascinating and potentially hazardous episode is continuing, but there is general agreement on the involvement of both magmatic and tectonic processes contributing to the unrest. Almost certainly, magma has accumulated in the crust beneath the caldera since 1980 and may have intruded to within a few kilometers of the surface beneath the caldera's south moat in 1983 and beneath Mammoth Mountain in 1989–1990 (Chapter 7).

Continuous magnetic field measurements have been made since 1983 at two sites within Long Valley Caldera and at a third reference station 26 km southeast of the caldera (Mueller and Johnston, 1998). Relative to the reference station, the magnetic field at Smokey Bear Flat in the western part of the caldera decreased by 8 nT from mid-1989 to 1996, while the field at Hot Creek Ranch in the southeast part of the caldera increased by more than 6 nT. The timing of these changes correlates with an acceleration in the rate of extension across the resurgent dome in late 1989, followed by a marked increase in the rate of seismic energy release beneath the caldera in early 1990. The magnitude and timing of the magnetic field changes are consistent with a simple piezomagnetic source model based on geodetic data. Frequent two-color EDM measurements show that the resurgent dome extended at an average rate of 3 ppm yr^{-1} from late 1989 through 1996. Both the pattern of line-length changes and the magnetic field observations can be modeled by inflation of a Mogi source located about 7 km beneath the center of the resurgent dome (Langbein *et al.*, 1993; Mueller and Johnston, 1998). Taken together, the seismic, EDM, microgravity, and magnetic data from Long Valley make a compelling case for renewed magmatic inflation beneath the resurgent dome starting in mid-1989.

CHAPTER 3

Continuous monitoring with *in situ* sensors
Daniel Dzurisin

The previous chapter described several ways to measure volcano deformation by making repeated network surveys. Frequent surveys of a well-designed network can serve to characterize the deformation field in space and time, albeit not completely nor continuously, thereby helping to constrain source models and anticipate future activity. However, this approach has two serious shortcomings during a crisis: (1) it repeatedly places survey personnel near a hazardous volcano, potentially in harm's way; and (2) repeated surveys cannot always keep pace with a rapidly evolving crisis, especially at night or during bad weather. Therefore, it is desirable to supplement periodic geodetic surveys with carefully sited continuous sensors that provide a steady stream of data in real time for analysis. Continuous monitoring also is important during non-crisis periods, because it can: (1) detect changes in the deformation pattern sooner than repeated surveys; (2) provide a dense time series of data as context for evaluating subtle changes in the pattern or rate of deformation; and (3) reduce the need for recurring field operations.

A disadvantage of *in situ* monitoring is that each sensor responds to its immediate surroundings, which may or may not represent the deformation field as a whole. If care is taken to couple sensors adequately to the ground and to insulate them from near-surface environmental effects, this shortcoming can be overcome in most cases. However, success is never guaranteed even with the most careful installations, so some type of redundancy is always desirable. For example, multiple sensors or sensor types might be installed near each other as a consistency check, or surveys might be conducted periodically near the stations to assess the reliability and fidelity of the continuous data. To optimize the design of a continuous sensor network for a par-

ticular situation, station locations should be chosen based on the results of repeated surveys or, in the absence of measured deformation, based on numerical models of the deformation field for plausible source types (Chapter 8).

This chapter describes five types of *in situ* deformation sensors: tiltmeters, strainmeters, continuous Global Positioning System (GPS) stations, gravimeters, and differential lake gauges, and discusses their use at selected volcanoes. Also included is a brief discussion of seismometers, for three reasons: (1) seismometers are used more than any other type of *in situ* sensor to monitor volcanoes; (2) the widespread use of broadband seismometers, borehole strainmeters, and continuous GPS instruments to record such phenomena as very long period (VLP) earthquakes and 'slow' earthquakes (e.g., Hill *et al.*, 2002b; Dragert *et al.*, 2001; Section 4.8.5) has blurred the traditional boundary between seismology and geodesy; and (3) to save the reader interested in a brief introduction to seismological concepts the task of looking elsewhere. Those seeking a more detailed treatment of modern volcano seismology might want to consult reviews of the subject by Chouet (1996b, 2003) and McNutt (1996, 2000a,b). Additional information on the use of GPS receivers and strainmeters to monitor volcano deformation is provided in Chapters 4 and 9, respectively.

3.1 SEISMOMETERS

No book about volcano monitoring would be complete without at least a brief discussion of seismometers and seismological principles. Volcano seismology and volcano geodesy have been closely linked throughout the modern era of geophysical instrumentation, which spans more than a century.

The linkage arose partly because early seismometers were sensitive to ground tilt (Section 1.4.1), modern tiltmeters and strainmeters are capable of recording long-period (LP) and VLP earthquakes, and broad-band seismometers respond to ground motions with periods as long as ~50 s. In recent years the two disciplines have become increasingly inter-twined as new discoveries reveal the full spectrum of Earth motions, including: (1) bradyseisms[1] with periods of months to decades, (2) slow earthquakes (days to weeks), (3) ultra long-, very long-, and long-period earthquakes (seconds to minutes), and (4) predominantly short-period earthquakes produced by brittle rock failure along faults (<0.1 to 1 second). The following treatment is basic and by no means complete, but it should suffice for most readers in lieu of consulting one or more of the review papers mentioned in the preceding paragraph.

3.1.1 A brief history of seismology

Aristotle (ca. 330 BCE) attributed earthquakes to winds blowing in underground caverns, and Chinese philosophers at about the same time suggested that ground shaking was caused by the blocking of a subtle essence (the *qi*). Chinese mathematician and philosopher Zhang Heng (or Chang Hêng) is credited with inventing the first seismo-scope, which is reported to have detected a four-hundred-mile distant earthquake that was not felt at the location of the instrument, in 132 CE (Dewey and Byerly, 1969; Needham, 1959). An early association between earthquakes and volcanoes developed during the scientific revolution in Europe between roughly 1500 and 1700. Familiarity with gunpowder led various writers to attribute both phenomena to explosions in the Earth resulting from combustion of pyrites or a reaction of iron with sulfur (Agnew, 2002).

The first recognition that ground shaking was caused by waves in the Earth propagating from a specific location came shortly after the destructive Lisbon earthquake of 1755, but a revolutionary 'new seismology' did not emerge for another century. The Meiji restoration of 1868 in Japan led to the estab-lishment of modern science in a very seismic region,

facilitated by foreign experts including John Milne, James A. Ewing, and Thomas Gray, who worked at prestigious Japanese universities at the invitation of the national government. Ewing and Gray used a horizontal pendulum[2] to produce the first good instrumental records of ground shaking, while Milne pioneered the field of global seismology. Agnew (2002) describes a 'classical' period in seis-mology that followed from 1920 to 1960, when ideas about wave propagation were refined and improved but there were no substantial technical advances. Interestingly, this period began shortly after Dr. Thomas A. Jaggar founded the Hawaiian Volcano Observatory (HVO) in 1912, and it ended shortly after Professor Kiyoo Mogi published his classic paper on volcano deformation (Mogi, 1958). Most volcanologists I know do not shy away from the fact that seismology often leads volcanol-ogy, especially in the theoretical arena. We tend to be pragmatists who welcome any source of insight into how real-world volcanoes work.

The modern era in seismology can be pegged to the emergence of plate tectonics theory starting in

---

[1] 'Bradyseism' refers to slow uplift or subsidence of the ground surface in response to inflation or deflation of a magma reservoir, or to pressurization or depressurization of a hydrothermal system. The best known example is the remarkable motion of the Phlegraean Fields caldera near Naples, Italy, which has persisted since Roman times.

[2] The horizontal pendulum was first described in 1832 by priest and scholar Lawrence (Lorenz) Hengler, who noted its utility as an astronomical instrument capable of measuring slight deviations from vertical. Zöllner (1869) was the first to suggest its use as a seismometer, and on 17 April 1889, von Rebeur-Paschwitz (1889) obtained one of the first known recordings of a distant earthquake using such a device. The earthquake had been felt in Japan about an hour before its surface waves were recorded in Germany. To achieve its remarkable sensitivity, the horizontal pendulum takes advantage of some basic physics. The more familiar vertical pendulum, consisting of a mass suspended on a string, oscillates with period $T = 2\pi\sqrt{L/g}$, where L is the length of the pendulum and g is acceleration due to gravity. For a rigid pendulum swinging in an inclined plane, the oscillation period is given by: $T = 2\pi\sqrt{L/g\sin\theta}$, where θ is the inclination from horizontal. Note that, as L approaches infinity or $g\sin\theta$ approaches zero, T tends toward infinity. Increasing L indefinitely has practical limitations, but dialing down the effect of gravity by decreasing θ turns out to be both easy and elegant. For an intuitive example, consider a gate swinging on a vertical hinge-post. If the post is exactly vertical, the gate swings freely through every position (ignoring friction). If the post is inclined slightly, the gate assumes an equilibrium position that is very sensitive to the orientation of the post. Once the gate is set into motion, it oscillates about the equilibrium position for a long time (θ is small, so T is large). Small changes in the orientation of the post produce large changes in the equilibrium position. In a horizontal-pendulum seismograph, a mass attached to one end of a rigid beam is free to oscillate in a horizontal plane about a vertical hinge near the other end of the beam. This arrangement is extremely sensitive to horizontal accelerations, including those caused by the passage of seismic waves. Add a means to record the swings of the mass at the pendulum's tip, and you have a functional seismograph.

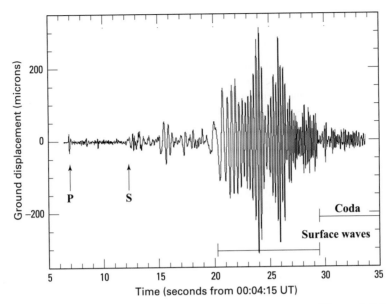

Figure 3.1. Seismogram showing vertical surface motion recorded at Kevo, Finland, beginning 500 seconds (8.3 minutes) after the start of the 17 October 1989, M_s 7.1 (M_w 6.9) Loma Prieta, California, earthquake near San Francisco. The distance from the epicenter to the seismometer is 71° (~7,880 km). Arrows indicate first arrivals of the P- and S-waves (body waves); bars show duration of surface waves and first ~5 s of the trailing coda. See text for discussion.

1960 (e.g., Vine and Matthews, 1963; Vine, 1966). Increasingly precise locations for sea-floor earthquakes placed most of them along mid-ocean ridges, focal-mechanism studies showed that oceanic fracture zones behave as transform faults connecting segments of the spreading ridge, and seismologists recognized that the distribution of many deep earthquakes, which theretofore had been enigmatic, was organized along subduction zones. As a historical note, Dr. Robert W. Decker and colleagues at HVO made the first electronic distance meter (EDM) measurements at Kīlauea and Mauna Loa Volcanoes in the 1960s, amidst the early excitement over plate tectonics. At the time I was a teenager fascinated by Solar System exploration, including the quest to land astronauts on the Moon. Another decade passed before I realized that Earth was the only planet I would ever visit, and volcanoes were among the most interesting things on Earth.

3.1.2 An introduction to seismic waves and earthquake types

Seismic waves are of two types: short-period (high-frequency) body waves that travel within the Earth and long-period (low-frequency) surface waves that travel along its surface. Body waves include longitudinal P-waves and transverse S-waves. P-waves travel faster than S-waves and arrive earlier; hence the designations P- (primary) and S- (second-

ary). P-waves involve compressional particle motion in the direction of wave propagation; S-waves involve shear (sideways) motion. P-waves propagate through liquids, including magma and partial melt, but S-waves disappear or are greatly attenuated.[3]

There are also two types of surface waves, Love waves and Rayleigh waves.[4] Surface waves travel slower than body waves but they attenuate less with distance. Surface waves from great earthquakes travel completely around the globe many times. They are responsible for most of the ground shaking produced by distant large earthquakes. Great earthquakes also excite free oscillations of the entire Earth, akin to the ringing of a giant bell, which can persist for days.

[3] Because liquids are largely incompressible and do not support shear stress, P-waves propagate through them but S-waves do not. Magma and hydrothermal fluids are compressible to varying degrees owing to the presence of gas in bubbles. As a result, S-waves and, to a lesser degree, P-waves are attenuated in magmatic and hydrothermal environments. Seismic tomography is a technique used to delineate the 3-D shapes of magma bodies and hydrothermal systems by mapping out S-wave attenuation along intersecting ray paths from a large number of earthquake sources to a dense array of seismometers.

[4] Love waves and Rayleigh waves travel along the surface of the solid Earth. They are distinct from T-waves, which are short-period ($\leq$1 s) acoustic waves that travel in the ocean at the speed of sound in water. A 'T-phase' is occasionally identified in the records of earthquakes in which a large part of the path from epicenter to station is across the deep ocean.

The first noticeable indication of an earthquake is often a sharp 'thud' signaling the arrival of the fast-moving compressional P-wave. This is followed by the S-wave and then the more gentle 'ground roll' caused by surface waves. P-waves, because they are the first to arrive, are used to locate earthquakes based on their differential travel times to an array of seismometers at different azimuths (directions) and distances from the earthquake's focus. S-waves and surface waves contribute to the concluding train of seismic waves, called the coda, which follows the primary wave (Figure 3.1). Codas from great earthquakes continue for hours while the slowly attenuating surface waves repeatedly circumnavigate the globe.

The frequencies of seismic waves range from as high as the audible range (greater than 20 Hz) to as low as the free oscillations of the Earth, with a longest period of 54 minutes (corresponding to a frequency of 0.0003 Hz). The amplitude range of seismic waves is also large. Measurable ground displacements from small to moderate earthquakes range from about 10^{-10} to 10^{-1} meters (0.1 nanometer (nm) to 1 decimeter). For comparison, the smallest displacement measured for geodetic purposes, to my knowledge, is $\sim$1 nm – the resolution of a differential transformer in a Sacks–Evertson borehole strainmeter (Section 9.1).[5] In the greatest earthquakes, the amplitude of the predominant P-waves can be several centimeters at periods of 2 to 5 s. Very close to the epicenters of great earthquakes, peak ground accelerations sometimes exceed that of gravity at high frequencies, and ground displacements can reach 1 m at low frequencies.

Reports of earthquake sounds are common, even though the dominant frequency of seismic waves is well below the audible range for most humans.[6] Most reports of a roar or train-like sound associated with moderate-to-large earthquakes un-

doubtedly refer to non-seismic noise sources such as objects crashing to the ground, breaking, or jostling together under the influence of seismic waves, rather than any sound generated by the waves themselves. Nonetheless, there are many credible reports of a boom or thud being heard close to the epicenter of small (M 2–4) earthquakes. I have heard such a noise dozens of times during swarms of earthquakes at Kīlauea and Mount St. Helens Volcanoes. Often the sound is followed almost immediately by perceptible ground shaking, but in some cases even shocks too small to be felt can be heard. Apparently, high-frequency ground vibrations couple to the atmosphere well enough to produce audible pressure pulses, at least within a few kilometers of the source. The sound of the Earth quivering like a drumhead in response to the beat of an earthquake is at once ominous and beguiling – a primal resonance that commands attention and instills a sense of vulnerability in even the most intrepid volcano watcher.

Any classification scheme for something as complicated and diverse as seismic events beneath volcanoes is fated to be revised as new information becomes available or new theories emerge to describe the causative processes. Nonetheless, a simple proposal by Japanese seismologist and volcanologist Takeshi Minakami (1960, 1961) has enjoyed considerable longevity. Minakami divided volcanic earthquakes into two types based on their signatures on seismograms and on their source depths. A-type earthquakes have a clear P-wave and S-wave and generally occur at depths of 1–10 km beneath a volcano. They are indistinguishable from normal shallow tectonic earthquakes. B-type earthquakes have emergent P-wave signatures on seismograms (i.e., gradual onsets), no distinct S-wave, and low-frequency content relative to tectonic earthquakes of the same magnitude. Minakami (1961) attributed the character of B-type earthquakes to an extremely shallow focal depth (<1 km). Poorly consolidated near-surface materials at most volcanoes attenuate high frequencies in P-waves and transmit S-waves very poorly, which could account for the characteristic signature of B-type events.

Recently the contrast between A-type and B-type earthquakes has been attributed to different source processes rather than to shallow path effects (e.g., Chouet, 1988, 1992, 1996a). Currently, A-type events are more often called volcano–tectonic (VT) or high-frequency (HF), and B-type events are called long-period (LP) or low-frequency (LF). VT earthquakes are attributed to brittle

---

[5] The wavelengths of ultraviolet radiation and visible light are short enough to be measured in nanometers. Most of the UV radiation that reaches the Earth from the Sun has wavelengths 100–400 nm, and visible light has wavelengths 400–700 nm.

[6] Numerous claims of pets or other animals acting strangely before a damaging earthquake remain unsubstantiated or unexplained, as do frequent earthquake 'predictions' by non-scientists based on a sense of impending calamity or some other indicator. Science is always open to discoveries that revise current understanding of the natural world, but for now predicting an earthquake before the arrival of the P-wave is mostly outside the scientific mainstream. Prediction is one arena in which volcanology leads seismology: At a time when successful eruption predictions are becoming more common, earthquake prediction remains an elusive research goal.

rock failure and, except for their occurrence near a volcano, are indistinguishable from normal tectonic earthquakes that occur elsewhere. LP events are attributed to resonance in a fluid-driven crack under choked-flow conditions (Chouet, 1988, 1996a; Chouet *et al.*, 1994), fluid pressurization processes such as bubble formation and collapse, and oscillation of a fluid-filled cavity (McNutt, 2000a,b). An alternative explanation has been proposed for deep LP (DLP) events that heralded the 1991 VEI[7] 5-6 eruption of Mount Pinatubo, Philippines, and which occur sporadically in the lower crust beneath some volcanoes in the Aleutian arc, for example. White (1996) and Power *et al.* (2004) attribute DLP earthquakes to injections of mixed phase (liquid + gas) basaltic fluids upward through cracks from the upper mantle or mid-to-lower crust into the roots of volcanoes.

Some volcanic earthquakes 'straddle the fence' between VT and LP types (i.e., they share attributes of both). They are called hybrid events and are thought to represent a combination of processes, such as an earthquake occurring near a fluid-filled crack and setting it into oscillation (McNutt, 2000a,b).

Another type of volcanic seismicity is tremor, which produces a continuous signal on seismographs with a duration of minutes to days or longer. The dominant frequencies of volcanic tremor are 1–5 Hz; 2–3 Hz is most common. This is similar to the frequency content of LP earthquakes, which leads many investigators to conclude that tremor is a series of closely spaced LP earthquakes. Two special types of volcanic tremor are harmonic tremor and spasmodic tremor. As the name implies, harmonic tremor consists primarily of a single-frequency sinusoid with smoothly varying amplitude; in some cases, a fundamental frequency is accompanied by its overtones (i.e., 'harmonics').[8] Spasmodic tremor is higher frequency, pulsating, and irregular in amplitude.

The origin of tremor has received considerable attention from volcano seismologists for decades. The observation that Hawai'ian eruptions are always accompanied by harmonic tremor suggests that magma flow through shallow conduits is involved in the source mechanism. The harmonic nature of some tremor can be modeled as resonance in a fluid-filled crack, akin to one of the mechanisms invoked to explain LP earthquakes (Chouet, 1988, 1992). Similarly, spasmodic tremor can be interpreted as unsteady flow in a volcanic conduit under choked-flow conditions. Although different mechanisms might operate in different situations, it seems appropriate to think of most tremor as one or more LP earthquakes, persisting either because the same fluid-filled source is repeatedly excited or because many similar sources are active in a small volume.

For additional discussion of types of volcanic earthquakes, terminology, and inferred source mechanisms, the reader is referred to, for example, Chouet (1996a, 2003) and McNutt (1996, 2000a,b).

3.1.3 Basic principles of seismometers

Although the modeling and interpretation of seismic waves is mathematically complex, seismic instrumentation is mechanically simple by comparison (e.g., Telford *et al.*, 1990, pp. 136–282). Geophones consist of four basic elements: (1) a frame securely affixed to the Earth, ideally embedded in solid bedrock deep enough to substantially attenuate surface noise; (2) an inertial mass suspended in the frame, called a proof mass, with a steady-state reference position established by springs (horizontal pendulum), gravity (vertical pendulum), or electrical forces (broadband seismometer); (3) a damper system to prevent long term resonant oscillations; and (4) a means of recording the motion of the mass relative to the frame. Passing seismic waves move the frame, while the mass tends to remain in a fixed position due to its inertia. The relative motion is sensed and recorded to produce a seismogram. Early seismometers used mechanical linkages to amplify the motion of the suspended mass. Modern instruments use electronic amplification of signals generated by position or motion sensors (Telford *et al.*, 1990, pp. 195–197).

The Mark Products L4C geophone is widely used for earthquake and volcano studies, and its design illustrates the basic elements of modern, short-

---

[7] Volcanic Explosivity Index (VEI) a measure of the size of volcanic eruptions akin to the Richter magnitude scale for earthquakes. The VEI is a 0-to-8 index of increasing explosivity, each interval representing an increase of about a factor of ten. It combines total volume of explosive products, eruptive cloud height, descriptive terms, and other measures (Newhall and Self, 1982).

[8] A 'harmonic' (noun) is a mode of vibration whose frequency is an integral multiple of a fundamental or base frequency; a series of oscillations is referred to as 'harmonic' (adjective) when each oscillation has a frequency that is an integral multiple of the fundamental frequency (i.e., each oscillation is a 'harmonic' of the fundamental frequency). Thus, the frequency content of harmonic tremor includes a fundamental frequency and its harmonics. If you're confused, complain to a seismologist.

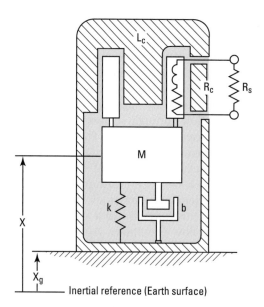

Figure 3.2. Mark Products vertical L4C geophone, from Bowden (2003). A proof mass M (kg) is suspended by a spring with stiffness k ($n \times m^{-1}$) and mechanically damped by dashpot b ($n \times m^{-1} \times s^{-1}$). The motion-sensing coil is mounted above the proof mass and is represented by inductance L_c in series with coil resistance R_c. See text for details.

period seismic sensors (Figure 3.2). The following description of the L4C closely follows Bowden (2003, pp. 1–2). The sensor produces a voltage proportional to the velocity difference ($dx/dt - dx_g/dt$) between its proof mass M and its housing. Voltage is generated in a pickup coil wound on the proof mass when a permanent magnet attached to the housing moves past the coil. The suspension system consists of a 1 kg mass suspended on soft springs with resonant frequency around 1 Hz. The system acts as a damped mechanical oscillator electromagnetically coupled to the output by the moving field of the permanent magnet. For ground motions at frequencies above mechanical resonance, the output signal follows the ground velocity, dx_g/dt. For frequencies below resonance, voltage falls off as proof mass motion starts to follow ground motion.

Modern seismometers come in two types, analog and digital, and three main varieties: short-period, long-period, and broadband. Most applications use either a single-component sensor oriented such that it responds to vertical motions, or a three-component sensor that responds along the vertical axis and two orthogonal horizontal axes. Analog seismometers are very sensitive to ground motion but they have a limited dynamic range (i.e., they 'clip' or go off-scale for most earthquakes large enough to be felt). The heart of a broadband seism-

ometer is a small mass confined in space by electrical forces. As the Earth moves, sophisticated electronics attempt to hold the mass steady through a feedback circuit. The force necessary to achieve this is proportional to ground acceleration ($F = ma$), which can be integrated to yield velocity and then output to a recording or telemetry system.

Digital broadband seismometers are more expensive, require more power, and are more difficult to deploy and maintain in the field than analog short-period instruments, which were the mainstay of volcano monitoring for several decades. However, the ability to record a broad range of frequencies, from about 0.02 Hz to 50 Hz (corresponding to periods of 50 s to 0.02 s), while staying on-scale through a wide range in earthquake induced ground motions, has made digital broadband seismometers increasingly popular among volcano seismologists during the past decade (e.g., McNutt, 1996). With records from an array of three-component, digital broadband instruments, seismologists can ogle, compare, and model the complete codas of thousands of earthquakes. Better them than me, but I am not a seismologist (a fact made clear to me on numerous occasions).

3.1.4 Current research topics in volcano seismology

Volcano seismology aims to understand the structure of magmatic and hydrothermal systems, the source mechanisms of volcanic seismicity, and the spatial and temporal patterns of magma transport through the crust. An overarching goal is to characterize magmatic systems well enough to enable the development of quantitative models capable of accounting for observed seismicity and of predicting future developments, including eruptions. Such an ambitious agenda requires that the scope of cutting-edge research be both broad and rapidly evolving. Consequently, rather than trying to paint a complete picture of the current state-of-the-art in volcano seismology, I have included just a snapshot of a few particularly vibrant topics in the field. The interested reader will find considerably more detail in the review article by Chouet (2003) and references therein.

Although determination of the seismic velocity structure of the Earth is a time-honored undertaking for seismologists, new instrumentation and approaches keep the work fresh and productive. For example, the 3-D velocity structure of a volcanic edifice can be imaged in unprecedented detail using high-resolution travel-time tomography. The technique uses crisscrossing ray paths from a large number of local earthquakes to a dense array of seismic stations to map local departures from an initial velocity model through comparison of predicted and observed arrival times. The result is an improved image of the 3-D velocity structure, typically with a spatial resolution of 1 km or better (Chouet, 2003).

Dawson et al. (1999) used tomographic inversion[9] of 4,695 P-wave and 3,195 S-wave arrivals from 206 earthquakes during a 20-day period in 1996 to derive high-resolution (0.5 km) 3-D P-wave velocity and V_P/V_S models of the Kīlauea Caldera region, Hawai'i.[10] The models delineate a zone of anomalously low P-wave velocities at depths of 1–4 km centered beneath the southeastern rim of the caldera, including two smaller zones of high V_P/V_S beneath the southern portion of the caldera and the upper East Rift Zone. The latter are interpreted as magma storage zones, consistent with many other studies that point to reservoirs in those areas.

The source mechanism of most VT earthquakes (brittle failure of rock) is well understood owing to their similarity to normal tectonic earthquakes, but the mechanisms of LP volcanic earthquakes and volcanic tremor remain fertile topics for research. The pioneering work by Aki et al. (1977), who interpreted volcanic tremor using fluid-driven crack models, inspired additional breakthroughs

by Chouet (1981, 1985, 1986, 1988, 1992, 1996a) and others during the ensuing two decades. The resulting theoretical framework, which draws heavily upon the dynamics of multiphase fluids, serves as a point of departure for most contemporary analyses.

One particularly promising facet of current research involves the use of dense, small-aperture arrays of short-period seismometers to delineate the source regions of LP earthquakes and tremor. As noted by Chouet (2003), this type of seismicity cannot be located by conventional means because the signals' first arrivals on seismograms are emergent, not impulsive (i.e., gradual, not sudden). To locate a seismic event, the relative timing of its first arrival at several stations must be known precisely (seismic waves travel at speeds of several km s^{-1}, so small timing errors have a big effect on the accuracy of the hypocenter location). This generally is not possible for emergent signals, whose first arrivals are obscured by unavoidable environmental and system noise (the slightly wavering background signal on an otherwise quiet seismogram). To circumvent this problem, Chouet and his colleagues deploy arrays of closely spaced seismometers around a source of LP events or tremor to act as seismic wave antennas.

One such array at Kīlauea Volcano, Hawai'i, consisted of 41 three-component sensors arranged in a semicircular spoked pattern with 50-m spacing along the spokes and 20° angular spacing between spokes. Each of several antennas (three in this case) provides a vector, called the slowness vector,[11] which represents the propagation direction and apparent velocity of seismic wave fronts across the array. The slowness vector data are combined with a model of the seismic velocity structure and inverted to produce a 3-D map of the source volume (Almendros et al., 2001a). Results for Kīlauea obtained during a swarm of LP events and tremor in February 1997 are very similar for both event types and point to a hydrothermal origin (Almendros et al., 2001b; Chouet, 2003). Those authors suggest that the swarm was a hydrothermal response to enhanced degassing associated with increased magma transport in Kīlauea's deeper magma conduit system.

---

[9] There are two general approaches to modeling geodetic and geophysical data, called *forward* and *inverse* methods (Chapter 8). Forward models use physical principles and known (or assumed) material properties to calculate a theoretical system response. For example, the surface-displacement field caused by a given pressure change in a Mogi-type source embedded at a given location and depth in a semi-infinite half-space with given elastic properties can be calculated using a forward model. An inverse model, on the other hand, starts with a generalized forward model and uses observations (data) with their associated uncertainties, plus known (or assumed) constraints called boundary conditions, to determine the best-fitting parameters for the model. This is accomplished by minimizing some model-to-data misfit criterion. For example, leveling or GPS data can be 'inverted' to determine the parameters of a Mogi source, including the optimal pressure change, location, and depth, that best fits the data in a least-squares sense. Inversion problems can be either linear or nonlinear, depending on the physics involved. Nonlinear inversions are non-unique, and require stronger constraints (e.g., more and better data, or independent knowledge of the system) to achieve a satisfactory result.

[10] V_P/V_S is the ratio of P-wave velocity to S-wave velocity. S-waves are more attenuated than P-waves by the presence of fractures, hydrothermal fluids, or partial melt, so the combination of relatively low P-wave velocity and high V_P/V_S is an indicator of pervasively fractured or fluid-rich zones within a volcanic edifice.

[11] Slowness is defined as the inverse of velocity for seismic waves; a large slowness corresponds to a low velocity. Most seismic tomography methods involve subdividing the medium into blocks and solving for slowness perturbations that cause predicted arrival times to match observed arrival times better than an initial model.

Another fruitful line of investigation uses moment tensor[12] inversion of VLP data (events with periods ranging from tens of seconds to a few minutes) recorded by broadband seismometers to infer magma conduit geometry and mass transport phenomena (Chouet, 2003). For example, Ohminato *et al.* (1998) analyzed broadband data acquired during a 4.5-hour period of rapid summit inflation and increasing seismic activity in the Kīlauea summit region on 1 February 1996. When the data are low-pass filtered and converted to displacement, they reveal a repetitive sawtooth signal with a rise time of 2–3 minutes and a drop time of 5–10 s.[13] What mechanism could account for mini inflation–deflation cycles with periods of just 2–3 minutes?

Ohminato *et al.* (1998) showed that the data are consistent with oscillations of a sub-horizontal crack or sill-like structure located about 1 km beneath the floor of Kīlauea Caldera. The volume change estimated from seismic moments of individual VLP events is $1–4 \times 10^3 \, \text{m}^3$, and integration over the duration of VLP activity yields a net volume budget on the order of $5 \times 10^5 \, \text{m}^3$. To explain the observations, the authors proposed a conceptual model of separated gas–liquid flow through a converging–diverging nozzle under choked conditions (Figure 3.3). During an inflation phase, magmatic gas accumulates upstream of the nozzle, building excess pressure and deforming the crack as a result. The gas slug is essentially stationary and the Mach number of the gas is $M_g \ll 1$. This phase coincides with the upgoing ramp in the sawtooth displacement signals. Eventually, pressure builds to the point that some of the gas is able to displace magmatic liquid and escape through the constriction in the nozzle.

The gas flow is choked ($M_g = 1$) in the nozzle and is supersonic ($M_g > 1$) immediately downstream of the nozzle. This phase corresponds to the rapid downdrop segment in the sawtooth displacement signals. Gas escape quickly lowers the pressure upstream of the nozzle, which again fills with liquid, thereby halting the gas flow and starting the cycle anew. In this fashion, $500{,}000 \, \text{m}^3$ of magma apparently chugged its way through a constriction in Kīlauea's shallow magma plumbing system during a 4.5-hour period, leaving behind as evidence some very long-period seismic fingerprints. That is an impressive story to extract from seemingly chaotic wiggles on broadband seismograms!

There are many other exciting research topics in contemporary volcano seismology, including precise relative relocation of swarm earthquakes to better delineate fault planes and other seismically active structures (e.g., Rubin *et al.*, 1998; Prejean *et al.*, 2002; Hill *et al.*, 2003; Got *et al.*, 2002; Got and Okubo, 2003; Battaglia *et al.*, 2003),[14] source mechanisms and implications of earthquake multiplets (e.g., Got *et al.*, 1994; Ramos *et al.*, 1996; Poupinet *et al.*, 1996; Got and Coutant, 1997),[15] spectral analysis of LP events to determine acoustic properties of magmatic and hydrothermal fluids (Kumagai and Chouet, 2000), and experimental modeling of the source dynamics of tremor (Lane *et al.*, 2001). Readers wanting to learn more about these topics might want to consult the original literature or a recent review article (e.g., Chouet, 2003).

Hopefully this section has conveyed a sense of the great breadth and diversity of contemporary research topics in volcano seismology. Now let us turn our attention back to geodesy and some additional tools of the ground deformation monitoring trade.

---

[12] The earthquake (or seismic) moment tensor is a symmetric, second-order (3×3) tensor that describes the nature of a fault's motion in terms of the amount of seismic moment release. Its components are 9 force couples, M_{ab}, where a and b correspond to the x, y, z axes, and $M_{ab} = M_{ba}$. Each force couple consists of two forces of magnitude f, separated by distance d, acting together in opposite directions; the magnitude of each component is the scalar moment, $f \times d$. Moment tensors corresponding to slip on a fault plane have a zero trace ($M_{xx} + M_{yy} + M_{zz} = 0$); a nonzero trace implies a volume change. The latter source type, sometimes referred to as a 'non-double--couple' source, is of particular interest in volcanology. Non-double--couple sources include explosions, implosions, and fluid injections into expanding cracks. For more information, the interested reader is referred to a seismology textbook such as Stein and Wysession (2003) or Aki and Richards (1980).

[13] Low-pass filtering the broadband signals removes higher frequency information that is superfluous for the analysis described here, but which can be analyzed separately to shed light on other volcanic processes.

[14] Relative relocation algorithms take advantage of the fact that errors in *absolute* hypocenter locations are largely attributable to sources, such as an imperfect seismic-velocity model, that affect tightly clustered events more or less equally (i.e., the errors are common-mode). The effect of common-mode errors can be reduced to the extent that hypocentral errors shrink by one to two orders of magnitude, revealing such structures as active fault planes and magma pathways (e.g., Prejean *et al.*, 2002).

[15] 'Earthquake multiplets' refers to a group of seismic events, usually occurring in a swarm, with very similar signatures (codas) on seismograms. In some cases, multiplet codas are virtually identical -- to the extent that the events appear to be clones of one another. At least some multiplets are thought to indicate repetitive, non-destructive excitation of the same source (e.g., opening and closing of a fluid-filled crack, bubble oscillation, or stick--slip motion across a resilient seismic patch on an otherwise freely slipping fault surface).

3.2 TILTMETERS

A review paper on the uses of strainmeters and tiltmeters published in 1986 began with a rather gloomy assessment: '*Despite steady effort over the last century, continuously recording tiltmeters and strainmeters have not yet been successful except for earth tide measurements*' (Agnew, 1986, p. 579). Since then, the situation has brightened considerably. The use of tiltmeters and strainmeters to monitor volcano deformation has become increasingly common, and there have been several striking successes. What's more, continuing design improvements are making these instruments more precise, durable, and affordable. This chapter takes a practical approach to monitoring volcanoes with tiltmeters and strainmeters. For a more thorough treatment of the principles involved, I recommend the review by Agnew (1986).

Broadly speaking, a tiltmeter is any device that can be used to measure changes in the local inclination of the Earth's surface or, in the case of borehole tiltmeters, in the orientation of the borehole. The physical means by which such changes are detected vary widely, and as a result tiltmeters are difficult to classify. One useful distinction that I adopt here from Agnew (1986) is between short- and long-base instruments. The former use a pendulum or bubble as their vertical reference, while the latter use the free surface of a liquid as a horizontal reference. Short-base tiltmeters generally are more portable and less expensive, making them better suited to most volcano crisis responses, although long-base tiltmeters are considerably more precise. Rather than describe the many types of tiltmeters with similar applications, I focus on three that have been used to monitor volcano deformation in a variety of settings worldwide.

3.2.1 Short-base bubble tiltmeters

The most widely used instrument for monitoring ground tilts near volcanoes is the bubble tiltmeter. As the name implies, the instrument's sensor includes a bubble in an electrolytic fluid enclosed in a small tube or disk, typically a few centimeters in size. In one common design, three wires are sealed into a bent glass tube to form electrodes (Figure 3.4). The tube is partially filled with an electrolyte (i.e., a fluid capable of conducting an electrical current). The electrolyte completely covers one of the wires, but includes a small bubble into which the other two wires penetrate. When the sensor tilts, movement of the bubble with respect to the wires causes a conductivity change that can be measured electronically. For small tilts, the magnitude of the tilt change is a linear function of the conductivity change. The tilt direction can be determined with two single-axis sensors, typically arranged perpendicular to one another, or a single biaxial sensor in which a bubble is free to move in two dimensions. Interested readers should consult the excellent article by Powell and Pheifer (2000), who provide a thorough discussion of electrolytic tilt sensors, including theory, design, applications, advantages, and limitations.

For Earth-tilt applications such as volcano monitoring, electrolytic bubble tilt sensors are usually mounted either in a platform housing or in a cylinder suitable for installation in a borehole (Figures 3.5–3.7). For engineering applications such as monitoring the structural integrity of dams or buildings, other types of housings sometimes are preferable (e.g., wall-mount or floor-mount designs). In either case, the sensor is arranged as one element in an electronic circuit designed to measure changes in fluid path conductivity within the bubble.

Westphal et al. (1983) used an alternating current Wheatstone bridge design with an inexpensive bubble sensor to construct expendable platform tiltmeters that were used to monitor and help predict dome-building eruptions at Mount St. Helens from 1981 to 1986 (Dzurisin et al., 1983). Commercial manufacturers offer similar designs that can be configured for the sensitivity, dynamic range, and mounting scheme required for a particular application. For most volcano-monitoring applications, 10^{-7} sensitivity (0.1 μrad) over 10^3 μrad of dynamic range is adequate and obtainable at reasonable cost. One microradian (μrad) corresponds to a vertical rise or fall of 1 mm over a horizontal distance of 1 km. Platform designs are preferred for crisis responses because they are relatively easy to transport and install. If time and conditions permit, it is advisable to emplace any tilt sensor at least several meters below ground level to mitigate the effects of temperature fluctuations and other environmental changes. At Kīlauea Volcano, Hawai'i, several bubble tiltmeters installed in boreholes up to 7 m deep exhibit sub-microradian sensitivity to short-term tilt changes and have proven to be very useful for tracking intrusive and faulting events (Cervelli et al., 2002a,b; 2003).

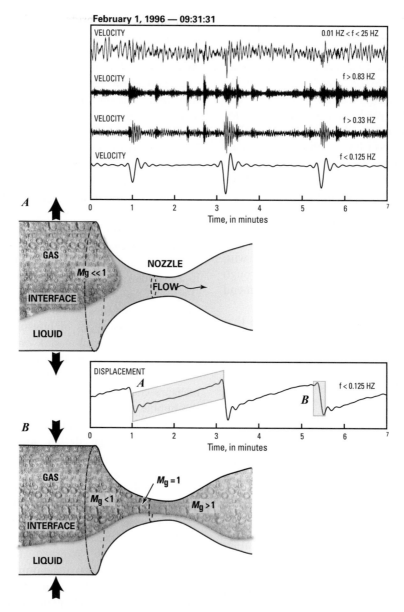

Figure 3.3. Broadband seismic record and associated filtered signals from Kīlauea Volcano, Hawai'i, during a volcanic crisis on 1 February 1996 that included rapid summit inflation, a swarm of VLP earthquakes, and tremor (reproduced from Chouet, 2003, p. 759). *Upper panel* shows the broadband vertical-velocity signal and the same signal filtered in three different frequency (f) bands to produce derivative velocity records for the same 7-minute time interval. Top trace shows the broadband signal ($0.01 < f < 25$ Hz), which is dominated by the oceanic wave-action microseism with periods in the range of 3 to 7 s. Second trace shows the signal after a high-pass filter has been applied ($f > 0.83$ Hz). The result is equivalent to a typical short-period record and shows a series of events superimposed on a background of tremor. Third trace also has a high-pass filter applied ($f > 0.33$ Hz); LP signals with a dominant frequency of about 0.4 Hz are enhanced in this record. Fourth trace shows the signal when a low-pass filter is applied ($f < 0.125$ Hz); in this case, a repetitive VLP signal consisting of pulses with a period of about 20 s ($f \sim 0.05$ Hz) is apparent. Fifth trace (*lower panel*) shows the corresponding displacement record (obtained by integrating the fourth trace above) with a low-pass filter applied ($f < 0.125$ Hz). The displacement record shows a repetitive sawtooth pattern with a rise time of 2–3 minutes and a drop time of 5–10 s. Also shown (lower left, A and B) is a conceptual model of separated gas–liquid flow through a converging–diverging nozzle under choked conditions (after Wallis, 1969, pp. 71–74), which was invoked by Ohminato *et al.* (1998) to explain repetitive VLP signals during the 1 February 1996 crisis at Kīlauea. (A) Inflationary periods lasting 2–3 minutes occur when flow of liquid magma through a constriction in the conduit system blocks the passage of a separate gas phase; these periods correspond to upgoing ramps in the sawtooth displacement record. (B) When the gas pressure upstream of the constriction reaches a critical value, some of the gas slug flows past the constriction, lowering the pressure and re-establishing the initial conditions; these periods correspond to downgoing ramps in the displacement record. The cycle repeats until the magma pressure drops below lithostatic, allowing the constriction to snap shut, and magma transport ceases.

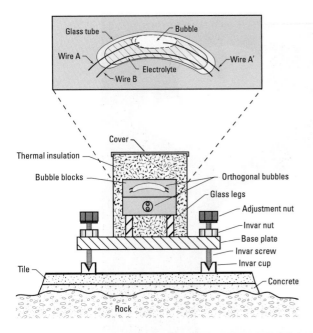

Figure 3.4. Schematic diagram of a typical platform tiltmeter used for volcano monitoring (Westphal *et al.*, 1983). Movement of a bubble (*inset*, dashed line) in an electrolytic fluid causes impedance changes across a glass tube that is part of an electronic impedance bridge. The impedance changes, which are proportional to ground tilt, are sampled periodically (e.g., at 10-minute intervals), converted to a digital signal, and telemetered to a data analysis facility. (*inset*) When the sensor tilts, movement of the bubble with respect to wires A and A causes a change in conductivity of the fluid path between wires A and B and between A and B, which can be measured electronically and related to the tilt in the direction of the long axis of the tube. Two sensors arranged perpendicular to one another measure two orthogonal tilt components, thus determining the resultant tilt vector. Platform tiltmeters can be installed on the surface, but a better approach is to isolate them from small-scale, near-surface effects by anchoring them securely in a tunnel or vault.

3.2.2 The Ideal-Aerosmith mercury capacitance tiltmeter

The Ideal-Aerosmith mercury capacitance tiltmeter near the summit of Kīlauea Volcano, Hawai'i, illustrates the utility of short-base electronic tiltmeters for monitoring ground deformation at active volcanoes. At each end of a 1-m-long tube, a pool of mercury is separated from a metal plate by an airspace gap, forming a parallel-plate capacitor.[16] A second tube filled with air, which is sealed from the atmosphere, also connects the pools. As the instrument tilts, mercury flows through the first tube from one pool into the other to maintain hydrostatic equilibrium, thereby changing the capacitances. The capacitance change is sensed electronically

with an electronic bridge circuit, converted to the equivalent tilt change, and recorded at the nearby United States Geological Survey (USGS) HVO. The instrument, which has short-term sensitivity of ~0.1 μrad, was installed in Uwekahuna Vault in 1965 and was still operational in 2006.

Records from the Uwekahuna Ideal-Aerosmith tiltmeter for the early parts of two long-lived eruptions along Kīlauea's East Rift Zone demonstrate the instrument's capability (Figure 3.8). At Mauna Ulu in 1969 and again at Pu'u 'Ō'ō in 1983–1984, eruptive episodes ranged in duration from hours to several days and were characterized by high-volume discharge of lava flows, vigorous lava fountaining, rapid summit subsidence, and strong harmonic tremor in the vent area. Recurring tilt patterns like those in Figure 3.8 led Dvorak and Okamura (1987) to propose a hydraulic model to explain variations in the tilt rate during summit subsidence events. They assumed that: (1) the magma flow rate is proportional to the pressure difference between the summit reservoir and a separate reservoir or series of reservoirs within the rift system, and (2) magmatic pressure is linearly related to volumetric strain. It follows that the magma flow rate and, hence, the summit tilt curve will follow an exponential decay. Dvorak and Okamura (1987) showed that the exponential curve represented by $\tau = 32 \times (1 - e^{-t/40})$, where τ represents the summit tilt value in microradians as a function of time t in days, fits the Ideal-Aerosmith record for episodes 2–9 of the Pu'u 'Ō'ō eruption remarkably well (Figure 3.9).

3.2.3 Long-base fluid tiltmeters

In situations where volcanic unrest may continue for several years, expected tilts are small, and adequate resources are available for monitoring, long-base fluid tiltmeters are preferable over short-base

---

[16] A capacitor stores energy in the electric field created between a pair of conductors on which equal but opposite electric charges have been placed; a simple design consists of two parallel, electrically conductive plates connected to a battery or other power source. For a *parallel-plate capacitor*, capacitance C is directly proportional to plate area and inversely proportional to plate separation. Furthermore, $C = Q/V$, where Q is electric charge and V is potential difference (voltage). A change in capacitance caused, for example, by a change in plate separation produces a corresponding change in Q/V, which can be measured electronically. In the Ideal-Aerosmith mercury capacitance tiltmeter, tilting of the ground surface produces a change in plate separation (capacitance) in an electronic sensing circuit.

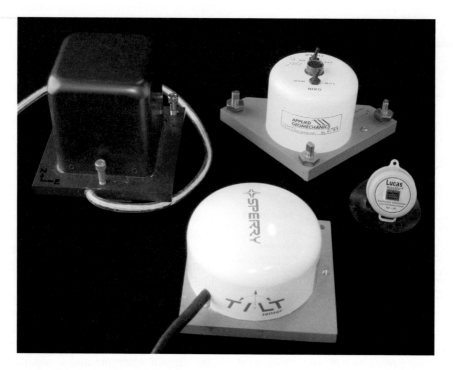

Figure 3.5. Four examples of platform tiltmeters suitable for volcano monitoring, including three from commercial suppliers and one prototype (*upper left*) that was designed and built specifically for use in the 1980 crater at Mount St. Helens (Westphal *et al.*, 1983; Dzurisin *et al.*, 1983). The 'Westphal tiltmeter' was an interim response to frequent dome-building episodes at Mount St. Helens during 1980--1986, which created a need for robust, relatively inexpensive, 'expendable' monitoring instruments. The small tiltmeter on the right (~5 cm in diameter) is designed for mounting on a vertical surface such as a rock wall; the other three are for use on a horizontal surface. Such designs are generally susceptible to thermal and other near-surface environmental effects that can be reduced substantially by installing them a few meters underground (e.g., in a vault or tunnel). USGS photograph by David E. Wieprecht.

designs. This is because long-base instruments generally are more precise and less susceptible to small-scale tilt fluctuations (i.e., 'noise' associated with localized ground movements that can be caused by differential settling, rain or snow loading, freeze/ thaw cycles, etc.) than their short-base counterparts (assuming careful installation of the long-base piers). The best long-base tiltmeters are hundreds of meters long and have sensitivity on the order of one part per billion (10^{-3} μrad). Such instruments easily resolve the solid Earth tide ($\sim 10^{-1}$ μrad), which can be computed accurately from theory and removed from the tiltmeter record. Because they are expensive and require elaborate installations and maintenance to isolate them from surface environmental effects, they are best suited to persistently active volcanoes where the threat to lives and property is high.

As is the case for short-base tiltmeters, long-base tiltmeters are available in a wide variety of designs. Most include one or two tubes at least 100-m long and either filled or partly filled with a liquid, usually water. Sensors at each end of the tube(s) measure the

height of the liquid surface with respect to some reference point tied to the ground. The liquid surface remains horizontal as the tube tilts, resulting in relative surface-height changes between the ends of the tube(s). These changes can be measured with various types of sensors, including a micrometer, pressure transducer, or interferometer, to determine the magnitude and, for biaxial designs, the direction of tilt (Agnew, 1986) (Figure 3.10).

Tubes that are completely filled with liquid are easier to install, because they don't have to be precisely level, but they are sensitive to slight environmental differences that are hard to eliminate along their entire length (e.g., temperature or pressure differences between the ends of the tube cause surface-height differences that are unrelated to ground tilt). An early implementation of this approach was the so-called wet tilt method developed by Eaton (1959) at the USGS HVO. In 1958, Eaton and the HVO staff installed 10 long-base watertube tiltmeters around the summit of Kīlauea Volcano (Yamashita, 1992). The installations consisted of three sunken concrete piers, each with a brass

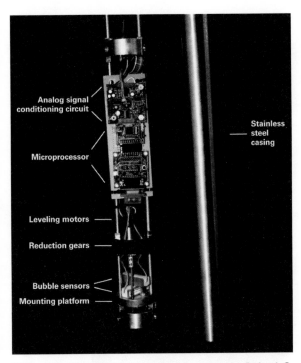

Figure 3.6. Dual-axis borehole tiltmeter design by Richard G. LaHusen, USGS. Analog and digital electronics are located downhole to avoid surface thermal effects. Leveling motors and reduction gears are operated by the microprocessor to re-level the mounting platform and its bubble sensors when the inclination exceeds full-scale range. Re-leveling can be accomplished either automatically or remotely via telemetry links. The tiltmeter package is housed in a submersible, 2¼ inch (55 mm) diameter stainless steel casing. Tiltmeter can be permanently grouted or temporarily set in sand within a 3-inch (75-mm) drill hole. USGS photograph by David E. Wieprecht.

container (pot) attached. The pots were machined from spent artillery shells, which were in abundant supply at a nearby military base. Each pair of pots was connected by a water hose and an air hose. The water hose was used to fill the pots and allow water to flow freely between them until equilibrium was reached. The air hose assured equal pressure between the pots. Micrometers were used to measure the height of the water in all three pots simultaneously, so the elevation differences between the piers could be calculated precisely.

Such installations were possible only where the elevation difference between the piers was no greater than 1.5 cm (the effective range of the micrometers) in an equilateral triangle with 20–50 m sides. The highest precision could be obtained only at night in fog or under heavy cloud cover, owing to thermal expansion or contraction of the water in the system during daylight hours. Because water was used as the medium to determine the elevation differences be-

tween the piers, and also because of the environmental conditions that were most conducive to good measurements, the term 'wet tilt' was applied to the method. Although no longer used at HVO, the simplicity of the wet tilt method was attractive to volcanologists elsewhere and it still is being used to monitor a few of the world's volcanoes. Another part of its continuing legacy is the term 'dry tilt,' which was coined at HVO to distinguish single-setup leveling (Section 2.5.2) from the wet tilt technique.

The shortcomings of water-filled tubes can be mitigated further by using a partly filled tube, which constitutes a miniature lake with an equipotential surface. Precision on the order of 1 nanoradian (10^{-9} rad or 10^{-3} μrad) can be achieved by burying such a device or installing it in a tunnel to insulate it from temperature fluctuations. Care must be taken to ensure that the tube is precisely level so a mini-lake can be maintained within it. This requires either a carefully constructed trench or an accessible tunnel, which can be hard to come by near an active volcano.

A biaxial long-base tiltmeter at the Long Valley Caldera in California uses two orthogonal tubes, 423 m and 449 m long, which are half-filled with water and buried at a mean depth of 1.5 m (Behr et al., 1992). The water level is measured at the ends of the tubes by laser interferometer transducers. Each transducer has a water-height resolution of approximately 0.25 μm, which corresponds to a tilt resolution of 0.6 nanoradians (0.0006 μrad). The ends of the tiltmeter are referenced to points 20-m deep by 3 borehole extensometers with displacement resolution of about 0.05 μm. After correction for atmospheric pressure variation, thermal expansion, and surface displacements, the instrument shows a long-term stability on the order of 0.1 μrad yr^{-1}. Such an elegant device is unaffordable or not feasible at most volcanoes, but in special cases its high precision and low drift rate can provide unique insights into magmatic processes. This is especially true if the strain rate is low but volcanic risk is high (e.g., relatively deep or gradual magma intrusion near a population center).

Another noteworthy long-base tiltmeter for volcano monitoring is installed in a tunnel 2.8 km northwest of the summit crater at Sakurajima Volcano in southern Japan (Kamo and Ishihara, 1989). The site was located at this distance to measure the maximum tilts expected from pressure changes in a magma source at 3–5 km depth beneath the crater, as revealed by repeated leveling surveys. To isolate the instrument from surface

Figure 3.7. Scientists install a shallow-borehole tiltmeter (~2-m) on the September 1984 lobe (foreground) of the 1980–1986 lava dome at Mount St. Helens, Washington. An active spine of the 2004–2006 lava dome, ~ 400 m distant, is visible near the upper left corner of the large photo, in front of the 1980 crater wall and rim on the skyline. View is to the south. A steel casing is hammered and cemented to the bottom of an 8 cm (3 inch) diamter drill hole (large photo). The tiltmeter is lowered to the bottom of the casing (*inset*, lower left) and secured with fine sand. The upper part of the casing is covered with a short section of PVC plastic pipe (*inset*, lower right); cables connect the tilt-meter to electrical power (solar panel and high-capacity rechargeable baterries) and a radio telemetry system (not shown). USGS photographs by Daniel Dzurisin, 5 August 2005.

environmental effects, a tunnel was constructed in the Harutayama lava dome specifically for this purpose, and a biaxial watertube tiltmeter and 3-component extensometer were installed in 1985. The watertube is 28 m long and 15 mm in diameter with water reservoirs at each end. Water levels are detected by a float in each reservoir and transformed into an electrical signal by a magnetic sensor. Inflationary radial tilt in the range from 0.01 μrad to 0.2 μrad is observed consistently for periods of 10 minutes to 7 hours prior to explosions in the summit crater. These episodes are followed by periods of deflation by approximately the same amount (Figure 3.11).

In an experiment to test an automated warning system based on the tiltmeter signal, scientists defined five stages of eruptive activity (ongoing erup-

tion, non-eruption, pre-eruption, warning, and critical) and used a computer to determine the stage in real time (Figure 3.11). During a 25-day experiment in 1985, there were 39 explosions. All but one occurred during pre-eruption or further advanced stages. The computer reported 59 critical stages, 38 of which correspond to eruptive events. The other 21 critical stages were not accompanied by significant eruptive activity. Most of these 'false alarms' were caused by temporary deflations associated with minor eruptions or gas emissions (Kamo and Ishihara, 1989). This was a pioneering experiment that led to significant improvements in the system, and now residents are accustomed to receiving accurate, automated, short-term warnings before most of Sakurajima's hazardous eruptions.

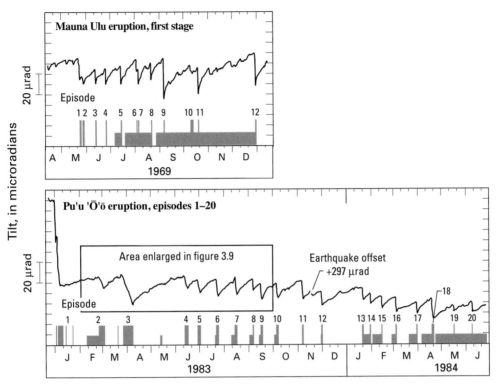

Figure 3.8. West–east tilt records from the Ideal-Aerosmith mercury capacitance tiltmeter, which has been housed in Uwekahuna Vault near the summit of Kīlauea Volcano since 1965, for parts of two long-lived eruptions along the East Rift Zone (Wolfe et al., 1987). Full-height bars represent eruptive episodes that were characterized by high-volume discharge of lava flows, vigorous lava fountaining, rapid summit subsidence, and strong harmonic tremor in the vent area. Half-height bars represent periods low-volume effusive activity at the vent during reposes between high-fountaining episodes.

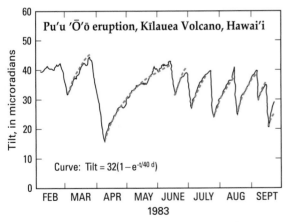

Figure 3.9. West–east tilt changes recorded from February to September 1983, which included episodes 2–9 of the Pu'u 'Ō'ō eruption along Kīlauea's East Rift Zone, by the Ideal-Aerosmith tiltmeter at Uwekahuna Vault in the summit area (Dvorak and Okamura, 1987). Periods of increasing tilt values correspond to summit uplift and magma accumulation beneath the summit area; periods of rapidly decreasing tilt values correspond to magma withdrawal from beneath the summit area and vigorous eruptive activity along the rift zone. The same exponential curve, for a maximum tilt change of 32 μrad and a time constant of 40 days, is fit empirically to all eight periods of summit uplift.

3.3 STRAINMETERS

Strain is loosely defined as a change in the volume (dilation) or shape (distortion) of a body as a result of stress, expressed as a dimensionless ratio: $\varepsilon = p_1/p_0$, where p_1 and p_0 represent final and initial states, respectively. Such a change implies relative displacements among the body's parts, and, in the most general case, these occur in three dimensions. Under the assumption of uniform strain throughout the body, we can write a strain tensor that relates the body's final state to its initial state by specifying the vector displacements of any part along three orthogonal axes. It follows that, for convenience, we can consider the linear strain l_1/l_0, surface strain s_1/s_0, or the most general case of volumetric strain v_1/v_0. In practice, most strainmeters are designed to measure either linear strain or volumetric strain. The first type is useful when strain is known to be localized and dominantly 1-D (e.g., across a narrow linear zone of surface extension or contraction), while the second type provides valuable information about spatially distributed strain changes caused by such processes as magmatic inflation or intrusion.

3.3.1 Linear strainmeters (extensometers)

Linear strainmeters are designed to measure the displacement between two separated points, either on the surface or in a borehole. The most common types of linear strainmeters are rod strainmeters, wire strainmeters, and laser strainmeters. Invar-rod and Invar-wire strainmeters used to monitor creep along the San Andreas Fault in central California have resolutions of 0.05 mm and 0.02 mm, respectively. Laser strainmeters are capable of fantastic precision, on the order of 10^{-10} to 10^{-13} (100 to 0.1 parts per trillion) (Agnew, 1986), but their complexity and high cost make them unsuitable for most volcano-monitoring applications. Because all designs are susceptible to environmental effects such as temperature fluctuations and loading by rain or snow, careful attention must be paid to isolating the strainmeter from such effects (e.g., by installing the instrument in a tunnel, burying it, or providing a sturdy insulated housing). Another important consideration is the means by which the strainmeter is attached securely to the ground. This is best and most easily accomplished by anchoring in bedrock, but where none is available, anchor rods driven to refusal may suffice. Deeply anchored piers isolated from surface movements are preferable to surface or shallow piers. Common sense, practicality, and budget are the best guides in this regard.

As the name implies, rod strainmeters use a solid rod to detect the relative displacement between two piers that are well anchored to the ground. One end of the rod is attached to one of the piers and the other is attached to a digital caliper or similar device, which is attached to the second pier. The rod transfers any relative displacement between the piers to the displacement sensor, which measures it; the result is recorded on site or telemetered to an analysis facility. If strain is distributed across a wide zone that can't be entirely spanned by the strainmeter,

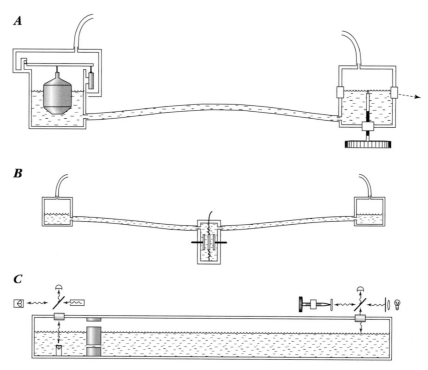

Figure 3.10. Three different designs of long-base tiltmeters (Agnew, 1986). For greater completeness, different types of sensors are shown at the two ends of each tiltmeter in A and C, though in reality such instruments are symmetric. (A) Pot-and-tube tiltmeter with (*left*) a typical design of float sensor and (*right*) a visually read micrometer. (B) Center-pressure tiltmeter with a pressure transducer to measure relative height changes. (C) Michelson-Gale tiltmeter, with (*left*) a laser interferometer and (*right*) a white-light interferometer used to measure the liquid level. A key feature of this design is that the tube is only half-filled with liquid, which ensures an equipotential surface. A biaxial instrument capable of determining tilt direction can be made by adding a second tube, usually perpendicular to the first, and a third sensor.

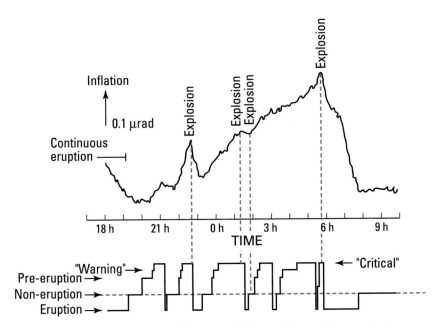

Figure 3.11. Radial tilt record for a 16-hour period on 18–19 December 1985, from a 28-m biaxial watertube tiltmeter in a tunnel near the summit crater of Sakurajima Volcano, Japan (Kamo and Ishihara, 1989). Also shown (*bottom*) are results of an experiment in which the tilt-meter record was used to automatically assign an activity stage in near-real time, according to a five-level scheme from 'ongoing eruption' (lowest alert) to 'critical.' See text for discussion.

wider spacing between the piers (i.e., using a longer rod) generally is a good idea, because the resulting signal at the displacement sensor is larger. This leads to a trade-off between portability and sensitivity. A longer rod captures more displacement, but is less portable than a shorter rod. For a given displacement sensor, sensitivity to small strains is greater with a longer rod. If displacement is known to be localized, (e.g., across a fault or crack), then a rod strainmeter long enough to span the feature is a good choice. Bilham (1989) used a ~1 cm diameter Invar rod, 6–7 m long, encased in a flexible lubricated tube to monitor creep at the Superstition Hills Fault in southern California. The equipment has sensitivity of 10 μm and therefore can detect a strain change of 1.7 μstrain over the instrument's length.

Where a longer reach is desired, a flexible wire sometimes provides a good alternative to a rigid rod. A wire stretched across a crack or deforming area, fixed to the ground at one end and attached to a displacement sensor at the other, can be very effective. For example, Iwatsubo *et al.* (1992b) used such an arrangement to monitor changes in the widths of radial cracks adjacent to a growing dacite dome at Mount St. Helens. Experience had shown that crack widening by several centimeters or more was likely in the days leading up to an extrusive episode, and that

monitoring stations near the dome were likely to be damaged or destroyed by repeated eruptive activity. Therefore, sophisticated strainmeter installations were impractical and not cost-effective. Instead, a relatively inexpensive ($500) commercial displacement meter with a stainless steel wire a few meters long was stretched between two 2-m steel fence posts driven into the ground and cemented on either side of a crack (Figure 1.22). The displacement meter used a potentiometer and battery to convert linear displacement to an electrical signal, which was digitized and telemetered to the USGS David A. Johnston Cascades Volcano Observatory (CVO), 70 km away in Vancouver, Washington. With a 3.8 m long wire, 12-bit analog-to-digital converter, and low-data-rate digital telemetry system (Murray, 1992), these homebuilt strainmeters could resolve crack displacements as small as ~1 mm – more than adequate for the task at hand. A wooden or metal conduit was used to protect the wire from heavy winter snow pack and other environmental hazards such as falling rocks and tephra. Starting about 6 days before an extrusive episode in May 1985, one such strainmeter recorded ~10 cm of extension across a radial crack in the crater floor near the base of the dome, which aided in a successful prediction of the event.

3.3.2 The Sacks–Evertson volumetric strainmeter

Most forms of solid-body deformation involve volumetric strain (i.e., an increase or decrease in volume (dilatational or compressional strain, respectively)). This type of strain is especially important in volcanology, because it commonly occurs as a result of magma movement, inflation or deflation of a magma reservoir, or pressure changes in a magma body or hydrothermal system. An elegant means to measure volumetric strain was first proposed by Benioff (1935), who suggested burying a large container of fluid with a small opening (akin to a sealed plastic bottle containing a narrow soda straw). Strains in the ground would change the volume of the container and force fluid to flow in or out of the straw. The narrower the straw, the farther a given volume of liquid (corresponding to a given change in volumetric strain) would move up or down inside the straw. By sensing the position of the liquid column in the straw, a precise measurement of volumetric strain could be obtained.

This principle was put into practice by I.S. Sacks and D.W. Evertson, who designed a remarkably sensitive instrument suitable for installation in a borehole (Sacks *et al.*, 1971; Figures 9.1 and 9.2). It is essential that such a precise instrument be installed in a borehole, preferably 200 m or more deep, to isolate it from spurious strains caused by environmental fluctuations near the surface. The Sacks–Evertson volumetric strainmeter and some of its applications to volcano monitoring are more fully described in Chapter 9, which deals specifically with borehole observations of strain and fluid pressure.

Sacks–Evertson strainmeters installed in California during the 1980s and 1990s, including four in the Long Valley region, have sensitivities on the order of 10^{-12} (one part per trillion) over periods of seconds to minutes. Instrumental drift and other factors (e.g., imperfect coupling to the host rock by means of an expanding grout, which creates strain as it hardens and cures) affect longer term signals to the extent that such minuscule strains cannot be resolved. Nonetheless, experience shows that the instruments are capable of resolving strain changes of 5×10^{-9} over periods of a few hours to a few days (Johnston *et al.*, 1994). Such high sensitivity means that interesting signals can be recorded much farther from the source than is possible with most other instruments. For volcano monitoring, this translates into safer and more sensitive monitoring stations

farther from the volcano – an important factor for working under hazardous conditions. On the other hand, the need to install this type of strainmeter in a relatively deep borehole, which is expensive and time-consuming to drill, makes it poorly suited for most volcano emergency responses.

Sacks–Evertson strainmeters have been installed near several volcanoes around the world, including Hekla Volcano, Iceland. Five of the instruments recorded remarkably similar sets of strain signals associated with explosive eruptions in 1991 and 2000. In the latter case, a warning based on the strain and seismic data was issued about 20 minutes before the eruption began (Agustsson *et al.*, 2000). This feat is all the more remarkable because four of the strainmeters are located more than 35 km from the volcano (Section 9.5.3). That's farther than the distance from the summit of Kīlauea to the summit of Mauna Loa, Hawai'i, and about halfway to Portland, Oregon, from Mount St. Helens, Washington! Tempering the excitement, however, is the fact that anomalous strain signals were recognized only about 30 minutes before the Hekla eruptions began – which leaves scant time for warning and response.

When and how magma accumulates before explosive eruptions, especially before the onset of precursory earthquakes, is a key question to be addressed by future volcano-monitoring efforts and theoretical studies. If magma accumulation beneath hazardous volcanoes can be detected months or even years before an eruption, reasonable steps can be taken well in advance to mitigate the eruption's social and economic impact. If, on the other hand, substantial volumes of magma can accumulate in the middle to upper crust and escape detection, even by the most sensitive instruments available, volcano hazards mitigation will continue to present a much greater challenge. Even in the first, more optimistic scenario, determining whether magma accumulation will culminate in a hazardous eruption or a relatively benign intrusion is likely to be extremely difficult for the foreseeable future.

Consider the dilemma posed by recent experiences at Redoubt Volcano in Alaska (1989), Hekla (1991), and Rabaul Caldera in Papua, New Guinea (1994). At Redoubt, more than two decades of repose ended suddenly with a large explosive eruption that began on 14 December 1989 (Brantley, 1990). In hindsight, volcanologists learned that a pilot flying near the volcano on 20 November had noticed a small white 'cloud' rising a few hundred meters above the summit crater and a strong sulfur odor downwind from the volcano. Additional pilot reports of small steam plumes were received on 3 December and 8 December. On 13 December, small earthquakes began to be recorded beneath the volcano at a rapidly increasing pace by a network of seismometers that included three geophones within 15 km of the summit (Power et al., 1994). Most of the earthquakes were located within 2 km of the surface and none was as large as M 3. The eruption began explosively just 24 hours after the first detection of unusual seismic activity. A collection of papers edited by Miller and Chouet (1994) provides a comprehensive account and analysis of the 1989–1990 eruption sequence at Redoubt.

The Redoubt experience left room for hope that a denser local seismic network or other monitoring techniques (e.g., continuous geodetic or geochemical measurements) might have provided a longer term warning. This optimistic view, however, seems to run contrary to experience at Hekla in 1991 and 2000. Had a similar eruption occurred in a populated area with very little warning, the consequences could have been severe. Hekla might be unusual in this regard, because there are many examples of volcanoes that inflate gradually over a long time and thus provide ample warning before eruptions. On the other hand, for the subset of historical eruptions with a paroxysmal event of Volcanic Explosivity Index (VEI) ≥ 3 ($>10^7$ m^3 of tephra, cloud column height >3 km, and described as 'severe' or 'violent'), in 45% of the cases the paroxysm came within the eruption's first day. Furthermore, half of the first-day paroxysms came within the eruption's first hour (Simkin and Siebert, 1984, pp. 112–113).

Rabaul Caldera is one example of a magmatic system that inflated gradually for many years before its 1994 eruption, but experience there offers scant comfort. Were it not for a comprehensive volcano hazards mitigation plan developed over more than two decades of sporadic unrest, and also for the resourcefulness of local residents who self-evacuated in accordance with the plan when the situation worsened to critical in just a matter of hours, thousands might have lost their lives in spite of an intensive monitoring effort. Here the problem was not a lack of intermediate-term precursors. On the contrary, repeated seismic swarms and broad uplift of the caldera floor began in 1971, increased dramatically in 1983, and continued to fluctuate in intensity until the early morning hours of 18 September 1994 (Newhall and Dzurisin, 1988, pp. 219–240). Then, in just 27 hours, the level of activity intensified abruptly from background (very low at the time) to simul-taneous eruptions at two sites along the caldera rim, Vulcan and Tavurvur Volcanoes, which had not been active since 1937–1943 (McKee et al., 1995; Blong and McKee, 1995; Lauer, 1995). Of course, longer term warning might have been possible if sensitive tiltmeters or strainmeters had been operating in the area, but there is no way to be sure. Sadly, there seems to be no guarantee of more than a few hours warning before some eruptions, even at well-monitored volcanoes.

In summary, volumetric strainmeters have the potential to significantly improve our ability to safely monitor crustal deformation caused by volcanic processes, especially if the instruments' size and installation cost continue to shrink. The two issues are related, because the cost of boring a hole is a significant fraction of the total installation cost, and small-diameter drill holes are much less expensive than large ones. Reducing the average installation cost would go a long way toward addressing two concerns that have delayed the instruments' widespread acceptance within the volcanological community. First is the potential for installation failures, especially failure to couple the strainmeter adequately to the wall of the borehole. Unfortunately, such flawed installations are permanent, rendering the instrument essentially useless. Of course, their adverse impact on a volcano-monitoring budget decreases in proportion to the shrinking installation cost. A second impediment to widespread use of volumetric strainmeters for volcano monitoring is the fact that a single strainmeter cannot provide directional information (unlike a tiltmeter, for example). This means that several strainmeters must be deployed in an array in order to locate and monitor a deformation source. Multiple-sensor arrays are always wise, but the relatively high cost of strainmeters has, in the past, caused most volcano observatories to opt for less expensive, less sensitive deformation sensors. If this obstacle can be overcome, perhaps through design changes that reduce cost while maintaining an adequate level of strain sensitivity, volumetric strainmeters likely will be deployed at many more volcanoes in the coming decades.

3.3.3 The Gladwin tensor strainmeter

Although it was designed primarily for mining and tectonic applications, the Gladwin tensor strainmeter (Gladwin, 1984; Gladwin and Hart, 1985) or some variation thereof might also see expanded use at volcanoes in the near future. Like the Sacks–

Evertson dilatometer, the tensor strainmeter is intended for installation in a borehole, preferably 200 m or more deep. Unlike the dilatometer, which measures volumetric strain (a scalar quantity), the tensor strainmeter determines the three independent components of the surface strain field, $\varepsilon_{xx}, \varepsilon_{yy}, \varepsilon_{xy}$.[17] For an isotropic material, the 3-D strain tensor generally has six independent components. Near the surface of the Earth, however, vertical stress and tractions are zero, so the strain field is completely described by three horizontal components.

The Gladwin tensor strainmeter uses three horizontal extensometers arranged vertically in a cylinder, each rotated 120° with respect to its neighbor, to measure the three independent components of horizontal strain. Each transducer consists of a three-plate capacitor. The plates are attached to the walls of the cylinder in such a way that deformation of the cylinder wall is transferred effectively to the gaps between plates. The resulting differential capacitance changes are measured with an electronic bridge circuit specially designed for high sensitivity and electronic stability. Strain sensitivity is on the order of 10^{-11} and the instrument's frequency response is flat from 0 Hz to more than 10 Hz. This means that the instrument is equally well suited to measuring gradual strain accumulation, such as might be caused by magmatic inflation, as well as much higher frequency strains associated with volcanic earthquakes and fluid flow, for example.

Tensor strainmeters of this type have been installed in mines since the late 1960s and near active faults since the early 1980s. Three were installed in 1986 along a creeping section of the San Andreas Fault near Parkfield in central California. Starting in mid-1993, two of those instruments detected a significant change in the accumulation rate of shear strain (0.5 and 1.0 microstrain per year). The third instrument also detected a change, but it is affected by localized deformation due to hydrology and, therefore, its data were discounted (Gwyther *et al.*, 1996). The strain-rate change, which was corroborated by two-color electronic distance meter (EDM) measurements, coincided with two other notable events. High rainfall in early 1993 ended a seven-year period of subnormal precipitation, and from late 1992 through late 1994 the Parkfield region had an increase in the number of magnitude 4–5 earthquakes relative to the preceding six years. Gwyther *et al.* (1996) and Langbein *et al.* (1999) recognized that the strain-rate change could have been induced by rainfall, but concluded that the more likely explanation is a change in the rate of aseismic slip on the San Andreas Fault at depth.

Tensor strainmeters are an integral part of the Plate Boundary Observatory (PBO), a state-of-the-art geodetic facility funded by the National Science Foundation (U.S.A.) to study deformation across the boundary zone between the Pacific and North American plates. Field operations commenced in 2004 and by the end of this decade PBO plans to install up to 175 Gladwin-type strainmeters and 875 continuous GPS (CGPS) stations in the western U.S.A., including instrument clusters at selected volcanoes in the Aleutians and Cascades, Yellowstone, and Long Valley (Section 4.10.3).

3.4 CONTINUOUS GPS

CGPS stations, which are set up to operate autonomously for long periods of time, have several advantages over repeated, campaign-style GPS surveys for volcano monitoring (Chapter 4). So-called 'permanent' CGPS stations include provisions for long-term power (typically a combination of batteries and solar panels at remote sites) and telemetry (radio, phone, or Internet). Therefore, they provide a continuous real-time data stream that can be analyzed either immediately using predicted satellite orbits or later using more precise orbits to obtain a record of 3-D surface displacement as a function of time.

An obvious advantage of this approach is the ability to track transitory deformation events such as the passage of a dike in near-real time. Once a network of CGPS stations is established, its maintenance requirements usually are lower in terms of field personnel than for repeated campaign-style surveys. Continuous records also offer an opportunity to recognize immediately such short-term deformation events as slow earthquakes, which might otherwise go unnoticed or be recognized only after the fact (Dragert *et al.*, 2001; Crescentini *et al.*, 1999; Linde *et al.*, 1996; Kawasaki *et al.*, 1995). Such

---

[17] For isotropic materials, the strain tensor is a symmetric second-rank tensor (i.e., a 3×3 matrix in which $\varepsilon_{ij} = \varepsilon_{ji}$ for $i, j = 1, 2, 3$) used to quantify the strain of an object undergoing a 3-D deformation. In general, six quantities are required: 3 diagonal terms to specify the relative changes in length along 3 orthogonal axes (principal strains $\varepsilon_{11}, \varepsilon_{22}$, and ε_{33}), and 3 off-diagonal terms to specify changes in the object's shape (shear strains $\varepsilon_{12} = \varepsilon_{21}, \varepsilon_{13} = \varepsilon_{31}$, and $\varepsilon_{23} = \varepsilon_{32}$). Volumetric strain is equal to the trace of the strain tensor: $\Delta V/V_0 = \varepsilon_{11} + \varepsilon_{22} + \varepsilon_{33}$. The Sacks–Evertson dilatometer measures the scalar quantity $\Delta V/V_0$, whereas the Gladwin tensor strainmeter determines $\varepsilon_{11}, \varepsilon_{22}$, and ε_{12} separately.

events have been observed recently for the first time at the Long Valley Caldera and Kīlauea Volcano by CGPS and borehole dilatometers. The mechanism of slow earthquakes at volcanoes is not yet understood, but their discovery lends credence to the idea that the closer we look at deforming volcanoes in space and time, the more we see.

A disadvantage of CGPS is that each receiver is tied to a single location rather than being available for re-location within a network of stations, as is the case for campaign-style surveys. Combined with the inevitable limitations on funding, this means that the spatial density of CGPS networks is often suboptimal (i.e., we can afford many fewer stations than we need). In addition, CGPS stations are subject to spurious environmental effects that plague all near-surface geodetic sensors. For CGPS, these include small movements induced by diurnal or seasonal temperature fluctuations, surface loading by precipitation, phase-center shifts caused by snow accumulation on the GPS antenna, and even such mundane annoyances as disturbance by animals, theft, or vandalism. On the other hand, the cost of GPS hardware and telemetry is continuing to decline as the utility of CGPS for a wide variety of applications becomes apparent. As a result, reasonably dense networks of 10–20 CGPS stations have been established at Kīlauea Volcano, Hawai'i (Segall et al., 2001; Owen et al., 2000b), the Long Valley-Inyo-Mono chain, California (Endo and Iwatsubo, 2000; Dixon et al., 1997; Webb et al., 1995), and several volcanoes in Japan. More are surely on the way. A preliminary instrument-siting plan for the Plate Boundary Observatory (PBO) in the western U.S.A. calls for clusters of 10–30 CGPS stations at several volcanoes in the Aleutian volcanic arc, Cascade Range, Long Valley-Inyo-Mono chain, and Yellowstone region (PBO Steering Committee, 1999). CGPS stations, borehole strainmeters and tiltmeters, improved interferometric synthetic-aperture radar (InSAR) coverage, and repeated geodetic surveys are essential components of a comprehensive volcano-monitoring program, which can be expected to yield fundamental new insights into how volcanoes operate (Dzurisin, 2003) (Chapter 11).

An excellent example of the utility of CGPS monitoring for revealing volcanic processes in unprecedented detail comes from Hawai'i, where Owen et al. (2000b) and Segall et al. (2001) analyzed CGPS data for the 30 January 1997 eruption at Napau Crater on the East Rift Zone of Kīlauea Volcano. The eruption occurred within a network of CGPS stations that recorded the temporal history of deformation during dike intrusion beginning ~8 hours prior to the onset of the eruption. Modeling results constrain the location, length, width, thickness, depth, strike, and dip of the dike. The authors concluded that: (1) magma was supplied to the eruption site from multiple sources, including the summit magma reservoir, a storage zone near Makaopuhi crater on the upper East Rift Zone, draining of the active lava pond at Pu'u 'Ō'ō, and perhaps elsewhere in the rift zone; (2) the intrusion was caused by continued deep rift zone inflation and slip on Kīlauea's south flank décollement, rather than by over-pressurization of stored magma; (3) the rate of dike propagation was limited by flow of magma into the dike; and (4) the volume of Kīlauea's summit magma reservoir is ~20 km^3, consistent with recent estimates from seismic tomography (Dawson et al., 1999). Such far-ranging and detailed insights likely would not have occurred without the CGPS network data.

To have learned this much about the internal structure and dynamics of Kīlauea from a single, relatively small event is very impressive, indeed. Furthermore, these results hint at even greater discoveries to come as growing networks of CGPS stations, strainmeters, and tiltmeters continuously monitor Kīlauea's ever-changing shape, and as increasingly detailed models reveal new aspects of its structure. Kīlauea is not alone in this regard, as many of the world's prominent volcanoes (e.g., Krafla, Iceland; Long Valley-Inyo-Mono chain, U.S.A.; Mount Etna, Italy; Rabaul, Papua New Guinea; Piton de la Fournaise, Réunion Island, Indian Ocean; Sakurajima, Japan; Soufriere Hills, Montserrat, West Indies; Taal; Philippines; Yellowstone, U.S.A.) are being examined with CGPS and other modern geodetic tools.

3.5 SOME CAUTIONS ABOUT NEAR-SURFACE DEFORMATION SENSORS

No geodetic sensor, however sophisticated in design, can provide useful information about volcanic processes unless it is well coupled to a representative piece of the Earth and isolated from unmonitored environmental effects. That might seem like a simple matter, but in fact it is a notoriously difficult issue to address with any consistency. This is especially true when time is short, as during most volcano emergencies, or where the surface consists of deeply weathered soil, fragmental debris, or intensely fractured rock, as is often the case near volcanoes. The most straightforward approach to mitigating this problem is to install all instruments in deep boreholes, far removed from insubstantial surface materials and capricious environmental effects. However, as emphasized by Agnew (1986, p. 596), '*The surface of the earth has two outstanding advantages as a place to put strainmeters and tiltmeters: it is accessible at low cost, and there is plenty of room.*' If for no other reason than this, surface installations will continue to dominate at most volcanoes for the foreseeable future.

One way to avoid the vagaries of near-surface installations is to take advantage of pre-existing tunnels, caves, or boreholes wherever possible. Agnew (1986) mentioned some advantages of this approach (e.g., reduced thermal effects), but also noted two disadvantages: (1) such openings may not be available where measurements are needed; and (2) instruments installed in such openings may not measure what is intended because the openings themselves distort the strain field. The ideal case, which can only be approximated in practice, is to bore a hole in competent bedrock well below the water table (preferably 100 m or more deep), case it with a watertight pipe, and couple a well-designed tiltmeter or strainmeter to the pipe.

Where such elaborate installations are not feasible (the general case), near-surface installations will suffice in most cases if simple precautions are observed to reduce temperature and precipitation effects. Burial of the sensor and any related electronics even by a few meters will help considerably, and usually can be accomplished in a single day's work. Attaching the sensor to several metal rods that have been driven to refusal in unconsolidated ground, coupled together, and partly encased in concrete is usually effort well spent. Finally, a sturdy housing to protect surface components of the installation (e.g., telemetry mast and antenna) from the worst of the elements is essential. Never underestimate the destructive power of nature (e.g., wind, heavy snow, or rime ice). Building a more robust monitoring station in good weather is always preferable to fixing a substandard one in a blizzard or rainstorm.

The following recommendations on the use of near-surface tiltmeters for volcano monitoring apply in part to other continuous sensors as well, and they still seem as pertinent as when they were first published (Dzurisin, 1992a). Out of necessity or oversight, I have violated these guidelines in the past

and may do so again in the future. Nonetheless, to the extent possible:

(1) **Be skeptical**. Resist the temptation to interpret tiltmeter data until adequate baseline information is available for each station. Remember that substantial ground tilt can be caused by site responses to installation of the tiltmeter, freeze/thaw cycles, heavy rainfall, groundwater changes, and other factors unrelated to volcanic activity.

(2) **Never believe a single tiltmeter**. Design a tiltmeter network with enough redundancy to compensate for unreliable stations that are likely to exist, regardless of the amount of effort expended during installation.

(3) **Never believe a single dataset**. Interpret tiltmeter data only within the context of other monitoring information and in light of the recent eruptive history of the volcano.

(4) **Be conscious of the social and economic impact of your work**. Impacts can be either positive (e.g., reduction of damage to life or property) or negative (e.g., hardships caused by unnecessary evacuation or decrease in property values), but in either case they are likely to be substantial.

3.6 CONTINUOUS GRAVIMETERS

As discussed in the previous chapter, gravimeters are sensitive to two processes that commonly occur at active volcanoes: (1) changes in the height of the ground surface, and (2) changes in the subsurface distribution of mass. Therefore, repeated gravity measurements at a volcano are a useful tool for studying both surface deformation and subsurface changes in mass or density that might be caused by magmatic processes. There are two ways to obtain such measurements: (1) by repeatedly surveying a network of bench marks using a portable gravimeter; and (2) by installing one or more continuous gravimeters, which produce a continuous record of the local gravitational acceleration, g. Good spatial and temporal coverage can be obtained by combining both approaches. Such information is especially valuable if independent measurements of surface height changes (e.g., from leveling or GPS) and groundwater levels (e.g., from monitored wells) are available, so their effects can be removed from the gravity data.

Despite their potential for illuminating magmatic

processes, few continuous gravimeters have been operated on volcanoes for extended periods of time. This is mainly because the instruments are relatively expensive to install and maintain, and the data obtained reflect the combined effects of surface height changes, groundwater fluctuations, instrumental drift, and environmental factors. Without other, independently determined monitoring data, these factors are difficult to separate and remove from the gravity data. Nonetheless, continuous gravimeters are likely to see increasing use at volcanoes, as they become more affordable and as more volcanoes are monitored continuously for surface-height changes and groundwater fluctuations. For a more thorough discussion of gravitational principles, instruments, and techniques, the reader might want to consult an applied geophysics textbook, such as Chapter 2 in Telford et al. (1990).

3.6.1 Absolute gravimeters

Continuous gravimeters can be classified further as either absolute gravimeters or relative gravimeters. Both types can provide useful information at volcanoes. One type of absolute gravimeter, the free-fall gravimeter, makes straightforward and very precise measurements of the time t it takes an object to fall a distance d in a vacuum. The measurements are then fit to the appropriate equation of motion, $d = gt^2/2$, to determine the local value of g (Faller, 1967). Typical free-fall gravimeters consist of four main components: (1) an evacuated chamber with a freely falling test mass, (2) a reference test mass that is held in isolation from nongravitational accelerations and other background noise, (3) a laser interferometer, and (4) an atomic clock. Laser light is directed onto the test masses, and the reflected light is combined with the laser reference to produce interference fringes. The optical signal is sent to a photo detector where the precise trajectory of the falling mass is sampled, resulting in many time-and-distance pairs that are fit to the motion equation to determine an absolute value for g (Niebauer and others, 1986).

The precision and accuracy of free-fall gravimeters were estimated to be about $10\,\mu\mathrm{Gal}$ until 1993, when a new generation of instrument called the FG5 became commercially available (Sasagawa et al., 1995). The FG5 uses an iodine stabilized He–Ne laser and a rubidium atomic clock as length and time standards, respectively. It measures sets of time t and falling distance d of a free-falling corner cube in a vacuum chamber. The FG5's claim of 1–$2\,\mu\mathrm{Gal}$

accuracy and precision has been verified by intra-comparison of FG5 instruments (Sasagawa et al., 1995; Niebauer et al., 1995) and by comparison of the FG5 to a superconducting gravimeter (Okubo et al., 1997; Francis et al., 1998). The theoretical free-air gravity gradient is $-3.086\,\mu\mathrm{Gal\,cm^{-1}}$ so this corresponds to 3–7 mm of height change in the absence of any mass change, which is adequate to detect vertical ground movements at many volcanoes.

Rise-and-fall absolute gravimeters are a more recent innovation that, as the name implies, make measurements during both the upward and downward parts of a free-fall trajectory. A test mass is thrown upward repeatedly in an evacuated chamber and its trajectory is tracked precisely using a laser interferometer and atomic clock (Brown et al., 1999). Rise-and-fall gravimeters have several advantages over drop-only designs. The rise-and-fall symmetry results in cancellation of many classes of errors, including air resistance and magnetic field errors. Frequency dependent errors in the system electronics also are eliminated (Marson and Faller, 1986). Rise-and-fall gravimeters are substantially smaller and faster to use than their drop-only counterparts. A drop-only gravimeter spends a relatively large amount of time gently lifting the test mass, letting it settle at the top, and then waiting for it to gain an acceptable initial velocity to begin measurements. In contrast, rise-and-fall gravimeters accelerate the test mass upward very quickly, measure on both sides of the trajectory, and finish at the initial starting point. In one experiment with a rise-and-fall gravimeter, a throw rate of $1\,\mathrm{s^{-1}}$ for 100 s resulted in a standard deviation in the measured value of g of $\pm1.5\,\mu\mathrm{Gal}$ (Brown et al., 1999).

3.6.2 Relative gravimeters – the magic of zero-length springs and superconductivity

Relative gravimeters measure differences in the acceleration due to gravity at different locations or times. In 1932, Lucien LaCoste, a graduate student at the University of Texas, and his faculty advisor, Dr. Arnold Romberg, developed a revolutionary design for portable instruments of this type. The elegance of the design, which still defines the state-of-the-art, merits a brief explanation. LaCoste and Romberg set out to develop a vertical long-period seismograph, akin to the horizontal pendulum seismograph used to record horizontal ground motions caused by earthquakes. In a horizontal pendulum seismograph, a mass attached to one end of a rigid beam is free to oscillate in a horizontal plane about a vertical hinge near the other end of the beam. If the axis of the hinge is perfectly vertical, the mass and beam form a horizontal pendulum with an infinite period. Any slight departure from verticality allows gravity to act as a damping force, decreasing the period of the pendulum and thus its sensitivity to horizontal accelerations.

LaCoste and Romberg wondered if a similar device could be constructed to measure vertical motions. This would require a mechanism to precisely counterbalance the force of gravity, so a mass attached to a beam with a horizontal hinge could be 'suspended' and allowed to oscillate in a vertical plane. A sketch of the required suspension is shown in Figure 3.12. The goal is to use a spring to precisely counterbalance the gravitational torque on a mass at the end of a beam. The gravitational torque is:

$$T_g = Wd \sin \theta \qquad (3.1)$$

where W is the weight of the mass, $W = mg$, and d is the distance from the center of the mass to the beam's hinge. The torque due to the spring is the product of the force exerted by the spring and the spring's lever arm, and can be written as (LaCoste and Romberg, 1998; Bomford, 1980, pp. 375–377):

$$T_s = -k \times (l - l_0) \times s = kl_0s - k \times \frac{b \sin \theta}{\sin \beta} \times a \sin \beta$$

$$= kl_0s - kab \sin \theta \qquad (3.2)$$

where s is the lever arm, k is the spring constant, l and l_0 are the lengths of the stretched and unstretched spring, respectively, and a, b, α, β, and θ are the distances and angles shown in Figure 3.12. The total torque on the mass is:

$$T = T_g + T_s = Wd \sin \theta + kl_0s - kab \sin \theta$$
$$= kl_0s + (Wd - kab) \times \sin \theta \qquad (3.3)$$

Equating the total torque to zero, we obtain:

$$kl_0s + (Wd - kab) \times \sin \theta = 0 \qquad (3.4)$$

The second term can be engineered to equal zero by selecting an appropriate mass and spring constant, then arranging the geometry correctly. On the other hand, eliminating the first term is more challenging. The spring constant k, and the spring's lever arm s, have finite values, so the only option available to LaCoste and Romberg was to design a 'zero-length' spring (i.e., a spring with $l_0 = 0$). At first glance, this seems impossible. How can we design a spring that

shrinks to zero length when no force is applied to it? LaCoste and Romberg realized they didn't have to. The trick is to engineer a spring that exerts an appropriate, non-zero force when the spring contracts to the point at which its coils touch one another. For such a spring, a plot of F versus l follows the relation $F = kl$, which passes through the origin when *extrapolated* to $l = 0$ (Figure 3.13). There are several ways to make a spring of this type. Modern LaCoste & Romberg gravimeters actually use a 'negative-length' spring in combination with fine wires to achieve the same effect (LaCoste and Romberg, 1998).

Modern LaCoste & Romberg gravimeters use this suspension together with a fine micrometer to adjust the extension of the spring and thus to counterbalance the gravitational torque to a precision of $1\,\mu$Gal or less (Figure 3.14). The accuracy of properly corrected measurements made with such a meter in survey mode is typically about $5\,\mu$Gal. For me, the graceful swings of the crosshair about the reading line when a LaCoste & Romberg gravimeter is perfectly adjusted in the local gravity field, especially at a bench mark within sight of an active volcano, elicit a feeling that borders on reverence. Such delicate mechanical balance provides a soothing counterpoint to the potential for volcanic mayhem.

In 2001, LaCoste & Romberg and Scintrex Ltd., merged to form a new company known as LaCoste & Romberg–Scintrex, Inc., which produces over 90% of the world's gravimeters. In 1989, Scintrex Ltd., developed a fused-quartz-based relative gravimeter, the CG-3, based on the Ph.D. thesis of another enterprising graduate student, Dr. Andrew Hugill. Today, the Scintrex CG-5 Autograv gravimeter, together with the LaCoste & Romberg Model G, Model D, and Graviton EG gravimeters, set the standards for precision and repeatability among portable gravimeters. Features include automatic noise rejection, self leveling and automatic tilt compensation, and automatic reading and internal data logging options.

Although portability is an important feature of zero-length-spring and fused quartz-sensor gravimeters, which are commonly used in survey mode to measure a network of gravity stations, an instrument of this type can also be operated at a single site for an extended period of time to produce a continuous record. The instrumental sensitivity is more than adequate to measure solid Earth tides, which can be removed from the record if desired to reveal residual signals caused by volcanic or tectonic processes, or by groundwater fluctuations. Unfortu-

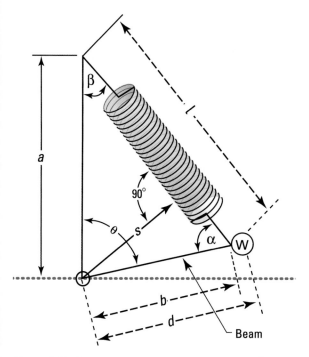

Figure 3.12. A suspension system for precisely counterbalancing the gravitational torque on a mass attached to a beam using a 'zero-length' spring (LaCoste and Romberg, 1998) (Figure 3.13). This ingenious mechanism is an essential component of LaCoste and Romberg gravimeters (Figure 3.14), which are suitable for volcano studies. See text, Section 3.6.2, for explanation of symbols.

nately, slow mechanical aging of the spring or other factors produce long-term drift that can be difficult to distinguish from other effects, including vertical surface movements and subsurface mass or density changes. Replacing the spring would only start the process over again. For applications where slowly varying, long-term signals might be important, as is the case for many volcanoes, other types of continuous gravimeters are a better choice.

One such design is the superconducting relative gravimeter, which balances the weight of a superconducting niobium sphere using a force produced by electrical current in a superconducting coil. Changes in the current needed to maintain the sphere at a null position are proportional to changes in the local gravity field (Francis et al., 1998). These instruments sample the gravity field every 1–10 s with a precision of <1 nGal (10^{-9} Gal) and an estimated accuracy on the order of 0.1 μGal. Because electrical losses in a superconducting material are virtually zero, this design can achieve extremely low long-term drift rates, on the order of 1 μGal yr^{-1} (Crossley et al., 1999). The best gravity-monitoring site would combine a superconducting relative

gravimeter with co-located measurements by an absolute gravimeter at regular intervals, a strategy that has been implemented at Mount Vesuvius, Italy (see below).

3.6.3   Gravity results from selected volcanoes

A recording LaCoste & Romberg Model D gravimeter was installed at Mount Vesuvius in 1988. The instrument is located at the Osservatorio Vesuviano on a concrete pillar in a 20 m deep artificial cave, where the daily temperature excursion is about 0.1°C and the annual excursion is about 2°C. The station also is equipped with temperature, pressure, and tilt sensors. High-precision absolute gravity measurements were made at a nearby site in 1986, 1994, and 1996. The absolute gravity decreased by about 60 μGal from 1986 to 1994, then changed very little from 1994 to 1996 (Berrino et al., 1997). EDM measurements of a 21-station, 60-line

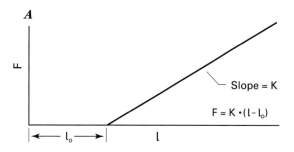

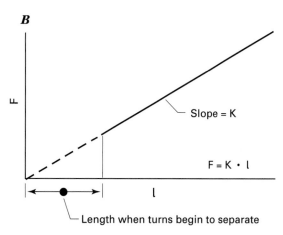

Figure 3.13. Relationship between force exerted by an ideal spring F and its length l. (A) For a typical spring, F is a linear function of l, and F approaches 0 as l approaches l_0, the unstretched length of the spring. (B) A 'zero-length' spring is designed such that it exerts a finite force $F = kl_0$ at its minimum length (when its coils touch each other) and such that F approaches 0 as l is extrapolated to 0 (LaCoste and Romberg, 1998).

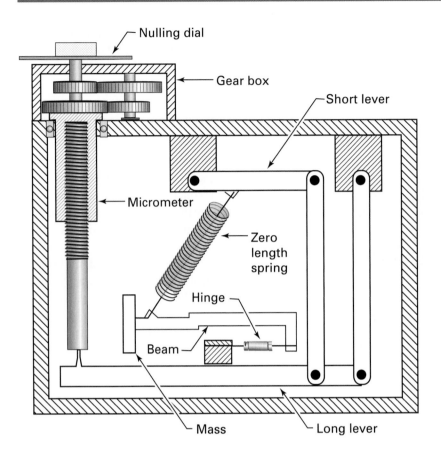

Figure 3.14. Schematic diagram showing the principal components of a LaCoste & Romberg gravimeter (LaCoste and Romberg, 1998).

trilateration network on the volcano revealed no significant ground deformation from 1975 to 1995 (Pingue *et al.*, 1998), which suggests that the observed gravity change was not caused by uplift. An analysis of the continuous gravity record revealed that the tidal amplification factor, which measures local gravitational response to solid Earth tides, increased by several per cent from 1988–1991 to 1994–1996. Berrino *et al.* (1997) suggested that both changes could have been caused by mass redistribution at depth (presumably withdrawal of material from beneath the gravity station with no associated surface deformation) or, alternatively, the increased amplification factor might indicate reduced crustal rigidity. Neither of the changes seemed to correspond with the intensity of local earthquake activity, which mostly fluctuated between about 10 and 100 earthquakes per month.

At Merapi Volcano, Indonesia, a recording gravimeter in the Babadan observatory has recorded long-term signals that correlate with seismic and volcanic activity (Jousset *et al.*, 2000). For example, the residual drift of the meter decreased steadily from November 1993 (when it was installed) to March 1994, then increased steadily until June 1994 when it stabilized. The trend reversal corresponded to a period of intense low-frequency earthquakes and nuées ardentes (i.e., dome-collapse pyroclastic flows) during March–April 1994. In addition, the gravimeter's admittance, a combination of the instrument's sensitivity (which is tilt dependent) and the mechanical response of the ground to tidal forces, decreased suddenly by almost 20% during the third week of April 1994, when a new phase of rapid dome extrusion began. Jousset *et al.* (2000) proposed a model in which extrusive activity was driven partly by internal pressure oscillations due to crystallization and degassing of magma in a shallow reservoir.

Continuous gravity results at Mount Etna, Sicily, are likewise intriguing (De Meyer *et al.*, 1995), although their interpretation is made difficult by the inability to sort out effects of ground deformation, magma movement or density changes, and groundwater fluctuations.

Taking full advantage of the potential of continuous gravimetry requires a substantial investment of funds, time, and effort. This is especially true because vertical surface displacements, groundwater fluctuations, and environmental changes also must be monitored continuously to remove their effects from the gravity signal. These difficulties aside, I believe this approach holds considerable promise for a significant advance in our understanding of volcanic processes, especially during non-eruptive periods when magmatic activity is hidden from direct observation.

3.7   DIFFERENTIAL LAKE GAUGING

Water-filled lakes are common features at volcanoes, and in some cases they provide a useful means to monitor ground deformation or surface volume changes related to eruptive activity. A lake on the flank of a volcano or near its base, for example, would exhibit differential water-level changes if the lakeshore tilted while the water surface remained horizontal. A crater lake at the summit of a volcano might not exhibit any differential tilt, but extrusion of a lava flow or dome onto the lake floor would raise the water level, while collapse of the floor would lower it. Both types of changes can be monitored by repeated surveys or continuously recording instruments.

If conditions are right, lake gauging can be a very useful volcano-monitoring technique. For example, periodic monitoring of water levels in a well-situated, 2 km diameter lake with an accuracy of ± 2 mm could provide information about cross-lake tilt to an accuracy of ± 1 µrad without having to deploy or maintain tiltmeters or other types of sensors. Alternatively, pressure transducers could be deployed in a lake to continuously sense water depth at several sites, thereby accumulating enough information to distinguish among the effects of wave action, seiche, filling and draining of the lake, and tilting of the lake basin. In practice, it is usually difficult to find a well-situated lake of sufficient size that does not freeze over in the winter and is not subject to violent wave action that can damage sensors. Nonetheless, if a suitable lake exists, consideration should be given to making use of it as a natural tiltmeter.

3.7.1   Monitoring active deformation at Lake Taupo, New Zealand

Lake Taupo occupies most of the 35 km diameter Taupo Caldera, which formed by collapse during eruption of the Taupo Pumice (90+ km^3 bulk vol-

ume) in 186 CE (Wilson *et al.*, 1984). Deformation of the area has been monitored since 1979 with a portable water-level gauge at sites around the shoreline and on two small islands (Otway and Sherburn, 1994). The standard deviation σ of lake-level measurements using this technique, based on the repeatability observed over short time periods, is $\pm[1.2 + (d \times 10^{-7})]$ mm, where d is the straight-line distance in kilometers between a reference site and some other measurement site. In the worst case (i.e., at the most distant site), $d \approx 40$ km and $\sigma \approx 5.2$ mm, which corresponds to a uniform tilt across the long axis of the lake of approximately 0.1 µrad.

From 1985 to 1990, vertical deformation measured with a portable water-level gauge at Lake Taupo and shallow seismicity there displayed two distinct patterns (Otway and Sherburn, 1994). In the Taupo Fault Belt north of the lake, subsidence at a maximum rate of 10 ± 1 mm yr^{-1}, which corresponds to inward tilt of more than 1.0 ± 0.1 µrad yr^{-1}, occurred steadily and nearly aseismically (Figure 3.15). In contrast, the central and southern parts of the lake tilted much more slowly to the southwest, but produced repeated small oscillations in height and frequent earthquake swarms. Short-term increases in the deformation rate preceded or accompanied four of the strongest swarms, but there was no apparent correlation between short-term deformation and individual seismic events.

Otway and Sherburn (1994) hypothesized that both patterns reflect responses, at different time-scales and depths, to underlying strain release. They noted that extension rates across the southern Taupo Fault Belt north of Lake Taupo averaged 18 ± 5 mm yr^{-1} for the 30-year period before 1986 (Darby and Williams, 1991), so aseismic sag in that area might indicate crustal thinning resulting from back-arc spreading accompanying subduction. Farther south, the deformation and seismicity patterns suggest a distinctive style of strain release, although the reason for the difference is not known.

3.7.2   Lake terraces as paleo-tiltmeters

Lake terraces, like bathtub rings, form horizontally at the margin of a body of water and, therefore, indicate a surface that was initially level. Any subsequent deformation of the lake margin is recorded by terraces as a departure from horizontality, which can be measured by surveying. Typically, terraces are not continuous along an entire lake margin, both

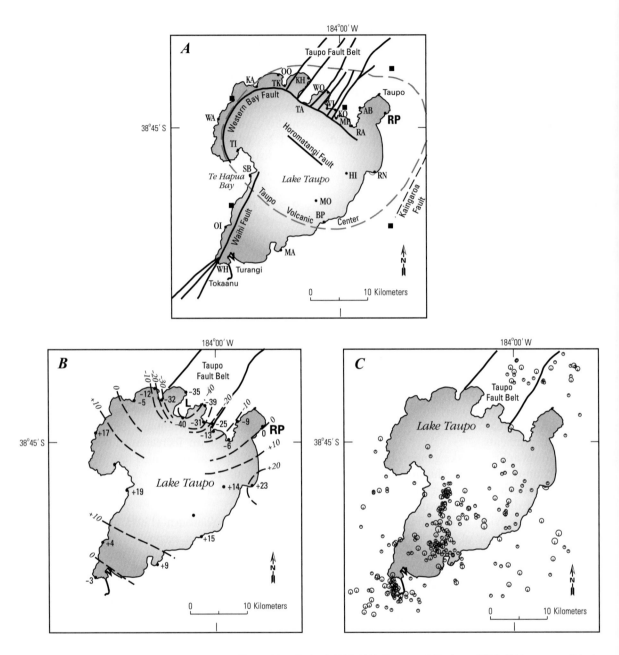

Figure 3.15. Deformation and seismicity near Lake Taupo, New Zealand, 1985–1990 (Otway and Sherburn, 1994). (A) Locations of lake leveling sites (dots), seismographs (squares), and main faults. (B) cumulative shoreline height changes, in mm relative to reference station RP, between 10 December 1985 and 6 December 1990, from repeated lake leveling surveys. Standard deviations are 3–7 mm and increase with distance from RP. (C) Epicenters of earthquakes of $M_L \geq 2.6$ (local magnitude, see Glossary entry 'earthquake magnitude' at the back of this book) between September 1985 and December 1990. Symbol size varies with earthquake magnitude.

because they do not form everywhere and because subsequent lake-level changes and wave action tend to remove terrace segments. Nonetheless, in some cases a deformation history can be pieced together through careful mapping, correlation, dating, and surveying of remnant terrace surfaces. Two such cases are Yellowstone Lake, Wyoming, which strad-

dles the southeast rim of Yellowstone Caldera, and Paulina Lake, Oregon, one of two lakes that occupy the summit caldera of Newberry volcano.

The shorelines of Yellowstone Lake record a rich Holocene deformation history, including two major cycles of uplift and subsidence that were each thousands of years long and tens of meters in amplitude

(Pierce *et al.*, 2002; Locke and Meyer, 1994; Hamilton and Bailey, 1988; Bailey and Hamilton, 1988; Meyer and Locke, 1986). Cyclic tilting of the lake basin alternately raised and lowered the lake outlet, causing high and low stands of the lake, respectively. The resulting shoreline terraces range from about 30 m below the present shoreline to about 20 m above it. Over the entire postglacial period (15,000 years), subsidence has balanced or slightly exceeded uplift, as shown by older shorelines that *descend* toward the caldera axis. These millennia-scale deformation cycles are akin to decadal-scale ups and downs of the caldera floor that have been revealed by repeated leveling, GPS, and InSAR observations (Dzurisin *et al.*, 1990, 1994, 1999; Wicks *et al.*, 1998) (Chapter 7).

Several processes might contribute to deformation at Yellowstone at different times and to varying degrees. These include magma intrusion or withdrawal, cooling of magma or hot rock, tectonic extension, and fluid pressure changes driven by magmatic volatiles and modulated by episodic cracking of a self-sealed layer in the deep hydrothermal system (Fournier, 1999; Dzurisin *et al.*, 1990). The hydrothermal mechanism is favored for both the historical and Holocene time periods, because it provides for both uplift and subsidence with little overall change (Pierce *et al.*, 2002; Dzurisin *et al.*, 1990).

At Newberry volcano in central Oregon, deformed lakeshore terraces and a nearby outlet stream preserve evidence for a large flood and subsequent uplift within the summit caldera. Newberry's 6 × 7 km caldera contains two lakes, Paulina Lake and East Lake, and has been the site of several eruptions during the past 10,000 years (MacLeod *et al.*, 1995). Along the channel of Paulina Creek, which drains Paulina Lake, a flood about 1,800 years ago swept away the regional Mazama Ash, deposited coarse gravel, eroded bedrock channels, and dropped lake level by as much as 1.5 m. Shoreline terraces that were abandoned when lake level dropped were subsequently tilted and displaced vertically upward by as much as 4–6 m (Jensen and Chitwood, 1996). It remains to be seen whether the flood or caldera deformation can be related to eruptions that produced rhyolitic pumice-fall deposits about 1,600 years ago or Big Obsidian Flow and associated ashflow deposits about 1,300 years ago (MacLeod *et al.*, 1995).

3.8 CONCLUDING REMARKS

This chapter has touched upon a wide variety of *in situ* sensors and techniques that volcanologists can draw upon to monitor deforming volcanoes. The best choices for a particular situation depend upon such things as the nature of the mission (hazards mitigation, curiosity driven research, or both), risk level (people, property, and major infrastructure), type of volcano (frequently active or long dormant, dominantly explosive or effusive), logistics (ease of access, year-round maintenance requirements), and available resources. No single type of sensor or combination of sensor types is appropriate for all situations, nor is a cookie-cutter approach to deformation monitoring likely to be successful. A monitoring program tailored to a specific set of circumstances and based on flexibility and innovation is usually the best alternative.

The GPS is likely to be a key component of any such program, for several reasons. GPS is flexible, precise, easy to use, universally accessible, and generally affordable. In the following chapter I examine this modern marvel and a few of its many applications to geodesy and volcano monitoring.

CHAPTER 4

The Global Positioning System: A multipurpose tool

Daniel Dzurisin

The Global Positioning System (GPS) is arguably the most versatile navigation, surveying, and geodetic tool ever devised. Some of its diverse applications include: (1) terrestrial, marine, and air navigation; (2) land surveying, cadastral mapping, and GIS support; and (3) geodynamic research into such topics as plate motion, tectonism, and volcanism. Although a relative newcomer to volcanology, GPS has quickly carved a niche for itself among mainstream volcano-monitoring techniques. Unlike most other geodetic techniques, GPS measures not just a single parameter pertaining to the relative positions of two points (e.g., height difference, line-length, or bearing), but rather the full 3-D position of each point in an absolute reference frame.[1] For the first time, a single technique can be used to monitor 3-D surface displacements remotely and continuously with millimeter accuracy. A shortcoming of GPS and all other point measurement techniques (i.e., the difficulty of obtaining data that are spatially dense enough to avoid under-sampling and thus aliasing localized deformation signals) can be addressed in many cases with deformation data obtained by interferometric synthetic-aperture radar (InSAR) studies (Chapter 5). GPS will not render classical geodetic techniques obsolete anytime soon, but its use will continue to increase as affordable equipment and data-analysis software tailored to the needs of the volcanological community become more widely available.

GPS usage has already become so widespread that the word GPS means different things to different users. For that reason, I need to specify how I will be using the term and what the relationships are among various components of the burgeoning Global Navigation Satellite System (GNSS), which includes GPS. In 1978, the US Department of Defense began launching a constellation of Navigation Satellite Time and Ranging (NAVSTAR) satellites to provide global positioning and timing information primarily to military users. Non-military users were provided access to the information with some restrictions, which were not overly burdensome (Section 4.3.2), and a multibillion-dollar civilian industry was born. The system was known initially as NAVSTAR GPS and that term persists in some official US military circles (e.g., the NAVSTAR Global Positioning System Joint Program Office, which manages the system). However, in popular usage and especially among civilian users (including geodesists), NAVSTAR GPS has become synonymous with GPS and the former term is seldom used.

The Soviet Union developed its own Global Navigation Satellite System (GLONASS) in parallel with GPS and for the same purposes (Section 4.2.3). Although GLONASS is no longer fully operational, satellite launches are occurring regularly again following a brief hiatus and the Russian government remains committed to re-establishing an operational constellation by 2008. Several manufacturers offer GLONASS-enabled GPS receivers[2] and GLONASS usage will likely increase when the system becomes operational again. I have included a

---

[1] The GPS reference frame is centered at the Earth's center of mass and includes three orthogonal axes that are defined by convention (Section 4.1.1).

[2] A GPS receiver is an electronic device that receives and processes signals from NAVSTAR or GLONASS satellites to calculate positioning information (e.g., latitude, longitude, elevation) directly or, alternatively, to produce a dataset that can be further processed elsewhere to obtain such information.

brief description of the Russian system that parallels a more thorough treatment of the US system.

Starting in 2005, the European Space Agency (ESA) plans to launch its own constellation of global navigation satellites. The European system will be called Galileo and is expected to become operational in 2008 (Section 4.2.4). Henceforth, I use the term GPS to refer specifically to the US Global Navigation Satellite System (i.e., NAVSTAR GPS), GLONASS for the Russian system, and Galileo for the European system. The term NAVSTAR is reserved for specific references to the US global navigation satellites. Readers interested in learning more about the Russian or European space programs in general will enjoy the books by Harvey (1996, 2001, 2003).

With GPS now a mainstay of global navigation, the revitalization of GLONASS underway, and the deployment of Galileo on the horizon, a need has arisen for a term to encompass all three systems and the concept they represent. The term Global Navigation Satellite System (GNSS) has emerged and its usage is growing. GNSS refers to any of, or the collective combination of, the operational space-based navigation systems. In 2006, this is generally taken to mean GPS and GLONASS, although GLONASS is not fully operational. Soon, Galileo will become a third component of GNSS. I use the term GNSS generically when the reference is not to a specific system but rather to any or all of them (e.g., 'GNSS applications will likely continue to grow at a rapid pace for the foreseeable future'). However, there are many instances when common practice dictates use of the term GPS instead of GNSS, even though the reference applies equally well to the US, Russian, and European systems. For example, phrases like GPS data processing, continuous GPS, and kinematic GPS are entrenched in the literature and in geodetic vernacular. Substituting GNSS for GPS in such cases would be confusing to most readers. Therefore, I also use the term GPS in this more generic sense when such usage conforms to common practice. The intended meaning should be clear from context.

A thorough description of GPS, GLONASS, and Galileo is beyond the scope of this chapter. So, too, is a technical treatment of data-processing techniques, a subject worthy of a book unto itself. Instead, the chapter includes brief, mostly nontechnical discussions of global navigation satellites and signal structure, data types ('observables' in GPS parlance), how GPS data are used to determine positions, surveying techniques and software, and

some applications to volcano geodesy. I have tried to avoid jargon and references to specific hardware or software wherever possible because techniques are changing rapidly and specific products become outdated quickly. I have also avoided the electrical engineering aspects of GPS because an understanding of those concepts is not necessary for most applications. Several equations are derived for illustrative purposes, but the discussion is mostly qualitative and hopefully intuitive. The unavoidable math notation might be tedious, but the mathematics itself is straightforward. My goal is to provide the reader with a conceptual understanding of how global satellite positioning works without resorting to anything more daunting than basic algebra and physics. Furthermore, the chapter is structured in such a way that the math in Sections 4.4 through 4.6 can be skipped without unduly interrupting the flow of the discussion. Many details are provided in footnotes to accommodate both casual and careful readers. More technical information is available from numerous sources, including books by Leick (1990), Logsdon (1992), Seeber (1993), Kaplan (1996), Hofmann-Wellenhof *et al.* (2001), and Misra and Enge (2001).

4.1 GLOBAL POSITIONING PRINCIPLES

4.1.1 Reference surfaces and coordinate systems: the geoid and ellipsoid

Before considering how satellites whizzing through space can be used to determine the position of a point on Earth to within a few millimeters, a brief explanation of reference surfaces and coordinate systems is in order (see also Section 2.2). Implicit in all geodetic coordinates is a reference system of some sort (e.g., an elevation of 100 m *above sea level*, or a latitude of 3.57 degrees *south of the equator*). For horizontal coordinates, a mathematical construct called the mean Earth ellipsoid (hereafter referred to as the ellipsoid) is widely used as the reference surface. The ellipsoid is a geometric figure centered at Earth's center of mass and defined in terms of Earth's equatorial and polar radii. It approximates the mean surface of the Earth, but differs in detail owing to topography and subsurface mass heterogeneity. If the spinning Earth consisted entirely of water, it would assume the shape of the ellipsoid (neglecting dynamic effects such as tides, water density variations, currents, and atmospheric fluctuations).

The real Earth is not homogeneous, of course, so

even if its surface were completely covered with water, the ocean surface would not assume the exact shape of the ellipsoid. Instead, it would be everywhere perpendicular to the local direction of gravity, which is affected by topography and the subsurface mass distribution. The resulting equipotential surface is called the geoid. Like the ellipsoid, the geoid approximates the shape of the real Earth but differs from it in detail. The difference between the geoid and ellipsoid can be as large as ±100 m (Figure 2.1).

These concepts might seem rather arcane, but they have very real implications for geodesists and GPS users in general. The mathematical simplicity of the ellipsoid makes it a good choice as the reference surface for horizontal coordinate systems such as latitude and longitude. We need only choose the origins from which to measure (e.g., the equator for latitude and a great circle passing through Greenwich, England, for longitude), slice the ellipsoid into equal angular increments, and project the resulting grid onto the Earth's surface. For vertical coordinates, the situation is more complicated. Until the advent of GPS, most techniques for measuring surface elevation relied on gravity in some way to determine the local vertical (i.e., the zenith). For example, when the bubble in an ideal spirit level is centered at its null position, the resulting sight line is perpendicular to the local vertical (i.e., to the direction of gravity). The complication arises from the fact that the mass distribution inside the Earth is non-uniform, and, therefore, the gravity field has irregularities, called gravity anomalies. Near such an anomaly, the geoid and ellipsoid are not parallel. The angular difference is called the vertical deflection or the geoid undulation (Figure 2.1). Levels are affected by gravity anomalies, which influence local vertical, so heights measured by leveling are intrinsically referenced to the geoid. Therefore, they differ from heights measured by GPS, which are referenced to the ellipsoid.

The difference between the geoid and ellipsoid has some interesting consequences in the real world. For example, imagine that two geodesists set out to determine the height difference between two coastal bench marks, one in San Francisco and the other in New York. Geodesist #1 organizes a leveling party in New York and heads west to San Francisco, while geodesist #2 sets up a GPS receiver in San Francisco and heads east to set up a second receiver in New York. When both are finished, they compare results and discover a 16 m discrepancy. Leveling shows that the east coast bench mark is 0.7 m *higher*

than the one on the west coast, but the GPS data show that the east coast mark is 15.4 m *lower*. Both are convinced that the other has made a mistake, so they repeat their surveys to resolve the dispute. Each claims victory upon obtaining essentially the same result as before. Who is right and why isn't sea level the same in New York as it is in San Francisco?

In fact, both measurements are correct. The discrepancy arises because the geoid undulation varies by ~16 m between San Francisco and New York. The New York bench mark is 15.4 m closer to the Earth's center of mass, but sea level in New York is 0.7 m higher than it is in San Francisco. The two statements are completely compatible. It's just that the shape of the ocean surface is contorted by Earth's irregular gravity field and by dynamic effects such as tides and currents in such a way that sea level does not equate to a fixed distance from the Earth's center of mass. As a result, height differences measured by leveling should not be compared to those measured by GPS without first making a correction for the geoid undulation. This requires that the gravity field be known over the area of interest.[3] For most applications, a better (i.e., more direct and accurate) approach is to compare leveling results to leveling results and GPS results to GPS results.

The coordinate system used for GPS (i.e., the GPS reference frame) is centered at the Earth's center of mass and includes three orthogonal axes that are defined by convention. The first axis passes through the intersection of the Greenwich meridian and Earth's equatorial plane. The third axis is defined as the average position of the Earth's rotation pole for the years 1900 to 1905. The second axis is orthogonal to the first and third axes. Heights determined by GPS are intrinsically ellipsoidal heights. The World Geodetic System 1984 (WGS 84) is used as the reference ellipsoid. Horizontal positions are expressed in terms of either latitude and longitude in degrees or in terms of Universal Transverse Mercator (UTM) coordinates in meters.

4.1.2 Point positioning and relative positioning

GPS positions can be specified either with respect to the Earth-centered coordinate system described above or with respect to some other point, which is taken as the origin of a local coordinate system. The first mode of positioning is called point positioning or absolute positioning, and the second mode is

---

[3] Software for this purpose is available from the National Geodetic Survey (http://www.ngs.noaa.gov).

called relative positioning or differential positioning. Point positioning establishes the coordinates of a point in an absolute reference frame (e.g., with respect to the center of mass of the Earth). Relative positioning establishes the coordinates of a point with respect to a known reference point or base by determining the vector between the two points, which is called a baseline.[4] Relative positioning generally is easier than point positioning, because the former does not require that local observations be tied to an absolute reference frame. Both approaches are suitable for volcano-monitoring applications, but relative positioning is more commonly used. This is because we are usually more interested in deformation within a local network of control points than we are in their absolute positions. It is important to ensure that at least one station, called the reference station, lies outside the deforming area or that its motion is known in some suitable reference frame, and also that the network spans the entire deforming area. GPS makes it possible to tie a local network to a global reference frame and thus to account for motions that extend beyond the network (e.g., to distinguish plate motion and regional tectonic strain from local volcano deformation). For precise geodetic applications, the local reference frame must be aligned with a global reference frame such as the International Terrestrial Reference Frame (ITRF) in order to express station displacements and velocities in absolute terms. The International Earth Rotation and Reference Systems Service (IERS) updates the ITRF regularly. ITRF2000 is the current frame used by the international geodetic community (Altamimi *et al.*, 2001, 2002).

The terms static positioning and kinematic positioning are used when the point to be measured is stationary or moving, respectively. Both of these approaches are useful for volcano monitoring. Network surveys are a form of static relative positioning in which the positions of control points are determined with respect to one another in a local reference frame. This type of surveying is a mainstay of classical geodesy (e.g., theodolite, EDM, and leveling surveys for horizontal and vertical control),

and it plays a prominent role in the modern world of GPS. GPS data requirements, surveying methods, and processing schemes vary between point and relative positioning modes and between static and kinematic modes. Each approach has its own advantages and disadvantages, and various hybrid modes (e.g., rapid static, pseudokinematic, and on-the-fly techniques) have been devised to meet specific needs (Section 4.7).

4.2 AN OVERVIEW OF GPS, GLONASS, AND GALILEO

GPS was designed by the US Department of Defense in the 1970s as a satellite-based navigation aid to provide global, 24-hour, all-weather position and timing information to military and (with some restrictions) civilian users. The system was designed to provide positioning accuracy of 16 m for military users and 100 m (deliberately degraded) for civilian users. Now, thanks to improved equipment and some changes to the system that were prompted by the overwhelming popularity of GPS among civilian users (Section 4.3.2), 5–10 m accuracy is routinely achievable by everyone. The development of GLONASS paralleled that of GPS. The two systems have similar designs and capabilities, and in some cases, they complement each other.[5] Geodetic applications of GPS became possible when civilian researchers recognized that positions could be determined to the millimeter level by processing standard GPS data in a special way (Section 4.6).

4.2.1 Who controls GPS?

GPS in the broadest sense consists of three components: (1) the space segment comprising the NAVSTAR satellite constellation; (2) the control segment (i.e., arrays of ground stations that control and update the satellites); and (3) the user segment, which includes such diverse civilian users as pilots, boaters, hikers, surveyors, and volcanologists. The GPS control segment consists of 5 monitoring stations (Hawai'i in the central Pacific Ocean, Kwajalein in the western Pacific Ocean, Ascension

[4] Here, baseline denotes a vector quantity (magnitude and direction) that specifies the relative positions of two points in 3-D space. However, the term is commonly used in electronic distance meter (EDM), GPS, and InSAR discussions to mean the scalar distance (magnitude only) between two points (e.g., a 10 km EDM or GPS baseline, or a 100-m baseline between the acquisition points of two SAR images). Elsewhere in the book, the meaning should be clear from context.

[5] Some GPS receivers are capable of tracking both NAVSTAR and GLONASS satellites, which increases the number of satellites available for positioning and potentially improves the results. However, GLONASS is not widely used by the worldwide geodetic community at this time – a situation that could change when the GLONASS constellation becomes fully operational again.

Island in the South Atlantic Ocean, Diego Garcia in the Indian Ocean, and Colorado Springs on mainland U.S.A.), 3 ground antennas (Ascension Island, Diego Garcia, Kwajalein), and a Master Control station at Schriever Air Force Base in Colorado. The system is managed by the NAVSTAR GPS Joint Program Office at the Space and Missile Systems Center, Los Angeles Air Force Base, California, and is operated by the 50th Space Wing at Schriever Air Force Base.

GLONASS is managed by the Russian Space Forces and operated by the Coordination Scientific Information Center (KNITs) of the Ministry of Defense of the Russian Federation. Its control segment includes the System Control Center near Moscow plus several Command Tracking Stations, Quantum Optical Tracking Stations (for satellite laser ranging), and other monitoring stations distributed widely across Russia.

4.2.2 NAVSTAR satellite constellation

The heart of GPS is a constellation of 24 operational NAVSTAR satellites plus 3 or more on-orbit spares deployed in 6 orbit planes (Figure 4.1). The satellites operate in circular orbits at an altitude of 20,200 km, with an inclination of 55 degrees and an orbital period of about 12 hours (Table 4.1). The constellation orbit was designed in such a way that the satellites repeat the same track and configuration over any point on the Earth's surface at the same sidereal time each day (i.e., about 4 minutes earlier each solar day). From any point on Earth, each satellite is visible for about 5 hours out of every 12, and at any time, there are 4 to 10 satellites above the horizon.

Five generations of the NAVSTAR satellite exist

---

[6] Eleven Block I NAVSTAR satellites designated Satellite Vehicle Number (or Space Vehicle Number, SVN) I through II were launched between 1978 and 1985 and used to test the system. Unlike their successors, Block I satellites had an orbital inclination of 63 degrees. None is still functioning. Block II and IIA satellites comprised the first operational series, including 9 Block II (SVNs 13 through 21) and 19 Block IIA (SVNs 22 through 40) satellites, with the last four tagged as replacements. Block II satellites were launched between 1989 and 1990, and Block IIA satellites between 1990 and 1997. Block IIR satellites (SVNs 41 through 62) were deployed starting in 1997 to replace Block II/IIA satellites as they reached the end of their service lives. Block IIR satellites include some dramatic improvements over their predecessors, including the ability to determine their own position by performing inter-satellite ranging with other Block IIR satellites. Current and future series include Block IIR-M (September 2005), Block IIF (2006), and Block III (2012) (see Section 4.3.1 for details).

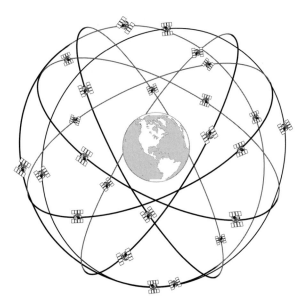

Figure 4.1. The NAVSTAR satellite constellation, which includes 24 operational satellites plus several spares (not shown) in six orbit planes, is the heart of GPS. See Table 4.1 for additional details.

or are planned: Block I, Block II/IIA, Block IIR, Block IIF, and Block III.[6] Initial operational capability (IOC) for GPS (i.e., 24 Block I, II, or IIA satellites) was declared in December 1993 and full operational capability (FOC; i.e., 24 Block II or IIA satellites operating satisfactorily) in July 1995. In February 2004, the operational GPS constellation consisted of 20 Block II/IIA and 9 Block IIR satellites. The most recent NAVSTAR launch, the first in a series of modernized Block IIR satellites (Block IIR-M1), was on 25 September 2005.

Some of the Block I satellites were equipped with quartz oscillators for clocks, but all Block II/IIA and Block IIR satellites contain two rubidium and two cesium oscillators for more precise timing. All GLONASS satellites operate with cesium clocks. The high positioning accuracy of GPS and GLONASS is due in part to the phenomenal accuracy of these atomic clocks. Typical drift rates of rubidium and cesium oscillators are 1 part in 10^{12} or 10^{13} (i.e., $\sim$1 second in 30,000 or 300,000 years, respectively).

4.2.3 GLONASS satellite constellation

The fully deployed GLONASS space segment comprises 24 satellites in three orbit planes whose ascending nodes are 120 degrees apart (Figure 4.2). Each satellite operates in a circular 19,100 km orbit at an inclination of 64.8 degrees and completes an orbit in 11 hours 15 minutes, which results in an 8-day repeat track (Table 4.1). The constellation was

Table 4.1. GPS and GLONASS comparison.

Satellites	GPS	GLONASS
Number of satellites	24 plus 3 or more on-orbit spares	24 planned, 13 operating (March 2006)
Number of orbital planes	6	3
Satellites per plane	4	8
Orbital inclination	55°	64.8°
Orbit altitude	20,180 km	19,130 km
Orbit period	11 hours, 58 minutes	11 hours, 15 minutes
Orbit repeat track	1 sidereal day	8 sidereal days
Signal structure		
C/A-code or S-code (L1)		
Code rate	1.023 MHz	0.511 MHz
Chip length	293 m	587 m
Selective availability	Discontinued May 2000	No
P-Code (L1 and L2)		
Code rate	10.23 MHz	5.11 MHz
Chip length	29.3 m	58.7 m
Selective availability	Discontinued May 2000	No
Anti-Spoofing	On since 31 January 1994	No
Signal separation	CDMA	FDMA
Carrier frequencies	L1 and L2	L1 and L2 bands: 24 frequencies spread over two bands at intervals of 0.5625 MHz
L1	1575.42 MHz	1602.5625–1615.5000 MHz
L2	1227.60 MHz	1246.4375–1256.5000 MHz
Approximate carrier wavelengths		
L1	19.0 cm	18.6–18.7 cm
L2	24.4 cm	23.9–24.0 cm
Almanac		
Duration	12.5 minutes	2.5 minutes
Capacity	37,500 bits	7,500 bits
General		
Time reference	GPS time is steered to UTC (USNO). The GPS epoch is 0000 UTC on 6 January 1980. GPS time is not adjusted for leap seconds and, as of 1 January 2006, leads UTC by 14 s.	Formerly UTC (SU, Soviet Union), now UTC (Russia)
Datum	WGS-84	PE-90 (also called PZ-90)
Initial operational capability	December 1993	September 1993
Full operational capability	July 1995	January 1996

designed so that there would be at least five GLONASS satellites visible worldwide at all times. The first launch in the experimental GLONASS series occurred in October 1982. IOC was achieved in September 1993 and FOC (i.e., 24 operating satellites) in January 1996.

The number of working GLONASS satellites has since declined and the system is no longer fully operational, following a hiatus in the normal launch schedule between December 1998 and October 2000. However, Russia began replenishing the constellation at the rate of three satellites per year starting in October 2000.[7] The December 2004 launch brought

[7] Russia launches three GLONASS satellites as a combined payload with a single Proton booster. In December 2005, it was agreed that India would share the development costs of the advanced GLONASS-K series and launch them from India. There were 13 GLONASS satellites operating in March 2006.

Figure 4.2. The Russian GLONASS satellite constellation, which when fully deployed includes 24 satellites in 3 orbit planes. FOC was achieved in January 1996, but the number of operating satellites has since declined. See Table 4.1 for additional details.

the number of operating satellites to 12 in 2 orbit planes. The Russian government has declared its intention to restore an operational constellation of 18 satellites, comprising advanced GLONASS-M and GLONASS-K series, by 2008 and to maintain it for the foreseeable future.

4.2.4   Galileo Global Navigation Satellite System

The ESA plans to launch a constellation of global-navigation satellites called Galileo, which will be interoperable with GPS and GLONASS.[8] FOC is expected in 2008. The fully deployed Galileo system will consist of 30 satellites (27 operational + 3 active spares) at 23,616 km altitude, positioned in three circular orbit planes at inclinations of 56 degrees with respect to the equatorial plane. Each of the satellites will have a rubidium clock and a hydrogen-maser clock on board.[9] A ground-based network of cesium clocks, which have better long-term stability than rubidium or hydrogen-maser clocks, will generate Galileo System Time (GST)

---

[8] The first Galileo satellite was launched from the Baikonur Cosmodrome in Kazakhstan on 28 December 2005. Plans call for up to four additional launches in 2006.

[9] Masers are atomic clocks with outstanding short-term stability. Current hydrogen masers at the US Naval Observatory (USNO) are designed to operate with maximum frequency instability of 3 parts in 10^{15} for periods of 10^3 to 10^4 s.

---

for synchronization of the satellite and ground-control clocks.

Galileo resembles GPS and GLONASS in many respects, but unlike those systems, it is intended entirely for civilian and commercial use. Galileo will transmit 10 signals in three frequency bands: 1164–1215 MHz, 1260–1300 MHz, and 1559–1591 MHz. The third band is already used by GPS for its L1 carrier signal (Section 4.3.1). Sharing the band on a non-interference basis will offer users simultaneous access to GPS and Galileo with minimal increases in receiver cost. In addition to supporting navigation and geodesy, Galileo will provide a global search and rescue function. Each satellite will be equipped with a transponder capable of transferring distress signals from specially designed transmitters to a rescue coordination center, which will initiate rescues. The system also will notify the user that his situation has been detected and help is on the way.

In hindsight, global positioning seems like an obvious opportunity for international cooperation. A single constellation of satellites would be more cost-effective and generally useful than separate national systems, and would benefit all nations with little or no incremental cost. Satellites, by their very nature, have the potential to serve as global resources. However, political barriers to cooperative international uses of space are more difficult to overcome than technological ones. There have been notable successes, including the Apollo–Soyuz Test Project in 1975, the International Space Station since 2000, and several international missions involving unmanned spacecraft. More progress in this regard would be welcome within the scientific community, particularly to geodesists in the areas of global positioning and InSAR (Chapter 5).

4.3   GPS SIGNAL STRUCTURE: WHAT DO THE SATELLITES BROADCAST?

If the NAVSTAR and GLONASS satellites are the heart of GNSS, its voice is the information-rich signals that the satellites broadcast continuously. Each satellite transmits a unique signal that can be deciphered by an electronic device called a receiver, which is attached to an antenna tuned to the appropriate frequencies (Figure 4.3). There are many types of receivers, from simple handheld units suitable for recreation to sophisticated geodetic models that provide millimeter precision. Signals from at least four satellites are required to compute a receiver's position. However, positioning accuracy improves dramatically whenever more than four satellites are visible. There is also a special class of time and frequency receivers that take advantage of the precise oscillators aboard GNSS satellites to generate extremely accurate timing signals, with a wide variety of applications. For example, such receivers provide a simple and cost-effective means to record precise timing information at remote seismometer stations.

Each GPS antenna has a 'phase center' that serves as an electronic bench mark; this is the precise point in space that is positioned by GPS observations.[10] For most geodetic applications, the antenna is set up directly over a physical bench mark and the height difference between the antenna and mark is measured to establish the position of the mark by way of the antenna's phase center. One notable exception is continuous GPS (CGPS), in which a receiver is operated at a fixed location for a long period and the phase center of its antenna serves as the mark. Most CGPS stations are tied to one or more nearby physical marks by surveying.

The next several sections include some technical jargon and tedious notation that I warned about earlier. The reader might wonder where the discussion is heading, or why it takes so many acronyms and Greek symbols to locate a GPS receiver. L1, L2, L5; pseudorandom noise codes; a W-key to unlock the Y-code; wide-lane, narrow-lane; single, double, and triple differences – yikes! If you start to bog down, I recommend either: (1) skim Sections 4.3 through 4.6, or skip them for now, and jump to the discussion of relative positioning techniques in Section 4.7 (less math, more field-oriented), or (2) better yet, bear with me and trust that the fog will lift later in the chapter. It might be comforting to know up-front that GPS data-processing strategies employing various data combinations and differencing techniques have received a lot of attention from some very bright people. Innovative solutions

have been worked out and implemented in firmware and software, so the reader can make good use of GPS without mastering the intricacies of Sections 4.3 through 4.6. You don't have to understand how GPS satellites and receivers work (Sections 4.3 and 4.4), how GPS software makes potentially crippling errors disappear from GPS data simply by differencing them away (Section 4.5), or the mathematics behind GPS positioning techniques (Section 4.6). Serious students of GPS will find the included material useful as background. Hopefully, there's enough cleverness involved to hold the attention of casual readers, too. Some people prefer to drive from point A to point B without understanding how an internal combustion engine works; others like to peek or even tinker under the hood before turning the crank. With GPS, as with automobiles, it's your choice.

4.3.1 L1 and L2 carrier signals, C/A-code, P-code, and Y-code

Each NAVSTAR satellite transmits signals on two carrier frequencies, 1575.42 MHz and 1227.60 MHz, which are called L1 and L2, respectively. The corresponding wavelengths are approximately 19.0 cm (L1) and 24.4 cm (L2). A military L3 signal is broadcast discontinuously at 1381.05 MHz, but it is not useful for positioning. The L1 and L2 carrier signals are modulated by binary pulse codes called pseudorandom noise (PRN) codes (i.e., noise-like but repeatable sequences of binary values 0 or 1, corresponding to +1 or −1 states of the code, respectively), which repeat at fixed intervals. Why, you might ask, did the designers of GPS allow the terms 'pseudo', 'random', 'noise', and 'code' anywhere near the L1 and L2 carrier signals? We'll see in Section 4.4 that there was method to their seeming madness: PRN-code modulation adds a distinctive layer of information to the carrier signals, which GPS receivers decipher to determine positions with remarkable accuracy. But first, there are a few more details of the GPS signal structure that we need to address.

Here's an example of how PRN-code modulation works. When the GPS coarse/acquisition (C/A-) code state is +1 (normal state, corresponding to binary value 0), the L1 carrier signal is unchanged. When the state of the C/A-code is −1 (mirror image state, binary value 1), the phase of the L1 signal is shifted by 180 degrees (Figure 4.4). Each 0 or 1 in a PRN code, or in some contexts the length of time required to transmit each 0 or 1, is called a chip. The

---

[10] The phase center of a GPS receiver antenna shifts slightly as the elevation angle of a satellite varies. Therefore, at any given time, differences in the elevation angles among satellites introduce fuzziness in the phase-center position. Phase-center stability is an important antenna design criterion, but even some geodetic-grade antennas are subject to phase-center variations of several millimeters. Fortunately, the effect can be removed from GPS data by carefully calibrating each antenna type. The National Geodetic Survey (NGS) maintains a database of antenna phase-shift functions, and most software packages use this information to model and remove the effect of antenna behavior.

Figure 4.3. (A) Trimble model 4000 SSE GPS receiver and antenna set up at bench mark HVO88-14 on Poli'ahu Cone near the summit of Mauna Kea Volcano, Hawai'i. Hualalai Volcano is visible in the distance. (B) Trimble choke-ring antenna set up at bench mark CVO 85-308 on the north flank of South Sister Volcano, Oregon. A Trimble model 4000 SSI GPS receiver, which is connected to the antenna by a cable on the right, is not visible in the photograph. Middle Sister (*left*) and North Sister (*right*) can be seen in the distance. (A) USGS photograph by Maurice K. Sako. (B) USGS photograph by David E. Wieprecht, 19 October 2001.

number of chips transmitted per second is called the chip rate. For the GPS C/A-code, each chip is 977 nanoseconds long and the chip rate is 1.023 MHz. The C/A-code is 1023 chips long, so it repeats every millisecond and each chip corresponds to a travel distance at the speed of light of 293 m.

The precise (P-) code is used to modulate both the L1 and L2 carrier signals from NAVSTAR satellites. The P-code is 2.3547×10^{14} chips long and it is broadcast at 10.23 megachips per second, so the repeat interval is 266.4 days and the chip length is 29.3 m. The total P-code length is partitioned into 38 unique segments and the code sequences are re-started at the beginning of each GPS week, so the effective P-code repeat interval is one week. At the discretion of the US Department of Defense, the P-code can be encrypted to form the Y-code. The algorithms that generate the C/A-code and P-code are available to receiver manufacturers, but the Y-code algorithm is classified.

Current GPS modernization plans call for the addition of a civilian carrier signal L5 at 1176.45 MHz and a new military code modulation structure called the M-code. Block IIR-M satellites (first launch on 25 September 2005) implement the M-code on L1 and L2, and the C/A-code on L2. The latter signal is called L2C and is analogous to the C/A-code on L1. Block IIF satellites will add L5 with M-code modulation starting in the fourth quarter of 2006. Combined with L1 and L2, L5 will offer

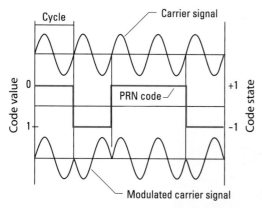

Figure 4.4. An example of binary biphase modulation of a carrier signal by a PRN code. Whenever the state of the PRN code is -1, corresponding to the binary value 1, the phase of the carrier signal is shifted by 180 degrees. When the code state is $+1$ (binary value 0), the signal is unchanged. The C/A-code or standard (S-) code is used to modulate the L1 carrier signals from NAVSTAR or GLO-NASS satellites, respectively. Both the L1 and L2 signals from both types of satellites are also modulated by a P-code. The NAVSTAR P-code can be encrypted with a secret W-key to form the Y-code. See text for discussion. Figure modified from Hofmann-Wellenhof et al. (2001, p. 73).

significant advantages for geodetic applications.[11] The first of the Block III satellites, with higher M-code signal strength, is to be launched in 2012. The entire constellation is expected to remain operational at least through 2030.

GLONASS signal structure is similar to that for NAVSTAR. Both types of satellites broadcast dual-frequency ranging signals modulated by PRN codes. The GLONASS S-code (standard) and P-code are analogous to the NAVSTAR C/A-code and P-code, respectively. GLONASS chip lengths are 587 m for S-code and 58.7 m for P-code (Table 4.1). Unlike the NAVSTAR satellites, each GLONASS satellite transmits on a separate frequency within two bands, which span from 1602.5625 MHz to 1615.5000 MHz (L1 band) and from 1246.4375 MHz to 1256.5000 MHz (L2 band). Whereas NAVSTAR uses the same frequencies for all satellites but different PRN codes for each satellite, GLONASS uses the same PRN codes

for all satellites but different frequencies in the L1 and L2 bands for each satellite. The first scheme is called Code Division Multiple Access (CDMA). The second is called Frequency Division Multiple Access (FDMA). NAVSTAR satellites are usually identified by their PRN codes, which are numbered from 1 through 32. GLONASS satellites are identified by their orbital slots, which are numbered sequentially from 1 through 24. In both cases, numbers are reassigned from decommissioned satellites to their replacements.

4.3.2 Selective availability and anti-spoofing

In the interest of national security, GPS was designed in such a way that positioning capability could be degraded by the US Department of Defense in two ways while the system remained fully operational for military users: selective availability (SA) and anti-spoofing (AS). SA was intended to deny most non-military users full accuracy by reducing the accuracy of the broadcast orbits and dithering[12] the satellite clock information. This had the effect of reducing the accuracy of point positions based on code pseudoranges (Section 4.4.1) from about 5 m to 50 m. The US government discontinued the use of SA in May 2000 for the foreseeable future.

AS guards against fake GPS transmissions by encrypting the P-code with a secret W-key to form the Y-code. Except for brief periods, AS has been turned on continuously since January 1994. Neither SA nor AS poses a serious problem for geodetic applications, including volcano monitoring. In the first case, improved orbits derived after the fact from tracking data can be substituted for real-time broadcast orbits and the satellite clocks can be eliminated or estimated as part of the relative positioning solution (Section 4.5.3). AS degrades the L2 phase observable (Section 4.4.2) for civilian users, because civilian receivers do not have access to the Y-code. In theory, this could lead to degraded geodetic solutions. However, the effect is small and comparable in magnitude to other noise sources, so its impact on geodesy has been minimal. As intended, AS mostly affects kinematic, rapid static, and other users employing short averaging times (<30 min). This could be a problem for some

---

[11] By forming linear combinations of the L1, L2, and L5 carrier-phase observables (ϕ_{L1}, ϕ_{L2}, and ϕ_{L5}; Section 4.5.1), we can form the so-called narrow-lane ($\phi_{L1} + \phi_{L2}$), medium-lane ($\phi_{L1} - \phi_{L5}$), wide-lane ($\phi_{L1} - \phi_{L2}$), and extra-wide-lane ($\phi_{L2} - \phi_{L5}$) observables. Each has certain advantages. For example, the extra-wide-lane ($\lambda_{ewl} = 5.861$ m) and wide-lane ($\lambda_{wl} = 0.862$ m) observables facilitate ambiguity resolution (Legat and Hofmann-Wellenhof, 2000; Misra and Enge, 2001, pp. 247–250), and other combinations are nearly free of ionospheric effects (Section 4.5.2).

[12] 'Dithering' refers to the introduction of digital noise. Initially, the US Department of Defense reduced the accuracy of broadcast orbits and dithered satellite clock information from NAVSTAR satellites to withhold full military capability from non-authorized users. The practice was discontinued in May 2000.

implementations of real-time GPS monitoring of volcanoes, but only during periods of extremely rapid deformation. Neither SA nor AS is implemented by GLONASS, and plans call for the Galileo signals to be unencrypted and fully accessible to all civilian users – a welcome sign of progress.

4.3.3 Navigation message

In addition to dual-frequency ranging signals, each GPS satellite broadcasts a navigation message that contains information about the satellite clock, satellite orbit, state of the ionosphere, constellation almanac, and satellite health status. The satellite clock information includes the satellite vehicle (SV) time (i.e., the time given by the satellite's on-board clock) and corrections for the clock's offset, drift, and drift rate relative to GPS time as determined by GPS controllers on the ground. The Keplerian orbital parameters describe the geometry and evolution of the satellite orbit as a function of time. This information is commonly referred to as the broadcast ephemeris or broadcast orbit to distinguish it from various types of improved orbits that are available from several sources.[13] The broadcast ionosphere model can be used to correct for ionospheric path-delay effects when computing positions (Section 4.4.2). The almanac is an approximate description of the satellites' orbits used to locate all visible satellites soon after the signal from any one satellite is acquired by a receiver. The health data provide information on the status of each satellite. A clean bill of health means that a satellite's data are reliable within operational guidelines. Lower health states serve as a flag to the user that the data might be compromised in some way. If desired, tracking data from satellites with impaired health states can be eliminated during data processing to improve the accuracy of the positioning solution.

4.4 OBSERVABLES: WHAT DO GPS RECEIVERS MEASURE?

How do receivers make use of the information contained in the signals from GPS satellites to determine the position of a point near the surface of the Earth? The answer lies in a receiver's ability to measure apparent distances, which are related by mathematical models to true distances (i.e., geometric ranges) from the receiver to several satellites.[14] Assuming for the moment that the satellite positions are known, it should be possible to determine the unknown position of a receiver by measuring at least four distances simultaneously, thus specifying three spatial coordinates for the phase center of the receiver's antenna. Algebraically, the requirement is for four independent equations to determine four primary unknowns (i.e., three receiver coordinates and the offset of the receiver clock from GPS time (Section 4.4.1)).

Actually it is considerably more complicated than that in practice, but the principle still applies. GPS receivers determine the travel times of satellite signals using the PRN codes and keep track of the number of received carrier-signal cycles as a function of time. A travel time multiplied by the speed of light equals an apparent distance, which is related to the true satellite-to-receiver range by a mathematical model that includes various error sources and unknowns (Sections 4.4.1 and 4.4.2). Likewise, a carrier-signal cycle count multiplied by the appropriate wavelength equals an apparent distance, which is related to the true range by another model. In this case, additional ambiguity arises from the fact that all carrier cycles look alike, but this isn't a show-stopper because it can be resolved as part of the modeling process (Section 4.4.2).

Ultimately, most of the intricacies of GPS

---

[13] Improved orbits incorporate information from a global network of GPS tracking stations; they include ultra-rapid orbits (formerly called predicted orbits; available in real time (no latency), 20-cm accuracy), rapid orbits (1 day latency, ~5-cm accuracy), and final orbits (13 day latency, 5-cm accuracy). Improved orbits are available from the International GPS Service (IGS, 2004) and from IGS analysis centers: Center for Orbit Determination in Europe, AIUB, Switzerland; European Space Operations Center, ESA, Germany; GeoForschungsZentrum, Germany; Jet Propulsion Laboratory, USA; National Oceanic and Atmospheric Administration/NGS, U.S.A. (NOAA, 2004); Natural Resources Canada, Canada; Scripps Orbit and Permanent Array Center, Scripps Institution of Oceanography (SOPAC, 2004); US Naval Observatory, U.S.A.; Massachusetts Institute of Technology, U.S.A.; and Geodetic Observatory Pecny, Czech Republic.

[14] The term 'range' in this context is synonymous with 'slant range,' which refers to the distance from the transmitting antenna aboard a GPS satellite to a receiving antenna on the ground. For an orbiting imaging-radar system (Chapter 5), slant range is the distance from the radar antenna to a point in the antenna footprint (i.e., the area on the ground illuminated by the radar beam). In the latter case, slant range should not be confused with ground range, which is the distance measured from the near-range line to a particular point in the footprint in the direction perpendicular to the flight path of the radar (Figures 5.5 and 5.6).

positioning can be understood in terms of two simple concepts: (1) the satellite ranging signals serve as electronic tape measures that can be used to measure apparent distances between receivers and satellites as a function of time, and (2) these distances can be related to receiver coordinates and to various other unknowns by equations (i.e., mathematical models), which can be solved for the receiver coordinates if the number of equations equals or exceeds the number of unknowns. The crux of the problem is to overtake and preferably to overwhelm the unknowns with data. Thereafter, what remains is simply arithmetic. The implementation of these concepts requires some mathematical and engineering wizardry that we'll touch on later, but those details can be mostly left to the experts and practitioners.

Here's where a little jargon is unavoidable if you want to read, speak, understand, and use GPS for geodetic purposes. The first type of measurement made by receivers is called *code pseudorange* or, in GPS shorthand, *pseudorange*. The second type is called *carrier phase* or simply *phase*. It's important to note that, in this context, phase refers to a cumulative carrier-signal cycle count, not to an angular measure that repeats every 2π radians.[15] Code pseudoranges are coarser than carrier phases because the chip lengths of the C/A-code and P-code (293 m and 29.3 m, respectively) are much longer than the wavelengths of the L1 and L2 carrier signals (19.0 cm and 24.4 cm, respectively). Code pseudoranges are used mainly for rough point positioning and navigation applications; carrier phases are used mainly for precise applications including geodesy.

4.4.1   Code pseudoranges

Imagine the simplest case of a receiver tracking the signals from a single GPS satellite. We want to determine the true range (i.e., straight-line distance or slant range) between the satellite and receiver as part of the receiver positioning solution. To formulate this problem correctly, we need to realize that there are three distinct time frames involved: GPS time, satellite time (also called SV time), and receiver time. For our purposes, we can think of GPS time as the product of an ideal clock (i.e., one that keeps

perfect time).[16] Satellite time and receiver time refer to the times kept by real clocks on the satellite and receiver, which are subject to small but significant departures from GPS time (i.e., clock errors or biases). The PRN code chips are generated by the satellite at precisely known instants according to the satellite clock, so the code signals are effectively time stamped (i.e., the receiver can 'read' the satellite clock time to determine when each chip was generated). The time of reception is determined from the receiver clock, and the apparent travel time multiplied by the speed of light is the code pseudorange.

The first step in determining a receiver's position from code pseudoranges is to relate pseudoranges to true ranges by way of known and measurable quantities. The following treatment follows that of Misra and Enge (2001, pp. 124–126). Let τ be the travel time associated with a specific code transition of the signal from a satellite, which arrives at a receiver at GPS time t. Let $t^s(t - \tau)$ be the transmit time determined from the satellite clock and $t_r(t)$ be the arrival time measured by the receiver clock. The superscript s and subscript r are used to identify terms associated with the satellite or receiver, respectively; quantities associated with both the satellite and receiver, such as the distance between them, are denoted with both a superscript and a subscript. The apparent range between the satellite and receiver (i.e., the code pseudorange $R_r^s(t)$), is related to the apparent travel time by:

$$R_r^s(t) = c \times [t_r(t) - t^s(t - \tau)] \qquad (4.1)$$

where c is the speed of light in a vacuum. The timescales of the receiver and satellite clocks are related to GPS time by:

$$t_r(t) = t + dt_r(t) \qquad (4.2)$$

$$t^s(t - \tau) = (t - \tau) + dt^s(t - \tau) \qquad (4.3)$$

where dt_r and dt^s are the receiver clock bias and the satellite clock bias, respectively (i.e., in this notation, the amounts by which the receiver and satellite clocks are advanced in relation to GPS time). Substituting these expressions into (4.1) and grouping terms, we obtain:

---

[15] This differs from InSAR and EDM usage, where phase is measured modulo 2π. According to number theory, m and n are equal modulo p, if p divides exactly into $m - n$. Therefore, $0 = 2\pi = 4\pi = 2N\pi$, modulo 2π, where N is any integer. Modulo-2π phase measurements have inherent ambiguity N which for InSAR and EDM measurements can be resolved by various means.

[16] The USNO establishes GPS time by steering it on a daily basis toward UTC (USNO), which is a paper timescale based on the average of an ensemble of approximately 50 cesium-beam frequency standards and a dozen hydrogen masers. System specifications call for GPS time to be within one microsecond of UTC (USNO), but during the past several years it has been within about 10 nanoseconds.

$$R_r^s(t) = c \times \tau + c \times [dt_r(t) - dt^s(t - \tau)] + \varepsilon_\rho(t) \tag{4.4}$$

where $\varepsilon_\rho(t)$ represents unmodeled effects, modeling error, and measurement error. The travel time multiplied by the speed of light can be modeled as:

$$c \times t = r_r^s(t, t - \tau) + \rho_{trop}(t) + \rho_{ion}(t) + \rho_{mp}(t)$$
$$+ \rho_{rel}(t) + \varepsilon_\rho(t) \tag{4.5}$$

where $r_r^s(t, t - \tau)$ is the geometric range between the receiver position at time t and the satellite position at time $(t - \tau)$; $\rho_{trop}(t)$ and $\rho_{ion}(t)$ are the delays, expressed as equivalent ranges, associated with transmission of the signal through the troposphere and ionosphere, respectively;[17] $\rho_{mp}(t)$ represents multipath errors,[18] and $\rho_{rel}(t)$ accounts for relativistic effects.[19] Depending on the application, some or

---

[17] GPS signals are delayed by propagation through both the ionosphere and the neutral atmosphere. Ionospheric delays, which occur when the signals interact with free electrons, are dispersive (i.e., they depend on the frequency of the signals), and therefore they can be corrected using dual-frequency observations (Section 4.4.2). Neutral-atmosphere delays can be divided into dry and wet delay terms. The dry delay depends on the mass of the atmosphere along the signal propagation path, and it can be calculated from surface-pressure measurements or from a meteorological model. The wet delay is caused by refraction of satellite signals owing to the presence of water vapor in the troposphere. It is nondispersive and must be modeled as a free parameter during data processing. The magnitude of the ionospheric delay depends on latitude, season, time of day, and level of solar activity. At zenith, expressed as an equivalent range, it is typically 1–10 m. Zenith values of dry and wet tropospheric delays are about 2.3 m and 0.1 to 0.3 m, respectively.

[18] Multipath errors arise when incoming signals are reflected from the ground surface or nearby objects into a receiving antenna, producing spurious phase data.

[19] The relativity term accounts for the fact that satellite and receiver clocks keep time differently. According to the Special Theory of Relativity, a clock traveling in a satellite at constant speed appears to lose time relative to an identical clock on the ground. Conversely, the General Theory of Relativity holds that a satellite clock appears to gain time relative to a clock on the ground owing to the difference in gravitational potentials. The net effect on a clock aboard a NAVSTAR satellite in a circular orbit around the Earth with a radius of 26,560 km would be to gain 38.4 microseconds per day relative to GPS time as defined on the ground. To compensate for this effect, the fundamental frequency of the nominally 10.23-MHz satellite clocks is set 0.00455 Hz lower. However, because the satellite orbits are not perfectly circular (the orbital eccentricity can be as high as 0.02), both the speed and gravitational potential change with the position of the satellite in its orbit. GPS receivers apply a relativistic correction to the satellite clock time that is based on the broadcast orbit and can vary between zero and about 45 nanoseconds (ns) (Misra and Enge, 2001, p. 92). For a thorough treatment of relativistic effects on GPS, see Ashby and Spilker (1996).

---

all of the delay, multipath, and relativistic terms can be lumped with the unmodeled-error term, $\varepsilon_\rho(t)$. Conversely, additional error sources can be split out and estimated; for example, some carrier-phase models estimate the receiver and satellite clock biases as part of the positioning solution. Dropping the explicit reference to the measurement epoch t and substituting from (4.5) into (4.4), we obtain:

$$R_r^s = r_r^s + c \times (dt_r - dt^s) + \rho_{trop} + \rho_{ion} + \rho_{mp}$$
$$+ \rho_{rel} + \varepsilon_\rho \tag{4.6}$$

Equation (4.6) is called the code pseudorange observable equation. The term pseudorange emphasizes the difference between the apparent satellite receiver distance R_r^s, which is subject to the error sources enumerated above, and the true range, r_r^s.

How does a receiver measure R_r^s? The signal from each NAVSTAR or GLONASS satellite contains enough information for a receiver to determine the travel time of the signal at some predetermined interval (typically 1–30 s), called the epoch interval, in the following way. A demodulator separates the arriving satellite signals into their constituent parts (i.e., two sinusoidal carrier signals (L1 and L2) and two binary PRN codes (C/A-code and P-code for GPS, S-code and P-code for GLONASS)). Digital signal generators use known algorithms for generating the PRN codes to produce exact replicas of the code signals. Because the patterns generated by the PRN codes repeat with known periods, it is relatively straightforward for a device called a digital correlator to slide the received code signals along the replica signals, searching for a match, in a procedure called cross correlation. A delay is applied to the clock controlling the receiver's code generator until a match is achieved. From this point on, the code generator stays in step with the incoming code and the receiver is said to be 'locked on' to the satellite. In this state, a reading of the delayed code clock is essentially a reading of the satellite clock. The offset of the code clock from the receiver clock (i.e., the travel time), multiplied by the speed of light in a vacuum, is the pseudorange R_r^s.

A good digital correlator can measure travel times with an accuracy of about 1% of the chip length, which corresponds to accuracy in pseudorange measurements of about 3 m for C/A-code or about 6 m for the GLONASS S-code. Because the C/A-code is used to modulate the L1 carrier signal and the P-code is used to modulate both the L1 and L2 carriers, there are three independent pseudorange observables, which we'll call $R_{L1,C/A}$, $R_{L1,P}$, and $R_{L2,P}$.

4.4.2   Carrier phase and carrier-beat phase

Like the PRN codes, the L1 and L2 carrier signals offer opportunities to measure apparent distances that can be related to true ranges and thus to determine a receiver's position. Imagine that at some instant we could determine accurately the number of full and fractional cycles in one of the carrier signals between a satellite and a receiver. We could then calculate the observed satellite–receiver range by multiplying the cycle count by the wavelength of the carrier signal. Receivers designed for geodetic applications are capable of measuring the fractional part of the carrier phase to better than 0.01 cycles, which corresponds to millimeter precision.

Unfortunately, there's a catch. There's no practical way to measure instantaneously the number of full carrier-signal cycles (i.e., the integer part of the phase) between a satellite and receiver. To understand this problem, imagine that a satellite starts transmitting a carrier signal at some arbitrary time t_0 and a receiver on the ground starts counting cycles as soon as it acquires the signal at time t_1. For the purposes of this discussion, all times are expressed in GPS time so the differences between satellite and receiver clocks do not matter. At some later time t_2, the satellite has moved along in its orbit and it continues to broadcast the carrier signal, which arrives at the receiver at time t_3. At this instant, the receiver has an accurate count of the number of integer and fractional cycles it received since t_1, but it has no means to determine how many cycles passed between t_0 and t_1 (i.e., during the interval between the transmit time and arrival time of the earliest part of the received signal). This is true regardless of how long the receiver tracks the satellite, because it can never determine what happened before it first acquired the signal. Because each cycle of the carrier signal looks like every other, there's no way for a digital correlator to match up portions of the received signal with an on-board replica to determine the travel time, as it does with the code signals. As a result, the satellite–receiver range measured in this way is ambiguous by an amount $N \times \lambda$, where λ is the radar wavelength and the integer N is variously called the initial ambiguity at first observation, the cycle ambiguity, the phase ambiguity, or the integer ambiguity. The value of N is different for each satellite–receiver pair, but is constant for a

particular satellite–receiver pair for as long as the receiver tracks the satellite continuously. If tracking is interrupted, a cycle slip[20] occurs and the cycle ambiguity is reinitialized (i.e., the receiver starts counting over again). Cycle slips can be identified and removed when carrier-phase data are processed.

Luckily, there's a way out of the $N \times \lambda$ conundrum by using a little algebra and physics. Just as they do with the PRN code sequences, GPS receivers produce replicas of the carrier signals, in this case using their on-board oscillators (clocks). The replica signals are not identical to the incoming satellite signals, owing to the Doppler shift produced by relative motion between the satellites and receiver and to slight differences among the oscillators. When the replica and incoming signals are differenced, what remains is a beat signal with its own characteristic phase. For the case of a receiver tracking a single satellite, and ignoring clock biases, measurement errors, and unmodeled effects for now, we can write the following expression for the beat phase in units of cycles (Misra and Enge, 2001, pp. 126–128):

$$\phi_r^s(t) = \phi_r(t) - \phi^s(t - \tau) + N \qquad (4.7)$$

where $\phi_r(t)$ is the phase of the receiver generated signal; $\phi^s(t - \tau)$ is the phase of the signal received from the satellite at time t, or the phase of the signal at the satellite at time $(t - \tau)$; τ is the travel time of the signal; and N is the integer ambiguity.[21] Phase is the product of frequency and time, so we can write:

$$\phi_r^s(t) = f \times \tau + N = \frac{r_r^s(t, t - \tau)}{\lambda} + N \qquad (4.8)$$

where f and λ are the frequency and wavelength, respectively, of the carrier signal, and $r_r^s(t, t - \tau)$ is the geometric range between the receiver position at time t and the satellite position at $(t - \tau)$. Introducing clock biases, propagation delays, multipath and relativistic effects, and unmodeled-error sources, and also dropping explicit reference to the measurement epoch as we did for the case of code pseudoranges, (4.8) can be rewritten as:

---

[20] A cycle slip results when the integer cycle count maintained by a GPS receiver is interrupted while tracking a satellite, introducing a cycle ambiguity. Cycle slips can be caused by obstructions that block the incoming satellite signal, by low signal-to-noise ratio due to such factors as noisy ionospheric conditions, multipath errors, or low satellite-elevation angle, and by incorrect signal processing by the receiver.

[21] Carrier-beat phase is the difference at time t between the phase of the replica signal generated by the receiver and the phase of the carrier signal received from the satellite. The latter term is the fractional part of the incoming cycle count referenced above.

$$\phi_r^s \times \lambda = r_r^s + c \times (dt_r - dt^s) + \phi_{trop} - \phi_{ion} + \phi_{mp}$$
$$+ \phi_{rel} + N \times \lambda + \varepsilon_\phi \qquad (4.9)$$

where in this case ϕ_{trop}, ϕ_{ion}, ϕ_{mp}, and ϕ_{rel} are propagation delays expressed in meters, and ϕ_r^s is in units of cycles. There are two carrier-beat phase observables corresponding to the two carrier signal frequencies, which we will call ϕ_{L1} and ϕ_{L2}.

The careful reader will have noticed the minus sign associated with the ionospheric delay term, which differs from (4.6) for code pseudoranges. Here's where we need a little physics. You might recall that the propagation speed of electromagnetic waves (including microwaves such as the GPS carrier signals) depends on the refractive index of the medium through which they travel, i.e.:

$$v = \frac{c}{n} \qquad (4.10)$$

where v is the propagation speed, c is the speed of light in a vacuum, and n is the refractive index. The ionosphere is dispersive, which means that its refractive index is a function of frequency. The approximate relationship is:

$$n = 1 \pm \frac{A \times N_e}{f^2} \qquad (4.11)$$

where A is a constant, N_e is the total electron density, and f is the frequency of the microwaves. The plus sign (+) applies to the group velocity of a modulated wave, which is the speed at which information such as the PRN codes propagate. In this case, $n > 1$, $v < c$, and the signals are delayed in the ionosphere. The minus sign in (4.11) applies to the phase velocity, which *increases* in the ionosphere because $n < 1$.[22] As a result, the phase of the carrier signal is *advanced* and a minus sign must accompany the ionospheric delay term in (4.9). Readers with an interest in Earth's ionosphere, its interaction with the solar wind, or other aspects of the joint NASA and ESA mission Ulysses to study the Sun will enjoy the book *The Heliosphere near Solar Minimum: The Ulysses Perspective* by Balogh *et al.* (2001).

Equation 4.9 for the carrier-beat phase observable is reminiscent of (4.6) for the code pseudorange observable. The only differences are the sign of the ionospheric delay term and the addition of a phase ambiguity term $N \times \lambda$. As we shall see shortly, there are several ways to resolve the phase ambiguity

---

[22] A phase velocity greater than the speed of light $v > c$ does not violate the theory of relativity or any other physical principle, because no information or energy travels at the phase velocity in this case.

(i.e., to determine N) or to reformulate the problem in such a way that the ambiguity term disappears.

4.5 DATA COMBINATIONS AND DIFFERENCES

All GPS observables are affected to varying degrees by the same error sources: satellite orbits, satellite clocks, receiver clocks, tropospheric and ionospheric propagation delays, multipath, relativistic effects, and receiver noise. For the case of relative positioning, which includes most geodetic applications, the resulting errors will exhibit some degree of correlation among the signals arriving at several receivers that are simultaneously tracking the same satellites. For example, the satellite signals arriving at two receivers located just a few meters apart will have taken essentially the same path through the atmosphere, so the ionospheric and tropospheric propagation delays will be nearly identical. If we were to subtract the two sets of observables, taking care to synchronize the datasets epoch by epoch using the receiver clocks, the resulting difference would be nearly free of propagation delay effects. Thus, by being clever in the way we form differences, we can remove or greatly reduce the effects of most error sources.

In some cases, there are advantages to summing GPS observables as well. The general case is a linear combination $n_1 X + n_2 Y$, where n_1 and n_2 are arbitrary numbers, and X and Y are GPS observables. The frequency of a linear combination is:
$$f = n_1 f_1 + n_2 f_2 \qquad (4.12)$$

and the wavelength of the combination is:

$$\lambda = \frac{c}{f} \qquad (4.13)$$

There are an infinite number of possible combinations from which to choose, but most are not very helpful. A few of the most useful choices are discussed below.

4.5.1 Wide-lane and narrow-lane combinations

Letting $f_1 =$ the L1 carrier frequency, $f_2 =$ the L2 carrier frequency, $n_1 = 1$, and $n_2 = \pm 1$, we can form the following linear combinations of the two carrier-beat phase observables:

$$\phi_{L1+L2} = \phi_{L1} + \phi_{L2} \qquad (4.14)$$
$$\phi_{L1-L2} = \phi_{L1} - \phi_{L2} \qquad (4.15)$$

From (4.12) and (4.13), the wavelengths of these combinations are $\lambda_{L1+L2} = 10.7$ cm and $\lambda_{L1-L2} = 86.2$ cm. Traditionally, ϕ_{L1+L2} is called the narrow lane and ϕ_{L1-L2} is called the wide lane. The wide-lane observable is especially useful for resolving integer ambiguities (Section 4.6.5).

4.5.2   The L3 combination

An especially useful combination of the carrier-beat phase observables, which is commonly called the L3 combination,[23] is:

$$\phi_{L1,L2} = \left(\phi_{L1} - \frac{f_{L2}}{f_{L1}}\phi_{L2}\right) \times \frac{f_{L1}^2}{f_{L1}^2 - f_{L2}^2} \qquad (4.16)$$

in cycles of L1 or, in cycles of L2:

$$\phi_{L2,L1} = \left(\frac{f_{L1}}{f_{L2}}\phi_{L1} - \phi_{L2}\right) \times \frac{f_{L2}^2}{f_{L1}^2 - f_{L2}^2} \qquad (4.17)$$

A similar formulation for the code pseudorange observable is:

$$R_{L1,L2} = \left(R_{L1} - \frac{f_{L2}^2}{f_{L1}^2}R_{L2}\right) \times \frac{f_{L1}^2}{f_{L1}^2 - f_{L2}^2} \qquad (4.18)$$

An important advantage of these combinations is that both are nearly free of ionospheric effects, so they can be used to estimate ionospheric delay. Such estimates based on code pseudorange measurements (4.18) are unambiguous but noisy, while estimates based on carrier-phase measurements (4.16) or (4.17) are precise but ambiguous in integer values (Hofmann-Wellenhof *et al.*, 2001, pp. 99–106; Misra and Enge, 2001, pp. 141–143).

4.5.3   Single differences

If we consider the general case of several receivers tracking several satellites through several epochs, we can envision forming differences between receivers, between satellites, and between epochs to eliminate common mode errors (Figure 4.5). Such differences are called single differences. Taking the process

further, we could take the difference between single differences to form a double difference, or the difference between double differences to form a triple difference.[24] Each of these formulations is useful for reducing the effects of various error sources on GPS data, as we shall see for a few of the most useful cases (e.g., Hofmann-Wellenhof *et al.*, 2001, pp. 190–192; Misra and Enge, 2001, pp. 218–227). For the following discussion, I have adopted the notation of Wells *et al.* (1987, Chapter 8), which is intended to be mnemonic:

- Δ denotes differences between two receivers (two vertices at the bottom).
- ∇ denotes differences between two satellites (two vertices at the top).
- δ denotes differences between two epochs.

First, let us consider two receivers tracking the same satellite simultaneously (Figure 4.6). Starting with a simplified version of (4.6) that ignores multipath, relativistic, and unmodeled-error sources including receiver noise, we can form the between-receiver single difference code pseudorange observable:

$$\Delta R_r^s = \Delta r_r^s + c \times \Delta dt_r + \Delta\rho_{trop} + \Delta\rho_{ion} \qquad (4.19)$$

where $\Delta\rho_{trop}$ and $\Delta\rho_{ion}$ are differential corrections for the tropospheric and ionospheric delays, respectively, and Δdt_r is a differential correction for the receiver clock errors. Note that the differential correction for satellite clock error Δdt^s drops out of (4.19) because there is only one satellite clock involved. Similarly, using (4.9) we can write the following expression for the between-receiver single difference carrier-beat phase observable:

$$\Delta\phi_r^s \times \lambda = \Delta r_r^s + c \times \Delta dt_r + \Delta\phi_{trop} - \Delta\phi_{ion}$$
$$+ \Delta N \times \lambda \qquad (4.20)$$

Again, the differential correction for satellite clock error drops out.

Between-receiver single differences greatly reduce the effects of errors associated with the satellites, including satellite clock errors and orbit errors. The effects of atmospheric propagation delays also are diminished if the baseline between receivers is short relative to the ~20,000 km altitude of the satellites. This is because the signals arriving at both receivers have traversed a similar path through the atmosphere, and, therefore, the differential corrections for ionospheric and tropospheric delays are small.

Another useful formulation, this for the case of a

---

[23] The L3 linear combination of the L1 and L2 carrier-beat phase observables should not be confused with the military L3 signal that is broadcast discontinuously at 1381.05 MHz by NAVSTAR satellites, or with the L5 carrier signal to be broadcast by Block IIF and Block III NAVSTAR satellites.

[24] Single, double, and triple differences can be thought of as successive linear data combinations in which $n_1 = 1$ and $n_2 = -1$, as described for the wide-lane combination. Whereas the wide lane involves two different observables (ϕ_{L1} and ϕ_{L2}), single differences involve the same observable from different receivers, different satellites, or different epochs.

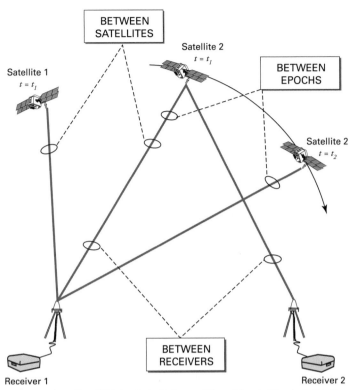

Figure 4.5. Many of the error sources that affect GPS signals, including satellite orbit, satellite clock, receiver clock, and atmospheric propagation delays, are correlated to some degree among the signals received at several stations simultaneously tracking the same several satellites. As a result, their effects can be removed or greatly reduced by taking differences between measurements in various ways. GPS measurements can be differenced between receivers, between satellites, and between epochs to form single differences (Figures 4.6–4.8). Single differences can be differenced to form double differences (Figures 4.9 and 4.10), and double differences can be differenced to form triple differences (Figure 4.11). Each approach has distinct advantages and disadvantages. Modified from Wells et al. (1987, Chapter 8).

single receiver tracking two satellites, is the between-satellite single difference (Figure 4.7). Using the same approach as above, the between-satellite single difference pseudorange observable is:

$$\nabla R_r^s = \nabla r_r^s - c \times \nabla dt^s + \nabla \rho_{trop} + \nabla \rho_{ion} \quad (4.21)$$

and the between-satellite single difference carrier-beat phase observable is:

$$\nabla \phi_r^s \times \lambda = \nabla r_r^s - c \times \nabla dt^s + \nabla \phi_{trop} - \nabla \phi_{ion}$$
$$+ \nabla N \times \lambda \quad (4.22)$$

These two formulations are free of receiver clock errors, because only one receiver is involved.

Lastly, the between-epoch single difference equations are (Figure 4.8):

$$\delta R_r^s = \delta r_r^s + c \times (\delta dt_r - \delta dt^s) + \delta \rho_{trop} + \delta \rho_{ion} \quad (4.23)$$

and

$$\delta \phi_r^s \times \lambda = \delta r_r^s + c \times (\delta dt_r - \delta dt^s) + \delta \phi_{trop} - \delta \phi_{ion} \quad (4.24)$$

Note that (4.24) does not include a differential integer ambiguity term, because the initial ambiguity does not change between epochs. Therefore, between-epoch single difference carrier-beat phase data are free of cycle ambiguities. The change in carrier-beat phase over a time interval corresponds to changes in both the receiver–satellite range and the receiver clock bias. Because such data are partly a measure of the frequency shift in the received signal caused by relative motion between the satellite and receiver, they are sometimes referred to as integrated Doppler data (Wells et al., 1987, Chapter 8; Hofmann-Wellenhof et al., 2001, pp. 90–91, 185–186; Misra and Enge, 2001, p. 128). Doppler data can be used to determine integer ambiguities in kinematic surveys (Remondi, 1991) or as an additional independent observable for point positioning (Ashjaee et al., 1989).

4.5.4 Double differences

The receiver-time double difference observable is the change from one epoch to the next in the between-

$$\Delta(*) = (*)_{receiver\,2} - (*)_{receiver\,1}$$
$$\Delta R_r^s = \Delta r_r^s + c \cdot \Delta dt_r + \Delta\rho_{trop} + \Delta\rho_{ion}$$
$$\Delta\phi_r^s \cdot \lambda = \Delta r_r^s + c \cdot \Delta dt_r + \Delta\phi_{trop} - \Delta\phi_{ion} + \Delta N \cdot \lambda$$

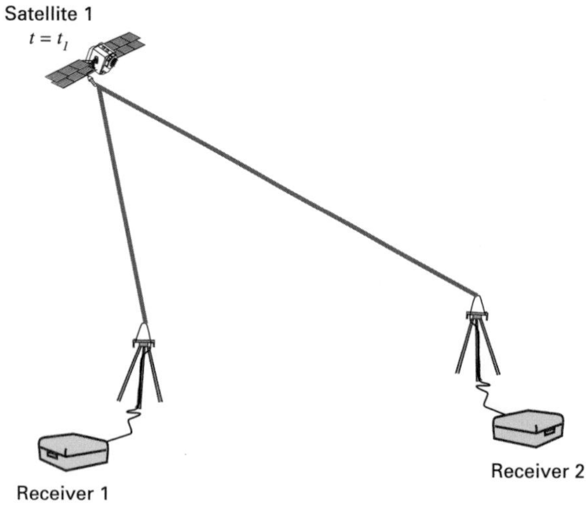

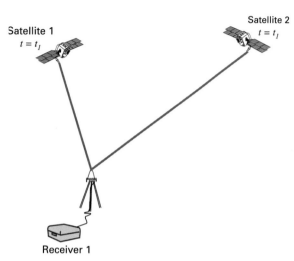

Figure 4.6. Between-receiver single differences are formed by differencing measurements from two receivers simultaneously tracking the same satellite, which reduces the effects of errors in the satellite orbit, satellite clock, and (for short baselines) atmospheric propagation delays. See text for discussion of the observation equations. Modified from Wells *et al.* (1987, Chapter 8).

$$\nabla(*) = (*)_{satellite\,2} - (*)_{satellite\,1}$$
$$\nabla R_r^s = \nabla r_r^s - c \cdot \nabla dt^s + \nabla\rho_{trop} + \nabla\rho_{ion}$$
$$\nabla\phi_r^s \cdot \lambda = \nabla r_r^s - c \cdot \nabla dt^s + \nabla\phi_{trop} - \nabla\phi_{ion} + \nabla N \cdot \lambda$$

Figure 4.7. Between-satellite single differences are formed by differencing the observations of two satellites as simultaneously recorded by a single receiver. The resulting data are free of receiver clock errors because only one receiver is involved. See text for discussion of the observation equations. Modified from Wells *et al.* (1987, Chapter 8).

receiver single difference for the same satellite (Figure 4.9). The equation for the receiver-time double difference carrier-beat phase observable is:

$$\delta\Delta\phi_r^s \times \lambda = \delta\Delta r_r^s + c \times \delta\Delta dt_r + \delta\Delta\phi_{trop} - \delta\Delta\phi_{ion} \qquad (4.25)$$

The same result is obtained by differencing the between-epoch single difference equations for two receivers (i.e., $\delta\Delta\phi_r^s = \Delta\delta\phi_r^s$). Note that the satellite dependent integer cycle ambiguities, which are present in $\Delta\phi_r^s$ as a differential term, are now absent. For this reason, receiver-time double differences are particularly useful for editing cycle slips. The corresponding double-difference equation for pseudoranges is:

$$\delta\Delta R_r^s = \delta\Delta r_r^s + c \times \delta\Delta dt_r + \delta\Delta\rho_{trop} + \delta\Delta\rho_{ion} \qquad (4.26)$$

Now consider two receivers and two satellites at the same epoch (Figure 4.10). There are three possible differencing operations (Wells *et al.*, 1987, Chapter 8): (1) two between-receiver single differences, each involving a different satellite; (2) two between-satellite single differences, each involving a different receiver; and (3) one receiver–satellite double difference, involving both receivers and both satellites. The receiver–satellite double difference can be constructed either by: (1) taking two between-receiver single difference observables, involving the same pair of receivers but different satellites, and differencing these between the two satellites; or (2) taking two between-satellite single difference observables, involving the same pair of satellites but different receivers, and differencing these between the two satellites. Both approaches produce identical results. The receiver–satellite double difference equation for pseudorange is:

$$\nabla\Delta R_r^s = \nabla\Delta r_r^s + \nabla\Delta\rho_{trop} + \nabla\Delta\rho_{ion} \qquad (4.27)$$

and the corresponding equation for carrier-beat phase is:

$$\nabla\Delta\phi_r^s \times \lambda = \nabla\Delta r_r^s + \nabla\Delta\phi_{trop} - \nabla\Delta\phi_{ion} + \nabla\Delta N \times \lambda \qquad (4.28)$$

Double differences greatly reduce the effects of receiver-clock errors, which contribute to both

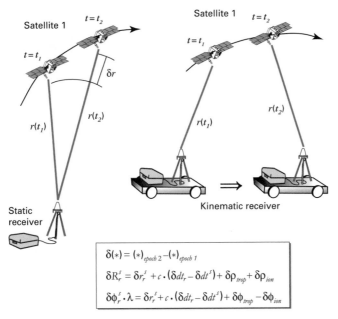

$$\delta(*) = (*)_{epoch\ 2} - (*)_{epoch\ 1}$$

$$\delta R_r^s = \delta r_r^s + c \cdot (\delta dt_r - \delta dt^s) + \delta\rho_{trop} + \delta\rho_{ion}$$

$$\delta\phi_r^s \cdot \lambda = \delta r_r^s + c \cdot (\delta dt_r - \delta dt^s) + \delta\phi_{trop} - \delta\phi_{ion}$$

Figure 4.8. Between-epoch (Doppler) single differences are formed by differencing the observations of one satellite made by one receiver, which can be stationary or moving, from one epoch to the next. The received frequency differs from the transmitted frequency owing to relative motion between the receiver and satellite, which produces a change in phase (code pseudorange and carrier phase) between epochs. Doppler data are free of any cycle ambiguity. See text for discussion of the observation equations. Modified from Wells *et al.* (1987, Chapter 8).

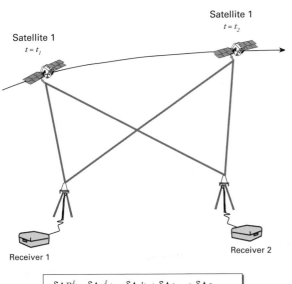

$$\delta\Delta R_r^s = \delta\Delta r_r^s + c \cdot \delta\Delta dt_r + \delta\Delta\rho_{trop} + \delta\Delta\rho_{ion}$$

$$\delta\Delta\phi_r^s \cdot \lambda = \delta\Delta r_r^s + c \cdot \delta\Delta dt_r + \delta\Delta\phi_{trop} - \delta\Delta\phi_{ion}$$

Figure 4.9. The receiver-time double difference is the change from one epoch to the next in the between-receiver single difference for the same satellite. Differencing the between-epoch single differences for two receivers simultaneously tracking the same satellite produces the same result. Satellite dependent cycle ambiguities, which are present in the between-receiver single differences, are absent from this formulation. See text for discussion of the observation equations. Modified from Wells *et al.* (1987, Chapter 8).

of the between-receiver single differences, and of errors associated with the misalignment between satellite clocks, which contribute to both of the between-satellite single differences (e.g., Wells *et al.*, 1987, Chapter 8; Hofmann-Wellenhof *et al.*, 2001, p. 191; Misra and Enge, 2000, pp. 223–225).

4.5.5 Triple differences

Finally, receiver-satellite-time triple differences measure the change in receiver–satellite double differences from one epoch to the next (Figure 4.11). The corresponding observation equations are:

$$\delta\nabla\Delta R_r^s = \delta\nabla\Delta r_r^s + \delta\nabla\Delta\rho_{trop} + \delta\nabla\Delta\rho_{ion} \quad (4.29)$$

and

$$\delta\nabla\Delta\phi_r^s \times \lambda = \delta\nabla\Delta r_r^s + \delta\nabla\Delta\phi_{trop} - \delta\nabla\Delta\phi_{ion} \quad (4.30)$$

The error sources that dropped out of the double difference observables are also absent from triple differences. In addition, for carrier-beat phase measurements, the initial cycle ambiguity terms cancel, which allows for easier automatic removal of cycle slips. However, position estimates based on triple differences tend to be less accurate than those based on double differences (Misra and Enge, 2001, pp. 225–226).

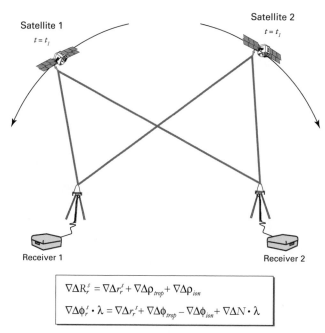

$$\nabla \Delta R_r^s = \nabla \Delta r_r^s + \nabla \Delta \rho_{trop} + \nabla \Delta \rho_{ion}$$

$$\nabla \Delta \phi_r^s \cdot \lambda = \nabla \Delta r_r^s + \nabla \Delta \phi_{trop} - \nabla \Delta \phi_{ion} + \nabla \Delta N \cdot \lambda$$

Figure 4.10. Receiver–satellite double differences can be constructed either by: (1) taking two between-receiver single differences, involving the same pair of receivers but different satellites, and differencing these between the two satellites; or (2) taking two between-satellite single differences, involving the same pair of satellites but different receivers, and differencing these between the two satellites. The two results are identical. Receiver–satellite double differences are free of receiver clock errors, which are present in between-receiver single differences, and of satellite clock errors, which are present in between-satellite single differences. See text for discussion of the observation equations. Modified from Wells *et al.* (1987, Chapter 8).

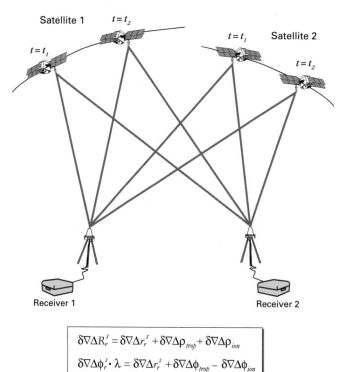

$$\delta \nabla \Delta R_r^s = \delta \nabla \Delta r_r^s + \delta \nabla \Delta \rho_{trop} + \delta \nabla \Delta \rho_{ion}$$

$$\delta \nabla \Delta \phi_r^s \cdot \lambda = \delta \nabla \Delta r_r^s + \delta \nabla \Delta \phi_{trop} - \delta \nabla \Delta \phi_{ion}$$

Figure 4.11. The receiver-satellite-time triple difference is the change in a receiver–satellite double difference from one epoch to the next. Errors that dropped out of the double differences (i.e., receiver clock and satellite clock errors) are also absent from the triple difference. In addition, for carrier-phase measurements, the cycle ambiguity terms cancel. See text for discussion of the observation equations. Modified from Wells *et al.* (1987, Chapter 8).

4.6 DOING THE MATH: TURNING DATA INTO POSITIONS

Let us explore the requirement that the number of pseudorange or phase measurements must exceed the number of unknowns, including three receiver coordinates and several nuisance parameters (e.g., cycle ambiguities, clocks errors, propagation delays), in order to obtain a unique positioning solution. The next two sections address this issue for point positioning using pseudoranges and carrier-beat phases to give the reader a feel for how things work. Sections 4.6.3 and 4.6.4 address static and kinematic relative positioning, respectively, where the power of differences becomes apparent. The treatment follows that of Hofmann-Wellenhof et al. (2001, Chapter 8.)

4.6.1 Point positioning with code pseudoranges

For the general case of several receivers simultaneously tracking several satellites, a simplified version of (4.6) for code pseudoranges that ignores multipath, relativistic, and unmodeled-error sources including receiver noise can be written as:

$$R_i^j = r_i^j + c \times dt_i - c \times dt^j + (\rho_{trop})_i^j + (\rho_{ion})_i^j$$
$$(4.31)$$

where R_i^j is the measured pseudorange between the ith receiver and the jth satellite; r_i^j is the true range (i.e., geometric distance) between the ith receiver at time t and the jth satellite at time $(t - \tau)$;[25] dt_i and dt^j represent the offsets from GPS time of the ith receiver and the jth satellite clocks, respectively; and $(\rho_{trop})_i^j$ and $(\rho_{ion})_i^j$ are the propagation delays in the troposphere and ionosphere, respectively, for each receiver–satellite pair.

The range r_i^j depends on the receiver coordinates at time t (X_i, Y_i, Z_i) and on the satellite coordinates at time $(t - \tau)$ (x^j, y^j, z^j) all expressed in the GPS reference frame:

$$r_i^j = \sqrt{(x^j - X_i)^2 + (y^j - Y_i)^2 + (z^j - Z_i)^2} \quad (4.32)$$

Substituting this expression into (4.31), we are faced with what seems like a long list of unknowns.

However, things are not as bad as they might seem. The satellite coordinates as a function of time can be derived from the broadcast orbit, usually to an accuracy of better than 1 m.[26] Also available from the navigation message are correction coefficients for the satellite clocks,[27] which can be used to determine dt^j, and an ionosphere model that can be used to estimate $(\rho_{ion})_i^j$, usually with an accuracy of 1–2 m.[28] The tropospheric delays, $(\rho_{trop})_i^j$, can be estimated from a combination of theory and atmospheric observations to a similar accuracy.[29] Thus, we are left with just four unknowns: three receiver coordinates X_i, Y_i, Z_i and the receiver clock bias dt_i. All of the unknowns are independent of the satellites, so by tracking four satellites simultaneously we can form and solve four independent equations for the four unknowns, thus determining the receiver's position to within a few meters. What seemed like magic is really just careful engineering plus a little algebra and physics!

Let us explore the implications of (4.31) a little further (Hofmann-Wellenhof et al., 2001, pp. 182–183). Let n_j be the number of satellites being tracked simultaneously and n_t be the number of epochs. The number of independent observation equations is thus $n_j \times n_t$. We'll ignore the two propagation delay terms for this analysis, keeping in mind that they can be determined independently

---

[25] Pseudoranges and true ranges are both referenced to the phase centers of the transmitting and receiving antennas. 'Satellite–receiver' range or pseudorange is shorthand for the range or pseudorange between phase centers. Similarly, 'satellite' and 'receiver' coordinates actually refer to the coordinates of the respective phase centers.

[26] The accuracy of the broadcast orbits has improved dramatically over time. Since the use of SA was discontinued in May 2000, the accuracy is typically better than 1 m except for short periods following satellite maneuvers.

[27] These take the form $dt^j(t) = a_0 + a_1(t - t_0) + a_2(t - t_0)^2$, where each a_n is a constant coefficient. Like the broadcast orbits, the timing coefficients are determined and updated periodically by the control segment.

[28] The electron density in the ionosphere varies diurnally by a factor of 10 at mid latitudes and by a factor of 5 over the 11-year solar cycle. As a result, ionospheric delays are typically 1–10 m but they sometimes exceed 30 m. In extreme cases, the broadcast ionosphere model might be in error by a factor of two. The effect on pseudorange positioning can be greatly reduced by using a dual-frequency receiver, because the ionospheric delay is dispersive (i.e., frequency dependent) – with dual-frequency data, ionospheric-delay effects can be estimated and removed.

[29] The magnitude of ρ_{trop} at sea level zenith is about 2.3 m in a dry atmosphere; precipitable water vapor typically contributes an additional 0.1 m to 0.3 m. The dry component of the tropospheric delay is directly proportional to atmospheric pressure and can be modeled very well. The wet component is more variable and difficult to account for. It can be measured directly with radiometers, Lidars, Fourier transform infrared spectrometers, and radiosondes (Solheim et al., 1999), or it can be included as a free parameter in a GPS positioning model.

from models of the ionosphere and troposphere, and, therefore, they do not introduce any additional unknowns into the point-positioning problem. We'll see that the problem quickly becomes over-determined, so we can solve for improved ionosphere and troposphere models as well. Moving the satellite clock bias to the left side of (4.31) (because it is known from the navigation message) and grouping terms, we obtain:

$$R_i^j + c \times dt^j = r_i^j + c \times dt_i \qquad (4.33)$$

The requirement that the number of observations must equal or exceed the number of unknowns can be expressed as:

$$n_j \times n_t \geq 3 + n_t \qquad (4.34)$$

where corresponding terms are written in the same order as in (4.33). If we let $n_t = 1$ (i.e., the single-epoch case), (4.34) becomes $n_j \geq 4$. This means that a receiver's position can be determined at every epoch whenever four or more satellites are being tracked simultaneously. This is the case even if the receiver is in motion, so kinematic point positioning using pseudoranges is possible. Handheld GPS receivers and those used for navigation applications (e.g., vehicle tracking) take this approach.

4.6.2   Point positioning with carrier-beat phases

The simplified observation equation for carrier-beat phase (Equation 4.9 without the multipath, relativistic, and unmodeled-error terms) for the case of several receivers simultaneously tracking several satellites can be written as:

$$\phi_i^j \times \lambda = r_i^j + c \times dt_r - c \times dt^s + \phi_{trop} - \phi_{ion}$$
$$+ N \times \lambda \qquad (4.35)$$

We assume that the satellite coordinates and the satellite clock biases are known from the navigation message or some other source, and we ignore the two propagation delay terms for now. Rearranging and grouping terms as before yields:

---

[30] This expression accounts for a rank deficiency encountered when attempting to solve simultaneous observation equations with the form of (4.36). The rank of a matrix is defined to be the maximum number of rows or columns that are linearly independent. A rank deficiency exists when the rank of a matrix is less than the number of rows or columns. In other words, the number of linearly independent equations is less than the number of unknowns, so a unique solution is not possible without additional constraints. Interested readers should consult Hofmann-Wellenhof et al. (2001, p. 184) and references therein.

$$\phi_i^j \times \lambda + c \times dt^j = r_i^j + c \times dt_r + N_i^j \times \lambda \quad (4.36)$$

The number of observations, as before, is $n_j \times n_t$, but the number of unknowns is increased by n_j because of the ambiguities. The requirement that the number of independent observations must equal or exceed the number of unknowns becomes (Hofmann-Wellenhof et al., 2001, pp. 183–185):[30]

$$n_j \times n_t \geq 3 + n_j + (n_t - 1) \qquad (4.37)$$

A solution for a single epoch (i.e., $n_t = 1$) is not possible unless we disregard the n_j integer ambiguities. In practice, this means that carrier-beat phase observations can be used for kinematic applications only if the integer ambiguities have been resolved by some other means. The simplest way to accomplish this is to preface a kinematic survey with a few minutes of static observations, which allows the integer ambiguities to be resolved before the receiver goes in motion. For example, tracking four satellites ($n_j = 4$) for a minimum of two epochs ($n_t \geq 2$) in theory would resolve the four integer ambiguities (i.e., from (4.37), $4 \times 2 \geq 3 + 4 + 1$). In practice, having to track four or more satellites for several epochs to resolve the ambiguities is not a serious constraint for most kinematic applications.

From (4.37), the required number of epochs for static point positioning with carrier-beat phases more generally is given by:

$$n_t \geq \frac{n_j + 2}{n_j - 1} \qquad (4.38)$$

The minimum number of satellites for a solution is $n_j = 2$, which leads to $n_t \geq 4$ observation epochs. Another integer solution pair is $n_j = 4$, $n_t \geq 2$, as noted above. For kinematic point positioning with carrier-beat phases when the ambiguities are not resolved with static observations, the requirement for a solution is:

$$n_j \times n_t \geq 3n_t + n_j + (n_t - 1) \qquad (4.39)$$

which yields the explicit relation (Hofmann-Wellenhof et al., 2001, p. 185):

$$n_t \geq \frac{n_j - 1}{n_j - 4} \qquad (4.40)$$

A solution is possible when, for example, five satellites are tracked for four or more epochs, or seven satellites are tracked for at least two epochs.

The static point positioning problem, which is of most interest for geodetic applications, is greatly over-determined for the typical case of $n_j > 4$ and $n_t \gg 3$. Therefore, we can solve not only for the integer ambiguities, but also for improved estimates of the satellite orbits, the satellite and receiver clock biases, and the ionospheric and tropospheric delays.[31] For observation periods of 24 hours or more, by using precise orbits and user-derived models of the ionosphere and troposphere, station coordinates can be determined from carrier-beat phase observations to a typical accuracy of 3 mm in northing (latitude), 5 mm in easting (longitude), and 10 mm in height, even for baselines hundreds to thousands of kilometers long. For shorter baselines, cancellation of common-mode errors yields further improvement by a factor of two or more in some cases, although difficulties in determining the wet component of the tropospheric delay still limit the typical accuracy of height measurements to about 10 mm. Continuously tracking GPS reference stations achieve the ultimate positioning accuracy. Repeatability of 1.9 mm, 2.4 mm, and 5.6 mm (northing, easting, and height, respectively) has been reported for representative CGPS reference stations (Herring, 1999).

4.6.3 Static relative positioning

A similar analysis can be done to determine the requirements for relative positioning (i.e., for specifying a baseline vector between two receivers as opposed to point positioning a single receiver) using carrier-beat phase single, double, and triple differences. The result is the same for single, double, and triple differences (Hofmann-Wellenhof et al., 2001, pp. 198–199):

$$n_t \geq \frac{n_j + 2}{n_j - 1} \qquad (4.41)$$

This is identical to the requirement for static point positioning (4.38), so solutions are again possible if $n_j = 2$ and $n_t \geq 4$, or if $n_j = 4$ and $n_t \geq 2$.

The details aren't important for our purpose. What matters is that point positioning is possible with even a small amount of data, and with more data the accuracy can be improved by estimating various error sources (e.g., ionospheric and tropo-

spheric-path delays, satellite and receiver clock biases) as part of the positioning solution.

4.6.4 Kinematic relative positioning

If we move one of the receivers while the other remains stationary at a known location, there are three unknown receiver coordinates for each epoch. Thus, the total number of unknowns is $3 \times n_t$ for n_t epochs. The requirements for a solution can be worked out in a manner similar to that used above for static relative positioning. The resulting relationships are (Hofmann-Wellenhof et al., 1998, pp. 200–201):[32]

Single difference: $\quad n_j \times n_t \geq 3 \times n_t + n_j + (n_t - 1)$

$$(4.42)$$

Double difference: $(n_j - 1) \times n_t \geq 3 \times n_t + (n_j - 1)$

$$(4.43)$$

Triple difference: $(n_j - 1) \times (n_t - 1) \geq 3 \times (n_t - 1)$

$$(4.44)$$

For kinematic positioning to be viable, we need a solution at every epoch. Unfortunately, none of these relationships can be satisfied for $n_t = 1$. The way out of this dilemma is to reduce the number of unknowns by omitting the integer ambiguities, which must be resolved in some other way. The result for all three cases is $n_j \geq 4$, which means that kinematic relative positioning is possible whenever two receivers can track the same four or more satellites simultaneously.

Ignoring the integer ambiguities is akin to requiring that we know the position of the roving receiver for a single epoch, which presupposes the ambiguities are known. We can meet this requirement in one of three ways (Hofmann-Wellenhof et al., 2001, pp. 201–202): (1) position the roving receiver over a known point, (2) keep the rover stationary long enough to obtain a static position, or (3) swap the antennas. The process, accomplished by whatever means, is called static initialization.

---

[31] Early in the development of GPS, users who were interested in obtaining the best possible results usually computed their own improved orbits in this way. With the establishment of a global GPS tracking network and of the means to easily disseminate improved orbits derived from the network, this is no longer necessary in most cases. Instead, users avail themselves of the ultra-rapid, rapid, and final orbits provided by the International GPS Service (IGS, 2004).

[32] The relationship for triple differences assumes that the coordinates of the roving receiver are known at some reference epoch, preferably at the starting point. The same requirement exists for single- and double-difference kinematic solutions, which are possible for $n_t = 1$ only if the integer ambiguities are known. This can be accomplished if the position of the rover is known at any reference epoch (Hofmann-Wellenhof et al., 2001, pp. 200–201).

Remondi (1984, 1986) devised the antenna swap, an ingenious ploy in which two receivers track at least four satellites for a few epochs at two nearby locations, then trade places while continuing to track. In this way, one can determine the starting baseline between the receivers very quickly. Thereafter, relative kinematic positioning is possible for as long as both receivers maintain continuous tracking. If the number of satellites being tracked drops below four, another antenna swap is required.

For some applications such as aircraft navigation, keeping the roving receiver stationary for even a brief period is untenable. Such applications require another form of ambiguity resolution called kinematic or on-the-fly (OTF) initialization. Several OTF methods have been devised, some of them proprietary. One scheme uses carrier-beat phase double differences, an initial carrier-smoothed code pseudorange solution, and a mathematical search technique supplemented by Doppler data to resolve ambiguities on-the-fly (Remondi, 1991).

4.6.5   Ambiguity resolution

To take full advantage of the high accuracy of the carrier-beat phase observable, the ambiguities inherent in phase measurements must be resolved to their correct integer values. The cleverness of GPS gurus is not limited to hardware engineers. Theoreticians and software developers have devised several strategies for ambiguity resolution, some that rely on mathematical brute force and others that strike me as particularly deft or ingenious. The topic is too narrow for a detailed discussion here. Instead, I briefly describe two common approaches used for precise geodetic applications and refer the interested reader to more thorough treatments by Hofmann-Wellenhof *et al.* (2001, pp. 213–248) and Misra and Enge (2001, pp. 227–247).

To resolve the integer ambiguities associated with dual-frequency, receiver–satellite, double-difference carrier-phase data (4.28), we can use the wide-lane data ϕ_{L1-L2} either alone or in conjunction with the ϕ_{L1} or ϕ_{L2} measurements. Because the wide-lane formulation has a significantly longer wavelength than ϕ_{L1} and ϕ_{L2}, the corresponding wide-lane ambiguities are more widely spaced and easier to estimate than the base-carrier ambiguities. Once the wide-lane ambiguities are determined, the base-carrier ambiguities can be calculated directly (Hofmann-Wellenhof *et al.*, 2001, pp. 216–218; Misra and Enge, 2001, pp. 230–233).

It is not always that easy in practice, because the wide-lane observable is not free of ionospheric influences that can be a problem for long baselines or under noisy ionosphere conditions. The ionosphere-free L3 combination $\phi_{L1,L2}$ (4.16 and 4.17) can be used instead, but this destroys the integer nature of the ambiguities. In this case, the ambiguities are estimated as real numbers and the resulting estimates are called float solutions. There are several integer ambiguity algorithms based on float solutions (Hwang, 1991; Hatch and Euler, 1994; Han and Rizos, 1997). These generally work well for long stretches of data without any break in signal tracking, as is usually the case for precise geodetic applications. Rapid or even single-epoch ambiguity resolution, which is required for navigation of fast-moving vehicles such as aircraft, presents a much bigger challenge. Fortunately, this situation will improve when three-frequency phase measurements become possible with the planned addition of the L5 carrier signal. Misra and Enge (2001, pp. 236–247) discuss several mathematical search techniques for identifying the best solution for the integer ambiguities based on a float solution.

Another approach that eliminates the main drawback of the wide-lane technique (i.e., ionospheric dependence) combines double-differenced, dual-frequency carrier phase and code pseudorange observations. A lengthy derivation produces the elegant result:

$$N_{L1-L2} = N_{L1} - N_{L2} = \phi_{L1-L2} - \frac{f_{L1} - f_{L2}}{f_{L1} + f_{L2}}$$
$$\times (R_{L1} + R_{L2}) \qquad (4.45)$$

where N_{L1}, N_{L2}, and N_{L1-L2} are the integer ambiguities for L1, L2, and the wide-lane combination, respectively; ϕ_{L1-L2} is the wide-lane phase observable; f_{L1} and f_{L2} are the frequencies of the L1 and L2 carrier signals, respectively; and R_{L1} and R_{L2} are the L1 and L2 code pseudorange observables, respectively. This equation allows for the determination of wide-lane ambiguities for each epoch and each site. The base-carrier ambiguities N_{L1} and N_{L2} can then be calculated using equations given by Hofmann-Wellenhof (2001, pp. 219–220). This approach is independent of the baseline length and of ionospheric effects, but multipath can still be a problem for very short observation times. This problem usually can be overcome by averaging over several epochs. If seven or more satellites can be tracked simultaneously, this technique is capable of single-epoch ambiguity resolution for short-baseline kinematic applications (Hatch, 1990).

4.7 RELATIVE POSITIONING TECHNIQUES

The requirements and strategies discussed in Sections 4.6.3 and 4.6.4 give rise to several relative positioning techniques that are in common use. These include static, stop-and-go kinematic, kinematic, pseudokinematic, rapid static, and real time kinematic (also called on-the-fly). In the following sections, I briefly describe field procedures for each technique and provide examples of how some have been used for volcano monitoring. Also included (Section 4.7.7) is a brief summary of which equipment and techniques are best suited to some common applications, to guide the reader through a bewildering array of hardware and field-procedure options.

4.7.1 Static GPS

As the name implies, static GPS observations are made with receivers that remain stationary for periods of minutes to days or longer. Static surveys are by far the most common technique used for high-precision geodetic applications. Typically, several dual-frequency receivers are deployed at fixed control points for a period of 24 to 48 hours. Thereafter, most or all of the receivers are moved to other points in the network for another observation session until all of the points have been occupied. This approach is sometimes referred to as campaign mode, which is a holdover from the early days when large GPS surveys in distant, often remote places with bulky, power-hungry receivers resembled a military campaign more than a geodetic survey. Happily, the situation has improved considerably in recent years with the introduction of lightweight, low-power receivers. When a static survey is complete, the carrier-phase and code pseudorange data are processed together with comparable data from several continuously tracking reference stations using one of several research or commercial software packages to obtain network station coordinates (Section 4.9). Observation times can be adjusted depending on the scale of the network (i.e., the baseline lengths) and on the required accuracy.

Positioning accuracy for static surveys increases nonlinearly with time because the data are subject to various error sources, some random or pseudorandom and others predictable with periods ranging from hours to years. To achieve centimeter accuracy for baselines a few tens of kilometers long, data must be collected for at least 5–10 minutes at each station. After several hours, horizontal accuracy typically

approaches 3 mm in northing (latitude), 5 mm in easting (longitude), and 10 mm in height. With data spanning 24 hours or more, improvement by another factor of two is possible. In part, this is because the satellite-constellation geometry repeats every sidereal day (i.e., the same satellites are visible from the same point at nearly the same time on successive days (about 4 minutes earlier each solar day)). When data are averaged over several days, the effect of shorter term errors related to satellite orbits is reduced. Similarly, the effect of multipath errors that arise when incoming GPS signals reflect from the ground surface or nearby objects into the receiving antenna, producing spurious phase data, is reduced when data are collected for several days. Multipath errors can be reduced also by selecting sites away from buildings and other reflectors, and by using a choke-ring antenna with a large ground plane that blocks access to the antenna by reflected signals.

The designs of static GPS networks are as diverse as the geodetic issues they address. For example, the NGS maintains the Federal Base Network (FBN), a nationwide network of permanently monumented stations spaced 100 km apart with higher density in areas subject to crustal motion. In cooperation with state and local survey organizations, NGS acquires GPS, leveling, and gravity observations at FBN stations in support of the National Spatial Reference System, a consistent coordinate system that defines latitude, longitude, height, scale, gravity, and orientation throughout the U.S.A. The FBN is designed to provide the following local accuracies with a 95% confidence level: 1 cm for latitudes and longitudes, 2 cm for ellipsoidal heights, 3 cm for orthometric heights, 50 μGal for gravity, and 1 mm yr^{-1} for crustal motion. Especially in the western U.S.A., where crustal deformation is concentrated, this network is a valuable resource that can be exploited by geodesists to investigate a wide range of research topics, including volcano deformation and volcano–tectonic interactions.

An example of a static GPS network designed specifically to investigate both tectonic and volcanic deformation is the 40-station network surrounding Mount St. Helens, Washington, which was established in 2000 by the USGS David A. Johnston Cascades Volcano Observatory (CVO) (Figure 4.12). The network is concentrated within 10 km of the volcano's summit to ensure its sensitivity to shallow magmatic sources, but it extends more than 30 km in all directions and covers a total area of more than 7,400 km^2. It spans the central

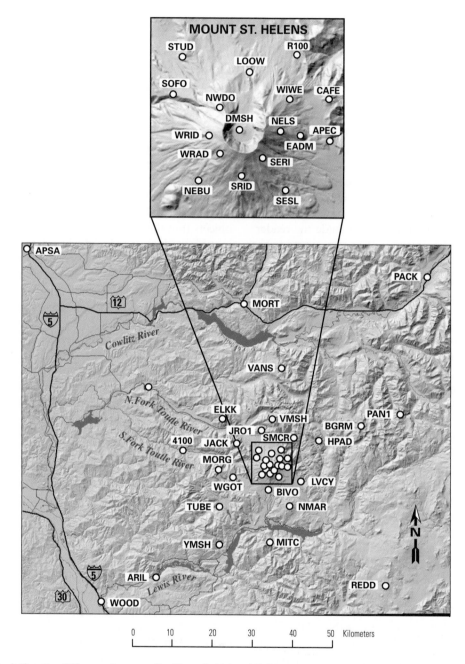

Figure 4.12. A 40-station GPS network surrounding Mount St. Helens, Washington, superimposed on a shaded-relief version of a USGS digital elevation model (DEM). The network was designed to capture both near-field and far-field deformation from magmatic sources beneath the volcano and also tectonic deformation within the St. Helens Seismic Zone, a 90 km long, north–northwest-trending feature centered near Mount St. Helens. The network was installed in 2000 and reobserved in 2003 and 2005. Additional surveys are planned every few years or as often as conditions warrant -- at least annually during the ongoing 2004--2005 dome-building eruption.

part of the St. Helens Seismic Zone, a 90 km long, north–northwest-trending zone of diffuse seismicity centered near Mount St. Helens that has generated several *M* 5 earthquakes in the last 50 years and is capable of producing a *M* 7 event (Weaver and Smith, 1983). More than half of the stations are

accessible easily by road or foot, so partial surveys are possible without a large investment in helicopter support.

There have been numerous demonstrations of the utility of static GPS surveys for monitoring tectonic deformation. For example, Figure 4.13 shows a

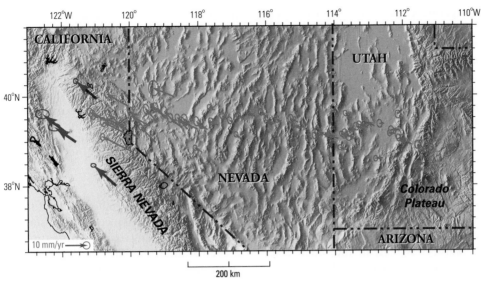

Figure 4.13. GPS station velocities in the Basin and Range province and Sierra Nevada block, western North America, relative to other stations on the stable North American plate lying to the east of the network shown here (Thatcher *et al.*, 1999). Red and blue arrows represent velocity vectors for stations in the Basin and Range province and on the stable Sierra Nevada block of the Pacific Plate, respectively. One standard deviation error ellipses are shown for each vector. Velocities were determined by static GPS surveys in 1992, 1996, and 1998. Deformation within the Basin and Range is concentrated near its western margin and, to a lesser degree, near its eastern margin. Repeated surveys of dense networks such as this are a cost-effective means for measuring steady deformation rates and for helping to site continuous GPS stations in areas of concentrated deformation.

linear array of 63 GPS stations that span about 800 km of the actively extending Basin and Range province between the relatively stable Sierra Nevada block on the west and the Colorado Plateau on the east. The stations were observed in campaign mode, most for 2 days or more, in 1992, 1996, and 1998 to determine station velocities relative to stable mid-continent North America (represented by a group of stations east of the array). Results show that contemporary deformation generally follows the pattern of Holocene faults, with extension concentrated in the westernmost ~200 km and easternmost ~100 km of the array, and with little internal deformation of the intervening ~500 km of the central Basin and Range (Thatcher *et al.*, 1999).

CGPS data from the same area have been used to support a different conclusion (i.e., that extensional strain is broadly distributed across the entire Basin and Range province). For example, Bennett *et al.* (1998, 1999) analyzed the first 13.5 months of data from the 18-station northern Basin and Range (NBAR) CGPS network and concluded that their results were consistent with uniform east–west extension across the entire province. Wernicke *et al.* (2000) came to a similar conclusion based on ~2.5 years of data from the 50-station Basin and Range Geodetic Network (BARGEN). This discrepancy undoubtedly will be resolved as more

data, both campaign and continuous, are acquired and analyzed. The important point here is that both techniques offer distinct advantages. GPS campaigns are more cost effective for determining long-term strain rates under the assumption that they are relatively constant through time, and they can be used to site CGPS stations efficiently in areas of concentrated deformation (or, depending on the application, to avoid such areas). CGPS networks, on the other hand, are better for detecting short-term variations in strain rates that might precede a slip event or, in a different setting, a magmatic intrusion or eruption.

4.7.2 Stop-and-go kinematic GPS

Stop-and-go kinematic positioning is similar to static relative positioning in that both methods require at least two receivers simultaneously tracking four or more satellites. A major difference is the amount of time required for a receiver to stay fixed in order to determine a point's position. At the start of a stop-and-go kinematic survey, the ambiguities are resolved using one of the static initialization techniques described in Section 4.6.4. A reference receiver stays fixed at a known position for the duration of the survey while a second, roving receiver collects static observations at a series of unknown positions for a period of time, usually for a few minutes. Both receivers must track at least the same four satellites continuously throughout the survey. If tracking is interrupted, the roving receiver must revisit a known position before continuing. With carrier-phase data, centimeter-scale accuracy can be achieved over baselines of several kilometers. The main advantage of stop-and-go surveys over static surveys is the shorter occupation time, which makes the stop-and-go technique well suited to many mapping and monitoring applications.

4.7.3 Kinematic GPS

The main difference between stop-and-go surveys and kinematic surveys is that, in the latter case, the roving receiver need not pause at the unknown positions. Once the ambiguities are resolved by some static initialization technique, the roving receiver is free to roam for as long as at least four common (with the reference station) satellites are tracked continuously. If tracking is interrupted, another static initialization must be performed. Single-frequency receivers are adequate for most kinematic surveys, which are best suited to navigation applications in open terrain. Meter-scale accuracy is achievable with code pseudorange data or centimeter-scale accuracy with carrier-phase data, typically at an epoch interval of 1 or 2 s. Sites of special interest can be identified as waypoints and relocated with a subsequent survey. This approach is very effective when integrated with a GIS to record accurate positions of features such as geologic contacts, sample localities, and monitoring stations together with ancillary information. For example, a volcanologist might use a kinematic survey to map quickly and accurately the margins of a new lava flow, or to record the locations of sampling stations along a traverse across the flow.

4.7.4 Pseudokinematic GPS

Pseudokinematic surveys are similar to stop-and-go kinematic surveys, except that satellite tracking is not required while the roving receiver moves between observation points. This is a big advantage in areas where obstructions such as trees, steep topography, or tall buildings inhibit satellite tracking. The field procedures are similar to those for a stop-and-go survey, except in a pseudokinematic survey each station is observed a second time (i.e., the survey is double-run) after a period of about an hour. The reference and roving receivers must track at least four common satellites during the first observation, and the same four or more satellites during the second observation. For this reason, the interval between occupations should not be much longer than an hour. Achievable accuracies are comparable with kinematic surveys (i.e., a few cm over baselines of several km). Usually, longer baselines are not practical owing to the requirement to return to each station within ~1 hour.

4.7.5 Rapid static GPS

Rapid static GPS is a combination of the stop-and-go kinematic, pseudokinematic, and static surveying methods. The roving receiver spends several minutes at each station, satellites need not be tracked between stations, and accuracies are comparable with those for static surveys. Rapid static surveying does not require reobservation of remote stations like pseudokinematic surveying. However, the rapid static technique does require the use of dual-frequency receivers with some means to compensate for AS. Recording carrier-phase data for 10–20 minutes at each station produces accuracies of the order of 1 cm for horizontal coordinates and a few cm for height. This is less accurate by roughly a factor of three than static observations of several hours duration, but for some applications, the trade-off is beneficial.

For example, at a volcano where road access is good and hazards are low, such as the summit area of Kīlauea Volcano, Hawai'i, one person could make rapid static observations at 2–3 dozen stations per day. Observing the same stations by leveling and EDM would require several days of work by three or four people. If the expected station displacements are large relative to the accuracy of the rapid static measurements, the savings in time and effort might be worthwhile.

4.7.6 Real time kinematic OTF GPS

The final step in automating the surveying process is to eliminate the post-processing task (i.e., computing station positions only after the survey is finished) by obtaining positions in the field as soon as the data are acquired. The technique is called precise OTF or real time kinematic (RTK) positioning. This breakthrough has been a boon to land surveyors, construction engineers, and others who require centimeter-scale positioning in real time. Numerous applications of OTF/RTK surveying include bench mark recovery, ship and aircraft navigation, and automated piloting of agricultural vehicles (hands-off aircraft landings and driverless tractors!).

RTK surveying is similar to kinematic surveying in that the roving receiver(s) can move continuously. Unlike kinematic surveying, RTK surveying requires the use of dual-frequency receivers and initialization can occur while a rover is moving. The main difference between kinematic and RTK surveying is that RTK requires a wireless communications link between the reference and roving receivers. The reference receiver continuously broadcasts a correction message, which is based on the difference between the reference receiver's known position and a point position computed at each epoch from carrier-phase observations. The rovers process this information together with their own carrier-phase data to compute their positions relative to the reference station epoch-by-epoch in real time. The accuracy of RTK surveys is about 10 mm + 1 ppm horizontal and 20 mm + 1 ppm vertical, which corresponds to ±20 mm horizontal and ±30 mm vertical for a 10-km baseline.

The rapidly growing demand for RTK positioning, especially by the airline industry, led to the development of the Satellite Based Augmentation System (SBAS). SBAS includes a set of geostationary satellites and ground stations that receive satellite navigation signals, ensure their integrity, and broadcast differential corrections in support of civilian applications. The locations of the geostationary satellites and ground stations are known precisely, so together they serve as reference stations for RTK surveys. Currently, SBAS comprises: (1) WAAS (Wide-Area Augmentation System) for North America; (2) EGNOS (European Geostationary Navigation Overlay Service) for Europe; and (3) MSAS (MTSAT Satellite-based Augmentation System, MTSAT being Multi-functional Transport Satellite) for Asia. China and India are developing similar systems to be called SNAS (Satellite Navigation Augmentation System) and GAGAN (GPS and Geo Augmented Navigation system), respectively.

In North America, SBAS-enabled receivers do not require any additional equipment to take advantage of WAAS, for which there are no setup or subscription fees. WAAS testing in September 2002 confirmed accuracy of 1–2 meters horizontal and 2–3 meters vertical throughout the majority of the continental U.S.A. and portions of Alaska. System upgrades are being developed with a goal of 1-m accuracy or better. In 2003, the Federal Aviation Administration (FAA) certified WAAS for use in low-altitude maneuvering and instrument approaches at thousands of airports and airstrips where there are no other precision landing capabilities. The FAA is evaluating the system for support of fully automated aircraft landings in the near future.

RTK surveying has many applications in volcanology, including flow mapping, grid layout for geophysical surveys, and deformation monitoring. In the latter case, a reference receiver is installed outside the deforming area and one or more 'roving' receivers are installed within it. Ground motion causes the rovers to move with respect to the reference station. If the relative motion is greater than a few centimeters, it can be tracked automatically in real time. In this case, it is preferable for data processing to be done at the reference station rather than at the rovers. If the reference station is Internet accessible, surveying parameters such as the epoch interval can be adjusted remotely, time series data can be served to the Worldwide Web, raw data files can be sent automatically by FTP for more accurate post-processing, and warning messages can be issued by e-mail whenever a predetermined displacement or displacement rate threshold is exceeded. One such system has been operating since 1998 at the Long Valley Caldera, California, with encouraging results (Endo and Iwatsubo, 2000).

4.7.7 Which type of GPS receiver and field procedures are right for the job?

The answer, of course, depends on the nature of the job and on the size of the budget. But other factors are important, too, such as: (1) the required accuracy (meter-scale, centimeter-scale, or millimeter-scale?); (2) logistics (e.g., How many receivers are available to be deployed? How many people are available to deploy them? Is the field area accessible on foot, by road, or helicopter?); (3) level of technical support (e.g., for data-telemetry or computer-networking solutions), timelines requirements (How soon are the results needed? Can the data-acquisition phase span several weeks, or must it be accomplished in a day?); and (4) level of commitment (Is GPS positioning a one-time requirement or a long-term research interest?). The following guidelines should help, but there is no substitute for hands-on experience in the field. In an operational sense, the best way to *learn* GPS is to *do* GPS. Developing an intuitive feel for how best to tackle a new GPS project is less daunting than it might seem from a purely theoretical perspective.

If your goal is to find your way out of the woods, or to pursue the geodetically inspired sport of geo-caching,[33] buy a simple handheld GPS receiver and enjoy. These cell-phone-size units implement kinematic point positioning using pseudoranges to provide accuracy of the order of 5–10 m, at an entry level price under $100. Upper end models cost 2–3 times more (2005 prices in US dollars), will upload and display topographic or street maps for virtually anywhere in the world, and store 1,000 or more user-specified or downloadable waypoints for quick retrieval. The ability to locate a seldom-visited bench mark in dense underbrush, navigate efficiently to a favorite vista or fishing hole, or find your vehicle at the end of a long field day – these are a few of the simple pleasures afforded by a handheld GPS receiver. Get one, and always know your place.

At the other end of the spectrum are state-of-the-art, research-grade receivers suitable for use in either continuous or campaign mode. They are available from a handful of manufacturers at prices upwards of several thousand dollars. Those designed primarily for use at CGPS stations are less fully featured than those intended for campaign use, to keep power consumption at a minimum. These are ideal for remote installations that rely on batteries and solar or wind power, but they work equally well at sites where commercial power is available. They also work well in campaign mode, especially for remote stations to which batteries and other equipment must be backpacked. High-end receivers designed mainly for campaign use tend to be more user-friendly and adaptable in the field, at a cost of greater power consumption. Fortunately, the cost of research-grade receivers of both types has plummeted during the past two decades, to the extent that large projects can afford dozens or even hundreds of the units. The choice of a specific model or configuration depends on many factors, including compatibility with legacy equipment, special requirements (e.g., maximum sample rate greater than 1 Hz, power consumption less than 3 watts),

and competitive pricing. The same is true for GPS data-analysis software (Section 4.9). There is no 'best' choice for every application, but there are many good alternatives from which to choose. Most users who employ GPS as a research tool become proficient with at least one commercial software package (typically supplied by the GPS receiver manufacturer), and at least one research/academic package (see Section 4.9.1). The choice is a matter of personal preference, influenced by specifics of the application as they relate to relative strengths and weaknesses of the software options.

Between handheld units for recreational use and high-end receivers for the most exacting geodetic applications, there exists a vast array of GPS equipment targeting various segments of the marketplace. From agriculture to construction, military to mining, there exists a GPS solution tailored to almost every conceivable need. Need to keep track of your taxicab fleet, or take a hands-off approach to bringing in the wheat? There are turnkey GPS systems for vehicle tracking/fleet management and, yes, even for automated control of agricultural machinery. In the latter case, the GPS 'operator' can interface with a GIS database that might include layers on topography, soil type, rainfall, chemical application rates, crop yield, etc., thereby optimizing field-management practices. It is not geodesy, but it is GPS, nonetheless.

Let us assume that you are an inexperienced but serious GPS user with a range of research and mapping applications. You have access to modern GPS equipment, but are not sure which of the many data-collection strategies described in previous sections is best for your application. Pseudokinematic, rapid static, RTK, stop-and-go kinematic – it can be confusing, even for surveying professionals. The good news is that there's plenty of help available – from GPS hardware manufacturers, GPS user groups, and the research literature, among other sources – and there's no penalty for experimentation. Start with a static survey, process the data by whatever means is convenient (usually commercial software supplied by the receiver manufacturer), then repeat the survey with shorter station occupations to determine if the rapid static technique is suitable for your needs. Is your field area relatively small ($<100\,\mathrm{km}^2$) with good access and sky view? Try a kinematic survey. If sky view obstructions are too much of a hassle, switch to pseudokinematic. You'll have to visit each station at least twice, but you will have first-hand knowledge of the reproducibility of your results. Before long, you will be comfortable

---

[33] The following description is from *http://www.geocaching.com/faq/*: 'Geocaching is an entertaining adventure game for GPS users. Participating in a cache hunt is a good way to take advantage of the wonderful features and capability of a GPS unit. The basic idea is to have individuals and organizations set up caches all over the world and share the locations of these caches on the Internet. GPS users can then use the location coordinates to find the caches. Once found, a cache may provide the visitor with a wide variety of rewards.' Readers wanting to participate are asked to be environmentally conscious and respect property rights. Collect only waypoints, leave only footprints.

using this versatile tool in one or several of its many configurations. Eventually, your research needs might lead you to the ultimate in GPS positioning: continuous, 24/7 tracking with CGPS.

4.8 CGPS NETWORKS

In the midst of a revolution, the unthinkable sometimes becomes possible or even commonplace in a very short period. Such has been the case with very large CGPS networks during the ongoing GPS revolution. The price of a state-of-the-art GPS receiver plummeted by roughly an order of magnitude from 1985 to 1995, while instrumental capabilities improved dramatically. At the same time, processing GPS data evolved from a time-consuming, labor-intensive chore to a mostly automated procedure that now can be accomplished in a matter of hours with little user intervention. Largely as a result of these advances, what was impractical only two decades ago (i.e., dedicating a receiver to continuously track the motion of a single point), has become commonplace. What is more, government agencies and research groups around the world are operating dozens, even hundreds, of such stations in dense networks and processing the resulting deluge of data (~2.5 Mbytes of 30-second-epoch RINEX[34] data/station/day) in near real time.

At dangerous volcanoes where repeated geodetic surveys would expose field crews to unacceptable risks, the emergence of CGPS as a practical monitoring tool is a welcome development. Together with tiltmeters and strainmeters, CGPS stations provide a means to track volcano deformation with excellent temporal resolution at enough sites to characterize the deformation field, study its causes, and anticipate hazardous outcomes.

Operating a CGPS network at a remote volcano poses some logistical difficulties, but usually these can be overcome with careful planning. For example, a typical dual-frequency GPS receiver consumes about twice as much power as a seismometer or tiltmeter, so more batteries and solar panels are required. Good sky visibility, which generally is not a factor when siting volcano-monitoring instruments, is critical for a good CGPS station. Whereas

seismic and tiltmeter stations can be mostly buried to protect them from the elements (except solar panels and telemetry antenna), the GPS antenna at a CGPS station must have an unobstructed view of the sky. This makes it susceptible to burial by snow, which degrades the positioning accuracy, and to damage from high winds or rime ice. Provision must be made for retrieving and processing the GPS data in a timely fashion, which can be accomplished by telephone dial-up where available, or by radio telemetry to an Internet accessible site for storage and remote access. None of these requirements is prohibitive, as evidenced by the proliferation of large CGPS networks for both navigation and geodetic applications. I discuss a few noteworthy examples below.

4.8.1 GEONET – The national GPS network of Japan

The Geographical Survey Institute of Japan operates the GPS Earth Observation Network (GEONET), a dense array of more than 1,000 CGPS stations distributed throughout Japan. The primary goal of the array is to monitor 3-D crustal deformation associated with large earthquakes and other tectonic processes. Several active volcanoes also are being monitored by stations in the nationwide array or by smaller, local CGPS arrays.

A subset of the nationwide array consisting of 100 stations became operational in 1994, four days before the M 8.1 Hokkaido–Toho–Oki earthquake of 4 October 1994. Twenty-one stations in the northern part of the array survived the shaking and provided continuous records of the event. No significant precursory deformation was noted at any of the stations during the week before the earthquake. At the time of the earthquake, however, the station closest to the epicenter (Nemuro, 170 km west) moved 44 cm eastward and subsided 10 cm (Tsuji et al., 1995) (Figure 4.14). Tsuji et al. (1995) noted that the nationwide CGPS network in Japan has three distinct advantages over conventional geodetic techniques: (1) GEONET provides information for a much larger area than local tiltmeter or strainmeter networks; (2) CGPS data provide continuous records of crustal motions; and (3) state-of-the-art GPS receivers and data-processing techniques are capable of horizontal precision on the order of a few millimeters. These points apply equally well to smaller arrays of CGPS stations that are being deployed at an ever-increasing number of volcanoes around the world.

---

[34] RINEX is an acronym for 'receiver independent exchange' format – an ASCII-based, receiver-independent GPS-data format designed to facilitate data exchange between different GPS receivers, software programs, and users. RINEX includes provisions for code pseudorange, carrier phase, and Doppler observations.

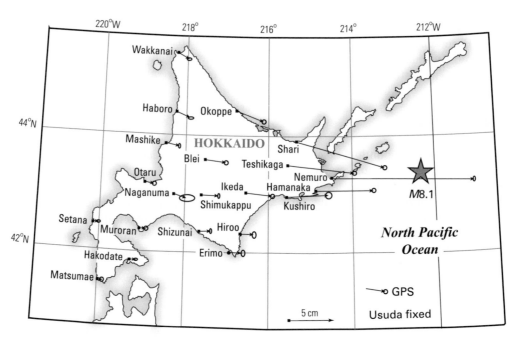

Figure 4.14. Horizontal displacement vectors for the 1994 Hokkaido–Toho–Oki *M* 8.1 earthquake from an array of CGPS stations (Tsuji *et al.*, 1995). Displacements are relative to Usuda, about 1,100 km southwest of the epicenter (star). Error ellipses represent 99% confidence. Nemuro moved 44 cm east and 10 cm down during the earthquake.

4.8.2   The US Continuously Operating Reference Station (CORS) network

In the U.S.A., the NGS coordinates a national network of continuously operating reference stations (CORS) in cooperation with various government, academic, commercial, and private organizations. The CORS network provides carrier-phase and code-pseudorange data to a large and diverse user group that includes surveyors, GIS professionals, engineers, and scientists. Standard positioning accuracy approaches a few centimeters relative to the National Spatial Reference System. Millimeter accuracy is achievable for many stations using improved orbits and appropriate data-processing techniques (Section 4.9). In January 2005, there were 667 CORS sites operating across all 50 states, parts of Central America, and the Caribbean.

4.8.3   SCIGN – The Southern California Integrated GPS Network

SCIGN is an array of more than 250 CGPS stations distributed throughout southern California with emphasis on the greater Los Angeles metropolitan area. SCIGN's objectives are: (1) to provide regional coverage for estimating earthquake potential throughout southern California; (2) to identify active blind thrust faults (i.e., buried faults that do not rupture all the way to the surface) and to test models of compressional tectonics in the Los Angeles area; (3) to measure local variations in strain rate that might reveal the mechanical properties of earthquake faults, and (4) in the event of an earthquake, to measure crustal deformation not detectable by seismographs, as well as the response of major faults to the regional change in strain. SCIGN data are processed daily and distributed by SOPAC, the Jet Propulsion Laboratory (JPL), and the USGS. Day-to-day repeatability is a few millimeters.

4.8.4   PANGA – The Pacific Northwest Geodetic Array

SCIGN's counterpart in the Pacific Northwest is PANGA, the Pacific Northwest Geodetic Array, which is operated by a consortium of universities and agencies in the U.S.A. and Canada. In January 2003, PANGA included the 18-station Western Canada Deformation Array (WCDA) (Dragert and Hyndman, 1995) plus about 40 additional stations distributed throughout Washington, Oregon, and northern California. The cooperative project aims to improve understanding of the regional kinematics, tectonics, volcanism, and hazards associated with the Cascadia subduction zone (Miller *et al.*, 1998).

4.8.5 The discovery of slow earthquakes in the Pacific Northwest

Among the more enigmatic and potentially important geodetic discoveries in recent memory is the detection of periodic slow slip events on the deep Cascadia subduction zone. These events were first recognized by Dragert *et al.* (2001), who reported a sudden shift from forearc contraction to extension at seven PANGA stations in southwestern British Columbia, Canada, and northwestern Washington State, U.S.A., starting in August 1999 (Figure 4.15). The duration of the event ranged from 6 to 15 days at individual stations, and the strain pulse migrated northwestward through the array at an average rate of about 6 km day^{-1} for 35 days, producing 2–4 mm of horizontal surface displacement. The authors dubbed the occurrence a 'silent slip event,' opening an exciting new avenue of research at the intersection between geodesy and seismology (Thatcher, 2001).

The unusual surface displacements are best fit by a model that includes 2.1 cm of down-dip slip along the subduction interface between about 30 km and 40 km depth, tapering linearly to 0 cm at about 25 km depth and migrating about 200 km northwestward during the course of the event. The 2.1-cm slip is comparable with about half a year of plate convergence at the average rate of about 4 cm yr^{-1}, and the total moment of the slip event is equivalent to a *M* 6.7 earthquake. Apparently, a large part of the subducting slab beneath the locked zone lurched downward about 2 cm over the course of a month, which is a disquieting thought for residents of a region that has produced repeated *M* 8–9 megathrust earthquakes (Atwater and Hemphill-Haley, 1997).

The story became even more intriguing when it was recognized that the 1999 slip event was just one in a series of similar events that occur with remarkable regularity. Miller *et al.* (2002) analyzed PANGA data for the period from 1992 to 2002 and discovered eight similar slip events with an average recurrence interval of 14.5 ± 1 months (Figure 4.16). Modeling suggests repeated creep episodes of a few cm each along the plate interface at depths of 30–50 km, in each case approximating the moment release of a *M* 6.7 earthquake.

The term 'silent slip' became a bit of a misnomer when it was discovered that the events are accom-

panied by tremor-like seismic signals, which Rogers and Dragert (2003) imaginatively called 'the chatter of silent slip.' Almost overnight, 'episodic tremor and slip' (ETS) activity or so-called 'slow earthquakes' became a hot topic for seismologists as well as geodesists. Shortly thereafter, Cervelli (2004) reported that, in early November 2000, the south flank of Kīlauea Volcano had lurched seaward about 10 centimeters over a period of nearly 36 hours – a Hawai'ian version of Cascadia's slow earthquakes.

What is going on beneath Seattle, Vancouver Island, and the Big Island of Hawai'i? How are these slip events similar (or dissimilar)? Why are part of the subducting slab in Cascadia and Kīlauea's south flank slipping episodically and what are the implications for earthquake hazards? Could such a slip event trigger the next megathrust earthquake in Cascadia or the next gigantic landslide in Hawai'i? Cervelli (2004) suggested that water may play an important role in both cases. In subduction zone settings like Cascadia, water squeezed out of hydrous minerals in the subducting slab at lithostatic pressure might temporarily wedge apart the megathrust, facilitating periodic slip events. At oceanic islands like Kīlauea, rainwater might seep into shallow faults at rates high enough to promote gravitational collapse.

The answers to these and many other questions might be forthcoming soon as researchers focus more attention on aseismic deformation transients along convergent plate margins and on oceanic islands. An important point to remember is that such events would not have been discovered without the PANGA network of CGPS stations in Cascadia or a similar network on the Big Island. In some situations, borehole strainmeters or tiltmeters might be a better choice owing to their superior strain sensitivity. Better yet is to co-locate all three types of sensors wherever possible, which is an approach being considered by the Plate Boundary Observatory (PBO) (Section 4.10.3). For studying transient deformation, especially when signals are small, continuous monitoring offers clear advantages over repeated GPS campaigns and other types of surveys.

4.8.6 Tracking deformation events at Kīlauea Volcano, Hawai'i, with CGPS

Kīlauea Volcano, Hawai'i, provides additional examples of the utility of CGPS for tracking short-term deformation events. Stanford University, the

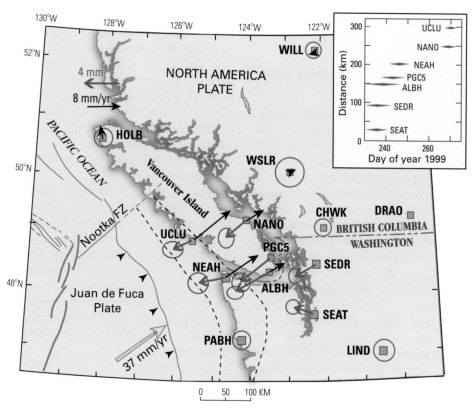

Figure 4.15. Evidence of a deep slip event on the Cascadia subduction interface captured by continuous GPS stations in the Western Canada Deformation Array (WCDA) and the Pacific Northwest Geodetic Array (PANGA), as reported by Dragert *et al.* (2001). Red arrows show displacements relative to station DRAO at the Dominion Radio Astrophysical Observatory south of Penticton, B.C., Canada, which occurred in August-September 1999. Black arrows show 3- to 6-year average station velocities with respect to DRAO. Error ellipses are double the 95% confidence limits derived from regression analysis. Two dashed lines show inferred downdip limits of locked and transition zones from model of Flück *et al.* (1997). Inset shows approximate time interval of transient signal at each station along a northwest-striking line. The temporary trend reversal is attributed to transient creep on the subducting slab downdip from the locked zone.

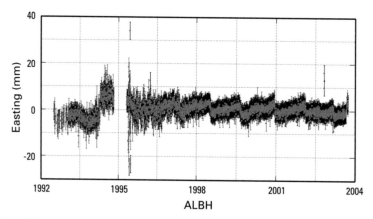

Figure 4.16. The periodicity (14.5 ± 1 month) of deep slip events along the Cascadia subduction zone is apparent in this time series plot of eastings for CGPS station ALBH, adapted from Melbourne *et al.* (2004). ALBH, one of 18 CGPS stations in the WCDA (Dragert and Hyndman, 1995), is located near the southeastern tip of Vancouver Island, near Victoria, British Columbia, Canada (Figure 4.15). Red data points represent the station's east coordinate (longitude) from daily 24-hour solutions. An average eastward velocity of 8.6 mm yr^{-1} relative to station DRAO (south of Penticton, British Columbia, Canada) has been removed to account for long-term tectonic motion. Annual and semi-annual sinusoid signals, and offsets in the data due to earthquakes or hardware upgrades, have also been removed. There have been 9 deep slip events recorded at ALBH since 1992, the most recent shown here in 2003. Each event produces 5–8 mm of southwestward motion at ALBH over a period of 1–3 weeks and has a moment magnitude equivalent to a $M \sim 6.7$ earthquake.

USGS Hawaiian Volcano Observatory (HVO), and the University of Hawaii operate a network of dual-frequency CGPS stations at Kīlauea, while UNAVCO, Inc., operates a network of single-frequency (L1) receivers along the volcano's East Rift Zone.[35] The Kīlauea CGPS networks have demonstrated their capability to track surface deformation caused by magmatic intrusions, eruptions, faulting, and south flank slip events. Accordingly, they have become an integral part of the volcano-monitoring program at HVO.

For example, CGPS data revealed deformation associated with the 30 January 1997 eruptive event at Kīlauea in unprecedented spatial and temporal detail (Owen et al., 2000b; Segall et al., 2001). The brief eruption at Napau Crater along the East Rift Zone drained the lava pond at Pu'u 'Ō'ō, located about 4 km downrift from Napau, and led to a two-month hiatus in the ongoing eruption that began in January 1983. CGPS data recorded summit deflation and rift zone extension, interpreted as the opening of a fracture, starting 8 hours before the onset of the 30 January 1997, eruption (Figures 4.17 and 4.18). The absence of precursory summit inflation led Owen et al. (2000b) to conclude that the event was caused by deep rift zone dilation and slip on the south flank décollement rather than by over-pressurization of the summit magma reservoir. Segall et al. (2001) used the CGPS data to model dike propagation in space and time and to explore conditions in the magma storage system that favor intrusions breaching the surface to feed eruptions instead of stalling in the subsurface to produce dikes.

Cervelli et al. (2002b) analyzed CGPS data together with several other geodetic datasets for the 12 September 1999 intrusion along the upper East Rift Zone and came to a similar conclusion (i.e., that the intrusion was caused by tensile failure in the upper part of the rift zone brought about by persistent deep rifting and seaward sliding of Kīlauea's south flank). CGPS stations, tiltmeters, and strainmeters also captured an aseismic slip event on the south flank in early November 2000, which might have been triggered by nearly 1 m of rainfall that fell on the area nine days before the slip event (Cervelli et al., 2002a).

These three examples serve to illustrate the analytical power of CGPS and other continuous-monitoring techniques for tracking deformation and anticipating eruptions at frequently active volcanoes such as Kīlauea. None of the CGPS analyses for Kīlauea were available in near-real time, but together they called attention to the potential value of timely CGPS information during a rapidly evolving deformation event. Currently, HVO produces daily and hourly solutions from CGPS data to track deformation at Kīlauea and Mauna Loa Volcanoes. Elsewhere, CGPS data are being processed in real time, epoch-by-epoch, to produce detailed records of 3-D surface displacements while they are occurring (e.g., Long Valley Caldera and the Three Sisters volcanic center) (Section 4.8.6).

4.8.7 Continuous, real time GPS network at the Long Valley Caldera

For most volcano-monitoring applications, especially when lives are at stake, timeliness is critical. During a rapidly developing crisis, daily GPS solutions might not be frequent enough for scientists to keep on top of the situation. Hourly or even minute-by-minute updates can be crucial for short-term eruption prediction and hazards mitigation. Toward that end, the USGS installed a network of 16 CGPS stations at the Long Valley Caldera, California, starting in 1998 (Endo and Iwatsubo, 2000) (Figure 4.19). Automated post-processing of the data produces daily 24-hour solutions with millimeter accuracy for all of the stations, and a subset of the stations are also tracked continuously in real time with sub-centimeter accuracy.

A key component of the real time system is 3D Tracker software[36] (Condor Earth Technologies, 2004), which updates the locations of six of the stations relative to a local base station (MWTP) every GPS epoch – in this case, every 5 s. The six stations telemeter carrier-phase and code pseudorange observations to a central recording site, where a delayed-state Kalman filter running on a PC processes the phase data as triple differences and the code data as double differences to obtain epoch-by-epoch solutions (Remondi and Brown, 2000; Rutledge et al., 2001). This strategy is free of errors introduced by satellite clocks, receiver clocks, atmospheric propagation delays, and initial cycle ambiguities (Section 4.5.5), which makes it robust

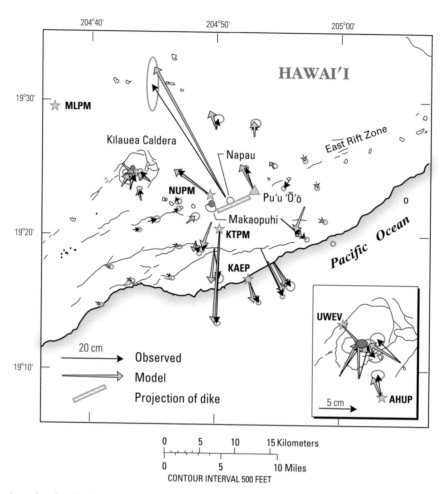

Figure 4.17. Horizontal surface displacements that accompanied the January 1997 eruption at Kīlauea Volcano, Hawai'i, from a combination of continuous and campaign GPS observations (Owen et al., 2000b; Segall et al., 2001). Black arrows with 95% confidence error ellipses represent observed displacements and gray arrows represent model predictions. Stars with four-letter station designations represent continuously recording GPS stations. Small shaded circles near Makaopuhi along the East Rift Zone and within Kīlauea Caldera mark the locations of inferred centers of deflation. Inset shows a close-up of the Kīlauea summit area. Thin lines represent faults, surface fractures, and eruptive fissures. See Figure 4.18 for a time series of the NUPM–KTPM baseline spanning the eruption site.

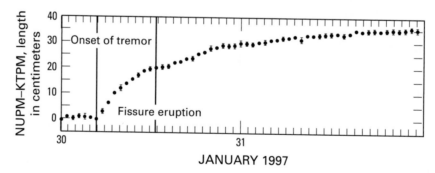

Figure 4.18. Extension across the East Rift Zone of Kīlauea Volcano, Hawai'i, as a function of time for the January 1997 eruption from Owen et al. (2000b). The baseline length between continuous GPS stations NUPM and KTPM (Figure 4.17) is plotted for a 48-hour period spanning the eruption. The data were post-processed, after the eruption, in hourly segments using the GIPSY–OASIS software in precise point-positioning mode (Zumberge et al., 1997). Error bars are two standard deviations. Vertical lines mark the onset of tremor, which was coincident with the start of rift zone extension, and the beginning of the eruption.

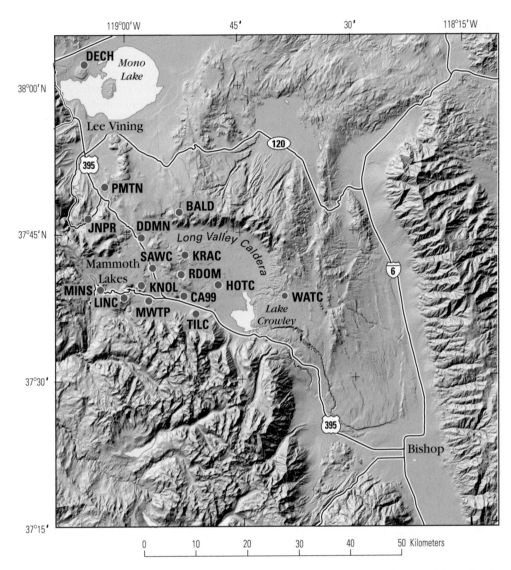

Figure 4.19. This network of 16 CGPS stations at the Long Valley Caldera, California, includes real time tracking of 6 stations relative to the local base station MWTP (red dots) using 3-D Tracker software from Condor Earth Technologies (2004). A delayed state Kalman filter running on a PC near the base station processes carrier-phase and code pseudorange observations as triple differences and double differences, respectively, and updates the positions of the other 4 stations every 5 s (Remondi and Brown, 2000; Rutledge *et al.*, 2001). Data from all 16 stations are also post-processed automatically to produce 24-hour solutions (Endo and Iwatsubo, 2000).

in terms of surviving cycle slips. It does not converge on a sub-centimeter position as quickly as alternative strategies that rely on double differences, but it provides nearly instantaneous motion detection once the solution converges. The software can display 3-D station displacements for comparison with other monitoring results through a graphical user interface at the base station, which is remotely accessible via the Internet.

The Long Valley real time CGPS network represents a significant advance in volcano-monitoring capability because it provides an essentially contin-

uous, automated geodetic data stream from several key sites with little or no user intervention. Although real time CGPS lacks the ultra-high precision obtainable with borehole instruments such as tiltmeters and strainmeters, or the millimeter accuracy achievable with post-processed GPS data, real time CGPS is ideal for volcano-monitoring applications, which generally place a premium on timeliness. Equally important is the combination of relatively low cost and easy-to-install equipment that makes real time CGPS an affordable and practical monitoring tool for volcanoes.

4.9 DATA PROCESSING

Just as there are many types of GPS receivers and GPS surveying techniques, there are many software packages available to process GPS data for various applications. GPS hardware and software are evolving quickly, so it would be pointless to discuss them here in detail. Instead, I'll mention a few popular software packages and direct the reader to GPS trade publications, GPS equipment manufacturers, and the Internet for more up-to-date information.

4.9.1 GPS software packages

GPS data-processing software can be grouped broadly as research/academic or commercial, depending on its source and intended use. For many applications, especially those involving baselines shorter than 10 km for which many common-mode errors can be ignored, both types provide essentially the same accuracy. Commercial packages generally are more user-friendly but less flexible. Research/academic packages have a steeper learning curve but are amenable to tinkering by users who require the best possible accuracy. Both types of software have been used successfully, at times concurrently, for volcano monitoring. Becoming proficient with more than one package for redundancy and error checking is a good idea in most cases.

Among the most widely used research/academic packages are Bernese, EPOS, GAMIT–GLOBK, and GIPSY–OASIS II. Each has loyal supporters and specific strengths that make it well suited to particular applications. All four packages deliver accurate results that usually agree to within a few millimeters (e.g., Simons W. J. F. *et al.*, 1999). Many researchers are becoming proficient with more than one of these packages to better address diverse applications. The idea that GPS data processing will evolve toward a single best or consensus package is probably naive, because GPS applications are so varied that different data-processing strategies are needed to meet the full range of user requirements. Any one processing scheme designed to address all possible applications would be unwieldy, and might not be optimized for the task at hand. Instead, we are likely to see continued development of a few research/academic packages and many commercial packages designed to serve specific user groups (e.g., geodesists, surveyors, outdoor enthusiasts).

Most GPS hardware manufacturers and many specialty software companies offer commercial software packages for processing GPS data. These packages typically are user friendly and offer a wide range of features, with options to process data from various types of static and kinematic surveys. In most cases, processing GPS data with software of this type is straightforward and produces excellent results. Many suppliers bundle their GPS data-processing software with utilities to translate among various data formats, perform network adjustments, and plot results in both time series and displacement-vector formats. A few companies offer specialized software that processes data automatically in real time to support such applications as precise navigation, monitoring critical structures such as dams, and notifying users of strong ground motion (Condor Earth Technologies, 2004). Processing GPS data in this way to obtain 3-D positions with centimeter-scale accuracy every few minutes will likely become much more common at volcanoes, especially during crises.

4.9.2 Precise point positioning

The chief advantage of relative positioning over point positioning is the strength of the solution. As shown in Section 4.6.2, solving simultaneously for the relative positions of all the points within a GPS network becomes a vastly over-determined problem with just a modest amount of data. Redundancy acts as a lever that moves such a network solution toward an accurate determination of the relative station positions, and when distant stations are included in the solution the entire network can be tied to a well-defined, external reference frame such as the ITRF. If adequate time and computing power are available, the relative positioning approach is preferable for most geodetic applications. It is computationally intensive, however, which can be a serious drawback for large networks that require timely processing. Point positioning requires substantially less computing power and offers nearly the same accuracy for relatively small networks (10–100 km in aperture). In this case, the absolute coordinates (e.g., latitude, longitude, and height) of each station are determined separately and independently, which is a much less daunting computational task than relative positioning, especially for a large number of stations.

A specialized implementation of point positioning commonly referred to as precise point positioning has gained wide acceptance in the geodetic community for its excellent precision and relatively low demand on computer resources (Zumberge *et al.*, 1997). An example is the GIPSY–OASIS II

(GOA II) software developed at NASA's JPL. Distinctive characteristics of GIPSY–OASIS II include: (1) the absence of double differencing as a means of eliminating satellite and receiver clock biases (instead, these are estimated explicitly as part of the positioning solution); and (2) simultaneous analysis of carrier-phase and code pseudorange data for ambiguity resolution. This approach makes it possible to use improved orbits and satellite clock solutions from a global analysis to point position one station at a time, which reduces dramatically the computation time required to analyze a GPS dataset. When solving for the positions of several stations simultaneously, the CPU requirement increases roughly as the cube of the number of stations. With precise point positioning, the CPU requirement instead increases only linearly with the number of stations. This permits the analysis of very large networks (such as GEONET in Japan) with ordinary workstations or PCs, with repeatability comparable to that obtained by double-difference methods for relative positioning. The trade-offs are debated among experts, but it seems clear that both approaches will play a role for the foreseeable future.

4.10 LOOKING TO THE FUTURE

Given the rapid pace of change since GPS and GLONASS became operational just a decade ago, any attempt to glimpse the future of global positioning is fraught with uncertainty. Nonetheless, current trends toward smaller, low-power receivers, automated data processing, and mega-CGPS networks are sure to continue. Below I speculate on three future developments that already are coming into focus. More are surely on the way.

4.10.1 Lightweight, low-power GPS receivers

Until recently, four factors have combined to limit the utility of GPS monitoring at volcanoes: (1) early receivers were expensive compared with most other types of geodetic instruments; (2) early receivers had high power requirements that were difficult to meet at remote sites; (3) the receivers and batteries were not easily portable; and (4) GPS data processing was time consuming and labor intensive. Fortunately, there has been considerable progress on all four fronts. Today, basic GPS receivers cost about the same as broadband seismometers (several thousand US dollars), consume comparable amounts of power (a few watts), and can be backpacked to remote sites

with relative ease. Furthermore, reliable data processing with minimal delay is virtually automatic.

On the cost issue, there have been two important developments. First, the proliferation of GPS applications has greatly expanded the customer base for manufacturers of GPS equipment, and consumer driven competition has dramatically lowered prices for state-of-the-art receivers. At the same time, technological advances have substantially improved the capabilities and reliability of modern receivers. Whereas a high-end receiver in the mid-1980s could track only 4–6 satellites simultaneously, cost more than $100,000, and consumed enough power to keep the user warm on a cold day, its counterpart in the 1990s could track 12 or more satellites at a cost of less than $20,000. By 2005, highly capable GPS receivers were available for less than $5,000 and required a mere 2–3 watts of power, which in many areas could be supplied indefinitely by a small solar panel.

Another significant step was taken in the late 1990s, when it was recognized that the relatively small horizontal extent of most volcano deformation meant that single-frequency (L1-only) receivers, which are much less expensive and power hungry than their dual-frequency counterparts, could be used for many volcano-monitoring applications. Most magma bodies of concern are less than 10 km deep and their geodetic footprints at the surface are of comparable size. Over such short distances, the ionosphere and troposphere are relatively uniform, so their differential effects on GPS signals across a typical volcano-monitoring network are small enough to be ignored. Therefore, single-frequency receivers are adequate to the task, especially if at least one dual-frequency receiver is included in the network to provide a tie to more distant stations. The resulting savings in money and power can be substantial. For example, in 1999 a full-featured GPS receiver with a choke-ring antenna cost about $15,000 and consumed several watts of power, while a no-frills, single-frequency receiver cost only a few hundred dollars, consumed about 100 milliwatts of power, and fit comfortably in a shirt pocket (LaHusen and Reid, 2000).[37]

---

[37] A choke-ring consists of concentric rings of metal around and below the ground plane of a GPS antenna for the purpose of reducing multipath effects. For comparison, the antennas shown in Figure 4.3 have a ground plane (4.3A) and a ground plane plus a choke ring (4.3B). The choke-ring design has become a *de facto* standard for precise geodetic applications, including most continuous GPS stations.

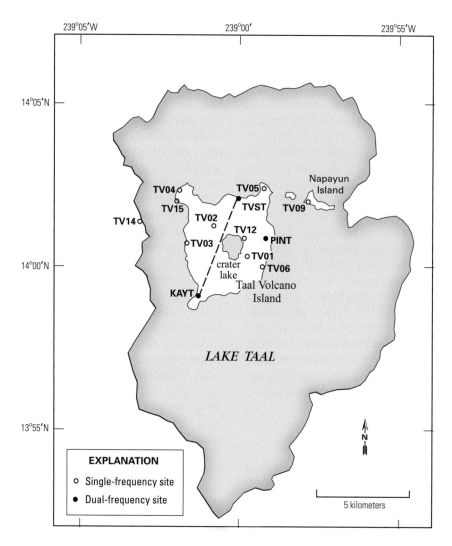

Figure 4.20. CGPS network at Taal Volcano, a basaltic–andesitic stratovolcano near the capital city of Manila on the island of Luzon in the Philippines, modified slightly from Bartel (2002). Data were radioed via several repeaters, not shown here, to a recording site in Quezon City, approximately 60 km north of Taal. Taal is one of the most active volcanoes in the Philippines and has produced some of its most powerful historical eruptions. Lake Taal nearly fills a 25 km × 30 km caldera. The large island in Lake Taal, which itself contains a crater lake, is known as Volcano Island. Data for the KAYT–TVST baseline (dashed line) in 2000–2001 are shown in Figure 4.21.

Using both single-frequency and dual-frequency receivers, Bartel (2002) tracked an episode of inflation at Taal Volcano in the Philippines from February 2000 to October 2000, which they attributed to magma intrusion into a reservoir located about 5 km beneath Volcano Island (Figures 4.20 and 4.21). Parts of the island rose as much as 120 mm with respect to the surrounding caldera rim and baselines extended more than 75 mm. Scatter was greater in the single-frequency data, especially in the north component, but the deformation was readily discernible in both datasets. For relatively small volcano-monitoring networks

where the expected surface displacements are large, single-frequency receivers offer an attractive trade-off between precision and savings, both in equipment cost and power consumption.

4.10.2   Automated GPS data processing

In the past two decades, the processing of GPS data for geodetic purposes has changed from a laborious and time-consuming chore for a few specialists to a routine activity accessible to everyone. Fully automated online GPS data-processing services are available free of charge from several sources, including

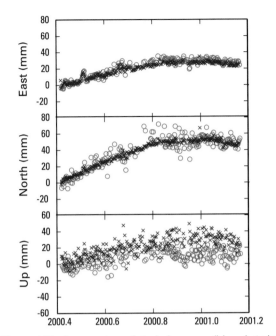

Figure 4.21. A comparison of single-frequency (L1, red circles) and dual-frequency (L1 and L2, blue crosses) processing of continuous GPS data for a ∼5 km baseline (KAYT–TVST; Figure 4.20) at Taal Volcano in the Philippines during a period of inflation in 2000 (Bartel, 2002). Greater scatter in the L1-only solutions is apparent in the east and especially in the north components of the coordinate solutions, but the same trend is discernible from either dataset. Greater separation between the single- and dual-frequency results for the up component (*bottom panel*) could be due to uncorrected path-delay effects in the single-frequency data. See Bartel (2002) for a description of the instrumentation and data processing.

anytime to anyone with a GPS receiver on their wrist – a geodesist's dream come true.

Human resources are always limited during a volcano crisis, so it is essential that the entire response team have easy access to all types of monitoring information. Software can enable users to integrate partly or fully automated GPS solutions with monitoring data from seismometers, tiltmeters, and other sensors in near-real time. For example, epoch-by-epoch GPS processing can be combined with hourly solutions to track rapidly moving stations during periods of intense deformation, and the results can be displayed together with other types of monitoring data. Later, the GPS data can be post-processed to achieve better accuracy (e.g., by using precise rather than predicted satellite orbits and by combining local and regional data to construct better models of the ionosphere and troposphere) for a more thorough analysis of all available datasets.

4.10.3 EarthScope and the PBO

The next quantum leap in our understanding of geodynamic processes at convergent plate margins, including volcanism, could come from a visionary project called EarthScope, conducted under the auspices of the US National Science Foundation (NSF). EarthScope, which comprises the United States Seismic Array (USArray), San Andreas Fault Observatory at Depth (SAFOD), PBO, and a dedicated radar interferometry satellite mission,[38] will study the structure and dynamics of the North American continent in unprecedented detail.

The goal of USArray is to image seismically the continuous range of length scales from whole Earth, through lithospheric and crustal, to local. The program consists of three major seismic components: (1) a transportable array of 400 portable, three-component broadband seismometers deployed on a uniform grid, which will be moved periodically to systematically cover the entire conterminous U.S.A. and Alaska; (2) a flexible component of 400 portable, three-component, short-period and broadband seismographs and 2000 single-channel high-frequency recorders for active and passive source studies; and (3) a permanent array of high-quality, three-component seismic stations, coordinated as

the JPL (Auto GIPSY) and the NGS (OPUS). I strongly recommend that the user acquire a basic understanding of GPS principles and data-processing techniques before taking such a 'black box' approach to data processing, but for the qualified user these services can supply all of the confidence afforded by meticulous, hands-on data processing with little of the pain or expense. As a quick check on data quality or on results from some other data-processing software, they are indispensable.

For volcanologists, the development of web-based automated GPS data-processing services has been revolutionary. Soon it may be possible to send GPS data continuously from remote volcanoes to a central processing facility and to receive high-quality solutions automatically within a matter of minutes. Alternatively, commercial software can be used for on-site, real time data processing and automated event notification (Endo and Iwatsubo, 2000). In the near future, precise-positioning information may be available anywhere,

---

[38] The fourth proposed component of EarthScope, a radar satellite dedicated to InSAR observations, had not yet been funded in January 2006. USArray, SAFOD, and PBO received initial funding in 2003.

part of the USGS's Advanced National Seismic System (ANSS), to provide a permanent reference array spanning the conterminous U.S.A. and Alaska. Also included is an array of 30 magneto-telluric sensors embedded within the transportable and permanent arrays, and 16 CGPS stations, closely integrated with PBO, to image continent-scale deformation.

SAFOD is a scientific drilling and down-hole monitoring program aimed at improving understanding of nucleation and rupture processes of earthquakes. A 4 km deep hole will be drilled through the San Andreas fault zone close to the hypocenter of the 1966 M 6 Parkfield earthquake, where slip is accommodated by a combination of small-to-moderate earthquakes and aseismic creep. The program will sample fault zone materials, measure a wide variety of fault zone properties, and monitor a creeping and seismically active fault zone at depth.

Of particular interest here is EarthScope's third component, the PBO, consisting of four major elements: (1) a backbone network of 100 CGPS stations that will cover western North America and Alaska at an average receiver spacing of 200 km and the eastern U.S.A. at an average receiver spacing of 500 km; (2) several dense instrument clusters comprising 775 CGPS stations and 175 tensor strainmeters along fault zones and magmatic centers in western North America and Alaska; (3) a pool of 100 portable GPS receivers for temporary deployments and rapid response; and (4) Geo-PBO, which will acquire extensive lidar imagery (i.e., high-resolution DEMs) to support geologic and paleo-seismic studies. Current plans call for PBO to operate a large fraction of the stations in preexisting CGPS networks, including SCIGN, PANGA, AKDA (Alaska Deformation Array), BARGEN, which includes the NBAR and Yucca Mountain sub-networks), BARD (Bay Area Regional Deformation Network), and EBRY (Eastern Basin–Range (Wasatch Front) and Yellowstone Hotspot (Yellowstone–Snake River Plain) Network). The magmatic systems component of PBO includes instrument clusters at selected volcanoes in the Aleutian and Cascade arcs, Yellowstone, and Long Valley.

Initial funding for EarthScope ($30 million) was secured in February 2003. Planned to span at least a decade, it is the most extensive and potentially rewarding Earth science initiative in recent decades. Volcano studies are a relatively small part of EarthScope, but the scientific return – direct and indirect – to volcanology from an undertaking of this magnitude is likely to be substantial.

In the 1980s, GPS was a complex technology accessible to just a few researchers, but it has since become a trusted and versatile tool for millions of users. GPS studies are underway at an ever-increasing number of the world's volcanoes, GPS equipment and data analysis tools are improving daily, and geodesy is becoming better integrated with seismology and many other aspects of volcanology – all of which sets the stage for further advances in our understanding of how, when, and why volcanoes deform.

CHAPTER 5

Interferometric synthetic-aperture radar (InSAR)

Daniel Dzurisin and Zhong Lu

Geodesists are, for the most part, a patient and hardworking lot. A day spent hiking to a distant peak, hours spent waiting for clouds to clear a line-of-sight between observation points, weeks spent moving methodically along a level line – such is the normal pulse of the geodetic profession. The fruits of such labors are all the more precious because they are so scarce. A good day spent with an electronic distance meter (EDM) or level typically produces fewer than a dozen data points. A year of tiltmeter output sampled at ten-minute intervals constitutes less than half a megabyte of data. All of the leveling data ever collected at Yellowstone Caldera fit comfortably on a single PC diskette! These quantities are trivial by modern data-storage standards, in spite of the considerable efforts expended to produce them.

Armed with a few hard-won data points, the geodesist must hope that they accurately characterize essential features of the deformation field, without introducing artifacts caused by sparse network coverage or bench mark instability. This is a problem at many volcanoes, where geodetic networks tend to be sparse relative to the complexity of the deformation field. Owing to difficult logistics, safety concerns, or limited resources, many networks are neither dense enough in proximal areas to show the details of shallow-seated deformation nor extensive enough distally to capture deformation from deeper sources (Chapter 11).

What a boon it would be to train a magical geodetic camera on a deforming volcano and take a picture of the entire deformation field, rather than trying to piece it together bench mark by bench mark! Imagine a technique capable of producing a detailed snapshot of the deformation field with

centimeter accuracy, over lateral dimensions of tens of kilometers, without a requirement for ground access to the field area, even at night or in bad weather. Too good to be true? Not necessarily, at least under favorable conditions that exist at many of the world's volcanoes. In fact, remotely sensed, remarkably detailed images of volcano deformation started appearing in some of the world's most prestigious research journals in the mid-1990s, creating a buzz among normally stodgy geodesists – and for good reason.

Under ideal conditions, radar images acquired by satellites can provide more information about a deforming volcano than even the most intensive ground-based geodetic surveys. For a geodesist, this is remote sensing at its very best! For example, the radar interferogram in Figure 5.1 reveals that a shallow-dipping dike intruded the southwest flank of Fernandina Volcano in the Galápagos Islands and fed a lava flow that reached the sea – a conclusion confirmed by the accompanying SPOT satellite image.[1] Closer examination and modeling reveal that the maximum surface uplift

---

[1] The art of interpreting interferograms might seem mysterious at first, but the basic principles are not difficult to understand. Think of the pattern of concentric color bands on Fernandina's southwest flank (Figure 5.I) as a contour map of surface deformation, with each band representing 2.83 cm of slant-range change in the direction of the SAR satellite (mostly uplift in this case). Greater the number of bands, greater the amount of deformation. In this example, there are about 27 bands, so the maximum range change is about 75 cm. The pattern is elongate in the northeast–southwest direction, which means that the source has similar geometry and orientation (i.e., a NE–SW-striking dike). There are many more color bands on the southeast side of the dike than on its northwest side, so the dike dips at a shallow angle to the southeast. (cont.)

was 0.75 m and that the dike was about 3.8 km long, 2.3 km high, 0.86 km thick, with a 34° dip to the southeast (Jónsson *et al.*, 1999). All without ever setting a bench mark or setting up a tripod! Let's take a closer look at this geodetic camera, shall we?

A breakthrough remote-sensing technique called interferometric synthetic-aperture radar (InSAR) burst on the scene with all the drama of a major earthquake.[2] The 8 July 1993 issue of *Nature* highlighted on its front cover a striking color image of the displacement field produced by the *M* 7.3 Landers earthquake, which struck about 150 km east of Los Angeles, California, on 28 June 1992 (Figure 5.2). In the accompanying article, Massonnet *et al.* (1993) introduced InSAR as a geodetic imaging technique and provided geodesists with a revolutionary new tool. Simply stated, InSAR applies interferometric image-processing techniques to two or more synthetic-aperture radar (SAR) images of the same area to measure ground-surface deformation during the time interval spanned by the image acquisitions. Ground-based InSAR had been used earlier to study the topography of the Moon (Shapiro *et al.*, 1972), and satellite interferometry had been used to detect swelling of the ground produced by selective watering of fields in California's Imperial Valley (Gabriel *et al.*, 1989). Not until the Landers interferogram made its splashy debut, however, did InSAR's potential for mapping ground deformation become widely recognized within the geodetic or volcanological communities.

Beneath the dazzle, InSAR rests on a foundation

(cont.) The fact that the dike vented to feed a lava flow is harder to see, but careful inspection reveals a chaotic zone that interrupts the color bands and extends to the sea in the lower left corner of the interferogram. To the trained eye this reveals a resurfacing event that can be interpreted as a new lava flow, which is clearly visible in the SPOT © 1995 optical image on the right.

[2] InSAR, like most technical fields, is well stocked with specialized and potentially confusing terminology. The meanings of many unfamiliar terms should become clear as the chapter unfolds. If the language becomes impenetrable, the reader is encouraged to consult the Glossary (at the end of this book) or one of the references listed at the end of the next paragraph. The term 'interferometric synthetic-aperture radar (InSAR)' is a case in point. 'Interferometric' refers to the interaction that occurs between sinusoidal waves when they encounter one another. You might recall from an introductory physics course that the resulting 'interference' can be either constructive or destructive (i.e., the waves either reinforce or diminish one another). Among the upshots of this phenomenon are multicolored oil slicks and radar interferograms. 'Synthetic-aperture' is an engineering term. Frustrated by practical impediments to building a *really big* radar antenna, enterprising engineers conjured up an imaginary one that works! Read on.

of classical electromagnetic theory and radar-engineering principles. A thorough treatment is beyond the scope of this chapter, but a brief explanation is in order before moving on to some recent applications of InSAR to volcanoes. More detailed information on SAR and InSAR is available in several books and review articles, including Elachi (1988), Curlander and McDonough (1991), Rodriguez and Martin (1992), Gens and Genderen (1996), Bamler and Hartl (1998), Henderson and Lewis (1998), Madsen and Zebker (1998), Massonnet and Feigl (1998), Bürgmann *et al.* (2000), Rosen *et al.* (2000), Zebker *et al.* (2000), Hensley *et al.* (2001), and Hanssen (2002).

5.1 RADAR PRINCIPLES AND TECHNIQUES

Radio and radar technology in various forms has been around for more than a century. In 1873, Scottish physicist and mathematician James Clerk Maxwell published his seminal work *A Treatise on Electricity and Magnetism*. Included in the work were four elegant partial differential equations describing the nature of electricity and magnetism – the now-famous Maxwell equations.[3] Among the far-reaching implications was the existence of electromagnetic waves traveling at the speed of light – the theoretical basis for the invention of radio and radar. In the latter part of the nineteenth century, German physicist Heinrich Rudolf Hertz expanded Maxwell's theory and performed a series of experiments that led to the development of the wireless telegraph and the radio. In 1904, Christian Hülsmeyer received a German patent for the *Telemobiloskop*, or Remote Object Viewing Device, which could detect ships at sea over ranges of up to 3 km. Sir Robert Watson-Watt successfully demonstrated the detection of an aircraft by a radio device in 1935, paving the way for

---

[3] The fundamental behavior of time-varying electric and magnetic fields is encompassed in the following four differential equations (cgs): (1) $\nabla \cdot E = 4\pi\rho$, (2) $\nabla \times E = -(1/c)(\partial B/\partial t)$, (3) $\nabla \cdot B = 0$, and (4) $\nabla \times B = (4\pi/c)J + (1/c)(\partial E/\partial t)$, where $\nabla \cdot$ is the divergence, $\nabla \times$ is the curl, E is the electric field, B is the magnetic field, ρ is the charge density, c is the speed of light, and J is the vector current density. Expressed in integral form, they are known as (1) Gauss's law, (2) Faraday's law, (3) the absence of magnetic monopoles, and (4) Ampere's law. Armed with these four equations and sufficient mathematical acumen to understand the meanings of 'divergence' and 'curl,' you could have invented radio, radar, the incandescent light bulb, GPS, and InSAR. It's not volcano science, but it's impressive nonetheless.

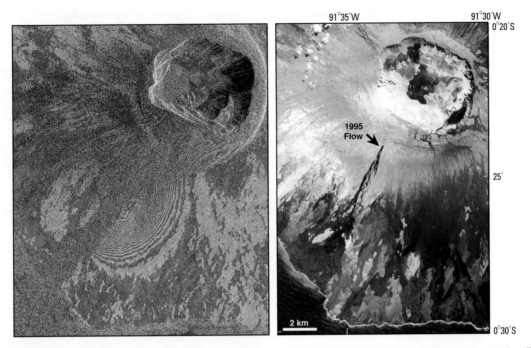

Figure 5.1. Radar interferogram (*left*) and SPOT false-color optical image (*right*) of Fernandina Volcano, Galápagos, showing the effects of an eruption on the volcano's southwest flank during January–April 1995. The interferogram was formed from ERS images acquired on 12 September 1992, and 30 September 1997 (Jónsson *et al.*, 1999). Concentric color bands near the center of the image on the left are interferometric fringes that represent a maximum of about 75 cm of slant-range decrease (2.38 cm per fringe), mostly surface uplift, caused by magmatic intrusion. The resulting lava flow, which appears chaotic in the interferogram owing to a lack of phase coherence, is clearly visible in the SPOT image acquired on 6 July 1995. White areas in the upper left of the SPOT image are meteorological clouds. ERS interferogram created by Harold Garbeil and provided by Peter Mouginis-Mark (both at University of Hawaii). SPOT data copyright SPOT Image Corporation 1995.

development in the UK of the first operational radar network, called Chain Home. The Chain Home, which became operational in 1937, played a key air-defense role during the Battle of Britain (1939–1941)[4].

The term *radar* is derived from '*ra*dio *d*etection *a*nd *r*anging', a phrase that hints at a few of radar's

[4] Here are some other interesting facts about radar excerpted with slight modification from the BBC online encyclopedia, h2g2 (http://www.bbc.co.uk/dna/h2g2/alabaster/A591545, *The History of Radar*). The oldest radar system was developed millions of years ago and is still used worldwide: the ultrasonic sensor of a bat. Bats emit a short 'cry' from their noses, receiving the echo with a set of two antennae (ears). A bat's echolocation system does not use electromagnetic waves, but the working principle is the same as that of a modern radar, with a chirped signal, target-tracking by Doppler estimation, pulse-repetition frequency (PRF) agility, terrain avoidance function, and fine-angle measurement based on the monopulse principle. The oldest radar warning device also was developed millions of years ago. Tiger moths (which frequently appear on bats' menus) are equipped with ears that can detect and jam the ultrasonic signal of a bat, and they have developed tactics to evade a bat's attack. Thus, electronic combat also came into being a long time ago. To my knowledge, *Homo sapiens* can lay sole claim to InSAR, however.

many applications, both practical and scientific. Radar systems make use of the microwave portion of the electromagnetic spectrum, at frequencies ranging from 1 billion to 1,000 billion cycles per second (1–1,000 GHz), which correspond to wavelengths from 30 cm to 0.3 mm. Short-wavelength radar signals are more sensitive to small range changes, but they do not penetrate clouds or vegetation as effectively as longer wavelength signals. This tradeoff requires a compromise between high resolution with less penetration at short wavelengths and low resolution with more penetration at long wavelengths. For imaging applications, the frequency range from 1 to 12 GHz, which encompasses X-band ($\lambda \sim 3$ cm), C-band ($\lambda \sim 5$ cm), and L-band ($\lambda \sim 20$ cm), is most useful (Table 5.1).

All radar systems employ a radio transmitter that sends out a beam of microwaves either continuously or in pulses. Two characteristics of radar that are important for volcano monitoring and many other applications are: (1) unlike optical and infrared systems that are inherently passive (i.e., they rely on natural reflected or radiated energy originating at the source), radar is an active sensor that provides its own illumination; and (2) owing to their longer wavelength, radar signals penetrate water clouds, diffuse ash clouds, and sparse to moderate vegetation better than visible light, enabling limited 'see-through' capability for objects that are opaque at optical wavelengths. Because radar is an active microwave system, it is equally effective in darkness and daylight, and during bad weather or good. This is a tremendous advantage for volcano monitoring, which requires round-the-clock operations during periods of unrest. Currently, this is possible only at volcanoes where an airborne radar could be deployed for an extended period of time – a rare luxury in most parts of the world. Nearly global coverage is available from a few operational SAR satellites, but their orbit repeat cycles are too long for intensive monitoring (Table 5.2). In the foreseeable future, though, daily surveillance of the world's volcanoes and other dynamic surface features such as faults and sea ice by a constellation of SAR satellites is an achievable goal.

Most radar systems can be classified as either *tracking* or *imaging* systems. Tracking radars determine the distance to an object based on the time required for a radar pulse moving at the speed of light to make the round trip from the transmitter to the object and back. If the object is moving with respect to the transmitter, its velocity can be determined from the frequency of the reflected signal, which differs from that of the transmitted signal as a result of the Doppler effect. If the receiver is arranged to reject echoes that have the same frequency as the transmitter and amplify only echoes with different frequencies, it shows only moving targets. Air traffic control radars and police speed-detectors, for example, work on this principle.

Imaging radar systems, as the name implies, are designed to record spatial reflectivity information from illuminated targets for display as a synthetic image, not unlike a photograph.[5] When microwaves strike a target, part of the energy is reflected back to its source where it can be received, amplified, and processed. The distance to and nature of the target are determined by the timing and character of the received signal. All targets do not reflect microwaves equally. The strength of the reflection depends on the size, shape, roughness, orientation, and dielectric

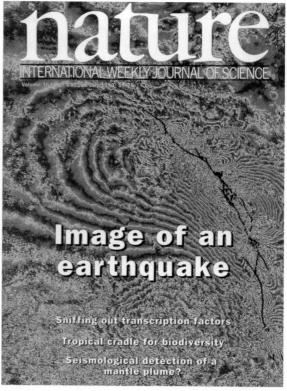

Figure 5.2. This historic satellite radar interferogram, which appeared on the cover of the 8 July 1993 issue of *Nature*, depicts in remarkable detail surface displacements caused by the M 7.3 Landers, California, earthquake on 28 June 1992 (Massonnet et al., 1993). Each color cycle from violet to red represents 2.8 cm of slant-range change. Black lines indicate the fault geometry as mapped in the field. Reprinted by permission from *Nature*, volume 364, no. 6433, 8 July 1993, copyright 1993 Macmillan Publishers Ltd.

constant (strongly influenced by moisture content) of the target. Metal objects are the best reflectors; wood and plastic produce weaker reflections. Turbulent seawater and lake ice are good reflectors; roads and highways are poor ones. Surfaces that are rough at the scale of the radar wavelength generally are brighter in radar images than smooth ones, because some of the roughness elements are oriented perpendicular to the incoming signal and reflect energy back toward the source. With smooth surfaces, most of the energy is deflected forward, away from the source, which causes them to appear dark. For this reason, a body of water appears dark on a calm day when its surface is smooth, and bright on a windy day when wave action causes a greater proportion of the incoming signal to be reflected back toward the radar.

To understand how a detailed image is assembled from a jumble of radar echoes, let us first consider the principles of real aperture radar. Then we will see

---

[5] We will see later that radar images also contain valuable information about surface material properties and satellite-to-ground range. The processed range data underlie the colorful fringes in radar interferograms (Section 5.2).

Table 5.1. Frequency and wavelength bands commonly used for imaging radar systems.

Frequency	Wavelength	Designation	Example
1–2 GHz	30–15 cm	L-Band	JERS-1, SIR-C, Seasat
2–4 GHz	15–7.5 cm	S-Band	NEXRAD weather radar (U.S.A.), Digital Audio Radio Satellite (DARS)
4–8 GHz	7.5–3.75 cm	C-Band	SIR-C, ERS-1, ERS-2 RADARSAT-1 Envisat
8–12 GHz	3.75–2.50 cm	X-Band	SIR-C/X-SAR

how a combination of physics, mathematics, and human ingenuity can transform a small radar antenna into a very large one that serves as a lens for our geodetic camera.

5.1.1 Real-aperture imaging radar systems

Unlike the cameras used to make vertical aerial photographs, imaging radar systems point to the side rather than straight down so the arrival path of the radar signal is oblique to the surface being imaged. This is necessary to better distinguish among targets located at different distances from the radar. Side-looking radar systems are classified as either *real-aperture* (called SLAR for side-looking airborne radar or SLR for side-looking radar) or *synthetic-aperture* radar (SAR).

SLR imaging systems use a long, straight antenna mounted on a moving platform, usually an aircraft or satellite, with the antenna's longitudinal axis parallel to the flight path (Figure 5.3). The antenna emits microwave pulses in a beam directed perpendicular to the flight path and obliquely downward toward the surface of the Earth. These pulses strike the surface along a narrow swath and are scattered, usually in many directions, including the direction back to the radar. The antenna captures a very small fraction of the energy it transmits, typically less than 0.000 1%, from the returning pulse echoes. The echoes arrive at the antenna at different times, depending mainly on the distance from the antenna to countless natural reflectors on the ground.[6] By keeping track of the arrival times,

an SLR profiles its distance from the ground in the direction perpendicular to the flight path (i.e., the cross-track or range direction). Motion of the aircraft or satellite advances the profile in the direction parallel to the flight path (i.e., the along-track or azimuth direction). By overlapping successive swaths, a large area on the ground can be mapped at a resolution of a few tens of meters (Curlander and McDonough, 1991; Madsen and Zebker, 1998; Hanssen, 2002). The resulting images contain information about surface slopes (i.e., topography) and other factors that influence radar reflectivity, including roughness and moisture content.

The dimensions of the antenna determine the spread of the radar beam and thus the size of its footprint on the ground. The angular beam width is $\beta_r \approx \lambda/D$ in the range direction and $\beta_a \approx \lambda/L$ in the azimuth direction, where D and L are the width and length of the antenna, respectively. The corresponding dimensions of the antenna footprint are:

$$W_r \approx \frac{\lambda R_m}{D \cos \theta_m} \tag{5.1}$$

and

$$W_a \approx \frac{\lambda R_m}{L} \tag{5.2}$$

where W_r is the width of the footprint in the range direction (i.e., the swath width), W_a is the width of the footprint in the azimuth direction, λ is the wavelength of the radar, R_m is the slant range from the antenna to the middle of the footprint, and θ_m is the incidence angle at the middle of the footprint (Figure 5.3).[7] For the European Remote-Sensing satellites (ERS-1 and ERS-2), $D = 1$ m, $L = 10$ m,

---

[6] Propagation delays in the ionosphere or troposphere also affect the round-trip travel times of the radar pulses. Inhomogeneities in the ionospheric electron-density or tropospheric water vapor concentration result in spatially variable delays that produce fringes in interferograms, which can be mistaken for ground deformation. Usually, such path-delay effects can be identified and discounted by comparing two or more interferograms of the same scene created from independent pairs of images (Section 5.2.8).

[7] The edges of the antenna footprint are not sharp, so its dimensions are somewhat arbitrary. The energy pattern in radar systems usually are considered to be constant between the half power (3 dB) angles, in which case the theoretical beam widths are $\beta_r = 0.886\lambda/D$ and $\beta_a = 0.886\lambda/L$. For ERS-1 and ERS-2, the beam widths defined in this way are $\beta_r = 2.870°$ and $\beta_a = 0.287°$ (Hanssen, 2002, p. 25).

Table 5.2. Satellite imaging radar systems.

Mission	Agency	Period of operation[1]	Orbit repeat cycle	Frequency	Wavelength	Incidence angle at swath center	Resolution[2]
ERS-1 and ERS-2	European Space Agency (ESA)	July 1991 to March 2000 (ERS-1) April 1995 to present (ERS-2)	35 days[3]	C-band 5.3 GHz	5.66 cm	23 degrees	30 m
JERS-1	National Aeronautics and Space Development Agency of Japan (NASDA) and Ministry of International Trade and Industry (MITI)	February 1992 to October 1998	44 days	L-band 1.275 GHz	23.5 cm	39 degrees	18 m
RADARSAT-1	Canadian Space Agency (CSA)	November 1995 to present	24 days	C-band 5.3 GHz	5.66 cm	10 to 59 degrees	8 m to 100 m
SIR-C/X-SAR	National Aeronautics and Space Administration (NASA), German Space Agency (DARA), and Italian Space Agency (ASI)	9–20 April 1994 (STS-59) 30 September to 11 October 1994 (STS-68)	N/A[4]	L-band 1.249 GHz C-band 5.298 GHz X-band 9.6 GHz	24.0 cm 5.66 cm 3.1 cm	17 to 63 degrees (L- and C-band) 54 degrees (X-band)	10 m to 200 m (30 m typical)

	Agency	Dates	Repeat cycle	Band/Frequency	Wavelength	Incidence angle	Resolution
SRTM	National Aeronautics and Space Administration (NASA)	11-22 February 2000 (Space Shuttle mission STS-99)	N/A	C-band 5.3 GHz / X-band 9.6 GHz	5.8 cm / 3.1 cm	17 to 63 degrees (L-band and C-band) / 54 degrees (X-band)	30 m
Envisat	European Space Agency (ESA)	March 2002 to present	35 days	C-band 5.331 GHz	5.63 cm	14 to 45 degrees	30 m
ALOS	Japan Aerospace Exploration Agency (JAXA), formerly National Aeronautics and Space Development Agency of Japan (NASDA)	Launched 24 January 2006	46 days	L-band 1.275 GHz	23.5 cm	8 to 60 degrees	10 m to 100 m

[1] Information was current in June 2005.

[2] The resolution of radar images differs in the range and azimuth directions and depends on the type of data processing (single-look or multi-look). A representative value is shown here for comparison purposes.

[3] To accomplish various mission objectives, the ERS-1 repeat cycle was shortened to 3 days (43 orbits) from 25 July 1991 to 1 April 1992 and from 23 December 1993 to 9 April 1994. From 10 April 1994 to 20 March 1995, the cycle was extended to 168 days (2,411 orbits) to support a radar altimetry mission. At other times for ERS-1 and throughout the mission of ERS-2, the repeat cycle was 35 days (501 orbits). From 17 August 1995 to 2 June 1996, ERS-1 and ERS-2 flew a tandem mission in which ERS-2 followed 30 minutes behind ERS-1 in the same orbital plane. As a result, the same swath on the ground was acquired by ERS-2 one day later than by ERS-1. The short repeat cycle resulted in good interferometric coherence in most areas, which facilitates the production of DEMs. ERS-1 was retired on 10 March 2000. ERS-2 was operational, with some limitations, in June 2005.

[4] Although repeat-pass interferometry was not a primary mission objective, more than one million square kilometers of repeat-pass data were obtained during STS-68.

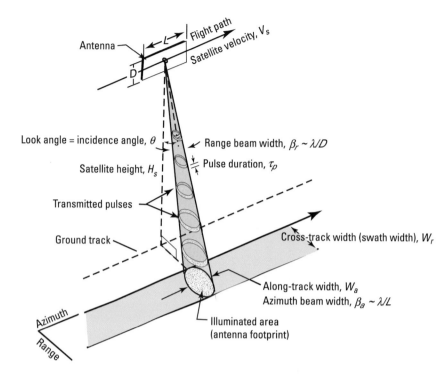

Figure 5.3. Scanning configuration for a right-looking radar, after Olmsted (1993). The dimensions of the antenna (D and L) combined with the radar wavelength (λ) determine the beam width in the range and azimuth directions: $\beta_r \sim \lambda/D$ and $\beta_a \sim \lambda/L$, respectively. The beam widths, in turn, determine the size of the antenna footprint on the ground.

$R_m \sim 850$ km, $\lambda = 5.66$ cm (C band), and $\theta \sim 23°$. The theoretical footprint dimensions are, therefore, $W_r \sim 50$ km and $W_a \sim 5$ km. In practice, the beam width in the range direction is broadened to achieve more even energy distribution across the full swath. This results in practical beam widths of $\beta_r = 5.4°$ and $\beta_a = 0.228°$ for ERS-1 and ERS-2, which correspond to $W_r = 102.5$ km and $W_a = 4.8$ km (Hanssen, 2002, p. 26).

5.1.2 Ground resolution of real-aperture imaging radars

The spatial resolution of an imaging radar system depends on different factors in the range and azimuth directions. Intuitively, the range resolution must depend both on the pulse duration and the incidence angle of the radar beam. In the first case, a shorter pulse duration provides a finer probe of the distance from antenna to ground, so range resolution should improve at shorter pulse durations. The situation is akin to measuring the width of this page using a ruler marked off either in millimeters or centimeters – the more-closely-spaced markings give better precision, for the same reason that short-duration radar pulses allow better range resolution than long-duration ones. The angle of incidence of the incoming radar signal must also play a

role, because a vertically directed radar would receive return echoes from all objects in the scene at nearly the same time, making it very difficult to distinguish among them. At the other extreme, distances across the ground swath for an SLR with a high angle of incidence (i.e., a grazing beam) would differ by a large amount so the range resolution would be good, but the return signal strength would be poor because most of the signal would be scattered forward away from the radar. The ERS satellites accommodate this tradeoff with a fixed incidence angle of $23°$ at the middle of the swath. The radar aboard the Canadian RADARSAT-1 satellite can be steered electronically to produce incidence angles from $10°$ to $59°$ at the middle of the swath.

Two objects can be distinguished in the range direction only if the leading edge of the pulse echo from the more distant object arrives at the antenna later than the trailing edge of the echo from the nearer object. In other words, the two slant ranges must differ by at least half of the pulse length for the objects to be resolvable (Figure 5.4):

$$\Delta R_s \geq \frac{c\tau_p}{2} \qquad (5.3)$$

where ΔR_s is the slant-range difference, c is the speed of light, and τ_p is the pulse duration. Otherwise, the

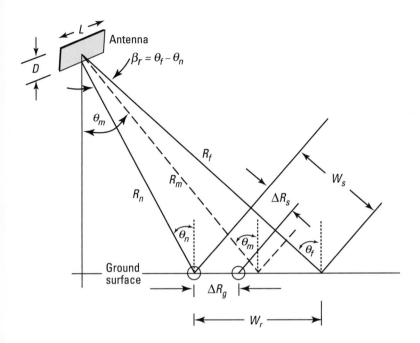

Figure 5.4. The swath width W_r for a real-aperture radar is determined by the wavelength λ, the antenna width D, the slant range at the middle of the antenna footprint R_m, and the incidence angle at the middle of the footprint θ_m. The ground-range resolution ΔR_g is related to the pulse duration τ_p and the incidence angle. R_n and R_f are the slant ranges at near range and far range, respectively, and β_r is the angular beam width in the range direction. See text for details. After Olmsted (1993).

echoes would arrive at the antenna at the end of their round trip separated by less than one pulse length and be indistinguishable. This constraint leads to the following expression for the theoretical ground-range resolution as a function of pulse duration and incidence angle:

$$\Delta R_g = \frac{c\tau_p}{2\sin\theta} \qquad (5.4)$$

where ΔR_g is the ground-range resolution and θ is the incidence angle at a particular ground range.

The best theoretical ground-range resolution would be produced by a combination of extremely short pulse duration and large incidence angle. For example, an X-band radar operating at a frequency of 10 GHz with a pulse duration of 10^{-8} s at an incidence angle of 60° would have a theoretical ground-range resolution of 1.7 m. Unfortunately, there is a catch. The quality of the radar image depends on the ratio of the return signal strength to the ambient system noise. The greater the signal-to-noise ratio, the more fidelity the radar image will have. To achieve meter-scale resolution over typical terrain, the pulse duration would have to be too short to generate enough energy to achieve an adequate signal-to-noise ratio. High-incidence angles pose the same problem. For the case of grazing incidence, forward scattering dominates and the return from back scattering is negligible. In other words, engineering constraints and the principles of wave scattering preclude the realization of a short-pulse-duration, high-incidence-angle SLR.

This difficulty can be overcome, however, by using a technique known as *range compression* together with a type of signal processing called *matched filtering*. A thorough description of these techniques can be found in Curlander and McDonough (1991). For our purposes, suffice it to say that the approach is based on transmitting and receiving not a single frequency but a spread of frequencies in the microwave range. The goal is to correlate specific outgoing and incoming pulses – a daunting task when one considers that a typical X-band radar operates at a frequency of 10 billion cycles per second with a pulse duration of ten one-billionths of a second! To solve this problem, radar engineers use some nifty mathematics and a special kind of pulse called a *chirp*. The frequency of a typical radar chirp increases linearly with time, say from 9.99 GHz to 10.01 GHz. In the auditory frequency range, such a pulse would sound like a chirp – hence the term. When the returned echo of a linearly frequency modulated pulse (a chirp) is correlated with the known transmitted signal, the autocorrelation function is nearly zero except for a narrow spike that corresponds to the round trip travel time of the pulse. The echo is thus 'compressed' into a form that can be readily identified with an electronic match filter tuned to the appropriate range of travel times.

With these techniques it is possible to achieve a ground-range resolution given by:

$$\Delta R_g = \frac{c}{2B\sin\theta} \qquad (5.5)$$

where B is the frequency bandwidth of the transmitted radar pulse. The range resolution can be improved by increasing the frequency bandwidth of the pulse or the incidence angle, as long as the signal-to-noise ratio remains high enough. Comparing the ground resolution without range compression (5.4) with that after range compression (5.5), we see that the improvement in ground-range resolution is equal to $\tau_p \times B$, which is commonly called the *pulse compression ratio*. For ERS-1 and ERS-2, $\tau_p = 37.1\,\mu s$, and $B = 15.5\,MHz$, so $\Delta R_g = 31\,m$ at near range ($\theta \sim 18°$), 25 m at mid-range, and 22 m at far range ($\theta \sim 26°$). The pulse compression ratio is about 575, so the use of a frequency modulated chirp waveform and the matched filtering technique equivalently 'compresses' the ERS-1, ERS-2 radar pulse from 37.1 μs into 64.5 ns. Nominally, the ground range resolution ΔR_g is slightly larger than the sampling spacing $c/(2f_s \sin \theta)$, where f_s is the sampling frequency in the range direction. For ERS-1 and ERS-2, $f_s = 18.96\,MHz$, so the sampling spacing is about 26 m at near range, 20 m at mid-range, and 18 m at far range.

We can see that the range resolution is controlled by the type of frequency modulated waveform and the way in which the returned signal is compressed. Both SLR and SAR systems resolve targets in the range direction in the same way. What distinguishes SLR from SAR is resolution in the azimuth direction (Curlander and McDonough, 1991).

In the azimuth direction, two objects at the same range can only be distinguished if they are not both illuminated by the beam at the same time. Therefore, the best azimuth resolution attainable with a real-aperture radar is, from (5.2) (Curlander and McDonough, 1991):

$$\Delta A_g = W_a = \frac{\lambda R_m}{L} \qquad (5.6)$$

where ΔA_g is the ground-azimuth resolution, akin to the ground-range resolution ΔR_g in (5.5). A 3-m-long antenna operating at a wavelength of 5.6 cm (C-band) on an aircraft flying 1,400 m above terrain with a beam incidence angle $\theta_m = 30°$ ($R_m \sim 1,600\,m$) has an azimuth resolution of about 30 m. The same C-band SLR operated from orbit at a slant range of 850 km from terrain would produce an azimuth resolution of about 16 km, which is much too coarse for most applications. To reduce this number to 30 m would require an antenna about 1.6 km long, which is not likely to be deployed in orbit anytime soon. The solution to

this dilemma lies in conjuring up a huge synthetic antenna by taking advantage of the along-track motion of a real antenna and using the Doppler effect to identify signals reflected from specific parts of the antenna footprint.

5.1.3 Synthetic-aperture radar

Carl A. Wiley (1965) was the first to realize that the frequency spread in the echo signal produced by the Doppler effect could be used to synthesize a much larger antenna aperture than is achievable with real-aperture radars (hence the term *synthetic-aperture radar*, or SAR) (Figure 5.5). This breakthrough led to dramatic improvement in the resolution of SLRs, paving the way for satellite-borne SARs with meter-scale resolution over swath widths of 50–150 km.[8]

Relative motion between an observer and a source of waves causes a frequency shift known as the Doppler effect. When a source emitting waves with frequency f approaches or recedes from the observer at velocity v, the observed frequency f' is:

$$f' = \sqrt{\frac{1 + (v/c)}{1 - (v/c)}} \times f \qquad (5.7)$$

where c is the speed of light and v is positive for an approaching source and negative for a receding source. Thus, an approaching source exhibits a higher frequency and a receding source a lower frequency than the frequency emitted by the source. This effect is familiar to anyone who has listened to a passing train whistle or emergency siren, which decreases in pitch as its velocity relative to the observer changes from positive (approaching) to negative (receding).

How can we take advantage of this classical principle of wave propagation to transform a small real antenna into a huge synthetic one? Notice from Figure 5.5 that targets in the radar illumination footprint which are located at the same range but different azimuths subtend slightly different angles with respect to the look direction of the antenna. As

---

[8] The resolution of SAR images differs in the range and azimuth directions and depends on how the raw data are processed (i.e., single-look or multi-look processing). For ERS-1 and ERS-2, the range and azimuth resolutions of single-look complex images are about 25 m and 5 m, respectively. The resolution of ERS multi-look images with 1 × 5 pixel averaging is about 30 m. Averaging several multi-look images has the desired effect of suppressing speckle, albeit at the expense of degraded spatial resolution. For more information about multi-look processing, see 'multi-look' and 'speckle' in the **Glossary**, or one of the books listed near the beginning of this chapter.

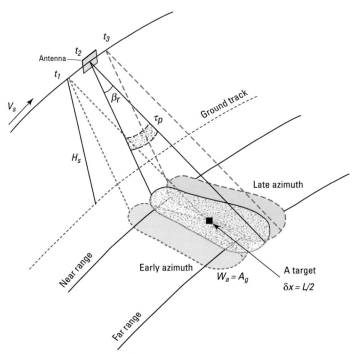

Figure 5.5. At any time t_2, a real-perture radar of length L illuminates a footprint on the ground of dimension $W_a = \lambda R_m/L$ in the azimuth direction. A synthetic-aperture radar takes advantage of the fact that the same target is illuminated continuously and remains in the footprint from time t_1 to t_3 as the radar travels along its flight path at velocity V_s. All of these echoes are used to synthesize a much longer antenna with an azimuth resolution improved from $A_g = W_a$ to $\delta x \sim L/2$ (Curlander and McDonough, 1991). For ERS-1 and ERS-2 ($L = 10$ m), the use of synthetic aperture improves the azimuth resolution by about three orders of magnitude, from $W_a \sim 4.8$ km to $\delta x \sim 5$ m.

a result, they have slightly different velocities at any given moment relative to the antenna. At the leading edge of the footprint the radar is approaching the target, while at the trailing edge it is receding. Therefore, the signal echoed from each target will have its frequency shifted a different amount from the transmitted frequency. Recall that the limitation on azimuth resolution for real-aperture radars derives from the requirement that, to be resolvable, two objects at the same range can not be within the beam's footprint at the same time. The Doppler effect provides a means of distinguishing two such objects, because the antenna will see a return from each that is frequency shifted by an amount that can be attributed to a specific azimuth within the antenna footprint. The procedure is known as *azimuth compression*. Through the wonders of range and azimuth compression, each of myriad echoes from resolution cells throughout the foot-

print can be assigned unique coordinates in both range and azimuth to produce a 'focused' SAR image (Figure 5.6).

The situation is complicated, however, by the fact that the round-trip travel time for pulses from the ERS-1 and ERS-2 radars ($R_m \sim 850$ km) is nearly 10 times the pulse repetition interval.[9] The SAR antenna switches between transmit and receive modes and back again after each pulse, but at any given instant the received signal is not from the last pulse transmitted, which has yet to reach the ground, but rather from an earlier pulse which has echoed from numerous resolution cells in the illuminated footprint. In this respect, the ERS-1 and ERS-2 radars are akin to a juggler who keeps 10 balls in the air by tossing #1, #2, #3 . . . #10, then catching #1, tossing #11, catching #2, tossing #12, etc. The feat is all the more remarkable when one considers that the ERS-1 and ERS-2 radars transmit and receive an average of 1,680 pulses *every second*! Still, it gets worse. Because the antenna footprint (100 km × 5 km) is much larger than the size of a resolution cell (20 m × 4 m), each sample of the return signal includes information from many different cells on the ground, and each cell contributes to many different samples. Rather than a melodious

---

[9] The pulse repetition interval (PRI) is the reciprocal of the pulse-repetition frequency (PRF), which for ERS-1 and ERS-2 is about 1,680 Hz. The time required for a pulse to make the SAR-to-surface round trip is $2R_n/c$, where R_n is the slant-range distance to near range and c is the speed of light. The number of pulses that could be transmitted before the echo of the first pulse is received is: $1 + 2R_n/c \cdot \text{PRF} = 10.3$.

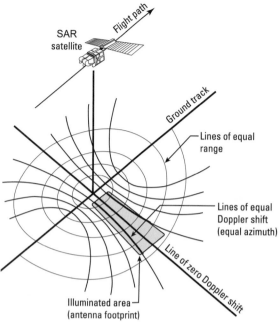

Figure 5.6. SAR images are focused by assigning range and azimuth coordinates to the echoes received from each resolution cell on the ground, which are delineated by the intersections of lines of equal range and lines of equal Doppler shift. The ERS-1 and ERS-2 satellites emit one radar pulse for each 4 m traveled along the flight path. The resulting single-look images have range and azimuth resolutions of 20–30 m and ~5 m respectively (Wright, 2000). Single-look images are resampled using 1 × 5 pixel averaging to produce standard ERS multi-look (see the Glossary at the back of this book) images, which have square pixels and ~30 m ground resolution in both range and azimuth.

sequence of well-defined chirps, the return signal is actually a cacophony of blurred echoes! How can a SAR data processor possibly make sense of such babble?

The following analogy is not perfect, but useful nonetheless. Imagine yourself standing at center stage in a very large auditorium, with 100 people seated across the front row. The room is quiet until suddenly all 100 people start talking simultaneously. From your central vantage point, you hear only a din of indistinct voices. Recalling your high school physics, you stroll to the side of the stage and then pace deliberately across the front while listening intently to the chatter. This time, you have the benefit of hearing a time history of each voice, including subtle frequency changes produced by your motion relative to the speakers. Distinct messages begin to emerge from the racket. Everyone continues to talk throughout your transit of the stage, but your motion allows you to distinguish 100 different voices as you pass by. In our SAR

analogy, you have just sorted the echoes from near-range pixels by azimuth.

Now imagine that every seat in the auditorium is occupied: 100 seats by 100 rows. Having all 10,000 people start to talk simultaneously would not be analogus, however, because a SAR is an active system (i.e., it initiates the action by transmitting a pulse to the ground and then listening for the echoes). So let us say that you instruct the audience to close their eyes and respond in kind each time they hear you clap your hands, which you do repeatedly as you pace across the stage. The speed of sound is finite, so people in the last row hear your first clap a short time after those in the first row, and the sound of their responses takes a correspondingly longer time to reach you. You hear a rolling series of handclaps from first row to last as the 'echoes' of your own clap are sorted by range. You can imagine the racket that would be created by the audience's responses to your repeated handclaps as you strolled across the stage, but in theory you could sort the sounds by seat and row (i.e., by azimuth and range) and attribute them to 10,000 separate sources (resolution cells). A SAR processor goes through an analogous procedure to create a focused SAR image, albeit with more mathematical rigor and considerably less panache!

Let us get back to the math. As the radar antenna moves along its flight path its velocity relative to points on the ground changes, inducing a corresponding Doppler shift in the frequency of the echoes (Curlander and McDonough, 1991, p. 17):

$$f_d = \frac{2 \times (v_s \sin \varphi)}{\lambda} \approx \frac{2v_s x}{\lambda R} \qquad (5.8)$$

where f_d is the Doppler shift, v_s is the antenna's velocity relative to a point on the ground, φ is the angle from a line perpendicular to the ground track and is defined on the plane constituted by the flight direction and the point on the ground, x is the along-track distance measured from the point of closest approach, R is the slant range, and the factor of 2 arises from the two-way travel time (Figure 5.7). Thus, the Doppler frequency of the echoes can be used to determine the along-track coordinate (azimuth) of a particular echo sample:

$$x = \frac{\lambda R f_d}{2 v_s} \qquad (5.9)$$

As the target passes through the antenna beam, it exhibits a constantly changing Doppler shift that is directly related to the position of the radar and the azimuth position of the target. For each azimuth

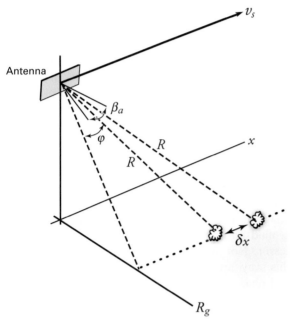

Figure 5.7. Azimuth resolution of a synthetic-aperture radar, after Curlander and McDonough (1991). v_s is the velocity of the satellite or aircraft, β_a is the angular beam width in azimuth, φ is measured from a line perpendicular to the ground track, R is the slant range, R_g is the ground range, x is the along-track distance measured from the point of closest approach, and δx is the azimuth resolution of the SAR.

position there is a unique Doppler history. So the echoes from multiple targets at the same range that are illuminated simultaneously by the beam can be discriminated on the basis of their Doppler shifts. The focused azimuth resolution δx of a SAR depends on the precision of the Doppler shift measurement:

$$\delta x = \left(\frac{\lambda R}{2v_s}\right) \times \delta f_d \qquad (5.10)$$

where δf_d, the measurement resolution of the Doppler shift, is nominally the inverse of the time span Δt, for which the target remains illuminated (Figure 5.5):

$$\Delta t = t_3 - t_1 = \frac{A_g}{v_s} = \frac{R\lambda}{Lv_s} \qquad (5.11)$$

Therefore, the azimuth resolution of a focused SAR image is approximately (Curlander and McDonough, 1991, p. 20):

$$\delta x = \left(\frac{\lambda R}{2v_s}\right) \times \left(\frac{Lv_s}{R\lambda}\right) = \frac{L}{2} \qquad (5.12)$$

This simple result seems surprising because it shows that the azimuth resolution of a SAR improves as the antenna size *decreases*, which is opposite to the

rule for a real-aperture radar. The smaller, the better would seem to be the rule in this case. This seems like good news, but again there is a catch. The same antenna must act both as a transmitter and as a receiver, so it must alternate between transmitting and listening modes. The effective gain of the antenna is proportional to the square of its real aperture, so reducing the antenna size to improve the azimuth resolution produces a tradeoff with the maximum obtainable signal-to-noise ratio (Henderson and Lewis, 1998). A very small antenna might produce excellent azimuth resolution, but with an unacceptably poor signal-to-noise ratio. The determining factor is the pulse repetition interval $\mathrm{PRI} = 1/\mathrm{PRF}$, where PRF is the pulse repetition frequency, because the difference in time between echoes from the near range $2R_n/c$ and far range $2R_f/c$ must be less than the time between pulses (Figure 5.4):

$$W_s = R_f - R_n \leq \frac{c}{2 \times \mathrm{PRF}} \qquad (5.13)$$

Otherwise, far-range echoes from one pulse would be confused with near-range echoes from the next. The maximum swath width is therefore:

$$W_g \approx \frac{c}{2 \times \mathrm{PRF} \sin \theta} \qquad (5.14)$$

Hence, large swath widths require small PRFs. For ERS-1 and ERS-2, $\mathrm{PRF} = 1{,}640{-}1{,}720\,\mathrm{Hz}$ and the maximum theoretical swath width is about 230 km.

To focus a SAR image in azimuth, a SAR processor must be able to relate an observed incremental phase change to a specific Doppler frequency. Curlander and McDonough (1991) show that this requires the bandwidth of the Doppler signal $B_d \approx (2\beta_a v_s/\lambda)$ to be less than the PRF. Recalling that $\beta_a \approx (\lambda/L)$ and $\delta x = (L/2)$, we can write:

$$\frac{2v_s}{L} = \frac{v_s}{\delta x} < \mathrm{PRF} \qquad (5.15)$$

So in the time between successive pulses $1/\mathrm{PRF}$ the antenna must travel a distance along its flight path of no more than half its own length $L/2$. Combining this result with that from the previous paragraph, we see that although the azimuth resolution δx is increased for smaller antennae, smaller antennae require larger pulse repetition frequencies, which reduce the swath width (Wright, 2000, p. 26). These relationships place a lower bound on the size of a practical SAR antenna. Most SAR systems designed for Earth orbit use an antenna 1–4 m wide and 10–15 m long, with a look angle in the range 10 to 60 degrees to illuminate a footprint of 50–150 km in range by 5–15 km in azimuth, producing a single-look ground resolution of 4–10 m in azimuth and 10–20 m in range (Table 5.2).

5.1.4 Characteristics of SAR images

A focused SAR image contains information about the SAR-to-surface range, the radar beam's two-way transit through the atmosphere, and its interaction with a myriad of small reflectors on the ground. Each illuminated pebble, tree, and fence post leaves a unique imprint on the reflected signal. In addition, because microwaves penetrate the surface for a distance that depends on their wavelength and on the dielectric constant of the surface material, the reflected signal also contains information about the shallow subsurface – primarily about the electrical conductivity, which in turn is a strong function of water content, but also about any buried reflectors such as a gravel layer, lost coins and keys, and so forth. When the signal scattered from a given resolution cell on the ground is received back at the SAR, it contains a microwave fingerprint of the cell encoded as amplitude (i.e., intensity or brightness) and phase information, which is mathematically distinctive.[10] This information can be recorded as a single complex number $Ae^{i\phi}$, with A representing the amplitude of the echo and ϕ representing its phase.[11]

For illustrative purposes, raw SAR data can be displayed as an image by assigning values to adjacent pixels equal to the real part and imaginary part of each successive echo sample, starting in the upper left corner of the image (i.e., pixel $1 = A_1 \cos \phi_1$, pixel $2 = A_1 \sin \phi_1$, pixel $3 = A_2 \cos \phi_2$, pixel $4 = A_2 \sin \phi_2$, etc.).[12] Each line in such an image corresponds to a single pulse echo. The average PRF for ERS-1 and ERS-2 is 1,680 Hz and the return signal is sampled at $f_s = 18.96$ MHz. The maximum number of samples per line in a raw ERS image is therefore: 18.96 MHz/1,680 Hz = 11,285 samples. In practice, the ERS-1 and ERS-2 radars spend only about half of the time between successive pulses in sampling the return signal, so the raw images contain 5,616 samples per line. Each raw image includes about 27,000 lines, which corresponds to 16 s of continuous data collection. At 16 bits/sample, each image comprises about 300 Mbytes. The samples in a raw image do not correspond to resolution cells on the ground until the data are focused by a SAR processor. Not surprisingly, an image constructed in this way appears random. When the amplitude and phase data are displayed separately, the result is only marginally better (Figure 5.8). This is because each echo contains information from many resolution cells on the ground, which are spread out in both range and azimuth. As a result, the information from individual cells is 'smeared' across the images. The unfocused phase image is entirely random, because each pixel includes contributions from many elementary reflectors in each of many cells, which combine to produce a random phase signature.

A SAR processor sharpens the unfocused image in range using the matched filter technique described earlier. The process is called range compression because information from each echo that had been spread from near range to far range is compressed into specific range bins. The resulting amplitude image resembles the target area, but still is unfocused in azimuth (Figure 5.9). The final step is to take advantage of the differing Doppler shifts among echo samples from the same range to compress the data in azimuth as well. A fully focused SAR amplitude image is similar in many ways to an aerial photograph. The dimensions on the ground of a standard, focused ERS-1 or ERS-2 image are about 110 km in range and about 110 km in azimuth.

There are some important differences between radar-amplitude images and vertical aerial photographs. Radar images are subject to foreshortening, layover, and shadowing. If the ground surface is level, the echo from a far-range reflector arrives at the SAR later than the echo from a near-range reflector. The time difference Δt is mapped correctly into a ground-range difference and the distance between the two reflectors is properly rendered in the radar image. However, if the surface slopes toward the SAR, Δt is reduced and the distance

---

[10] It is important to note that the term 'phase' is used here to denote a dimensionless number from 0 to 2π (phase is modulo 2π for a sinusoidal wave), which differs from the usage in Chapter 4 for GPS.

[11] A complex number has both a real part and an imaginary part. In this case, $Ae^{i\phi} = A \cos \phi + iA \sin \phi$, where A is the amplitude of the signal scattered back to the SAR antenna from a given resolution cell on the ground, ϕ is the corresponding phase, $A \cos \phi$ is the real part of $Ae^{i\phi}$, and $A \sin \phi$ is the imaginary part. Thus, both the amplitude and phase of the return signal are recorded in the complex number $Ae^{i\phi}$.

[12] For ERS-1 and ERS-2, the raw data are stored as 16 bit integers with bits 1–8 (byte 1) corresponding to the real part of the complex-valued echo and bits 9–16 (byte 2) corresponding to the imaginary part. Bytes 1 and 2 are sometimes called the I-channel and Q-channel (in-phase, quadrature), respectively. I suspect that SAR engineers call this sort of ingenuity 'fun.' I'm just glad there are real volcanoes on which to train their ingenious, synthetic, partly imaginary, geodetic camera!

MOUNT PEULIK VOLCANO, ALASKA

Raw SAR Data

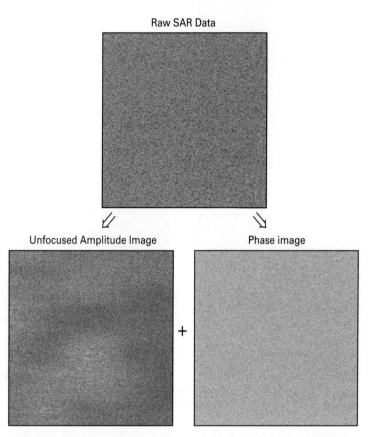

Unfocused Amplitude Image   Phase image

+

Figure 5.8. Raw SAR data consist of amplitude and phase information from time-sequential samples of the echo signal produced by reflection of transmitted pulses from the ground surface. This information can be represented by a complex number $Ae^{i\phi}$, with A representing the amplitude of the echo and ϕ representing its phase. See text for details. Until the data are focused by a SAR processor, the samples do not correspond to resolution cells on the ground, because each cell contributes to many samples and each sample includes information from many cells. Here, raw SAR data are displayed as an image by assigning the real and imaginary parts of the complex-valued signal from each sample ($A\cos\phi$ and $A\sin\phi$, respectively) to two adjacent pixels. The resulting image (*top*) appears random. When the amplitude and phase data are separated, the result is unfocused amplitude and phase images (*bottom left and right*, respectively). The amplitude image does not resemble the target area because each echo contains information from many resolution cells on the ground, which are spread out in range and azimuth. The random nature of the phase image reflects the fact that each resolution cell includes a unique montage of radar reflectors (e.g., pebbles, shrubs, fence posts), each of which contributes a different phase shift to the echo signal. These data for Mount Peulik Volcano, Alaska, were acquired by ERS-1 on 4 October 1995. North is approximately at the top.

between the reflectors is artificially compressed in the ground-range direction. This type of spatial distortion is known as foreshortening, and the feature appears to lean toward the radar in the image. If the surface slope is equal to the incidence angle, the incoming wavefronts are parallel to the surface so they arrive at both reflectors simultaneously. In this case, Δt is zero and the two reflectors appear on top of each other in the image. Steeper slopes cause layover, which occurs when the echo from the top of a slope arrives at the antenna before the echo from the bottom. The top gets mapped on the wrong side of the bottom. This is

a particular problem for systems with a small incidence angle, such as ERS-1 and ERS-2, which produce layover wherever the surface slope exceeds about 23° (18° at near range or 26° at far range). Shadowing occurs when intervening topography blocks the line-of-sight from the radar antenna to the ground. This is more common for systems with large incidence angles, including RADARSAT-1 in some of its imaging modes. The effects of foreshortening can be removed if the surface topography is known, but the information from areas of layover or shadowing cannot be recovered.

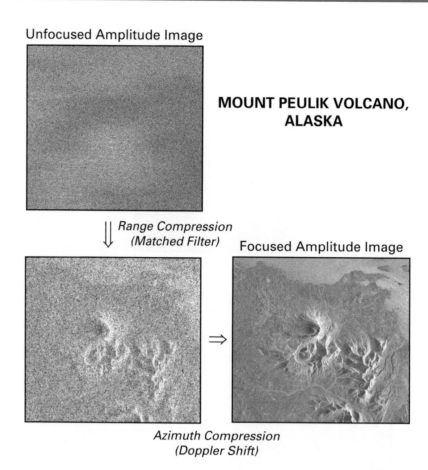

Unfocused Amplitude Image

MOUNT PEULIK VOLCANO, ALASKA

Range Compression
(Matched Filter)

Focused Amplitude Image

⇒

Azimuth Compression
(Doppler Shift)

Figure 5.9. A SAR processor focuses the image in range using a matched filter, which sorts the echo signal into range bins in a process called range compression. The result is an image that resembles the target area but is not yet focused in azimuth (*middle*). Motion of the SAR with respect to the target area causes the echo signals to be Doppler shifted by an amount that depends on azimuth, which allows the SAR processor to focus the image in azimuth as well. The resulting focused image (*right*) resembles an aerial photograph. Foreshortening in the direction of the incoming radar signal, from the right in this case, is apparent on steep east-facing slopes. North is approximately at the top.

5.2 PRINCIPLES OF SAR INTEROMETRY

We have seen that it is possible to map the surface of the Earth with a SAR from an aircraft or satellite at meter-scale resolution, but to produce a digital elevation model (DEM) or to image centimeter-scale surface displacements requires some additional cleverness and mathematical manipulation. First, we will need a pair of overlapping radar images taken from slightly different vantage points, analogous to a stereo pair of aerial photographs used to produce a topographic map (Chapter 6). Overlapping radar images can be obtained in two ways. First, an aircraft or satellite can be equipped with two spatially separated antennas. One of the antennas serves as both transmitter and receiver, while the other serves as a second receiver. The images produced by the antennas are similar but subtly different owing to the different viewing geometries.

The National Aeronautics and Space Administration (NASA) used this single-pass, dual antenna approach for its Shuttle Radar Topography Mission (SRTM), which mapped approximately 80% of Earth's land surface between 60°N and 56°S during an 11-day flight in February 2000 (Farr and Kobrick, 2000; Williams, 2002). For SRTM, the SIR-C/X-SAR radar system was augmented by secondary C-band and X-band receive (slave) antennas mounted at the tip of a 60-m boom that could be extended from Space Shuttle Endeavor's cargo bay to form a single-pass interferometer (Bamler *et al.*, 1996; Jordan *et al.*, 1996; Farr and Kobrick, 2000). In this configuration, Endeavor/SRTM was the largest rigid structure ever flown in space – a record that will not be eclipsed until the International Space Station grows to its planned 108-m wingspan.[13]

Another way to obtain overlapping radar images is to observe the same land area at least twice with the same radar by returning to nearly the same vantage point at different times. The European Space Agency (ESA) used this repeat-pass approach with a C-band radar system aboard its ERS-1 and ERS-2 satellites. From August 1995 to May 1996, these satellites flew a unique tandem mission in which ERS-2 followed approximately 30 minutes behind ERS-1 in the same orbital plane. Both orbits had a 35-day repeat cycle, which meant that once every 35 days, the same land area was observed first by ERS-1 and then, one day later, by ERS-2. Any surface changes that might have occurred during the one-day observation interval generally were small, so coherence was maintained in most areas and the tandem images are well suited to making DEMs (Section 5.2.5). The vertical accuracy of DEMs derived from ERS tandem pairs can be ≤ 10 m across an entire 100 km × 100 km image and, under favorable conditions, ≤ 2 m for local areas.

5.2.1 Co-registration of overlapping radar images

The next step in producing an interferogram is to co-register the two radar images so that we can difference (subtract) the phase information from corresponding pixels. Usually, this is possible if the image-acquisition points are separated by less than about 1 km for ERS-1 or ERS-2. The separation of image-acquisition points, or the distance between the SAR trajectories at the times of the two image acquisitions, is called the baseline. For longer baselines, differences caused by topography and viewing geometry can be too severe for coherence to persist. In such cases, the images are said to be spatially decorrelated. If the spatial coherence is

good, we can use both the amplitude and phase information to co-register the images. The two sets of phase data are random but in a similar way, so they are helpful if they are spatially coherent. Otherwise, the amplitude data alone are generally sufficient. Think of the amplitude and phase data as two layers of a single image. These layers are perfectly co-registered for each image, but corresponding pixels in the two images are offset from one another owing to the different viewing geometries and topography. Conceptually, we would like to slide the two images over one another and distort them to achieve the best possible alignment of corresponding pixels.

In practice this procedure is accomplished by a computer, which uses cross correlation to determine the offsets between corresponding pixels. In one co-registration algorithm, a Fast Fourier Transform (FFT) is first run on a down-sampled version of each single-look complex (SLC) image.[14] The FFT of the first image is multiplied by the conjugate of the FFT of the second image to compute the correlation matrix in the frequency domain, and an inverse FFT is run on the product to return to the spatial domain (Figure 5.10). The location of the peak in the correlation matrix approximates the overall offset between the images. This result is refined by selecting corresponding, full-sampled portions from near the center of each image, using the approximate offset determined earlier, and repeating the procedure. Finally, the images are divided into sections called *chips*[15] and the correlation procedure is repeated on corresponding chips to determine the offsets between each of them as a function of position in the image. By fitting a polynomial to the results (second-order is sufficient in most cases), co-registration at the sub-pixel level can be achieved throughout the entire image. The second image is then re-sampled into the viewing geometry of the first so the pixels in both images correspond precisely.

---

[13] SRTM's primary objective was to produce a worldwide DEM with 1 arc-second (~30 m) spatial resolution, 16 m absolute height accuracy, and 10 m relative height accuracy (Farr and Kobrick, 2000). Plans to distribute global SRTM data were sharply curtailed after 11 September 2001 for reasons of US national security. SRTM data with 30 m spatial resolution are currently available for the US, and with 90 m resolution for all of North America and South America. Data for other continents will follow. Portions of the 30 m dataset outside of the US are available for scientific applications by special request. SRTM DEM data can be accessed at *http://seamless.usgs.gov/*.

[14] The FFT is an algorithm for computing the Fourier transform of a set of discrete data values, thereby converting data from the time domain to the frequency domain. To reduce the computing time required for the FFT when co-registering SAR images, the images are down-sampled in both range and azimuth directions. For example, the resolution cell size for ERS-1 and ERS-2 single-look images is 20–30 m in range and ~5 m in azimuth. Using every second column and every tenth row in the SLC images greatly reduces the time required for the FFT.

[15] This usage of the term chip differs from that in the Global Positioning System (GPS) realm, where it refers to part of a binary sequence called a pseudorandom-noise code that is used to modulate GPS carrier signals (Section 4.3.1).

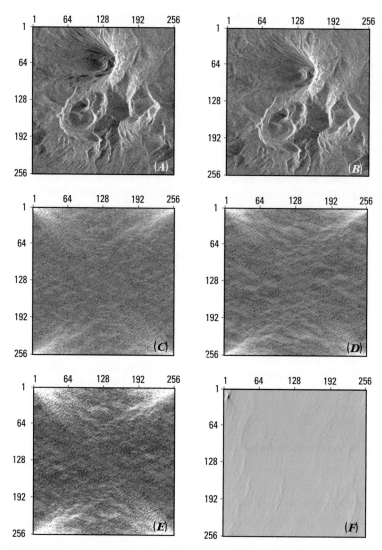

Figure 5.10. A Fast Fourier Transform (FFT) algorithm for co-registering SAR images. (A) Image 1 in the spatial domain; (B) Image 2 in the spatial domain; (C) FFT of Image 1 in the frequency domain (unit of 1/256); (D) FFT of Image 2 in the frequency domain (unit of 1/256); (E) product of the FFT of Image 1 with the conjugate of the FFT of Image 2 (in the frequency domain with a unit of 1/256); (F) inverse FFT of the product in (E). The coordinates of the peak near the upper left corner in (F) indicate the horizontal and vertical offsets in the spatial domain beween Image 1 and Image 2; in this case, 8 and 15 pixels, respectively. Readers interested in digital signal processing might want to consult an introductory textbook on the subject, such as Lyons (2004).

If the preceding paragraph was a painful reminder of how much math you have forgotten since college, do not worry. The algorithms have all been worked out and the details can be fairly transparent to users who prefer it that way. Alternatively, interested readers might want to consult an introductory textbook on digital signal processing, such as Lyons (2004), to learn how to commute between time and frequency domains by way of such conveyances as the FFT, inverse FFT, and complex conjugate of FFT. Here, let us forge ahead to the fun part – processing the pictures from our geodetic camera.

5.2.2 Creating the interferogram

When two images have been successfully co-registered, we can produce an interferogram simply by differencing the phase values of corresponding pixels (Figure 5.11). How can the difference between two random images be anything other than random? The random character of the phase images arises from the fact that the phase value associated with each resolution cell on the ground is determined in large part by the properties of numerous elementary reflectors within the cell, which combine to produce a unique phase signature.

MOUNT PEULIK VOLCANO, ALASKA

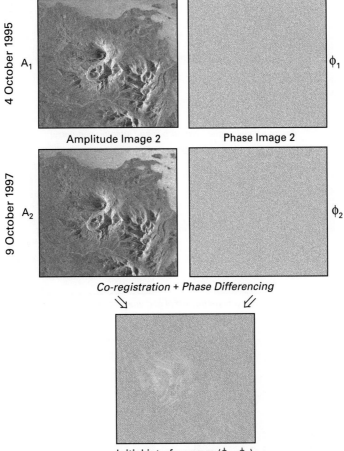

Figure 5.11. Information from two SAR images of the same target area (*top and middle*) acquired at different times and from slightly different vantage points (perpendicular baseline $b = 35$ m, altitude of ambiguity $h_a = 279$ m) can be combined to produce an interferogram by differencing the phase values after the images have been co-registered (*bottom*) . The resulting interferogram contains fringes produced by the differing viewing geometries (orbital fringes), topography (topographic fringes), any path delays present in the images, and surface displacements (deformation fringes). These images of Mount Peulik Volcano, Alaska, were acquired by ERS-1 on 4 October 1995 (*top*) and ERS-2 on 9 October 1997 (*middle*). North is approximately at the top. On the ground, the dimensions of the area shown are about 28 km from left to right by 25 km from top to bottom.

If these properties do not change appreciably between the acquisition times of the first and second images, their random contributions vanish in the difference image (i.e., the interferogram) and we are left with nonrandom, useful information.

An interferogram produced in this way includes contributions from the different viewing geometries of the two parent images, topography, path delays owing to different atmospheric conditions, noise, and any range changes caused by surface deformation during the interval spanned by the image acquisitions. It is this last bit of information that we would like to tease out of the data. First, though, we have to

remove or account for the other sources of phase differences in the interferogram.

5.2.3 Removing the effects of viewing geometry and topography

The difference in viewing geometry between two images acquired from slightly different vantage points produces a regular pattern of phase differences between the images. If the target area were perfectly flat, these differences would manifest themselves in the interferogram as a series of nearly parallel bands called orbital fringes (Figure 5.12).

MOUNT PEULIK VOLCANO, ALASKA

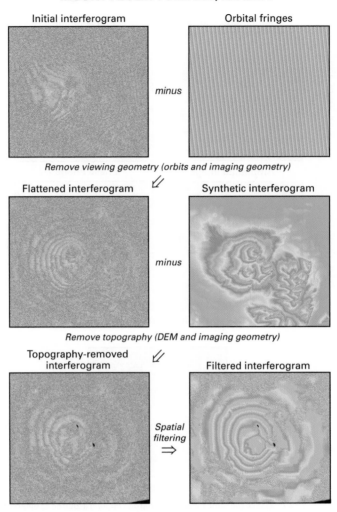

Figure 5.12. Orbital fringes can be computed from the known trajectory and imaging geometry of the SAR and removed from the initial interferogram by subtraction (*top*) to produce a flattened interferogram. Similarly, topographic fringes can be removed by subtracting a synthetic interferogram (*middle*) based on the known topography (Section 5.2.3) to produce a topography removed interferogram, which includes fringes produced by surface displacements (deformation fringes) and any path delays present in the images, plus noise. Spatial filtering suppresses the noise and accentuates the fringes (*lower right*).

These can be computed and removed based on the known trajectory and imaging geometry of the SAR, along with the small additional effect of Earth's curvature. The result is called a flattened interferogram.

Surface topography also contributes to the phase images, which is helpful if our goal is to produce a DEM but superfluous if our real interest is ground deformation. In the latter case, the effect of topography can be removed from a flattened interferogram by first constructing a synthetic interferogram based on known topography (i.e., on a DEM). A useful concept in this regard is the altitude of ambiguity h_a which is the amount of surface

height difference that would produce exactly one topographic fringe:

$$h_a = \frac{H\lambda \tan\theta}{2b} \qquad (5.16)$$

where H is the SAR altitude (about 800 km for most ERS-1 and ERS-2 orbits), λ is the wavelength of the SAR, θ is the incidence angle, and b is the perpendicular component of the baseline with respect to the incidence angle θ. For example, $h_a = 10$ m means that, for a given image pair, each 10 m of surface height difference would produce one topographic fringe in the interferogram, which would resemble a topographic map with a contour interval of 10 m.

Given h_a for a particular image pair, it is relatively straightforward to produce a synthetic interferogram that simulates topographic fringes.[16] Subtracting the synthetic interferogram from the observed flattened interferogram, we are left with a topography removed interferogram that contains only the effects of noise, path delays, and (most importantly) surface deformation (Figure 5.12).

Any errors in the DEM used to produce the synthetic interferogram will remain in the topography removed interferogram, so for deformation studies it is important to choose image pairs with a large altitude of ambiguity (i.e., short baseline). An interferogram formed from such pairs will have low sensitivity to DEM errors but full sensitivity to surface displacements. For example, a 30 m error in the DEM would result in 3 residual fringes in a topography removed interferogram if $h_a = 10$ m, but only one-third of a fringe if $h_a = 90$ m. Conversely, if the goal is to produce a DEM from the interferogram, a small h_a value (long baseline) is desirable for better sensitivity to topography. For ERS-1 and ERS-2, a useful rule of thumb is $h_a(\mathrm{m}) \approx 9,400/b$ (m).

A topography removed interferogram can be spatially filtered to reduce the effect of short-wavelength noise, and displayed over an amplitude image to facilitate visual correlation of deformation fringes with surface features (Figure 5.13(A) and (B)). A final step to enhance the utility of the interferogram is to rectify (warp) the image to a geographical coordinate system, so it can be displayed over a shaded relief image derived from the DEM or with other georeferenced map features (Figure 5.13(C) and (D)). The flattened, topography removed, filtered interferogram in Figure 5.13, for example, shows 6 concentric fringes that represent about 17 cm of range decrease (mostly uplift) centered on the southwest flank of Mount Peulik Volcano, Alaska. The time spans of several interferograms include an uplift episode that occurred between October 1996 and September 1998

---

[16] The altitude of ambiguity for a given viewing geometry h_a is analogus to the contour interval for a topographic map (i.e., it is the surface height difference encompassed by one interferometric fringe or by successive topographic contours, respectively). A synthetic interferogram can be derived from known topography (e.g., a DEM) by first contouring the height data using a contour interval equal to h_a, then assigning modulo-2π phase values to the cells between successive contours. If a suitable DEM is not available, one can be produced from a flattened interferogram of the same area, using an independent image pair (Section 5.2.5).

– a period that was aseismic at the volcano, but included an intense earthquake swarm near Becharof Lake, 30 km to the northwest, starting in May 1998 (Lu et al., 2002a).

The ability to georeference SAR images and interferograms (i.e., to assign geographical coordinates in a known reference frame to each pixel in an image) opens the door to the wonderful world of geographic information systems (GISs). All geodetic and ground-deformation data are inherently geospatial, because they pertain to specific locations on the surface of the Earth; for example, the ellipsoidal height at point A, height difference from point B to point C, or change in baseline vector from point D to point E. All forms of geodetic and deformation data, including SAR images and interferograms, can and should coexist within a GIS with data pertaining to topography, geology, geophysics, geochemistry, hazard zones, infrastructure, land-management jurisdictions – all of the information pertinent to monitoring volcanoes, studying volcanic processes, assessing volcano hazards, and communicating effectively with colleagues, public officials, and the general public.

Housed in a robust and accessible geospatial database, SAR and InSAR data can be put to many uses beyond volcano monitoring – limited mainly by one's imagination – from map-making to hazards analyses and source inversions. As repeated SAR coverage of the globe increases and InSAR monitoring of the world's volcanoes becomes routine, the need to store and manipulate large geospatial datasets will only become more acute. The tool of choice for the foreseeable future will be GIS, and that suits volcano geodesy quite well.

5.2.4   Two-pass, three-pass, and four-pass interferometry

The method described above is called two-pass interferometry because it requires two overlapping radar images plus a DEM (Massonnet and Feigl, 2000). The two-pass method has the advantage of requiring only two spatially coherent radar images, but it also requires an accurate DEM from some other source (e.g., photogrammetry, digitized contours from a topographic map, SRTM). Typically, a DEM with a cell size of 30–90 m is adequate for this purpose. This is a drawback in many parts of the world where high-resolution DEMs are not yet available. Alternatively, two or more SAR images can be used to generate a DEM (Section 5.2.5),

MOUNT PEULIK VOLCANO, ALASKA

Filtered interferogram

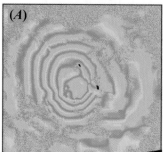

Filtered interferogram over amplitude image

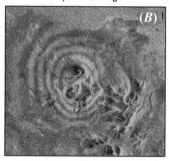

Transformation to geographical coordinates

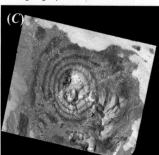

Filtered, transformed interferogram over shaded relief from DEM

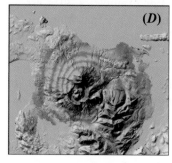

Figure 5.13. Displaying an interferogram over an amplitude image facilitates visual correlation of fringes with surface features (**B**). Images **A** and **B** are displayed in the SAR viewing geometry. Because the orbit and imaging characteristics of the SAR are known, the interferogram and amplitude image can be georeferenced (i.e., rectified to geographical coordinates) for consistency with standard map products (**C**) (Williams, 2002), and the rectified interferogram can be displayed over a shaded relief image produced from a digital elevation model (**D**). Each fringe corresponds to 2.8 cm of range change (mostly uplift in this case). The concentric pattern indicates ∼17 cm of uplift centered on the southwest flank of Mount Peulik Volcano, Alaska, which must have occurred sometime during the interval spanned by the interferogram (October 1995–October 1997). Other interferograms for overlapping time periods bracket the inflation episode, which was aseismic at the volcano, between October 1996 and September 1998 (Lu et al., 2002a). The dimensions of the area shown are about 28 km from left to right by 25 km from top to bottom.

which then can be used to remove topographic effects from other interferograms of the same area. This technique requires only radar data, but the requirement for multiple, spatially coherent images of the same scene is a disadvantage for areas where coverage is relatively sparse (e.g., near the equator) and where the number of useful images is limited further by seasonal rainfall, snowfall, vegetation, or other factors that destroy coherence.

In three-pass interferometry, three SAR images are used to form two topographic interferograms; one is used to predict and remove the topographic fringe pattern from the other (Gabriel *et al.*, 1989; Zebker *et al.*, 1994). Usually, one image is selected as the reference image (call it *A*) and two interferograms are formed between the reference image and the other two images (*B* and *C*). In other

words, one interferogram is formed from *A* and *B*, and another is formed from *A* and *C*. Because the two interferograms are referenced to the same geometry, extra registration and re-sampling procedures are not necessary.

Four-pass interferometry is similar to three-pass, with the exception that different reference geometries are used to generate the two interferograms (i.e., the two interferograms do not share a common image). Consequently, one of the two interferograms needs to be re-sampled to the geometry of the other, and the topographic contribution in one of the interferograms needs to be removed using the other. The three-pass or four-pass method works well when no suitable DEM is available, but each requires extra image-processing steps relative to the two-pass method. Given the considerable number-crunching power of modern

PCs and workstations, this is more of a bother than a real impediment. The two-pass method is more straightforward, but requires a DEM from some other source. Nonetheless, the two-pass method is likely to become the technique of choice in most cases when the entire worldwide DEM from SRTM becomes widely available.

5.2.5 DEMs derived from InSAR

Because the observed SAR-to-target range depends mainly on the known vehicle track and the unknown topography along the ground track, data from a SAR can be used to create a DEM of the terrain it illuminates. The procedure is identical to the one described above for producing a topography removed interferogram, except in this case we would like to eliminate any contribution to the interferogram from surface deformation so we can extract just the topographic information. For this reason, it is advantageous to choose an image pair that spans the shortest possible time interval. For ERS-1 and ERS-2, tandem pairs with one-day separation are ideal for this purpose. If a tandem pair for the area of interest is not available, a small multiple of the 35-day repeat cycle for each satellite is usually adequate, except in areas of very rapid surface change or deformation. Relatively long baselines correspond to the smallest altitudes of ambiguity and therefore to the best sensitivity to topography. Baselines between 1,000 m and 300 m, which correspond to $h_a \sim 10$ m to $h_a \sim 30$ m, respectively, are suitable for DEM production. For baselines longer than $\sim 1,000$ m, the difference in viewing geometry between two images usually is too severe for the images to be spatially coherent.

To produce a DEM from a flattened interferogram, we need to keep track of the integer number of phase cycles between points using a procedure called phase unwrapping. The phase of the interferogram, which is directly related to the topography through the altitude of ambiguity, is only measured modulo 2π (i.e., values range from 0 to 2π and then repeat). To calculate the elevation of each point, it is necessary to establish an elevation at some point in the interferogram and to add the correct integer number of phase cycles to each phase measurement. The 'known' elevation should be taken from the best source available (e.g., a spot elevation on a topographic map or a surveying result for the summit of a prominent peak in the area). The problem of solving

this 2π ambiguity is called phase unwrapping. Several approaches to this problem have been proposed (e.g., Goldstein *et al.*, 1988; Zebker and Lu, 1998; Costantini, 1998). Phase unwrapping is relatively easy in areas of high coherence. It is a matter of counting fringes from a 'known' starting point to other points in the image, then multiplying by the altitude of ambiguity (for a flattened interferogram to calculate elevation difference) or by the radar half-wavelength (for a topography removed interferogram to calculate apparent range-change). The procedure is akin to counting contour lines on a topographic map from a labeled contour to some other point, then multiplying by the contour interval to determine the elevation difference. However, where decorrelation or layover is a problem, it can be very difficult to unwrap an interferogram correctly. In such cases, it is sometimes necessary to fill holes in the unwrapped interferogram with lower resolution data from an existing DEM.

We have seen that a good DEM is an essential component of our geodetic camera, but there are other important applications, too. For example, the information contained in a DEM can be represented as a shaded-relief image, which simulates an oblique aerial photograph. The shaded-relief image in Figure 5.14 was produced from an InSAR-derived DEM; it shows in considerable detail a large debris avalanche deposit from Socompa Volcano in north Chile. The 40-km-long deposit was emplaced about 7,000 years ago when the northwestern flank of the volcano collapsed (Wadge *et al.*, 1995). Unlike a photograph, a shaded-relief image is 'illuminated' by simulated lighting that can be manipulated to accentuate features of particular interest (e.g., low-angle lighting from the north or south in an equatorial area to accentuate east–west trending ridges or valleys). Thus, shaded-relief images can reveal surface morphology that might otherwise be obscured in heavily vegetated areas. Combined with an empirical model of how lahars or lava flows interact with topography, a high-resolution DEM can be used to produce inundation-hazard maps quickly, efficiently, and with relatively little work in the field area (Iverson *et al.*, 1998). In areas of rapid geomorphic change, sequential DEMs can provide quantitative information about the sites and rates of such processes as erosion and deposition by rockfalls (Mills, 1992), lava-dome growth (Mills and Keating, 1992), eruption volume estimation (Lu *et al.*, 2003b), or glacier evolution (Schilling *et al.*, 2002, 2004).

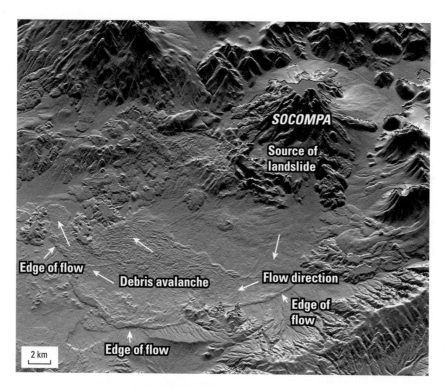

Figure 5.14. Shaded-relief image of Socompa Volcano and a prehistoric debris avalanche deposit in north Chile. The image was created from a DEM, which was produced by radar interferometry using a tandem pair of images from the ERS-1 and ERS-2 satellites. Image produced by Harold Garbeil and provided by Peter Mouginis-Mark, both at the University of Hawaii.

5.2.6 Lidar, InSAR, and photogrammetry – a potent remote-sensing triad

Additional sources of DEMs for InSAR and other applications include SRTM, digitization of conventional topographic maps, photogrammetry (Chapter 6), and lidar surveys. *Lidar*, an acronym for **li**ght **d**etection **a**nd **r**anging, is similar to radar except that a lidar system emits laser-light pulses instead of microwave chirps. The most widespread scientific application of lidar is in the field of atmospheric remote sensing (see, e.g., Weitkamp, 2005), but an aircraft equipped with lidar, GPS, and an inertial navigation system can also be used to map surface elevations with decimeter-scale accuracy, even in heavily vegetated terrain. How is this possible, given that: (1) most lidars operate at near-infrared wavelengths that are 10^4 to 10^5 times *shorter* than the microwave wavelengths used for radars (Figure 6.1), and (2) electromagnetic signal penetration generally *decreases* with wavelength?

The key to lidar's success in this regard is persistence. Lidar systems emit thousands of pulses per second, so each resolution cell on the surface is illuminated repeatedly. The size of the resolution cell is much larger than the slant-range resolution of the lidar (typically a few meters versus ~0.1 m), which means that the return signal potentially includes contributions from multiple reflectors in

each cell that can be resolved in slant-range. The lidar system determines round-trip travel times for pulses arriving from each cell. The earliest returns correspond to the shortest slant range between the lidar and any reflector in a particular cell. In densely forested areas, for example, the first returns to arrive are from treetops; later ones are from branches, leaves, or other reflectors at successively lower levels in the canopy. Eventually, if all goes well, at least a few pulses make it to the ground surface and back to the lidar. These tardiest returns from each cell are used to produce a DEM, because they effectively 'strip away' any vegetation and reveal the topography of the underlying ground surface.

It should come as no surprise that the foregoing description is grossly oversimplified – if it were that easy, lidar surveys wouldn't be so expensive.[17] In the real world, several complications arise. Steep, uneven terrain is an obvious problem. If the

---

[17] The cost of lidar data varies greatly depending on factors such as the size and location of the project, horizontal postings (point density), vertical accuracy requirements, and type of data product requested. The standard product is (x, y, z) point data; derivative products including DEMs, digital terrain models (DTMs), and digital contours add to the cost. In 2005, the average cost for point data was $400 to $800 per square kilometer for 2- to 3-meter postings. Production of a DEM with 1-m postings in steep terrain could increase the cost by an order of magnitude or more.

surface height within a resolution cell varies by more than the slant-range resolution of the lidar, the return times of signals scattered from different parts of the cell will vary accordingly. The upshot is added uncertainty in the cell-height estimate. Patches of extremely dense vegetation present a similar problem. If none of the returns from a given resolution cell penetrates the canopy completely (i.e., the lidar cannot see the ground for the trees), the height estimate for that cell is biased upward.

Another fact that should not be surprising is that lidar engineers and digital-signal-processing experts have come up with clever ways to suppress the undesirable characteristics of raw lidar data, and to extract remarkably accurate approximations to the topography of the bare Earth. Sub-decimeter accuracy with 1-m postings is within the capability of modern lidar systems, and lidar coverage is expanding rapidly. Among lidar's recent successes in solid Earth science is delineation of fault scarps beneath heavy vegetation in the active Seattle fault zone, which strikes through downtown Seattle in the densely populated Puget Lowland of western Washington State (Harding and Berghoff, 2000; Haugerud *et al.*, 2003).

Lidar, InSAR, and photogrammetry are complementary remote-sensing approaches to DEM generation that, when integrated, can yield a robust and widely useful geospatial data product. InSAR provides geometric fidelity, lidar supplies highly accurate 'point-level' data, and photogrammetry provides realistic terrain visualization (Althausen *et al.*, 2004).

5.2.7 Range-change resolution of InSAR

To better appreciate the potential of InSAR, let us quickly review some essential points from the preceding sections. Recall that the phase value for each pixel in a radar image is determined by many factors, including a phase shift caused by the radar beam's interaction with surface materials. The radar wavelength is small compared with the size of a resolution cell on the ground, which typically contains hundreds of individual reflectors with differing complex reflection coefficients, each of which causes a different phase rotation or delay. The complex returns from all of these reflectors are summed to give the phase and amplitude contribution of the cell as a whole. The instantaneous SAR footprint is large compared with the cell size, so many cells contribute to each pulse echo. The echoes from many cells arrive simultaneously at the SAR, which samples the echo signal between successive chirps and passes the time-and-frequency information for each sample to a SAR processor for deciphering. The processor unscrambles the raw signal and assembles a focused SLC image with pixels that correspond to specific resolution cells on the ground. Because the phase value for each pixel depends on a unique montage of reflectors in the corresponding resolution cell, the phase image is random.

Here is where things seemingly get 'magical'. When two such random images of the same scene are co-registered and differenced, the randomness disappears (assuming phase coherence has been maintained) and distinctive patterns emerge with information about viewing geometry, surface topography, and ground deformation. When the effects of viewing geometry and topography are removed, the remaining pattern of phase differences in some cases reveals subtle ground deformation and thus the locations and shapes of the causative sources buried kilometers beneath the surface. It is not really magic, but the analogy to pulling a rabbit out of a hat seems apt. Let us analyze this trick more carefully before examining some snapshots of deforming volcanoes.

We have seen that our geodetic camera is capable of imaging topography with ~10 m resolution, but how sharp is it when the subject matter is much more subtle (e.g., millimeter-scale surface displacements)? Consider what would happen if part of the target area moved toward or away from the SAR by exactly half of the radar wavelength, measured along the SAR-to-target (i.e., slant-range) direction, during the time between two image acquisitions.[18] The round-trip distance would have changed by one full wavelength, which means that signals reflected from the same resolution cell in the two images (after corrections for viewing geometry and topography) would be exactly in phase back at the SAR. The phase difference between corresponding pixels in the two images, and, therefore, the corresponding data value in the interferogram formed from the two images, would be zero (or, equivalently, 2π). The same would be true if the surface moved by $n\lambda/2$, where n is any integer. For cells that moved by an intermediate amount, $(n\lambda/2) \leq \Delta R \leq ((n+1)\lambda/2)$,

---

[18] The directionality of the SAR-to-target vector is commonly denoted by a unit vector (i.e., a vector of unit length pointing in the direction of the radar beam). The range vector is the product of this unit vector and the scalar range (i.e., the line-of-sight distance from SAR to target (Chapter 8)).

the data values in the interferogram would range from $0 < \Delta\phi < 2\pi$.

Now imagine that we use the color red to depict those pixels in the interferogram where $\Delta\phi = 0$ and violet for those areas where $\Delta\phi$ is slightly greater than zero. Let us also assign a continuous spectrum of colors from violet to red to pixels with intermediate phase differences (i.e., $0 \le \Delta\phi \le 2\pi$). Because the phase of a sinusoidal wave is modulo 2π (i.e., the waveform repeats every 2π or $360°$), we would not know a priori whether a given violet pixel had moved $\Delta R = 0, \lambda/2, 3\lambda/2$, etc. If we assumed that most of the scene did not move at all and counted the number of violet-to-red color cycles (i.e., fringes) that formed a closed pattern, we could deduce the total amount of movement in much the same way that we interpret the contours on a topographic map.

Imagine that a volcano subsided in a concentric pattern centered at the summit by a maximum amount corresponding to several radar wavelengths (Figure 5.15). The resulting interferogram would show a concentric, repeating pattern of colors from red-to-green-to-violet in the direction of the summit. Each complete color cycle would represent differential movement of the ground surface by $\lambda/2$, and we could count the number of cycles (i.e., fringes) to determine the total amount of movement. If, instead, the surface had moved *toward* the SAR (i.e., uplift), the color pattern would be the reverse, from violet-to-green-to-red toward the summit (Figure 5.16).

For a C-band SAR operating at a wavelength of 5.6 cm, each fringe in the interferogram corresponds to a range change of $\lambda/2 = 2.83$ cm. By interpolation, we could in theory measure surface displacements corresponding to a small fraction of a single fringe, perhaps as little as 1 mm. In practice, various sources of image noise, mostly phase delays caused by inhomogeneity in the troposphere and ionosphere, limit the range-change accuracy of C-band interferometry to 1–10 mm under favorable conditions. Combined with spatial coverage from orbiting SARs that is typically on the order of 100 km × 100 km for each scene, this means that our dream of a geodetic camera is not so far-fetched as it might have seemed at first thought.

A crucial assumption in the foregoing discussion is that nothing in the target area changed during the time between successive radar images other than the surface height. This is not necessarily true over time-scales of days to years, which are of interest for volcano monitoring. Over that period of time,

vegetation might grow or die, rainfall might change the soil moisture, snow might accumulate or melt, or the surface might be affected by erosion or deposition. If such non-deformation changes are large with respect to the radar wavelength, our attempt to co-register the images and form a useful interferogram might fail. In this case, the images are said to lack coherence or be temporally decorrelated, and our geodetic camera would be hopelessly out of focus.

5.2.8 Coping with decorrelation and atmospheric-delay anomalies

The main limitations of repeat-pass SAR interferometry (i.e., temporal decorrelation and artifacts caused by atmospheric-delay anomalies), are subjects of intensive study. By understanding these phenomena better, researchers hope to mitigate their effects and thus make InSAR more accurate and widely applicable. Not all of the factors that contribute to temporal decorrelation are well understood, but it's clear that most are inescapable (i.e., soils erode and deposit, trees move in the wind, ice forms and melts, etc.). The best means available to extend the period over which successive radar images of the same area maintain coherence are to: (1) work in lava or other rocky terrains, or in urban areas, where temporal decorrelation generally is less problematic (not always possible if your target is a specific volcano), or (2) use a longer wavelength SAR. The penetration depth of a radar pulse increases with wavelength, so a longer wavelength pulse interacts with scatterers that are buried more deeply and is therefore less susceptible to changes occurring at the surface. In vegetated areas, a long-wavelength pulse is more likely to reach the ground, which increases the likelihood of obtaining coherent images on successive passes. A short-wavelength pulse is more likely to interact with branches, leaves, grasses, etc., that are much more fickle as scatterers. As a result, a target area that is virtually inaccessible to C-band interferometry might yield good results at L-band. Alternatively, an area might stay coherent for a matter of days at X-band, months at C-band, and years at L-band. Whereas most InSAR studies of volcanoes to date have relied on C-band images from ERS-1 and ERS-2, the fact that so many of the world's volcanoes are vegetated means that L-band is a better choice for future volcano–InSAR missions. This is true in spite of the fact that L-band's longer wavelength produces correspondingly less range-change resolution, a

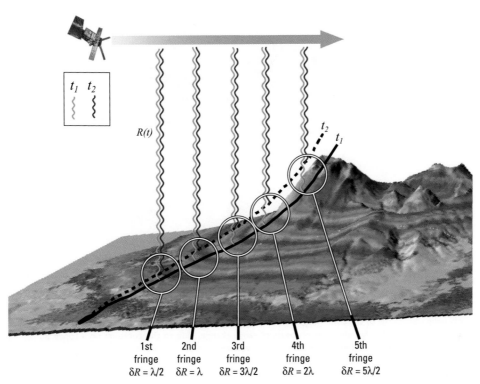

1st fringe $\delta R = \lambda/2$ | 2nd fringe $\delta R = \lambda$ | 3rd fringe $\delta R = 3\lambda/2$ | 4th fringe $\delta R = 2\lambda$ | 5th fringe $\delta R = 5\lambda/2$

Figure 5.15. An inflating volcano (or any other landform) produces a pattern of concentric fringes in a radar interferogram from which the effects of viewing geometry and topography have been removed (Massonnet, 1997) . If the volcano's height profile changes from the solid line to the dashed line between the acquisition times, t_1 and t_2, of two radar images, the range $R(t)$ from the SAR to the surface will decrease by half the radar wavelength ($\delta R = \lambda/2$) in some areas, by $\delta R = \lambda$ in other areas, $\delta R = 3\lambda/2$ in others, etc. When the images are differenced, each half-wavelength change produces one complete interference fringe, which is portrayed as a spectrum of colors from red to violet (or violet to red, the choice is arbitrary) in the interferogram. The result is a concentric pattern of fringes around the volcano, akin to contours of range change with a contour interval of $\lambda/2$. Point-source subsidence would produce a similar pattern of fringes but with the opposite color sequence (i.e., from violet to red in the direction of increasing subsidence) (Figure 5.16). The satellite passes at times t_1 and t_2 need not be sequential; they could be separated by up to a few years if the surface retains interferometric coherence.

disadvantage that is outweighed by the prospect of better coherence in vegetated areas.

Even when temporal decorrelation is not a problem, atmospheric-delay anomalies can cause artifacts in radar interferograms that complicate their interpretation. Delay anomalies originate in the troposphere or ionosphere and are caused by inhomogeneities in water content, temperature, pressure, or electron density. They cannot be avoided entirely, but often their effects can be recognized and dealt with. For example, Massonnet and Feigl (1998) noted that interferograms made from night-time scenes show fewer atmospheric artifacts than those from daytime scenes, presumably because the atmosphere generally is more stable at night. If they are available, night-time images are therefore preferable.

How can path-delay anomalies be recognized and distinguished from surface deformation if both phenomena produce fringes in coherent interfero-

grams? Fortunately, this is relatively straightforward if more than one interferogram is available for the same target area and approximately the same time period. By comparing multiple interferograms of the same scene, several types of potential artifacts can be recognized and discounted. For example, imagine that we form 3 coherent interferograms, AB, BC, and AC, from radar images A, B, and C. For simplicity, let us assume that the data-acquisition geometry is such that each of the interferograms has the same low sensitivity to any errors in the DEM (i.e., small b, large h_a). Assume also that image B was acquired during a period of heavy thunderstorm activity in the target area. The concentration of raindrops in a storm cell can interact with the radar beam to alter the phase of the echo signal and thus produce a path-delay anomaly, which appears as a closed pattern of fringes in interferograms AB and BC. The anomalies will have similar shapes and sizes but opposite signs,

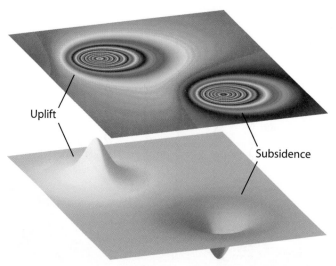

Figure 5.16. In a topography-removed interferogram, each fringe can be represented by a color band that spans the spectrum from red to violet, or *vice versa*. Each fringe corresponds to a change in the SAR-to-surface range by $\lambda/2$, where λ is the wavelength of the radar. For C-band radars like those aboard ERS-1, ERS-2, Radarsat, and Envisat, each fringe corresponds to 2.8 cm of range change. In this case, decreasing range (i.e., mostly uplift) is denoted by the color progression red--yellow--green--blue--violet; the opposite progression denotes increasing range (subsidence). Fringes can be thought of as contours of range change, akin to the contours on a topographic map. The fringe patterns corresponding to an axisymmetric source (e.g., a sphere or vertical pipe, see Chapter 8) are not circular because the radar is side-looking and therefore sensitive to both vertical and horizontal surface displacements. The fringes are more closely spaced where the vertical and horizontal components of motion are additive in the range direction (i.e., where the displacements are upward and toward the radar or downward and away). Accordingly, the patterns shown here are appropriate for a right looking SAR. Graphic provided by Charles W. Wicks, Jr., USGS.

which suggests they originated in the common image B. Interferogram AC will be free of the anomaly, because the storm effects only appear in image B. How can we be sure that the anomalies in AB and BC were caused by the atmosphere and not by ground deformation?

If we are fortunate enough to have another image of the study area, D, acquired at approximately the same time as B but not during the period of storm activity, we can form interferograms AD, BD, and DC. The storm-induced anomaly will appear only in BD, tying its origin more closely to image B. If the anomaly appears exclusively in interferograms formed from image B, we have good reason to discount it as a tropospheric propagation-delay effect. This strategy can fail if delay anomalies are persistent or recurring, as might be the case near volcanoes with large topographic relief. Tall volcanoes can influence local weather and set up quasi-steady patterns of water vapor concentration both horizontally and vertically (e.g., windward clouds and leeward clear skies, or summit cap clouds and sunny flanks). Such quasi-steady tropospheric patterns might be especially troublesome at large volcanoes such as Mount Etna, Italy (Beauducel *et al.*, 2000) and Mauna Loa, Hawai'i.

Emardson *et al.* (2003) pointed out that propagation delays arising in the ionosphere and troposphere might have different small-scale variability and decorrelation times, and that the effect of the ionosphere could be mitigated in the future by deploying a dual-frequency or split-band radar system. For the neutral atmosphere (troposphere), they suggested stacking (i.e., averaging) successive interferograms as a reasonable approach to increasing the signal-to-noise ratio. Stacking takes advantage of the fact that any geodetic signal is likely to persist through several interferograms, whereas tropospheric noise is more nearly random over time-scales longer than a day (with some exceptions as noted above). Stacking interferograms tends to enhance the geodetic signal and average out the noise.

The standard deviation of the neutral-atmosphere-induced noise σ is approximated by (Emardson *et al.*, 2003):

$$\sigma = cL^{\alpha} + kH \qquad (5.17)$$

where L and H are the differential length and height in kilometers and, for southern California, typical values of c, α, and k (determined empirically) are 2.5, 0.5, and 4.8, respectively. Using Mount Shasta as an example, $L \sim 20$ km and $H \sim 3$ km, so $\sigma \sim 26$ mm or ~ 1 fringe at C-band. Emardson *et al.* (2003) contend that the value of α is largely site-independent, but the value of c depends on the variability of the atmospheric water vapor content at the site of interest. Experience with Aleutian volcanoes suggests that σ might be as large as 2–3 fringes in some cases (Lu *et al.*, 2000c). The greater the atmospheric noise level, the greater the need for multiple interferograms. Emardson *et al.* (2003) estimated that resolving a deformation rate of 1 mm yr^{-1} over a distance of 10 km with radar images acquired every 7 days would take 2.2 years, and that resolving the same rate over 100 km would take 4.8 years. Thus, a small constellation of radar satellites dedicated to interferometry could track deformation from magmatic and tectonic sources worldwide with a precision rivaling that of other geodetic techniques.

Now consider the case of anomalies that appear in interferograms *AB*, *BC*, and *AC* at the same location but with differing amplitudes. No single image could be responsible for all three anomalies, but they all could be topographic artifacts rather than deformation signals. This is especially true if the altitudes of ambiguity are small with respect to the accuracy of the DEM. For example, if *AB* has an altitude of ambiguity twice as large as *BC* and the anomaly in *BC* has twice as many fringes as the one in *AB*, there is good reason to be suspicious. In this case, we could try to find a fourth image to form other interferograms with greater altitudes of ambiguity.

Failing this, we could employ another trick of the InSAR trade. By adding (or subtracting) co-registered interferograms with altitudes of ambiguity of opposite (or same) sign, we can synthesize an interferogram from two ghost images that were never acquired! This is a useful technique to increase the effective altitude of ambiguity and thus to produce an interferogram with less sensitivity to topography. The same technique can be used to decrease the effective h_a and thus to increase the topographic sensitivity. This can be helpful if the desired product is a DEM rather than a topography removed interferogram (Massonnet and Feigl, 1998).

In summary, three factors that can limit InSAR's usefulness for geodetic studies are temporal decorrelation, path-delay anomalies, and topographic artifacts caused by DEM errors. Decorrelation is less of a problem in lava or other rocky terrains, and in urban areas, owing to relatively sparse vegetation and the presence of permanent scatterers such as blocky lava, boulders, pavement, or buildings. Elsewhere, including at sparsely vegetated volcanoes where deformation rates are high, useful interferograms sometimes can be produced from images separated by a relatively short time interval (e.g., the 35-day repeat cycle for ERS-1 and ERS-2). Another approach is to select images acquired during the same season of successive years, when decorrelation caused by snow or rainfall is likely to be at a minimum (Lu and Freymueller, 1998). At several Aleutian volcanoes, C-band coherence has been observed to persist for 3–5 summers in some areas (Lu *et al.*, 2002b,c). Future L-band SAR missions could greatly extend InSAR's applicability by reducing the problem posed by temporal decorrelation at shorter wavelengths. Artifacts caused by path-delay anomalies or DEM errors can be identified in a relatively straightforward way if several interferograms are available for the same area. InSAR is fast becoming an operational tool suitable for use at many of the world's volcanoes, and its utility will surely grow as next-generation InSAR missions are realized.

5.2.9 Volcano InSAR studies: a growing list of success stories

Fortunately, experience has shown that many real-world surfaces stay correlated for periods of months to a few years, long enough for most volcano-monitoring purposes, even in areas of sparse-to-moderate vegetation. For example, coherent interferograms over periods of 1 to 4 years have been produced for Mount Etna, Italy (Massonnet *et al.*, 1995; Lanari and others *et al.*, 1998); Long Valley Caldera, California (Thatcher and Massonnet, 1997); Yellowstone Caldera, Wyoming (Wicks *et al.*, 1998), Okmok (Lu *et al.*, 1998), Akutan (Lu *et al.*, 2000a, 2005), Westdahl (Lu *et al.*, 2000b, 2004), Makushin (Lu *et al.*, 2002c), and Kiska (Lu *et al.*, 2002d) Volcanoes, Alaska; Piton de la Fournasie, La Réunion, Indian Ocean (Sigmundsson *et al.*, 1999); several volcanoes in the Galápagos Islands (Jonsson *et al.*, 1999; Amelung *et al.*, 2000) and in Chile (Pritchard and Simons, 2002); and for a growing list of other volcanoes worldwide (e.g., Zebker *et al.*, 2000). These volcanoes span a wide range of climatic conditions and vegetative cover, suggesting that many more of the world's volcanoes are amenable to study using InSAR, especially at L-band wavelengths.

However, there have been unsuccessful attempts as well, especially at volcanoes with dense vegetation, rugged topography, or permanent snow-and-ice cover (Zebker *et al.*, 2000). The volcanoes of the Cascade Range in the western USA present difficult challenges in this regard, as do many volcanoes in tropical or subtropical regions. Even at the volcanoes where InSAR has been successful, it generally has been necessary to use only images acquired in the same season to avoid temporal decorrelation caused by changes in the land surface or snow cover. Together with the 35-day orbital repeat cycle of the ERS-1 and ERS-2 satellites, this essentially limits the useful repeat cycle for InSAR monitoring at most volcanoes to periods of either 1–3 months during the summer or to one or more years (e.g., successive summers). Nonetheless, there have been some striking successes that bode well for the future of InSAR studies at volcanoes. A few of these are discussed below. Several others are described in excellent reviews of the subject by Massonnet and Feigl (1998) and Zebker *et al.* (2000).

5.3   EXAMPLES OF INTERFEROMETRIC SAR APPLIED TO VOLCANOES

5.3.1   Mount Etna

Spurred by the remarkable success of InSAR in capturing coseismic displacements from the 1992 Landers earthquake (Massonnet *et al.*, 1993), InSAR specialists soon turned their attention to another class of promising targets – volcanoes. The first positive results were for Mount Etna during the course of an effusive eruption from 14 December 1991 to 31 March 1993. Massonnet *et al.* (1995) produced a series of interferograms from ERS-1 images that revealed progressive subsidence of the edifice from 17 May 1992 to 24 October 1993. A series of best-fit models based on sequential interferograms showed that the volume of subsidence increased linearly with time until the end of the eruption, when the subsidence also stopped. The overall best fit was obtained from a deflation source located 2 ± 0.5 km east of the summit at 16 km depth.

The InSAR results for Mount Etna are supported by an abundance of observational data. During the latter part of the 1991–1993 eruption, ERS-1 acquired 13 images of Mount Etna during ascending orbits (i.e., traveling north) and 16 images during descending orbits. Only images from the same family of orbits can be combined to form interferograms. The ascending images give 78 potential interfero-

metric combinations, and the descending images give 120 independent combinations. Of these, Massonnet *et al.* (1995) selected 32 ascending and 60 descending combinations with favorable data-acquisition geometries to minimize any residual topographic errors resulting from the DEM. About 30 of the 92 favorable combinations were coherent enough to be analyzed and only 12 were used for modeling. An important lesson here is that no single interferogram is likely to be definitive. It should be standard procedure to analyze as many potential image pairs as possible to overcome problems caused by temporal decorrelation, path-delay anomalies, or DEM errors.

Following the initial InSAR study of Mount Etna, Lanari *et al.* (1998) produced additional interferograms of the volcano for several intervals between September 1992 and October 1996. The 1992–1993 interferograms revealed a subsidence source similar to that inferred from the earlier study but at shallower depth (9 km versus 16 km). Following the end of an effusive eruption in March 1993, a period of uplift ensued until the start of another eruption in July 1995. The uplift source was deeper (>11 km), which is consistent with the idea that deflation at shallow levels is followed by inflation at deeper levels as the volcano recharges with magma from below. Sequential interferograms showed steady inflation following the end of the 1991–1993 eruption, which accelerated in the months preceding the next outbreak in July 1995. This was the first InSAR study to detect a reversal from deflation to inflation and the first to track pre-eruptive inflation.

The foregoing interpretations were questioned by Beauducel *et al.* (2000), who argued that tropospheric effects could account for most of the reported ground deformation during 1992–1998. They estimated tropospheric delays in 238 interferograms of -2.7 to $+3.0$ (± 1.2) fringes, and suggested that any volume changes in the Mount Etna magma reservoir were within the observational uncertainty. The volcano's imposing size (3,300 m) undoubtedly is a factor, because the edifice extends through a large range in atmospheric conditions and influences local weather. It is not uncommon for delay anomalies to produce 2–3 fringes in interferograms of Aleutian volcanoes (Lu *et al.*, 2000c), which are subject to notoriously poor weather. Better understanding of path delays would improve InSAR's reliability for volcano monitoring, especially at tall volcanoes capable of influencing local weather patterns.

5.3.2 Long Valley Caldera, California

Several striking successes followed on the heels of the initial InSAR studies of Mount Etna. At the Long Valley Caldera in eastern California, InSAR was used to map uplift of the resurgent dome and surrounding caldera floor in more detail than had been possible with repeated geodetic surveys (Chapter 7). A multifaceted episode of tectonic and magmatic unrest at Long Valley began in October 1978 with the occurrence of a M 5.7 earthquake about 15 km southeast of the caldera rim. This was followed on 25–27 May 1980, by a swarm-like sequence of four M 6 and thousands of smaller earthquakes near the south caldera rim, and by the onset of inflation of the resurgent dome (Hill *et al.*, 1985a,b). Dozens of small earthquake swarms occurred during the ensuing two decades, including especially strong swarms beneath the south moat in January 1983 (Savage and Cockerham, 1984) and December 1997–January 1998 that may have been accompanied by intrusions. In 1989, a swarm of small earthquakes beneath Mammoth Mountain that included dozens of long-period events probably marked another intrusion (Hill *et al.*, 1990). The swarm was preceded by more than 2 months of increased extensional strain within the caldera (Langbein *et al.*, 1993) and was followed several months later by the start of tree-kill caused by increased emission of carbon dioxide (Farrar *et al.*, 1995).

Repeated leveling and two-color EDM surveys at Long Valley revealed persistent uplift and extension across the caldera since 1980 at rates that varied with time (Chapter 7). Maximum uplift of the resurgent dome from 1975 to 1997 was more than 70 cm, and from mid-1983 to mid-1998 a two-color EDM line across the resurgent dome extended by nearly 40 cm at rates that sometimes exceeded $10 \, \text{cm} \, \text{yr}^{-1}$. Most models of the 1980–1998 deformation have in common a spherical or ellipsoidal source 5–8 km beneath the resurgent dome and a fault or dike beneath the south moat. Some studies also include a second, deeper ellipsoidal source beneath the south moat or a dike beneath Mammoth Mountain (Langbein *et al.*, 1995, 1993, 1987a,b; Langbein, 1989; Savage, 1988; Savage and Cockerham, 1984; Rundle and Whitcomb, 1986).

Thatcher and Massonnet (1997) were first to study the Long Valley region with InSAR. They constructed several interferograms from ERS-1 and ERS-2 images, and obtained coherent results with image pairs spanning 2-year and 4-year intervals. The interferograms revealed a pattern of surface deformation centered in the caldera and generally consistent with that deduced from repeated leveling and two-color EDM surveys. Synthetic interferograms based on the ellipsoidal-inclusion models of Langbein *et al.* (1995), which were developed from the leveling and EDM results, showed good agreement with the observed interferograms. This demonstrated that InSAR could produce essentially the same result as field-intensive geodetic surveys, with much better spatial resolution (albeit with less time resolution in this case).

Simons *et al.* (1999) combined sequential InSAR, continuous GPS, and two-color EDM measurements at Long Valley to constrain time-dependent parameters for various deformation sources needed to explain the observed deformation pattern. The InSAR dataset included ERS-1 and ERS-2 images acquired at sub-yearly intervals from 1992 to 1999. Topographic effects were removed from the interferograms using a DEM with 5 m cell size derived from the NASA/JPL airborne SAR instrument (TopSAR). Inversion[19] of the InSAR data suggested a spherical or ellipsoidal source 9 ± 1 km beneath the resurgent dome, deeper than the best-fit model of Langbein *et al.* (1995) based on inversion of two-color EDM and leveling data (5.5 to 7 km depth). The discrepancy might reflect the much greater spatial aperture of the interferograms, which capture far-field deformation from a deeper source more effectively than the less extensive EDM and leveling networks.

Fialko *et al.* (2001b) came to a similar conclusion for the 1997–1998 inflation episode at Long Valley. Their modeling of ERS-1 and ERS-2 interferograms, which collectively span the time interval from 6 June 1996 to 12 July 1998, showed that the observed uplift pattern (~11 cm maximum) could be explained by an inflating sill at ~12 km depth or an elongate prolate spheroid (i.e., a 'pluton-like body') at ~8 km depth beneath the resurgent dome. Joint inversion of the InSAR and two-color EDM results suggested that the inferred magma body is a steeply dipping prolate spheroid with a depth of 7–9 km beneath the dome and an aspect ratio greater than 2 : 1. Although the exact nature of

---

[19] Inversion is a mathematical procedure for determining best-fit model parameters from a set of observations, starting with a generalized forward model and known or assumed boundary conditions. For related information, see *inversion (numerical modeling)* in the Glossary (at the back of this book), and Chapter 8.

the deformation source(s) might be obscured by such complications as crustal heterogeneity and deviations from linear elasticity, the InSAR results at Long Valley have helped to constrain the range of viable models and thus have contributed to a better understanding of unrest at one of the most intensively monitored calderas on Earth.

5.3.3 Yellowstone Caldera, Wyoming

Encouraged by the success of InSAR at Long Valley, two colleagues and I applied the same approach to the Yellowstone Caldera, where leveling surveys had shown both uplift and subsidence for several decades. There have been at least seven M 6 earthquakes and one M 7 earthquake in the Yellowstone region during historical time, and swarms of smaller earthquakes are commonplace. The vigor of Yellowstone's hydrothermal system, epitomized by the regular spouting of Old Faithful geyser, is well known throughout the world. Less obvious to the casual visitor, but equally fascinating, are the ups and downs of the caldera floor as revealed by repeated leveling and GPS surveys (Chapter 7). Comparison of leveling surveys in 1923 and 1975–1977 first detected uplift within the caldera that continued through 1984 at an average rate of 1–2 cm yr^{-1}. When the uplift paused during 1984–1985, parts of the caldera floor had risen more than 90 cm since 1923 (Dzurisin and Yamashita, 1987). Annual leveling surveys thereafter showed subsidence at rates of 1–3 cm yr^{-1} through 1995, raising the question of when uplift would resume (Dzurisin et al., 1990, 1994). InSAR provided the answer, and some surprises too.

As expected from the leveling results, interferograms formed from ERS-1 and ERS-2 images show a pattern of subsidence within the caldera during 1992–1993 and 1993–1995 (Wicks et al., 1998) (Figures 5.17 and 5.18). From August 1992 to June 1993, more than 3 cm of subsidence accumulated in the northeast part of the caldera near the Sour Creek resurgent dome (Figure 5.18(A)). Subsidence continued from June 1993 to August 1995 (more than 4 cm), but surprisingly the center of deformation shifted to the Mallard Lake dome in the southwest part of the caldera (Figure 5.18(B)). Interferograms for the entire 1992–1995 interval show about 6 cm of subsidence centered between the two domes, consistent with leveling surveys in the northeast part of the caldera in 1992, 1993, and 1995 (Dzurisin et al., 1999). This was the first time that a shift in the deformation center had been

recognized at Yellowstone. But the surprises continued.

An interferogram for the period from August 1995 to September 1996 shows that the Sour Creek dome rose about 2 cm while the Mallard Lake dome was still subsiding slightly (Figure 5.18(C)). The renewal of uplift following a decade of subsidence was confirmed by an interferogram for the period from July 1995 to June 1997, which shows over 3 cm of uplift extending throughout the central part of the caldera (Figure 5.18(D)), and by leveling in 1998. Whereas leveling surveys spanning more than seven decades had failed to detect any movement of the deformation center at Yellowstone or a change from subsidence to uplift, InSAR accomplished both in very short order. Even before the implications of these discoveries had fully settled in, we were treated to yet another surprise.

Figure 5.19 shows Yellowstone interferograms for the periods from September 1996 to September 2000 and from August 2000 to September 2001. An elliptical uplift centered along the north caldera rim about 10 km south of Norris Geyser Basin is beautifully portrayed in the 1996–2000 image, which spans part of the July 1995 to June 1997 period of renewed uplift in the caldera (Figure 5.18(D)). The Norris uplift extends about 30 km east–northeast from Madison Junction to Canyon Junction, and 40 km north–northwest from the center of the caldera. It is tucked neatly between the two resurgent domes, which had been identified as separate deformation centers just a few years earlier. The 2000–2001 interferogram shows that the Norris uplift continued at an average rate of $\sim$3 cm yr^{-1} (i.e., $\sim$1 fringe yr^{-1}) while the central part of the caldera floor, including both resurgent domes, subsided at a comparable rate (Wicks et al., 2002c). More recent InSAR results indicate that the 2000–2001 pattern persisted through summer 2002 but had stopped by summer 2003.[20] In less than a decade, three distinct deformation sources have been active at Yellowstone, and two of the three have produced both

---

[20] The year-to-year range changes are small enough to have been caused by tropospheric-path delays, as discussed by Beauducel et al. (2000) for Mount Etna, but three lines of evidence point instead to ground deformation as the cause: (1) the fringes around Norris Geyser Basin appear in several independent interferograms spanning similar time periods, (2) the Norris area is relatively flat so there is no reason to suspect a strong topographic influence on tropospheric water vapor concentration, and (3) the uplift inferred from InSAR was corroborated by comparison of leveling results from 1987 and 2004 (CVO, unpublished data).

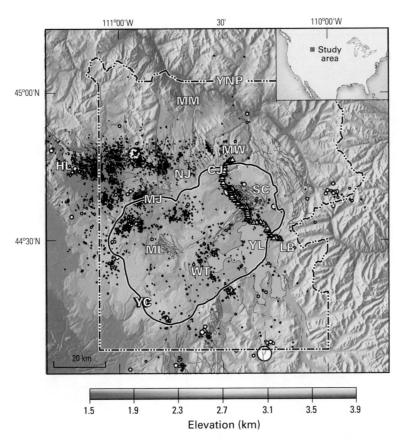

Figure 5.17. Topography of the Yellowstone region derived from USGS DEMs, after Wicks et al. (1998). YNP (dash-dot line), Yellowstone National Park boundary; YC (solid black line), Yellowstone Caldera boundary; MM, Mammoth; HL, Hebgen Lake; MW, Mount Washburn; CJ, Canyon Junction; MJ, Madison Junction; NJ, Norris Junction; SC, Sour Creek resurgent dome; ML, Mallard Lake resurgent dome; WT, West Thumb; LB, Lake Butte; YL, Yellowstone Lake. String of white-filled black triangles running northwest from LB to MW represents bench marks along a level line that was measured annually from 1983 to 1998, except 1994, 1996, and 1997 (Dzurisin et al., 1994). Red lines show faults mapped (solid) and inferred (dashed) by Christiansen (1984). Earthquake epicenters from the University of Utah Seismographic Stations' Yellowstone National Park Earthquake Catalogs for 1983--1995 are shown as white-filled black circles and black dots, with symbol size proportional to earthquake magnitude. The three largest earthquakes shown are M 5.1 (28 August 1995) on the southern boundary of YNP, M 4.9 (26 March 1994) ~20 km east of HL, and M 4.8 (24 September 1994) on the west edge of SC along the level line; the smallest earthquakes shown are M ~0.0.

uplift and subsidence. Thanks to InSAR, we now know that the dynamics of unrest are far more complicated than we anticipated!

No comprehensive model has been proposed to explain all of the InSAR results at Yellowstone, although the essential ingredients have been identified from earlier work. There is abundant geophysical evidence for the existence of a rhyolitic magma body in the crust beneath Yellowstone that is sustained by basaltic intrusions from the mantle (Smith and Braile, 1994; Smith and Siegel, 2000; Christiansen, 2001). The existence of a vigorous hydrothermal system is evident from the countless thermal features that grace Yellowstone National Park. It, too, requires basaltic heat input to have persisted since Yellowstone's most recent eruption ~70,000 years ago (Fournier and

Pitt, 1985; Dzurisin et al., 1990). The hydrothermal system consists of a deep zone in which pore-fluid pressure is near lithostatic and a shallow zone in which pore pressure is hydrostatic (Fournier and Pitt, 1985; Fournier, 1991, 1999).[21] The two zones are separated by an impermeable, self-sealed layer created by mineral deposition and quasi-plastic flow at a depth of about 5 km. Pore-fluid pressure in the shallow zone is always less than lithostatic because a

---

[21] Lithostatic pressure refers to the vertical pressure at a point in the Earth's crust caused by the weight of the overlying column of rock or soil. Pore fluids in the lithostatically pressured zone are isolated from the surface by impermeable rock or a self-sealed zone. Hydrostatic pressure is due to the weight of groundwater at higher levels in the system. In highly fractured or porous rocks where groundwater flows freely, the total pressure approximates the hydrostatic pressure in this upper zone.

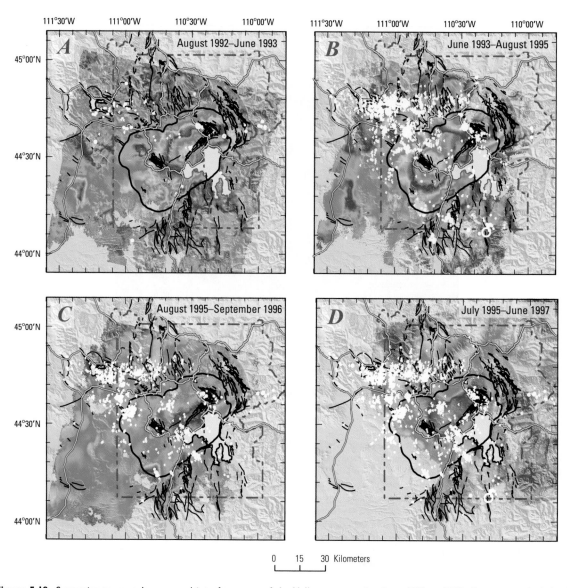

Figure 5.18. Successive topography-removed interferograms of the Yellowstone region from 1992 to 1997, shown over shaded relief (Wicks *et al.*, 1998; Dzurisin *et al.*, 1999; Wicks *et al.*, 2002c). Range of colors from violet to red corresponds to 2.8 cm (one C-band fringe) of range change (mostly uplift or subsidence). Dashed line, Yellowstone National Park boundary; heavy black line, Yellowstone Caldera; fine black lines, faults; yellow lines, roads; white circles, earthquake epicenters (sized by magnitude) for each time interval. (A) From August 1992 to June 1993, subsidence was centered in the northeast part of the Yellowstone Caldera near the Sour Creek resurgent dome (Figure 5.17). (B) The subsidence center shifted to the southwest part of the caldera near the Mallard Lake resurgent dome from June 1993 to August 1995. (C) From August 1995 to September 1996, a small amount of subsidence persisted near the Mallard Lake dome, but the Sour Creek dome began to rise slightly. (D) Uplift spread throughout the central part of the caldera from July 1995 to June 1997. Note that the color progression inside the caldera is reversed between D (uplift) and either A or B (subsidence).

complex system of microfractures and other fluid passageways connects it to the surface. For this reason, pressure changes in the hydrostatically pressured zone cannot be the cause of widespread uplift or subsidence. Beneath the self-sealed layer the deep hydrothermal system is near lithostatic pressure, so any change in pore-fluid pressure has the potential to raise or lower the ground surface.

Based on the interferograms shown in Figure 5.18, Wicks *et al.* (1998) proposed that uplift and subsidence inside the caldera are caused by pressure changes in two interacting fluid reservoirs, one under each of the resurgent domes at 8 ± 4 km depth. Inflation and deflation are regulated by flow through conduits that connect the reservoirs to each other and to a deeper magmatic pressure

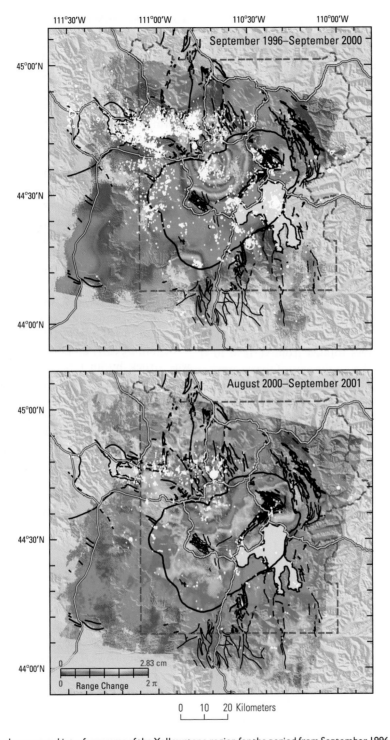

Figure 5.19. Topography-removed interferograms of the Yellowstone region for the period from September 1996 to September 2000 (*top*) and from August 2000 to September 2001 (*bottom*), showing a large uplifted area centered along the north caldera rim ~10° south of Norris Geyser Basin (NJ, Figure 5.17) and, for the later period, subsidence of the central part of the Yellowstone Caldera. White circles and dots represent earthquakes that occurred during the periods spanned by the interferograms, from the University of Utah Seismographic Stations' Yellowstone National Park Earthquake Catalogs. The caldera rim and Yellowstone National Park boundary are shown with heavy solid and dashed black lines, respectively. Mapped and inferred faults are shown with thinner black lines. Yellow lines represent roads. Uplift in the Norris area began in 1997 or 1998 and continued at an average rate of ~3 cm yr^{-1} (~1 fringe yr^{-1}) at least through 2001. Starting in 2000, the uplift was accompanied by subsidence of the central part of the caldera, including the Mallard Lake and Sour Creek resurgent domes (Wicks *et al.*, 2002c).

source. According to their model, uplift occurs when part of the deep hydrothermal system becomes over-pressured by magmatic volatiles. Subsidence ensues when the self-sealed layer eventually ruptures, releasing volatiles and depressurizing the deep hydrothermal system. Self-sealing and re-pressurization allow the process to repeat indefinitely.

Waite and Smith (2002) proposed a similar mechanism to explain the 1985–1986 earthquake swarm near Madison Junction, which was associated with the onset of subsidence inside the caldera. The activity was noteworthy for several reasons: (1) it was the most energetic earthquake swarm ever recorded at Yellowstone;[22] (2) seismic activity migrated at an average rate of 113 m day^{-1} to the northwest, away from the caldera rim, during the first month of the swarm, and thereafter it migrated gradually from depths of 2–5 km to more than 10 km; (3) the dominant focal mechanisms for the early part of the swarm were oblique-normal strike-slip instead of more typical normal-faulting mechanisms; (4) the maximum principal stress rotated from vertical to horizontal and subparallel to the axis of the swarm; and (5) the swarm was accompanied by increased geyser activity, changes in the clarity and temperature of hot springs, formation of new fumaroles and mud caldrons, and two small steam explosions (Dzurisin *et al.*, 1994). Waite and Smith (2002) attributed these characteristics of the swarm to lateral intrusion of magmatic or hydrothermal fluid from beneath the caldera floor into the Hebgen Lake fault zone.

The 1997–2002 uplift near Norris (12.5 cm maximum) seems to bleed westward toward the Hebgen Lake fault zone along a corridor marked by repeated earthquake swarms (Figure 5.19), which is consistent with the idea that the fault zone is connected in some way to a fluid reservoir beneath the caldera. Whether the fluid is hydrothermal or magmatic remains a mystery. The base of the hydrothermal system is thought to lie at a depth of ~5 km (Fournier, 1991, 1999), whereas the deformation source depths based on interferometry are 4–12 km beneath the resurgent domes (Wicks *et al.*, 1998) and 11–16 km beneath Norris (Wicks *et al.*, 2002c). At least in the latter case, the involvement of Yellowstone's magmatic system seems inescapable.

5.3.4   Akutan Volcano, Alaska

Earthquake swarms are a common occurrence near volcanoes, but often whether the primary cause of the seismicity is magmatic or tectonic cannot be determined. InSAR can help to resolve this ambiguity, as illustrated by the March 1996 earthquake swarm beneath Akutan Island, Alaska. Akutan Volcano, which is situated on the western half of the island, is one of the most active volcanoes in the Aleutian arc with at least 27 separate eruptive episodes since 1790 (Miller *et al.*, 1998). An eruption in 1948 produced measurable amounts of ash in Akutan village, 13 km east of the volcano's summit. In 1978, airline pilots observed large incandescent bombs, some of them 'as big as a car', ejected more than 100 m above the summit and lava flowed down the volcano's north flank. The most recent eruptive activity was a series of small (VEI ~1)[23] steam and ash emissions during March through May 1992 and possibly into 1993.

Beginning early on 11 March 1996, Akutan Island was struck by an intense earthquake swarm that lasted about 11 hours. The swarm consisted of more than 80 earthquakes greater than M 3.5; the largest was M 5.1. A second, more vigorous swarm on 14 March lasted about 19 hours and included more than 120 earthquakes greater than M 3.5. Thousands of smaller earthquakes also occurred during these periods of strong activity. During peak activity, residents of Akutan Island reported feeling continuous low-level ground shaking punctuated by individual shocks about once per minute! The shaking caused considerable anxiety for the people of Akutan, many of whom left the island voluntarily in fear of an eruption. However, no eruption ensued and the level of earthquake activity gradually returned to normal during the next several months.

In July 1996, scientists from the Alaska Volcano Observatory (AVO) discovered fresh ground cracks that extended discontinuously across Akutan Island from near Lava Point on the northwest, past the summit of Akutan Volcano, nearly to the southeast end of the island (Figure 5.20). The cracks broke snowfields and their sides

---

[22] Excluding the aftershock sequences of the 1959 M_s 7.5 Hebgen Lake earthquake and the 1975 M 6.1 Yellowstone Park earthquake.

[23] Volcanic Explosivity Index (VEI): a measure of the size of volcanic eruptions akin to the Richter magnitude scale for earthquakes. The VEI is a 0-to-8 index of increasing explosivity, each interval representing an increase of about a factor of ten. It combines total volume of explosive products, eruptive cloud height, descriptive terms, and other measures (Newhall and Self, 1982).

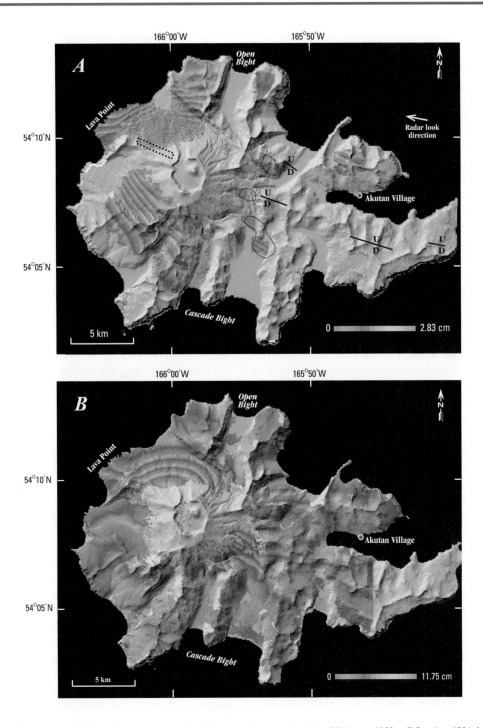

Figure 5.20. (A) Topography-removed interferogram of Akutan Island for the period from 20 August 1993 to 7 October 1996, from C-band images acquired by the ERS-1 satellite (Lu *et al.*, 2000a). The breached summit caldera and active cinder cone of Akutan Volcano are visible slightly left of center in the shaded relief image. Dotted rectangle in upper left surrounds a 250-m wide zone of ground cracks that formed on the volcano's northwest flank during an intense earthquake swarm in March 1996. Four lines on the eastern part of the island represent normal faults that reactivated during the swarm, with the upthrown (U) and downthrown (D) sides, respectively. There are more than 21 fringes on the northwest flank of the volcano and more than 8 fringes on the southwest flank, representing ~60 cm and ~23 cm of uplift relative to the shoreline, respectively. Coherence on the eastern two-thirds of the island is patchy, but fringes near Akutan Village and within three areas enclosed by dotted lines indicate downwarping along an axis roughly coincident with the ground cracks and reactivated faults. (B) Coherence is much better in this interferogram produced from L-band images acquired by the JERS-1 satellite on 28 October 1994 and 22 June 1997 (Lu *et al.*, 2005). In this case, lower resolution in range of L-band compared to C-band (11.76 cm per fringe versus 2.83 cm per fringe, respectively) is more than offset by improved coherence at L-band.

showed little evidence of erosion, suggesting they had formed recently – almost surely during the March 1996 swarm. The most extensive cracks occurred in a zone roughly 250 m wide and 3 km long between Lava Point and the summit. Local graben structures with net vertical displacements of 30–80 cm suggested that the cracks formed in response to tumescence of the volcano's northwest flank. The cracks on the east side of the island were only a few centimeters wide and apparently formed by activation of Holocene normal faults (Richter *et al.*, 1998). Was the March 1996 activity a tectonic earthquake swarm, a failed eruption, or both?

To address this question, Lu *et al.* (2000a) produced several interferograms of Akutan from ERS-1 and ERS-2 images using the two-pass technique and a United States Geological Survey (USGS) DEM. Figure 5.20 shows one of these for the period from 20 August 1993 to 7 October 1996, which brackets the March 1996 earthquake swarm. There are more than 21 fringes on the northwest flank of Akutan Volcano, representing ~60 cm of uplift of the upper part of the volcano relative to the northern shoreline. Similarly, there are more than 8 fringes representing ~23 cm of uplift on the volcano's southwest flank. This result was verified by forming an independent interferogram with images acquired on 4 June 1995 and 9 July 1997. The second interferogram showed essentially the same fringe pattern as the first, so any contribution from path-delay anomalies was negligible. Instead, the fringes map out ground deformation that must have occurred during the period of overlap in the interferograms (i.e., between June 1995 and October 1996), almost surely during the March 1996 earthquake swarm. Coherence on the eastern part of the island is limited to a few small patches, but the fringes in those areas are sufficient to reveal a consistent pattern of deformation.

Several features of the Akutan interferograms are noteworthy. First, the sense of motion was different on the western and eastern parts of the island. Whereas the western part of the island including Akutan Volcano moved upward, the eastern part (i.e., east of a line from Open Bight on the north coast to Cascade Bight on the south coast) moved downward. Second, the fringe pattern is asymmetric on the western flank of the volcano. The fringes are denser (i.e., deformation is more concentrated) north of the ground cracks than they are to the south. The uplift axis aligns with the zone of fresh ground cracks observed on the volcano's northwest flank. Third, fringes in several patches

of coherence on the eastern part of the island indicate subsidence along an axis roughly parallel to young normal faults in the area, some of which were reactivated during the swarm. The fringe pattern in this area suggests broad downwarping along a northwest–southeast axis rather than slippage along normal faults, most of which step down to the south.

A best-fit model of the Akutan interferograms includes at least two deformation sources: (1) an east–west-trending, north-dipping dike that intruded to within 500 m of the surface beneath the northwest flank of the volcano to account for concentrated, asymmetric deformation there; and (2) a Mogi-type inflation source ~13 km beneath the northwest flank to account for broad uplift of the volcanic edifice (Lu *et al.*, 2000a). The addition of one or more contracting sources to account for downwarping does not significantly improve the overall fit, although some type of subsidence mechanism clearly is required. Subsidence might have occurred when magma moved out of deep storage within a fault system beneath the island to intrude the northwest flank of the volcano, or it might reflect some other change in the fault system before, during, or after the intrusion (e.g., stress relaxation or pore fluid depressurization).

Lu *et al.* (2005) came to a similar conclusion based on comparison and modeling of 43 interferograms from C-band ERS images and 1 interferogram from L-band JERS images, which collectively span 1992–2002. Notably, the L-band interferogram has greater coherence than the others, especially in areas with loose surface material or thick vegetation. This bodes well for the future of L-band radar interferometry at volcanoes, starting with the successful launch on 24 January 2006 of the L-band Advanced Land Observing Satellite (ALOS) (Section 5.3.7).

For our purposes, the details of the Akutan model are less important than the first-order interpretation that a magma body beneath the volcano inflated and fed a dike that propagated to within 500 m of the surface during the March 1996 earthquake swarm. Tectonic influences probably played an important role as well, but the magmatic character of the event is clear from the InSAR results. Consider how difficult logistically it would have been to obtain a similar result from EDM, GPS, or leveling results. The number of bench marks required to capture details of the deformation pattern shown in Figure 5.20 would have been prohibitive, and access to the island is difficult even during brief Aleutian summers. Placed in a broader context,

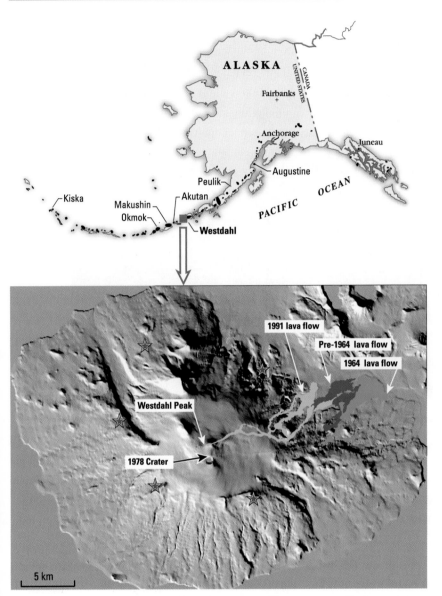

Figure 5.21. (*top*) Map of Alaska including the 2500-km-long Aleutian volcanic arc. Small red square and arrow indicate the area shown below. (*bottom*) Shaded-relief image of Westdahl Volcano. The 1978 crater, Westdahl Peak (1991--1992 vent), and three recent lava flows are labeled. Red stars represent seismic stations near the volcano. Figure slightly modified from Lu *et al.* (2003a).

the Akutan InSAR results suggest that intrusions beneath arc volcanoes are more common than previously thought, because the associated earthquake swarms easily could be mistaken for tectonic activity. Furthermore, as the following two cases illustrate, some volcanoes inflate without any accompanying unusual seismicity, so their activity would likely escape detection entirely were it not for InSAR.

5.3.5 Westdahl Volcano, Alaska

Unlike Akutan, Westdahl Volcano has been seismically quiet since its last eruption in 1991–1992. Nonetheless, InSAR studies by Lu *et al.* (2000b, 2003a) revealed an unmistakable pattern of progres-

sive inflation at Westdahl from 1992 to 1998. This result is noteworthy because it confirms that magma sometimes accumulates and moves beneath volcanoes without causing earthquakes. This raises the possibility of longer term warnings of future eruptions, based on geodetic detection of aseismic inflation, than would be possible otherwise.

Westdahl Volcano is located on Unimak Island in the eastern Aleutian arc, about 85 km southwest of the tip of the Alaska Peninsula and 100 km northeast of Akutan Island (Figure 5.21). Known eruptions of Westdahl occurred in 1964, 1978, and 1991–1992. The 1991–1992 eruption, from a fissure through ice, produced a lava flow that extended about 7 km down the northeast flank, debris flows that reached the sea 18 km from the vent, and ash plumes to 7 km altitude

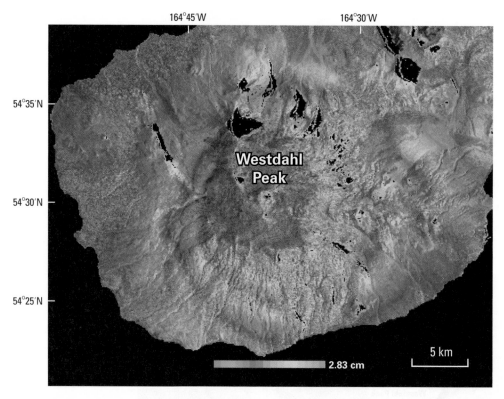

Figure 5.22. Topography-removed interferogram showing inflation at Westdahl Volcano, Alaska, from 21 September 1993 to 9 October 1998 (Lu *et al.*, 2000b). Range of colors in the color bar corresponds to 2.83 cm of range change between the ground and ERS satellite. Earthquake activity was at a very low, background level throughout this period.

(Miller *et al.*, 1998, pp. 45–46). The AVO installed a five-station seismic network at Westdahl Volcano in July 1998. Local seismic activity remained at a very low background level of about 5 earthquakes per year through 2001.

Motivated primarily by eruptions at nearby Shishaldin Volcano in 1995–1996 and early 1999, Lu *et al.* (2000b, 2003a) produced interferograms for Unimak Island that span the interval from 1992 to 1998 and made a surprising discovery. Unheralded by unusual activity of any kind, Westdahl Volcano began to re-inflate almost immediately after the end of its most recent eruption in January 1992 (Figure 5.22). Several independent interferograms reveal a consistent pattern of uplift centered beneath the summit area. The best-fit model is a Mogi-type source located about 8 km beneath the summit area that inflated by 0.05 km^3 from 1993 to 1998.

The Westdahl result is important for two reasons. First, it was the first time that InSAR revealed inflation of an otherwise quiescent volcano for several years *after* an eruption, and presumably *before* a future eruption. This raises the possibility that InSAR can be used to prospect for volcanic systems where magma is accumulating before other signs of unrest are detected or recognized. If so, scientists could take advantage of the advance information to intensify their monitoring efforts at inflating volcanoes and provide longer term warnings of future eruptions. Second, the inflation sources beneath Westdahl and Akutan are relatively deep (8–13 km), which means that InSAR can be used to identify deforming magma bodies before they rupture toward the surface to feed shallow intrusions or eruptions.

Local seismic activity at Westdahl began to increase gradually in 2002. The pace quickened in 2003, and a small earthquake swarm occurred beneath the volcano on 7 January 2004. Whether the seismicity is directly precursory to the next eruption remains to be seen.

5.3.6 Three Sisters volcanic center, Oregon

Another striking example of aseismic crustal uplift was provided by an InSAR study of the Three Sisters volcanic center in central Oregon, U.S.A.

Three Sisters is a long-lived center of basaltic to rhyolitic volcanism that has produced five large cones of Quaternary age: North Sister, Middle Sister, South Sister, Broken Top, and Mount Bachelor (Figure 2.33). South Sister, the youngest of the five cones, has erupted lavas ranging in composition from basaltic andesite through rhyolite. Most, if not all, of the volcano is probably of late Pleistocene age. The youngest eruptions occurred at a series of vents on the south and northeast flanks of the volcano that erupted rhyolite tephra, lava flows, and domes between about 2,200 and 2,000 years ago (Taylor *et al.*, 1987; Scott and Gardner, 1990; Scott *et al.*, 2001).

In the spring of 2001, my good friend and USGS colleague, Chuck Wicks, decided to study the Three Sisters area with InSAR, mainly because its eruptive history and setting make it a promising target. Geologically young eruptive products suggest that an active magmatic system may still exist beneath the area, and its eastern part is sparsely vegetated and therefore promising for InSAR. Coincidentally, the area is also one of Chuck's favorite backpacking haunts, so he was familiar with the terrain and curious whether it might be deforming. There had been no reports of unusual activity in the area and, in fact, central Oregon is known for its low level of seismic activity relative to other parts of Cascadia.

Imagine our surprise, then, when some of the first interferograms revealed a striking bullseye pattern of uplift centered about 5 km west of the summit of South Sister. An area about 20 km in diameter had been uplifted as much as 11 cm sometime between August 1996 and October 2000. An elastic point-source model indicates a volume increase by $0.023 \pm 0.003 \, \mathrm{km}^3$ located $6.5 \pm 0.4 \, \mathrm{km}$ beneath the surface, presumably a magmatic intrusion (Wicks *et al.*, 2001). The deformation field extends beyond the margins of EDM and tilt-leveling networks that were established in 1985 and re-measured in 1986 in anticipation of future activity at South Sister (Yamashita and Doukas, 1987; Iwatsubo *et al.*, 1988). Re-observation of these networks during summer 2001 (the former by GPS) corroborated the InSAR results and showed that there had been little or no deformation prior to the time spanned by the interferograms. Subsequent analysis of additional interferograms revealed that the deformation episode started in 1997 or 1998 and that cumulative uplift had reached 14 cm by September 2001 (Figure 5.23).

At the time of this writing (May 2005), the uplift is continuing at a steady rate of about $3 \, \mathrm{cm \, yr}^{-1}$ and

the area is producing only a few small earthquakes ($M < 3$) per year. The USGS Volcano Hazards Program, through its David A. Johnston Cascades Volcano Observatory (CVO), has enhanced its monitoring of the area in cooperation with the US Forest Service, which manages the Three Sisters Wilderness Area. Both agencies have taken steps to keep the public informed about the activity and its implications for volcano hazards.

5.3.7 The future of volcano InSAR

In the coming years, there are sure to be successful applications of InSAR at many other volcanoes. Just as there was a period of uncertainty over the long-term availability of GPS to civilian users, some doubts are now being expressed over future satellite missions suitable for InSAR. In May 2005, JERS-1 and ERS-1 had been decommissioned after exceeding their design lifetimes. ERS-2 and RADARSAT-1 were still functioning after exceeding their design lifetimes, but problems aboard ERS-2 were limiting its usefulness for InSAR applications. Envisat, the European Space Agency's follow-on to ERS-1 and ERS-2, had completed its third full year in orbit and was functioning normally. The Japan Aerospace Exploration Agency (JAXA, formerly the National Space Development Agency of Japan, NASDA) sucessfully launched its follow-on to JERS-1, called the Advanced Land Observing Satellite (ALOS), on 24 January 2006. Also in 2006, the Canadian Space Agency plans to launch RADARSAT-2. Unfortunately, there seems to be no US InSAR mission on the horizon in the near future, although several proposals have been advanced within NASA.

In the book's final chapter, I try to envisage the near-term future of volcano geodesy, including a much-expanded role for InSAR. With a small constellation of SAR satellites dedicated to interferometry at multiple wavelengths (X-band, C-band, L-band), and with on-board GPS and careful orbit control to reduce spatial decorrelation, it should be possible in the foreseeable future to: (1) monitor most of the world's volcanoes with L-band interferometry (for greater penetration of vegetation) at repeat cycles of a few days or less; (2) resolve the problems currently posed by atmospheric-delay anomalies and lack of coherence; (3) routinely detect magma accumulation or hydrothermal pressurization before the onset of shallow seismicity or other indications of unrest; and (4) identify key volcanoes for intensified monitoring with con-

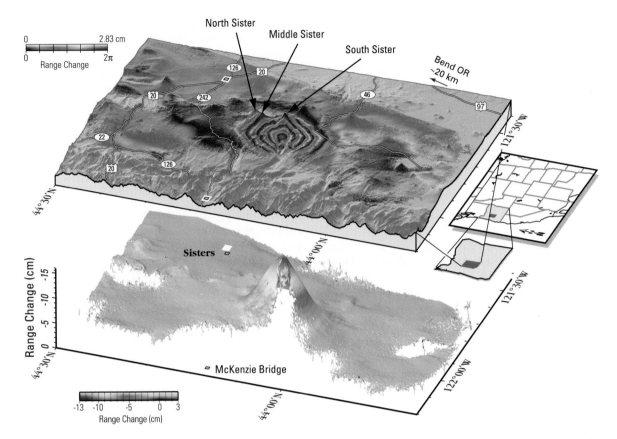

Figure 5.23. Wrapped (*top*) and unwrapped (*bottom*) interferograms of the Three Sisters area (red rectangle in inset), central Oregon Cascade Range, for the period September 1995 to September 2001 (Wicks *et al.*, 2002a,b). Approximately five fringes centered 5 km west of South Sister Volcano correspond to ~14 cm of surface movement toward the satellite (mostly uplift) during this period. Analysis of additional time sequential interferograms shows that the uplift started in 1997 or 1998 and proceeded at a steady rate of 3--4 cm yr^{-1} at least through summer 2002. Data from repeated leveling and GPS campaigns, and also from two continuous GPS stations in the deforming area, show that uplift was continuing in May 2005. Modeling indicates that the uplift is caused by inflation of a source 5--7 km deep, presumably a magma body, at a rate of ~0.006 km^3 yr^{-1}.

tinuous GPS, strainmeters, tiltmeters, dense seismic networks, and volcanic gas sensors. Short-term uncertainty notwithstanding, InSAR has established itself as a powerful geodetic tool for studying ground deformation at volcanoes and elsewhere. Given its proven capabilities and vast potential, this geodetic camera is likely to remain in service for a long time to come.

In the meantime, there will be a continuing need in volcanology for other geodetic tools, including cam-

paign and continuous GPS, leveling, photogrammetry, tiltmeters, strainmeters, and gravimeters. InSAR provides an exciting means to identify and study deforming volcanoes worldwide, but there are more riddles left to solve than any single technique or sensor can handle. The best strategy is to attack the unknown on multiple fronts, with as much weaponry as can be brought to bear. Volcanologists often find themselves outgunned, but seldom daunted, in their struggle to understand how volcanoes work.

CHAPTER 6

Photogrammetry

Ren A. Thompson and Steve P. Schilling

6.1 INTRODUCTION

Since its conception over a century ago, photogrammetry has been a mathematically rigorous cartographic tool for extracting geodetic information from stereo photography. Defined by the American Society of Photogrammetry as '*the science or art of obtaining reliable measurements by means of photography*' (Whitmore, 1952), photogrammetry provides the foundation for many standard cartographic datasets that serve as tools in the volcanologist's kit, including topographic maps, orthophotographs, digital elevation datasets, geologic maps, and infrastructure maps. With the advent of high-speed microprocessors, real-time GPS systems (Chapter 4), and the growing availability of satellite-imaging systems, photogrammetry has evolved from dedicated photogrammetric mapping venues to portable modular applications that can be readily integrated with other systems to provide near real-time, 3-D mapping and monitoring capability over large areas.

Despite recent technological developments, the fundamental principles of photogrammetry and the basic requirements for developing successful applications remain unchanged. As noted in the Preface and Chapter 1, volcanologists charged with assessing unrest at volcanoes are often faced with the challenge of implementing field-based monitoring systems that require innovative and often unconventional solutions to complex problems. Understanding the basic principles of photogrammetry enables them to tailor photogrammetric tools to a broad range of applications and geologic situations.

Wolf (1983, p. 159) provides the following useful introduction to the phenomenon that lies at the heart of photogrammetry (i.e., *stereoscopic parallax*). Hold up a finger in front of your face and alternately blink your eyes while focusing on the finger. This is *not* a test of mental or physical acuity – please bear with us for a moment. Your finger will appear to move from side to side with respect to objects beyond it, such as the pages of this book. The closer the finger is held to the eyes, the greater will be the apparent shift. No one is looking – try it!

Conceptually, photogrammetry is the mathematical formulation of the procedure that your eye–brain combination uses to determine the distance from your eyes to your finger or any other object. The shift in apparent position is related to the angular separation between your eyes at the distance in question, which is called the *parallactic angle*.[1] The average person, with an eye separation of about 66 mm, is capable of discerning parallactic-angle changes of about 3 seconds of arc. This corresponds to distinguishing range differences of about 4 m at 100 m – pathetic by raptor standards, but mostly adequate for *Homo sapiens*.

6.2 HISTORICAL PERSPECTIVE

The application of geologic photogrammetry to volcano geodesy has its roots in the development

---

[1] The parallactic angle, also known as the convergence angle, is formed by the intersection of the left eye's line of sight with that of the right eye. The closer this point of intersection is to the eyes, the larger the parallactic angle. As the eyes scan overlapping areas between a stereo image pair, the brain receives a continuous 3-D impression of the ground. This is caused by the brain constantly perceiving the changing parallactic angles of an infinite number of image points making up the terrain. The perceived 3-D model is known as a stereoscopic model (Section 6.3.6).

of modern photography and cartography. The popularity of photography and stereoscopy in the second half of the 19th century provided the impetus to apply this emerging technology to cartographic applications. In 1855, Nadar (Gaspard Félix Tournachon) obtained the first aerial photographs from a balloon about 80 m above the ground. Aerial photography was used during the American Civil War (1861–1865) and Franco–Prussian War (1870–1871) for purposes of military reconnaissance. Concurrently, the effects of lens distortion in cartographic applications became known, and techniques were devised to both reduce distortion and mitigate its consequences. However, not until George Eastman developed the nitrocellulose film base in 1885 and the Wright brothers pioneered the airplane in 1903 did the two critical elements of early photogrammetry materialize in a way that enabled collection of stereo imagery over large areas suitable forcartographic applications. Consequently, photographic film became the first remote-sensing medium utilized in the production of maps and for other applications in geology (Ray, 1960; Sabins, 1999).

The first half of the 20th century witnessed the development of a number of stereoplotting instruments (i.e., instruments to view adjacent overlapping photographs and make measurements) designed to facilitate the quantitative production of maps and geodetic measurements from stereo aerial photographs, along with development of cameras specifically designed for aerial photography. The derivation of rigorous mathematical formulas constraining the resection, orientation, rectification, and scaling of imagery in the 1920s and 1930s quickly led to the development of both analog and analytical instrumentation for stereoscopic mapping during the remainder of the 20th century. Subsequent commercialization of the photogrammetry industry evolved around the ever-increasing demand for photography and instrumentation capable of higher accuracy and output. By the 1960s, topographic mapping based on photogrammetry was employed worldwide for national-scale programs. Geologic applications including geologic mapping, geomorphic change studies, and landslide and volcano monitoring became common by the mid-1970s through 1980s. Fully automated digital-photogrammetric systems were introduced in the 1990s, and today the range of possibilities is limited only by the availability of high-quality stereo photography and geodetic control with which to orient the imagery to real-world coordinate systems. Even this last require-

ment is easing, as both aerial and handheld cameras with on-board Global Positioning System (GPS) receivers (Section 4.3) can alleviate or minimize the need for external ground control in many situations. The growing advantage to the volcano community is that photogrammetric applications which, in the past, took days to weeks to execute, are now possible in near real-time and with a relatively modest investment in equipment.

6.3   PHOTOGRAMMETRY FUNDAMENTALS

6.3.1   Introduction

Photogrammetric applications in geology can be classified as either aerial or terrestrial, based on whether the photography was collected from aerial or ground-based cameras. Traditionally, both approaches have focused on extraction of geologic information from photographs collected specifically for the purpose of reconstructing a distortion-free stereoscopic terrain model – one of several applications of direct interest to volcanologists. The geometric requirements for collection of photography in both approaches are identical, although aerial photography is more widely available and instrumentation required for extraction of quantitative data from aerial photography is far more accessible than that required for terrestrial imagery. Additionally, aerial photography is widely available covering a broad range of nominal scales and is well suited to rigorous geologic and topographic mapping, as well as reconnaissance or 3-D terrain analysis over large areas. Generally, terrestrial applications are more aerially restricted and site-specific. The basic elements of aerial photogrammetry are outlined below; terrestrial photogrammetry is addressed in Section 6.6. More detailed treatments of these subjects can be found in Wolf and Dewitt (2000), American Society for Photogrammetry and Remote Sensing (1997), and references therein.

Nearly all imagery collected for photogrammetric analysis is based on electromagnetic energy with wavelengths within the visible (0.4 to 0.7 mm), ultraviolet (0.3 to 0.4 mm), and near infrared (0.7 to 1.2 mm) portions of the spectrum (Figure 6.1). If an object reflects all wavelengths of visible light, it will appear white to the human eye or film emulsion; conversely, if it absorbs all wavelengths it will appear black. If an object or surface absorbs all visible blue

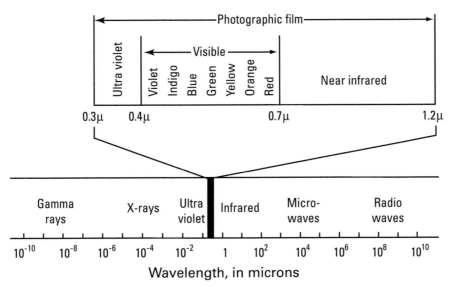

Figure 6.1. Simplified classification of the electromagnetic spectrum by wavelength. Most photographic and photogrammetric applications make use of the visible, near infrared, and ultra violet part of the spectrum (expanded portion).

and red wavelengths it will appear green, and likewise for other wavelength combinations within the visible spectrum. Film emulsions used in photogrammetry have been optimized for sensitivity to certain parts of the electromagnetic spectrum depending on the intended application. For example, black and white emulsions are sensitive to blue and ultraviolet energy; color or panchromatic emulsions are sensitive to red, green, and blue wavelengths. Color infrared emulsions record images in the near-infrared part of the spectrum, beyond what is visible to the human eye. By far, the most common applications in geology utilize conventional black and white emulsions because of their wide availability, low cost, and high resolution. True-color emulsions are reserved for applications in which rendering natural scenes as accurately as possible is a high priority.

6.3.2 Aerial cameras

Cameras used for collection of aerial photographic images are classified as either *film* or *digital*. Film cameras record images by focusing reflected electromagnetic radiation, predominantly visible light rays, on silver-halide film emulsions. Digital cameras employ *charge coupled devices* (CCDs) to measure the intensity of light exposed on a matrix of picture elements (pixels). Each element builds up an electric charge proportional to the incident light energy received and represents this as a tonal value for each pixel in a digital photograph. Digital-

imaging technologies are relatively new in the field of photogrammetry and acquisition of digital images collected with digital cameras is rare. Far more commonly, film-based aerial photographs are digitally scanned as needed for digital applications. A discussion of digital-scanning applications is presented in Section 6.4.

Film cameras can be subdivided further into fixed-frame and scanning varieties. Fixed-frame cameras are more common and expose a single image frame on film with each opening of the camera shutter. Scanning cameras typically employ a rotating lens element to focus collimated light rays along predefined paths or rows of the film plane, progressively exposing the entire image after multiple passes. They come in many different geometric configurations and are largely restricted to special purpose applications. The single lens, fixed-frame aerial camera remains the mainstay of geologic and cartographic applications in photogrammetry.

In addition to being extremely precise and expensive instruments, all single lens, fixed-frame cameras used in aerial photography, regardless of manufacturer, share several common attributes. Modern camera lenses employed in the manufacture of aerial cameras represent the epitome of the optical lens-makers craft. Each is characterized by extremely high resolving power and minimal to virtually non-existent lens distortions – two features that are critical to using aerial photography for quantitative, 3-D measurements. Since the lens-to-object distance is large relative to the typical focal length of the lens, fixing the focus at infinity enables optimum optical performance and simplifies lens construction and calibration.[2]

To capitalize on the optical integrity of the lens, the lens-mounting system and camera body must be rigid and maintain precise geometry such that the lens axis is as nearly perpendicular as possible to the film plane in the camera body. As film-based photographic emulsions are inherently flexible, all modern aerial cameras utilize some form of film stabilization system. The most common type is a vacuum system in the film chamber that activates immediately prior to release of the camera shutter, pulling the film perfectly flat against the film backing plate. The vacuum is released prior to advancing the film to the next frame using a motorized, high-capacity film magazine capable of holding enough film rolls to capture hundreds of exposures.

6.3.3   Format, focal length, and field of view

Stereoplotters, instruments that are used to orient overlapping aerial photographs, are designed to work with imagery at the scale of the negative produced by the camera/lens combination. Consequently, most aerial cameras utilize large film formats, typically 230 mm (9 in) square. The focal length of an aerial camera using the 230 mm-square format determines the field of view recorded in each image. By far, the most common focal length used today is 152 mm (6 in), with an angular field of view of 94°. This is considered a wide-angle lens and yields suitably large image footprints for many applications. The relationship between angular field of view, lens focal length, and film format is:

$$\theta = 2 \tan^{-1}(d/2f) \qquad (6.1)$$

where θ is the lens angular field of view, d is the

diagonal distance across the film frame, and f is the lens focal length. For a 230 mm film camera using a 152 mm lens, the angular field of view is:

$$\theta = 2 \tan^{-1}\left[\frac{\sqrt{(230^2 + 230^2)}}{2(152)}\right] = 94°$$

Aerial camera lenses are classified on the basis of their angular field of view as super wide-angle ($\theta \geq 100°$), wide-angle ($\theta = 75–100°$), or normal ($\theta \leq 75°$). Less common, but still used, are lenses with focal lengths of 89 mm (3.5 in), 210 mm (8.25 in), and 305 mm (12 in), which have angular fields of view of 123°, 76°, and 56°, respectively. Examination of (6.1) reveals the inverse relationship of focal length to field of view (i.e., as the focal length increases, the field of view decreases). Consequently, the image footprint decreases with increasing focal length for the same flight height. Lenses with focal lengths significantly shorter or longer than 152 mm are used mostly for unusually low- or high-altitude photo missions, respectively.

6.3.4   Photo collection and scale

Aerial photography for mapping and monitoring applications is typically classified as vertical or oblique, depending on the orientation of the camera relative to the ground surface. In the case of vertical aerial photography, the goal is to maintain an orthogonal orientation of the camera-lens axis relative to the ground surface. For oblique aerial photography, the camera axis can be oriented at high to low angles relative to the ground surface (i.e., near-vertical to near-horizontal), depending on the application. An important consideration for aerial photo collection in volcanic terrain is the need for combining vertical and oblique aerial photography to optimize image resolution in areas of high relief or intense shadowing resulting from less than ideal sun angles. Partial obscuring of the ground surface by clouds or volcanic fume can also be a problem. Often these difficulties can be circumvented by combining hand-held photography, typically collected from a light aircraft or helicopter, with conventional vertical aerial photography. In either instance, strips of photos are collected at the same altitude, with each successive photo overlapping its predecessor by approximately 60%. Adjacent parallel strips are collected while maintaining approximately 20–30% side overlap with the previous strip (Figure 6.2). In this manner, large areas can be collected in a single

---

[2] Careful calibration of aerial cameras and lenses is essential to achieving the best possible photogrammetric precision. Clarke and Fryer (1998) provide a thorough discussion of the development of calibration methods and models, which includes the following introduction (p. 58): 'A camera consists of a image plane and a lens which provides a transformation between object space and image space. This transformation cannot be described perfectly by a perspective transformation because of distortions which occur between points on the object and the location of the images of those points. These distortions can be modelled. However, the model may only be an approximation to the real relationship. How closely the model conforms to reality will depend on the model and how well the model's parameters can be estimated. Choosing parameters which are both necessary and sufficient has taxed those involved in the process of lens calibration for as long as lenses have been used to make precise measurements.'

A

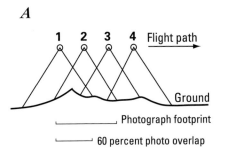

1 2 3 4 Flight path

Ground

Photograph footprint

60 percent photo overlap

B

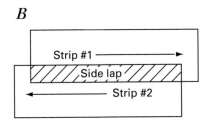

Strip #1

Side lap

Strip #2

Figure 6.2. Geometric representation of aerial photography collected with 60% overlap in sequential photos (A) and 20–30% overlap for adjacent photo strips (B).

flight while maintaining the photo geometry necessary for stereo rectification.

The nominal *photo scale*, or ratio of the unit length measured on the photo to the equivalent distance represented on the ground, is a function of the height of the camera platform above the ground surface and the focal length of the camera lens:

$$S = f/H = 1/(H/f) \qquad (6.2)$$

where S is the scale, f is the focal length, and H is the camera flight height. Both f and H are expressed in the same units, typically feet, meters, or millimeters. H represents the nominal camera height above the ground, usually determined by the aircraft altimeter. For example, photos taken with a 152 mm lens at an altitude of 3,648 m would have a scale represented by: $Scale(S) = 1/(3,648/0.152) = 1/24,000$, more commonly written 1 : 24,000. Figure 6.3 shows aerial photos flown over Mount St. Helens on 25 September 2000, at nominal scales of 1 : 24,000 and 1 : 12,000, using the same camera with a 152 mm focal length lens flown at nominal altitudes of 3,648 m and 1,824 m above the 1980–1986 lava dome surface, respectively.

The horizontal scale of aerial photographs can vary significantly across the field of view, depending on the vertical relief of the ground surface. Any departure from a flat surface effectively changes the camera flight height (H) in (6.2), which affects the scale in the photographs. For the photo shown in Figure 6.3, for example, point A on the crater floor of Mount St. Helens lies at an elevation of approximately 2,000 m above sea level and point B on the crater rim is nearly 450 m higher. As a result, the scale in the photo varies from 1 : 24,000 at A to 1 : 21,039 at B (assuming a nominal flight height of 3,648 m above the 1980–1986 dome or 5,648 m above sea level). Commonly, aerial photographs flown for both commercial and geologic purposes include the average photo scale and the precise focal length of the camera lens on the film negatives for each project.

Most aerial photography flown for purposes of monitoring deformation or other surface change at active volcanoes is specifically targeted at resolving objects at a given photo scale. Typically, this might be a control-point tower or air-photo panel, a series of visually identified landmarks used as control points, or even a geomorphic landform such as a lava dome or flow. Owing to the variation in scale that can occur across a single image in regions of high terrain relief, substantial care must be taken to ensure sufficient stereo overlap at the required scale for image interpretation. Examples of the potential for omission of critical data are illustrated in Figure 6.4.

6.3.5 Relief displacement

Relief displacement occurs in all vertical aerial photographs as a result of the perpendicular orientation of the camera axis relative to the ground surface. This results in objects, trees for example, that appear to lean away from the *principal point*, or optical center of the photograph. The amount of relief displacement in any given photo is directly proportional to the height of the object and its distance from the principal point of the photo, and is inversely proportional to the flight height above the object. The practical consequence of this phenomenon is that areas of high relief, photographed at relatively low altitude, are typically subject to large amounts of relief displacement. This relief displacement, coupled with variable scale typically observed in these photographs, results in images that require rectification before they can be used to make quantitative measurements. This type of rectification differs significantly from mathematical transformations or digital rubber-sheeting techniques that serve to locally adjust photo-image coordinates to accommodate local cartographic ground control. Such methods work reasonably well for photographs with

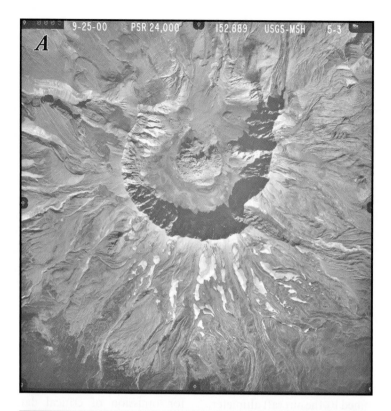

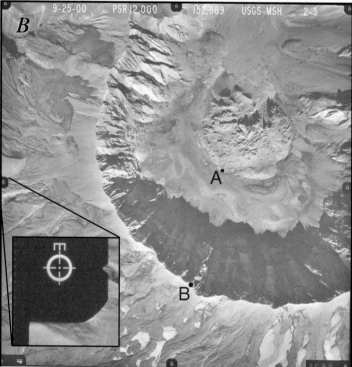

Figure 6.3. Aerial photographs of the crater and lava dome of Mount St. Helens taken in September 2000 at nominal scales of 1 : 24,000 (A) and 1 : 12,000 (B), the latter reproduced at 50% of original scale. See text for discussion of points 'A' and 'B' in the bottom image. Note camera focal length (152.889 mm) at top of each photo. Inset in bottom image shows enlarged view of left center fiducial mark. Similar marks occur in all four corners and at the center of each edge of all aerial photographs. The exact configuration and style of fiducial marks varies from camera to camera.

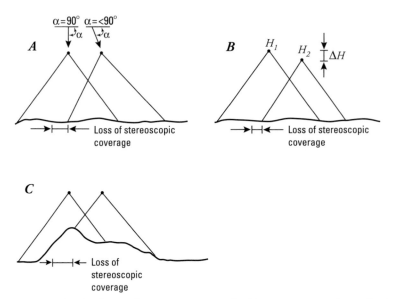

Figure 6.4. Circumstances preventing the degree of stereo coverage required for photogrammetric rectification include unanticipated camera axis rotation such that $\alpha < 90°$ (A), fluctuation in altitude where $(H_1 - H_2) = \Delta H$ (B), or severe variation in topographic relief between adjacent photos (C).

minimal scale variation and low relief, but produce unacceptable levels of distortion when applied to photographs of high- or variable-relief volcanic terrain. In such cases, preparation of an *orthophoto*, or rectified photographic image, is needed to extract quantitative 2-D image information, and stereo-model orientation is required to obtain 3-D information. Both of these techniques require photogrammetric instrumentation for proper orientation of aerial photographs prior to analysis.

6.3.6   Orientation

A photogrammetric instrument called a stereoplotter is used to reconstruct a distortion-free stereoscopic model of the ground surface. In such a model, points, lines, and surfaces can be constructed in 3-D space and projected in 2-D as needed for map or ortho-photo preparation. Modern orthophoto production requires digital photogrammetric instrumentation (Section 6.4), but relies fundamentally on techniques developed for analog, or opto-mechanical stereoplotters, for image rectification.

Any two overlapping photographs of an aerial strip can be viewed in stereo, providing a 3-D visual image of the ground surface. Such a pair of photos is referred to as a *stereo model*. There are typically many stereomodels depending on the length and number of flight strips associated with the photo mission. Each image of the stereo model can be thought of as a series of bundled light rays

representing the fixed geometry of the reflected electromagnetic radiation from the ground surface relative to the camera in the aircraft. For the overlapping 60% of adjacent photographs, two such bundles of light rays are represented for each model. For each stereo model, a 3-D image of the terrain can be reconstructed from the two photographs by knowing the relative position and rotation of the camera lens for each photograph. To create the stereo model, modern analytical and softcopy stereoplotters use mathematics to reconstruct the geometry of the original light paths relative to the camera, taking into account known quantities such as photo-flight parameters and characteristics of the aerial camera system. For our purposes, the geometric constraints involved are better illustrated by reference to mechanical (analog) stereoplotters.

Photogrammetric *orientation* of stereo images is basically the same for both aerial and terrestrial photography, and is divided into three steps: interior orientation, relative orientation, and absolute orientation. All three orientations must be performed to establish photogrammetric and geodetic rectification.

Interior orientation involves establishing the geometric configuration of the camera relative to the film image. To re-establish the path of light for any point on the film image, the projection center and focal length of the camera/lens combination used to collect the photographs must be known. These parameters are readily available in the form of a calibration report for aerial cameras used in most aerial applications. Often the camera focal length is recorded as a notation directly on the film image, and information about the projection center can be derived indirectly from the aerial photos. *Fiducial marks* are recorded on the border of each photo, typically one on each side and one in each of the four corners (Figure 6.3(B)). The intersection of lines connecting opposite fiducial marks is called the *center of collimation*. This is an important point of reference for photogrammetric measurements. For an ideal camera/lens combination, the center of collimation is coincident with the photo principal point (i.e., the intersection of the film plane and the lens projection center). Hereafter, we assume this to be the case, and refer simply to the principal point. Fiducial marks are characteristic of the aerial camera; the lens projection center and focal length are determined by repeated calibration throughout the life of the camera.

Relative orientation re-establishes the photo positions for the stereo model and, consequently, the ray paths for adjacent photos established during photo collection. One way to accomplish this is to physically position the photographs next to each other and re-establish the ray paths for each photograph relative to the same points in space. Assuming the position of one photograph is fixed in space, the second photograph has six degrees of freedom with respect to the first. Three of these parameters are translations along the x, y, and z axes, and the other three are rotations about the same axes (Figure 6.5). The translation bx is along the flight line, by is orthogonal to the flight line, and bz is along the vertical axis. The rotations ω, ϕ, and κ, are around the x, y, and z axes, respectively.

For any two photographs for which the ray paths have been reconstructed, a translation in the direction of flight bx influences scale changes between photos but not the intersection of light rays from adjacent photos; therefore, such translations are not considered in the orientation procedure. However, all five of the remaining translations are treated as unknowns and must be determined such that five pairs of light rays intersect when positioning the second photo relative to the first. A displacement in the y-axis direction by results in non-intersection of light rays for all five pairs of light rays for adjacent photos. This by translation, or y-parallax, is removed during the relative-orientation process by making iterative and diminishing adjustments to by, bx, ω, ϕ, and κ for one photo relative to the other. In opto-mechanical stereoplotters, these

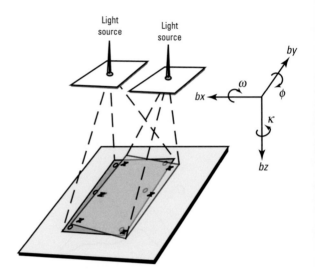

Figure 6.5. Diagrammatic representation of the three translations and rotations about the x, y, and z axes relative to the projected view of the overlap area on a single stereo pair.

adjustments are made physically by moving of one photo relative to the other using both translations parallel to, and rotations about the x, y, and z axes. In analytical and softplotter systems, the same problem is solved mathematically. In both cases, the solution converges toward the same relative orientation that existed when the photos were acquired.

The implementation of translations for removing y-parallax varies from instrument to instrument, but follows the same basic procedure (Figure 6.6). The five pairs of light paths for adjacent photos intersect at points 1 through 5 for a properly oriented stereo model. Although the actual positions of the points in the model are arbitrary, the relative position and the numbering system used have become industry standards. A relative orientation starts by removing the y-parallax in point 1 by means of a by translation.

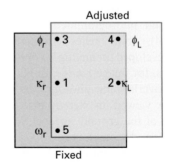

Figure 6.6. Effect of iterative adjustment of translations and rotations on one photo in a stereo pair relative to the fixed photo during a relative orientation. See text for discussion.

Next, the parallax in point 2 is removed without adversely affecting the orientation of point 1. A rotation about the z-axis, κ, centered at point 1 will remove the parallax from point 2. The y-parallax in point 3 is removed by a translation along the z-axis, bz. This introduces an x-parallax at point 2, which can be removed later by proper scaling, but does not introduce y-parallax at points 1 and 2. The y-parallax in point 4 is removed with a translation ϕ, again not introducing any significant new parallax to points 1, 2, and 3. Finally, to remove the y-parallax in point 5 we introduce a translation ω about the axis of the flight line. This translation introduces y-parallax in previously adjusted points 1 through 4, which must be revisited. The iterative procedure continues until parallax is minimized at points 1 through 5 and, consequently, at all points in the stereo model. The impact of adjusting one parameter on the relative position of points within the model is illustrated in Figure 6.7.

Precise 3-D measurements can be made from the stereo model once a suitable relative orientation has been established. However, the model is not oriented and scaled to a geodetic coordinate system (i.e., to the real world) until ground-control information is included. This is done mechanically for analog systems and mathematically for analytical and softcopy instruments.

Absolute orientation allows the model, cleared of all distortion, to be precisely oriented in space so that 3-D coordinate measurements (x, y, z) in the model can be transformed to ground coordinates in a geodetic coordinate system. The absolute orientation, like the relative orientation, is an iterative process involving solution for seven unknown parameters: scale (S), three linear translations (x_o, y_o, z_o), and three angular rotations (ω around the x axis, ϕ around the y axis, and κ around the z axis).

Solving for these seven parameters requires seven independent pieces of information about the precise location of the aerial camera. The most sophisticated aerial camera systems have inertial platforms that record, at the instant of exposure, the camera rotations (ω, ϕ, and κ) and precise camera coordinates using sophisticated GPS systems. Together with the flight altitude, this theoretically provides all the information necessary to solve for the seven unknowns. Unfortunately, these systems lack sufficient precision or accuracy to provide a real-time solution completely independent of external GPS ground control. Nonetheless, they can reduce the amount of ground control required to solve the

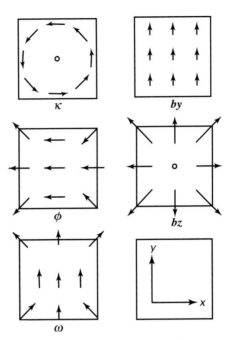

Figure 6.7. Iterative adjustment sequence and the impact of adjusting each parameter on the relative positions of points during a relative orientation of a stereo model. See text for discussion.

external orientation by nearly an order of magnitude. Lacking information about camera orientation and position, an exterior orientation can be performed using ground control points (i.e., points of reference visible in the model that have known geodetic control positions).

The minimum amount of ground control required to orient a stereo model is two horizontal points (x and y coordinates) and three elevation points (z coordinates). These need not be the same points, but the horizontal and elevation requirements can be combined such that two of the required elevation points can be horizontal control points as well. The distribution of control points within the area of stereo overlap determines the accuracy of the orientation outside the area encompassed by the control points. For best results, elevation points should form a large triangle and the (x, y) coordinate pairs should be separated as far as possible (Figure 6.8).

A standard spatial coordinate transformation is used in analytical and softcopy stereoplotter systems to perform the absolute orientation. This method can be illustrated geometrically and described relative to the physical parameters required for the orientation. This method only works for models having minimal tilts, typically less than 1°. This requirement is met for the majority of aerial photographs, making these photos well suited to analysis

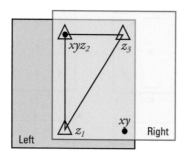

Figure 6.8. Relative positions of representative control points necessary to rectify a single stereo model during an exterior orientation. x and y are horizontal coordinates; z is the vertical coordinate.

with analog instruments. On the other hand, most hand-held aerial and terrestrial photographs have variable tilts exceeding several degrees, which requires spatial coordinate transformations that can be accomplished with analytical and softcopy photogrammetric systems.

The model scale is calculated by comparing the distance L on the ground to the same distance l in the model. This is expressed as:

$$l = \sqrt{(x_2 - x_1)^2 + (y_2 - y_1)^2} \qquad (6.3)$$

where the x and y coordinates are known horizontal control positions. The scale for the absolute orientation is:

$$S = l/L \qquad (6.4)$$

The rotation about the z-axis (κ) is determined by comparing the azimuth between the two plane ground control points and the azimuth between the same points measured in the model. The translation is calculated as:

$$\kappa = Az_t - Az_m \qquad (6.5)$$

where Az_t and Az_m are the azimuths of the ground control points on the Earth and the photo model, respectively. Once the scale and the κ translation are determined, the ground distances between the three elevation points are determined by multiplying the model distances by the scale. Subsequently, the ω and ϕ translations are determined in a similar way.

For purposes of illustration, consider the case in which two of the three elevation points (z_1 and z_2) lie along a line in the model y-axis, and the third point (z_3) forms a line with z_2 along the model x-axis (Figure 6.8). The translation ω is determined by:

$$\omega = \arctan((z_{1m} - z_{1t})/l_{12}) \qquad (6.6)$$

The translation ϕ is likewise determined from elevation points z_2 and z_3 by:

$$\phi = \arctan((z_{3m} - z_{3t})/l_{23}) \qquad (6.7)$$

Obviously, real-world situations would dictate more complicated geometry for the location of control points, but the geometric solutions remain similar, albeit more complex.

After determination of the scale and three rotations (ω, ϕ, and κ), the linear transformations (x_o, y_o, z_o) are determined simply by subtracting the rotated-and-scaled model coordinates from the known ground coordinates for the horizontal control points. Once the absolute orientation of the model is completed, points from one model can be transferred to an adjacent photo and used as control in the orientation of the second model. This generally works well, as long as the control points being transferred are easily identified in the adjacent photo. Two of the ground control points required to orient the second model can be transferred from the first model, requiring only one additional elevation control point for absolute orientation. In practice, using three or more control points in the overlapping part of the model ensures sufficient control for statistical analysis of error associated with the orientation. The procedure can be continued for orientation of an entire strip of models with the understanding that any orientation or control-point errors present early in the process will be propagated throughout the length of the strip.

6.3.7 Photogrammetric accuracy

Photogrammetric accuracy based on a properly oriented model is determined primarily by the quality of the instrumentation used and the quality of the photographic film. Individual points can typically be collected with an accuracy of about 10 microns at the film scale on analog stereoplotters, and down to 3 microns on high-end analytical stereoplotters. This translates to ground units by multiplying the instrument accuracy by the scale of the photographs. In practical application, only the most experienced operators are capable of realizing accuracies close to those possible with modern instruments. However, even inexperienced users are routinely capable of making measurements with accuracies of 20 microns at the film scale, or ~0.5 m on the ground, using 1 : 24,000-scale photography. This is adequate for most geologic mapping applications, but might not be sufficient for deformation monitoring.

It is important to distinguish relative accuracy from absolute accuracy in photogrammetric projects. Relative accuracy is the accuracy of measuring two points relative to each other in a single stereo model. In this case, the previous discussion of accuracy applies. Absolute accuracy refers to the accuracy of measured model points relative to a geodetic coordinate system. The degree to which the absolute accuracy approaches the relative accuracy is determined by the quality of the ground control used in the absolute orientation of the stereo model.

In most photogrammetric applications in geology, the quality of the ground control is the weak link in the orientation process. In many remote areas of the world, high-quality geodetic ground control is limited in extent and availability. Existing topographic maps often provide the only ground control for orientation purposes, unless allowances are made for collection of GPS or conventionally surveyed control. When GPS or survey control is unavailable, cartographic features on the topographic map such as road or stream intersections that can also be identified in the photographs are used for the absolute orientation. Unfortunately, the cartographic uncertainties associated with the original map production might be unknown, which severely limits the absolute accuracy of the photogrammetric project. If full photogrammetric accuracy is required, ground control measurements must be accurate to the same order as the relative or theoretical accuracy of the photogrammetric instrumentation.

6.4 INSTRUMENTATION AND DATA TYPES

Photogrammetric instrumentation falls into three classes, which share common requirements for orientation of stereo aerial photographs, including interior, relative, and absolute orientation for quantitative image rectification. To a degree, all three also share common ergonomic controls for data collection. Beyond adherence to the fundamental principals of photogrammetry, the three classes of instrumentation vary dramatically in capability and utility for geodetic application at volcanoes. Brief descriptions of each class are provided below with an emphasis on fundamental characteristics, advantages, and limitations inherent in each design. For a more thorough treatment of the historical context, mechanical construction, and

theoretical design characteristics of stereoplotters, the reader is referred to Wolf and Dewitt (2000) and references therein.

6.4.1 Analog stereoplotters

Analog stereoplotters are characterized by coupled optical and mechanical systems designed to replicate precise geometries linking object points on the ground to the orientation of stereo aerial photographs at the precise moment of acquisition. With properly oriented photographs, these instruments enable 3-D acquisition of coordinate data, facilitated either by digital encoders corresponding to the x, y, and z coordinate axes or by projection, typically through use of a mechanical pantograph, onto 2-D base materials. Historically, the primary application of analog instruments was for topographic mapping, but these instruments were easily adaptable for geologic map compilation. Largely displaced by analytical and softcopy stereoplotters for topographic mapping, analog instrumentation is still commonly used for geologic applications.

A stereo pair of diapositives (positive images on a transparent medium, usually polyester film or glass) or paper prints is oriented so light projected through, or reflected from, the images produces bundled light-ray paths such that intersecting light rays of the same object point on the left and right images form a stereo model (Figure 6.9). This viewing geometry corresponds to the inverse of the angular geometry experienced during the in-flight acquisition of the same stereo pair. Analog stereoplotters developed largely during the 1960s and 1970s shared common characteristics, including an optical viewing system, a measuring and tracing system, and a reference or tracing table (Figure 6.10). The mechanical carriage and optical-viewing systems of analog stereoplotters are designed to facilitate the range of motions necessary to perform interior and relative orientations of individual stereo models (Section 6.3), while maintaining an optical-viewing path that provides sufficient magnification for photo interpretation and visual confirmation of stereo superposition. Most analog systems employ: (1) overhead carriage-mounted photographs; (2) high-quality binocular viewing systems directed upward toward the illuminated left and right images; and (3) mechanical, hand-operated tracing stands for simultaneously moving the optical train through the region of stereo overlap on the photos and for adjusting

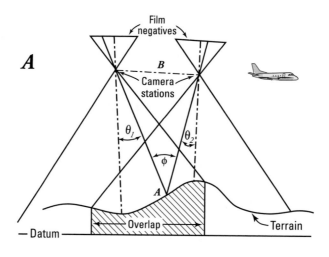

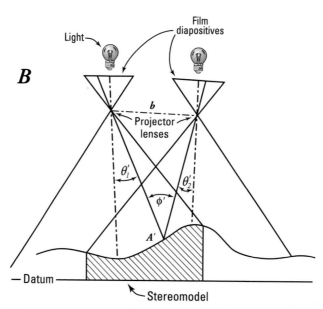

Figure 6.9. Geometric configuration of camera stations and aerial photograph negatives relative to the ground surface (A), and equivalent geometry established in an analog stereoplotter for diapositive images (B). The stereoplotter orientation ensures that original camera positions, distances (B, b), and ray path angles (ϕ, θ_1, θ_2) and ($\phi', \theta'_1, \theta'_2$) are related by the model solution (modified from Wolf and Dewitt, 2000).

the relative positions of the photos to maintain stereo viewing. A floating mark, or narrow beam of projected light, is superimposed on the center of the optical path for both left and right images. A rotating wheel on the tracing stand is adjusted until identical points on each photo, representing a single position on the ground, occupy the same relative positions as when the photos were collected. At this point, the left and right projected light sources form a single light path superimposed on a 3-D view of the ground surface. If proper stereo orientation is not achieved, parallax offset of common object points on the photos will result in separation of left and right image dots, which appear to float above the ground or rest beneath it relative to the stereo view of the ground surface. Only when the dots are superimposed and visually appear to sit precisely on the ground surface can 3-D coordinates be extracted from the model. Additional adjustment to the mechanical carriage is made to level the model relative to the tracing stand or geographic plane of the model.

Stereo models were originally scaled to ground coordinates by physically linking the photo carriage to a pantograph arm, which carried a pencil and was

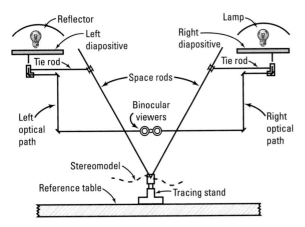

Figure 6.10. Sketch of a typical configuration of an analog stereo-plotter, including the optical-viewing system, measuring and tracing system, and reference or tracing table (modified from Wolf and Dewitt, 2000).

used to translate motion of the instrument to projected lines drawn on base materials. Relative movements of the mechanical carriage were translated to the appropriate degree of motion of the pantograph arm by a series of gears in combinations that corresponded to the scale of the geographic base map. Most analog systems still in use today have been retrofit with digital encoders to translate the x, y, and z motions of the mechanical carriage into digital streams of coordinate triplets that are passed to digital-mapping applications. Because of the mechanical limitations of these systems, they are only suitable for work with aerial photographs collected with minimal camera-axis rotation. Although limited in the range of application, these instruments are extremely precise, have excellent optics, and, after interior, relative, and absolute orientations are complete, are easy for a geologist to operate. Additionally, stereo models can usually be oriented with no knowledge of the original camera beyond its focal length. This advantage makes these instruments particularly well suited to analysis of a large database of archived imagery that might lack accompanying camera-calibration reports.

6.4.2 Analytical stereoplotters

Analytical stereoplotters became readily available during the 1980s with the advent of low-cost micro-computers, digital encoders, and servo controllers. These stereoplotters are designed around an analytical solution to a photogrammetric orientation. The fundamental characteristics of these instruments are an ability to precisely drive to defined x and y coordinate positions on left and right image carriages,

and to accurately measure photo coordinates on the rectified image pair. This is accomplished by controlling image-plate motions with high-precision servo motors linked to electronic operator controls for positioning the camera plates and the floating mark. Optical systems include high-power binocular viewing and variable light source controls, which help to maximize image measurement capabilities. Analytical stereoplotters remain the highest precision photogrammetric instruments available, often claiming precision tolerances within three microns on the film plane.

Analytical stereoplotters compute a mathematical model for the interior, relative, and absolute orientation of a stereo model and translate operator controls to photo-plate motions based on the orientation. This facilitates automated data collection, especially for digital elevation model (DEM) collection, by allowing automated grid positioning. Grid spacing is predetermined and the stereoplotter drives the floating mark to the proper x, y coordinate position on the photo. The operator uses the stereo-viewing system to place the floating mark on the ground, fixing the z coordinate in space based on the photogrammetric-model orientation. The ability to drive the instrument also aids in data collection for features such as geologic contacts in mapping applications, by enabling the operator to edit vector line work while the instrument drives the floating mark to the node position indicated by the mapping software.

The calculation of the z coordinate in analytical systems and softcopy systems requires that two conditions be satisfied: (1) collinearity for a single photograph, and (2) coplanarity for a stereo pair. These conditions are mathematically represented by two non-linear equations, one for the x coordinate and one for the y coordinate, based on the parameters of the interior, relative, and exterior orientations. The equations can be transformed to a series of linear expressions that are solved simultaneously and iteratively to determine the 'best-fit' z coordinate for any x, y pair. This procedure is represented graphically in Figures 6.11 and 6.12. Collinearity requires that the exposure station, any object point, and associated photo image point must all lie along a straight line in 3-D space. In the example shown in Figure 6.11, L, a, and A all lie along a straight line. Coplanarity requires that two exposure stations of a stereo pair, object point, and the associated image points on the two photos must all lie in a common plane. In Figure 6.12, points L_1, L_2, a_1, a_2, and A all lie in the same plane, enabling

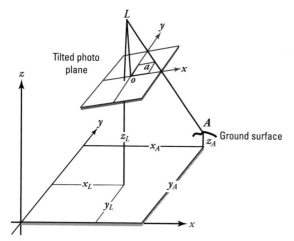

Figure 6.11. Illustration of the collinearity condition for any point A on the ground surface and its equivalent image point on an aerial photograph (modified from Wolf and Dewitt, 2000). See text for discussion.

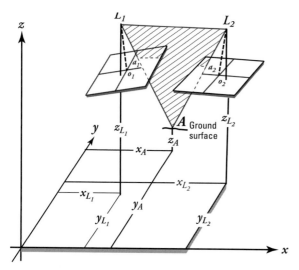

Figure 6.12. The condition of coplanarity for any point A on the ground and its equivalent image points a_1 and a_2 in a stereo model (modified from Wolf and Dewitt, 2000). See text for discussion.

iterative determination of the z coordinate for any object point in the stereo model. These calculations are performed in real time as the operator drives the analytical stereoplotter and positions the floating mark on the ground surface in the model.

6.4.3 Softcopy stereoplotters

Softcopy photogrammetric systems are essentially analytical stereoplotters that utilize digital images and software for translation of images, instead of mechanical control of image movement. Softcopy systems evolved from analytical stereoplotters as high-precision film scanners and high-speed com-

puting platforms were developed, enabling creation and manipulation of high-resolution digital imagery in real time (American Society of Photogrammetry and Remote Sensing, 1996a, 1998; Graham *et al.*, 1997; Gwinner *et al.*, 2000). Modern photogrammetric scanners are capable of distortion-free, 7-micron scans resulting in image files in excess of a gigabyte each, with sufficient resolution to rival measurement capability of many analytical stereoplotters, despite a somewhat degraded viewing system. Although softcopy systems offer all of the advantages of conventional analytical stereoplotters, with the possible exception of the viewing system, its capacity for automated measurement based on image matching algorithms has catapulted this technology to the forefront of modern photogrammetry. The simplification of hardware requirements for softcopy systems has been accompanied by a dramatic increase in the complexity of associated software, leading to fierce market competition for what is, in effect, a software tool with minimal hardware requirements relative to those for analog and analytical stereoplotters.

6.4.4 Display systems

The basic hardware requirements for a softcopy system include a stereo-viewing system, high-speed central processing unit (CPU), and several hundred gigabytes of data-storage capacity. The two most common types of viewing systems both employ large cathode ray tube (CRT) monitors with 120-MHz screen refresh rates alternating views of the left and right photo image of a stereo model pair 120 times per second. The most comfortable viewing system employs an active polarizing filter fitted to the CRT screen and a lightweight pair of passive polarized glasses for the operator to wear. The active polarizing filter alternates the polarizing orientation at the same refresh rate employed by the CRT for displaying left and right images of the model pair. The glasses have left and right lenses with opposite polarizations, corresponding to those used by the active screen. When the computer displays the left image, the active polarizing screen blocks the view from the right eye. This alternates with display of the right image, blocked to view by the left eye, at a rate of 120 times per second; too high to be perceived by the human brain. The alternating projection of the overlapping images provides visualization of the area of overlap in stereo. An alternative method employs the same alternating display of left and right images on the CRT, but uses an infrared signal keyed to the screen refresh rate that controls active polarizing lenses in headgear that alternates blocking the left and right images from view.

In both systems, the images are either panned relative to a fixed floating mark in the center of the screen, or the floating mark is allowed to track sequentially across the fixed stereo model. The floating mark, the ground surface, and any point or line data collected are viewed in 3-D, enabling immediate evaluation of adherence to the ground surface. Zooming is accomplished by sequentially displaying smaller regions of the digital stereo model, while increasing the displayed resolution to maintain a fixed image dimension. This is implemented in real time by maintaining multiple copies, typically up to seven, of the same image at successively higher resolutions. Maintaining multiple image copies not only aids zoom and pan operations, but also provides a means to apply image matching algorithms during the orientation procedure.

6.4.5 Computer-assisted orientation

Orientation of softcopy imagery involves the same steps required when using analog or analytical systems, with the advantage of additional image-matching capabilities inherent in all digital photogrammetric systems. Interior and relative orientations are greatly assisted by utilizing *image pyramids*, or linked image layers representing successively higher resolution versions of a given image area for each scanned photograph. Having multiple copies of the same image at different resolutions enables rapid identification of fiducial marks for interior orientations and of *pass points* (i.e., points common in three successive images used in relative orientations).

For orientations, fiducial marks are recognized automatically by matching image pixel patterns to templates identifying styles of marks utilized by different camera manufacturers. Alternatively, unknown fiducial patterns can be 'learned' by most systems and recognized in subsequent photographs. Fiducial marks are measured using pattern recognition for successively higher resolution scans and by digitally constraining the floating point to occupy the center of the marks. After fiducial marks are measured, the principal point of the digital photos is calculated by resecting lines connecting individual fiducial marks.

Combined with knowledge of the focal length and lens distortion parameters of the interior orienta-

tion, a relative orientation can be performed by identifying pass points common to both photos of a stereo pair or three photos in multi-model strips. Pass points are identified by using pattern recognition and image-matching algorithms that initially search low-resolution photos for coarse object patterns common to left and right images of the stereo model. Then the algorithms successively zoom in to higher resolution images, refining the pattern search to occupy x and y coordinates of common pixels. Analog and analytical stereoplotters typically require a minimum of three manually determined points to constrain a relative orientation. Usually, more points are collected to provide sufficient redundancy in an analytical solution; this can be a tedious manual process. Softcopy stereoplotters typically calculate many (up to hundreds) pass points for use in the relative orientations, thereby minimizing the impact of errors associated with any given point. Automatically generated pass points with high residual errors associated with the orientation can be removed from the solution, and the relative orientation can then be recalculated.

Despite the automation inherent in interior and relative orientations (Hannah, 1989; Kersten and Haering, 1997), absolute orientation requires operator input to visually identify ground control with known positional coordinates in both the left and right photos. In conventional photogrammetry, these are indicated in the photos by GPS or survey control points marked with white or black plastic panels (photo targets). After the x and y coordinates of these control points have been measured monoscopically in the left and right images, an analytical solution to the exterior orientation is performed. When the errors associated with control point measurements have been determined, control points can be re-measured, added, or deleted to optimize the best-fit solution. Upon completion of the exterior orientation, stereo models are viewed in three dimensions, point and line data are collected and viewed in three dimensions, and derivative products such as digital elevation models and orthophotos are produced.

6.4.6 Digital elevation models

Digital elevation models (DEMs) are mathematical representations of topography based on x, y, z coordinate sampling of the ground surface at regular grid intervals. These datasets can be used to produce topographic maps or detailed mathematical models of the ground surface. They provide critical quantitative data to volcanologists monitoring geomorphic change at active volcanoes or conducting predictive modeling of volcanic hazards such as dome growth (Robertson *et al.*, 2000) or lahar runout (Iverson *et al.*, 1998).

DEMs can be collected on analytical stereoplotters by forcing the instrument to occupy x and y coordinate positions in a stereo model as the operator places the floating mark on the ground to determine the z position at each grid node. This produces a very accurate DEM, but it is time consuming and leads to operator fatigue for large grids with many nodal points (perhaps thousands). Softcopy stereoplotters utilize the same pattern-recognition and image-matching algorithms employed in the determination of pass points to calculate x, y, and z coordinates for nodal points on grids, with spacing determined by the operator. This requires solving the collinearity and coplanarity equations for every node on the grid. The process is automatic with most systems and requires no input from the operator during the calculation of the model.

An accurate representation of the ground surface, especially in areas of high relief, usually requires additional information for calculation of accurate DEMs. *Breaklines*, or x, y, z coordinate strings representing the tops of ridges or the bottoms of valleys, and individual points representing peaks or objects of interest, are collected manually and provided as input to the surface-generating (DEM or triangulated irregular network, TIN) software. During the creation of the DEM or TIN, the manually collected points force the grid surface to follow these important inflections in the topography. The creation of a high-quality DEM typically is an iterative process in which successively higher resolution models are produced by successively decreasing the grid size of each model iteration. This forces the DEM calculations to conform to points generated in the previous, lower resolution model. Performing the calculations in this fashion minimizes the amount of editing required, because a percentage of points in any given calculation will be incorrectly placed either above or below the ground surface as a result of incorrect image correlation.

6.4.7 Orthophotos

Digital orthophotos are photogrammetrically rectified photo images derived from aerial photographs and a DEM. Orthophotos have the distinct advantages for field applications of having both the image qualities of aerial photographs and the cartographic

registration necessary for mapping and measurement purposes. The critical requirements for the creation of digital orthophotos are a high-quality DEM and digital aerial photographs having known exterior-orientation parameters. The process is based on the application of the collinearity condition for any point of the DEM to relate x and y position coordinates to a given pixel, or series of pixels, in the associated photograph. Repetitive application of the collinearity condition for every point of the DEM results in population of the x and y matrix with corresponding image pixels, resulting in the orthophoto image (Hohle, 1996).

6.4.8 Satellite imagery

The rapidly increasing availability of commercial satellite imagery, especially in the visible and radar portions of the electromagnetic spectrum (Theodossious and Dowman, 1990; Weston, 1990), has generated considerable interest in the potential for photogrammetric exploitation of this data for three reasons. First, commercial optical sensors have an image resolution of ~ 1 m and regional radar imagery is now available with better than 50 m resolution. Such resolutions are suitable for many mapping applications and the satellite images have the advantage of a significantly larger image footprint for any given scene. A second reason is that satellites provide predictable and repeatable image coverage, which makes them ideal platforms for monitoring applications. Radar waves can penetrate clouds and, to a degree, airborne dust, so satellite radar imaging holds promise for 'seeing' through clouds and ash during an eruption, when other imaging systems are less effective. Third, softcopy stereoplotters are readily adapted to accommodate digital imagery of all types, which makes them a particularly flexible tool.

Most imaging satellites use linear-array sensors, which acquire images along scan lines that are perpendicular to the path of the satellite. Because satellites are highly stable platforms with predictable flight paths and imaging geometry, stereo images suitable for photogrammetric rectification are routinely collected by such systems. The details of the scanning geometries associated with commercial satellites are routinely made available to photogrammetry software vendors, with hopes that they will incorporate scanning imagery into the mainstream of softcopy systems. Photogrammetric applications of satellite imagery are beyond the scope of this chapter, but the interested reader is referred to Weston (1990) and Theodossious and Dowman (1990) for a description of photogrammetric rectification using commercial SPOT satellite imagery. Utilization of higher resolution IKONOS and QuickBird imagery and RADARSAT datasets is becoming more common, and several commercial software vendors already offer orientation packages with satellite-specific geometric solutions for interior, relative, and absolute orientations.

6.5 AEROTRIANGULATION

Aerotriangulation is the general term for the method used to bridge control between photos within a strip of stereo models or a block of multi-model strips. Exterior orientation of single stereo models requires a minimum of five control points, two of which have x, y coordinates and three of which have elevation or z-coordinate control. As discussed above, these can be combined such that two horizontal control points with vertical control serve as two of the three z-control points, reducing the total number of points required to three. These photogrammetric solutions for single-model stereo pairs are relatively easy to perform on analytical or softcopy stereoplotters, but they require a high density of control points for each model solution. For small projects, control can be passed along the strip from model to model, but this introduces photogrammetric error that tends to increase the farther the solution is carried down the strip. The only way to avoid this problem is to perform a strip or block adjustment based on aerotriangulation that incorporates control from geodetic markers, large enough to be visible in the photos, strategically placed at ground survey points in the study area. This is usually accomplished with panels made of cloth, plastic, wood, or similar material positioned in a symmetrical pattern centered on bench marks or other survey points prior to the photo flight.

Geologic projects designed around mapping or monitoring applications based on aerial or terrestrial photography comprise multiple models or multiple strips of photos and require complex analytical orientation solutions. Fortunately, the mathematical derivations of these solutions, though complex, are well documented (Wolf and DeWitt, 2000, and references therein) and are available as part of orientation software packages on most analytical photogrammetric systems. There are numerous aerotriangulation schemes for orienting strips or blocks of photographs. Most are optimized for particular types of images or numbers of photos included in the orientation. For our purposes, these schemes can be discussed in generic terms by describing the mechanical steps an operator performs when doing an aerotriangulation adjustment (sometimes referred to as a 'bundle adjustment').

The general procedure involves the same steps required to perform any standard photogrammetric orientation, including: (1) interior and relative orientation of individual stereo models; (2) connection of adjacent models to form continuous strips or multiple strips; and (3) absolute orientation, tying the adjusted model to ground-control coordinates. Basically, aerotriangulation uses condition equations that represent the unknown elements of absolute orientation of each photo in the strip or block in terms of camera parameters, internal photo coordinates, and external ground coordinates. These relationships are expressed as a series of over-determined, non-linear expressions with many unknowns. The expressions are transformed by Taylor's theorem[3] into a series of linear equations that can be solved simultaneously once initial boundary conditions are established. In the same sense that orientation of stereo models on an analog stereoplotter involves iterative reduction of error about any geometric variable on the instrument, aerotriangulation is a numerical solution that minimizes photogrammetric error across the entire strip or block of stereo models. The most common form of equations employed for this purpose describe the condition of collinearity which, as we have seen, is essential to analytical and softcopy photogrammetric orientation of single models.

The most immediate benefits of aerotriangulation in geodetic studies at volcanoes are: (1) ground control can be minimized across the project area, (2) the aerotriangulated solution provides independent evaluation of the geodetic ground control, and (3) the final solution provides for full-field 3-D geodetic control.

A diagrammatic representation of interior pass points and exterior ground control for a typical strip and block of photos is depicted in Figure 6.13. The total number of ground control points necessary for orientation of the entire strip or

---

[3] In calculus, Taylor's theorem, named after the mathematician Brook Taylor, who stated it in 1712, gives the approximation of a differentiable function near a point by a polynomial whose coefficients depend only on the derivatives of the function at that point. The most basic example is the approximation of the exponential function near $x = 0$:
$e^x \approx 1 + x + (x^2/2!) + (x^3/3!) + \cdots (x^N/N!)$.

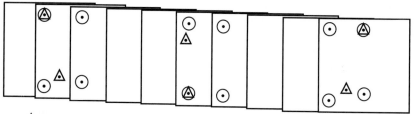

△ Horizontal control point

⊙ Vertical control point

Ⓐ Horizontal and vertical control point

Figure 6.13. Control point configuration for aerotriangulation of a single-strip project consisting of 9 stereo models and 10 photos (from Wolf and Dewitt, 2000). The diagram represents an ideal situation in which control is available in the center of the strip; however, the aerotriangulation can also be performed with control only on both ends. See text for details.

block of photographs is reduced by nearly an order of magnitude relative to that required for the corresponding series of individual stereo models. This is only possible because of the strength of the relative orientation, which is based on the availability of photo identifiable pass points common to adjacent images. The positions of pass points can be measured independently on each photo in analytical stereoplotters, or they can be determined automatically using image-matching algorithms in digital softcopy systems. The great advantage of softcopy systems in this regard is that many pass point measurements can be made in a short period of time, increasing redundancy in the simultaneous equations that are solved for the aerotriangulation, and consequently reducing the error associated with the final solution. Upon convergence, an aerotriangulated solution provides an indication of residual errors associated with any given control point in the exterior orientation, and also a mean residual error associated with the entire orientation. These control point errors can be compared with the errors associated with the original geodetic control point measurements. Large deviations that exceed the residual errors or high residual errors associated with the aerotriangulated solutions indicate a photogrammetric bust in the control point data, which can be attributed to errors in the original control point measurements. In some circumstances, weighting control points relative to the first-order errors associated with individual control point measurements can reduce the mean residual error for the project as a whole. Control points with the lowest first-order measurement errors are given preference over less well-determined ground control in the final solution. Typically, this is not a problem with modern GPS ground control, but it can be an issue when using conventional survey control. Upon completion of an aerotriangulated solution, every point within any stereo model encompassed by the solution has an associated error equal to the mean residual error of the solution.

6.6 TERRESTRIAL PHOTOGRAMMETRY

Terrestrial photogrammetry refers to photogrammetric analysis of photos taken with cameras at ground locations. Historically, a distinction has been drawn between photographs taken from less than approximately 300 m and those taken at greater distances. The former is typically referred to as close-range photogrammetry. Applications in volcanology encompass a wide range of camera-to-object distances, and commonly employ low-altitude aerial platforms such as helicopters or light aircraft in addition to camera stations on the ground. As the distinction is photogrammetrically irrelevant, we have opted to include all non-conventional aerial photography in this discussion under the heading terrestrial photogrammetry.

Two classes of cameras are used in terrestrial photogrammetry. Metric cameras are those designed specifically for photogrammetric use; non-metric cameras include all others. This distinction is made regardless of the type of camera (i.e., a film-based or CCD digital imaging camera). Metric cameras have factory calibrated lenses, reliable construction, vacuum backs for film stabilization, and permanent fiducial marks in the film plane for accurate and repeatable determination of the principal point. Geologists most often use less sophisticated non-metric cameras, typically either (1) 35-mm or medium-format film cameras, or (2) point-and-shoot or single-lens reflex (SLR) digital cameras, which serve double duty as standard field cameras. Only one 35 mm and one medium-format film camera in this category have vacuum backs for film stabilization, but nearly all can be fitted with high-quality, single-focal length lenses suitable for photogrammetric work. Digital cameras, by design, do not require vacuum backs. High-quality, interchangeable lenses designed specifically for use with digital SLRs are widely available, and, in some cases, lenses designed for use with film cameras can be used with digital camera bodies.[4] In either case, the lens should be calibrated for photogrammetric applications.

Lens calibration is essential when using non-metric cameras in photogrammetric applications. Calibrating lenses provides four pieces of information critical to performing interior orientations with terrestrial photographs. The first two are determined by photographing an object of known dimensions, usually a 3-D grid, from multiple angles and inverting the interior orientation solution to solve for the unknown lens parameters. Determination of the second two parameters is more subjective and requires knowledge of both the lens and camera body.

A camera's calibrated focal length represents the distance from the *nodal point* of the lens to the principal point of the photograph. For aerial cameras, this is equivalent to the optical focal length of the lens, but the same is not necessarily true for hand-held, non-metric cameras. Unlike metric cameras, non-metric cameras utilize standard photographic lenses, which commonly produce noticeable *radial lens distortion*. This is observed in standard photographs when horizontal or vertical objects near the outer edges of the photograph appear curved, either inward or outward. This distortion tends to be symmetric and radially distributed outward from the principal point of the lens (Figure 6.14). Modern, high-quality, wide-angle to normal focal length photographic lenses with low to moderate radial distortion are easily calibrated and lend themselves well to terrestrial photogrammetry. These can be either conventional film camera lenses

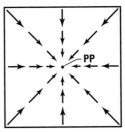

Figure 6.14. Radial distortion pattern characteristic of small- and medium-format camera lenses most commonly used in terrestrial photogrammetry. Arrow tails represent positions of points in object space; heads represent corresponding, distorted positions in image space. Object points nearer to the edges and corners of the lens' field of view are displaced to a greater degree in image space than those nearer to the principal point (PP).

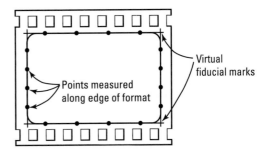

Figure 6.15. Asymmetric or non-square corners on small- and medium-format film positives require extrapolation of photo edges to establish fiducial marks (from Wolf and Dewitt, 2000).

or newer versions designed specially for use with digital SLRs.

Non-metric, small- and medium-format cameras have no fiducial marks visible in the image plane. *Fiducial marks* provide the x and y coordinates of the edges or corners of the film image. These marks are necessary for calculation of the principal point of the image, usually the intersection of obliquely opposing corner marks, and for establishing the position of the images themselves. The corners of image frames can be used, but they are rarely square (Figure 6.15). In most cases, the frame edges forming a given corner can be extrapolated to the point where they intersect, and this point can be used in lieu of corner fiducial marks. It is important to note that the geometry of the image corner is directly related to the film mask of the camera body in use. Hence, calibration usually refers to lens and camera combinations in non-metric applications.

A critical uncertainty in terrestrial photogrammetry using non-metric cameras is the effect of film-plane instability. Interior orientation of small- and medium-format images assumes the film plane is flat across the entire field of view. Cameras with vacuum

---

[4] When a lens designed for use with a 35-mm film camera is used with a digital SLR, the focal length of the lens must be multiplied by a factor that depends on the size of the camera's image sensor. If the image sensor is full-frame (i.e., the same size as 35-mm film), then the multiplier is 1. However, only a few digital SLRs use full-frame image sensors. Most are smaller and, consequently, the multiplier is greater than 1. For example, a conventional 55-mm lens, when used with a digital SLR with a typical focal-length multiplier of 1.6, has an effective focal length of ~88 mm − and a correspondingly narrower field of view. Two salient points to consider when choosing a camera lens combination for photogrammetric applications are: (1) the imaging system's field-of-view and resolution should be consistent with other mission parameters and requirements (e.g., flight altitude, desired DEM accuracy); and (2) for best results, the lens should exhibit minimal radial distortion and must be calibrated.

film backs actuated at the moment of shutter release can approach this condition and reduce this uncertainty essentially to zero. Alternatively, digital CCD-based cameras with a fixed geometry can be used, thereby eliminating this uncertainty altogether. Non-metric film cameras, on the other hand, suffer from variations in film flatness by up to an order of magnitude across the film plane, typically ranging from zero to ten microns. Such deviations are comparable to the measurement capability of many analytical instruments, and therefore they introduce unknown errors to a photogrammetric solution. However, these errors might be insignificant for many geologic applications.

Collecting suitable terrestrial imagery can be as simple as photographing handheld stereo pairs from the ground or air along linear traverses, or as complex as rigorously establishing GPS control for each camera position and monitoring three camera-axis rotations for each photograph (Dueholm, 1992; Dueholm and Pedersen, 1992; American Society of Photogrammetry and Remote Sensing, 1996b). The more information that is available for camera position and orientation, the less ground control is necessary for exterior photogrammetric orientation. Nearly all ground control established for terrestrial applications in volcanology comes from one or more of the following: (1) GPS ground control established at visibly identifiable control points within the terrestrial imagery; (2) visually identifiable control point targets whose positions can be determined by conventional survey techniques; or (3) transferred coordinate data from photogrammetrically rectified aerial photographs. External GPS control or survey control is minimal or not readily available in many volcanically active regions; in such cases, absolute orientation is limited to control transferred from existing aerial photography. For many applications, this provides adequate geodetic control to establish a ground-coordinate datum from which relative geomorphic changes, such as those associated with dome growth, can be rigorously modeled based on relative, as opposed to absolute, position changes.

Terrestrial photogrammetry associated with geodetic monitoring is typified by single strips of overlapping photographs obtained at an oblique angle to the horizontal datum, either looking down into volcanic craters and steep canyons, or up at crater walls and mountain fronts. These situations are characterized by large scale changes across the photograph or strips of photographs, which can result in loss of stereo coverage in parts of individual stereo pairs or across the strip. This problem usually stems from unanticipated variation in the geometry of the camera orientation relative to the ground, or failure to accommodate sharp changes in relief with corresponding changes in base length of the photography. Loss of stereo coverage can be mitigated by varying the photographic interval across the length of the photo mission, in such a way as to accommodate changes in topographic relief. Inadvertent variation in camera orientation can often be avoided by maintaining a distinct line-of-sight relative to the horizon in the camera's viewfinder, and by tracking the progress of a visually distinct image object through the viewfinder. Tripping the shutter after the object of interest moves 40% of the way across the field of view yields 60% stereo coverage in the resultant image. In some cases, multiple strips might be flown at different scales to ensure proper stereo coverage.

The United States Geological Survey (USGS) David A. Johnston Cascades Volcano Observatory (CVO) is using both conventional 9 × 9 inch vertical aerial photographs and small-format oblique stereo photographs to track the dome-building eruption that began at Mount St. Helens in September–October 2004 (Dzurisin *et al.*, 2005). The oblique photos are useful for extending stereo coverage into areas that are obscured in the vertical photos by unavoidable clouds, deep shadows, or the eruption plume.

6.7 APPLICATION TO MOUNT ST. HELENS

The catastrophic eruption of Mount St. Helens on 18 May 1980, involved or initiated geologic processes that produced geomorphic changes amenable to both short- and long-term photogrammetric monitoring. Post-18 May activity included smaller explosive eruptions during summer 1980, and episodic lava dome growth through October 1986 (Swanson and Holcomb, 1990). Contemporaneously, frequent rock avalanches from the steep and unstable crater walls resulted in their rapid retreat and in the formation of large talus cones (Mills, 1992). Within a few years, snow, ice, and rockfall debris accumulating on the crater floor formed a new glacier, the youngest in the Cascade Range (Schilling *et al.*, 2002, 2004). Although some aspects of the activity, particularly dome growth at Mount St. Helens and more recently at active volcanoes elsewhere (Jordan and Kieffer, 1981; Zlot-nicki and others, 1990; Achilli *et al.*, 1998; Baldi *et al.*, 2000), have been analyzed quantitatively using aerial photogrammetry, few active volcanoes are as well characterized through archival aerial photographs as Mount St. Helens. This archival database, spanning two decades and counting, enables quantitative characterization and reconstruction of all aspects of geomorphic change and re-glaciation at the volcano.

Determination of volumetric changes associated with geomorphic change at Mount St. Helens requires systematic retrospective photogrammetric aerotriangulation of aerial photography of the volcano's crater and flanks, followed by DEM extraction from the stereo photography, and determination of volumetric change as a function of time based on DEM grid calculations. The ability to make these determinations in real time using softcopy photogrammetric systems has largely gone unrealized, but it will likely influence future monitoring efforts at Mount St. Helens and elsewhere. For example, Robertson *et al.* (2000) used photogrammetric analysis to develop predictive models of dome growth during the eruption of the Soufrière Hills Volcano, Montserrat, that began in 1995. In addition, geologic mapping of the largely inaccessible Mount St. Helens crater walls based on terrestrial photogrammetry affords geologists a rare opportunity to reconstruct the 3-D geologic record of a young Cascades stratovolcano (Singer *et al.*, 1997; Dungan *et al.*, 2001).

During the years following the 1980 eruption of Mount St. Helens, aerial photographs were used to construct large-scale contour maps of the growing dome (Swanson and Holcomb, 1990), and to estimate the volume of snow and rock accumulation within the crater moat (Mills, 1992; Mills and Keating, 1992). Mills (1992) described the crater walls as one of the most rapidly eroding places on Earth. Frequent and nearly ubiquitous rockfalls spalled off the steep, unstable crater walls and were, in part, incorporated in the growing lava dome. In 2005, rockfalls were still nearly continuous from the crater walls in summer and fall. In winter and spring, the walls are stabilized by moisture trapped mainly as snow and ice – a condition that has prevailed since the crater formed in 1980. As the moat continued to fill with rock, snow, and ice throughout the 1990s, crevasses began to form in the crater fill, indicating the presence of a new glacier. As the glacier grew, so did the concerns of public officials, planners, and emergency managers, who were aware that any new eruptive activity posed a

threat of hazardous lahars or debris flows in the North Fork Toutle River valley. Photogrammetry provided an affordable means to address these concerns and to estimate the volume of talus, snow, and ice in the crater.

Determining the change in volume of geomorphic features in the crater is based in part on a high-resolution, high-accuracy DEM created from 1 : 12,000 scale aerial photographs flown in September 2000. Photogrammetric ground control was based on a network of GPS control targets in the crater and on the outer flanks of the volcano. Precise aerotriangulation of four strips of aerial photographs enabled autocorrelation of hundreds of thousands of x, y, z coordinates and extraction of a TIN surface using commercial softcopy photogrammetric software. The resulting topographic model can be used to calculate volume changes by differencing with models for earlier epochs, and it provides an accurate and well-characterized baseline for analysis of future geomorphic changes.

Roving GPS receivers were used to occupy control points in the study area, which were paneled with standard aerial photography targets. GPS observations were collected for approximately 1 hour at each point and differentially corrected using data from the continuous GPS station JRO1 at the Johnston Ridge Observatory, 9 km north of the crater. Accuracy associated with the resulting aerotriangulation for the entire volcano is on the order of a few centimeters, based on both GPS and photogrammetric uncertainties. High precision resulting from the internal orientation enabled highly accurate placement of TIN surface points and breaklines. The detailed TIN surface for the crater derived from year 2000 aerial photographs was used in combination with the elevation of TIN points and interpolated elevations from TIN facets to create a DEM. The configuration of control points and examples of overlapping aerial photographs are shown in Figure 6.16.

The high degree of accuracy and precision associated with the year 2000 model can be applied to archival models while collecting precise TIN surfaces (i.e., control points visually identified in the 2000 model that fall outside of the region of change can be identified in the archival photography and assigned x, y, z coordinate positions determined from the 2000 aerotriangulation). These new control points in the archival photography are treated as standard ground control points in the subsequent aerotriangulation of the archival strips of aerial photography. In this manner any number of

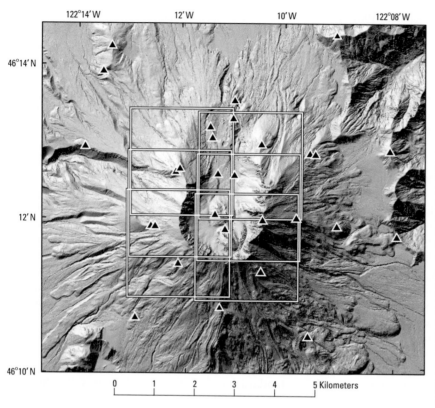

Figure 6.16. Example of aerial photo coverage used to produce a series of DEMs for Mount St. Helens from 1980 to 2005, including the September 2000 DEM discussed in the text. White squares represent the extent of single aerial photographs, overlapping about 60% along north–south flight lines with about 30% sidelap between adjacent flight lines. Black triangles represent control points included in photogrammetric analyses. Shaded relief depiction is based on a November 2003 DEM from lidar and an April 2005 DEM from photogrammetery. Graphic created by Sarah K. Thompson and Steve P. Schilling, USGS.

archival datasets can be used for quantitative extraction of geodetic data without having had the benefit of original ground control. The same approach can be used to visually transfer control into future sets of aerial photographs, even if the original control points are no longer visible.

Even though we have not yet constructed TIN surfaces from previous years of crater aerial photographs, as a preliminary estimate we generated DEMs from 1:24,000 scale digital contours constructed from aerial photographs taken in 1980 (published USGS topographic map) and 1989 (unpublished map data). The vertical accuracy of the DEMs created from contours is derived from the accuracy of the contours. National Map Accuracy Standards specify that the vertical accuracy for contours is one-half of a contour interval (i.e., 20 feet or about 6 meters in this case). The DEM elevation values are taken from the contours or interpolated between contours using an algorithm with eight-way, inverse-distance weighting. The algorithm uses, with appropriate weighting, the distances from each element in the DEM to the first contour encountered in each of the four cardinal and four intermediate directions (N, NE, E, SE, S...) to compute an elevation for each element.

Closer contours are assigned greater weight. The Mount St. Helens DEMs span two decades (vintage October 1980, September 1989, and October 2000) and share a common projection (Universal Transverse Mercator, zone 10), horizontal datum (NAD27), geographic location, and 10-m cell size. Each of the three surfaces is shown in gray shaded relief and colored perspective view in Figure 6.17. In the shaded-relief view, white lines sketched from the 1989 DEM mark the crater rim and outline the 1980–1986 lava dome; an intervening line separates areas of erosion from areas of deposition as determined by subtracting the 1980 DEM from the 2000 DEM.

We calculated volume changes by subtracting one DEM surface from a second one, after ensuring proper alignment of grid cells from one surface to the next, using ArcINFO geographic information system (GIS) software. The absolute values of all resulting negative cells (erosion or subsidence) and positive cells (deposition or uplift) were summed separately and multiplied by the area of a single cell ($100\,m^2$). Following a similar methodology established by Mills (1992), we took the total volume summed from positive values and subtracted the volume of the 1980–1986 lava dome. Then we

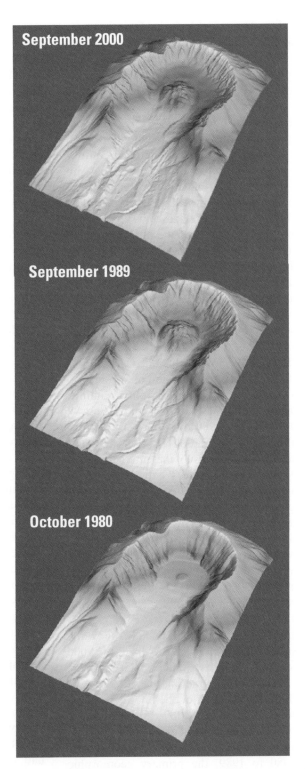

Figure 6.17. Shaded relief and perspective color contoured depictions (left and right, respectively) of three DEMs, showing surface morphology for the crater and upper flanks of Mount St. Helens for October 1980, September 1989, and September 2000. North is toward the bottom in the shaded relief images, and toward the bottom-left in the color contoured images. Yellow lines in the shaded relief images were sketched from the 1989 DEM; they mark the extent of the crater rim, the dome, and the intervening boundary between areas of erosion and deposition. Note progressive development of erosional channels on crater floor north of lava dome. The lava dome grew episodically from 1980 to 1986. Thereafter, the base of the dome was buried progressively by a combination of rockfall debris from the crater walls, annual snow, and perennial ice that together form the newest glacier in the Cascade Range (Schilling et al., 2002, 2004).

subtracted the volume of rock eroded from the crater walls, assuming a closed system. The remaining volume is an estimate of the volume of snow and ice in the crater.

Dome growth at Mount St. Helens began in June 1980 and ceased in October 1986.[5] Aerial photographs were collected frequently during this six-year time period for purposes of visual monitoring, and some were used to construct detailed contour maps of areas within the crater (Swanson and Holcomb, 1990). To estimate the volume of the 1980–1986 dome, we subtracted the 1980 DEM from the 1989 DEM. Our result, $0.09 \, km^3$, differs slightly from the $0.08 \, km^3$ estimate of Swanson and Holcomb (1990), but it agrees well with the calculations of Mills (1992), who suggested the difference is due to Swanson and Holcomb's (1990) exclusion of crater wall debris that was incorporated into the dome as it grew.

The results of our calculations are depicted in Figure 6.18. The three images on the left (Figure 6.18(A)–(C)) show the magnitude and location of both negative values (erosion or subsidence) and positive values (deposition or uplift) obtained by differencing the DEMs for 2000 and 1980 (A), 2000 and 1989 (B), and 1989 and 1980 (C). The three images on the right (Figure 6.18(D)–(F)) show perspective views from northeast of the same surfaces as in (A)–(C), plus isopach surfaces of the net elevation changes for the same three intervals (floating surfaces). Each row in the figure shows three portrayals of the differences created by subtracting DEM surfaces of different vintages. For example, images in the top row show the results of subtracting the 1980 DEM from the 2000 DEM draped over a shaded relief image of the 2000 DEM ((A) and (D)), and also draped over an isopach surface of the net elevation changes during that interval ((D), floating). Similarly, the middle two images ((B) and (E)) show the result of subtracting the 1989 DEM from the 2000 DEM, and the bottom two images ((C) and (F)) show the results of subtracting the 1980 DEM from the 1989 DEM. Each of the images is color-coded: warm colors (red, yellow) for erosion or subsidence and cooler colors (green, blue, purple) for deposition or uplift.

From 1980 to 1989, the primary geomorphic changes within the crater were (Figure 6.18(C) and (F)): (1) growth of the lava dome from 1980 to 1986; (2) erosion of the steep crater walls by mass

wasting; and (3) accumulation of talus, snow, and ice in the moat on the west, south, and east sides of the dome. During the ensuing decade (Figure 6.18(B) and (E)), the dome subsided slightly, probably as a result of cooling and compaction, while material continued to be shed from the crater walls and to accumulate in the moat. By 2000, the deepest fill in the moat was over 240 m thick between the south crater wall and the 1980–1986 dome (Figure 6.18(A) and (D)), and the total volume of snow and ice stored within the crater was approximately $80 \times 10^6 \, m^3$ ($0.08 \, km^3$) (Schilling et al., 2002, 2004).

In addition to creating sequential DEMs that capture volumetric changes in the crater through time, we have used terrestrial photogrammetry to support detailed 3-D geologic mapping of the crater walls. The resulting maps document multiple eruptive and erosion cycles during the last millennium at Mount St. Helens. For the latter application, visually identifiable control, based on the year 2000 aerotriangulated model, was used to establish control in oblique aerial strips of stereo photographs collected from a helicopter flying north–south and east–west flight lines in the crater. The general configuration of the data collection for this terrestrial photogrammetric study is depicted in Figure 6.19(A). The resulting stereo photographs from a segment of the west crater wall are shown in Figure 6.19(B), along with an example of the minimum control requirements typically associated with a short aerotriangulated strip. In this case, prominent outcrops and geologic contacts were established as visually identifiable objects in both the aerial photographs and the oblique strips. The rigorous treatment of oblique photography in photogrammetric studies such as this enables precise determination and analysis of 3-D geologic contacts, which record past eruptive and erosive events but typically are not exposed at mid-latitude stratovolcanoes of the northern hemisphere.

Starting in late September 2004, another prolonged episode of dome growth at Mount St. Helens dramatically altered the topographic configuration of the 1980 crater. By March 2005, more than $40 \times 10^6 \, m^3$ of gas-poor, crystal-rich dacite lava had extruded onto the crater floor immediately south of the 1980–1986 dome. The new glacier in the crater was intensely deformed and eventually cut in two by the encroaching dome, resulting in spectacular crevassing and rapid advance of the east arm of the glacier (Figure 6.20) .

Not surprisingly, photogrammetry is playing a

---

[5] Dome growth resumed in September 2004 and was continuing when this chapter went to press in January 2006.

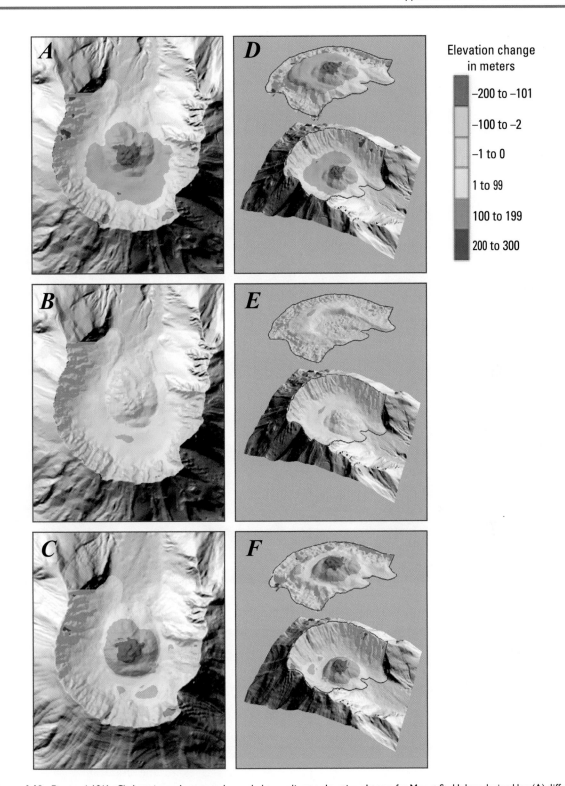

Figure 6.18. Figures 6.18(A–C) show isopach maps, color-coded according to elevation change, for Mount St. Helens derived by: (A) differencing 2000 and 1980 DEMs and draping on the 2000 DEM; (B) differencing 2000 and 1989 DEMs and draping on the 2000 DEM; (C) differencing 1989 and 1980 DEMs and draping on the 1989 DEM. Figures 6.18(D–F) show perspective views from northeast of the same color-coded isopach maps draped over the same shaded relief maps as in (A–C). Above each shaded relief map in (D–F) is a shaded isopach surface color-coded according to thickness (i.e., the floating surfaces represent elevation *changes* during 1980–2000 (D), 1989–2000 (E), and 1980–1989 (F). Negative isopach values indicate erosion, subsidence, or both. Positive isopach values indicate deposition, uplift, or both. See text for discussion. Cell size for the DEMs is 10 meters.

Figure 6.19. (A) Oblique view southward into Mount St. Helens crater showing approximate orientation of N–S flight line used to collect oblique stereo photography of the east and west crater walls. This photograph was acquired on standard 230-mm (9-inch) film with a metric camera and a relatively long focal length (305-mm, 12-inch) lens (noted at top-right edge of photo), which provided a 56° angular field of view. White panels (1, 2, 3) show approximate coverage of oblique photos shown in (B), which were aquired with a small-format (35-mm) non-metric camera and a 28-mm wide-angle lens. (B) Small format (24 mm × 36 mm) stereo photographs of Mount St. Helens' west crater wall with relative positions shown in (A). White dots indicate minimum control point configuration used to orient the oblique photo strip. Coordinate locations of control points were transferred from September 2000 aerial photography (1 : 12,000 scale).

2004-2006 lava dome

Deformed glacier

1980-1986 lava dome

Figure 6.20. The 1980 crater at Mount St. Helens as it appeared on 11 March 2005. View is to the southwest; north is to the lower right in the photo. A lava dome that grew episodically from 1980 to 1986 had been surrounded on three sides and partly buried by a combination of rockfall talus from the crater walls and perennial ice, which persisted owing to the shadowing effect of the steep crater walls and to insulation by annual (summer) talus layers. The rock-and-ice glacier, which was thickest south of the 1980–1986 dome, moved northward around the dome's east and west flanks. The 2004–2005 lava dome issued from a southward-inclined vent near the south margin of the older dome, severely disrupting the thickest part of the glacier on the south crater floor. The east arm of the glacier, which was pushed eastward and upward by the growing dome, became deeply crevassed and its northward motion accelerated dramatically. Dark material on the east arm of the glacier and east crater wall is airfall ash lofted by a small explosive event on 8 March 2005. USGS photograph by Steve P. Schilling.

key role in documenting and interpreting this latest activity. Both conventional 9 × 9-inch vertical aerial photographs and small-format oblique stereo photographs of the growing dome are being acquired at 2–4 week intervals, the latter from a helicopter using a digital SLR camera with a calibrated lens. The oblique photos are particularly useful for extending stereo coverage into areas that are obscured in some of the vertical photos by clouds, deep shadows, or a persistent eruption plume. High-resolution (2-m cell size) DEMs are being produced from the photos within several days to a few weeks, depending on the urgency of the situation and area of interest. The entire 1980 crater is included in each set of vertical or oblique photos, but the area of interest for DEM construction varies. Should the area of interest increase in the future (e.g., as the dome grows or more of the glacier is deformed), the DEMs for past epochs can be

extended using a digital photo archive – an efficient and cost-effective approach. The resulting archive of digital photographs and 3-D terrain models is likely to be the most detailed ever assembled during a dome-building eruption, and it constitutes a valuable resource for current and future investigators.

Clearly, photogrammetry has earned its place among other geodetic techniques in the modern volcanologist's tool kit. As digital SLRs become more affordable and widely available, and softcopy stereoplotters become more generally accessible, the analytical power of photogrammetry will extend to many of the world's deforming volcanoes. Great care still will be required, both in the field and on the stereoplotter, to produce optimal results. But the potential exists for legions of geologists, seismologists, geochemists, and others – already armed with digital cameras – to serve the cause of volcano geodesy. Now, *that's* exciting!

CHAPTER 7

Lessons from deforming volcanoes

Whereas previous chapters are focused primarily on geodetic techniques, this one addresses specific applications of those techniques at a few well-studied volcanoes. Mount St. Helens, Kīlauea, Yellowstone, and Long Valley illustrate the tremendous diversity in terms of style, duration, dimensions, and mechanisms of ground movements. That is not to say that these four examples encompass the full range of deformation observed at volcanoes worldwide, or that other examples would not serve equally well. Rather, I chose these four because I know them personally, having worked at each for at least several years, and because they have been studied thoroughly with a variety of monitoring techniques, including geodesy. Therefore, the geodetic results can be placed in a rich context of other information to help interpret the processes responsible for ground deformation and other symptoms of unrest.

Geodesy provides tools for scientists to study processes as diverse as dome growth and edifice instability, over length scales of meters to kilometers, for periods of minutes to decades, across a wide spectrum of magmatic and tectonic settings. Most volcanoes, even during periods of no visible activity, are dynamic landforms that respond to magmatic, tectonic, or hydrothermal processes. Ground deformation provides a useful window into those processes that can be explored through an integrated program of geodetic monitoring. The next four sections highlight a few of the things that have been learned from geodetic investigations at four very different volcanic systems: Mount St. Helens (stratovolcano), Kīlauea (intraplate shield volcano), Yellowstone, and Long Valley (silicic calderas).

7.1 MOUNT ST. HELENS – EDIFICE INSTABILITY AND DOME GROWTH

In hindsight, the remarkable series of events at Mount St. Helens from 1980 to 1986 was not unprecedented in that volcano's relatively brief eruptive history or in the geologic records at other volcanoes, even during historical time. Nonetheless, the 1980–1986 activity at Mount St. Helens arguably had a disproportionate impact on volcanology and on the public's perceptions of volcano hazards in the USA. As a direct consequence of the Mount St. Helens experience, important strides were made during the 1980s in volcano monitoring, eruption prediction, and hazards assessment techniques. These gains have since been applied and extended during responses to numerous volcano emergencies around the world.

Geodetic measurements played an important role at Mount St. Helens, both before and after the 18 May 1980 eruption. During the pre-climactic phase, repeated EDM measurements of the famous 'bulge' tracked the growth of a cryptodome and increasing instability of the volcano's north flank. Following the explosive events of 1980, more than a dozen successful predictions of extrusive dome-growth episodes were issued on the basis of repeating patterns of ground deformation and seismicity. The following account highlights a few of the geodetic techniques used to study the 1980–1986 activity to illustrate how volcanologists were able to adapt their monitoring strategy to rapidly changing conditions at the volcano.

7.1.1 Precursory activity: the north flank 'bulge'

At 3:47 p.m. PST on 20 March 1980, an M 4.2 earthquake shook the snow-covered slopes of Mount St. Helens in southwest Washington. In hindsight, the quake was a wake-up call. More than a century of volcanic slumber was about to come to an explosive end. Many residents of the area were unaware that their 'mountain', which had not erupted in anyone's memory, was actually a volcano. But the mountain's true character was clearly recorded by its young and plentiful eruptive products, which tagged Mount St. Helens as the most active and explosive volcano in the Cascade Range. Five years earlier, United States Geological Survey (USGS) geologists Dwight R. 'Rocky' Crandell and Donal R. 'Don' Mullineaux, who had studied the volcano for nearly two decades, forecast that it would erupt again '... *within the next hundred years, perhaps even before the end of this century*' (Crandell *et al.*, 1975; Crandell and Mullineaux, 1978, p. C25).

The shallow earthquake swarm that began on 20 March 1980, and intensified during the next several days, was the first recognized precursor to the renewed volcanic activity that Rocky and Don had forecast. An electronic distance meter (EDM) network had been established on the volcano's flanks in 1972, but the bench marks lay deeply buried under snow. When the endpoints of one of the EDM lines on the east side of the volcano (Smith Creek Butte and East Dome, located 10.5 km and 3 km, respectively, from the volcano's summit) were dug out and the line length remeasured in April 1980, it was discovered that there had been virtually no change since 1972 (Figure 7.1). No tiltmeters or other continuously recording instruments to measure ground deformation had been installed at the long-dormant volcano, so any geodetic precursors to the first earthquakes, if they occurred, went unnoticed.

Within a week, however, all that was needed to recognize the profound topographic and structural changes occurring at the volcano was a keen eye. A new crater formed during the first of many phreatic eruptions and the summit area was cut by an east–west-trending system of fractures. Most telling of all, the area north of the fractures appeared to be bulging outward and possibly upward, destabilizing the volcano's north flank (Christiansen and Peterson, 1981; Lipman *et al.*, 1981a) (Figure 7.2). USGS geologists who rushed to the scene had studied

mainly the active volcanoes of Hawai'i or the deposits of long-quiescent volcanoes in the Cascade Range. At Mount St. Helens, they were confronted with the unfamiliar task of monitoring a reawakening stratovolcano. Those responsible for measuring ground deformation used traditional geodetic techniques and developed a few new ones to cope with the rapidly evolving situation.

The first priority was to ascertain whether the fractures and other obvious surface disturbances in the snow-covered summit area were caused primarily by earthquake-induced ground shaking or by intense deformation of the volcano. As is often the case during volcanic emergencies, circumstances conspired to make this task difficult. Phreatic eruptions starting on 27 March and poor weather in early April prevented geodetic measurements on the upper flanks of the volcano. Unable to pursue their first priority, scientists instead did what was possible under the circumstances. They deployed two continuously recording platform tiltmeters on the lower flanks of the volcano and three more around its base. In addition, they used the mostly ice-covered surface of Spirit Lake, about 10 km north of the summit (Figure 7.1), as a large 'natural' tiltmeter and established two single-setup leveling triangles on the volcano's north flank.

Water level measurements at Spirit Lake were simple, easy to make even in bad weather, and important for determining the spatial extent and magnitude of any ground deformation that might have been occurring (Lipman *et al.*, 1981a). Inexpensive wooden yardsticks were attached to fixed objects such as tree stumps or dock piers at six sites around the lakeshore where open water was present. Water levels were read in rapid succession, typically about 20 minutes, to reduce seiche effects. Through mid-April, while ice on most of the lake served to dampen wave action, water levels could be read with a precision of 1/16 of an inch (1.6 mm). Thereafter, precise measurements could only be made at times when the lake surface was relatively calm. Any tilt of the lake basin would have been manifest as a change in differential water levels around the lakeshore. The size of the lake and precision of the measurements combined to provide a detection threshold of about 1 part in 500,000, equivalent to a tilt of 2 microradians (μrad), at a distance of 9–12 km from the volcano's summit.

The differential water level measurements at Spirit Lake revealed no significant tilts during the first critical weeks of the unrest, which meant that any deformation source was either relatively small or

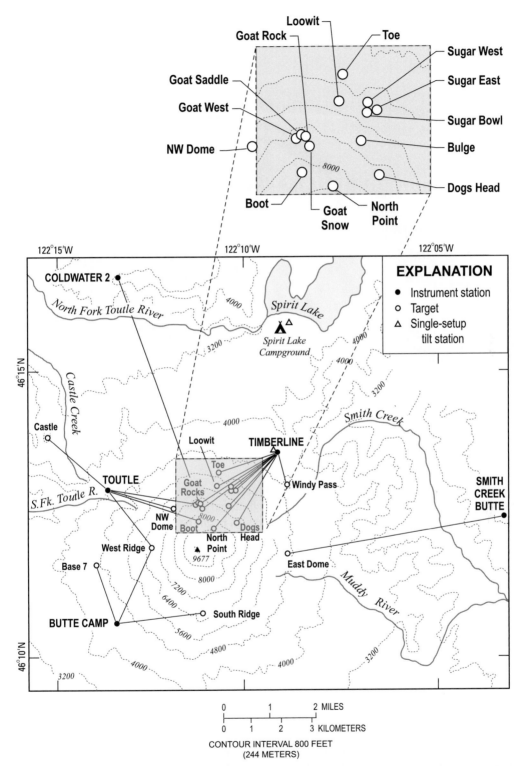

Figure 7.1. Stations used for measuring ground tilt and surface displacements at Mount St. Helens during April–May 1980 (Lipman *et al.*, 1981a). Six water-level stations (not shown) along the shore of Spirit Lake and a single-setup leveling array at Spirit Lake Campground (triangle), all 9–12 km from the summit, detected no significant tilt. Solid dots and open circles connected by lines represent places where vertical angles and slope distances were measured repeatedly using an EDM and theodolite, revealing a bulge centered near Goat Rock on the volcano's north flank. Instrument stations are shown as dots and labeled in uppercase, targets as circles and in lowercase. A single-setup leveling array at Timberline (triangle), 4 km from the summit, showed progressive tilt down to the northeast (i.e., summit uplift) and short-term, large tilt changes directed toward or away from the bulge (Figure 7.3).

Figure 7.2. Oblique aerial photograph taken on 7 April 1980, looking south at Mount St. Helens. Mount Hood, 100 km distant, is visible on the horizon at upper left. A thin layer of volcanic ash mantles fresh snow near the summit. A series of phreatic eruptions that began on 27 March 1980 deposited ash and formed the summit explosion crater seen here, which by this date had grown in diameter to about 500 m W–E and 300 m N–S. Dark streaks on the steep north flank are from fresh rockfalls. The intensely cracked 'blister' below and north of summit crater shows distension of the north flank at this early date. This feature enlarged gradually during the ensuing weeks. By 24 April, it became known as 'the bulge.'
USGS photograph by Richard B. Waitt.

shallow. In the crisis atmosphere that prevailed, this was vitally important information. If deformation was occurring beyond the disturbed summit area, it did not extend as far as Spirit Lake. Those responsible for monitoring the volcano moved closer.

Two single-setup leveling arrays (sometimes called 'dry tilt' arrays; see Yamashita, 1981, 1992; Section 2.5.2) were established in late March and early April in the parking lots of the Spirit Lake and Timberline campgrounds (Figure 7.1). The Spirit Lake array was occupied only once, because water level measurements at nearby Spirit Lake detected no significant tilting that far from the volcano. The Timberline station, on the other hand, was measured seven times between 30 March and 30 April. The results showed a generally consistent tilt of about 2μrad day^{-1} down to the northeast (i.e., away from the summit of the volcano). The source was confirmed in various ways to be a bulge growing on the volcano's north flank.

In addition to progressive tilting away from the bulge, the measurements at Timberline revealed a remarkable pattern of short-term inflationary and deflationary tilt cycles. The first clue that such deformation might be occurring came from poor closures of the tilt-leveling array. After eliminating other possible sources of error, observers began measuring a 40-m line every few minutes for several hours. The measurements revealed that significant tilt changes were occurring over periods of a few minutes. Thereafter, a small (14 m) triangular array was used to reduce measurement errors and the time required for closure. Results showed both inflationary and deflationary tilts of as much as 50μrad hr^{-1} directed away from or toward the center of the bulge, respectively (Figure 7.3). On 10 April, the bullseye bubble used for rough leveling of the instrument moved off level while the observers watched (Lipman et al., 1981a, p. 146)!

I had a similar experience while making microgravity measurements at East Dome on the volcano's east flank. To my amazement, the meter's level bubbles moved well off center in a direction suggesting inflation several seconds before the start of a phreatic eruption in the summit area. By the time the

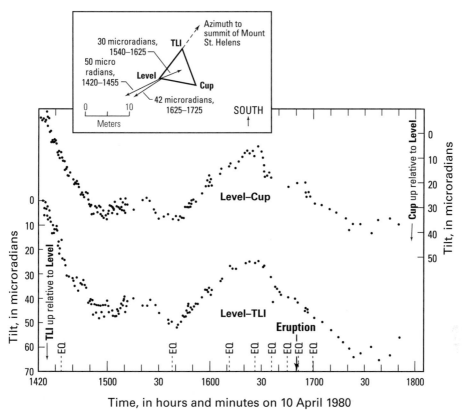

Figure 7.3. Short-term tilt fluctuations measured by single-setup leveling at Timberline, on the northeast flank of Mount St. Helens, from 14:20–17:50 PST on 10 April 1980 (Lipman *et al.* 1981a). Tilt vectors (*inset*) derived from elevation changes at TLI and Cup relative to Level. Vectors point toward or away from the center of a bulge growing on the volcano's north flank. Occurrences of felt earthquakes (EQ) and a phreatic eruption (Eruption) are also noted on the time axis. No consistent relationship was recognized among tilt changes, earthquakes, and eruptions.

eruption waned a few minutes later, the inflation episode was over and the bubbles were centered again. It was an adrenaline-charged experience that I'll never forget. Everyone who works near active volcanoes develops a personal threshold of acceptable risk. Maybe we were lucky that day on East Dome and on many subsequent occasions, but we were never capricious. Each of us assessed the risks and rewards as we understood them and made deliberate decisions about how to proceed within the boundaries established by those in charge. We did not always agree, but we respected each other's right to decide and act accordingly.

The source of the short-term tilt fluctuations at Mount St. Helens remains unclear even to this day. One plausible explanation is that they reflected pressurization–depressurization cycles in a shallow hydrothermal system driven by the intrusion of a cryptodome. The timing of inflation immediately before some of the phreatic eruptions, followed by deflation during the eruptions, supports

this hypothesis. On the other hand, this pattern was not observed consistently and the orientation of short-term tilt vectors toward or away from the center of the bulge, rather than the summit area, suggests that the tilts might instead have been caused by jerky movement of magma intruding beneath the bulge (Lipman *et al.*, 1981a, p. 146).

Five continuously recording tiltmeters were installed around Mount St. Helens during late April and early May 1980 at distances of 3–15 km from the summit (Dvorak *et al.*, 1981). The biaxial, platform-type instruments, with ~0.1-μrad resolution, were secured with expansion bolts to cast-concrete base plates that were cemented to rock outcrops. Surface installations are quick and can be useful during a crisis or where ground tilts are large and rapid, but they are subject to large diurnal fluctuations and other noise sources that limit their effectiveness in most situations (Dzurisin, 1992a). At Mount St. Helens, they provided important information about ground deformation on a continuous

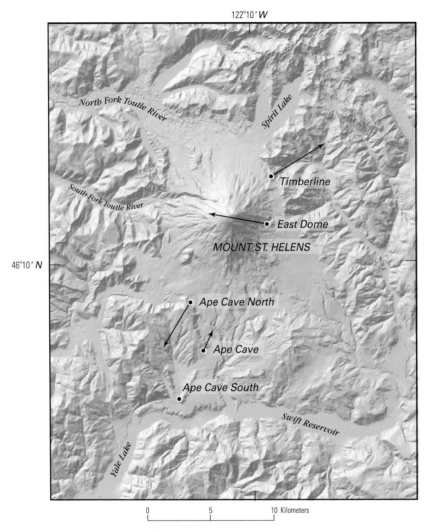

Figure 7.4. Locations of electronic tiltmeters around Mount St. Helens prior to the landslide and eruption of 18 May 1980, and net ground-tilt vectors for the period from 6 May to 18 May 1980 (Dvorak *et al.*, 1981). Tilt changes are represented by arrows that point away from an uplift source or toward a subsidence source. In this case, the Timberline and Ape Cave North stations tilted away from the summit area, while the East Dome station tilted toward it. Tilt changes at the Ape Cave and Ape Cave South stations were small enough to be considered not statistically significant. Dvorak *et al.* (1981) interpreted these trends as a general inflation of the volcano plus formation of a graben across the summit area, the latter to explain the East Dome result.

basis, including times when other geodetic observations were impossible owing to darkness or poor weather.

Two of the three tiltmeters within 6 km of the summit recorded net inflationary trends (i.e., points closer to the volcano moved up relative to points farther away) from the time they were installed until the onset of the eruption on 18 May (Figure 7.4). The third, at East Dome on the volcano's east flank, recorded an apparent deflationary trend of approximately the same magnitude (10–15 μrad). Two tiltmeters located more than 6 km from the summit recorded only minor tilt changes. In hindsight, Dvorak *et al.* (1981) attributed this tilt pattern to a general inflation of the volcano plus formation of an east–west-trending graben across the summit area. Unfortunately, the data are too sparse to convincingly test this idea or distinguish among alternative models. Imagine the frustration felt by scientists in the midst of the crisis, when the volcano was often shrouded in clouds and the tiltmeters were telling an ambiguous story.

The tiltmeter results at Mount St. Helens emphasize the need for as much geodetic information as possible during a volcanic crisis (Chapter 11). The

Figure 7.5. Timberline theodolite and EDM station on 1 May 1980 (Lipman *et al.*, 1981a). Named features are sites of fixed reflectors. Dashed line indicates approximate limits of the deforming bulge and of rock subsequently removed during the 18 May debris avalanche and lateral volcanic blast.
USGS photograph by P.W. Lipman.

surface deformation field is likely to be complex in both space and time, especially if more than one deformation source is involved (e.g., an inflating magma reservoir, opening dike, pressurizing or depressurizing hydrothermal system, or slipping faults). To distinguish among these possibilities with any real confidence, the spatial and temporal evolution of the deformation field must be known in considerable detail. This is especially difficult during a rapidly evolving crisis, when hazards to field personnel and limited resources must also be considered. Scientists facing this dilemma at Mount St. Helens during the spring of 1980 did what they could under trying circumstances, with generally good results.

The most revealing geodetic dataset during the two-month precursory period from 20 March to 18 May came from repeated measurements of horizontal angles, vertical angles, and slope distances among an array of bench marks on the flanks of the volcano and around its base (Lipman *et al.*, 1981a). Every 1–3 days, weather permitting, theodolite and EDM

measurements were made from 5 instrument stations around the volcano to as many as 19 fixed targets on the edifice (Figures 7.1 and 7.5). The data were combined to calculate displacement vectors for the targets relative to the instrument stations.

The geodetic results were alarming. From 23 April through the early morning of 18 May, several targets located within an elliptical area about $1.5\,km \times 2.0\,km$ in size were observed to move steadily northward at rates of 1.5–$2.5\ m\,day^{-1}$ (Figures 7.6–7.9). The movements were subhorizontal, indicating that the deforming area was not simply sliding downslope under the influence of gravity, but rather was being forced laterally outward by intrusion of a cryptodome (Figure 7.10). Stations outside the rapidly deforming area, including Dogs Head, East Dome, South Ridge, and West Ridge, were remarkably stable. Especially notable were the results from East Dome, which was first measured from Smith Creek Butte in 1972. No significant changes were observed during re-occupations of that line on 10

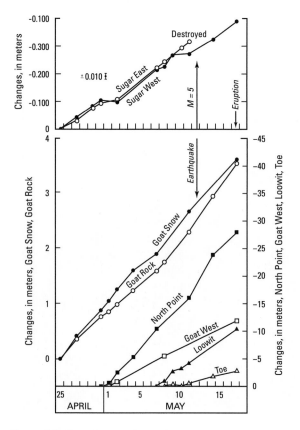

Figure 7.6. Slope-distance changes between two instrument stations and several targets on the bulge measured with an EDM, 25 April to 18 May 1980 (Lipman et al., 1981a). Goat West was measured from Toutle, all others from Timberline. See Figure 7.1 for locations.

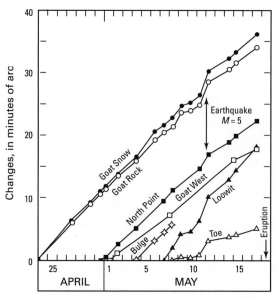

Figure 7.7. Horizontal-angle changes to several points on the bulge measured by theodolite, 23 April to 18 May 1980 (Lipman et al., 1981a). Goat West and Bulge targets were measured from Toutle, all other targets from Timberline. See Figure 7.1 for instrument and target locations.

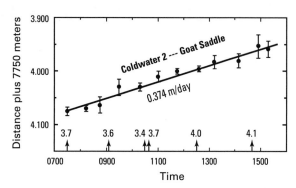

Figure 7.8. Changes in slope distance between Coldwater 2 and Goat Saddle (Figure 7.1) measured with an EDM for approximately 8 hours on 4 May 1980 (Lipman et al., 1981a). Error bars indicate spread among measurements at three different frequencies. Arrows indicate times of earthquakes under Mount St. Helens with magnitudes greater than 3.5. The deformation rate increased from about $0.4\,\text{m}\,\text{day}^{-1}$ on 4 May to an average of $1.4\,\text{m}\,\text{day}^{-1}$ between 4 May and 16 May, and then slowed to about $0.5\,\text{m}\,\text{day}^{-1}$ until the landslide and eruption on 18 May.

April, 25 April, or even after the 18 May eruption (Lipman et al., 1981a). The significance of this null result is explored further in Chapter 11.

By mid-May 1980, the increasing instability of the north flank was apparent to all concerned, prompting the question: What would be the outcome, and when? The geologic record at Mount St. Helens included deposits from numerous plinian eruptions in the past few thousand years, so a vertically directed explosive eruption seemed the most likely scenario. The volcano had also produced several young dacite domes, including an amalgamation of domes called the Summit Dome that comprised the upper third of the edifice. Given the compelling evidence for a cryptodome actively growing beneath the bulge, a relatively non-explosive dome-building eruption was also a distinct possibility. Outcomes regarded as less likely included the extrusion of a lava flow, like those exposed in several places on the volcano's flanks, or a debris avalanche triggered by failure of the north flank. Ironically, no evidence for

previous large debris avalanches at Mount St. Helens was exposed at the volcano until the crater that formed on 18 May 1980, was eroded by subsequent lahars and seasonal stream flow. Only then were deposits from two earlier debris avalanches exposed for study (Hausback and Swanson, 1990). Of course, another possible scenario was that the 1980 unrest would simply stop

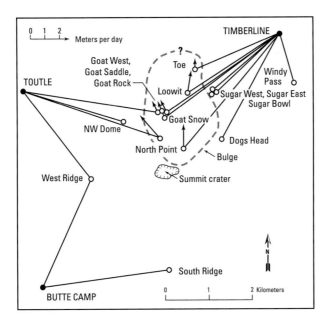

Figure 7.9. Typical daily displacement vectors in the area of the bulge on the north flank of Mount St. Helens during the first half of May 1980 (Lipman *et al.*, 1981a). Instrument stations are in uppercase, targets in lowercase. Targets without vectors showed displacements too small to plot. Dashed line represents approximate limit of the bulge.

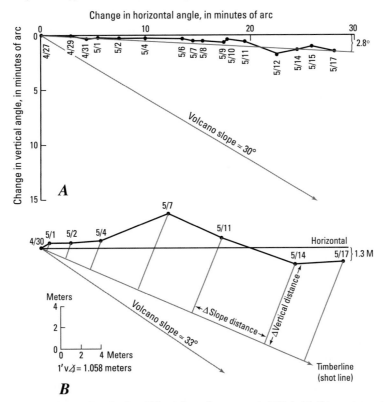

Figure 7.10. Vertical changes at stations Goat Rock and North Point (Lipman *et al.*, 1981a). (A) Changes in vertical angle versus changes in horizontal angle for Goat Rock from 27 April to 17 May. Average local slope of the volcano near the target is about 30°, much steeper than the displacement vector. (B) Vertical angle versus slope distance for the North Point target from 30 April to 17 May. For each measurement interval, the change in slope distance is scaled along the shot line, and the change in vertical angle is converted into a length (1 minute vertical angle = 1.058 m for the 3.65 km line) and plotted perpendicular to the shot line. Again, the average volcano slope is much steeper than the displacement vector. Therefore, points on the bulging north flank were being shoved outward, in hindsight by a growing cryptodome, not merely sliding downslope under the influence of gravity.

and the volcano would resume its century long slumber.

Aware as they were of these diverse possibilities and of the increasing hazard to observers near the volcano, scientists wrestled with the conflicting requirements for personal safety and better information about what was happening. No single source of geodetic data was sufficient to adequately monitor the rapidly evolving situation, so in addition to the continuously recording tiltmeters, microgravity and EDM–theodolite surveys were continued through 9 May and the morning of 18 May, respectively. EDM measurements from Coldwater 2 to Goat Rock on 16 and 17 May, and at 6:53 a.m. PDT on 18 May showed that the deformation rate slowed from an average of $1.4 \, \text{m day}^{-1}$ from 4 May to 16 May to $0.5 \, \text{m day}^{-1}$ from 16 May to the last measurement an hour and a half before the north flank failed catastrophically at 8:32 a.m. PDT on 18 May.

Faced with the same situation today, observers might try to install continuously recording tiltmeters, strainmeters, and Global Positioning System (GPS) receivers on the bulge as a substitute for geodetic surveys that require field personnel to go repeatedly in harm's way. At Mount St. Helens in 1980, the window during which such installations would have been possible in relative safety, as determined by weather and activity at the volcano, was very narrow. Less than two weeks elapsed between the onset of unusual seismicity beneath the volcano (15 March) and the first phreatic eruption (27 March). Theory suggests that deformation of the bulge might have accelerated rapidly in the final minutes before failure of the north flank, but there is no evidence for this in the distal tiltmeter records. The use of radar interferometry to study the bulge would have been hindered by coherence loss caused by the spring snow pack and by the relatively long orbital repeat cycles of present-day radar satellites. Even in hindsight and with the benefit two more decades of geodetic sophistication and experience, obtaining this critical piece of information without exposing field crews to substantial risk would have been difficult at best. A solution to a similar dilemma emerged in 2004 – in the form of a 'spider' – during another period of intense surface deformation at Mount St. Helens (Section 11.3.4).

A lesson to be learned from the 1980 experience is that geodetic monitoring of dangerous volcanoes is best accomplished by installing continuously recording instruments at key sites as early and robustly as possible, in order to minimize subsequent visits when conditions might be more hazardous. If the unrest persists for several months, aerial photogrammetry or satellite radar interferometry can sometimes provide additional information without exposing field crews to the restless volcano. On the other hand, even as continuous monitoring and remote-sensing techniques continue to improve, trained observers are essential to provide context, analysis, and an assessment of the hazards.

7.1.2 Monitoring and predicting the growth of a lava dome

Another example that illustrates the importance of adapting geodetic monitoring techniques to changing volcanic conditions is the case of episodic dome growth at Mount St. Helens from 1980 to 1986. Following the catastrophic debris avalanche and eruption on 18 May, five smaller explosive eruptions occurred between 25 May and 18 October 1980 (Christiansen and Peterson, 1981). Dacite domes were emplaced on the floor of the 18 May crater in the waning stages of three of those eruptions, but the first two domes were mostly destroyed by subsequent explosive eruptions. The October 1980 dome survived as the core of the 1980s dome, which developed during a series of exogenous and endogenous growth episodes that ended in October 1986 (Swanson et al., 1987).[1] Each of these episodes was successfully predicted tens of minutes to three weeks in advance, on the basis of a recurring pattern of seismicity and ground deformation in the crater and eventually on the dome itself.

A key ingredient for each of the successful predictions at Mount St. Helens was seismic monitoring that allowed the recognition of three main types of seismicity: (1) tectonic-like earthquakes with focal depths greater than 4 km beneath the volcano or at any depth away from the volcano, which produced high-frequency impulsive arrivals (i.e., volcano–tectonic (VT) earthquakes); (2) earthquakes at depths of less than 3 km beneath the dome, which produced medium- to low-frequency arrivals (including long-period (LP) earthquakes); and (3) surface events including rockfalls and gas bursts from the dome with complicated signatures and generally emergent onsets (Malone et al., 1983).

The second and third types of seismicity were most

---

[1] A prolonged dome-building eruption at Mount St. Helens that began in September–October 2004 produced another dacite dome immediately south of the 1980s dome (Dzurisin et al., 2005). The eruption was continuing in March 2006.

diagnostic of impending dome-growth episodes over periods of a few days to weeks. Typically, both the number of shallow earthquakes and the rate of seismic energy release increased progressively for several days to four weeks before dome-building episodes. A sudden pronounced increase in the occurrence of shallow quakes, usually a few hours before the onset of exogenous dome growth, was a reliable short-term indicator of impending activity. After magma reached the surface, shallow earthquakes essentially stopped and surface events dominated the seismic records. In some cases, when the onset of exogenous growth was unobserved owing to darkness or poor weather, the marked change from intense shallow earthquake activity to mostly rockfall activity was the only basis for concluding that magma had reached the surface and an eruption was underway (Malone *et al.*, 1983).

The pattern of seismic activity varied from episode to episode, partly as a result of the changing morphology of the dome. Endogenous growth became more prevalent as the dome grew larger; thus, in some cases, rockfalls from the dome increased dramatically several days before magma eventually appeared at the surface. Another factor that affected the timing and intensity of rockfall activity was the changing stability of the outer parts of the dome. Growth episodes that began when parts of the dome were gravitationally unstable, as was the case for some extrusive lobes that were perched on steep slopes and for large cliffs that formed on the dome as a result of intense deformation, were characterized by earlier and more intense rockfalls.

Now let us consider the geodetic data, which takes several forms. Electronic tiltmeters placed on the crater floor near the dome or on the dome itself were one source of useful information. Near-surface installations with detection thresholds of a few to several tens of microradians were adequate to monitor tilt changes within a few hundred meters of the dome starting several weeks before each growth episode (Dzurisin *et al.*, 1983) (Figure 7.11). Tilting was generally outward from the dome and started to accelerate rapidly a few hours to several days before the start of exogenous growth. The tilt direction often was affected by nearby cracks or faults, but typically reversed sharply from outward (i.e., away from the center of the dome) to inward several minutes to hours before magma reached the surface (Figure 7.12). These reversals may have reflected the combined effects of depressurization within the dome and increased surface loading as

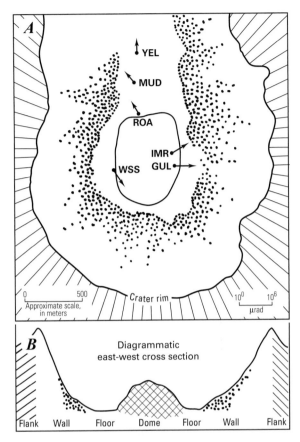

Figure 7.11. Sketch map of the 1980 crater and lava dome at Mount St. Helens (A) showing the locations of electronic tiltmeters from May 1981 to May 1982, and diagrammatic east–west cross section (B) (Dzurisin *et al.*, 1983). Net tilt vectors for periods that culminated in eruptions (i.e., dome growth episodes) are shown on a logarithmic scale to accommodate very large changes at some stations. Not all tiltmeters operated simultaneously; all were eventually destroyed by rockfalls or eruptive activity. Vectors for several periods are shown together to illustrate the typical pre-eruption pattern of tilting outward from the dome. Tilt data are for the following intervals: IMR, 29 May to 23 June 1981; GUL, 3 July to 6 September 1981; YEL, MUD, and ROA, 13 October to 31 December 1981; WSS, 29 April to 15 May 1982.

magma broke through to the surface and moved onto the exterior of the dome.

One of the more effective geodetic monitoring instruments used to predict dome-building eruptions at Mount St. Helens was decidedly low-tech (i.e., a common steel tape measure). Observers noted that the crater floor surrounding the dome often became cracked and wrinkled several weeks before dome-building eruptions (Figure 2.15). The cracks, which were mostly radial to the dome, were caused by uplift and extension of the crater floor as magma rose in the feeder conduit beneath the dome. The 'wrinkles' were shallow thrust faults that formed because the crater floor was being shoved outward against the

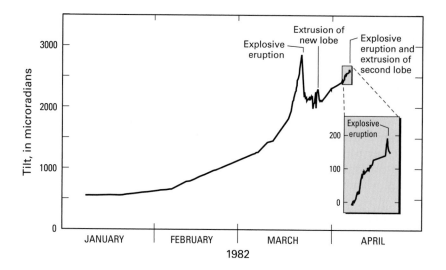

Figure 7.12. Tiltmeter data from station ROA (see Figure 7.11) for the period January–April 1982, which included an explosive eruption and lahar on 19 March, extrusion of a small lobe onto the lava dome during 20–24 March, and smaller explosions and extrusive of a second lobe during 5–10 April (Dzurisin *et al.*, 1983). Detectable radial uplift began in mid-January and accelerated sharply on 16 March. Rapid subsidence started 30 minutes before the 19 March explosion. Uplift resumed in late March and again reversed to subsidence 36 hours before extrusion resumed on 5 April (*inset*).

relatively rigid crater walls. To monitor the development of these features, observers needed to measure relative movements of up to several meters over timescales of several weeks. Risking expensive sensors was futile, because they were likely to be damaged or buried by frequent rockfalls from the growing dome or encircling crater walls. Experience showed that the survivability of equipment left in this dynamic, hazardous environment was measured in weeks to a few months at best.

The solution was both simple and elegant. Short sections of steel rebar or fence posts were driven into the crater floor to serve as markers, and the distances among them were measured periodically with a steel tape. A typical monitoring station consisted of one or two markers on each side of a developing crack or thrust fault (Figures 2.13–2.15). Taping required only two observers, who could visit several stations on foot in just a few hours. As a crack widened, the taped distance across it increased, usually in a regular and accelerating pattern. Conversely, as the upper plate of a thrust over-rode the lower, the distance between points on opposite sides of the fault decreased. Extrapolation of these trends helped to define a predictive window during which the eruption was expected to begin, in some cases as much as 2–3 weeks in advance (Figure 2.16).

Similar data were obtained using a different technique that partly avoided the problem caused by thick accumulations of snow on the crater floor during the winter. In this case, slope distances between a few instrument stations on the crater floor and several fixed targets on the flanks of the dome were measured repeatedly with an EDM. As the dome swelled prior to the start of exogenous

growth, the data revealed a pattern of slope-distance contraction very similar to that already described for thrust faults on the crater floor (Figure 7.13). Based primarily on the combination of seismic, tiltmeter, thrust-fault, and slope-distance data, the onset of each exogenous growth episode at Mount St. Helens from December 1980 to October 1986 was successfully predicted a few days to three weeks in advance (Figure 7.14). In some cases, the prediction window was narrowed to just a few days (several days before the start of the eruption).

My point in describing the geodetic techniques used to predict dome-building eruptions at Mount

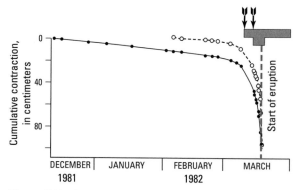

Figure 7.13. Cumulative contraction of slope distances between two points on the lava dome and a relatively stable point on the crater floor before the eruption of March–April 1982 (Swanson *et al.*, 1983). Left arrow indicates the date on which the USGS David A. Johnston Cascades Volcano Observatory (CVO) first issued an eruption prediction, and upper shaded box indicates the time window in which the eruption was predicted to occur (i.e., one to three weeks after the prediction was issued). An updated prediction with a shorter window was issued 3 days after the initial prediction, as indicated by the right arrow and lower box. Dashed line represents the start of the eruption on 19 March 1982.

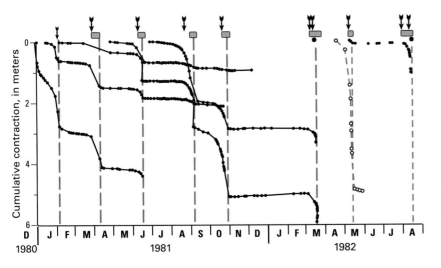

Figure 7.14. Summary of displacement data for thrust faults (solid circles) and a slope distance from the crater floor to the flank of the dome (open circles). Also shown is the timing of predictions and dome-building eruptions from late December 1980 to August 1982 (Swanson et al., 1983). Symbols are as in Figure 7.13.

St. Helens is not to imply that exactly the same approach should be applied at other volcanoes with growing domes. In fact, success at Mount St. Helens depended on a favorable combination of uncommon factors (i.e., a repetitive process acting on viscous magma beneath a relatively open, easily accessible vent area, where substantial resources could be committed to volcano monitoring (Swanson et al., 1983). Rather, I have included this example to illustrate the importance of frequent visits to a restless volcano by trained observers, and of adapting an effective monitoring strategy to the specific situation at hand. Considerable standardization of monitoring equipment and strategy is not only possible but also desirable, as evidenced by the experience of the USGS–USAID (US Agency for International Development) Volcano Disaster Assistance Program (VDAP) during numerous volcano-emergency responses since the program's inception in 1986 (Murray et al., 1996a). Nonetheless, experience also shows that searching out special opportunities, such as the rapidly developing thrust faults at Mount St. Helens, and then finding innovative ways to take advantage of those opportunities, is sometimes key to a more effective monitoring strategy. In the words of a scientist who played a key role in monitoring and predicting eruptions at Mount St. Helens during the 1980s: '*Field observations go hand in hand with more sophisticated equipment and techniques to form a complete system for monitoring volcanoes. Monitoring programs should explicitly include provisions for geologic field observations and instill in field workers, scientists and technicians alike, the need to be flexible and clever in designing simple experiments and measurements to*

test important field observations on the spot.' (Swanson, 1992, p. 219).

Mount St. Helens had more lessons to teach during an 18-year eruptive hiatus from October 1986 to September 2004, and especially during an extended period of new dome growth that began in October 2004. The story through early 2006 is told in Section 11.3.4.

7.2 KĪLAUEA VOLCANO, HAWAI'I – FLANK INSTABILITY AND GIGANTIC LANDSLIDES

7.2.1 The volcano's mobile south flank: Historical activity

The largest earthquake (M 7.2) in more than a century shook the Big Island of Hawai'i at 4:48 a.m. HST on 29 November 1975 (Tilling et al., 1976; Ando, 1979; Lipman et al., 1985). Most residents were accustomed to feeling small earthquakes during normal background activity at Kīlauea and Mauna Loa volcanoes, and many also had experienced moderate shocks (M 5–6.5) that occur beneath the island once every decade or so. But that morning's shaking was much worse than anything in recent memory. Not since 1868 had the ground heaved so hard and so long. In both 1868 and 1975, a local tsunami, meter-scale subsidence of the island's south coast, and an eruption of Kīlauea accompanied the intense shaking. What could possibly explain this recurring pattern of large earthquakes, coastal subsidence, and eruptions?

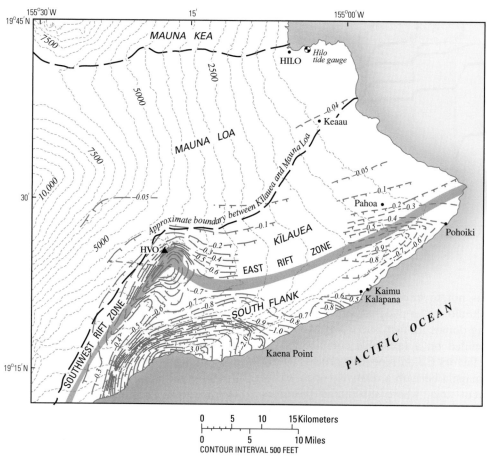

Figure 7.15. Interpreted contours of elevation changes (meters) associated with the *M* 7.2 Kalapana earthquake of 29 November 1975, based on repeated leveling surveys and coastal subsidence measurements (Lipman *et al.*, 1985). Rift zones and summit caldera are indicated by shaded pattern. Hachures show direction of relative movement.

Thanks in large part to a program of geodetic measurements by staff members of the USGS Hawaiian Volcano Observatory (HVO) dating back to 1912, we now understand that events like those in 1868 and 1975 occur when part of the gravitationally unstable south flank of Kīlauea or Mauna Loa slips and lurches toward the sea. Repeated measurements, especially EDM and leveling surveys, reveal a striking pattern of downward and seaward movements by up to several meters during the 1975 event (Figures 7.15 and 7.16). The sudden movements produce a tsunami where the coastline plunges seaward and extension across the summit area, which facilitates movement of stored magma toward the surface. Geodesy has played a key role in advancing our understanding of such events in Hawai'i and at many other island volcanoes around the world.

Even a cursory examination of the topography and structure of the Big Island of Hawai'i reveals the potentially unstable nature of Kīlauea's south flank, especially in light of the recent discovery of extensive

landslide deposits surrounding many of the Hawaiian Islands (Figure 7.17). Dominating the skyline to the northwest is the massive edifice of Mauna Loa Volcano, the largest on the island. The buttressing effect of giant Mauna Loa on Kīlauea, its diminutive neighbor to the southeast, has long been recognized. Separating the two volcanoes is the Kaoiki seismic zone (Figure 7.18(B)) and fault system, a group of normal faults downthrown mainly to the southeast. Running sub-parallel to the boundary between the volcanoes are Kīlauea's two rift zones, the East Rift Zone and Southwest Rift Zone. Southeast and downslope from the rift zones is the Hilina fault system (Figures 7.18(A) and 7.25), another group of young normal faults downthrown mainly to the southeast. Seen from this perspective, the south flank appears to be clinging precariously to the rest of the island, separated from its bulk by two active fault systems and two active rift zones. The combined forces of gravity and magmatic intrusion along the rift zones have the effect of pulling and

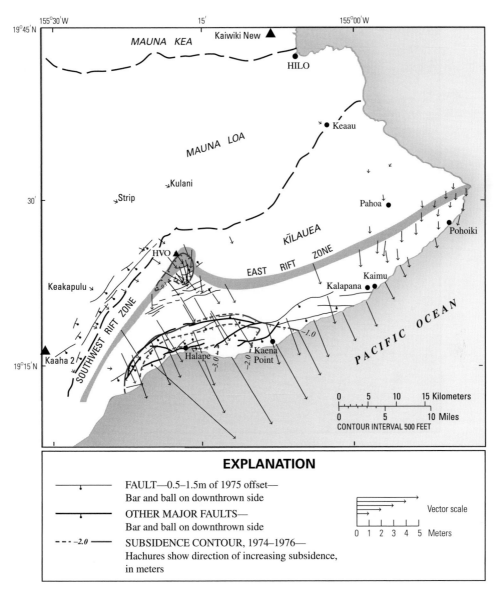

Figure 7.16. Horizontal displacements associated with the 1975 M 7.2 Kalapana earthquake derived from trilateration surveys in fall 1974 and spring 1976 (Lipman et al., 1985). Arrow tails are at marks; lengths of arrows indicate amounts of displacement. Stations Kaaha 2 and Kaiwiki New (triangles) were assumed fixed. Solid dots (e.g., Hilo, Keaau) mark named geographic features or towns.

shoving the south flank toward the sea. Not surprisingly, seismic and geodetic data confirm the mobility of the south flank and suggest the possibility of future catastrophic collapse. Confirmation that such collapses have occurred in the recent geologic past involved contributions from geodesy, geology, and undersea exploration (e.g., Takahashi et al., 2002) over the course of two decades, and makes for a compelling geologic detective story.

A strong clue that the south flank of Kīlauea is, in fact, moving seaward comes from a striking pattern of earthquakes beneath it. In plan view, the epicenters of most earthquakes at Kīlauea in the depth range 2–4 km correspond closely to the seismically active portions of the volcano's two rift zones (Figure 7.18(A)). Most of these earthquakes occur during swarms caused by lateral intrusion of basaltic magma along a rift zone from a reservoir beneath the summit area. At depths of 5–10 km, the pattern is distinctly different: most of these earthquakes occur not along the rift zones but beneath the volcano's south flank (Figure 7.18(B)). Viewed in cross section, the significance of this pattern is more apparent (Figure 7.19(A) and (B)). Earthquakes in the depth range 10–40 km clearly outline a vertical conduit system that delivers basaltic magma from

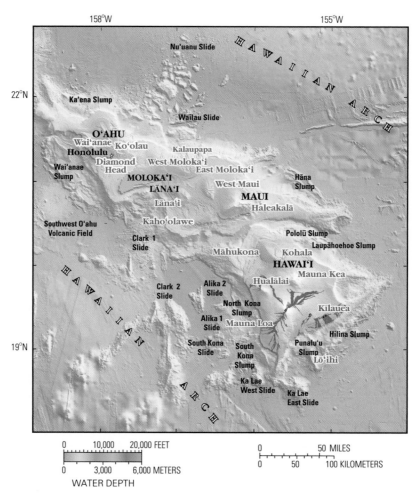

Figure 7.17. Shaded relief and bathymetry of the major Hawaiian Islands from Eakins *et al.* (2003). The islands are surrounded by deposits from gigantic landslides that originate on the unstable flanks of Hawai'i's volcanoes. Volcanic activity is now focused near the southeast tip of the island chain (i.e., at Kīlauea and Mauna Loa on the Big Island of Hawai'i and at nearby Lō'ihi). The steep submarine slopes of Mauna Loa and Kīlauea have been the source of several huge slides, which are partly responsible for the hummocky nature of the seafloor southwest and southeast of the Big Island. Historical lava flows are shown in red. North is at the top. Seafloor bathymetry derived in part from high-resolution multibeam-sonar surveys carried out by the Japan Marine Science and Technology Center (JAMSTEC), Yokosuka, Japan (*http:// www.jamstec.go.jp/*). Additional sources for submarine bathymetry and subaerial topography: USGS, Menlo Park, California (*http://walrus.-wr.usgs.gov/infobank/*); Monterey Bay Aquarium Research Institute, Monterey, California (*http://www.mbari.org/data/mapping/hawaii/index.htm*); University of Hawaii, School of Ocean and Earth Science and Technology, Honolulu, Hawaii (*http://www.soest.hawaii.edu/HMRG/*); National Geophysical Data Center, Boulder, Colorado (*http://www.ngdc.noaa.gov/mgg/bathymetry/relief.html*); Scripps Institution of Oceanography, San Diego, California (*http://sioexplorer.ucsd.edu/*); US Army Corps of Engineers, Mobile, Alabama (*http://shoals.sa-m.usace.army.mil/default.htm*); Global seafloor topography (predicted bathymetry) (*http://topex.ucsd.edu/marine.topo/mar.topo.html*).

the upper mantle to Kīlauea's summit reservoir. Also discernible are the two rift zones, which are seismically active in the 2–4 km depth range. By far the greatest number of earthquakes, and an even greater proportion of the seismic energy release, occurs at 5–10 km depth beneath the south flank. Focal mechanisms for many of the south flank earthquakes are consistent with slip on a sub-horizontal fault, which suggests seaward sliding along a shallow detachment fault, or décollement.

More direct evidence for the mobility of Kīlauea's south flank is available from repeated leveling and EDM surveys that span several decades. Comparison of leveling traverses across the historically active part of the East Rift Zone, for example, reveals the classic deformation pattern produced by intrusion of a near-vertical dike (i.e., broad uplift on either side of the dike trace punctuated by a sharp zone of subsidence directly above the dike) (Figure 7.20). Another effect of repeated intrusions along the rift zone is compression of the adjacent south flank, as recorded by repeated EDM

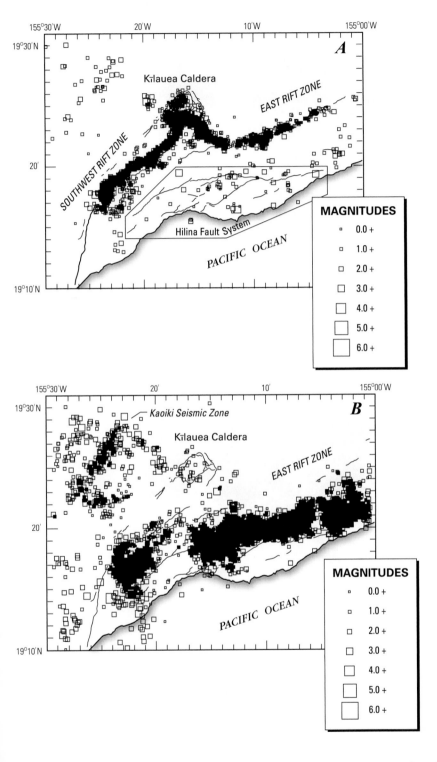

Figure 7.18. (A) Earthquake epicenters (squares) at Kīlauea Volcano in the depth range 3–4 km during 1970–1983. (B) Epicenters in the 6–7 km range for the same period. The shallower earthquakes outline the active rift zones, while the preponderance of deeper events occur beneath the volcano's mobile south flank.

From Klein *et al.* (1987).

measurements (Swanson *et al.*, 1976) (Figure 7.21). The north flank of Kīlauea is buttressed against Mauna Loa and, therefore, moves very little in response to intrusions into the rift zones. The south flank, on the other hand, is un-buttressed and also steeper than the north flank. As a result, it moves seaward under the combined influences of rift zone intrusions and gravitational stress. This motion is apparent in displacement vectors derived from repeated triangulation and trilateration surveys (Figure 7.22). With the geodetic data in mind, the significance of the strong concentration of

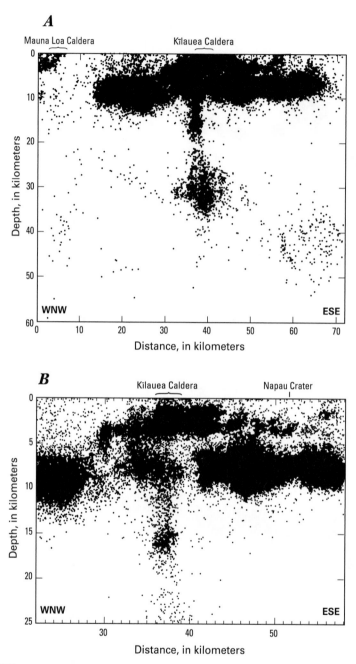

Figure 7.19. (A) WNW–ESE cross section through Kīlauea Volcano showing earthquake locations for 1970–1983. The vertical distribution of epicenters from about 40 km to 10 km depth outlines the conduit system that delivers magma from a mantle source to a reservoir beneath the summit region. Most of the earthquakes concentrated in the depth range 5–10 km occur beneath the volcano's mobile south flank. (B) Enlargement of (A) showing the concentrations of south flank earthquakes at 5–10 km depth and another at 2–4 km depth beneath the summit region and along the East Rift Zone. Napau Crater is a collapse feature (pit crater) along the rift zone. From Klein *et al.* (1987).

earthquakes 5–10 km beneath the south flank is clear. As the flank moves seaward, it grinds over a basal décollement and deforms internally, producing thousands of small earthquakes. About once per century, the flank lurches seaward in a large event like those in 1868 and 1975.

In a pioneering paper, Swanson *et al.* (1976) reported seaward displacements of Kīlauea's south flank as large as 4.4 m from 1914 to 1970 and 2.3 m from 1958 to 1970. They noted that the direction of the displacements was similar to that of maximum stress axes derived from focal mechan-

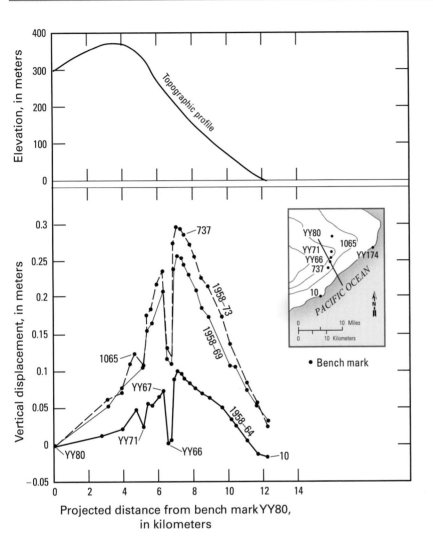

Figure 7.20. Vertical displacement and topographic profiles across the lower East Rift Zone of Kīlauea Volcano between 1958 and 1973 (Swanson *et al.*, 1976). The elevation of bench mark YY80 was held fixed. (*Inset*) The location of leveling route, key bench marks, and line of topographic profile. Contour interval of inset map is 150 m.

isms of south flank earthquakes and concluded (p. 1): '*We anticipate a subsidence event in the not too distant future, possibly similar to the damaging events of 1823 and 1868.*' The November 1975 *M* 7.2 Kalapana earthquake occurred while their paper was in press.

A major change in Kīlauea's behavior occurred on 3 January 1983, when the volcano entered an extended period of almost continuous eruption from vents along the East Rift Zone (Wolfe, 1988). By September 2002, 2.3 km^3 of lava had covered 110 km^2 and added 220 hectares (2.2 km^2 or ∼540 acres) to Kīlauea's south shore – ranking the Pu'u 'Ō'ō-Kūpaianaha eruption as the longest and largest rift zone eruption of Kīlauea Volcano in more than 600 years (Heliker *et al.*, 2003; Heliker and Brantley, 2002; USGS Hawaiian Volcano Observatory, 2002). In the process, lava flows destroyed 189 buildings, many of them houses, and buried 13 km of highway with as much as

25 m of lava. A wealth of geodetic information reveals that the eruption was accompanied by steady subsidence of the summit area and rift zones and by continued seaward motion of the south flank. From 1976 to 1996, the summit widened by more than 250 cm and subsided more than 200 cm, while the adjacent south flank rose more than 50 cm (Delaney *et al.*, 1993, 1998). Summit widening slowed from about 25 cm yr^{-1} in 1976 to about 4 cm yr^{-1} in 1983, at the beginning of the Pu'u 'Ō'ō-Kūpaianaha eruption. Likewise, the average subsidence rate along the upper East Rift Zone slowed from ∼9 cm yr^{-1} during 1976–1983 to about 4 cm yr^{-1} during 1983–1996. Horizontal motions across the East Rift Zone and south flank were correspondingly large. Delaney *et al.* (1998, p. 18,003) concluded that: '*Because the magnitudes of these contractions and extensions [across the subaerial south flank] are much less than the extension across the rift system, the subaerial south*

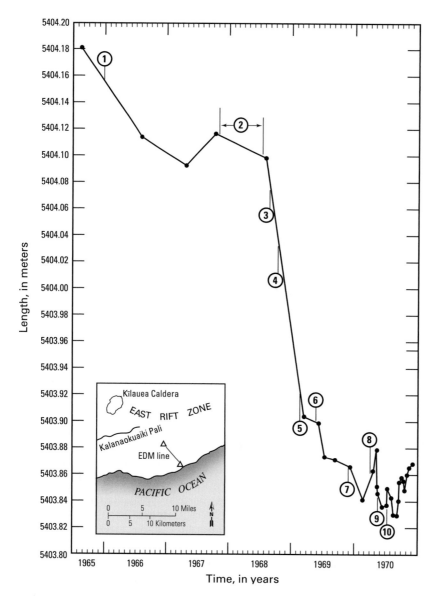

Figure 7.21. Contraction of a 5.4-km-long EDM line (*inset*) on the south flank of Kīlauea between August 1965 and December 1970. Major magmatic and structural events (circled numbers) during this period were: (1) December 1965 eruption and ground cracking; (2) November 1967–July 1968 summit eruption; (3) August 1968 eruption; (4) October 1968 eruption; (5) February 1969 eruption; (6) beginning of May 1969–October 1971 Mauna Ulu eruption; (7) new fissure north of Alae Crater, December 1969; (8) new fissure and cracks in and west of Aloi Crater, April 1970; (9) intrusion and cracking in southern part of Kīlauea Caldera, May 1970; and (10) new fissure east of Mauna Ulu, July 1970. See table 1 in Swanson *et al.* (1976) for additional information. Contraction is attributed to forceful intrusions along the East Rift Zone.

Modified from Swanson *et al.* (1976).

flank is apparently sliding seaward on its basal décollement more than it is accumulating horizontal strains within the overlying volcanic pile. Kilauea suffers from gravitational spreading made even more unstable by accumulation of magma along the rift system at depths in excess of about 4–5 km in the presence of hot rock incapable of withstanding deviatoric stresses.'[2]

Repeated GPS surveys and a growing network of continuous GPS stations have also tracked contin-

ued south flank motion during the Pu'u 'Ō'ō-Kūpaianaha eruption. For example, Owen *et al.* (1995, 2000a) analyzed repeated GPS surveys from 1990 to 1996 and showed that the south flank moved seaward at an average rate of $\sim 8\,\mathrm{cm\,yr^{-1}}$. This is substantially slower than immediately after the 1975 earthquake, but still amazingly fast by most geodetic and geologic standards. While the south flank slipped seaward, the summit subsided $\sim 8\,\mathrm{cm\,yr^{-1}}$, the upper East Rift Zone subsided a few $\mathrm{cm\,yr^{-1}}$, and the south coast rose $1–2\,\mathrm{cm\,yr^{-1}}$. These movements are modeled very well by deep opening along the East Rift Zone, slip along a décollement near the base of the volcano, and deflation beneath the summit area (Figures 7.23 and 7.24). It is not yet known

---

[2] The term 'deviatoric stress' is quoted here from the original literature. The current author is aware of the article by Engelder (1994) entitled 'Deviatoric stressitis: A virus infecting the earth science community,' which points out widespread misuse of the term and attempts to set the record straight. In this case, the intended meaning seems clear.

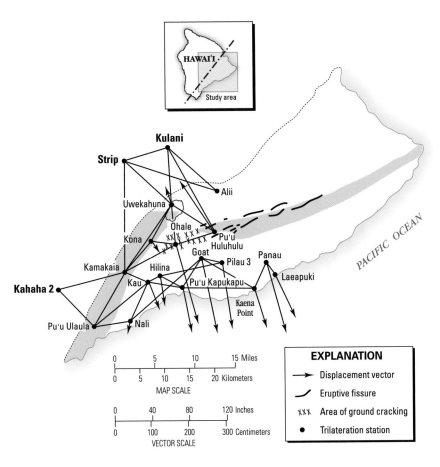

Figure 7.22. Horizontal ground displacements (arrows) at Kīlauea Volcano for the period 1961–1970, which included several eruptions along the East Rift Zone (Swanson et al., 1976). Displacements were derived from comparison of a triangulation survey in 1961 and a trilateration survey in 1970. Stations in bold were held fixed. Shading represents the approximate extents of the Southwest Rift Zone and active part of the East Rift Zone. Also shown are areas of ground cracking and eruptive fissures during the survey interval.

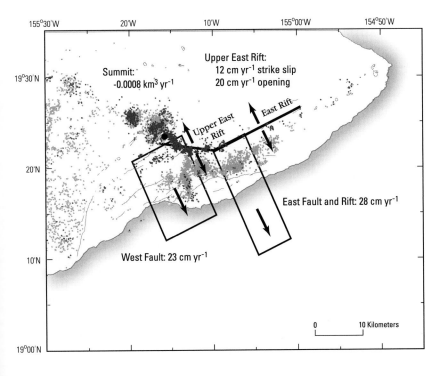

Figure 7.23. Elements of a best-fit model for deformation of Kīlauea Volcano measured by repeated GPS surveys from 1990 to 1996 (Owen et al., 2000a). Rapid seaward motion of the south flank was accompanied by subsidence in the summit area and along the upper East Rift Zone, and by uplift of the south coast (Figure 7.24). These movements are consistent with deep opening along the rift zone (opposing arrows), slip along a décollement near the base of the volcano (arrows inside rectangles, which are the surface projections of two model dislocations), and deflation beneath the summit area (large solid circle). Light shaded dots represent earthquakes that occurred between 5 and 12 km depth; dark shaded dots represent earthquakes that occurred between 0 and 5 km depth.

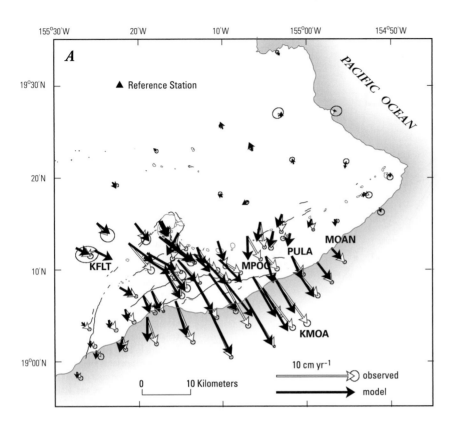

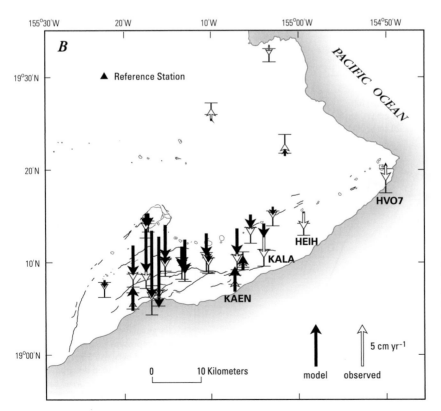

Figure 7.24. Model and observed horizontal (A) and vertical (B) displacement rates at Kīlauea Volcano based on repeated GPS surveys from 1990 to 1996 (Owen *et al.*, 2000a). Uncertainty in observed horizontal rates given by circle at arrow head (A). These results make a compelling case for deep opening along the rift zone, slip along a décollement near the base of the volcano, and deflation of the summit area.

how this motion is partitioned among after-slip from large earthquakes, steady gravitational spreading, and the ongoing Pu'u 'Ō'ō-Kūpaianaha eruption. One thing is crystal clear, though: Kīlauea's south flank has been and still is slipping into the sea at a remarkably high rate. What might be the implications of such rapid motion over geologic timescales?

7.2.2   Colossal prehistoric landslides and sea waves

We know that small, routine movements of the south flank are marked by earthquakes in the *M* 3–4 range and accumulate at a rate of about 1 meter per decade. Single earthquakes in the *M* 7–8 range have produced coastal displacements of several meters three times in the past 200 years. Could it be that 'mega-landslides' occur on the south flank with even longer recurrence intervals? The answer is yes, and the evidence is irrefutable.

The pattern of south flank movements revealed by repeated geodetic surveys led Swanson *et al.* (1976) to interpret the seaward-facing scarps of the Hilina fault system as headwalls of huge landslide blocks

that are reactivated when cumulative lateral and upward displacement of the south flank renders the un-buttressed and over-steepened parts of the south flank unstable (Figures 7.25 and 7.26). This interpretation gained wide acceptance and posed some obvious questions. How large were the submarine landslide blocks hidden off the south coast of Hawai'i? Were the historical subsidence events of 1823, 1868, and 1975 typical of this process, or had much larger events occurred in the prehistoric past?

More than a decade earlier, Moore (1964) had correctly inferred, based on bathymetric evidence, the presence of two giant submarine landslides on the Hawaiian Ridge adjacent to the islands of Oahu and Molokai. In what turned out to be a particularly prescient discussion of the origin of these remarkable features, Moore concluded (p. D97): '*Because the Hawaiian Ridge is one of the earth's steepest and youngest major topographic features, it is a region favorable to large-scale landsliding.*' Earlier, I referred to the discovery of huge landslides off the south flank of Kīlauea as a compelling geologic detective story. Moore's 1964 paper meant that

Figure 7.25. Aerial view of several *en echelon* faults of the Hilina fault system on Kīlauea Volcano, Hawai'i. Pu'u Kapukapu is the nearest and most prominent fault scarp; Hilina Pali is the most distant. These historically active scarps are interpreted as headwalls of giant landslide blocks, most of which are submarine. This USGS photograph, taken by D.A. Swanson in 1971 before a *M* 7.2 earthquake in November 1975 caused coastal areas to subside more than 3 m, along these and other faults, drowning the coconut grove near the small island in the center foreground (circle).

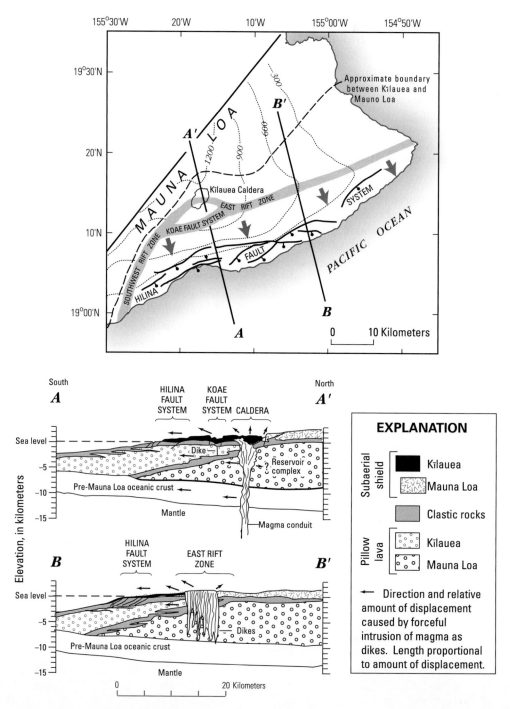

Figure 7.26. Index map and cross sections of Kīlauea illustrating the role of south flank landslides in seaward growth of the shield (Duffield *et al.*, 1982) . The mobile south flank is bounded by the East Rift Zone, the Koae fault system, and the Southwest Rift Zone. Magma intruded as dikes into the rift zones from a summit reservoir wedges the south flank seaward and upward. The north flank is buttressed by much larger Mauna Loa and therefore is relatively stable. Contours in meters. Dashed line approximates the contact between Mauna Loa and Kīlauea lavas.

the case had been cracked, but the full story was yet to be told.

During 1976 and 1978, the research vessel *S.P. Lee*, operated by the USGS, studied areas of inferred

submarine landslide deposits off the coasts of both Kīlauea and Mauna Loa. The voyage, which was inspired by the work of Moore (1964) and Swanson *et al.* (1976), produced a dramatic discovery: echo sounding and seismic reflection profiling revealed the presence of slump and slide features extending as far as 80 km offshore! These were the largest submarine landslides known at the time.[3] Their discovery confirmed the landslide origin of the Hilina fault system on the island's southeast flank, and the Kealakekua and Kahuku faults on the southwest flank. However, a more complete understanding of the morphology, chronology, and impact of Hawai'i's giant landslides would take at least another decade.

The next bit of evidence came not from Hawai'i itself, but from the nearby island of Lanai. There, a widespread gravel deposit containing limestone derived from coral reefs was thought to have been deposited along ancient marine strandlines that formed during worldwide high stands of the sea. However, Moore and Moore (1984) used dated submerged coral reefs and tide gauge measurements to demonstrate that the island was sinking so fast under the weight of the Hawaiian Ridge that former high stands of the sea now lay below sea level. If the limestone-bearing gravel, found at a maximum height of 326 m above sea level, was deposited while the sea was essentially at its modern level, how were bits of coral reef carried to such a height?

The authors came to an astounding conclusion: The gravel was deposited by the surge of a giant ocean wave that swept several hundred meters up the flanks of Lanai and nearby islands about 100,000 years ago. Marine material in the deposit was ripped up from the littoral and sublittoral zone and mixed with basaltic debris as the wave swept inland, then deposited high above sea level as the wave receded. What could have caused such a tremendous wave? Its great run-up suggests it was not a tsunami caused by a submarine earthquake, because the largest historical tsunami recorded in Hawai'i reached only 17 m above sea level in 1946. Moore and Moore (1984, p. 1314) reasoned: '*Either the impact of a meteorite on the sea surface or a shallow submarine volcanic explosion could have generated the Hulupoe*

wave. We believe, however, that a more likely explanation is a rapid downslope movement of a sub-sea landslide on the Hawaiian Ridge, which is among the steepest and highest landforms on Earth...We infer that rapid movement of a submarine slide near Lanai displaced seawater forming a wave that rushed up onto the islands, carrying with it rock and reef debris from the near-shore shelf and beach.' It was time to take a closer look at the seafloor around Hawai'i.

In 1986, the research vessel *M.V. Farnella* surveyed the offshore west flank of Mauna Loa using the sidescan sonar system GLORIA, echo sounding, and seismic-reflection profiling (e.g., Moore *et al.*, 1989). The results were striking. The GLORIA images revealed that the Alika slide off the southwest coast of Hawai'i was even larger and morphologically more complex than previously thought (Figure 7.27). Its hummocky surface resembles that of subaerial debris avalanches such as the famous 1980 deposit at Mount St. Helens. Rather than a prolonged or repetitive sequence of slump or creep events comparable in magnitude to the November 1975 event, '*...the Alika slide represents several geologically rapid events involving mass flowage*' (Lipman *et al.*, 1988, p. 4285). In other words, the slide consists of far-traveled debris from several catastrophic mass movements in which part of the west flank of Hawai'i literally slumped into the sea. What is more, the Alika slide is associated with less rapidly emplaced gravitational slump and slide features that occupy virtually the entire submarine west slope of Hawai'i (Lipman *et al.*, 1990, 2002).

Subsequent GLORIA data revealed similar large flowage deposits off the southeast slope of Hawai'i that head in the area of the Hilina fault system. Finally, a conjecture based on a structural interpretation and supported by both geodetic data and the occurrence of a large earthquake was verified by direct observation. The recurrence interval for catastrophic debris avalanches in Hawai'i seems to be on the order of 100,000 years, while that for events on the scale of November 1975 is approximately 100 years. It is especially remarkable that this fascinating story was pieced together both before and after the 1980 debris avalanche at Mount St. Helens, which so vividly demonstrated the importance of flank failures in the development of many stratovolcanoes. Thanks to decades of persistent detective work, we now know that the same is also true for the gently sloping shield volcanoes of Hawai'i.

---

[3] Other enormous submarine landslide deposits were discovered soon thereafter, including the 5,600 km³ Storegga Slide off the western coast of Norway (Bugge, 1983), which left evidence of a tsunami along the eastern coast of Scotland about 7,000 years ago (Dawson *et al.*, 1988; Long *et al.*, 1989). Masson (1996) and Masson *et al.* (2002) identified several large submarine landslides and associated deposits offshore the western Canary Islands, including the large ~15 ka El Golfo debris avalanche off the flank of El Hierro Volcano, and a 1,000 km³ slide offshore the Oratava and Icod valleys on Tenerife from Las Cañadas Volcano (Watts and Masson, 1995).

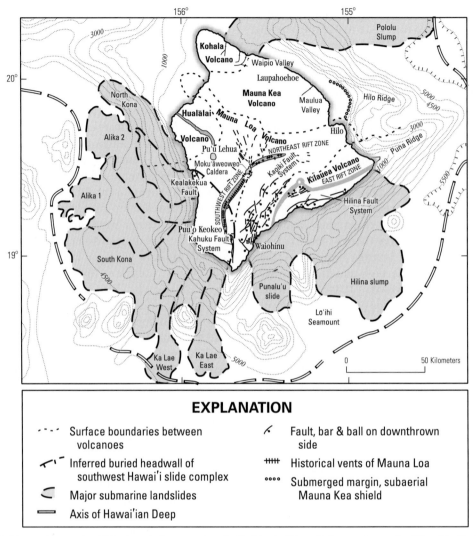

Figure 7.27. Major submarine landslides (stippled) surrounding the Big Island of Hawai'i. Contour interval is 500 m. Modified from Lipman *et al.* (1990, 2002).

7.3   YELLOWSTONE – THE UPS AND DOWNS OF A RESTLESS CALDERA

On 1 March 1872, the US President Ulysses S. Grant set aside 2.2 million acres (8,900 km²) of wilderness for '... *the benefit and enjoyment of the people*' and declared Yellowstone to be the world's first national park. The event followed several million years of persistent and at times frenetic groundwork by tectonic, volcanic, and hydrothermal processes that formed Yellowstone's spectacular landscape and continue to shape it today. Even so, the region's geologic heritage is mostly lost on millions of visitors each year, who fail to make a connection between the Park's renowned hydrothermal features and the world-class magmatic system that lies hidden below the surface. Fortunately, this is changing

for the better as more is learned about Yellowstone's tumultuous geologic past, including evidence for dramatic ground movements that continue to the present day.

7.3.1   Tectonic setting and eruptive history

The scenic Yellowstone Plateau lies along the northeastward extension of the eastern Snake River Plain – a region of young extensional tectonics and basalt–rhyolite volcanism in the western USA (Figure 7.28). The Yellowstone Plateau is generally acknowledged to be a 'hotspot' that marks the location of either a buoyant plume originating in the mantle (e.g., Pierce and Morgan, 1992; Smith and Braile, 1994), or a melting anomaly that reflects feedback between upper-mantle convection and regional lithospheric

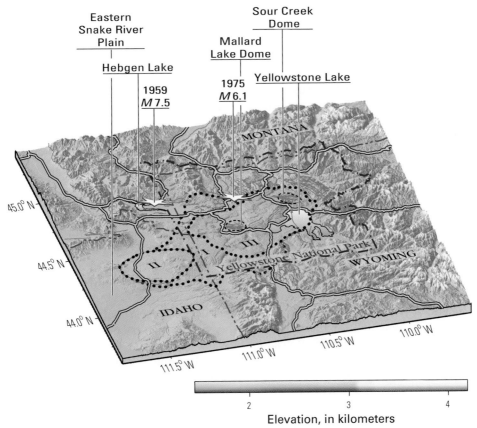

Figure 7.28. Eastern Snake River Plain and Yellowstone Plateau with outlines (heavy dotted line) of three youngest calderas: (I) 2.0 Ma Huckleberry Ridge; (II) 1.3 Ma Mesa Falls; and (III) 0.64 Ma Lava Creek. Also shown, outlines of the Mallard Lake and Sour Creek resurgent domes (light dashed line). White stars indicate epicenters of 1959 M_s 7.5 Hebgen Lake earthquake and 1975 M 6.1 Yellowstone Park earthquake.
Graphic created by C. Wicks, USGS.

tectonics (Christiansen, 2001; Christiansen *et al.*, 2002). The 700-km by 90-km northeast-trending Snake River Plain–Yellowstone Plateau volcanic province marks the southwestward track of the North American plate over the plume or melting anomaly during the past ∼16 million years. According to Morgan and McIntosh (2005, p. 288): '*Passage of the North American plate over the melting anomaly at a particular point in time and space was accompanied by uplift, regional extension, massive explosive eruptions, and caldera subsidence, and followed by basaltic volcanism and general subsidence.*'

For the past 2 million years, explosive rhyolitic volcanism has been focused at the Yellowstone Plateau, including Yellowstone National Park. The dominantly extensional tectonic regime at Yellowstone reflects the influence of northeast–southwest Basin and Range extension in the region. Yellowstone seismicity is characterized by swarms of $M < 3$ earthquakes within the 0.64 Ma

caldera (see below) and between the caldera and the eastern end of the 44-km-long rupture of the M_s 7.5 Hebgen Lake earthquake (Waite and Smith, 2004) (Section 7.3.3).

For the interested reader, I recommend the book *Windows into the Earth* by Smith and Siegel (2000), which tells the geologic story of Yellowstone and Grand Teton National Parks in vivid detail. For a more technical treatment of Yellowstone geology, USGS Professional Paper 729-G, *The Quaternary and Pliocene Yellowstone Plateau volcanic field of Wyoming, Idaho, and Montana* (Christiansen, 2001), is the definitive work.

Three times during the past 2 million years – 2.0, 1.3, and 0.64 million years ago – large rhyolite magma bodies formed in the upper crust beneath Yellowstone, eventually fueling immense explosive eruptions that resulted in caldera collapses (Christiansen, 1984, 2001). The current Yellowstone Caldera, 45 km wide by 75 km long, formed about 640,000 years ago during the eruption of 1,000 km^3

of rhyolitic ash flows constituting the widespread Lava Creek Tuff. Soon thereafter, structural resurgence formed the Sour Creek and Mallard Lake resurgent domes, and rhyolitic volcanism resumed within the caldera. Renewed doming in the western part of the caldera culminated with extrusion of $1,000\,km^3$ of intra-caldera rhyolite flows that virtually buried the Yellowstone Caldera between 150,000 and 75,000 years ago. Abundant geophysical evidence exists for the presence of partial melt beneath the caldera (e.g., Benz and Smith, 1984; Miller and Smith, 1999; Husen et al., 2004) and the consensus among those who have studied the area is that the Yellowstone magmatic system will almost surely erupt again. We just don't know when.

7.3.2   Results of repeated leveling surveys

In view of Yellowstone's explosive past, characterizing the current state of the Yellowstone magmatic system and assessing its potential for future eruptions has been a priority of the USGS Volcano Hazards Program for several decades. The USGS initiated a program of detailed geologic mapping and exploration of Yellowstone's hydrothermal system in the 1960s, and since then numerous investigators from around the world have studied the Park's geology, geophysics, geochemistry, and hydrology. In cooperation with the USGS, the University of Utah operates a regional seismic network to record Yellowstone's frequent and sometimes damaging earthquakes. In 2001, the USGS, University of Utah, and National Park Service formed a partnership to create the Yellowstone Volcano Observatory (YVO).

From a geodetic perspective, an important milestone was reached during 1975–1977 when a 1923 leveling survey throughout Yellowstone National Park was repeated for the first time. The resulting discovery of rapid uplift within the caldera set the stage for a more vigorous leveling effort starting in 1983. The Yellowstone leveling network consists of approximately 380 km of interconnected loops and spurs along all major roads and one back-country trail in Yellowstone National Park. Most of the network was measured in 1923, in increments from 1975 to 1977 (hereafter called the 1976 survey), and in 1987; partial surveys were conducted in 1936, 1941, 1955, 1960, and most years from 1983 to 2000 (see below).

During the interval between the 1923 and 1976 surveys, the central part of the Yellowstone Caldera rose as much as $726\pm21\,mm$ with respect to K12

1923, a reference bench mark located about 8 km outside the east caldera rim (Pelton and Smith, 1979, 1982; Figure 7.29). The maximum uplift measured was at bench mark B11 1923, near LeHardys Rapids (variant names, Le Hardys Rapids, LeHardy's Rapids) at the base of the Sour Creek resurgent dome in the eastern part of the caldera. By the time of the next complete survey in 1987, bench mark DA3 1934 near LeHardys Rapids had risen an additional $115\pm5\,mm$ with respect to 36MDC 1976 (near K12 1923). The average uplift rate from 1976 to 1987 was $10\pm1\,mm\,yr^{-1}$, apparently less than from 1923 to 1976 ($14\pm1\,mm\,yr^{-1}$). However, we know from annual surveys starting in 1983 that uplift stopped during 1984–1985 and subsidence began during 1985–1986, so the average uplift rate for 1976–1987 underestimates the actual rate for the early part of that period.

Each year from 1983 to 1998, except 1994, 1996, and 1997, we measured part or the entire leveling traverse between Lake Butte and Mount Washburn to first-order, class II standards (Chapter 2). We chose this route because it is approximately perpendicular to the uplift axis determined by the earlier surveys, and because it includes the area of maximum uplift near LeHardys Rapids. Leveling at Yellowstone in late summer, when the season's highest temperatures and heaviest traffic have receded and the Park starts to hunker down for another brisk Rocky Mountain winter, is pure delight for a volcanologist interested in ground deformation. Over a period of several days, we survey from Lake Butte, a high point on the caldera rim overlooking the glistening expanse of Yellowstone Lake, down into the caldera and across its floor to Mount Washburn on the opposite rim. Along the way, we skirt the shore of Yellowstone Lake, pass by the drowned hydrothermal explosion craters of Indian Pond and Mary Bay, cross Fishing Bridge at the lake outlet, and follow the Yellowstone River through herds of bison in Hayden Valley to the brightly colored canyon that inspired Yellowstone's name. From Canyon Junction, we climb the flank of Mount Washburn to a final bench mark overlooking Washburn Hot Springs, one of countless areas in the Park where hot water and magmatic gases seep out of the ground along fractures that mark the caldera boundary. A year later, we do it all over again, and then compare the two surveys to see how much the ground surface has risen or fallen while we were gone.

During the 16 years spanned by our surveys, vertical displacements of Canyon Junction and Mount

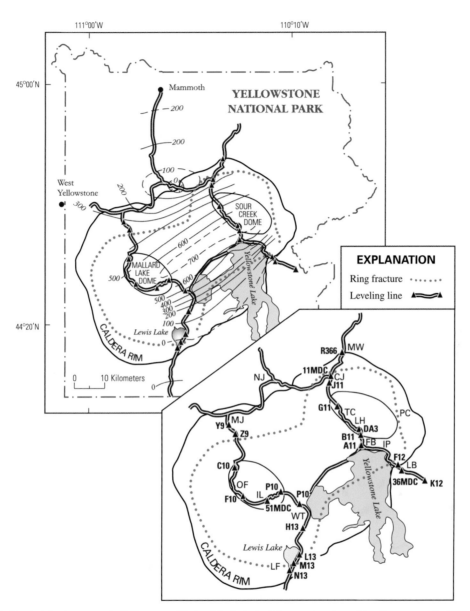

Figure 7.29. Vertical displacements in Yellowstone National Park derived from comparison of leveling surveys in 1923 and 1975–1977. Uplift contours in millimeters are from Pelton and Smith (1982). Localities (Lake Butte to Mount Washburn): LB, Lake Butte; IP, Indian Pond; FB, Fishing Bridge; LH, LeHardys Rapids; TC, Trout Creek; CJ, Canyon Junction; MW, Mount Washburn; (Lewis Falls to Madison Junction): LF, Lewis Falls; WT, West Thumb; IL, Isa Lake; OF, Old Faithful; MJ, Madison Junction; (other): NJ, Norris Junction; PC, Pelican Cone. Triangles represent bench marks (Lake Butte to Mount Washburn): K12 1923, 36 MDC 1976, F12 1923, B11 1923, DA3 1934, G11 1923, J11 1923, 11MDC 1976, R366 1987; (Lewis Falls to Madison Junction): N13 1923; M13 1923, H13 1923, P10 1923, M10 1923, 51MDC 1976, F10 1923, C10 1923, Z9 1923, Y9 1923.

Washburn relative to Lake Butte were generally less than two standard deviations of the measurements. So the caldera rim has been relatively stable while virtually the entire caldera floor has moved up and down (Figure 7.30). Between the 1976 and 1984 surveys, DA3 1934 rose 177 ± 5 mm with respect to 36MDC 1976, at an average rate of 22 ± 1 mm yr^{-1}. To our surprise, the displacement at DA3 1934 with respect to 36MDC 1976 during

1984–1985 was only -2 ± 5 mm, and the largest displacement measured anywhere along the traverse, -7 ± 5 mm, was hardly significant. In hindsight, the uplift rate near LeHardys Rapids had probably dropped below its historical average of 14–23 mm yr^{-1} by the time of our 1984 survey, and it was essentially zero during 1984–1985. We wondered how long it would be before uplift resumed.

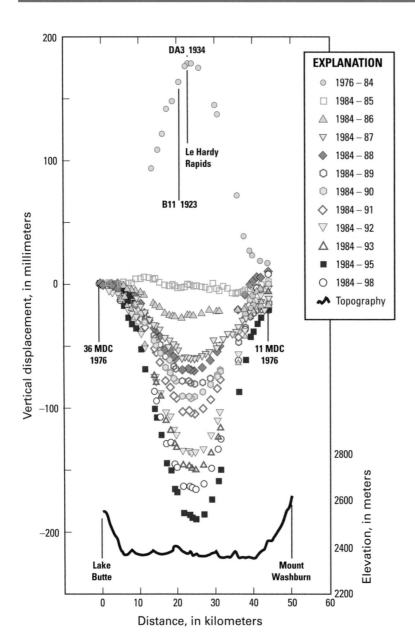

Figure 7.30. Vertical displacement profiles across the Yellowstone Caldera between two points on the caldera rim, Lake Butte and Mount Washburn, from repeated leveling surveys between 1976 and 1998. Uplift that began before 1976 stopped during 1984–1985, and the caldera floor progressively subsided during 1985–1995. Renewed uplift starting in 1995 was discovered with interferometric synthetic-aperture radar (InSAR) and confirmed by leveling.

We were surprised again when the caldera floor began to subside during 1985–1986 and continued to do so for the next decade (Dzurisin *et al.*, 1990, 1994) (Figures 7.30 and 7.31). Some of the annual displacements are less than the analytical uncertainty in the measurements, but the net displacement at DA3 1934 from 1985 to 1995, -189 ± 5 mm (-17 ± 1 mm yr^{-1}), is unequivocal. While subsidence persisted for more than a decade, its rate varied with time. For example, subsidence was relatively rapid during 1985–1986 (-25 ± 1 mm yr^{-1}), 1986–1987 (-34 ± 1 mm yr^{-1}), and 1991–1992 (-32 ± 1 mm yr^{-1}). But during 1988–1991, the annual rate was only 9–13 ± 1 mm yr^{-1}.

The shape of the 1985–1995 subsidence profile mirrors that of the 1976–1984 uplift profile, and the average maximum displacement rates for the two intervals also are similar (-19 ± 1 mm yr^{-1} and 22 ± 1 mm yr^{-1}, respectively). This suggests that the sources for uplift and subsidence have similar locations and dynamics, although the leveling data alone do not constrain the depth or geometry of the sources very tightly. A more constrained solution is obtained by modeling simultaneously the vertical displacements from leveling surveys and horizontal displacements from repeated GPS surveys (Meertens and Smith, 1991; Meertens *et al.*, 1992, 1993; Vasco *et al.*, 1990; Section 7.3.4).

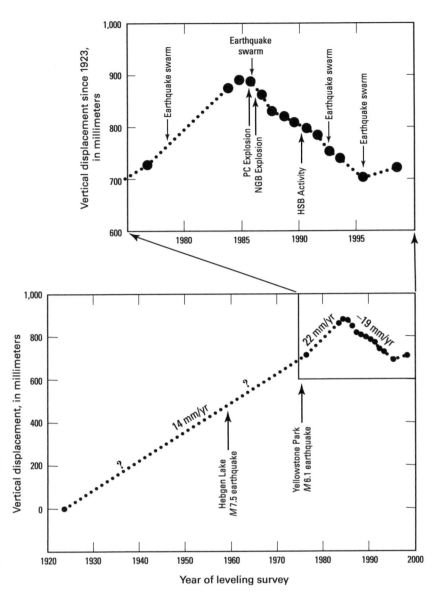

Figure 7.31. History of vertical displacements at bench mark B11 1923 near LeHardys Rapids as measured by leveling surveys from 1923 to 1998. Displacements are relative to bench marks 1923 or 36 MDC 1976, both located outside the caldera near its east rim. Arrows mark times of noteworthy events discussed in the text. Dotted lines represent one possible time history for the deformation, although the actual time dependence between surveys is unknown. PC, Pelican Cone; NGB, Norris Geyser Basin; HSB, Hot Spring Basin.

The history of vertical displacements at Yellowstone as measured by leveling surveys is summarized in Figure 7.31, which shows the elevation of bench mark B11 1923 as a function of time from 1923 to 1998. B11 1923 was chosen because, among the marks that survived from 1923 to 1998, it is closest to the area of maximum surface displacement. It is located about 2 km south of LeHardys Rapids and records about 95% of the maximum displacement measured at DA3 1934. The closely spaced data points for the years from 1983 to 1998 are satisfying, especially to someone who enjoyed carrying a leveling rod across the caldera for most of those surveys, but long intervals between the earlier surveys pose an obvious question.

7.3.3   What happened between leveling surveys?

The largest earthquake ever recorded in the Yellowstone region, the M_s 7.5[4] Hebgen Lake earthquake, occurred on 18 August 1959, about 25 km north–northwest of the northwest rim of the Yellowstone

---

[4] M_s refers to surface-wave magnitude, which is based on the amplitude of Rayleigh surface waves measured at a period near 20 s. Depending on how they were determined, earthquake magnitudes are expressed as local magnitude (M_L), body magnitude (M_b, based on the amplitude of P body-waves), or moment magnitude (M_w, based on the moment of the earthquake). Unless specified otherwise, earthquake magnitude M refers to local magnitude.

Caldera and 65 km west–northwest of LeHardys Rapids. Within 24 hours, six aftershocks of M 5.5 to 6.3 had occurred in an east–west-trending zone nearly 100 km long centered on the main shock. Another major earthquake, the M 6.1 Yellowstone Park event on 30 June 1975, was centered approximately 5 km beneath the north caldera rim near Norris Junction, 20 km northwest of LeHardys Rapids (Pitt *et al.*, 1979). All of these large earthquakes occurred during the 53-year interval between the 1923 and 1976 leveling surveys (Figure 7.31). Any effects they may have had on the short-term vertical displacement rate are undocumented.

No earthquakes larger than M 5.5 have occurred in the Yellowstone region since the 1976 leveling survey, but there have been many swarms of smaller earthquakes within or near the caldera. Those in 1978, 1985, 1992, and 1995 (Figure 7.31) are especially noteworthy. The May–November 1978 swarm occurred at depths of 1–5 km beneath the Mud Volcano hydrothermal area, about 5 km northwest of LeHardys Rapids. At least 8 earthquakes in the swarm were larger than M 2.5, and the largest was M 3.1. At its peak the swarm produced more than 100 events per hour. The earthquakes were followed by increased thermal activity in the Mud Volcano area starting in December 1978. Pitt and Hutchinson (1982, p. 2762) concluded that the earthquakes '*expanded pre-existing fracture systems, permitting increased fluid flow from depths of several kilometers.*' The net vertical surface displacement near LeHardys Rapids from 1976 to 1984 was 177 ± 5 mm. Any deformation that might have accompanied the 1978 swarm is otherwise undocumented.

The most intense earthquake swarm ever recorded in the Yellowstone region occurred between West Yellowstone and Madison Junction, just outside the northwest rim of the caldera, starting on 7 October 1985. Twenty-eight events larger than M 3.5 (maximum 4.9) occurred during the first three months of activity, which persisted into 1986. These earthquakes, and many others in the vicinity that occurred before and afterward, are attributed to stress release in relatively cold, brittle crust outside the caldera. Inside, the crust below ~5 km is too hot to fracture; instead, it deforms in ductile fashion without producing many earthquakes (Smith and Arabasz, 1991).

A third interesting swarm occurred about 3 km south of LeHardys Rapids on 20 July 1992. It included 8 earthquakes larger than M 1.8 (maximum 4.6), all within 4 km of the surface.

These earthquakes occurred directly beneath the level line that we measured each year from 1983 to 1993, and their occurrence coincides with an increase in subsidence rate detected by the 1991 and 1992 surveys (from -11 ± 5 mm yr^{-1} during 1990–1991 to -32 ± 5 mm yr^{-1} during 1991–1992).

The second most intense earthquake swarm in the Yellowstone region since the 1959 M_s 7.5 Hebgen Lake earthquake occurred during June–July 1995 near Madison Junction, along the northwest caldera boundary. It comprised over 560 locatable earthquakes, including over 170 on 4 July alone, with a maximum magnitude of 3.1.

In addition to earthquake swarms, there have been some dramatic changes in Yellowstone's hydrothermal features since we began our annual leveling surveys in 1983. For example, in 1985, 1986, and 1990, notable hydrothermal events occurred at three widely separated locations along the caldera rim. In 1985, probably in early July, an explosion near Pelican Cone, about 5 km east of the east caldera rim, killed mature trees, formed a crater 5 m × 2 m × 2 m deep, and gave rise to a new superheated fumarole.[5] In January 1986, a larger explosion in mature forest 3 km west of Norris Junction threw debris 35 m laterally, knocked down trees, and formed a crater 10 m × 15 m × 5 m deep. Starting in early 1990 and continuing through the end of 1993, increasing ground temperatures in a part of the Hot Springs Basin thermal area near the head of Astringent Creek killed trees and led to the emergence of another superheated fumarole and a vigorous new mud volcano (Hutchinson, 1993). The active area is about 9 km east–northeast of LeHardys Rapids.

Changes in hydrothermal activity are common at Yellowstone, but the magnitude of these three events is unusual in recent history. The first occurred about 3 months before the start of the 1985 swarm and during the interval between the 1984 and 1985 leveling surveys, which showed that uplift had stopped (Figures 7.30 and 7.31). The second explosion occurred about three months after the start of the 1985 swarm and during the interval between the 1985 and 1986 surveys, which detected the onset of subsidence. The 1990–1993 activity near the head of Astringent Creek apparently started

---

[5] The activity in a remote part of Yellowstone National Park went unnoticed at the time, except possibly for a false report of smoke from the area on 3 July 1985. The new feature was discovered on 5 October 1986 by National Park Service geologists Roderick (Rick) Hutchinson and C. Craig-Hunter (Hutchinson, 1992).

before and continued after a period of relatively rapid subsidence during 1991–1992.

An episode of surface uplift centered near Norris Geyser Basin along the north caldera rim began in 1997, as revealed by radar interferometry (Wicks *et al.*, 2002c, 2003) (Chapter 5). The uplift affected an area of 30 km by 40 km and was accompanied by increased thermal activity, including the formation of a 75-m line of fumaroles just north of Norris Geyser Basin, an eruption of Porkchop Geyser, which had been dormant since 1989, and five eruptions of Steamboat Geyser. Major eruptions of Steamboat usually occur about once per decade and can reach heights of 90 m, prompting the US National Park Service to celebrate Steamboat as 'the world's tallest active geyser.' Modeling indicates the source depth for the Norris uplift is 10–12 km beneath the surface, which places it well below the inferred base of the hydrothermal system. The most likely cause is a magmatic intrusion near the intersection of the caldera ring fault system and the Norris–Mammoth corridor, a north–south-trending system of faults, volcanic vents, and active hydrothermal features between Norris Geyser Basin and Mammoth Hot Springs. Such intrusions undoubtedly power Yellowstone's vigorous hydrothermal system and might have escaped detection before the advent of InSAR. The Norris uplift paused during 2002–2003 (Wicks *et al.*, 2005). When and where the next episode of uplift or subsidence will occur is anyone's guess.

7.3.4 Causes of uplift and subsidence

The association of earthquake swarms, hydrothermal activity, and changes in the rate of caldera floor uplift or subsidence suggests that these three processes are connected. Although the details are still unknown, there is general agreement on a conceptual model that accounts for most aspects of the uplift, subsidence, and relative timing between seismic and hydrothermal events described above. Before discussing the model, though, it is useful to consider several potential deformation mechanisms and the role that each might play over various timescales.

Over timescales of 10^5–10^6 years, the ultimate cause of uplift and other forms of unrest at Yellowstone almost surely is episodic intrusion of new basaltic magma from the mantle into the crust beneath the caldera. Continued heat transfer to the crust by basaltic magma is the only plausible explanation for the longevity and vigor of Yellowstone's spectacular hydrothermal system (Fournier and Pitt, 1985). We know from measurements of the chloride flux from rivers draining Yellowstone that the integrated convective heat flux is about 4×10^{16} cal yr^{-1}. Withdrawing heat at that rate would completely crystallize and cool about 0.1 km^3 of rhyolite magma each year. Since the last magmatic eruption at Yellowstone about 70,000 years ago, this corresponds to a lens of cold rock about 3 km thick beneath the entire caldera. So, if it were not being reheated periodically from below, Yellowstone's hydrothermal system would be literally stone cold by now. Instead, it is vigorously active and has not shown any signs of slowing down for at least the past 50,000 years. The heat that sustains it must come from episodic intrusions of basalt into the base of the silicic magma system, and such intrusions likely cause uplift.

There is plenty of other evidence for basaltic intrusions beneath Yellowstone throughout its history. Basalt has erupted all around the Yellowstone Caldera, both before and since the caldera-forming Lava Creek eruption 640,000 years ago, and basaltic lava flows have flooded the older part of the Yellowstone magmatic system west of the caldera (Christiansen, 1984, 2001). In addition, a preponderance of geophysical and geochemical evidence indicates that a body of partly molten rhyolite still exists beneath the Yellowstone Caldera, presumably remnant from the last caldera-forming eruption (Christiansen, 1984, 2001; Smith and Braile, 1994; Smith and Siegel, 2000). The rhyolitic magma probably acts as a density shadow for basaltic magma (i.e., basalt does not rise through rhyolite by buoyant convection because basalt is denser than rhyolite). Instead, basalt tends to underplate rhyolite beneath the caldera, thereby heating and sustaining it in a partly molten state. In the process, the magmatic system pressurizes, overlying rocks are shoved upward, and the ground surface rises.

Intrusions of silicic magma into the upper crustal rhyolite body are also likely. Basalt that rises from the upper mantle becomes less buoyant in the lower crust, because the crust is less dense than the mantle. As a result, basalt tends to pond near the base of the crust. Partial melting of the lower crust by heat derived from ponded basalt generates silicic melts, which are buoyant enough to rise to the level of the sub-caldera rhyolite body. Inflation of the magma body causes surface uplift.

In summary, magmatic heat input is absolutely required to sustain Yellowstone's hydrothermal system over timescales longer than about 10^5

years, and there is ample evidence that basaltic magma has repeatedly intruded the crust in the vicinity of the Yellowstone Caldera since its formation 640,000 years ago. Accumulation of basalt, both near the base of the crust and beneath a sub-caldera zone of silicic partial melt, is the principal cause of uplift over such long timescales. It might also contribute to uplift over much shorter timescales, possibly even during historical time, but there are other mechanisms to consider before drawing that conclusion.

Another likely mechanism for historical uplift at Yellowstone is pressurization of the deep hydrothermal system, most likely as a consequence of rhyolitic magma crystallizing near the top of the magmatic system. Fournier and Pitt (1985) proposed that Yellowstone's hydrothermal system consists of a deep zone in which pore-fluid pressure is near lithostatic and a shallow zone in which pore pressure is hydrostatic. According to their model, the two zones are separated by an impermeable, self-sealing layer created by mineral deposition and plastic flow at a depth of about 5 km. Dzurisin et al. (1990) noted that the thermal energy carried to the surface at Yellowstone by convecting thermal water $(4 \times 10^{16} \text{ cal yr}^{-1})$ could be derived entirely from crystallization of $0.2 \text{ km}^3 \text{ yr}^{-1}$ of rhyolitic magma. Calculations by Fournier (1989) showed that if the magmatic fluid liberated upon crystallization of $0.2 \text{ km}^3 \text{ yr}^{-1}$ of rhyolitic magma initially containing 2 wt% water were trapped at lithostatic pressure, the net volume change at depth would be more than adequate to account for historical uplift rates measured by leveling surveys.

This mechanism is particularly appealing because it can account not only for Yellowstone's high convective heat flux and historical uplift rates, but also for episodic subsidence. If a self-sealed layer within the hydrothermal system ruptured during an earthquake swarm, the resulting depressurization and fluid loss would cause the overlying surface to subside. Subsidence would extend far beyond the epicentral area if, as suggested by Fournier and Pitt (1985), a state of hydraulic equilibrium prevails throughout the deep hydrothermal system. Thus, rupturing of the self-sealed layer *anywhere* might cause the entire caldera floor to subside. The 1978 and 1992 earthquake swarms might have been hydrofracturing events that released pressure and fluids from the deep hydrothermal system, allowing the caldera floor to subside. No subsidence was measured in the first case, presumably because

uplift dominated between leveling surveys in 1976 and 1984.

The 1985 earthquake swarm near West Yellowstone seems enigmatic, because it occurred outside the caldera and might be explained by tectonic strain release in the vicinity of the Hebgen Lake fault zone. However, the swarm coincided with the onset of subsidence within the caldera and with two small hydrothermal explosions near the caldera rim. A possible link between the 1985 epicentral area and the sub-caldera hydrothermal system is suggested by results of regional trilateration surveys. Savage et al. (1993) showed that post-seismic strain accumulation in the epicentral area of the 1959 Hebgen Lake earthquake can be modeled as a 20-km-wide zone of extension approximately parallel to the 1959 rupture trace, extending at least 100 km S75°E to the vicinity of the Sour Creek resurgent dome (Figure 7.32). The hypothesized zone of extension is marked by: (1) a broad band of seismicity during the 1973–1988 interval; (2) the epicenters of the 1959 Hebgen Lake earthquake and the 1975 Yellowstone Park earthquake; (3) the epicentral areas of the 1985, 1978, and 1992 earthquake swarms; and (4) the sites of small hydrothermal explosions in 1985 and 1986, plus the emergence of a vigorous new mud volcano starting in early 1990. It passes a few kilometers south of Norris Geyser Basin, near the center of uplift detected by radar interferometry starting in 1997. Savage et al. (1993) noted that the zone cuts across the trend of major regional structures such as the Madison and Gallatin ranges, and that the evidence for extending the zone into the caldera is weak. However, if the zone *does* extend to the Sour Creek dome or beyond, it might provide a passageway for fluids to migrate from beneath the caldera to the 1985 swarm area, thereby depressurizing the hydrothermal system, inducing caldera subsidence, and accounting for the alignment of seismic and hydrothermal events noted above. At other times, its intersection with the caldera ring fracture system or with faults of the Norris–Mammoth corridor might be a preferred site for magmatic intrusion and surface uplift (e.g., uplift centered near Norris during 1997–2002).

This idea received additional support from Waite and Smith (2002), who noted that the 1985 earthquake swarm had several unusual characteristics indicative of interaction between seismicity and hydrothermal/magmatic activity. These include: (1) the swarm was roughly coincident with a reversal from uplift to subsidence inside the caldera; (2) swarm hypocenters occupied a nearly vertical north-

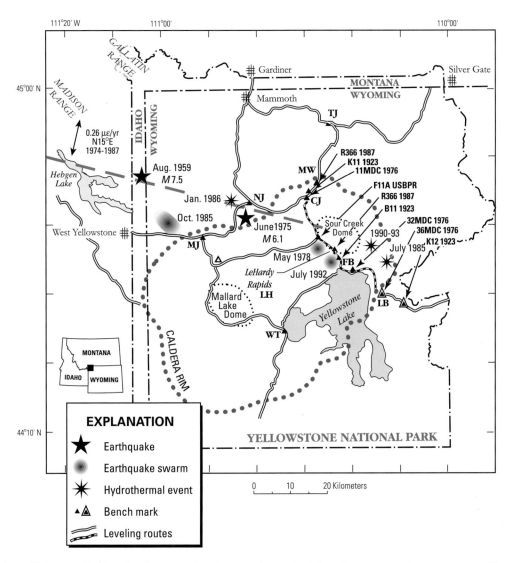

Figure 7.32. Yellowstone Caldera, leveling network, and a zone of extension inferred from trilateration measurements (Dzurisin et al., 1994). Level line from Lake Butte (LB) to Canyon Junction (CJ) via LeHardys Rapids (LH) and Fishing Bridge (FB), shown as a short-dashed line with solid triangles representing key bench marks, was measured each year from 1983 to 1998 except for 1994, 1996, and 1997. Bold, long-dashed line represents a 20-km-wide zone of extension hypothesized by Savage et al. (1993), who described ground deformation in the Hebgen Lake area between 1974 and 1987 as roughly a uniaxial, 0.266 ± 0.014 microstrain/year, N15°E ± 1° extension that extends southeastward into the Yellowstone Caldera, possibly as far as the Sour Creek resurgent dome. Caldera rim (bold dotted outline) and resurgent domes (fine dotted outlines) after Christiansen (1984). MW, Mount Washburn; TJ, Tower Junction; NJ, Norris Junction; MJ, Madison Junction; WT, West Thumb. Reference bench marks 36MDC 1976 and K12 1923 are represented by solid triangles inside larger, open triangles.

west-trending zone, and during the first month of activity, the pattern of epicenters migrated laterally away from the caldera at an average rate of 150 m day^{-1}; (3) the dominant focal mechanisms of the swarm were oblique-normal to strike-slip, contrasting with the normal-faulting mechanisms typical of the region; and (4) the maximum principal stress axis averaged for the swarm events was rotated 90 degrees from that of the normal background seismicity, from vertical to horizontal with a

trend of 30 degrees from the strike of the plane defined by the swarm. Although Waite and Smith (2002) did not offer a unique explanation for the 1985 swarm, they proposed that the rate of lateral migration of the activity along a steeply dipping plane and the orientations of the principal stress axes are consistent with models of migration of magmatic or hydrothermal fluids. In their words (p. 2190 or ESE 1-14): '*The most likely scenario involves the rupture of a self-sealed hydrothermal layer and subsequent migration of hydrothermal fluid through a preexisting fracture zone out of the caldera.'*

Three recent models based on ground deformation data suggest that the spatial patterns of uplift and subsidence are consistent with a source located at the depth and with the geometry of Yellowstone's deep hydrothermal system. Dzurisin *et al.* (1990) concluded that 1984–1987 leveling and trilateration data from the eastern part of the caldera were best fit by contraction of a horizontal tabular body located 10 ± 5 km beneath the Sour Creek resurgent dome. For the 1923–1976 uplift period, an inversion of the leveling data by Vasco *et al.* (1990) showed that the largest volume expansions were 3–6 km beneath each of Yellowstone's two resurgent domes. Meertens *et al.* (1992, 1993) showed that regional GPS data for 1987–1991 (a period when the caldera floor was subsiding) are best explained by deflation of two sub-horizontal, tabular bodies centered 3–6 km beneath the caldera floor, superimposed on a regional tectonic strain signal. They attributed the deflation to movement of hydrothermal or magmatic fluids.

It is useful to visualize these hypothetical deformation sources in the context of what is known about Yellowstone's upper crustal structure from seismic data. For example, Miller and Smith (1999) used first-arrival times from 7,942 local earthquakes and 16 controlled-source explosions (for which locations and origin times are known precisely) to model the 3-D P and S velocity structure beneath the caldera. They showed that a caldera-wide 15% decrease from regional P velocities at depths of 6 to 12 km is coincident with a -60 mGal Bouguer gravity anomaly, and they attributed both anomalies to a hot, subsolidus, granitic batholith with quasi-plastic rheology. Localized 30% reductions from regional seismic velocities and higher V_p/V_s ratios 8 km beneath the resurgent domes were interpreted as reflecting the presence of partial melts and vestigial magma systems. These latter anomalies correlate reasonably well with the two sources of deformation inferred by Vasco *et al.* (1990), Meertens *et al.* (1993), and Wicks *et al.* (1998).

Both the geodetic and seismic data point to the shallowest part of Yellowstone's magmatic system, or the deepest part of its hydrothermal system, as the source of contemporary crustal deformation. The data are not sufficiently precise to distinguish between these two sources, and in fact both may be involved, contributing to varying degrees at different times. Although the geodetic models differ in detail, all are consistent with the idea that (1) caldera uplift from 1923 to 1984 was caused at least partly by pressurization of the hydrothermal system below a self-sealed layer about 5 km deep, and (2) subsidence since 1984 was caused primarily by rupturing of the layer during shallow earthquake swarms. In addition, the regional GPS data indicate that sagging of the caldera floor in response to regional crustal extension contributed to the subsidence measured since 1984. Surface uplift centered near Norris Geyser Basin during 1997–2002 might be evidence for episodic intrusion of magma into the crust beneath the caldera. Such intrusions *must* occur from time to time to sustain Yellowstone's vigorous hydrothermal system over long timescales. Deeper intrusions might be aseismic and difficult to detect geodetically without using continuous GPS (CGPS) or InSAR. So it is possible that magmatic intrusions also contributed to the uplift measured at Yellowstone between 1923 and 1984.

The evidence from temporal gravity changes is consistent with this idea but equivocal. Smith *et al.* (1989) noted that the gradient determined at more than 100 precision-gravity and leveling stations in Yellowstone for 1977–1987, a period that was dominated by surface uplift, was significantly less than the theoretical free-air gradient ($-1.7\,\mu$Gal cm^{-1} *versus* $-3.086\,\mu$Gal cm^{-1}). This suggests a net mass increase beneath the caldera during this period (Section 2.7.1), presumably an intrusion of hydrothermal fluid or magma, but the density of the intrusion is poorly constrained. Arnet *et al.* (1997) extended the observations through 1994 and came to a similar conclusion. During a period of uplift from 1977 to 1983, the gravity field decreased across the caldera by as much as $-60 \pm 12\,\mu$Gal. The observed gradient could not have been caused solely by pressurization of the deep hydrothermal system, without any significant mass increase. The authors (p. 2744) attributed the changes to '...*widespread hydrothermal fluid movement, which furthermore is related to input by magma.*' From 1986 to 1993, while the caldera floor subsided, the gravity field increased by as much as $60 \pm 12\,\mu$Gal and the $(\Delta g/\Delta h)$ gradient was near the theoretical free-air gradient. This indicates that subsidence was not accompanied by a measurable mass change, which Arnet *et al.* (1997, p. 2744) attributed to '...*depressurization of the deep hydrothermal system as a result of fracturing and volatile loss to the shallow hydrothermal system or, less likely, reduced input of brine to the deep system.*' They preferred the first mechanism because it more easily explains the relatively abrupt change from uplift to subsidence between 1984 and 1986 (Dzurisin *et al.*, 1990).

The June–July 1995 earthquake swarm near Madison Junction remains enigmatic, because it coincided with the resumption of uplift that was first detected with InSAR and later verified by leveling. None of the models described above easily accounts for the fact that the most intense swarm ever recorded in the Yellowstone region (starting in October 1985) coincided with the beginning of subsidence, while the second most intense swarm (June–July 1995) coincided with the resumption of uplift. Clearly, we still have much to learn from Earth's largest restless caldera.

7.3.5 Spatiotemporal changes in deformation revealed by InSAR

As described in Chapter 5, the most recent advance in our understanding of ground deformation at Yellowstone came not from leveling surveys, but from InSAR observations starting in 1992. Radar interferograms for the periods 1992–1993 and 1993–1995 agree reasonably well with leveling results along the traverse from Lake Butte to Mount Washburn (Wicks *et al.*, 1998). However, the InSAR data also show that the center of subsidence shifted from the northeast part of the caldera, near the Sour Creek resurgent dome, during 1992–1993, to the southwest part, near the Mallard Lake dome, during 1993–1995. What is more, while the Mallard Lake dome was still subsiding, the Sour Creek dome began to rise during 1995–1996 for the first time since 1983–1984. Uplift was occurring throughout most of the caldera by 1997, but by 2000 the central part of the caldera floor was subsiding again, while a broad area centered near Norris was rising (Wicks *et al.*, 2005).

The rapidity of the change from uplift to subsidence within the caldera before and after 1984–1985, and from subsidence to uplift again from 1993–1995 to 1995–1996 suggests that the process responsible for these displacements is relatively shallow-seated and reversible. The same argument applies to the rapid lateral migration of the center of subsidence from 1992–1993 to 1993–1995. Thus, it seems implausible that changes of this magnitude could occur so quickly and reversibly if they are primarily driven by deep-seated magmatic or regional tectonic processes. Rather, the most likely source of historical surface displacements within the Yellowstone Caldera is the deep hydrothermal system, which

apparently is capable of pressurizing and depressurizing over timescales of years to decades. Because uplift of the Sour Creek dome began while the Mallard Lake dome was still subsiding and the Mallard Lake dome caught up within the next year, I imagine some sort of sluggish fluid connections that convey pressure changes within the lowermost part of the hydrothermal system and perhaps the uppermost part of the magmatic system.

The fact that uplift and subsidence occur at approximately the same rate, when averaged over several years, suggests that the rate of pressure change is regulated in some way, perhaps by the complexity of the fluid pathways. Like an inflatable raft with a small inlet/outlet port, Yellowstone's hydrothermal system may inflate and deflate – essentially 'breathe' – only very slowly in most cases. Catastrophic ruptures are probably associated with large hydrothermal explosions, which are responsible for several craters and associated ejecta deposits throughout the Yellowstone Caldera. At least some of these formed when glacially dammed lakes suddenly drained, abruptly decreasing the confining pressure on the shallow hydrothermal system and triggering explosions (Muffler *et al.*, 1971).[6] Even if the hydrothermal system is responsible for most of the vertical surface displacements observed at Yellowstone in recent decades, over longer timescales magmatic intrusions undeniably play an important role as well. The recent uplift episode near Norris Geyser Basin would seem to be a case in point.

7.4 LONG VALLEY CALDERA AND THE MONO-INYO VOLCANIC CHAIN: TWO DECADES OF UNREST (AND STILL COUNTING?)

7.4.1 Eruptive history and recent unrest

Among the most detailed and long-term geodetic datasets available for any volcanic system on Earth is that for the Long Valley Caldera and nearby

---

[6] Muffler *et al.* (1971) identified ten hydrothermal explosion craters in Yellowstone National Park ranging in diameter from a few tens of feet (~10 m) to about 5,000 feet (1,500 m). Geologic relations at the Pocket Basin crater, in the western part of the Park about 7 miles (11 km) north of Old Faithful (Figure 7.29), indicate that the explosion there was triggered by an abrupt decrease in confining pressure when an ice-dammed lake suddenly drained during the waning stages of early Pinedale Glaciation about 15,000 years ago. The authors suggested that most of the other explosion craters in the Park could have formed in the same manner.

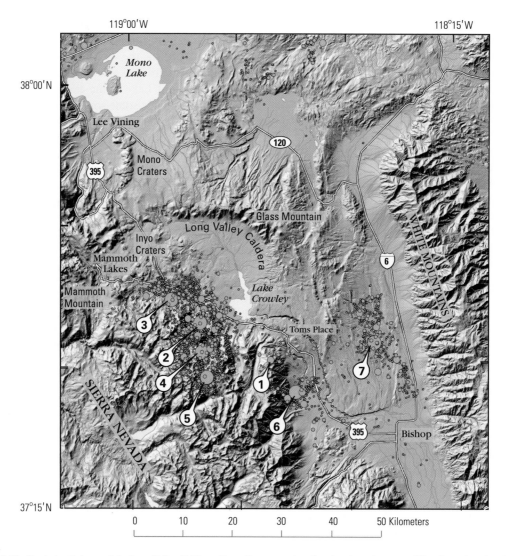

Figure 7.33. Shaded-relief map of the Long Valley Caldera–Mono Craters region showing the epicenters of $M \geq 3$ earthquakes (orange circles) for 1978–1999. Circle sizes represent earthquake magnitude in four steps from $M = 3.0$ to $M = 6.0$. Earthquakes discussed in the text include: (1) 4 October 1978 M 5.8 Wheeler Crest earthquake; (2–5) 25–27 May M ~6 Long Valley Caldera and Sierra Nevada earthquakes; (6) 23 November 1984 M 6.0 Round Valley earthquake; and (7) 21 July 1986 M 6.4 Chalfant Valley earthquake. Figure modified slightly from Hill *et al.* (2002a).

Mono-Inyo Craters volcanic chain in eastern California (Figure 7.33). The Long Valley Caldera is a $32 \, \text{km} \times 17 \, \text{km}$ elliptically shaped depression that formed about 730,000 years ago when the roof of a large crustal magma reservoir ruptured catastrophically and collapsed, resulting in the expulsion of $600 \, \text{km}^3$ of rhyolitic magma (Bailey *et al.*, 1976; Bailey, 1989, 2004). The resulting ash flow deposits, known as the Bishop Tuff, inundated $1,500 \, \text{km}^2$ of the surrounding countryside and accumulated locally to a thickness approaching $200 \, \text{m}$ (Figure 7.34). Some ash flows spilled westward over the crest of the Sierra Nevada, and within the caldera they locally ponded to a thickness of

more than $1,500 \, \text{m}$. Plinian ash clouds associated with the eruption drifted thousands of kilometers downwind, depositing an ash layer that is still recognizable as far east as Kansas and Nebraska, more than $1,000 \, \text{km}$ away (Izett *et al.*, 1970; Izett, 1982). Thinner Bishop Ash deposits have been recognized in southwest California (Merriam and Bischoff, 1975) and in Pacific Ocean seafloor cores (Sarna-Wojcicki *et al.*, 1987).

Following collapse, eruptions continued within the 2- to 3-km-deep Long Valley Caldera for a few tens of thousands of years. Early postcaldera pyroclastic eruptions formed a thick sequence of intracaldera bedded tuffs that were followed by

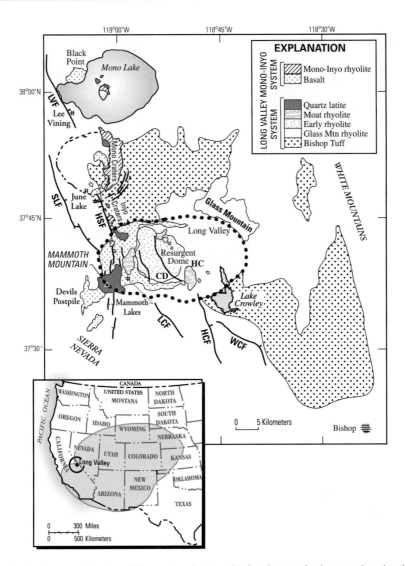

Figure 7.34. Simplified geologic map of the Long Valley region showing the distribution of volcanic rocks related to the Long Valley Caldera magmatic system and the younger Mono-Inyo magmatic system, after Bailey *et al.* (1976) and Bailey (2004). The Bishop Tuff is a welded ash flow deposit from the caldera-forming eruption 730,000 years ago. It is mostly buried by younger deposits in the vicinity of the caldera. (*Inset*) The approximate distribution of airfall ash from the caldera-forming eruption. LVF, Lee Vining fault; HSF, Hartley Springs fault; LCF, Laurel Creek fault; HCF, Hilton Creek fault; SLF, Silver Lake fault; WCF, Wheeler Crest fault (also known as Round Valley fault); CD, Casa Diablo; HC, Hot Creek.

extrusion of hot, fluid obsidian flows known collectively as the early rhyolites. Contemporaneously, renewed magma pressure at depth uplifted, arched, and faulted the central part of the caldera floor, forming a resurgent dome about 10 km in diameter and 500 m high. Early rhyolites exposed in the dome are tilted radially outward as much as 30 degrees, attesting to the intensity of deformation caused by magmatic resurgence and structural adjustments to caldera formation. Eruption of the early rhyolites and formation of the resurgent dome occurred rapidly, within 100,000 years or less after caldera collapse (Bailey, 1989).

During the rise of the resurgent dome and eruption of the early rhyolites, a large lake filled the caldera. Continued rise of the resurgent dome eventually raised the lake surface to the level of the low southeast caldera rim, where erosive overflow subsequently cut the spectacular Owens River Gorge (Figure 7.35). Glaciers flowing into the lake from the High Sierra Nevada generated debris-laden icebergs that drifted across the lake, depositing large erratics of Sierran granite on the flanks of the resurgent dome, which stood as an island until the lake drained between 100,000 and 50,000 years ago (Bailey, 1989).

Figure 7.35. The Owens River Gorge in eastern California, which was cut when resurgent doming within the Long Valley Caldera raised the surface of an intracaldera lake to the level of the low southeast caldera rim (Bailey, 1989). USGS photograph by David E. Wieprecht.

After a pause of about 100,000 years, crystal-rich rhyolite began erupting from three groups of vents in the caldera moat peripheral to the resurgent dome. The moat rhyolites erupted at about 200,000-yr intervals – approximately 500,000, 300,000, and 100,000 years ago (Figure 7.34). Bailey and others (1976) speculated that the coarse crystallinity, presence of hydrous minerals, and relatively high vesicularity of the moat rhyolites may reflect their origin as residual magma from the caldera-forming eruption. Cooling and crystallization of residual magma concentrated volatiles and increased pressure in the magma reservoir to the extent that the ring-fracture system was repeatedly reactivated and served as a conduit to the surface for the moat rhyolite magmas.

The younger Mono-Inyo Craters volcanic chain is localized along a narrow, north-trending fissure system that extends 45 km northward from south of Mammoth Mountain through the west moat of Long Valley Caldera to the north shore of Mono Lake (Figure 7.36). This system began by erupting basalt and andesite from vents in and around the west moat of Long Valley Caldera, including those at Devil's Postpile National Monument, about 300,000 years ago. Between 200,000 and 50,000 years ago, extrusion of at least 12 rhyolite domes and flows built Mammoth Mountain, which straddles the southwest caldera rim and hosts a popular ski resort (Figure 7.37) (Bailey, 1989).

Rhyolites began erupting to form the Mono-Inyo Craters about 40,000 years ago, first at the Mono Craters and later at the Inyo Craters. The Mono Craters comprise an arcuate chain of 30 or more overlapping rhyolite domes, flows, and craters, apparently fed from the margin of a sub-circular pluton that may still be partly molten. The Inyo Craters form a discontinuous 10-km-long line of rhyolite dome-flows and craters within and immediately north of the caldera's west moat. During the past 3,000 years, the Mono-Inyo Craters have erupted at intervals of 700 to 250 years, the most recent eruptions being from Panum Crater and the Inyo Craters 650–550 years ago (Miller, 1985; Sieh and Bursik, 1986), and from Paoha Island about 250 years ago (Stine, 1990) (Figure 7.36). Research drilling beneath the Mono Craters and detailed stratigraphic studies of associated tephra deposits have shown that the eruption 650–550 years ago produced five separate dome-flows and several phreatic explosion craters, all within a span of probably no more than a few years, or possibly months, and all originating from a single dike (Eichelberger *et al.*, 1985).

Against this backdrop of frequent eruptions in the recent geologic past, a period of relative seismic quiescence that followed a sequence of M 5.5 to 6.0 earthquakes in 1941 near Tom's Place ended on 4 October 1978, when the M 5.8 Wheeler Crest earthquake struck midway between the towns of Bishop and Mammoth Lakes (Figure 7.33). Clusters of small to moderate ($M > 4$) earthquakes generally migrated northwestward toward the caldera until, on 25–27 May 1980, four M ~6 earthquakes and an intense swarm of smaller events shook the region. Two more comparably sized events struck the region on 23 November 1984 (M 6.0 Round Valley earthquake) and 21 July

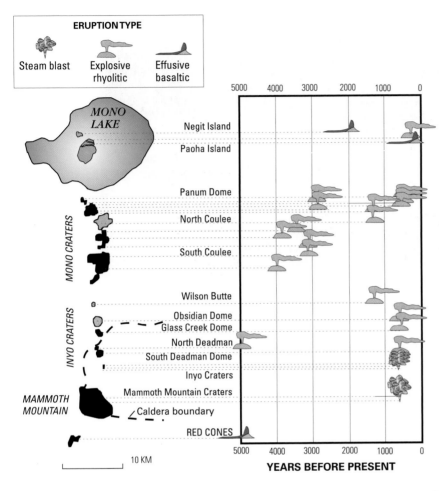

ERUPTION TYPE

Steam blast   Explosive   Effusive
               rhyolitic   basaltic

Figure 7.36. Eruptive history of the Mono-Inyo Craters volcanic chain for the past 5,000 years, from Hill *et al.* (2002a). The Mono Craters consists of 30 or more overlapping rhyolite domes, flows, and craters. An eruption 650–550 years ago formed Panum Dome and the Inyo Craters to the south. The most recent eruption, from vents on the floor of Mono Lake, formed Paoha Island about 250 years ago. Dashed line shows part of the caldera boundary.

1986 (*M* 6.4 Chalfant Valley earthquake). Frequent earthquake swarms, notably in 1983 (south moat), 1989–1991 (Mammoth Mountain), 1997–1998 (south moat), and 1998–1999 (Sierra Nevada) were accompanied by about 80 cm of uplift of the resurgent dome, changes in the shallow hydrothermal system, and increased emission of magmatic CO_2 in the vicinity of Mammoth Mountain (Hill, 1984, 1996; Hill *et al.*, 1990, 2002a, 2002b, 2003).

Because any recurrence of eruptive activity in the Long Valley area would pose a substantial threat to the community of Mammoth Lakes, to regional communications and transportation infrastructure, and to commercial air traffic (e.g., Miller *et al.*, 1982), the USGS conducts an intensive monitoring program through its Long Valley Observatory (LVO). As a result, there exists a detailed record of the unrest that has persisted for more than two decades, including repeated leveling, EDM, and GPS surveys, plus continuous data from arrays of tilt-meters, strainmeters, magnetometers, and GPS stations (Hill and Prejean, 2005). Some of the geodetic

datasets are discussed individually in other chapters. Here, I briefly review the leveling, trilateration, and GPS results to illustrate the effectiveness of a thorough geodetic program for studying volcanic unrest, especially in conjunction with other monitoring techniques integrated in an observatory environment.

The May 1980 earthquake swarm at Long Valley, which ranks among the most intense episodes of non-eruptive unrest ever recorded at an active magmatic system, prompted immediate re-surveys of regional leveling and trilateration networks to help determine the cause of the activity and its implications for earthquake and volcano hazards.[7] The results showed unequivocally that

---

[7] The intense earthquake swarm at Long Valley, including four *M* 6 shocks during 25–27 May 1980, came little more than a week after the catastrophic eruption of Mount St. Helens on 18 May. There is no plausible connection between the two events, separated as they were by 1,000 km, and in hindsight their timing was almost surely coincidental. For those of us struggling to cope with Mount St. Helens at the time, however, the news from Long Valley was unsettling – an eerie reminder of how little we understood about volcanoes and how they behave.

Figure 7.37. Mammoth Mountain as seen looking west from EDM station CASA near the south edge of the resurgent dome at Long Valley Caldera. Mammoth Mountain rises steeply above the Town of Mammoth Lakes, which is visible near the center of the photograph. Jagged peaks to the right of Mammoth Mountain are the Minarets of the Sierra Nevada. The major road leading to Mammoth Lakes is California Highway 203, which intersects US Highway 395 in the lower left portion of the photograph. Dead vegetation immediately northeast of the intersection was killed by recent changes in the Casa Diablo geothermal area, which is tapped by three geothermal power plants with a combined rating of 40 Megawatts (*lower right*).
USGS photograph by Elliot T. Endo, 14 August 2003.

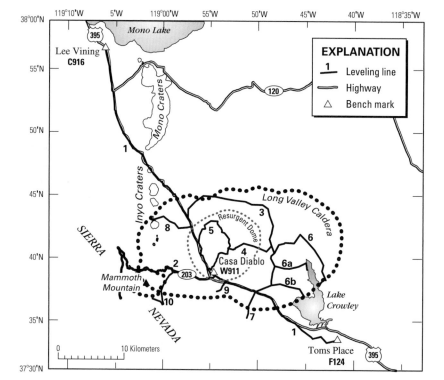

Figure 7.38. Leveling network (heavy solid lines) used to monitor vertical displacements within and around Long Valley Caldera (Hill *et al.*, 2002a). Numbers identify individual leveling loops or lines following the convention used by Savage *et al.* (1987). Bench marks C916 near Lee Vining and F124 near Toms Place are reference marks. Bench mark W911 near Casa Diablo is close to the point of maximum uplift along Highway 395. See text for discussion. The configurations of other types of monitoring networks at Long Valley (e.g., seismic stations, borehole strainmeters and tiltmeters, CGPS stations, CO_2 and hydrologic monitoring sites) are shown elsewhere by Hill *et al.* (2002a) and Dzurisin (2003).

the long-dormant magmatic system beneath Long Valley Caldera was involved in the unrest. Subsequent activity and ongoing investigations have confirmed the potential for renewed eruptive activity, and the recent geologic record points to the Mono-Inyo Craters volcanic chain as the most likely site of the next eruption in the region (Miller *et al.*, 1982; Hill *et al.*, 1985a,b).

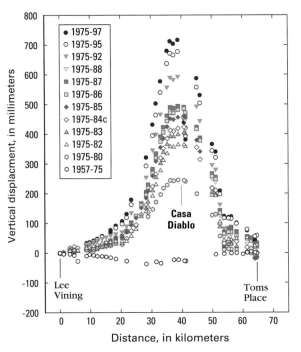

Figure 7.39. Results of first-order leveling surveys across Long Valley Caldera along Highway 395 between Lee Vining and Toms Place from 1957 to 1997. Following an intense earthquake swarm that included four M 6 events on 25--27 May 1980, leveling in October 1980 revealed nearly 25 cm of uplift centered at the resurgent dome near Casa Diablo since the previous survey in 1975. Uplift continued at a gradually declining rate through October 1989, when two-color EDM measurements revealed an increase in the extensional strain rate about 2 months prior to the start of a prolonged earthquake swarm beneath Mammoth Mountain on the caldera's southwest rim. Subsequent leveling surveys in 1992, 1995, and 1997 also show the effects of renewed caldera inflation. The 1984 leveling survey was contaminated by a systematic error that produced a large apparent tilt between Lee Vining and Toms Place, which has been removed from the data shown here (1984c). The dome subsided about 2 cm from early 1999 through the end of 2001, but that change was largely offset by renewed uplift starting in early 2002. Uplift stopped again in early 2003. Since then, the dome has shown only minor fluctuations in height and remains roughly 80 cm higher than in the late 1970s (current as of January 2006).

7.4.2 Leveling results: tracking caldera inflation in space and time

The first clear evidence that magma was involved in the unrest at Long Valley came from re-surveys in 1980 of a first-order leveling traverse along US Highway 395, which crosses the caldera's west moat and resurgent dome, and of a regional trilateration network designed primarily to measure tectonic strain (Savage and Clark, 1982). The Highway 395 leveling traverse between Lee Vining and Toms Place was measured partly or entirely in 1932, 1957, 1975, 1980, annually from

1982 to 1988, 1992, 1995, and 1997. No significant vertical displacements occurred from 1932 to 1975, but from 1975 to 1997 (probably starting sometime after July 1979, see below), a broad area centered on the resurgent dome rose cumulatively almost 80 cm (Figures 7.38 and 7.39).

When the traverse was first remeasured 5 months after the May 1980 earthquake swarm, bench mark W911 near Casa Diablo had risen about 25 cm since the previous survey in 1975. It rose an additional 12 cm between surveys in 1980 and 1982, and continued to rise at a slowly declining rate through 1988 (Figures 7.39 and 7.40). Two-color EDM measurements made every few days detected an increase in the rate of extension across the resurgent dome starting in October 1989 (Langbein et al., 1993), and a corresponding increase in the uplift rate was measured by leveling surveys along Highway 395 in 1992, 1995, and 1997.

More extensive leveling surveys along a network of roads mostly within the Long Valley Caldera (Figure 7.38) were conducted in 1975, then annually from 1982 to 1986, 1988 and 1992. Because the network is sufficiently dense, vertical displacements

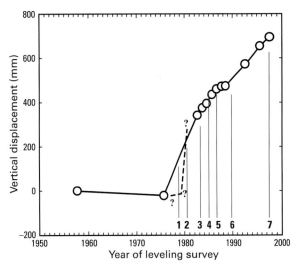

Figure 7.40. Progressive uplift of bench mark W911 near Casa Diablo with respect to bench mark C916 near Lee Vining from leveling surveys in 1957, 1975, 1982–1988, 1992, 1995, and 1997. A 1980 survey did not include C916, but W911 rose 250 mm with respect to F124 near Toms Place from 1975 to 1980 (dotted line). See Figure 7.38 for bench mark locations. (1) 4 October 1978 M 5.8 Wheeler Crest earthquake; (2) 25–27 May 1980 earthquake swarm including four M ∼6 events; (3) January 1983 south moat earthquake swarm including two M 5 earthquakes; (4) 23 November 1984 M 6.0 Round Valley earthquake; (5) 21 July 1986 M 6.4 Chalfant Valley earthquake; (6) 1989–1991 Mammoth Mountain earthquake swarm and renewed caldera inflation; and (7) 1997–1998 south moat earthquake swarm and renewed caldera inflation.

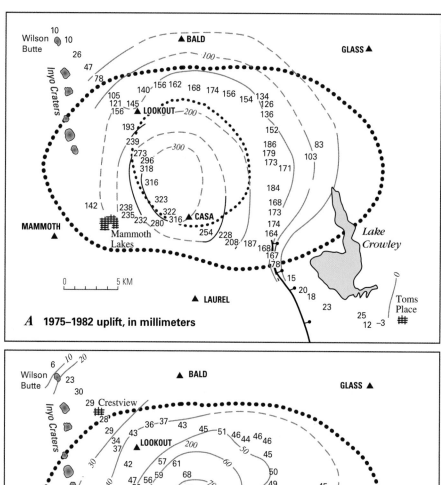

Figure 7.41. Contour maps showing uplift within Long Valley Caldera for the following periods: (A) 1975–1982; (B) 1982–1983; (C) 1983–1985; and (D) 1982–1985 (Savage *et al.*, 1987). Numbers show amounts of uplift, in millimeters, relative to bench mark F124 near Toms Place, which was assumed to have remained fixed. C916 near Lee Vining, not F124, was assumed to have remained fixed for Figures 7.39 and 7.40. Resulting differences are small compared with the amount of uplift within the caldera. The apparent westward shift of the center of uplift during 1983–1985 could be an artifact of relative rod-scale error. Triangles represent trilateration stations (see also Figure 7.43).

between surveys can be contoured to delineate both the spatial extent of the displacement field and any changes in its shape as a function of time (Figure 7.41). Such information provides a strong constraint on deformation-source models, especially in conjunction with contemporaneous data on horizontal displacements (e.g., Castle *et al.*, 1984; Savage *et al.*, 1987; Langbein *et al.*, 1995; Langbein, 2003).

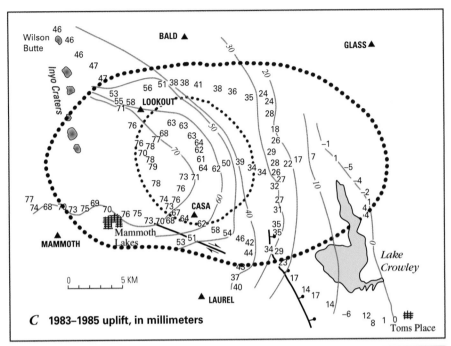

C **1983–1985 uplift, in millimeters**

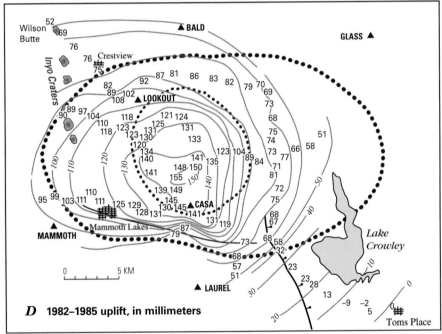

D **1982–1985 uplift, in millimeters**

7.4.3 Regional and intracaldera trilateration surveys

Horizontal deformation at Long Valley has been measured with two types of precise EDM surveys: regional trilateration using a Geodolite starting in 1972, and intracaldera trilateration with a two-color EDM starting in 1983. Both approaches have yielded important insights into the processes responsible for unrest. Regional surveys in 1972, 1973, 1976, 1979, and 1980 revealed that a distinctive pattern of outward-radial displacements centered at Long Valley Caldera developed sometime between

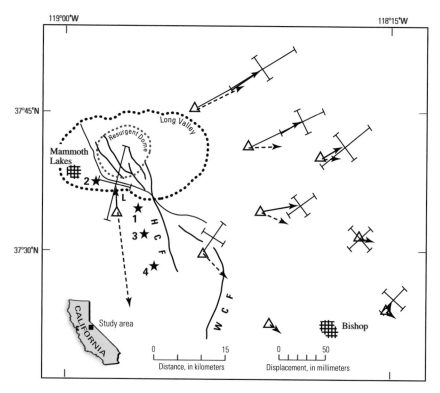

Figure 7.42. Horizontal displacement vectors determined by trilateration surveys near Long Valley Caldera in July 1979 and September 1980 (Savage and Clark, 1982). Dotted ellipse, Long Valley Caldera rim; smaller dotted circle, resurgent dome; triangles, trilateration stations; solid arrows, observed displacements with error bars showing principal axes of 95% confidence ellipse for each displacement vector; dashed arrows, model displacements predicted by inflation of a spherical magma body located 10 km beneath the caldera; stars, epicenters of four $M{\sim}6$ earthquakes on 25–27 May 1980, numbered in order of occurrence; HCF, Hilton Creek fault; WCF, Wheeler Crest fault (also known as Round Valley fault, e.g., Bailey, 2004).

July 1979 and September 1980 (Figure 7.42). Recall that leveling surveys along Highway 395 detected the beginning of uplift sometime between 1975 and October 1980. Savage and Clark (1982) concluded that the most likely cause of extension and uplift was inflation of a magma body located about 10 km beneath the caldera starting sometime after July 1979, just before or during the May 1980 earthquake sequence.

The same pattern of uplift and radial extension continued for several years at a gradually declining rate, as documented by annual leveling and Geodolite surveys from 1982 to 1986 (Figures 7.39, 7.40, 7.43–7.45). Savage et al. (1987) concluded: (1) the principal deformation sources were inflation of a magma body beneath the resurgent dome and right-lateral strike-slip on a vertical fault in the south moat of the caldera; (2) the inflation rate was roughly constant at $0.02\,\mathrm{km}^3\,\mathrm{yr}^{-1}$, but the slip rate on the south moat fault decreased substantially after the January 1983 earthquake swarm; and (3) there is evidence for a source of dilatation (possibly dike intrusion) beneath the south moat in 1983 and less certain evidence for a deep source (possibly magmatic inflation beneath Mammoth Mountain) in the western caldera during 1983–1985. The latter possibility was strengthened by the occurrence of a

prolonged earthquake swarm beneath Mammoth Mountain in 1989–1990, which Hill et al. (1990) attributed to an intrusion of basalt into the upper crust.

By the early 1980s, regional leveling and trilateration surveys at Long Valley clearly showed that the locus of ground deformation was inside the caldera, beneath the resurgent dome and south moat. To better study the source area, more detailed and frequent geodetic measurements were required. Following a particularly intense earthquake swarm and possible dike intrusion beneath the south moat in January 1983, a dense trilateration network was established in the south moat in June 1983 and expanded by 1985 to cover the resurgent dome and western half of the caldera (Langbein, 1989, 2003) (Figure 7.46). A two-color EDM with a precision of $0.1-0.2\,\mathrm{ppm}$ (1–2 mm for a 10-km baseline) is used to measure several of the 42 baselines every few days, weather permitting, and the rest of the network monthly or yearly. The resulting record might be the most detailed and precise long-term, time series geodetic dataset ever obtained at a restless volcano.

The two-color EDM results clearly show that extensional strain rates in the south moat gradually decreased from as high as $5\,\mathrm{ppm}\,\mathrm{yr}^{-1}$ several months

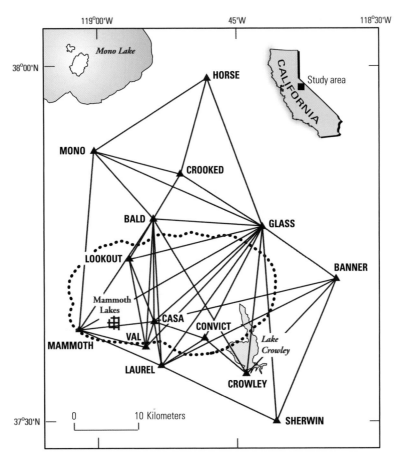

Figure 7.43. Long Valley regional trilateration network (Savage *et al.*, 1987).

after the January 1983 earthquake swarm to near zero in mid-1989 (Langbein *et al.*, 1993; Langbein, 2003). Then, starting in October 1989, a remarkable thing happened. Extension rates increased to as high as 9 ppm yr^{-1} over a period of a few weeks, while the level of earthquake activity beneath the south moat and elsewhere in the caldera remained low (Figure 7.47). About 2 weeks later, a subtle increase in the occurrence of small earthquakes beneath Mammoth Mountain was noted. This area had been relatively quiet since the beginning of unrest more than a decade earlier. Earthquake activity beneath the caldera picked up several weeks later, in December 1989, and continued at an elevated level through 1991. The two-color EDM results clearly foretold increases in earthquake activity beneath Mammoth Mountain and the caldera, but what processes were at work to cause the earthquakes?

There is compelling evidence that seismicity beneath Mammoth Mountain during 1989–1990 was caused by an intrusion of magma into the upper crust. Renewed inflation beneath the

caldera starting several weeks later probably was caused by a separate intrusion into the reservoir located about 6 km beneath the resurgent dome. Evidence for an intrusion beneath Mammoth Mountain includes (Hill, 1996):

(1) a dike-like distribution of hypocenters at depths of 6–9 km with a north–northeast strike essentially perpendicular to the *T*-axes of the earthquake focal mechanisms (Hill *et al.*, 1990);

(2) deformation in the vicinity of Mammoth Mountain consistent with a north–northeast-striking dike extending to within 2 km of the surface beneath Mammoth Mountain (Langbein *et al.*, 1993);

(3) frequent spasmodic bursts suggesting rapid-fire brittle failure driven by transient surges in local fluid pressure (Hill *et al.*, 1990);

(4) an increase in ^{3}He/^{4}He ratios from fumaroles on Mammoth Mountain detected in late 1989 (Sorey *et al.*, 1993);

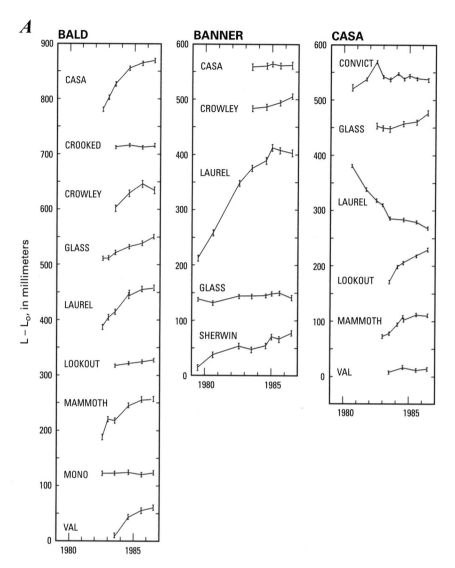

Figure 7.44. Line-length changes in the Long Valley trilateration network measured by annual Geodolite surveys, 1979 to 1986 (Savage *et al.*, 1987). (A) Lines measured from instrument stations BALD, BANNER, and CASA. (B, *opposite*) Lines measured from instrument stations CONVICT, CROOKED, CROWLEY, GLASS, HORSE, LAUREL, LOOKOUT, and MAMMOTH. See Figure 7.43 for station locations. Measured length less a constant nominal length is plotted for each line as a function of time. Error bars represent one standard deviation.

(5) a persistent sequence of LP volcanic earthquakes beneath the southeast flank of Mammoth Mountain at depths between 10 km and 30 km; and

(6) accelerated outgassing of CO_2, almost surely of magmatic origin, around the flanks of Mammoth Mountain first noticed when trees began dying in the area during 1990 (Farrar *et al.*, 1995; McGee and Gerlach, 1998b) (Figure 7.48).

Not to be outdone by Mammoth Mountain, Long Valley Caldera caused a brief stir with another earth-

quake swarm beneath its south moat during January–April 1996. This was followed by a period of major concern starting in July 1997 when south moat seismicity and resurgent dome swelling began to increase exponentially. As was the case prior to the 1989 Mammoth Mountain swarm, the deformation rate within the caldera began to increase as much as 2 months prior to the onset of earthquake swarm activity in July 1997 (Figure 7.49). By mid-November, the extension rate across the resurgent dome had reached $\sim$36 cm yr^{-1} ($\sim$1 mm day^{-1}). Seismicity peaked on 22 November at $\sim$1,000 $M \geq 1.5$ earth-

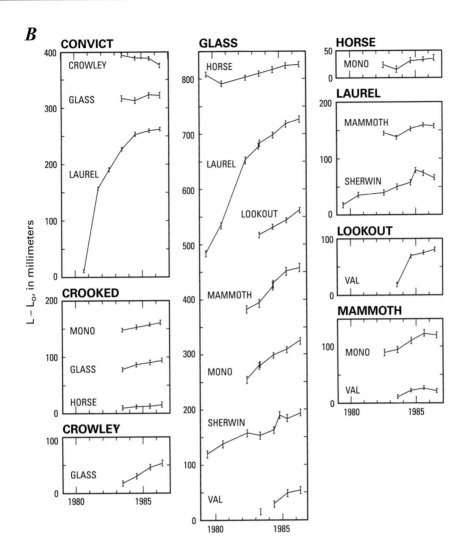

quakes day^{-1}, including $M = 4.6$, 4.9, and 4.6 events. The swarm earthquakes were concentrated in a 1-km-wide, east–west-striking lineation beneath the western part of the south moat (Prejean *et al.*, 2002; Hill *et al.*, 2003). The extension rate began to decay in early December 1997, but earthquake activity surged again at the end of December and early January 1998. By the time activity dropped to background levels in late March 1998, the resurgent dome had extended by an additional 10 cm and the cumulative number of $M \geq 1.5$ earthquakes exceeded 6,000, including 8 $M \geq 4.0$ events. Hill *et al.* (2003) concluded that the 1997–1998 episode was caused by intrusion of magma or magmatic brine beneath the south moat and that the associated earthquake swarm was largely driven by the intrusion process.

Another period of heightened activity in the region began on 9 June 1998, with a M 5.2 earthquake at the southern margin of the caldera just 2 km west of the point where the surface trace of the Hilton Creek fault intersects the caldera boundary. A typical aftershock sequence was interrupted by a second M 5.2 earthquake on 15 July, followed by another aftershock sequence. The third and largest earthquake in the series, a M 5.6 event on 15 May 1999, also was followed by an aftershock sequence and, like the earlier M 5.2 events, it was not accompanied by any response from the caldera. Hill *et al.* (2003, p. 190) concluded that '... *this sequence of M 5 earthquakes has a decidedly "tectonic" character*' and '... *the earthquakes within this Sierra Nevada sequence appear to be high-frequency, brittle-failure events with no evidence for active fluid involvement in the source process.*' Clearly, not all earthquake swarms near volcanoes have

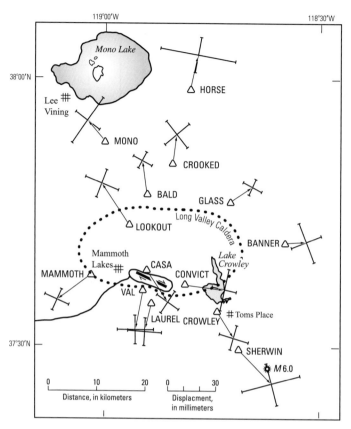

Figure 7.45. Station displacements from 1983 to 1985 determined by annual trilateration measurements (Savage *et al.*, 1987). Error bars represent the principal axes of the 95% confidence ellipse for each displacement vector. Also shown are the trace of an inferred fault in the south moat of the caldera (strike-slip symbol) and the epicenter of the 23 November 1984, *M* 6.0 Round Valley earthquake (sunburst symbol, lower right).

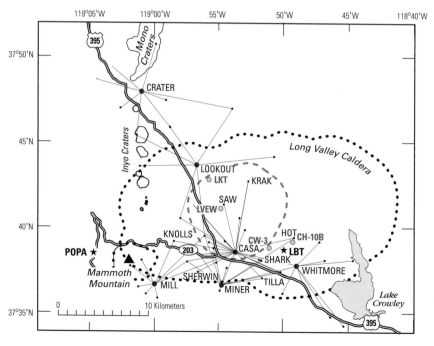

Figure 7.46. Two-color EDM network at Long Valley Caldera and Mono-Inyo Craters (Hill *et al.*, 2002a). Large black dots, instrument sites; small black dots, reflector sites. Baselines radiating from CASA are measured several times per week, weather conditions permitting; others are measured monthly or annually. Stars represent the Devils Postpile borehole strainmeter (POPA) and a long-base (0.5 km) Michelson long-base tiltmeter (LBT). Shaded circles mark the locations of four wells that have exhibited water level changes in response to earthquakes (CH-10B, CW-3, LKT, and LVEW).

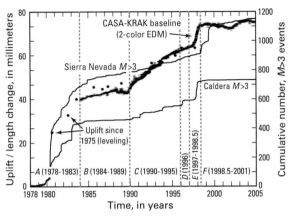

Figure 7.47. History of earthquake activity and swelling of Long Valley's resurgent dome from 1978 through 2004, updated from Hill *et al.* (2002a). Thin lines show the cumulative number of $M \geq 3$ earthquakes within Long Valley Caldera (Caldera $M \geq 3$) and within the Sierra Nevada block to the south (Sierra Nevada $M \geq 3$). Filled circles represent the uplift history of a bench mark near the center of the resurgent dome (i.e., near the intersection of lines 4 and 5 in Figure 7.38) from leveling surveys in 1980, 1982, 1983, 1985–1988, 1992, 1995, and 1997. Thick line shows the amount of extension of an 8-km-long baseline between CASA and KRAK (Figure 7.46) since 1983 based on frequent two-color EDM measurements.

the same origin, and careful detective work is needed to decipher their implications for volcanic and earthquake hazards.

7.4.4 Repeated and continuous GPS measurements

GPS has also played an important role in monitoring unrest at Long Valley – both repeated surveys of a regional network and continuous measurements at several sites within the caldera. An intriguing result was reported by Marshall *et al.* (1997), who showed that annual GPS surveys from 1990 to 1994 of a network spanning both Long Valley Caldera and the Mono-Inyo volcanic chain revealed not only caldera inflation ($0.007 \, \text{km}^3 \, \text{yr}^{-1}$) and Basin and Range extension ($0.15 \pm 0.03 \, \text{ppm} \, \text{yr}^{-1}$ oriented $N66^\circ W \pm 1^\circ$), as expected, but also the possibility of dike intrusion beneath the Mono Craters. Their best-fit model included as much as $6 \, \text{cm} \, \text{yr}^{-1}$ of opening along a north–south trending dike starting at 10 km depth beneath the Mono Craters chain.

CGPS measurements have been made at station CASA on the south side of the resurgent dome since early 1993 and at KRAK on the north side of the dome since late 1994 (Dixon *et al.*, 1997). The 3-D displacement vectors for these two stations originate at a source $5.8 \pm 1.6 \, \text{km}$ beneath the resurgent dome,

Figure 7.48. Horseshoe Lake tree-kill area, one of several zones of tree mortality around Mammoth Mountain, California, caused by magmatic carbon dioxide leaking from depth along faults and fractures and accumulating in the shallow soil layer. Soil CO_2 concentrations exceed 90% in some locations. (*top*) View looking north--northwest at Horseshoe Lake (foreground) and Mammoth Mountain (skyline); yellow square outlines area shown in middle and bottom photographs. (*middle*) View looking southwest at dead trees and parking area near Horseshoe Lake; CO_2 concentration is monitored continuously both in air inside the small building and in soil outside. (*bottom*) View looking southeast at the same CO_2 monitoring site with dead trees in foreground.
USGS photographs by Michael P. Doukas, August 1999 (*top*) and Kenneth A. McGee, 23 September 1999 (*middle and bottom*).

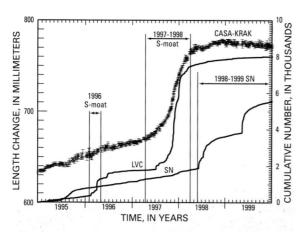

Figure 7.49. Time history of Long Valley resurgent dome inflation for 1995–1999, as represented by: (1) extension of the CASA–KRAK baseline; (2) the cumulative number of $M \geq 1.5$ earthquakes within Long Valley Caldera (LVC); and (3) the cumulative number of $M \geq 1.5$ earthquakes in the adjacent Sierra Nevada (SN) block. Vertical lines bracket the January–April 1996 earthquake swarm in the south moat (1996 S-moat), the July 1997–March 1998 south-moat seismic and deformation episode (1997–1998 S-moat), and the June 1998–May 1999 sequence of three $M > 5$ earthquakes and their aftershocks in the Sierra Nevada immediately south of the caldera (1998–1999 SN). The cumulative seismic moment releases of LVC and SN earthquakes during this time period were $M \sim 5.4$ and $M \sim 5.8$, respectively. Figure modified from Hill et al. (2003). Figure 7.47 shows similar information for 1978–2004.

in good agreement with the models of Langbein et al. (1995) and Langbein (2003) based on leveling and two-color EDM observations. Sixteen CGPS stations were operating in the Long Valley area in 2004. A subset of these include real-time, epoch-by-epoch automated data processing that promises to greatly expand the role played by CGPS in volcano monitoring (Endo and Iwatsubo, 2000).

As many as 30 additional CGPS stations and 15 borehole strainmeters will be deployed in the Long Valley area as part of the Plate Boundary Observatory (PBO). PBO is a component of the ambitious EarthScope project, which received initial funding in 2003. EarthScope is designed to investigate the structure and evolution of the North American continent and the physical processes controlling earthquakes and volcanic eruptions. In addition to PBO, the major components of EarthScope are USArray (United States Seismic Array), SAFOD (San Andreas Fault Observatory at Depth), and InSAR. Additional information about this major Earth science initiative, which promises a quantum leap in our understanding of continental structure, earthquakes, and volcanoes, can be found in reports by the US National Research Council

(2001a,b) and in the scientific literature as the project unfolds.

7.4.5 Temporal gravity changes

A compelling case for magma intrusion beneath the resurgent dome at Long Valley can be made from the deformation data alone, but additional confirmation is available from repeated gravity surveys. Recall from Chapter 2 that the gravitational acceleration at any point on the Earth's surface depends on both surface elevation and subsurface mass distribution. It follows that, if elevation changes are measured independently and their effects are removed, any residual gravity changes can be attributed to subsurface mass changes. A complication arises, however, if part of the mass changes are caused by changes in groundwater level. If these, too, are measured and their effects are removed from the gravity data, what remains is a direct and rather elegant measure of additional subsurface mass changes.

The results of such an analysis for the Long Valley Caldera were initially presented by Battaglia et al. (1999) and later refined by Battaglia et al. (2003a,b). After correcting for the effects of uplift and water table fluctuations, Battaglia et al. (2003b) found a positive residual gravity change of up to $59 \pm 19\,\mu$Gal centered on the resurgent dome for the period from 1982 to 1999. The data are fit well by a vertical prolate ellipsoid centered 5.9 km beneath the resurgent dome with a volume change of $0.105\,\mathrm{km}^3$ to $0.187\,\mathrm{km}^3$ and a density of $1,180\,\mathrm{kg\,m}^{-3}$ to $2,330\,\mathrm{kg\,m}^{-3}$. This result excludes *in situ* thermal expansion or pressurization of the hydrothermal system as the primary cause of surface uplift, and confirms the intrusion of magma beneath the caldera since the onset of unrest in mid-1978. In a cautionary note, Battaglia et al. (2003b) stressed the importance of using an accurate source geometry for studies of this type. Relative to the best-fit prolate ellipsoid, a point source biased the source depth estimate by 2.9 km (a 33% increase), the volume change by $0.019\,\mathrm{km}^3$ (a 14% increase), and the density estimate by $1,200\,\mathrm{kg\,m}^{-3}$ (a 40% increase). Even so, the results are unequivocal. Magma, not water, is primarily responsible for surface uplift and other symptoms of unrest at Long Valley.

A similar study at Yellowstone would be very useful for distinguishing between magmatic and hydrothermal models for surface deformation there (Section 7.3.4). Leveling, GPS, and micro-gravity observations have been made repeatedly

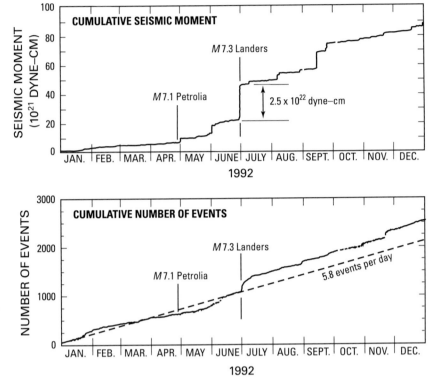

Figure 7.50. (*top*) Cumulative seismic moment for earthquakes near Long Valley Caldera during 1992. Double arrow indicates the cumulative seismic moment for seismicity triggered by the 28 June 1992, M 7.3 Landers, California, earthquake. (*bottom*) Cumulative number of earthquakes during 1992 for the same area. Dashed line indicates the average seismicity rate of 5.8 events per day. Vertical lines mark the times of occurrence of the M 7.1 Petrolia (Cape Mendocino) and the M 7.3 Landers earthquakes (Hill *et al.*, 1995).

at Yellowstone (e.g., Smith *et al.*, 1989; Vasco *et al.*, 1990; Arnet *et al.*, 1997), but the absence of wells suitable for making water-table measurements makes it difficult to constrain the density of fluid involved in historical uplift-subsidence cycles. However, achieving such a constraint would be diagnostic: A source density $\geq 2.3 \times 10^3 \, \text{kg m}^{-3}$ at a depth $>5 \, \text{km}$ would favor the magmatic model, whereas a source density of $\sim 1 \times 10^3 \, \text{kg m}^{-3}$ at 3–5 km depth would favor the hydrothermal model. Like Long Valley, Yellowstone is a target for focused study by EarthScope, so exciting new results are likely to be forthcoming soon.

7.4.6 Borehole strainmeter and long-base tiltmeter results: implications of triggered seismicity

What could an earthquake 450 km away from Long Valley possibly tell us about the state of the caldera and the nature of recent unrest? The answer is plenty, and the story might have broad implications for the poorly understood link between regional tectonism and magmatism. The surprising response of Long Valley and 14 other sites scattered across much of the western USA to the M 7.3 Landers, California, earthquake of 28 June 1992, forced scien-

tists to reconsider the possibility of long-distance interactions between large earthquakes and magmatic systems. Although the jury is still out, a plausible and particularly intriguing possibility is that large earthquakes can trigger episodic recharge of the deep roots of crustal magmatic systems, either by liquefying a partially crystallized magma body or by inducing deep magmatic intrusion (Hill *et al.*, 1995). This intriguing hypothesis suggests that intensive monitoring of magmatic systems can pay unexpected dividends, especially when the Earth conducts a natural experiment at the scale of the Landers earthquake.

Long Valley responded to the Landers mainshock with a surge in seismicity and a transient strain pulse that began while the S-wave coda and crustal surface waves were still passing through the caldera. The triggered seismicity comprised more than 200 events that occupied the entire seismogenic volume beneath the caldera; the cumulative moment was $2.5 \times 10^{22} \, \text{dynes cm}^{-1}$, or the equivalent to a single M 3.8 earthquake (Figure 7.50). The POPA borehole strainmeter at the western base of Mammoth Mountain recorded a transient strain pulse that reached peak amplitude of just over 0.2 microstrain (2×10^{-7}) in 6 days, then decayed over the next 10–30 days (Figure 7.51). The seismic slip

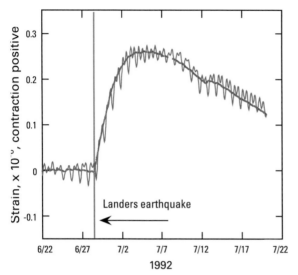

Figure 7.51. Strain transient at Long Valley Caldera triggered by the M 7.3 Landers, California, earthquake of 28 June 1992, as recorded by the POPA dilatational strainmeter (see Figure 7.46 for location). Data before (blue) and after (red) removal of tidal and atmospheric-pressure responses are shown. Modified from Roeloffs et al. (2003), after Johnston et al. (1995) and Hill et al. (1995).

represented by the cumulative moment of the triggered seismicity was too small by 2 orders of magnitude to produce the transient strain pulse recorded at POPA (Hill et al., 1995).

How was the Landers earthquake able to trigger a response at Long Valley over a distance of 450 km? The observed coseismic static dilatational strain in the vicinity of the caldera from the Landers mainshock was a 0.006 microstrain (~0.003 bar) compressional step, which is about an order of magnitude smaller than daily tidal stress fluctuations and therefore not a viable mechanism for the triggered seismicity. Peak dynamic stresses,[8] however, were about 3 bars – plausibly large enough to trigger one or more responses at Long Valley (Hill et al., 1993):

(1) Linde et al. (1994) suggested that the dynamic waves from Landers triggered a transient pressure increase in one or more magma bodies by advective overpressure (i.e., bubble formation and ascent), which in turn triggered the observed seismicity and strain.
(2) Johnston et al. (1995) proposed that the dynamic strains from Landers ruptured com-

partments of super hydrostatic fluid pressures, which are commonly encountered in volcanic and geothermal regions (Fournier, 1991). They suggested that the result was an upward surge of fluids in the crust beneath the caldera by hydraulic fracturing.
(3) Anderson et al. (1994) and Bodin and Gomberg (1994) hypothesized that the shear pulse from Landers initiated aseismic slip on midcrustal faults beneath the caldera, and the associated deformation induced brittle failure at shallower depths.
(4) Hill et al. (1995, p. 13,000) reasoned that: '*A magma body with a relatively small melt fraction transmits shear waves from local earthquakes and thus behaves as a solid under small, high-frequency strains. The large, low-frequency strains associated with the shear wave pulse and crustal surface waves from the Landers mainshock may have partially liquefied such a body, thereby releasing some of the differential stress supported by the solid phase. Alternatively, the large dynamic stresses from Landers may have induced a magmatic intrusion by disrupting the cohesive strength of an incipient or partially healed dike adjacent to a crustal or upper mantle magma source.*' According to the authors, either process, occurring at a depth of about 60 km beneath the caldera, satisfies all of the observational constraints.

All of these models explain the essential features of Long Valley's surprising response to the Landers earthquake, but one may be more generally applicable than the others. Most of the seismicity triggered by Landers was concentrated along the margins of the Basin and Range (Hill et al., 1993), where deep zones of basaltic magma are likely drawn into the lower crust as part of the mass balance accompanying crustal extension (Lachenbruch et al., 1976). The relaxing magma body or dike intrusion model (i.e., hypothesis 4 above) thus can account for triggered seismicity at many widely dispersed and seemingly diverse sites, as occurred in Landers' wake. As noted by Hill et al. (1995, p. 13,002): '*If this model is correct, it suggests that significant influx of basaltic magma into the deep roots of crustal magmatic systems occurs episodically in response to large, regional earthquakes. This in turn offers a specific link between regional tectonism and magmatism.*' It is a link being forged, at least in part, by practitioners of volcano geodesy.

---

[8] Peak dynamic stresses accompanied the S-wave (secondary shear wave) and the crustal Love and Rayleigh waves, which have periods of 10 to 20 seconds and wavelengths of crustal dimensions (15–50 km) (Hill et al., 1995).

7.4.7   Water-level changes induced by distant earthquakes: evidence for stimulated upward movement of magma or hydrothermal fluid

Triggered seismicity is not the only response to distant earthquakes that has been observed at Long Valley. Roeloffs *et al.* (2003) reported that water levels in five wells, including the 3-km-deep Long Valley Exploratory Well and four others open to formations as deep as 300 m, responded to 16 earthquakes, both local and distant, from 1989 to 1999 (Chapter 9). Not all of the wells responded to every earthquake, but the responses mostly were consistent from event to event. Water levels in three of the wells always dropped following an earthquake, while in one well the water level always rose. The water-level changes had repeatable time histories from event to event and they generally increased with earthquake magnitude and proximity.

For example, the 1992 *M* 7.3 Landers earthquake produced water-level changes in all three wells being monitored at that time (Figure 7.52). A gradual water-level drop of ~0.4 m at the LKT well is the largest earthquake-induced change observed there so far (January 2006). The largest seismic event in Long Valley triggered by the Landers earthquake was *M* 3.7, which is smaller than any of the other local events that produced water-level drops at LKT. So it is unlikely that the post-Landers drop was caused by local triggered seismicity. Instead, Roeloffs *et al.* (2003, p. 283) attributed it to '...*seismic waves from the Landers earthquake itself or to aseismic deformation triggered by those waves.*' They noted that contractional strain that could account for an observed water-level rise at CW-3 is inconsistent with two-color EDM observations. Instead, they (p. 269) attributed the rise to '...*diffusion of elevated fluid pressure localized in the south moat thermal aquifer.*'

Based on careful analysis of the entire dataset for 1989–1999, Roeloffs *et al.* (2003) concluded that earthquake-induced water-level changes in Long Valley could be accounted for by accelerated inflation of the resurgent dome plus localized fluid-pressure increases due to thermal pressurization in the south moat. Inflation of the dome produces extensional strain near the surface, which accounts for water-level drops observed at most of the wells, and thermal pressurization accounts for the observed water-level rises at CW-3. But how could distant earthquakes trigger accelerated

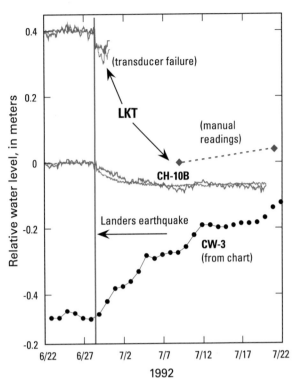

Figure 7.52. Water-level changes in three wells induced by the 1992 *M* 7.3 Landers earthquake, from Roeloffs *et al.* (2003). Linear trends have been subtracted from the data to minimize water-level variations in the 20 days before the earthquake. For LKT and CH-10B, data before (blue) and after (red) removal of tidal and atmospheric pressure responses are shown. See Figure 7.46 for locations of the wells. The maximum water-level decline at LKT is undetermined because the transducer failed shortly after the earthquake, but a lower bound, based on manual measurements, is 0.396 m.

inflation of the resurgent dome or thermal pressurization in the south moat? Roeloffs *et al.* (2003) noted that most of the locations in the western USA where the Landers earthquake triggered seismicity were geothermal areas (Hill *et al.*, 1993). Roeloffs *et al.* (2003, p. 301) concluded: '*Features at Long Valley that might be preferentially affected by seismic waves include fluid-filled fissures at low effective stress, recently active faults, magma in subsurface reservoirs, dikes or pipes, and young seals in hydrothermal systems. Hot magmatic fluids in the subsurface alter and weaken the rocks at their edges. These weakened rocks could be loosened or cracked by dynamic strains, clearing pathways for fluid movement ... This study provides observational evidence that seismic waves can stimulate hydrothermal systems, that such stimulation is likely related to remote triggering of earthquakes, and that the processes involved can be illuminated by hydrologic monitoring in active volcanic areas.*'

7.4.8 Long Valley summary

With additional leveling and two-color EDM data through 1992, Langbein *et al.* (1995) refined earlier models of Long Valley deformation to include: (1) an ellipsoidal source 5.5 km beneath the resurgent dome; (2) an ellipsoidal source or pipe 10–20 km beneath the locus of earthquake activity in the south moat; (3) a northwest-trending, northeast-dipping dike beneath the southwest part of the caldera; and (4) a dike from 2 to 12 km beneath Mammoth Mountain for the period of the 1989–1990 earthquake swarm. Regional GPS observations suggest another possible deformation source (i.e., dike inflation beneath the Mono-Inyo volcanic chain). With the benefit of several years' more data and additional modeling, Langbein (2003, p. 265–266) concluded: '... *inflation at about 6–7 km beneath the resurgent dome explains most of the deformation detected during two periods of unrest, the 1989–1992 and the 1997–1998 episodes. However, an additional inflation source appears to be required and the data suggest that this poorly resolved source is located in the mid-crust (12–20 km) beneath the south moat* ...' In their introduction to a special issue of the *Journal of Volcanology and Geothermal Research* dedicated to Long Valley, Sorey *et al.* (2003, p. 171) concluded: '*The studies described in this volume and in previous publications lead to the conclusion that both deformation and seismicity in the Long Valley region occur in response to the more fundamental process of magmatic intrusion into the crust. Indeed, episodes of accelerated deformation generally precede increases in earthquake activity by several weeks to months.*' That is music to a volcano geodesist's ears, and ample justification for continued intensive study of this fascinating magmatic–tectonic system.

The picture that emerges from more than two decades of detailed seismic, geodetic, geophysical, and geochemical monitoring at Long Valley is one of complex, space- and time-varying interactions among tectonic, magmatic, and hydrothermal processes. A response plan for volcano hazards in the Long Valley and Mono Craters region has been prepared and will be updated as necessary (Hill *et al.*, 2002a). Regardless of the outcome, the value of an ambitious program of geodetic measurements for monitoring unrest at large silicic calderas has been convincingly demonstrated, as it was several decades earlier for the active basaltic shields of Hawaii.

From basalt to rhyolite, intraplate to convergent margin, persistently active to long dormant, gently effusive to hyper-explosive – the four magmatic systems discussed in this chapter are remarkably diverse. So, too, are the pertinent geodetic datasets. To impose some order, and in hopes of learning more about what makes volcanoes tick, many volcanologists turn to numerical models as interpretive tools. Models are generally better behaved, more easily accessible, and less hazardous than the volcanoes they are intended to emulate. There are simple models, complicated models, forward models, inverse models – a model for every circumstance. Models are the finishing tools in the volcanologist's tool kit. A good one can help turn data into ideas, and ideas into understanding. In the next chapter, a modeling craftsman discusses the tools of his trade and displays a few of his wares. For anyone with a quantitative bent or, like me, with a longing for one, it is a show worth seeing.

CHAPTER 8

Analytical volcano deformation source models

Michael Lisowski

8.1   INTRODUCTION

Forces applied to solids cause deformation, and forces applied to liquids cause flow.

(Fung, 1977)

Primary volcanic landforms are created by the ascent and eruption of magma. The ascending magma displaces and interacts with surrounding rock and fluids as it creates new pathways, flows through cracks or conduits, vesiculates, and accumulates in underground reservoirs. The formation of new pathways and pressure changes within existing conduits and reservoirs stress and deform the surrounding rock. Eruption products load the crust. The pattern and rate of surface deformation around volcanoes reflect the tectonic and volcanic processes transmitted to the surface through the mechanical properties of the crust.

Mathematical models, based on solid and fluid mechanics, have been developed to approximate deformation from tectonic and volcanic activity. Knowledge of the concepts and limitations of continuum mechanics is helpful to understanding this chapter. The models predict surface deformation from forces acting, or displacements occurring, within the Earth. These subterranean forces or displacements are referred to as sources of deformation. Quantitative estimates of their location, geometry, and dynamics are inferred by comparing or fitting surface observations to the predictions from these idealized mathematical models.

We do not derive the equations that relate forces or displacements at the source to deformation at the surface, but we do provide an overview of the methods and references for such derivations. These source models are mathematical abstractions and, as a result, this chapter is filled with equations

that may be daunting, but we use plots and tables to describe important characteristics of the predicted deformation.

Volcanic deformation sources include inflating, deflating, and growing bodies of various shapes and sizes, which are collectively known as volumetric sources. The opening or closing of a cavity or crack is distinct from the typical tectonic source, such as a strike–slip or dip–slip fault, where the two sides of a fault slide by one another. Of course, there are composite sources that include both tensile and shear movements, such as a leaky transform fault. Volumetric sources grow and shrink through the movement of fluids and, in some cases, include both sources and sinks. For example, magma filling a growing dike is drawn from (deflates) an adjacent magma chamber.

Observed surface deformation can be fit to the predictions of the source models. The modeling of surface deformation, however, does not provide a unique description of the source causing the deformation. Even with a perfect description of the surface deformation, we could find many different ways to account for it. Model assumptions, simplifications, and data uncertainty further complicate interpretation. Nevertheless, much can be learned from non-unique modeling of sparse and imprecise data.

We are interested in predicting or fitting geodetic data: station displacements, line length changes, tilt, and strain. We limit our discussion to the slow static changes that occur over long periods of time and permanent offsets associated with volcanic or tectonic events. We do not discuss oscillatory, high-frequency ground motions, such as the dynamic strains that accompany earthquakes, even though they sometimes excite volcanic systems. Earth scientists do not like to label any change as static, so they often refer to these very low frequency ground movements as quasi-static ground deformation.

This chapter is more elementary than a previous discussion on modeling ground deformation in volcanic areas by De Natale and Pingue (1996), although there is some overlap. Both chapters include a short introduction to the theory of elementary strain sources in an elastic medium, a summary of spheroidal pressure sources, and a discussion of the ambiguities inherent in modeling surface deformation. DeNatale and Pingue extend their discussion to modifications needed to make the elastic half-space models more realistic. These include inversion techniques for non-uniform pressure distributions and crack openings, and the effects of inhomogeneity, plastic rheologies, and structural discontinuities. The simplifications that make analytical models tractable can, particularly in the case of structural discontinuities, result in misleading volcanological interpretations.

All equations, calculations, and most figures in this chapter are included in a *Mathematica* notebook, although some of the chapter figures are modified versions of those created in the notebook. *Mathematica* is one of several mathematics software packages capable of symbolic mathematics. *Mathematica* allows entry of equations in familiar typeset forms, rather than the more cryptic inline expressions typical of programming languages such as FORTRAN or C. In general, equations for surface displacements for each model are entered directly into *Mathematica* and derivatives, such as tilt and strain, are calculated directly. The *Mathematica* notebook that forms the basis of this chapter and a free reader to access the notebook are included in the DVD that accompanies this book.

8.2 THE ELASTIC HALF-SPACE: A FIRST APPROXIMATION OF THE EARTH

8.2.I Properties of an isotropic linearly elastic solid

The mathematical source models we discuss represent the Earth's crust as an ideal semi-infinite elastic body, known as an elastic half-space. The half-space has one planar surface bounding a continuum that extends infinitely in all other directions. The half-space is materially homogeneous and mechanically isotropic (i.e., mechanical properties do not vary with direction), and it obeys Hooke's law, which specifies a linear relation between displacements (strains) at any point in the body and applied forces (stresses). A body with these characteristics is called an isotropic linearly elastic solid. In short-term laboratory tests, rocks behave like linear elastic solids for strains less than about 1% (10,000 ppm), particularly at low temperatures. Over long time periods or at high temperatures, a non-linear rheology is more appropriate for the crust. Elastic half-space models often neglect many characteristics of the real Earth, but they provide good approximations of deformation resulting from infinitesimal, short-term phenomena on the surface of, or within, the shallow crust.

More realistic Earth models have been developed to account for curvature, topography, gravity, vertical layering, lateral inhomogeneity, and time-varying material properties. We expect that continued advances in computing power and improvements in geodetic data quality and station density will eventually lead to widespread use of these more realistic models. We limit our discussion, however, to the simplest analytical elastic half-space models, partly because they are convenient but also because modeling buried sources is inherently ambiguous. We include a short discussion of topographic corrections. Even with these simple models, there are many trade-offs between model parameters. Our discussion emphasizes the assumptions inherent in the models and how the predicted deformation is related to model geometry, location, and pressure change or displacement. Our goal is to understand both the characteristics and limitations of these simple analytical models.

8.2.2 Elastic constants

The constitutive equations for an isotropic, linearly elastic solid need only two independent elastic constants to describe the relationship between stress and strain. Generally, volcanic source models use Poisson's ratio ν and the modulus of elasticity in shear G. The shear modulus, also called the rigidity modulus or the second Lamé constant, is often represented by μ. We use G to avoid confusion with the prefix used to represent 1 ppm (e.g., μstrain). Okada (1985) and Okada (1992) use both Lamé's constants λ and G, and:

$$\lambda = \frac{2G\nu}{1 - 2\nu} \qquad (8.1)$$

What is the physical significance of G and ν? G relates shear stress to strain providing a material's rigidity or 'stiffness' under shear and has units of pressure. A typical value for intact crustal rock is in the range of 10,000 MPa to 30,000 MPa. Fractured rock and sediments have lower values of G, on the order of 100s to 1,000s of MPa. Poisson's ratio ν is the ratio of lateral unit strain to longitudinal unit strain in a body that has been stressed longitudinally within its elastic limit. Imagine a sample of rock being squeezed along its longitudinal axis (e.g., a cylinder confined at both ends). The sample contracts longitudinally and expands laterally under the stress. Poisson's ratio is simply the ratio of lateral expansion to longitudinal contraction. Laboratory experiments on intact rocks yield values of ν in the range of 0.15 to 0.30, and it is common to assume $\nu = 0.25$ (or, equivalently, $\lambda = G$).

The values of the elastic constants cannot usually be resolved independently of other model parameters and, therefore, typical values are assumed based on the average composition of the area being studied. An independent estimate of the values of the elastic parameters can be obtained from the velocity of seismic waves. Given rock with an average density of ρ, the seismic compressional velocity V_p, and shear-wave velocity V_s:

$$G = \rho V_s^2 \qquad \lambda = \rho V_p^2 - 2G \qquad (8.2)$$

8.3   NOTATION

8.3.1   Coordinate system and displacements

We use a right-handed Cartesian local coordinate system with the spatial coordinates given by (x, y, z) and with z vertical and positive up. In most cases, the local coordinate system is centered on the free surface $(z = 0)$ directly above the center of the source, which is at a depth $-d$, and the x-axis is aligned with the strike of the source. Displacements are denoted by (u, v, w). Many of the models discussed were derived in a local coordinate system with z positive down. Williams and Wadge (2000) discuss the operations needed to transform equations between the different local coordinate systems used to derive the expressions.

8.3.2   Stress and strain

Stresses are denoted by σ_{ij} and the stress tensor consists of:

$$\begin{pmatrix} \sigma_{xx} & \sigma_{xy} & \sigma_{xz} \\ \sigma_{yx} & \sigma_{yy} & \sigma_{yz} \\ \sigma_{zx} & \sigma_{zy} & \sigma_{zz} \end{pmatrix} \qquad (8.3)$$

We assume that displacements are so small that higher order terms of the partial derivatives are negligible. The strains ε_{ij} (extension reckoned positive) are then given by Cauchy's infinitesimal strain tensor (Fung, 1977):

$$\varepsilon_{ij} = \frac{1}{2}\left(\frac{\partial u_j}{\partial x_i} + \frac{\partial u_i}{\partial x_j}\right) \qquad (8.4)$$

For an isotropic elastic solid, the stress is a linear function of strain as given by Hooke's law:

$$\sigma_{ij} = \lambda \varepsilon_{\alpha\alpha}\delta_{ij} + 2G\varepsilon_{ij} \qquad (8.5)$$

where $\varepsilon_{\alpha\alpha} = \varepsilon_{xx} + \varepsilon_{yy} + \varepsilon_{zz}$, $\delta_{ij} = 1$ for $i = j$ or zero otherwise, and $\sigma_{yx} = \sigma_{xy}$, $\sigma_{zx} = \sigma_{xz}$, $\sigma_{zy} = \sigma_{yz}$ (giving six independent stress components). Equation (8.5) illustrates the linear relationship between stress and strain inherent in the elastic assumption.

For most of the problems, the planar surface of the elastic half-space (x, y-plane, at $z = 0$) is free of out-of-plane stress $(\sigma_{xz} = \sigma_{yz} = \sigma_{zz} = 0)$ and is called the free surface. With these stress conditions at the surface, the six independent strain components are reduced to just three:

$$\begin{pmatrix} \varepsilon_{xx} & \varepsilon_{xy} & 0 \\ \varepsilon_{xy} & \varepsilon_{yy} & 0 \\ 0 & 0 & \varepsilon_{zz} \end{pmatrix}_{z=0} = \begin{pmatrix} \dfrac{\partial u}{\partial x} & \dfrac{\partial u}{\partial y} & 0 \\ \dfrac{\partial v}{\partial x} & \dfrac{\partial v}{\partial y} & 0 \\ 0 & 0 & \dfrac{\partial w}{\partial z} \end{pmatrix}$$

$$= \begin{pmatrix} \dfrac{1}{2G(1+\nu)}\left(\sigma_{xx} - \nu\sigma_{yy}\right) & \dfrac{1}{2G}\sigma_{xy} & 0 \\ \dfrac{1}{2G}\sigma_{yx} & \dfrac{1}{2G(1+\nu)}\left(\sigma_{yy} - \nu\sigma_{xx}\right) & 0 \\ 0 & 0 & \dfrac{-\nu}{2G(1+\nu)}\left(\sigma_{xx} + \sigma_{yy}\right) \end{pmatrix} \qquad (8.6)$$

with $\varepsilon_{xy} = \varepsilon_{yx}$ and $\varepsilon_{zz} = (-\nu/1 - \nu)(\varepsilon_{xx} + \varepsilon_{yy})$. The vertical strain (ε_{zz}) is a function of Poisson's ratio and the areal dilatation (fractional change in area $\Delta_{\text{area}} = \varepsilon_{xx} + \varepsilon_{yy}$), as is the volumetric dilatation (fractional change in volume):

$$\Delta_{\text{volume}} = \varepsilon_{xx} + \varepsilon_{yy} + \varepsilon_{zz} = \frac{1 - 2\nu}{1 - \nu}\left(\varepsilon_{xx} + \varepsilon_{yy}\right)$$

$$= \frac{1 - 2\nu}{1 - \nu}\Delta_{\text{area}} \qquad (8.7)$$

Some models produce deformation that is axisymmetric about the vertical axis. For these models, the surface strain in a cylindrical polar coordinate system $(r, \theta, z$, with u_r the radial displacement) is reduced to two independent components, the radial (ε_{rr}) and the tangential ($\varepsilon_{\theta\theta}$) strain:

$$\varepsilon_{rr} = \frac{\partial u_r}{\partial r} \qquad \varepsilon_{\theta\theta} = \frac{u_r}{r} \qquad (8.8)$$

(Note that, in general, $\varepsilon_{\theta\theta} = (1/r)(\partial u_\theta/\partial\theta) + (u_r/r)$ but in axisymmetric deformation there is no θ dependence and all derivatives with respect to θ are equal to zero.) The vertical strain, $\varepsilon_{zz} = (-\nu/1 - \nu)(\varepsilon_{rr} + \varepsilon_{\theta\theta})$, and volumetric dilatation, $\Delta_{\text{volume}} = (1 - 2\nu/1 - \nu)\Delta_{\text{area}}$, are again functions of Poisson's ratio and the areal dilatation, $\Delta_{\text{area}} = \varepsilon_{rr} + \varepsilon_{\theta\theta}$.

Strain is non-dimensional and is expressed in units of μstrain (ppm) with extension reckoned positive.

8.3.3 Tilt

Tilt is defined simply as the inclination to the vertical or horizontal, and can be treated as a 2-D vector, which is resolved into components (ω_x, ω_y). The tilt components are given by:

$$\begin{pmatrix} \omega_x \\ \omega_y \end{pmatrix} = \begin{pmatrix} \dfrac{\partial u}{\partial z} \\ \dfrac{\partial v}{\partial z} \end{pmatrix} = \begin{pmatrix} -\dfrac{\partial w}{\partial x} \\ -\dfrac{\partial w}{\partial y} \end{pmatrix} \qquad (8.9)$$

We define ω_x to be positive when the vertical displacement w decreases in the $+x$ direction, and ω_y to be positive when w decreases in the $+y$ direction. Note that with our sign convention the component slopes $((\partial w/\partial x), (\partial w/\partial y))$ of an elevation change profile are equal to the negative of the corresponding tilt components. Tilt is non-dimensional and usually given in units of μradians (ppm).

There are two independent ways to measure tilt. A borehole tiltmeter records the rotation of a vertical bar or the variation of the horizontal displacements with the depth $((\partial u/\partial z), (\partial v/\partial z))$. The leveling of a tilt array (Chapter 3) gives the variation of the vertical displacement as a function of the horizontal coordinates $(-(\partial w/\partial x), -(\partial w/\partial y))$. A more complete discussion of tilt in which the local vertical is defined relative to the gravitational potential is given by Agnew (1986).

For models with axisymmetric deformation about the vertical axis, the surface tilt in a cylindrical coordinate system ($+z$ is up) is given by a single radial component:

$$(\omega_r) = \frac{\partial u_r}{\partial z} = -\frac{\partial w}{\partial r} \qquad (8.10)$$

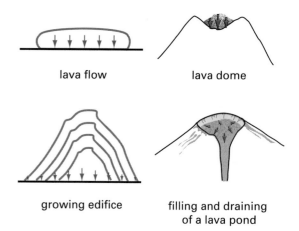

lava flow   lava dome

growing edifice   filling and draining of a lava pond

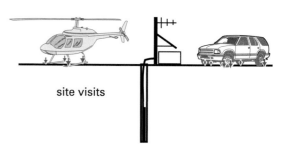

site visits

Figure 8.1. Examples of surface loads in volcanic areas.

8.4 SURFACE LOADS

Eruption products load the surface of the Earth. Lava flows, edifice or dome growth, and even filling of a lava pond will compress the elastic crust that supports the load (Figure 8.1). Likewise, removal of these products in a landslide, blast, or by drainback or flow will allow the crust to rebound. Some types of loads are periodic. For example, the gravitational attraction of the Moon and Sun produce the Earth tides and cause the movement of water that results in the ocean tides. Atmospheric pressure changes, rain, and accumulation and melting of snow are examples of environmental loads. Continental ice sheets are so large and long-lasting that they warp the entire lithosphere, causing flow of the viscous asthenosphere and displacements that vary over time (e.g., Nakiboglu and Lambeck, 1982; Iwasaki and Matsu'ura, 1982; Spada, 2003).

8.4.1 Deformation from point, uniform disk, and uniform rectangular surface loads

An estimate of the crustal deformation produced by the weight of a surface load is obtained by solving

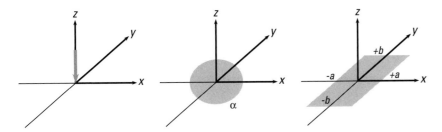

Figure 8.2. Elementary surface loads include point force (*left*), disk of radius α (*middle*), and rectangle of length a and width b (*right*) of uniform applied pressure.

Boussinesq's problem of the response of a non-gravitating, elastic half-space to a point force acting normal to its free surface at the origin. Component strains are calculated from the stress components and then integrated to give the displacements. The surface displacements ($z = 0$) from a normal force $F_z = -P_z\pi\alpha^2$ applied at the origin are given by:

$$\begin{pmatrix} u \\ v \\ w \end{pmatrix} = \frac{F_z(1-2\nu)}{4\pi G} \begin{pmatrix} \dfrac{x}{r^2} \\ \dfrac{y}{r^2} \\ \dfrac{2(1-\nu)}{(1-2\nu)r} \end{pmatrix} \quad (8.11)$$

where $r = \sqrt{x^2 + y^2}$ is the distance from the origin, $-P_z$ is the pressure, α is the radius of pressure source, ν is Poisson's ratio, and G is Lame's shear modulus. The force from a surface load is directed in the $-z$ direction (Figure 8.2). The displacements are axisymmetric, singular near the origin, and they scale with (P_z/G) and decreasing ν (Figure 8.3). A derivation and equations for the complete internal displacement field from a point force applied at the surface of an elastic half-space can be found in Farrell (1972).

For pratical applications, the weight of a real load is distributed over an area (Figure 8.2) and solutions for a disk- or rectangular-shaped load are derived by integrating a point load over a finite surface area. For a disk-shaped load that applies a constant pressure ($-P_z = (F/\pi\alpha^2)$), the displacement of the free surface is given by (Farrell, 1972):

$$\begin{pmatrix} u \\ v \\ w \end{pmatrix} = \frac{-P_z(1-2\nu)}{4G\alpha} \begin{pmatrix} rx \\ ry \\ \dfrac{4\alpha^2(1-\nu)\,{}_2F_1\left(\frac{1}{2}, -\frac{1}{2}; 1; \frac{r^2}{\alpha^2}\right)}{1-2\nu} \end{pmatrix} \quad \text{for } 0 < r < \alpha$$

and

$$\begin{pmatrix} u \\ v \\ w \end{pmatrix} = \frac{-P_z\alpha^2(1-2\nu)}{4G} \begin{pmatrix} \dfrac{x}{r^2} \\ \dfrac{y}{r^2} \\ \dfrac{2(1-\nu)\,{}_2F_1\left(\frac{1}{2}, \frac{1}{2}; 2; \frac{\alpha^2}{r^2}\right)}{(1-2\nu)r} \end{pmatrix} \quad \text{for } r > \alpha$$

$$(8.12)$$

where α is the radius of the disk, $-P_z$ is pressure applied by the load ($-P_z = \rho hg$, with ρ the average density, h the height of the cylinder, g the attraction exerted by the Earth on the load), and ${}_2F_1(\alpha, \beta; \gamma, x)$ is the hypergeometric function. The displacements are axisymmetric with an inflection in the horizontal and vertical displacements at the edge of the disk (Figure 8.4) . For $r \gg \alpha$, an approximate solution for the vertical displacement is obtained by replacing the hypergeometric function, ${}_2F_1(\frac{1}{2},\frac{1}{2}; 2; (\alpha^2/r^2))$, with the leading terms in its power series expansion (Farrell, 1972):

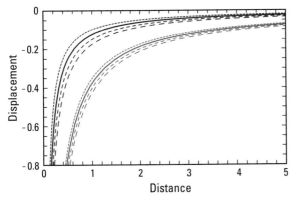

Figure 8.3. Profiles of axisymmetric horizontal (blue) and vertical (red) displacements from a point force $F_z = -P_z\pi\alpha^2$ applied normal to the surface of an elastic half-space as a function of radial distance from the force (see (8.11)). The displacements are normalized by the ratio of $-P_z/G$, and curves for Poisson's ratio ν of 0.30, 0.25 (heavy line), 0.20, and 0.15 are shown. The magnitudes of the displacements increase rapidly and become singular near the origin.

$$w = \frac{-P_z(1-2\nu)}{4G} \left(\frac{4(1-\nu)\alpha\left(1 + \dfrac{\alpha^2}{8r^2}\right)}{(1-2\nu)r} \right) \quad \text{for } r \gg \alpha$$

$$(8.13)$$

The displacements from an equivalent point load are

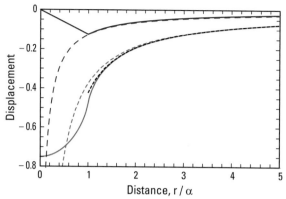

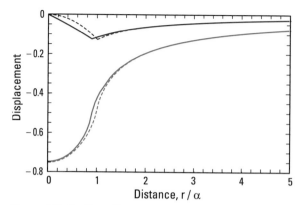

Figure 8.4. Profiles of horizontal (solid blue) and vertical (solid red) surface displacements from a circular load of radius α applying a constant pressure $-P_z$ normal to the surface of an elastic half-space with Poisson's ratio $\nu = 0.25$. An approximate solution for vertical displacement when $r \gg \alpha$ is shown with a black dashed curve, and that for an equivalent point load by dashed blue and red curves. The displacements are normalized by the strain $(-P_z/G)$ and the distance is in disk radii.

Figure 8.5. Profiles of horizontal (solid blue) and vertical (solid red) surface displacements from a uniform unit square load with a side length of $\alpha\sqrt{\pi}$ compared with that from a disk load of radius α (dashed curves). The loads apply a constant pressure $-P_z$ normal to the surface of an elastic half-space with Poisson's ratio $\nu = 0.25$. The displacements are normalized by the strain $(-P_z/G)$ and the distance is given in units of load radii.

similar to those from a disk at distances more than 2 radii from the center of the disk (Figure 8.4).

The surface displacements from a uniform load distributed over a rectangular area $-a < x < a, -b < y < b$ are found by integrating a point force over a rectangular area (Love, 1929) and are given by:

$$u = -\frac{P_z}{G}\frac{(1-2\nu)}{8\pi}\int_{-a}^{a}\int_{-b}^{b}\frac{x-x'}{(x-x')^2 + (y-y')^2}\,dy'\,dx'$$

$$v = -\frac{P_z}{G}\frac{(1-2\nu)}{8\pi}\int_{-a}^{a}\int_{-b}^{b}\frac{y-y'}{(x-x')^2 + (y-y')^2}\,dy'\,dx'$$

$$w = -\frac{P_z}{G}\frac{1-\nu}{2\pi}\int_{-a}^{a}\int_{-b}^{b}\frac{1}{\sqrt{(x-x')^2 + (y-y')^2}}\,dy'\,dx'$$

(8.14)

The resulting equations are lengthy and are included in the accompanying *Mathematica* notebook. A complete derivation and equations for the internal displacement field in the half-space are given in Becker and Bevis (2004). Component surface strains and tilts can be derived by taking the derivatives of the displacement components as given in (8.6) and (8.9). A comparison of equivalent disk and rectangular loads shows small differences close to the load from the different geometries, but essentially the same deformation at distances greater than two source radii or one side length from the center of the load (Figure 8.5).

Deformation from irregularly shaped areas and non-uniform loads is estimated by the linear super-position of a mosaic of disk or rectangular loads.

Example 1: load deformation from filling a lava pond

Pu'u 'Ō'ō cone is an active vent for the ongoing eruption at Kīlauea Volcano in Hawai'i. The cone is about 300 m high and has developed a crater that sometimes fills with lava. Increased lava flux or plugging of leaks on the side of the cone cause the lava pond to fill. As the pond fills, increased pressure on the floor and walls of the pond causes the ground around Pu'u 'Ō'ō to subside and move inward. When the pond drains, the deformation is recovered. As a first approximation, we estimate the deformation from increased pressure on the floor of the pond and leave the problem of accounting for the increased pressure on the pond's wall for Example 5. We assume that the pond has a cylindrical shape with a radius of 100 m and that it fills at the rate of 400,000 m³ day⁻¹, which will produce a 12.7 m day⁻¹ increase in the level of magma in the pond. The pressure on the floor of the pond $-P_z$ increases at a rate of $\rho g \dot{h}$ per day, where ρ is the density of the lava, g is the gravitational attraction, and $\dot{h} = (dH/dt)$ the daily increase in the level of lava in the pond. The deformation scales with the ratio of $-P_z/G$, and we expect the shear modulus (G) for Pu'u 'Ō'ō, which consists of agglutinated spatter and lava flows, to be less than that of intact crustal rocks. Using $G = 10$ GPa, the horizontal and vertical deformation (Figure 8.6) is less than

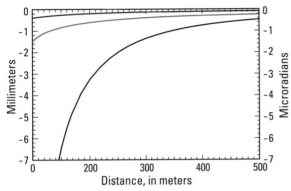

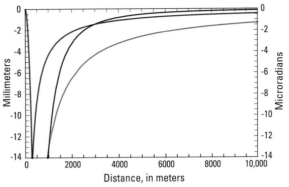

Figure 8.6. Profiles of horizontal (blue) and vertical (red) surface displacements and radial tilt (black in μradians) from the filling of a 100 m radius lava pond with 400,000 m^3 of 2,500 kg m^{-3} molten lava. The displacements will decrease with G, the shear modulus (we use 10 GPa), and with Poisson's ratio (we use 0.25). The displacements are less than 1 mm, except on the edge of the pond, but the tilt is $> 1\ \mu$radian out to 350 m from the center of the pond.

Figure 8.8. Axisymmetric horizontal (blue) and vertical (red) surface displacements and radial till (black) from the 3.8 MPa average load applied by a cylindrical representation of the new dome. The actual displacements could vary by a factor of two or more because of uncertainty in the shear modulus.

1 mm day^{-1}, except at the edge of the pond. The tilt, however, is several μradians near the edge of the pond and a well placed borehole tiltmeter could detect the pond filling.

Example 2: subsidence and inward collapse from growth of a lava dome

The landslide and lateral blast of the 18 May 1980, eruption of Mount St. Helens, Washington, left a large crater, which has been partially filled during the 1981 to 1986 dome-building eruptions and during the eruption of an adjacent dome starting in October 2004. The recent eruption has produced both near- and far-field subsidence and inward collapse toward the crater. How much of the recent deformation results from the cumulative load of the 2004–2005 dome (Figure 8.7)?

The dacite dome extruded between 11 October 2004, and 21 April 2005, is about 580 m in diameter with a maximum height of 381 m above the debris cover on the crater floor, and it has a volume of about 50×10^6 m^3. We use a cylinder with a radius of 290 m, height of 189 m, and average density of 2,200 kg m^{-3} to represent the new dome. The average pressure ($\rho g h$) over the base of this cylinder is 3.8 MPa and calculated load displacements assuming a shear modulus of 10 GPa are shown in Figure 8.8. The magnitudes of the displacements vary inversely with the strength of the material supporting the load and to a lesser extent with Poisson's ratio.

8.5 POINT FORCES, PIPES, AND SPHEROIDAL PRESSURE SOURCES

The response of an elastic half-space to an embedded single force (Mindlin, 1936; Press, 1965) can be

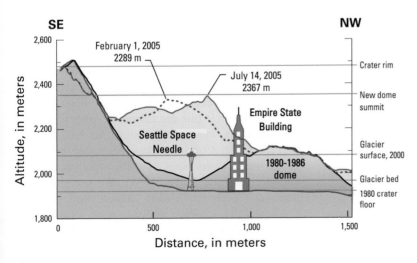

Figure 8.7. Topographic profiles along a N–S axis through Mount St. Helens' crater. Profiles show geometry of new lava dome relative to south crater rim, 1980 crater floor, 1980–1986 lava dome, and 2000 glacier surface.

decomposed into displacements from two infinite medium (full-space) terms, a surface deformation term, and a term that varies with depth (Okada, 1992).

Internal sources of deformation are represented by force dipoles, a pair of forces that are equal in magnitude and opposite in sense. Tensional forces are represented by force dipoles that share a line of action (vector dipoles), producing no net force. Shear forces are represented by a force dipole (couple) that does not share a line of action. Further, shear couples are paired (double couple) so that their torques cancel. All forces act at a point and have an infinite displacement potential (are mathematically singular) at their origin. To obtain the displacements for a point source in an elastic half-space, the equation for displacements from a single force are differentiated to obtain expressions for elementary strain sources. These elementary strain nuclei are combined using Volterra's formula (Steketee, 1958) to compute theoretical displacements for combinations of elementary strain sources known as point dislocation sources.

We are primarily concerned with point sources used to construct analytical models of displacements from tensile faults and pressurized cavities, but also show examples associated with shear faulting (Figure 8.9). These point sources are built up as the superposition of the strain equivalent of orthogonal double couples or vector dipoles. The point-inflation source or center of dilatation consists of three orthogonal vector dipoles of equal magnitude. The strike, dip, and tensile point sources are used to

represent motion at a point on a plane. The strike and dip point sources are composed of a pair of orthogonal shear couples of equal magnitude, with one plane parallel and the other plane normal. The tensile source consists of three orthogonal dipoles, with the plane normal dipole being about 3 times larger than the plane parallel dipoles (the relative magnitude depends on the elastic properties of the half-space). The tensile source can also be thought of as the superposition of a dipole and a point dilatation, with the plane normal dipole being twice the strength of the point dilatation dipoles. Shear motion in any direction on the plane can be represented by a combination of the strike and dip point sources, while a volumetric point source can be constructed from the combination of tensile point sources of varying strengths on three orthogonal planes. For example, a point inflation can be constructed as the sum of three orthogonal tensile point sources of equal strength.

8.5.1 Spheroidal cavities and pipes: model elements for inflating and deflating magma chambers and vertical conduits

A simple model of the eruption cycle of a volcano includes two principal elements, a magma chamber fed from below and a conduit to allow magma to escape to the surface (Figure 8.10). New magma input, differentiation of magma in the chamber, or exsolution and trapping of gasses will build up pressure in the chamber and conduit. If the pressure is sufficient to force the plug out or crack through the rock around the conduit, an eruption will occur and continue until the pressure decays to lithostatic levels or until the conduit is blocked. Between eruptions, magma remaining in the conduit cools and solidifies into a plug.

We are interested in estimating the surface deformation produced by pressure changes in these elements during the eruption cycle. Magma chambers are idealized as fluid-pressurized ellipsoidal cavities in an elastic half-space (Mogi, 1958; Davis *et al.*, 1974; Davis, 1986; Yang *et al.*, 1988; Fialko *et al.*, 2001a). We limit our discussion to spheroidal cavities that are symmetric about the vertical axis (Figure 8.11). Conduits are represented by fluid-pressurized closed pipes or dislocating open pipes (Bonaccorso and Davis, 1999). We use the same pressure increment in each element and sum the predicted deformation to obtain an estimate of the total deformation. Interactions between the sources are ignored.

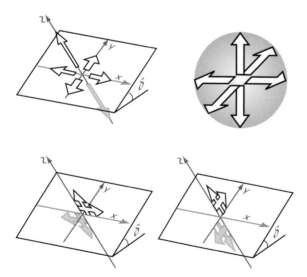

Figure 8.9. Four types of buried point dislocation sources: tensile, dilatational, strike slip, and dip slip (After Okada, 1992).

Plugged conduit with magma input from depth

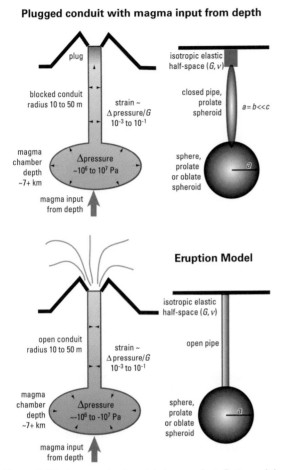

Figure 8.10. An example of model elements for inflating and deflating (erupting) composite volcanoes. Magma input from depth increases pressure in the magma chamber, represented by a spheroidal cavity, and the blocked conduit, represented by a closed pipe. Once the path to the surface is clear, an open pipe is used to represent the conduit.

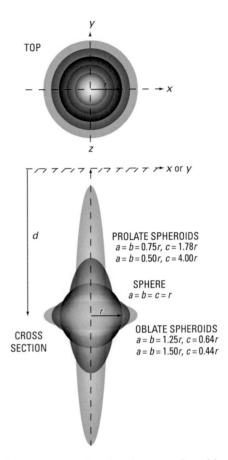

Figure 8.11. Some examples of equal volume spheroidal cavities of depth d. All have a circular cross section when viewed along their vertical axis. The equatorial semi-axis are a and b while the vertical axis is c. In this figure the c axis is vertical, but it can dip at different angles. A prolate (elongate) spheroid has minor axes that are equal and that are less than the major axis ($a = b < c$). An oblate (flattened) spheroid has major axes that are equal and that are larger than the minor axis ($a = b > c$). A degenerate prolate spheroid ($a = b \ll c$) is a closed vertical pipe and the degenerate oblate spheroid ($a = b \gg c$) is a horizontal disk.

This deformation depends on the shape and size of the source, the increment of pressure, and the elastic properties of the medium. For any particular source depth and geometry, the surface deformation scales with the ratio of the cavity pressure change to the half-space elastic modulus ($\Delta P/G$), and to a lesser extent with Poisson's ratio. The elastic modulus can vary by several orders of magnitude, producing an equal uncertainty in the pressure change estimated from the fits of a model to surface deformation.

What range of pressure change is physically consistent with our assumption of elastic behavior? The effective shear modulus in volcanic areas is estimated to be around 1 GPa (Davis *et al.*, 1974; Rubin and Pollard, 1988; Bonnaccorso, 1996), although it can be much less in unconsolidated near-surface materials. Models of volcanic eruptions suggest pressures of about 1 MPa for moderately explosive eruptions and 10 MPa for large explosive eruptions (Wilson, 1980). The corresponding strain ($\Delta P/G$) ranges from 10^{-3} to 10^{-1}, and the higher value is beyond the typical elastic limit of crustal rocks. The corresponding eruptive volumes range from 10^5 for moderate to 10^{10} m^3 for large explosive eruptions, with rare colossal events of greater than 10^{12} m^3 (Newhall and Self, 1982). The largest eruptions probably represent the collapse of a magma chamber. The size of the magma chamber is the key factor between the pressure and volume change. Large eruptions that are within the elastic limit of the crust require large magma chambers. Unfortunately, our elastic models provide weak or no constraints on the size of the cavity.

8.5.2 Point pressure source

It is the writer's opinion that the assumed spherical origin seems to harmonize well with the idea of the magma reservoir under the earth's surface.

(Mogi, 1958)

The point pressure or point dilatation source is often called the 'Mogi model' after Kiyoo Mogi, who concluded that geodetically measured elevation change and horizontal displacements associated with eruptions in Japan and Hawaii resulted from inflation and deflation of magma bodies within the volcanoes (Mogi, 1958). Mogi did not derive nor was he the first to use the mathematical expressions relating surface displacements to hydrostatic inflation or deflation of a small sphere (center of dilatation) buried in an elastic half-space. Anderson (1936) first solved this problem, and other solutions to similar problems published before Mogi's classic paper include those obtained by Mindlin and Cheng (1950), McCann and Wilts (1951), Sen (1951), and Yamakawa (1955). Mogi referenced the derivation by Yamakawa (1955) in his classic study of volcano deformation. Despite the many simplifications inherent in this formulation, it remains the best-known and most widely used method to model surface deformation from a deflating or inflating magma chamber.

The method used to obtain the analytical solution for a point pressure source begins with exact expressions for the effects of a hydrostatically pressurized cavity in a uniform, isotropic elastic full-space (e.g., Love (1927)). The stress-free surface of an elastic half-space (see (8.6)) is constructed by introducing an image of the original source into the full-space on the opposite side of a plane (Figure 8.12) to cancel out-of-plane shear stresses. The combined normal stress on the plane from the source and its image is then cancelled by a similar distribution of opposing normal stress. The resulting plane satisfies the boundary conditions for the free surface of an elastic half-space. A complete derivation can be found in McCann and Wilts (1951) and a clear discussion of the theory and compact expressions for deformation anywhere in the half-space are given by Okada (1992). This solution ignores higher order corrections for stresses reflected back on the source cavity by its image, and so forth. A more complete, but still approximate, solution for a finite-sized pressurized spherical cavity is given by McTigue (1987). The point pressure source is found to be the limiting case as $(\alpha/d) \to 0$ of a finite-sized pressurized cavity.

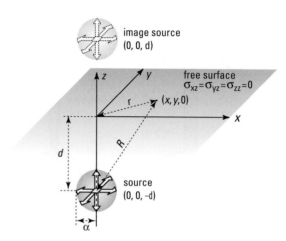

Figure 8.12. Coordinate system and geometric relationships used to derive surface deformation from an embedded point pressure source.

Essential characteristics of the 'Mogi' point pressure source are discussed next, followed by a more complete discussion of finite pressurized spherical cavities.

Surface displacements

The surface displacements produced by hydrostatic pressure change within a spherical cavity embedded in an elastic half-space with radius much smaller than its depth $(\alpha \ll d)$ are given by:

$$\begin{pmatrix} u \\ v \\ w \end{pmatrix} = \alpha^3 \Delta P \frac{(1-\nu)}{G} \begin{pmatrix} \dfrac{x}{R^3} \\ \dfrac{y}{R^3} \\ \dfrac{d}{R^3} \end{pmatrix} \qquad (8.15)$$

where u, v, w are displacements at the point $(x, y, 0)$, the center of the cavity is at $(0, 0, -d)$, and $R = \sqrt{x^2 + y^2 + d^2}$ is the radial distance from the center of the cavity to a point on the free surface. Equation (8.15) is often further simplified by setting $\nu = \frac{1}{4}$. The scaling coefficient, representing the power of the source, lumps together the pressure change ΔP in the cavity, its radius, the shear modulus G, and Poisson's ratio ν of the half-space. Their individual contributions, therefore, cannot be separated. That is, a small pressure change in a large cavity will produce the same surface deformation as a large pressure change in a small cavity provided the radius of the cavity is much smaller than its depth. The elastic constants are usually arbitrarily set to values appropriate for the problem.

The displacements are axisymmetric with the vertical displacement peaking directly above the source,

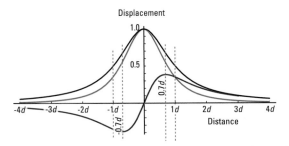

Figure 8.13. Profiles of axisymmetric displacements (vertical (red), horizontal (blue), magnitude of the total displacement (black)) for a point pressure source (8.15) as a function of horizontal distance in source depths. The displacements are normalized by $\alpha^3 \Delta P((1-\nu)/Gd^2)$, the power of the source multiplied by the inverse square of its depth. The dashed vertical lines mark the maximum horizontal displacement $(\pm d/\sqrt{2})$ and the distance $(\pm 1d)$ where horizontal displacement becomes larger than the vertical displacement.

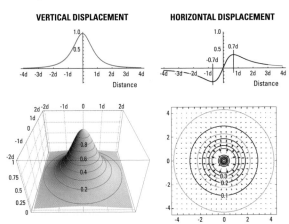

Figure 8.14. Details of the vertical and horizontal displacements for an inflating point pressure source. Axes are scaled the same as in Figure 8.13 and contours are labeled in the same units.

while the horizontal displacement obtains its maximum value at $\pm(1/\sqrt{2})d \simeq 0.7d$ (Figures 8.13 and 8.14). The maximum horizontal displacement is $(2/3\sqrt{3}) \simeq 38.5\%$ of the maximum vertical displacement, and it decays more slowly from its peak value becoming larger at $d > 1$. The horizontal displacements point away from the source (Figure 8.14), making the relative displacement between points on opposite sides of the source up to twice as large as that at a single point.

The surface displacement vector is radial to source (Figure 8.15) and its magnitude:

$$U_R = \sqrt{u^2 + v^2 + w^2} = \alpha^3 \Delta P \frac{(1-\nu)}{G} \frac{1}{R^2} \quad (8.16)$$

varies with the inverse square of the distance from the center of the buried cavity. The location of the source (its center and depth), therefore, can be defined by measurement of two well-chosen

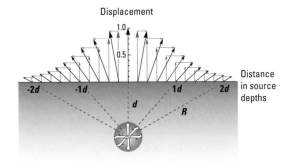

Figure 8.15. Surface-displacement vectors and their components from inflation of a point pressure source, also known as a Mogi source, located below the origin at a depth d. All displacements are axisymmetric and the surface displacement (black arrows) is radial to the center of the buried source with a magnitude that varies as $(1/R^2)$, where R is the distance from the source to the point on the surface. The displacements are normalized by the power of the source and inverse square of its depth with distance given in units of source depth.

surface displacement vectors. The ratio $(w/U_r)r$, where $U_r = \sqrt{u^2 + v^2}$, equals the source depth.

Substituting the measured displacements, estimated source depth, and assuming some values for the source radius α and elastic constants ν and G, provides the pressure increment ΔP producing the deformation. Further, even though it seems paradoxical to talk about the volume change associated with a point dilatation, we can use ΔP to estimate an equivalent cavity volume change $(\Delta V_{\text{cavity}})$. A pressure increment ΔP on the inner surface of a spherical cavity (area $= 4\pi\alpha^2$) will increase the cavity radius by $\Delta\alpha$, where (McTigue, 1987, equation 11):

$$\Delta\alpha = \frac{1}{4}\frac{\Delta P}{G}\alpha \quad (8.17)$$

The volume increase in a sphere from an incremental increase in its radius $\simeq \Delta\alpha 4\pi\alpha^2$ (from a Taylor series expansion of $\frac{4}{3}\pi(\Delta\alpha + \alpha)^3$), and:

$$\Delta V_{\text{cavity}} \simeq \frac{\Delta P}{G}\pi\alpha^3 \quad (8.18)$$

A radius of 1 km is usually assumed.

This volume change considers only the mechanical properties of the surrounding half-space and is not equivalent to the injection volume ΔV_{magma} of a compressible fluid, such as magma (Johnson et al., 2000). The ΔV_{cavity} is a measure of work done on the half-space by the increment of pressure, and when scaled by the elastic constants is equivalent to the seismic moment.

The strains and tilts are derived from the displacements using relations given in (8.6), (8.8), and (8.9), and we examine their characteristics in the next section. The requirement that the source radius be much smaller than its depth, and the ambiguity between pressure and volume change are resolved by using a finite spherical source.

8.5.3   Finite spherical pressure source

Magma bodies often are not 'deep.' Implicit in the representation of a magma body by a point dilatation is the assumption that the characteristic dimension of the body is very small compared to its depth.

(McTigue, 1987)

In the point pressure source approximation (8.15), the radius of the source cannot be separated from the pressure change. This means that we can only obtain estimates of the depth, location, and power of the source. Ideally, we would also like to know the size of the cavity being pressurized and the increment of pressure. McTigue (1987) derived expressions to approximate the deformation from a pressurized finite spherical cavity by applying higher order corrections for stresses reflected back on the source by its image.

Surface displacements

At the free surface of an elastic half-space the displacements are given by

$$\begin{pmatrix} u \\ v \\ w \end{pmatrix} = \left(\alpha^3 \Delta P \frac{(1-\nu)}{G} \left(\left(1 + \left(\frac{\alpha}{d} \right)^3 \right. \right. \right.$$

$$\left. \left. \left. \times \left(\frac{(1+\nu)}{2(-7+5\nu)} + \frac{15d^2(-2+\nu)}{4R^2(-7+5\nu)} \right) \right) \right) \begin{pmatrix} \dfrac{x}{R^3} \\ \dfrac{y}{R^3} \\ \dfrac{d}{R^3} \end{pmatrix}$$

(8.19)

where all symbols are the same as for the point pressure source (8.15). The first term in the scaling coefficient is that of a point pressure source, while the second term scales this coefficient by the corrections for a finite-sized cavity. The terms of the corrections have a common factor of the cube of the ratio of cavity radius to source depth $(\alpha/d)^3$, and, as a result, are very small except when the radius of the cavity is similar to its depth $(0.2 \lesssim (\alpha/d) < 1)$. If we set $\nu = \frac{1}{4}$, the corrections diminish the $(1/R^3)$

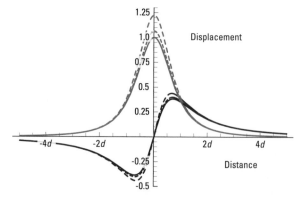

Figure 8.16. Profiles of horizontal (blue) and vertical (red) surface displacements for a finite spherical pressure source (8.19). The displacements are normalized by the product of the power of the equivalent point pressure source, and the inverse square of its depth, and the distance is in source depths. The solid curves are for $(\alpha/d) \to 0$, equivalent to a point pressure source, and the dashed curves for $(\alpha/d) = 0.4, 0.6$.

displacement components by $(15/184)(\alpha/d)^3$ but they increase the $(1/R^5)$ displacement components by $(315/368)(\alpha/d)^3$, resulting in increased deformation out to distances of about 2 source depths (Figure 8.16). To a first approximation, the maximum displacements for a finite sphere are about $1 + (\alpha/d)^3$ times those for a point sphere. The pattern of deformation, however, is very similar and it would be difficult to distinguish a finite source from a slightly shallower point source.

Although the near-field surface deformation for a finite sphere increases as $(\alpha/d) \to 1$, the displacement vectors remain radial to the center of the buried source for all values of (α/d) (Figure 8.17). As with the point source, if we know the ratio of vertical to horizontal displacement at a point, then we can estimate the depth of a spherical source by taking the product of this ratio and the approximate horizontal distance to the source.

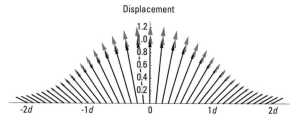

Figure 8.17. The surface displacements are radial to the source for spherical sources of all sizes as shown by vectors for $(\alpha/d) = 0.01$ (black), 0.4 (red), 0.6 (green), which become more 'peaked' over the source as $(\alpha/d) \to 1$. The vectors are normalized by the product of the power of the equivalant point pressure source and the inverse square of the source depth, and the distance is in source depths.

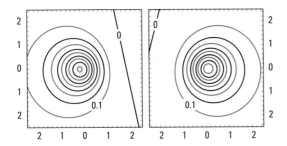

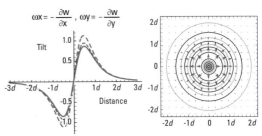

Figure 8.18. Contours of predicted unwrapped InSAR range change for a spherical source with maximum vertical displacement of 1 m (0.1-m contours) as a function of distance in source depths. The plot on the left corresponds to ascending (unit vector $= (-0.479, -0.101, 0.872)$) and that on the right to descending (unit vector $= (0.349, -0.090, 0.933)$) ERS ranges.

Figure 8.19. Surface tilt from inflation of a sphere at depth d, with profiles of tilt along the x–y plane (*left*) and contoured tilt vectors on the x–y plane (*right*). Tilt profiles for $(\alpha/d) = 0.01$ (solid), 0.4, 0.6 (dashed) are shown. The tilt is normalized by the product of the power of the source with the inverse cube of the source depth ($(3\Delta P\alpha^3/4Gd^3)$ for $\nu = \frac{1}{4}$), and the distance is given in source depths.

InSAR: distance change in the look direction of the satellite

Interferometric synthetic-aperture radar (InSAR) (Chapter 5) provides a measure of the distance change in the look direction of the satellite, which is inclined to the vertical. There are ascending and descending satellite paths, each with a look direction defined by a unit vector (i, j, k). Interferograms are usually displayed as fringes of range change equal to $\frac{1}{2}$ of the radar wavelength (the two-way distance is measured), but these fringes can be added (unwrapped) to give the total range change. The range change predicted by a model is equal to the dot product of the predicted surface displacements and InSAR unit vector (Figure 8.18). The inclined look direction of the satellite captures components of the vertical and horizontal displacements, resulting in a skewed and offset pattern of range change.

Surface tilt and strain components

Surface tilts are calculated from elevation changes measured by leveling or Global Positioning System (GPS), or they are measured instrumentally by a tiltmeter. Likewise, surface strains are calculated from line length or horizontal position changes, or they are measured instrumentally by a strainmeter. The surface tilts and strains represent the variation in the displacement components with horizontal distance from the source and are, therefore, unitless. They are computed by taking the partial derivatives of the component displacements with respect to the spatial coordinates (8.6). Tilts are expressed in μradians or the equivalent ppm (10^{-6}). The tilt components are reckoned positive when the vertical displacement decreases in the $+x$ or $+y$ direction. Strains (8.6, 8.7, 8.8) are expressed in μstrain or ppm, with extension reckoned positive.

Using the relationships given in (8.9), the surface tilt components are obtained by taking the negative of the partial derivative of the vertical displacement (w) with respect to the x and the y coordinate. For example, the surface tilt components for a point pressure source (8.15) are given by:

$$\begin{pmatrix} \omega_x \\ \omega_y \end{pmatrix}_{z=0} = \begin{pmatrix} -\dfrac{\partial w}{\partial x} \\ -\dfrac{\partial w}{\partial y} \end{pmatrix} = \alpha^3 \Delta P \dfrac{(1-\nu)}{G} \begin{pmatrix} \dfrac{-3dx}{R^5} \\ \dfrac{-3dy}{R^5} \end{pmatrix} \tag{8.20}$$

The expressions for spherical pressure source tilt components are more complicated and are derived in the accompanying *Mathematica* notebook.

The tilt for a spherical source is axisymmetric, with the magnitude increasing from zero directly above the source to a maximum value at $\pm\frac{1}{2}d$, and then decreasing to less than 10% of the maximum value at a distance of $\pm 2.2d$ (Figure 8.19). The surface tilt, like the surface displacement, is more peaked as $(\alpha/d) \to 1$, but the effects are confined to within about two source depths.

The surface strain for spherical sources, which produce axisymmetric deformation, is simplified to two independent components in a cylindrical polar coordinate system. The radial strain gives the variation in the radial displacement (u_r) with radial distance ($r = \sqrt{x^2 + y^2}$) and the tangential strain, which has no θ dependence, is simply the ratio of the radial displacement to the radial distance (8.8). Plots of strain are shown in Figure 8.20.

The magnitude of the strain is maximum over the source, and the radial strain becomes negative (compressive) at $\pm(d/\sqrt{2})$. The vertical and volumetric strains, which are scalar functions of the radial and

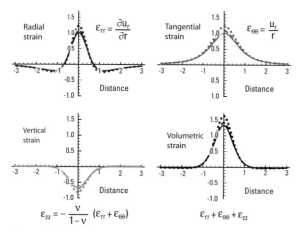

$$\varepsilon_{zz} = -\frac{\nu}{1-\nu}(\varepsilon_{rr} + \varepsilon_{\theta\theta}) \qquad \varepsilon_{rr} + \varepsilon_{\theta\theta} + \varepsilon_{zz}$$

Figure 8.20. Profiles of surface strain from inflation of a sphere at a depth d. Strain profiles for $(\alpha/d) = 0.01$ (solid), 0.4, 0.6 (dashed) are shown in each plot. The strain is normalized by the product of the power of the source and the inverse cube of the source depth $((3\Delta P\alpha^3/4Gd^3)$ for $\nu = 0.25)$ and the distance is in source depths.

tangential strains, have sign reversals at $\pm d\sqrt{2}$. The strains become more peaked within about one source depth as $\alpha/d \to 1$, and the zero crossings shift slightly.

8.5.4 Closed pipe: a model for a plugged conduit or a cigar-shaped magma chamber

Our simple model of the eruption cycle (Figure 8.10) includes a conduit to transport magma from the chamber to the surface. During repose, the magma in the conduit cools, forms a plug, and pressure in the magmatic system can increase. Bonaccorso and Davis (1999) formulated a model for a closed vertical conduit using a hydrostatically pressurized degenerate prolate spheroid embedded in an elastic half-space. Their solution included a term corresponding to the integral of a line of dilatations (point pressure sources) and another term for a line of vertical double forces. Walsh and Decker (1971) proposed the line of dilatations model to model surface deformation from vertically elongate magma bodies, but this model is only appropriate for phenomena such as cooling that produce a uniform contraction.

The distribution of surface deformation from inflation of a closed pipe is quite different from that previously described for a sphere. There are two important factors to consider: (1) most conduits are quite small relative to magma chambers, and (2) the near-field deformation from an elongate embedded source varies with the value of Poisson's ratio.

Closed pipe surface displacements

Expanding Bonaccorso and Davis (1999, equations 7a and 7b), the displacements for a closed pipe are given by:

$$u = \frac{\alpha^2 \Delta P}{4G}\left(\frac{c_1^3}{R_1^3} + \frac{2c_1(-3+5\nu)}{R_1} + \frac{5c_2^3(1-2\nu) - 2c_2r^2(-3+5\nu)}{R_2^3}\right)\frac{x}{r^2}$$

$$v = \frac{\alpha^2 \Delta P}{4G}\left(\frac{c_1^3}{R_1^3} + \frac{2c_1(-3+5\nu)}{R_1} + \frac{5c_2^3(1-2\nu) - 2c_2r^2(-3+5\nu)}{R_2^3}\right)\frac{y}{r^2}$$

$$w = -\frac{\alpha^2 \Delta P}{4G}\left(\frac{c_1^2}{R_1^3} + \frac{2(-2+5\nu)}{R_1} + \frac{c_2^2(3-10\nu) - 2r^2(-2+5\nu)}{R_2^3}\right)$$

$$(8.21)$$

where c_1 is the depth of the top of the pipe, c_2 the bottom of the pipe, $r = \sqrt{x^2 + y^2}$ the horizontal distance from the source, and $R_1 = \sqrt{r^2 + c_1^2}$ and $R_2 = \sqrt{r^2 + c_2^2}$ the distances from a point on the surface to the top and bottom of the pipe.

The closed pipe generates proportionally more horizontal displacement than a spherical source, the vertical and horizontal displacements decay more slowly, and a profile of vertical displacement has a characteristic dimple over the pipe (Figure 8.21). The dimple results from Poissonian

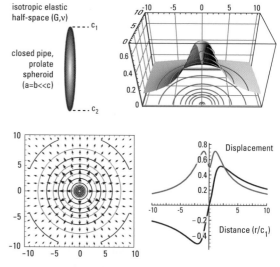

Figure 8.21. Closed pipe model and predicted surface displacements. (*top left*) Pressurized degenerate prolate spheroid used to represent a blocked conduit or elongate magma chamber with its minor axes much smaller than its major axis. (*top right*) Cross section of contoured vertical displacements showing central dimple. (*bottom left*) Contoured horizontal vector displacements. (*bottom right*) Profiles of horizontal (blue) and vertical (red) displacements. The displacements are normalized by the power of the source multiplied by the inverse of the top depth $(\alpha^2\Delta P/4Gc_1)$, and the distance is in top depths (r/c_1).

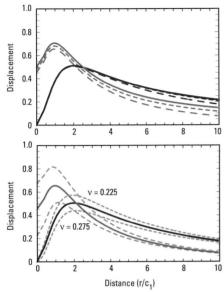

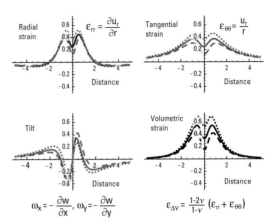

$$\omega_x = -\frac{\partial w}{\partial x}, \quad \omega_y = -\frac{\partial w}{\partial y} \qquad \varepsilon_{\Delta v} = \frac{1-2v}{1-v}\left(\varepsilon_{rr} + \varepsilon_{\theta\theta}\right)$$

Figure 8.23. Profiles of surface strain and tilt from inflation of an infinitely long vertical closed pipe with its top at depth c_1. Strain profiles for $v = 0.25$ (solid), 0.225, 0.275 (dashed gray) are shown in each plot. The strain is normalized by the product of the power of the source and the inverse square of the top depth $(\alpha^2 \Delta P / 4Gc_1^2)$, and the distance is in top depths (r/c_1).

Figure 8.22. Profiles of vertical (red) and horizontal (blue) surface displacements for a closed pipe of radius α with its top at a depth c_1 and bottom at depth c_2. The displacements are normalized by the power of the source multiplied by the inverse of the top depth $(\alpha^2 \Delta P / 4Gc_1)$, and the distance is in top depths (r/c_1). Solid curves on top are for $c_2 \to \infty$ and dashed curves for $c_2 = 10c_1$ and $20c_1$ with $v = 0.25$. Solid curves on bottom are for $c_2 \to \infty$ and $v = 0.25$, and gray dashed curves for $v = 0.225$ and 0.275.

contraction (i.e., necking of surface layer overlying pipe) in the vertical direction in response to horizontal expansion of the pipe. For Poisson's ratio $v = \frac{1}{4}$, the maximum vertical displacement is at a distance equal to the depth of the top of the pipe (c_1), and the maximum horizontal displacement is at $2c_1$ and is up to 77% of the maximum vertical displacement (for $c_2 \to \infty$). Deformation varies slightly with the vertical extent of the pipe. Shorter pipes produce less near-field vertical displacement and less far-field horizontal and vertical displacements (Figure 8.22). The amount of necking in the material above the pipe scales with Poisson's ratio, and there is a strong dependence between near-field vertical deformation and Poisson's ratio (v) (Figure 8.22). A 10% variation in v results in a 50% change in the vertical displacement directly above the pipe, which decreases with distance. The horizontal displacements are affected less, with the maximum values changing by about 14%.

8.5.5   Closed pipe tilt and strain components

The surface strain and tilts for a closed pipe (Figure 8.23) are more varied and decay more slowly with distance than those for a sphere. Relatively small changes in v produce relatively large

changes in the near-field strain and tilt, but these effects are concentrated at distances of less than $2c_1$. The radial strain becomes negative at around $2c_1$ (compared with $1d$ for a sphere), and the volumetric strain stays positive (zero crossing for sphere at $d\sqrt{2}$). Although a closed pipe is used to represent a plugged conduit, it is a degenerate prolate spheroid and could also represent a deep magma chamber with the minor axes being much smaller than the major axis (e.g., the prolate spheroid with $a = b = 0.5$ and $c = 4$ in Figure 8.11). The tilt and strain components are derived from the displacements (8.21) using the relationships given in Section 8.3.2.

Example 3: combined inflation of a magma chamber and plugged conduit

Our simple model of an inflating volcano includes a coupled magma chamber and conduit (Figure 8.10). Given realistic dimensions, what are their relative contributions to the surface deformation around an inflating volcano? We assume a spherical magma chamber of radius of 1 km and depth of 7 km, and a conduit of radius 50 m that extends from the top of the magma chamber (6 km) to within 1 km of the surface: the uppermost 1 km of the conduit is assumed to be plugged. We apply the same pressure increase to the magma chamber and conduit, but do not account for the effect of one source upon the other.

Although the conduit has greater vertical extent than the magma chamber, the surface deformation from a pressure increase in the system is dominated

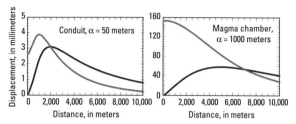

Figure 8.24. Individual contributions to surface deformation from a conduit (*left*) and magma chamber (*right*) subject to a strain $(\Delta P/G) = 0.01$ and $\nu = 0.25$.

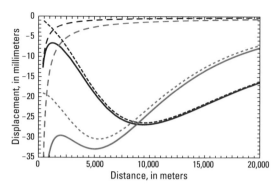

Figure 8.25. Profile of estimated cumulative surface displacement during the 2004 to 2005 lava dome building eruption of Mount St. Helens, assuming that 50 Mm³ eruptive volume was lost from an elongate magma chamber with a top at 6,500 m, bottom at 11,500, and a radius of 500 m -- with no magma resupply. The solid curve is the sum of the load deformation (wide dashed) and the deflation deformation (narrow dashed). The predicted load deformation is for $G = 10^{10}$ Pa.

by that from the magma chamber (Figure 8.24). A conduit radius of 50 m is at the upper end of the 10 to 50 m radius for a typical volcanic conduit (Bonaccorso and Davis, 1999). There are cases, however, where the deformation near a narrow conduit can be much larger. If the shallow portion of the conduit is within a weak layer of fill, then the effective modulus is lower and the deformation greater (Chadwick *et al.*, 1988).

Example 4: closed pipe model approximation of deformation from the deflation of an elongate magma chamber – Mount St. Helens 2004–2005 dome building eruption

Eruption of a new dacite dome in the crater at Mount St. Helens began in October 2004. By 21 April 2005, it grew to about 580 m in diameter, a maximum height of 381 m above the debris cover on the crater floor, and a volume of about $50 \times 10^6 \, \mathrm{m}^3$ (Figure 8.7). If the growth of the dome depleted an equal volume of magma from a buried chamber, what cumulative deformation is predicted for the deflation that occurred to this time?

The magma chamber geometry is approximated from seismic data. Based on the location of an earthquake-free zone in 1980 (Scandone and Malone, 1985), magma is believed to exist between depths of 6.5 to 11.5 km. The radius of this chamber is between 0.1 and 0.75 km. The lower bound is constrained by rock-equivalent volumes erupted in 1980 (Pallister *et al.*, 1992) and the upper bound by the diameter of the aseismic zone (a little less than 2 km in 1980 and a little more than 1 km in 1987–1992). A small chamber diameter is also indicated by Lees and Crossen (1989) tomographic study that did not identify a low-velocity zone within the 0.25 km resolution of the method. We assume a radius of 500 m, which gives the magma chamber a total volume of 5.2 km³. The chamber is connected to the surface by a conduit with an upper diameter of 25 m or less, based on the

observed size of the conduit exposed in the crater following the 18 May 1980 landslide and lateral blast (Chadwick *et al.*, 1988).

Given the above dimensions for a magma chamber and dome volume, the strain $(\Delta P/G) = 0.01$. The predicted surface displacements using (8.21) with $\nu = 0.25$ are shown in Figure 8.25. The growth of the dome applies a load to the floor of the crater, and, as seen in Example 2, the load deformation will increase the magnitude of the near-field deformation caused by magma chamber deflation.

8.5.6 Open pipe: a composite model for the filling of an open conduit

Once an eruption starts, the conduit is more or less open to the surface. During explosive eruptions rapid exsolution of gases causes the magma to expand and erupt. As the gases escape the level of magma in the conduit will subside. During effusive eruptions new vents form or old vents become blocked, and the level of the magma in the conduit changes accordingly. Bonaccorso and Davis (1999) presented a dislocation model for surface deformation from magma rising in a conduit that is open at the top. A constant cylindrical dislocation is used to model portions of the conduit subject to a uniform pressure change. In portions of the conduit that refill, the dislocation of the wall is proportional to the linearly increasing pressure (Figure 8.26). This model does not account for the shear force exerted on the magma walls from the upward movement of viscous magma.

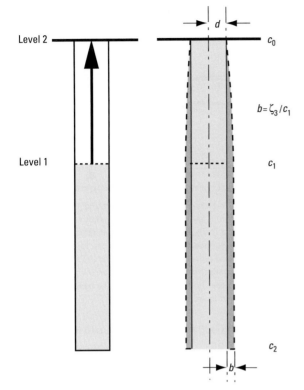

Figure 8.26. Geometry of a dislocating open pipe model used to represent the filling of a vertical conduit that is open on the top (after Bonaccorso and Davis (1999)). The dislocation of the conduit wall varies with the pressure change. In newly filled portions of the conduit, the pressure increases linearly. Elsewhere the increase is equal to rise in the magma head.

Open pipe surface displacements

Expanding integral equations given by Bonaccorso and Davis (1999), the surface displacements produced by a wall displacement of b between the depths of c_2 and c_1 in a cylinder of radius α are given by:

$$u = \frac{b\alpha}{2}\left(\frac{c_1^3}{R_1^3} - \frac{2c_1(1+\nu)}{R_1} + \frac{c_2^3(1+2\nu)+2c_2r^2(1+\nu)}{R_2^3}\right)\frac{x}{r^2}$$

$$v = \frac{b\alpha}{2}\left(\frac{c_1^3}{R_1^3} - \frac{2c_1(1+\nu)}{R_1} + \frac{c_2^3(1+2\nu)+2c_2r^2(1+\nu)}{R_2^3}\right)\frac{y}{r^2}$$

$$w = -\frac{b\alpha}{2}\left(\frac{c_1^2}{R_1^3} - \frac{2\nu}{R_1} + \frac{-c_2^2 + 2R_2^2\nu}{R_2^3}\right)$$

(8.22)

where all symbols are as in (8.21). For a dislocation

of the cylinder's wall that increases linearly with depth from c_1 to c_0, the displacements are:

$$u = \frac{b\alpha}{2}\left(-\frac{c_0^2}{R_0^3} + \frac{2\nu}{R_0} + \frac{c_1^2 - 2(c_1^2 + r^2)\nu}{R_1^3}\right)\frac{x}{c_1}$$

$$v = \frac{b\alpha}{2}\left(-\frac{c_0^2}{R_0^3} + \frac{2\nu}{R_0} + \frac{c_1^2 - 2(c_1^2 + r^2)\nu}{R_1^3}\right)\frac{y}{c_1}$$

$$w = -\frac{b\alpha}{2}\left(\frac{c_0^3}{R_0^3} - \frac{c_1^3}{R_1^3} + \frac{c_1(-1+2\nu)}{R_1} + \frac{c_0(1-2\nu)}{R_0}\right.$$

$$\left. + (-1+2\nu)\ln(c_0+R_0) - (-1+2\nu)\ln(c_1+R_1)\right)\frac{1}{c_1}$$

(8.23)

where $R_0 = \sqrt{c_0^2 + x^2 + y^2}$, and all other symbols are as before. These solutions are superimposed to obtain a complete solution for deformation from magma filling a conduit. To relate the dislocation model to the other pressure dependent models, the outward dislocation b equals the strain times the radius of the cylinder as given by Landau and Lifshitz (1975):

$$b = \frac{\Delta P}{G}\alpha$$

(8.24)

Horizontal surface displacements for the top and bottom open pipe components are larger than the vertical surface displacements. The combined near-field $(< 2c_1)$ surface displacement (Figure 8.27, bottom left) is dominated by the dislocation in

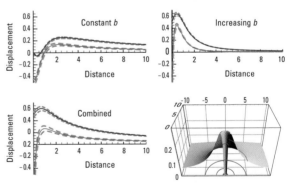

Figure 8.27. Profiles of vertical (red) and horizontal (blue) surface displacements from filling an open pipe. The top left profile is for the case of a constant pressure change in the lower section of the conduit with its top at depth c_1, the top right profile is for filling the top portion of the conduit from c_1 to the surface, and the bottom left profile is for the combined effect of filling a conduit from c_1 to the surface, and the bottom right a cut-away contour plot of the vertical displacement. The displacements are in units of the power of the source multiplied by the inverse of the top depth $((b\alpha/2c_1) = (\Delta P\alpha^2/2Gc_1))$ and the distance is in units of top depths (r/c_1).

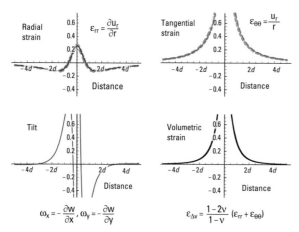

$$\omega_x = -\frac{\partial w}{\partial x}, \ \omega_y = -\frac{\partial w}{\partial y} \qquad \varepsilon_{\Delta v} = \frac{1-2v}{1-v}(\varepsilon_{rr} + \varepsilon_{\theta\theta})$$

Figure 8.28. Profiles of surface strain and tilt from filling a vertical open pipe from a depth c_1 to the surface. Strain profiles for $v = 0.25$ (solid), 0.225, 0.275 (dashed gray) are shown in each plot. The strain is normalized by the product of the power of the source and the inverse square of the top depth $((b\alpha/2c_1^2) = (\Delta P\alpha^2/2Gc_1^2))$, and the distance is in top depths (r/c_1).

the upper section of the conduit, whereas the far-field displacement is similar to that for the lower part of the conduit alone. The maximum positive displacements occur along a circular ridge at a distance of 1 to $2c_1$, and the magnitude is a large fraction of the dislocation b. Except in the region very close to the conduit, the magnitude of the horizontal displacement is greater than the vertical. In contrast with the closed pipe, displacements and strains for an open pipe are only slightly altered by small changes in Poisson's ratio.

Open pipe tilt and strain components

We examine strains and tilts for the combined open pipe model only. The large gradients in the displacements near the edge of an open pipe produce large near-field strains and tilts, except in the radial strain (Figure 8.28).

Example 5: surface deformation from filling a lava pond

To demonstrate the deformation produced by filling an open pipe, we complete Example 1 by adding the deformation from increased pressure on the walls of the conduit and pond to that already determined for the vertical load on the floor of the pond. The conduit feeding the pond is narrow (radius of 25 m) and extends to a depth of 300 m. The pressure on the walls of the lava pond (radius 100 m) increases linearly as it fills. As in Example 1, the pond fills at

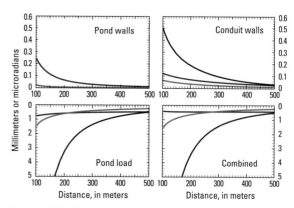

Figure 8.29. Filling a lava pond loads the crust and pressurizes the walls of the conduit and pond. Profiles of predicted horizontal (blue) and vertical (red) displacements and tilt (black) are shown in each plot. The surface displacements from the load imposed by the lava in the pond were previously calculated in Example 1. The pressure on the walls is constant within the narrow portion of the conduit and increases linearly in the pond. The contributions are summed in the combined plot.

the rate of $400,000 \, \text{m}^3 \, \text{day}^{-1}$, which causes the level to rise $12.7 \, \text{m} \, \text{day}^{-1}$.

The deformation contributed by increased pressure on the conduit and pond walls is opposite in sense and much smaller than that from the load applied to the floor of the pond (Figure 8.29). Their combined effect decreases the horizontal displacements to near zero, and slightly decreases the vertical displacement and tilt.

8.5.7   Sill-like magma chambers

Sill-like magma intrusions or chambers are represented by finite rectangular tensile dislocations (Davis, 1983; Yang and Davis, 1986), pressurized oblate spheroids (Davis, 1986), or finite pressurized horizontal circular cracks (Fialko et al., 2001a). Earlier work with 2-D fracture mechanics models (Pollard and Holzhausen, 1979) proved difficult to extend to 3-D geometries. The displacements from the approximate solutions of Davis (1983) and Sun (1969) lead to significant error when the crack radius is similar to or exceeds its depth. The exact expressions of Fialko et al. (2001a) are appropriate for a crack with a radius that is up to 5 times larger than its depth. In their review of related publications, Fialko et al. (2001a) found the zone of surface displacements from a finite-element model of a sill in Dieterich and Decker (1975) to be too large, while horizontal displacements were too small (see Fialko et al., 2001a, figure 3).

We use a horizontal point tensile dislocation to illustrate the characteristics of surface deformation

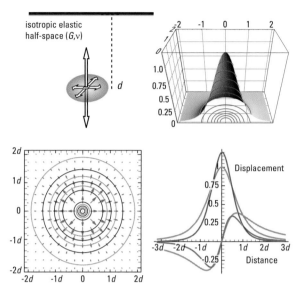

Figure 8.30. Surface deformation from a pressurized sill-like magma body. (*top left*) Pressurized degenerate oblate spheroid (horizontal point crack) used to represent a sill that is deep relative to its radius. (*top right*) Cross section of contoured vertical displacements. (*bottom left*) Contoured horizontal vector displacements. (*bottom right*) Profiles of horizontal (blue) and vertical (red) displacements compared with those from an equivalent spherical pressure source. The displacements are normalized by the power of the equivalent spherical source multiplied by the inverse square of the depth ($\alpha^3 \Delta P/4Gd^2$), and the distance is in source depths (r/d).

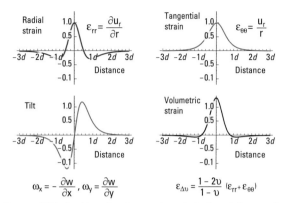

$$\omega_x = -\frac{\partial w}{\partial x}, \; \omega_y = \frac{\partial w}{\partial y} \qquad \varepsilon_{\Delta v} = \frac{1-2v}{1-v}(\varepsilon_{rr} + \varepsilon_{\theta\theta})$$

Figure 8.31. Profiles of surface strain and tilt from inflation of a horizontal point crack at depth d. The strain is normalized by the power of the source multiplied by the inverse cube of the source depth ($(3\Delta P\alpha^3/4Gd^3)$ for $v = 0.25$), and the distance is in source depths (r/d).

produced by a deep pressurized sill-like magma body embedded in an elastic half-space (Figure 8.30, top left). The point tensile dislocation corresponds to a degenerate oblate spheroid (Davis, 1986) with the lengths of the major axes much less than the length of the minor axis. This model, while approximate, is adequate when the radius of the sill is much less than its depth. The free surface displacements are given by (simplified from the equations of Okada (1992)):

$$\begin{pmatrix} u \\ v \\ w \end{pmatrix} = \left(\frac{3M_0}{2G\pi}\right) \begin{pmatrix} \dfrac{xd^2}{R^5} \\[2mm] \dfrac{yd^2}{R^5} \\[2mm] \dfrac{d^3}{R^5} \end{pmatrix} \qquad (8.25)$$

where M_0 is the moment, equivalent to the amount of opening multiplied by the area and G. A point pressure source can be constructed from three orthoganol point cracks, each consisting of an isotropic intensity of $(\lambda/G)M_0$ and a uniaxial intensity of $2M_0$ (Okada, 1992). For $\lambda = G$ (i.e., $v = \frac{1}{4}$), the intensity $5M_{0\,\mathrm{crack}} = M_{0\,\mathrm{point}}$. This scaling factor is used in the comparisons of predicted displacements from point cracks and point pressure sources.

The profiles of surface displacements from sill-like magma bodies are similar to those from a corresponding spherical source (Figure 8.30), but the maximum vertical displacement is about 20% greater and the maximum horizontal displacement about 10% smaller. The zone of deformation is smaller, with the maximum horizontal displacement occurring at a distance of $d/2$.

The volume of surface uplift or subsidence (integral of vertical displacements) for a pressurized sill is equal to the cavity volume change (Delaney and McTigue, 1994; Fialko *et al.*, 2001a). This result contrasts with that for a spherical source, for which the surface volume change is $2(1-v) = \frac{3}{2}$ (for $v = 0.25$) times the cavity volume change (Delaney and McTigue, 1994). This is discussed in more detail in Section 8.8.

The strain and tilt profiles (Figure 8.31) are similar in character to those from a spherical source (Figures 8.19 and 8.20) but, like the displacements, are confined to a smaller region around the source. The magnitude of strain is maximum over the source, with the radial strain becoming negative at $\pm d/2$. The volumetric strain has a sign reversal at $\pm\frac{3}{4}d$.

8.6 DIPPING POINT AND FINITE RECTANGULAR TENSION CRACKS

The horizontal point dislocation introduced in Section 8.5.6 to approximate a deep sill-like magma chamber is a special case of a more general class of deformation source models used to represent tabular intrusions such as dikes or sills. These elongate planar or gently curved faults are filled with magmatic fluids during formation. The lateral extent of the fracture tends to be orders of magnitude larger than the amount of opening. Dikes grow through some combination of regional extension and fluid pressure (hydrofracture), and they tend to initiate at depth and propagate toward the surface (Pollard and Holzhausen, 1979).

Davis (1983) demonstrated that a dipping rectangular tensile dislocation buried in an elastic half-space provides a computationally efficient method to model deformation associated with a tabular tension crack. Although real cracks tend to be elliptical in shape with a variable amount of opening, the deformation predicted by the rectangular dislocation with a constant amount of opening is similar to that predicted by the pressure driven 2-D elastic crack models of Pollard and Holzhausen (1979). Differences in the predicted deformation are small and are only apparent near the fracture (Davis, 1983). Both models satisfy the free-surface boundary condition, but the dislocation model ignores higher order corrections for stresses reflected back on the source by its image.

We use the solutions of Okada (1992) to calculate surface deformation from dipping point and finite rectangular tensile dislocations. The equations are complicated and are given in the accompanying *Mathematica* notebook. Simplified equations for the vertical displacements for the special cases of $\delta = 0°$ and $90°$ are given in Davis (1983) and Maruyama (1964). The local coordinate system for a point dislocation (Figure 8.32) is fault centered, with the crack strike parallel to the x-axis and its dip at an angle δ from horizontal. The local coordinate system for the rectangular dislocation is

centered on the lower edge of the fault plane, which strikes parallel to the x axis and dips at an angle δ. L is the fault's half length and W is its down-dip width. To compare results from equivalent point and finite cracks, the depth of the fault center ($W \sin\theta$ shallower) and its y offset ($W \cos\theta$) are used to transform the rectangular model to fault-centered local coordinates. Transformation of results from a local coordinate system to a global coordinate system is discussed in Williams and Wadge (2000).

Dipping point crack surface displacements

A point crack approximates the deformation from a tabular fracture that is shorter than it is deep, and is used to show how deformation across the center of the crack varies with the crack's dip angle (Figure 8.33). At a shallow dip ($\delta \approx 0°$) the deformation is axisymmetric, as shown previously in Figure 8.30. As the dip becomes steeper, the maximum uplift decreases and its center moves away from the origin in the down-dip ($-y$) direction, and a depression develops along the up-dip ($+y$) projection of the crack on the surface. The vertical deformation becomes symmetric across the crack again at $\delta = 90°$, with a depression centered over the crack that is flanked by two lobes of uplift. Horizontal displacements across the crack become concentrated in the down-dip ($-y$) direction with increasing dip, becoming equal in magnitude again when $\delta \to 90°$.

The deformation along strike will vary with the length of the crack, although this is only apparent when the length is greater than the depth to its top (Davis, 1983). For a point crack, vertical and horizontal displacements (Figure 8.34) decrease with

Point tensile crack   Rectangular tensile dislocation

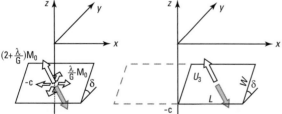

Figure 8.32. Local coordinate systems of Okada (1992) for point and finite rectangular tension cracks. The cracks strike along the x axis and dip at an angle δ from horizontal. The point crack is centered below the origin at a depth $-c$. The tensile moment (intensity) is composed of isotropic and axial components. The depth and center of a rectangular crack is defined by the center of its lower edge. The fault has a half-length of L (i.e., includes the dashed area), a down-dip width of W, and uniform opening of U_3.

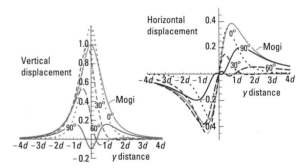

Figure 8.33. Profiles of normalized vertical (*left*) and horizontal (*right*) displacements across a point tension crack with dips of $0°$, $30°$, $60°$, and $90°$. The distance is in source depths. A profile for a point source with equivalent moment is shown in gray for comparison.

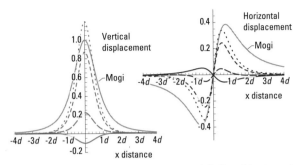

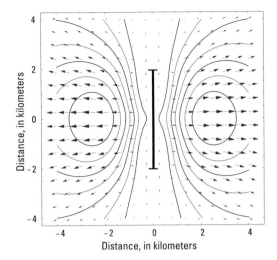

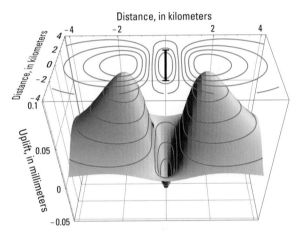

Figure 8.34. Profiles of normalized vertical (*left*) and horizontal (*right*) displacements along a point tension crack with dips of 0°, 30°, 60°, and 90° (see Figure 8.33 for key profiles). The distance is in source depths. A profiles for a point source with equivalent moment is shown in gray for comparison.

increasing dip angle, and reverse at steep dips $(\delta \gtrsim 70°)$. The inward horizontal displacement near the end of the fault is the elastic response to the outward motion across the crack, and it occurs regardless of the fault's length.

Rectangular tensile dislocation surface displacements

The pattern of vertical and horizontal deformation around a finite vertical rectangular dike that is four times as long as the depth to its top edge is shown in Figure 8.35. The fault is flanked by two lobes of maximum displacements, the vertical maxima at a distance about 20% less than depth to the center of the fault and the horizontal maxima at a distance 25% greater than this depth. The ends of the fault are associated with a strong gradient in the vertical and horizontal displacements, and a fanning of the horizontal displacement. The inward motion along strike, near the fault ends, is small but characteristic for steeply dipping extensional fractures.

How does the deformation vary with fault width? Figure 8.36 shows displacement profiles of deformation across a fault with increasing depth while the top of the fault (1 km) and the amount of opening (1 m) are kept constant. The intruded volume and, therefore, the moment will increase as the width increases. There is little change in the profiles near the fault, but farther out deformation increases fault width. Nevertheless, because the profiles are similar, it would be difficult to distinguish an increase in fault width from an increase in the amount of slip.

Simulated InSAR range change

InSAR measures range change in the inclined look direction of the satellite, which is described by a unit vector. The predicted range change (an artificial

Figure 8.35. Contoured 3-D vertical displacement surface and contoured horizontal vector displacements for a vertical rectangular opening dislocation with half-length $L = 2$ km, width $W = 2$ km, bottom edge depth 3 km (top edge 1 mm), and slip $b = 1$ m. The surface projection of the fault is shown with a solid black line.

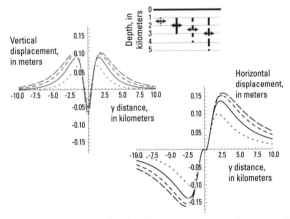

Figure 8.36. Profiles of surface displacements across the center of a vertical opening dislocation with half-length $L = 2$ km, top edge at 1 km, width $W = 1, 2, 3, 4$ km, and slip $b = 1$ m. Inset keys profile line type to fault width.

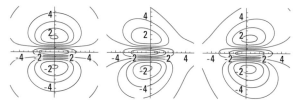

Figure 8.37. Predicted unwrapped InSAR range change for a vertical tensile dislocation with half-length $L = 2$ km, top edge at 1 km, width $W = 3$ km, and opening $b = 1$ m (corresponds to solid line profiles in Figure 8.36). Contours are 0.02 m and subsidence is shown with dashed contours. The plot on the left is modeled uplift (unit vector $= (0, 0, 1)$), the middle plot is simulated ascending ERS $((-0.479, -0.101, 0.872))$ range change, and the right plot is simulated descending ERS $((0.349, -0.090, 0.933))$ range change.

unwrapped interferogram) is obtained by taking the dot product of the modeled displacement vector and the InSAR unit vector. Figure 8.37 compares the pattern of vertical deformation from the intrusion of a vertical dike with the pattern of predicted range change for particular ascending and descending unwrapped interferograms. The mix of horizontal and vertical displacements viewed by the satellite produces an asymmetric pattern of range change.

8.7 GRAVITY CHANGE

Surficial gravity change associated with the intrusion of magma results from three processes: source volume and density change, surface uplift, and density change from deformation of the rock around the source (Hagiwara, 1977). As demonstrated by Hagiwara (1977), Rundle (1978), Walsh and Rice (1979), and Savage (1984), geometrical inflation or deflation of a point source of dilatation results in no surficial gravity change. There is, however, gravity change when there is an associated mass flux. It is instructive to review this result in terms of the three causative processes and considering mass flux (Johnson, 1995).

For a point pressure (Mogi) source a volume change $\Delta V_{\text{chamber}}$ from injection of magma with density ρ_0 will result in a gravity change of:

$$\delta g_0 = \gamma(\rho_0 - \rho_c)\Delta V_{\text{chamber}}\frac{d}{R^3} \qquad (8.26)$$

where ρ_c is the crustal density, $\gamma = 6.67 \times 10^{-11}\ \text{Nm}^2\,\text{kg}^{-2}$, d is the depth to the center of the source, and $R = \sqrt{x^2 + y^2 + d^2}$ is the distance from the center of the source to a point on the surface of the half-space. This component of gravity change will increase when

the mass $((\rho_0 - \rho_c)\Delta V_{\text{chamber}})$ increases and decrease when mass is lost.

The resulting surface uplift will produce a gravity change of:

$$\delta g_1 = 2\gamma\rho_c(1 - \nu)\Delta V_{\text{chamber}}\frac{d}{R^3} \qquad (8.27)$$

which is often called the 'free-air' correction.

Further, elastic deformation of the crust alters its density, resulting in a gravity change given by:

$$\delta g_2 = -\gamma\rho_c(1 - 2\nu)\Delta V_{\text{chamber}}\frac{d}{R^3} \qquad (8.28)$$

The sum of the three components is zero only if $\rho_0 = 0$ (i.e., if there is no inflow or outflow of material).

8.8 RELATIONSHIP BETWEEN SUBSURFACE AND SURFACE VOLUME CHANGES

Another useful relationship for conceptualizing ground deformation at volcanoes is that between subsurface and surface volume changes. Ideally, we would like to relate the volume of surface uplift or subsidence to the volume of magma intruded or withdrawn from an underlying magma chamber. Delaney and McTigue (1994) derived expressions for this relationship for a spherical chamber, sill, and dike. For a spherical chamber:

$$\frac{\Delta V_{\text{surface}}}{\Delta V_{\text{chamber}}} = 2(1 - \nu) \qquad (8.29)$$

For incompressible materials, $\nu = \frac{1}{2}$ and $\Delta V_{\text{surface}} = \Delta V_{\text{chamber}}$. Otherwise, $\nu < \frac{1}{2}$ and $\Delta V_{\text{surface}} > \Delta V_{\text{chamber}}$. For $\nu = \frac{1}{4}$, the change in surface volume is 1.5 times the change in chamber volume, which might seem counterintuitive. Indeed, Delaney and McTigue (1994) noted that uniform expansion of a spherical chamber in an infinite elastic space gives rise to no mean, or dilatational, strain, so the two volumes are equal. When a free surface is included in the problem, however, a bowl-shaped zone of dilation develops above the magma chamber, and the uplift volume exceeds the expansion volume by the factor $2(1 - \nu)$.

For sheet-like chambers that are small compared with their depth, Delaney and McTigue (1994) showed that:

$$\frac{\Delta V_{\text{surface}}}{\Delta V_{\text{chamber}}} = 1 - \frac{(1 - 2\nu)}{2}\sin^2\delta \qquad (8.30)$$

where δ is the dip of the sheet such that $\delta = 0°$ is a sill and $\delta = 90°$ is a dike. For an inflating sill, $(\Delta V_{surface}/\Delta V_{chamber}) = 1$, independent of Poisson's ratio. An inflating dike, on the other hand, compresses the host rock such that $(\Delta V_{surface}/\Delta V_{chamber}) = 1 - ((1 - 2\nu)/2)$. For $\nu = \frac{1}{4}$, the uplift volume caused by an inflating dike is $\frac{3}{4}$ of the volume expansion of the dike.

The foregoing discussion ignores any differences in elastic parameters between rock and magma, which modify the relationship between subsurface and surface volume changes. Johnson *et al.* (2000) addressed this issue for basaltic magma in a spherical magma chamber and showed that:

$$\frac{\Delta V_{surface}}{\Delta V_{magma}} = \frac{2(1 - \nu)}{1 + \dfrac{4G_{rock}}{3K_{magma}}} \quad (8.31)$$

where ΔV_{magma} is the volume of magma that enters or exits the chamber, ν is Poisson's ratio, G_{rock} is the shear modulus of the host rock, and K_{magma} is the effective bulk modulus of magma in the chamber. K_{magma} is defined as follows:

$$\left.\begin{array}{l} K_{magma} = K \text{ for } N < N_s \text{ and} \\[2em] K_{magma} = \dfrac{K}{1 + \dfrac{KN\rho_{magma}RT}{P^2\omega}} \text{ for } N > N_s \end{array}\right\} \quad (8.32)$$

where K is the gas-free melt bulk modulus, N is the total weight fraction of CO_2 dissolved and exsolved in the magma, N_s is the limiting amount of CO_2 dissolved in the melt at saturation, ρ_{magma} is the bulk-magma density, $R = 8.314\,m^3\,Pa/mol\,°K$ is the gas constant, T is absolute temperature, P is average pressure (assumed to be lithostatic), and $\omega = 0.044\,kg$ is the molar mass of CO_2. The effect of other gases was ignored because, at depths of a few km, CO_2 is the volatile most likely to be present in magma as a gas phase (Gerlach and Graeber, 1985), and thus to affect the value of K_{magma}.

As $(G_{rock}/K_{magma}) \rightarrow 0$, the result of Johnson *et al.* (2000) reproduces that of Delaney and McTigue (1994) for an incompressible magma. In the latter case, $(\Delta V_{surface}/\Delta V_{magma}) = 2(1 - \nu)$ for all values of ν, and $(\Delta V_{surface}/\Delta V_{magma}) = 1.5$ for $\nu = \frac{1}{4}$. For a compressible magma, on the other hand, $(\Delta V_{surface}/\Delta V_{magma}) < 1$ if $(G_{rock}/K_{magma}) > (3(1 - 2\nu)/4)$, and if $(G_{rock}/K_{magma}) > \frac{3}{8}$ for $\nu = \frac{1}{4}$. For the particular case of $(G_{rock}/K_{magma}) = \frac{3}{8}$ and $\nu = \frac{1}{4}$, the net volume change due to crustal dilation is equal in value with opposite sign to the net volume change due to dilatation of magma in the chamber. Only in

Table 8.1. Nomenclature used to represent volume change.

Symbol	Definition
$\Delta V_{surface}$	Integral of surficial vertical displacement
$\Delta V_{chamber}$	Source chamber volume change (cavity volume change)
ΔV_{magma}	Volume of intruded magma
$\Delta V_{compression}$	Net volume change of stored magma due to pressure change in chamber $\Delta V_{compression} = \Delta V_{magma} - \Delta V_{chamber}$

this special situation is $\Delta V_{surface}$ identical to ΔV_{magma} (Johnson *et al.* 2000).

The above discussion ignores magma volume changes resulting from gas expansion as magma depressurizes during transport to the surface.

8.9   TOPOGRAPHIC CORRECTIONS TO MODELED DEFORMATION

The interaction of a source with the irregular surface is significant only when the burial depth is of the same order as the characteristic horizontal scale of the topography. For vertical displacements the interaction is strongest when the horizontal strain and topography are correlated.

(McTigue and Segall, 1988)

The planar free surface of an elastic half-space is a poor match to the high relief of volcanic regions. What distortions are introduced by topography to the deformation from a buried magmatic source? What is the optimal reference surface for source depth in areas of high relief? How important are topographic corrections and what are the implications if they are ignored?

Unlike many other of the simplifying model assumptions, such as material homogeneity and isotropy, topography is well known and, with a suitable method, could be used to estimate model corrections. Topographic relief imparts a slope to the surface and it alters the distance from the surface to the source. We have shown that surface deformation varies with some power of the inverse distance from the source, so we might expect topography to be more of a factor for shallow sources where this distance is more variable.

The efficacy of applying topographic corrections to analytical deformation models has been studied by comparison of topographically corrected analytical models with predictions from numerical models (finite element and boundary element), where topography can be incorporated directly into the model. These studies demonstrate that topographic corrections for the typical composite volcano are of the same order as differences generated by different source geometries. Effects can also be significant for shield volcanoes, even when the average slope of the edifice is as small as $10°$.

8.9.1 Reference elevation model

The simplest topographic correction consists of an offset of the free surface by a constant amount, which is called the reference elevation (Figure 8.38). The source depth d is then taken relative to the reference elevation rather than sea level or some other assumed vertical datum. This has no effect on the computed deformation, but it does provide a realistic reference surface for the source depth in areas of high topography. The optimal value of the reference elevation varies, however, depending on the topography, the source depth, and the deformation component of interest (Williams and Wadge, 1998). Clearly, this method works best in areas of low relief.

For prominent volcanoes with a topographic profile similar to Mount Etna, Cayol and Cornet (1998) found the summit to be the best reference elevation. They used a numerical boundary element method (Cayol and Cornet, 1997) to compute surface deformation from a spherical pressure source (radius α) at a depth of 5α for the case of axisymmetric volcanoes (radius 6α) with average flank slopes of $0°$, $10°$, $20°$, and $30°$. They found that the topography reduces the near-field displacements (within about 1 source depth for the vertical and 2 source depths for the horizontal) by as much as 50%, but it has little effect in the far-field. They then inverted the surface displacements predicted by the numerical model with a point pressure source (8.15) to obtain the best-fitting depth and volume change. They found the best-fitting magma chamber depth to be relative to the summit, but the inversion overestimates the volume change by 10% to 50%, with the bias increasing with the average slope.

Without constructing a numerical model to test for the optimal reference elevation, Williams and Wadge (2000) provide the following guidelines: use a reference elevation somewhere between the average and the maximum for shallow sources and decrease this value with increasing source depth.

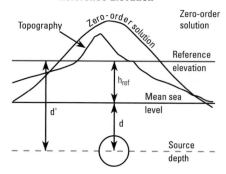

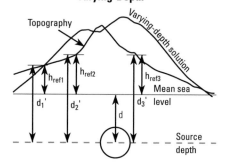

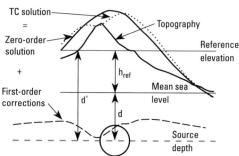

Figure 8.38. Graphical representation of three methods used to correct analytical source models for topography (after Williams and Wadge (2000)). The reference elevation method adds a constant depth to the source to correct for average topography but does not account for effects from varying elevation. The varying depth model (Williams and Wadge, 1998) adjusts the reference elevation for each point and provides reasonably accurate vertical displacements and tilts. The topographically corrected method (Williams and Wadge, 2000) begins with the zero-order reference elevation solution and then adds first order corrections based on the ground slope.

8.9.2 Varying depth model

The varying depth model corrects for the changing distance between the source and the surface by varying the magma chamber depth with topography (Figure 8.38). To test the method, Williams and Wadge (1998) compared results from the modified analytical models of a spherical pressure source with those from a 3-D finite element model that included topography. The topography used in the calculations was a digital elevation model (DEM) of Mount Etna, and they compared results for several source depths. They found that the varying depth model captured both the reduction in deformation close to the summit and the asymmetry in the deformation from non-uniform topography. In particular, the vertical deformation and tilt, its spatial derivative, nearly replicated the deformation predicted by the numerical model. The horizontal deformation, however, was more closely matched using a reference elevation model with a reference surface of 2 km. They concluded that, at least for the tested case of a spherical pressure source, topography significantly reduces displacement magnitudes primarily by increasing the distance of the surface from the source. Inversions for source parameters that ignore topography may overestimate the source depth and mislocate its center.

8.9.3 Topographically corrected model

Williams and Wadge (2000) developed the topographically corrected method to include arbitrary topography in elastic half-space source models. Their method extends the small-slope perturbation method of McTigue and Segall (1988) to three dimensions and is applicable to most analytical source models. Extensive testing with a tilted triaxial ellipsoid source model yielded excellent agreement between the analytical and numerical models within the constraints of the small-slope approximation.

The application of this method is complex, requiring free-surface displacements, derivatives, and curvatures for the elastic half-space source model and a decimated DEM. The DEM and source model depth are first corrected by an arbitrary reference elevation. The choice of reference elevation is important but not critical to the accuracy of the result. The zero-order displacement field (no topography) is then calculated over the mesh defined by the DEM. The first-order topographic corrections, given as convolutions to the zero-order solution, are then computed using Fourier transforms. This technique can be extended to handle several deformation sources. The method, although complex, provides a computationally efficient way to accurately represent topographic effects with an analytical model.

The importance of including topographic correc-

tions was demonstrated by comparing the misfits of finite element models to a spherical and ellipsoidal source with those from a simple analytical model with a reference elevation. In some cases, the fit of the finite element spherical source to the deformation from an elliptical source was better than that of the analytical model to that source. In regions with high relief, topographic corrections may be more important to fitting the data than variations in the source geometry.

8.10 INVERSION OF SOURCE PARAMETERS FROM DEFORMATION DATA

Observations of ground deformation are fit to model predictions to estimate the geometry and power of volcanic sources. The predictions of models having different geometries and depths can be similar (e.g., Dieterich and Decker, 1975; Fialko *et al.*, 2001a), and finding the best model requires a systematic search through likely values of the source parameters for each appropriate model, techniques to estimate uncertainty in the derived model parameters, and methods to compare results from different types of models. While trial-and-error estimates can fit major features of the deformation, non-linear inversions are needed to find the optimal model and to determine the uncertainty in the model parameters.

8.10.1 Non-linear inversion and model parameter error estimates

The following discussion is summarized from Cervelli *et al.* (2001) and Murray *et al.* (1996c). The forward problem linking observed deformation to model predictions can be expressed by the matrix equation:

$$\mathbf{O} = \mathbf{G}(\mathbf{m}) + \epsilon \qquad (8.33)$$

where **O** is the observed deformation, **G** is the function relating the source geometry with **m** components (e.g., location, depth, size, slip) to the deformation at a particular point, and ϵ is the vector of observation error. The relation between the deformation and the source geometry is not linear, and an optimized search of **m** is used to find the model that minimizes the misfit between the data and the model. The misfit is usually given by the sum of the weighted residuals squared, which is then normalized by the degrees of freedom (number of data less the number of free parameters). Statistical tests can be used to evaluate the significance of the misfit, but the results of these tests depend upon assumptions about the distribution of the misfit and assumptions used in formulating the source model.

A simple grid or random search for optimal model parameters is easily implemented, but computationally inefficient. Even a coarse search of parameter space involves a very large number of permutations, which grows geometrically with the number of model parameters. Random searches are more efficient at locating minima, but random or grid searches might miss the global minimum. Multiple iterations over finer grids can be used to narrow the search for the global minimum. Derivative-based searching algorithms (e.g., Marquardt, 1963; Gill et al., 1981) quickly find a local misfit minimum, but finding the global minimum depends on choosing the right starting point for the search. Several methods, collectively known as Monte Carlo algorithms, combine the efficiency of a derivative-based search with a wide-ranging search of the source parameters. (Cervelli et al., 2001) used synthetic geodetic data to compare the results of two Monte Carlo optimization techniques, simulated annealing (Metropolis, 1953), and random cost (Berg, 1993). They found random cost easier to use but simulated annealing, which requires a separate calculation to optimize the 'cooling schedule', to be more efficient. Given sufficient time both converged on the global minimum. They recommend using a Monte Carlo search to find a best-fit model, and take that result to refine the model with a derivative-based search.

A range of model values will provide an acceptable fit to the observations, and estimates of the uncertainties in and correlations between the derived source parameters are important to interpreting the modeling results. Some part of the misfit results from simplifications in source model. De Natale and Pingue (1996) showed inhomogeneity, different rheologies, and, especially, structural discontinuities affect ground deformation in volcanic areas. Unless explicitly included in the source model, such effects become part of the unmodeled noise.

The non-linear relationship between the model geometry and the data complicates the estimation of confidence intervals for individual model parameters. The bootstrap method (Efron and Tibshirani, 1993) can be used to rigorously estimate confidence intervals by providing a distribution of best-fit values from a large number (>1,000) of resampled

sets of the observed data. Care must be excercised to obtain a statistically valid resample of the geodetic data (Cervelli et al., 2001). The resampled best-fit solutions can be quickly generated with the efficient derivative-based parameter searches, although there is a chance that some searches will lead to a local rather than the global minimum. Confidence intervals are then derived by examining the range of source parameters within a chosen percentage of the solution distribution, excluding the highest and lowest values. The uncertainties are given by individual upper and lower limits of the source parameters within the selected distributions. Scatter plots between source parameters reveal whether some parameters are highly correlated and, therefore, individually poorly determined. Statistical tests, such as the F ratio test, that assume that the observations are normally distributed and that the minimization function can be linearized in the vicinity of the global minimum, tend to slightly underestimate confidence regions (Arnadóttir and Segall, 1994). In general, the range of acceptable model parameter values that fit within a high confidence limit is surprisingly large.

8.10.2 Choosing the best source model

Not only can a range of source parameters for a particular model adequately fit the data, but there may be several candidate source models. The model with the smallest misfit, after adjusting for the number of free model parameters, is usually preferred. The misfit, however, depends on how the data are weighted. Some types of data, such as leveling and InSAR, can be very numerous and may detect or be most sensitive to only one component of deformation. Measurements of horizontal and vertical deformation are needed to distinguish between some source models (Dieterich and Decker, 1975), and a relative weighting scheme is sometimes used to prevent one data type from having an overwhelming influence on the misfit.

Standard statistical tests have been used to justify the choice of one model over another. Cervelli et al. (2001) caution that the results of such tests should be regarded skeptically. The same reasoning applies to models with multiple elements. Lacking other data to guide a decision, the preferred model is the simplest one that fits the data within the expected error.

CHAPTER 9

Borehole observations of continuous strain and fluid pressure

Evelyn A. Roeloffs and Alan T. Linde

Strain is expansion, contraction, or distortion of the volcanic edifice and surrounding crust. As a result of magma movement, volcanoes may undergo enormous strain prior to and during eruption. Global Positioning System (GPS) observations can in principle be used to determine strain by taking the difference between two nearby observations and dividing by the distance between them. Two GPS stations 1 km apart, each providing displacement information accurate to the nearest millimeter, could detect strain as small as $2\,\mathrm{mm\,km^{-1}}$, or 2×10^{-6}. It is possible, however, to measure strains at least three orders of magnitude smaller using borehole strainmeters. In fact, it is even possible to measure strains as small as 10^{-8} using observations of groundwater levels in boreholes.

Since volcanic strain may be very large, techniques capable of detecting minute amounts of deformation might seem like overkill. There are, however, numerous advantages to high-resolution strain measurements for studying volcanic deformation. Their high sensitivity allows these instruments to be deployed farther from the volcano. Borehole strainmeters operate automatically, with little field maintenance to bring observers into danger, and can continue to transmit data throughout, and after, an eruption. High strain sensitivity is achieved without long observation periods, allowing sampling at up to 200 Hz. Detecting the tiny surface deformation resulting from magma movements at several kilometers depth provides information about the physics of volcanic eruptions, and as the examples will show, borehole strain and water level data enhance predictive capability because aseismic deformation can precede the onset of pre-eruptive volcanic seismicity.

In this chapter, we describe the design and capabilities of borehole strainmeters, as well as ways of interpreting water level measurements in terms of crustal strain. Data-analysis procedures are outlined. Finally, we summarize examples of strain and fluid-pressure changes recorded during volcanic unrest.

9.1 BOREHOLE STRAINMETER DESIGN AND CAPABILITIES

Detecting strain due to volcanic activity (or any tectonic activity) is challenging because the Earth's surface is an extremely noisy environment. Fortunately, large strain changes due primarily to weather and cultural effects diminish with depth below the surface. Tremendous improvements in the signal-to-noise ratio can be realized by installing sensors at depths of about 200 m below the surface. The most feasible and cost-effective way of achieving this is to install the sensor deep in a small-diameter (typically less than 15 cm) borehole.

The borehole installation is shown schematically in Figure 9.1. The highest quality data come from instruments installed below active aquifers and in competent rock. When the hole is complete, a quantity of expansive cement, or grout, is lowered in a container to the bottom of the hole. The container opens and is withdrawn, depositing the cement at the bottom in a way that avoids separation of sand and cement. The strainmeter is then lowered into the hole and sinks through the grout that completely surrounds and covers the instrument. As the grout hardens, it expands and locks the strainmeter to the surrounding rock. Because the strainmeter is

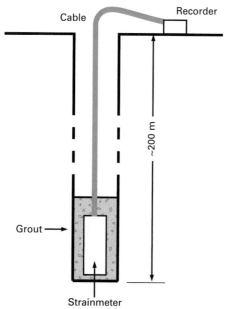

Figure 9.1. Schematic diagram of a borehole strainmeter installation.

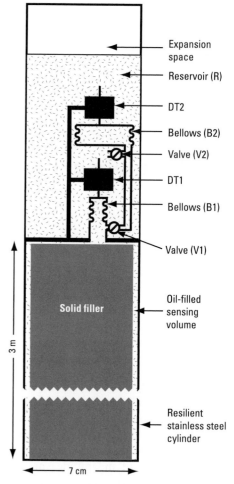

Figure 9.2. Diagram of a Sacks--Evertson two-stage borehole dilatometer; see text for discussion.

resilient, it will then track subsequent deformation of the rock wall.

The borehole strainmeters that have so far provided almost all the strain data relating to volcanic activity have been Sacks–Evertson instruments, often referred to as 'dilatometers' (Sacks *et al.*, 1971). See Agnew (1986) for descriptions of other types of strain instrumentation. The strain-sensing chamber of the strainmeter (Figure 9.2) is a long ($\sim$3 m) cylindrical stainless steel cylinder (7 cm diameter) effectively filled with a liquid, typically silicone oil. Both horizontal and vertical strain modify this cylinder's volume, causing oil to move in or out of the volume. Attached to the sensing chamber is a small bellows (12.7 mm in diameter) whose length changes in proportion to the amount of oil entering or leaving the chamber. The position of the top of the bellows is measured by means of a differential-transformer-based displacement transducer (DT1) with a resolution of about 1 nanometer (1 nm $= 10^{-9}$ m). The geometric ratio between the sensing unit and the bellows provides hydraulic amplification of about 40,000 and, since hydraulic amplifiers are effectively noise free, the instrument has a resolution of about 10^{-12} in strain. The bellows has limited extension capabilities (about 5 mm) so a hydraulic valve, controlled by surface electronics, is used to allow a reset of the sensing hydraulics. The dynamic range of the instrument is about 140 dB (more than a 24-bit analog-to-digital (A/D) converter). Since 1998 this

design has been modified to incorporate a second, less sensitive bellows-displacement transducer-valve assembly (DT2). The exit port of the primary valve is connected to a second, larger diameter (5 cm) bellows that is also monitored with a DT. A second valve allows this bellows to be equilibrated with an unstressed oil reservoir. No data are lost during the reset procedure since only one valve is open at a time. During normal operation, both valves are closed so that the second bellows responds to changes in volume of an unstressed constant mass of oil (i.e., the second DT2 output voltage is a measure of temperature changes with resolution better than $10^{-4}\,^{\circ}$C). The instrument is connected to surface electronics by a cable that provides power for the displacement transducers as well as the various signals.

The electronic unit provides a regulated voltage supply and monitors the output signals to implement opening and closing of the valves. The

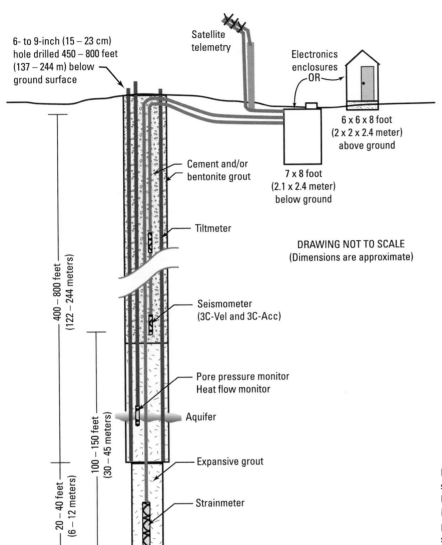

6- to 9-inch (15 – 23 cm) hole drilled 450 – 800 feet (137 – 244 m) below ground surface

Satellite telemetry

Electronics enclosures
OR

6 x 6 x 8 foot (2 x 2 x 2.4 meter) above ground

7 x 8 foot (2.1 x 2.4 meter) below ground

Cement and/or bentonite grout

Tiltmeter

DRAWING NOT TO SCALE (Dimensions are approximate)

Seismometer (3C-Vel and 3C-Acc)

Pore pressure monitor
Heat flow monitor

Aquifer

Expansive grout

Strainmeter

400 – 800 feet (122 – 244 meters)

100 – 150 feet (30 – 45 meters)

20 – 40 feet (6 – 12 meters)

Figure 9.3. Schematic diagram showing a well-instrumented borehole-monitoring station, including a strainmeter, pore-pressure and heat-flow monitors, seismometer, tiltmeter, telemetry, and electronics enclosure.

displacement transducer output signals are analog voltages that are connected to A/D converters. Changes in atmospheric pressure also deform the near-surface rock, so it is necessary also to record barometric changes with resolution of at least 1 hectopascal (1 hPa = 1 millibar). An inexpensive GPS receiver is incorporated to provide absolute time information. The data can be recorded on site as well as transmitted to a central recording location. The total on-site power consumption is small enough that solar cells or wind generators are able to power remote installations. The strainmeter frequency response is flat from 0 Hz to more than 10 Hz so that sampling at about 50 times per second with 24-bit resolution is needed for complete recording of the instrument output.

Drilling suitable boreholes is relatively expensive and thus measures need to be taken to ensure that the data return more than justifies the initial expense. The majority of the borehole strainmeters installed during the last 15 years, using recommended procedures, are still operating, and some borehole strainmeters have been in continuous operation for more than 20 years. Strainmeter boreholes also provide very quiet sites for seismometers and, in many cases, downhole seismometers are installed with the strainmeters. Tiltmeters can be installed in the same holes and it is also possible to include sensing of pore pressure at depth. Thus, the borehole can provide a number of different data streams for a time period that is more than adequate to justify the costs. Figure 9.3 shows a scheme for installing all of these types of instruments in a single borehole. The data obtained from such installations provide critical information for understanding volcanic activity.

Although each strainmeter can be calibrated before installation, in practice the instrument must be calibrated *in situ* by analyzing its response to Earth tides. The tidal calibration is necessary because the strainmeter output is a function of the elastic properties of the instrument as well as of the formation in which it is installed. Tidal calibration is usually accomplished by determining the amplitude of the M_2 tidal constituent in the strainmeter data and equating this to the calculated amplitude, computed from the site's latitude and longitude, with modifications for ocean loads. Software for estimating the amplitudes of tidal constituents from hourly data is described by Ishiguro (1981), and ocean load corrections can be computed using the program GOTIC2 described by Matsumoto *et al.* (2001). Earth-tidal variations recorded by strainmeters are generally closely in phase with calculated Earth-tidal strains, so that the strainmeter data can be expressed in terms of strain using only a multiplicative factor. Strainmeter data can usually be adequately corrected for atmospheric-pressure effects using a single coefficient of proportionality. The instrument output is proportional to a combination of areal and vertical strain. Since, compared with the tectonic sources being studied, the strainmeters are very close to the Earth's surface, it can be assumed that vertical stress vanishes, so that vertical and areal strain are proportional. Under this assumption, the relative weightings of strainmeter sensitivity to vertical versus areal strain are unimportant, and the instrument is thought of as measuring either areal or volumetric strain.

Borehole strainmeters yield very sensitive measurements of time variations in strain, but do not provide useful information about absolute strain rates. Although the strainmeter has constant response to strain changes over an extremely broad frequency band (0 Hz to >10 Hz), the data from a borehole site will generally be most reliable over a smaller band. Initial transients due to curing of the expansive grout (time constant of months) and to viscoelastic relaxation of stress around the newly drilled borehole (time constant typically of a year or more) perturb the record, albeit with removable exponential trends. Tectonic deformation at periods of several months may be masked by seasonal effects of precipitation and aquifer recharge, as well as possible instrument drift. At periods longer than about a year, the data may reflect instability of

the geological environment in the immediate vicinity of the hole rather than signals of tectonic interest. Such noise characteristics can vary greatly at different sites and special care should be taken in the interpretation of signals in the seasonal period range and at very long periods. In general, however, the data are very reliable at periods less than a few months and may well provide important data at periods at least as long as a year after careful analysis in conjunction with other types of data.

During volcanic activity that involves magma movement, deformation around the volcano generally results from a compound source. To determine the characteristics of the sources (e.g., a deflating magma reservoir combined with a propagating dike), it is necessary to measure the strains at a number of sites around the volcano. Ideally, there would be sites covering all azimuths and at a variety of distances from the volcano, from about 2 km to several times the reservoir depth. Financial support and logistical considerations limit both the number of sites and where they can feasibly be located. Site selection is strongly influenced by knowledge of the geological structure because the best quality data come from strainmeters installed in competent rock. With a network of about 5 well-distributed borehole sites, the strain information can place strong constraints on source models. Since dilatational strain from subsurface magma chambers and dikes is a strong function of the source depth (undergoing a change in sign at distances dependent on that depth), continuous strain data allow upward dike propagation to be tracked with time, which has potential for the prediction of eruptions (see below).

9.2 GROUNDWATER LEVEL AS A VOLUMETRIC STRAIN INDICATOR

Crustal deformation can produce fluid-pressure fluctuations in subsurface formations. These pressure fluctuations can be measured directly in a borehole sealed from the atmosphere, but it is usually easier to measure groundwater level in an open well, which acts as a manometer. The variation in groundwater level Δh is related to the pressure variation Δp as $\Delta p = \rho g \Delta h$, where g is the acceleration due to gravity and ρ is the density of fluid. Here we will use the terms groundwater level and fluid-pressure change interchangeably.

Groundwater level variations in a borehole can be recorded using a pressure transducer suspended below the water surface in an open well. If feasible,

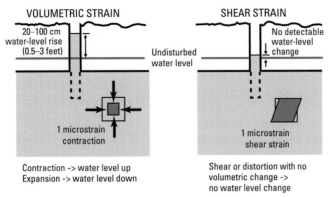

Figure 9.4. Diagram illustrating the water-level changes produced in a porous formation by volumetric strain as opposed to shear strain.

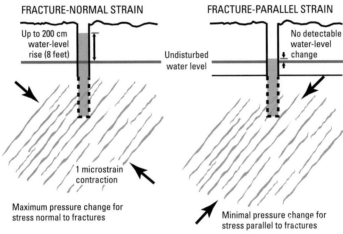

Figure 9.5. Diagram illustrating the water level response of a fracture-dominated formation to fracture–normal strain as opposed to fracture--parallel strain.

a device called a 'packer' can be installed to seal the fluid pressure from the atmosphere, in which case the transducer is placed below the packer, while a power and instrumentation cable passes through it. Compared with a borehole strainmeter, the transducer is many times less expensive, and in some cases existing boreholes or wells can be instrumented. The low cost and simplicity of the data collection, however, are offset by lower sensitivity to strain and greater complexity of data analysis. Like borehole strainmeters, groundwater level variations also are of limited value at long periods, with the upper limit of the period of usefulness sometimes difficult to determine. Nevertheless, water level observations have yielded some intriguing examples of pre-eruptive crustal deformation as well as crustal deformation accompanying earthquake swarms.

9.2.1 Water levels and crustal strain

Why should water levels in wells respond to crustal strain? The reason is that deforming a rock changes the volume of pore space, which in turn causes the pressure in the fluid to change. Such strain-induced changes generally dissipate with time due to fluid flow.

The ratio of water level change to strain depends critically on certain properties of the formation being monitored. For many rocks in which void space is distributed approximately uniformly in the form of pores, changes of pore volume occur primarily with rock deformation that changes the overall volume of the rock (Figure 9.4). That is, shear deformation with no associated volume change has little effect on groundwater level. For a rock in which void space is dominated by fractures with a preferred orientation, fluid pressure in the fractures is most greatly affected by deformation that is normal to the plane of the fractures (Figure 9.5). Other features that affect fluid-pressure response to strain are the elastic properties of the rock, and the degree of connection to the water table.

The mathematical formulation for the coupling between the elastic deformation of a porous rock and the pressure of fluids in the pores was first developed by Biot (1956, 1941), with a useful revisiting by Rice and Cleary (1976). Wang (2000) provides a comprehensive summary. Of chief importance to understanding water level fluctuations caused by crustal deformation is that, in the absence of fluid flow, fluid-pressure changes are proportional to changes in average stress $\Delta\sigma_{kk}$ or, equivalently, in volumetric strain $\Delta\varepsilon_{kk}$ according to:

$$\Delta p = -B\Delta\sigma_{kk} = -(BK_u)\Delta\varepsilon_{kk} \qquad (9.1)$$

In (9.1), the minus signs are needed because pore pressure decreases with positive (extensional) stress or strain. B, known as Skempton's coefficient, is a dimensionless material property with a value between 0 and 1, and K_u is the porous material's undrained bulk modulus. There are few measurements of Skempton's coefficient from laboratory work, but those that exist confirm this range of values. Moreover, Skempton's coefficient and the undrained bulk modulus can also be inferred from other rock properties: rock bulk modulus, porosity, and fluid and grain compressibility. These calculations, observations of Earth tides in wells, and laboratory studies suggest that the coefficient of proportionality between water level changes and volumetric strain is in the range of 30–100 cm per microstrain (1 microstrain $= 10^{-6}$) for rocks whose void space is in the form of pores.

In a rock where void space is dominated by fractures, the sensitivity of water pressure to strain can be greater than in a porous rock, but is not as easily calculated based on rock properties because the coefficient depends on fracture compliance. Bower (1983) has shown how to relate fracture parameters to observed Earth tide response. Observed coefficients of water level change in response to strain in fracture-dominated formations are as high as 2 m per microstrain (e.g., Woodcock and Roeloffs, 1996).

Fluid pressure changes in wells that respond to Earth tides can be expressed in units of strain by the same technique used for calibrating borehole strainmeters. In water level data, however, tidal variations may lead or lag actual tidal strain due to time-dependent flow in the well–aquifer system (Roeloffs, 1996). Poorly confined aquifers may also have frequency-dependent responses to atmospheric pressure (Rojstaczer, 1988; Rojstaczer and Agnew, 1989). If these effects are significant, then they can be taken into account in converting water level measurements to units of strain.

9.2.2 Effects of groundwater flow

Subsurface fluid pressure of course also varies in response to rainfall, pumping, or any other factor changing the mass of fluid in the system. When fluid mass per unit volume of material is not constant, (9.1) generalizes to:

$$\Delta p = BK_u - \Delta\varepsilon_{kk} + \left[\frac{1}{1 - K/K_s}\frac{m - m_0}{\rho_0}\right] \qquad (9.2)$$

where K is the (drained) bulk modulus of the material, K_s is the bulk modulus of the solid grains, and $(m - m_0)/\rho_0$ is the change in fluid mass per unit volume, divided by the fluid density in a reference state. Equation (9.2) shows that fluid mass changes are coupled not only to fluid pressure changes, but also to volumetric strain changes. One implication of this coupling is that when measurements of strain exhibit time variations, it is generally necessary to evaluate climatic or artificial changes in subsurface fluid pressure as a causative factor rather than attributing the deformation to tectonic causes.

Fluid pressure changes in response to strain occur instantaneously with deformation. In fact, the water level variation would have the same time history as the deformation were it not for the ability of water to flow. Flow causes spatial variations in pressure to equilibrate with each other over a timescale governed by the material's hydraulic diffusivity. In particular, a sudden, localized change of pore pressure induced by a tectonic event such as an earthquake or rapid intrusion will spread and dissipate with time, causing time-dependent fluid pressure changes in locations not originally affected by the tectonic event. Although in principle the pore pressure field is always coupled to elastic deformation, in practice the time-dependent fluid pressure field can often be adequately modeled using a groundwater model that solves the (uncoupled) diffusion equation of groundwater flow.

An important factor that limits the usefulness of water level monitoring is the flow path from the depth where pressure is sensed to the water table (Figure 9.6). When strain is applied far below the water table, the pressure in the fluid changes because the volume available for it to occupy has changed. The strain may also affect the rock at the depth of the water table. However, at the water table, fluid has the option of moving up or down into unoccupied pore space. There is a very large difference between the amount by which the water table must change, and the equivalent pressure head induced by a pressure change at depth. For example, a water pressure

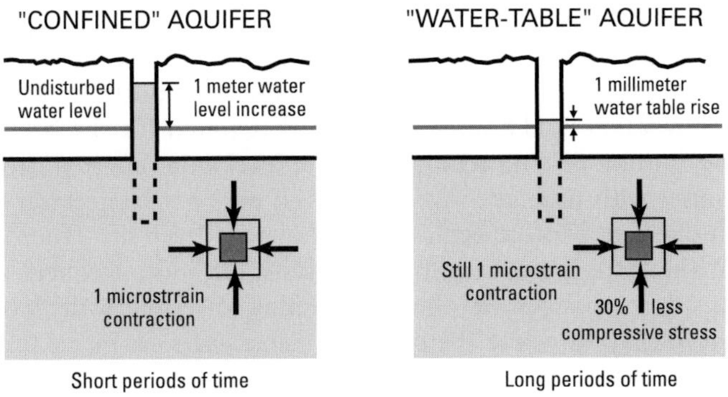

"CONFINED" AQUIFER

"WATER-TABLE" AQUIFER

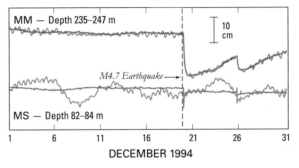

Figure 9.6. Diagram illustrating the greater response of fluid pressure to strain in a confined aquifer as compared with a water table aquifer.

rise of 0.01 MPa (1.45 psi) increases the water level in a well by 1 m. But the compressibility of water is 0.435 GPa^{-1}, so a fluid volume decrease of only 0.000 44% can relieve the pressure increase, if fluid can escape from the system. In a 100 m thick aquifer with 10% porosity, removal of 0.000044 cubic meters from each 1-m square area of aquifer would relieve the pressure increase while causing a barely detectable water table drop of only 0.44 mm. Thus, if the strain is applied instantaneously, and remains in force, then the fluid pressure at depth will instantaneously change, while that at the water table will not. Upward or downward flow will occur in response to this induced gradient, with the time to equilibrate increasing as the square of the depth of measurement below the water table, and decreasing in inverse proportion to the hydraulic diffusivity. Since hydraulic diffusivity varies over many orders of magnitude, this timescale also varies tremendously. Figure 9.7 shows co-seismic water level changes at two different depths in the same borehole that recover at different rates due to the greater thickness of material over the deeper sensor. Figure 9.8 shows the range of times over which strain-induced pressure would be expected to persist. This range of timescales is one basis for distinguishing 'confined' from 'unconfined' aquifers, for purposes of characterizing their responses to strain. Although there is no specific cut-off point, a well in which a pressure disturbance caused by strain can persist for a month can usefully be thought of as 'confined', while a well in which a pressure disturbance caused by strain equilibrates within a matter of hours is behaving like an unconfined aquifer and is unlikely to provide useful information about tectonic strain. To achieve a sufficient degree of confinement, the observation well needs to be open to the formation at some

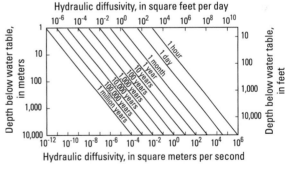

Figure 9.7. Raw (blue) and filtered (red) water level records from different depths in the Middle Mountain well near Parkfield, California, showing water level drops caused by a M 4.7 earthquake. Note rapid recovery in shallow interval (MS) and more sustained pressure change in deeper interval (MM).

Figure 9.8. Range of times over which strain-induced pressure would be expected to persist, as a function of depth beneath the water table and vertical hydraulic diffusivity. The graph can be viewed as giving the elapsed time after the imposition of a strain step such that the response has fallen to one-tenth of its initial value. Modified from Roeloffs (1996).

depth below the water table, ideally 100 m or more. Aquifer pumping or slug tests (e.g., Moench, 1985) can be used to determine the hydrologic properties of monitoring wells, and in some cases can provide information on the connection between the monitored formation and the water table.

Segall *et al.* (2003) have shown that fluid flow can be expected to modify the output of a borehole strainmeter, if the strain varies with time at a rate that is slow compared with the flow. More specifically, this effect could limit the response of strainmeters to low-frequency strain changes, if the strainmeter is installed in a formation where there is a high-diffusivity path to the water table.

9.2.3   Thermal pressurization

Where hot magma or magmatic gases are in proximity to pore fluid, the heated fluid will undergo thermal expansion that tends to increase fluid pressure (Delaney, 1982; Bonafede, 1990). Typical thermal expansion coefficients for water under hydrostatic pressure in a region with a normal geothermal gradient range from $1.7 \times 10^{-4}{}^{\circ}\mathrm{K}^{-1}$ at 0.1 km depth to $9.3 \times 10^{-4}{}^{\circ}\mathrm{K}^{-1}$ at 5 km depth (Delaney, 1982). Pressures induced by heating of pore fluid tend to dissipate by fluid flow, so high hydraulic diffusivity can work against thermally induced pressure. During periods of volcanic unrest, intrusions typically occur over timescales of hours to days, and are particularly effective at raising fluid pressures, because flow cannot proceed rapidly enough to dissipate them. Bonafede (1990) estimated that, assuming a spherical source and typical rock properties, the pore pressure field generated by a $10{}^{\circ}\mathrm{C}$ increase of source temperature is comparable with that induced by a 10 MPa increase in source pressure.

9.2.4   Data collection requirements

As the examples will show, some volcanic eruptions have been preceded or accompanied by extremely large water level variations that were observed without special equipment. The simplicity of these observations is appealing and unequivocally demonstrates that groundwater levels change in response to volcanic deformation. Much more can be learned, however, from high-resolution water level measurements in well-characterized aquifers.

Water level data that are to be analyzed for tidal and barometric response need to be collected using sensors with a resolution of 1 mm of water or better. If the ratio of water level change to strain is 100 cm per microstrain, then sensor resolution of 1 mm corresponds to strain resolution of 1 nanostrain

(1 nanostrain = 10^{-9}). In practice, the smallest strains that can be observed in water level data are about 10 nanostrain. The same pressure resolution is required for the barometric data as for the water level data, but in a network of closely spaced sites, it may not be necessary to operate a separate barometer at each site. Hourly data are ideal for tidal analysis, but more frequent sampling is helpful for detecting co-seismic water level changes. Most pressure transducers are prone to long-term drift, so it is worth visiting the site periodically to make manual measurements of water levels with a steel tape. Submersible pressure transducers are available to measure either absolute pressure, or pressure relative to the atmosphere (called a 'gauge' or 'vented' transducer). Either type of transducer can be used provided a stable, high-resolution barometer is available to allow the atmospheric pressure correction. An advantage of vented transducers is that an instrument with a smaller overall range can be used, since atmospheric pressure is not included in this range. However, special attention must be paid to preventing moisture from entering the vent tubes of such sensors and destroying the transducer. Meticulous splicing, storage in a dry environment, and maintenance of a desiccant pack over the vent tube's surface termination are required to avoid problems.

Water levels respond to rainfall and aquifer recharge. To date, the ability to model aquifer response to precipitation is limited, but nevertheless it is critical to make adequate measurements of precipitation. A well with a high sensitivity to strain can exhibit small water level changes due to the weight of heavy rainfall or snow accumulation on the Earth's surface. Aquifer pressure can usually be expected to change slowly in response to variations in the rate of rainfall, or, more generally, in the rate of surface infiltration. On or near a volcano, where weather varies dramatically with elevation, it is especially important to have nearby precipitation sensors and to be able to distinguish snow from rain in the data record. Daily records of precipitation are adequate for most purposes, but at remote sites a heated rain gauge sampled at the same interval as the water level sensors is ideal.

9.3   PROCESSING AND ANALYZING CONTINUOUS STRAIN AND WATER LEVEL DATA

Continuous deformation time series – strain or water level – accumulate at typical sampling

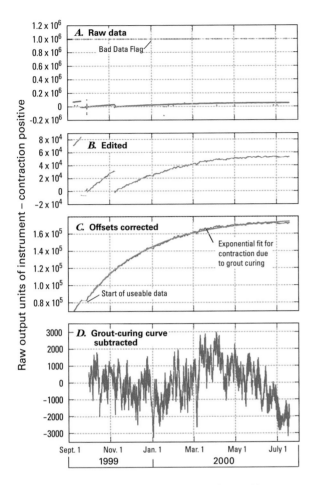

Figure 9.9. The first year of data from the Big Springs dilatometer at Long Valley Caldera, California. (A) Raw data as obtained from incoming telemetry. (B) Data with outliers removed. (C) Data corrected for offsets due to valve resets, with exponential curve fit superimposed. (D) Data after removal of exponential curve to account for contraction due to grout curing; see text for details. Data source: USGS Menlo Park.

intervals of minutes to an hour. The basic procedures for analyzing continuous deformation data are similar for strain, water level, and high-resolution tiltmeters.

Figure 9.9 illustrates these procedures using data from the Big Springs dilatometer, installed during the summer of 1999 on the north rim of the Long Valley Caldera in eastern California. Figure 9.9(A) shows the raw data as downloaded directly from incoming telemetry. A large number of points are equal to the 'bad' data flag 999999, and there are many additional outliers. Figure 9.9(B) shows the same dataset on an expanded scale after all outliers have been identified and replaced with the bad data flag. Large offsets in the first three months of the record are due to valve openings that bring the instrument back into range as it is compressed by

curing of the expansive grout. In Figure 9.9(C), these offsets have been corrected for, resulting in a smoothly rising curve, except for a short initial period when field adjustments were being made. Figure 9.9(C) also shows a curve of the form $A_1 + A_2 \exp[C(t - t_0)]$, which was fit to the daily averages of the strain values to approximate the contraction imposed by the curing grout. In Figure 9.9(D), the data record is shown after subtracting the grout-curing curve. This time series now displays, among other possible signals, the strain variations due to tectonic processes, but it no longer contains absolute strain rate information. Comparing the 25-fold differences in the spans of the y-axes in Figures 9.9(C) and (D) underscores a fundamental feature of borehole–strainmeter data: because the instrument has such high resolution, it is not possible to discern all signals of tectonic interest until instrument resets and grout-curing effects have been removed from the data.

Figure 9.10 shows one month of data from the Big Springs dilatometer after removing strains due to Earth tides and atmospheric pressure loading; an

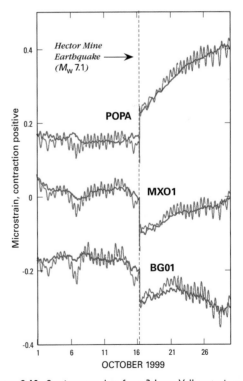

Figure 9.10. Strainmeter data from 3 Long Valley strainmeters, showing offset associated with the 16 October 1999, Hector Mine, California, earthquake (M_w 7.1, 416 km from Long Valley Caldera). Blue curves show data after editing, offsetting, and removal of long-term trend; red curves show data after removing effects of Earth tide and atmospheric pressure using harmonic analysis and linear regression.

offset due to the M_w 7.1 Hector Mine earthquake is well recorded. The tidal analysis also yields calibration factors that can be used to convert the strainmeter output to approximate units of strain.

The data may contain variations that are non-tectonic in origin. Rainfall can affect water level observations by infiltration to the aquifer. For strainmeters and wells with high strain sensitivity, however, it is more common to record strain caused by the weight of the rain falling on the Earth's surface. Strain and water level variations induced by rainfall may be removed, but most commonly are left in the data and merely identified by comparison with a nearby record of rainfall.

Seasonal signals may also appear in both water level and strain data. Subsurface fluid pressure is known to vary seasonally in response to seasonal changes in the rates of rainfall and/or evapotranspiration. Strainmeter data are affected by the coupling between these fluid pressure changes and deformation, as well as by the weight of precipitation or snow on the Earth's surface. These effects can be difficult to model, so it is important to use caution in ascribing tectonic significance to seasonal changes in strain or fluid pressure.

9.4 VOLUMETRIC STRAIN FIELDS OF IDEALIZED VOLCANIC SOURCES

Mathematically, strain is a dimensionless tensor obtained by differentiating the displacement vector with respect to each coordinate direction (Chapter 8). In this section we describe the volumetric strain fields of a center of dilatation, a closed cylindrical source, and a vertical dike, all for the case of a homogeneous, isotropic elastic half-space with Poisson's ratio ν and shear modulus G.

9.4.1 Center of dilatation

The surface uplift due to a spherical cavity with a prescribed internal pressure change ΔP is the mathematical solution used most frequently to model measured volcanic displacements (Figure 9.11). Davis (1986) gives the expressions for the horizontal and vertical surface displacements, as well as a brief history of the solution. The solution is nominally only valid for a cavity whose radius a is much smaller than its depth d, but McTigue (1987) has shown that a more exact solution differs little. The common term 'Mogi'

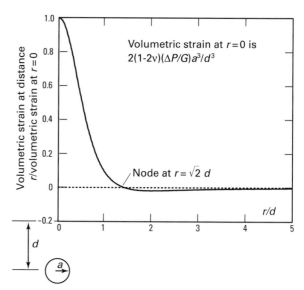

Figure 9.11. Diagram showing assumed configuration of a subsurface inflation source (center of dilatation), and plot of normalized volumetric strain versus dimensionless distance. See text for discussion.

source refers to this solution with the Lamé constants equal (i.e., $\nu = 0.25$).

The volumetric strain at the surface can be obtained by differentiating the radial displacement u_r (Equation 1a of Davis (1986)) to obtain $\varepsilon_{rr} = \partial u_r / \partial r$ and $\varepsilon_{\theta\theta} = u_r / r$. These two strain components are sufficient to obtain volumetric strain because at the free surface, the vertical stress must vanish, allowing the vertical strain ε_{zz} to be expressed in terms of the horizontal strains as:

$$\varepsilon_{zz} = -\frac{\nu}{1-\nu}(\varepsilon_{rr} + \varepsilon_{\theta\theta}) \qquad (9.3)$$

The result is:

$$\varepsilon_{kk} = (1-2\nu)\frac{\Delta P}{G}a^3 \frac{2}{(r^2+d^2)^{3/2}} - \frac{3r^2}{(r^2+d^2)^{5/2}} \qquad (9.4)$$

Equation (9.4) is plotted in Figure 9.11 in non-dimensional form. Like the solution for the displacements, the volumetric strain field is proportional to the ratio of pressure change to the shear modulus of the half-space. The largest strain changes are directly above the source, and are extensional when source pressure increases. Unlike the displacement field, however, the strain field changes sign at $r = \sqrt{2}d$. Beyond this distance, the greatest contractional value is approximately 2% of the maximum strain and occurs at a distance equal to twice the source depth. In practice, only two parameters can be obtained by fitting (9.4) to a set of strain observations: the source depth and an estimate of $(1 - 2\nu)(\Delta P/G)a^3$, the source-volume change. If there is independent information about the size of the source and the elastic constants, then an estimate of the pressure change can be made. Alternatively, the coefficient can be expressed in terms of the volume change ΔV of the source by assuming that:

$$\Delta P = \frac{2G(1+\nu)}{3(1-2\nu)}\frac{\Delta V}{V} \quad (9.5)$$

Combining (9.5) with $V = (4/3)\pi a^3$ leads to the following alternate form of (9.4):

$$\varepsilon_{kk} = 0.5(1+\nu)\frac{\Delta V}{\pi}\frac{2}{(r^2+d^2)^{3/2}} - \frac{3r^2}{(r^2+d^2)^{5/2}}. \quad (9.6)$$

For a concrete example, consider the signals that might have been recorded had continuous strain instruments been in place at the time of the 1980 eruption of Mount St. Helens. Scandone and Malone (1985) infer that the magma chamber at Mount St. Helens is located at a depth of about 9 km (Figure 9.12), and that approximately 0.2 km³ of material was erupted on 18 May 1980.[1] Assuming a spherical source geometry, a borehole strainmeter installed 18 km from the summit would likely have survived the eruption and would have recorded a contractional signal of about 1 microstrain.

The strain field for the center of dilatation can also be computed using the equations or computer code given by Okada (1992).

9.4.2 Vertical conduit models

Bonaccorso and Davis (1999) present analytic expressions for the surface displacements and tilts

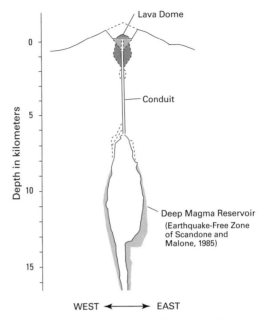

Figure 9.12. Cross-sectional view of Mount St. Helens, showing inferred locations and shapes of shallow- and deep-magma reservoirs.

due to inflation of narrow vertical pipes. These models can be applied to situations where a narrow conduit connects a subsurface magma chamber to the surface or where the magma chamber itself is elongated in the vertical direction. The strains can be obtained by differentiating the published expressions numerically, or by symbolically differentiating them using software such as Mathematica.

The difference between the volumetric strain fields of the closed pipe source and the center of dilatation is most pronounced for radial distances less than one-half the source depth. The closed pipe source has maximum strain at about this radial distance, with smaller strain closer to the source. As an example, consider a more realistic vertically elongated model of the deep magma reservoir at Mount St. Helens (Figure 9.12), based on the zone observed by Scandone and Malone (1985) to have been devoid of earthquakes during the 1980 eruptions. Figure 9.13 shows volumetric strain from an assumed pressure change of 100 MPa, corresponding to about 10% of the pressure change in the main 1980 eruption. For both models, strains readily measured by borehole strainmeters would be expected within 10 km of the volcano, although distinguishing between these two reservoir geometries would clearly require more than one instrument. Periods of increased seismicity at Mount St. Helens since eruptive activity stopped in 1986

[1] The 9-km depth estimate is based on a model of surface-deformation data for June–November 1980, which show cumulative subsidence associated with five lesser explosive eruptions that followed the paroxysmal event of 18 May 1980. On that basis, plus the distribution earthquake hypocenters associated with the eruptions, Scandone and Malone (1985) proposed a model in which 1980 magma was supplied from depths of 7–14 km. According to the model, a mean source depth of 7 km would correspond to a source-volume change of 0.15 km³, while a mean depth of 9 km would correspond to a source-volume change of 0.24 km³ (Scandone and Malone, 1985, p. 254). The net volume of products erupted during the 6 explosive eruptions between 18 May 1980, and 16 October 1980, was estimated to be about 0.23 km³, including 0.19 km³ on 18 May (Scandone and Malone, 1985, p. 241, reporting values from Lipman et al., 1981b).

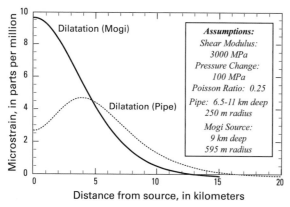

Figure 9.13. Volumetric strain fields from a pressure change in the deep magma reservoir under Mount St. Helens, assuming either a spherical (Mogi, 1958) or closed-pipe (Bonaccorso and Davis, 1999) geometry.

could reflect pressure changes in the deep reservoir (Moran, 1994), and these might produce deformation that would be detectable with borehole strainmeters.

9.4.3 Dike intrusion

The strain field of an inflating dike can be modeled as an opening-mode rectangular dislocation in an elastic half-space using the codes of Okada (1992). Figure 9.14 shows an example for a dike that is 5 km long and extends from 5 to 10 km beneath the surface. An interesting feature of the dike's strain field is the area of contraction directly over the dike. This feature, which is not present for a dike that reaches the surface, is surrounded by an area of extension. Unlike the center of dilatation or the closed cylindrical pipe, the maximum contraction produced by the dike is almost as great as the maximum extension. The width of the area of contraction decreases as the upper limit of the dike nears the surface; a strainmeter placed in this area of contraction will record a reversal in the strain field if the dike propagates far enough upward. To simulate strain from a propagating dike, the strain field can simply be calculated for a succession of positions of the source. To simulate the pore-pressure field of a propagating source is somewhat more complicated because the diffusing pressure fields must also be superimposed. Elsworth and Voight (1992) present a partially analytic solution in a moving coordinate system for pore pressure induced by advance of a dike intruding at a steady rate into a full space. They show that the solution closely resembles time-dependent pore-pressure changes recorded in a well several

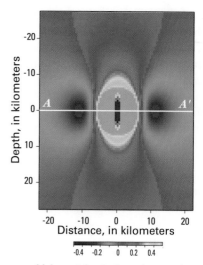

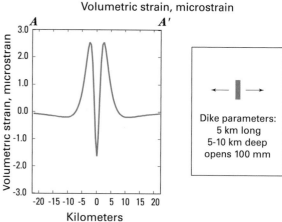

Figure 9.14. Volumetric strain field of a vertical dike, calculated using the computer code of Okada (1992).

kilometers from two dike intrusion events in Krafla, Iceland, in 1977 and 1978.

9.5 EXAMPLES

Water level changes preceding eruptions have on occasion been so large that they have been observed without special instrumentation. Newhall *et al.* (2001) review numerous examples of groundwater level changes associated with volcanic unrest and eruptions. For example, groundwater levels in wells on the upper slopes of Mayon Volcano, the Philippines, dropped as much as 2–3 m beginning several months before an eruption in 1993. These water level drops were attributed at least in part to inflation of the volcano accompanying subsurface upward movement of magma. Continuous water level data are now being collected to infer the magnitude of the extensional strain and help constrain the influence of rainfall on the observations.

Below we discuss case histories for which actual strain and fluid pressure data have been recorded. Currently, there are detailed strain datasets for two eruptions of Hekla Volcano in Iceland and fluid pressure datasets for an eruption of Mount Usu in Japan. Crustal deformation was measured prior to all three of these eruptions. In addition, these sensitive techniques have detected crustal deformation accompanying earthquake swarms on the Izu Peninsula in Japan, on the Juan de Fuca Ridge in the western Pacific Ocean, and in the Long Valley Caldera, California. These deformation records strongly support the hypothesis that earthquake swarms in volcanic areas are a manifestation of the dominantly aseismic upward movement of magma in the form of dikes, which may or may not breach the surface to pose eruptive hazards.

9.5.1   Izu Peninsula, Japan

The Izu Peninsula in Japan is a volcanic area with ongoing uplift and frequent seismic swarms (Okada and Yamamoto, 1991). A submarine eruption off Ito City in 1989 was preceded by earthquake swarms and rapid strain and tilt, as well as groundwater outflow at four wells in Ito City, two of which have been monitored continuously since then. The strain, tilt, and deformation data from the 1989 eruption have been interpreted to indicate intrusion of subvertical dikes beneath a shear fault (Okada and Yamamoto, 1991). Between 1995 and 1998, several seismic swarms have been accompanied not only by water level changes (Ohno et al., 1999; Koizumi et al., 1999), but also by deformation recorded using borehole strainmeters, tiltmeters, and continuous GPS (CGPS). The water level changes accompanying the seismic swarms exceed those that would be expected in response to individual co-seismic strain steps due to swarm earthquakes. According to Ohno et al. (1999) and Koizumi et al. (1999), the sizes and directions of water level changes are consistent with their being proportional to the volumetric strain field calculated from the dike intrusion models.

9.5.2   Long Valley Caldera, California: stimulation by distant earthquakes

Groundwater levels in several wells, and discharges at several springs and creeks, have been monitored

since the early 1980s at Long Valley Caldera, a potential volcanic hazard in eastern California. Tectonic and volcanic activity at Long Valley is monitored by a seismic network, periodic leveling surveys, a two-color electronic distance meter (EDM) network, three borehole volumetric strainmeters, a long-baseline tiltmeter, and, more recently, survey-mode and CGPS (Chapter 7). Although the volcano is judged unlikely to erupt anytime soon, it has exhibited a high rate of seismic activity and uplift in recent decades. Seismic swarms and accelerated uplift are closely watched, and local officials and the population are kept advised concerning volcanic hazards in case of renewed activity.

The intensive monitoring has revealed that local and distant earthquakes can induce persistent strain and water level changes at Long Valley that resemble those recorded during seismic swarms on the Izu Peninsula. The best-known example is the 1992 Landers, California, earthquake (M_w 7.3), whose effects in Long Valley, more then 400 km from the epicenter, included triggered microseismicity, a transient contractional signal on a borehole strainmeter, and water level changes (Hill et al., 1995; Johnston et al., 1995; Roeloffs et al., 1995) (Figures 9.15 and 9.16). Other distant earthquakes that produced water level changes at Long Valley, but did not trigger seismicity, also have been

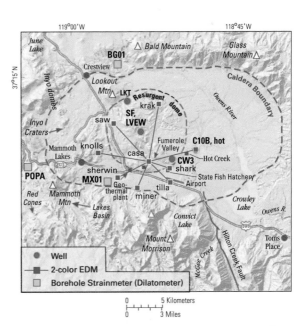

Figure 9.15. Map of the Long Valley Caldera area showing locations of the wells, strainmeters, and the two-color EDM network. The Hilton Creek Fault, Mammoth Mountain, and the resurgent dome are also shown.

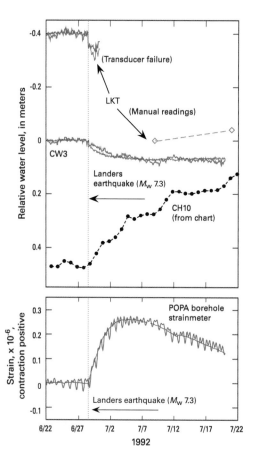

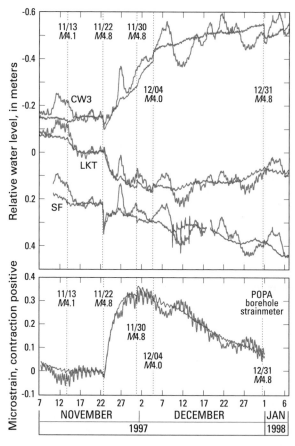

Figure 9.16. Earthquake-induced water level and strain changes at Long Valley Caldera caused by the 28 June 1992 Landers, California, earthquake (M_w 7.3), 446 km from Long Valley Caldera. Data with tidal and barometrically induced fluctuations removed (red) are superimposed on raw data (blue). A linear trend has been subtracted from the data so as to minimize the water-level variation in the 20 days before the earthquake.

Figure 9.17. Earthquake-induced water level and strain changes at Long Valley associated with earthquakes in Long Valley Caldera, November–December 1997. Data with tidal and barometrically induced fluctuations removed (black) are superimposed on raw data (blue). A linear trend has been subtracted from the data so as to minimize the water level variation in the 20 days before the earthquake. Dotted lines mark occurrences of $M \geq 4$ earthquakes.

accompanied by strain-rate changes (Roeloffs *et al.*, 1998).

During an intensive earthquake swarm in late 1997, strain and water level changes closely resembling those induced by the Landers earthquake began simultaneously with one particular M 4.9 local earthquake on 22 November 1997 (Roeloffs *et al.*, 2003) (Figure 9.17). This earthquake's focal mechanism contained a large component of opening mode displacement (Dreger *et al.*, 2000), and it was followed by upward migration of seismicity over the next several days (Hill *et al.*, 2003). These features of the 1997 activity strongly suggest upward movement of material into a shallow subvertical dike in the caldera's south moat, beginning with the November 22 earthquake. The similarity of the time histories of the water level and strain responses following this earthquake, to those following the 1992 Landers

earthquake, leads to the hypothesis that incremental propagation or inflation of dikes can be stimulated at Long Valley by distant earthquakes.

9.5.3 Eruptions of Hekla, Iceland, in 1991 and 2000

A small network of Sacks–Evertson strainmeters was installed in southern Iceland in 1979 with the objective of capturing strain changes associated with an expected moderate earthquake in the south Iceland seismic zone (Figure 9.18). The volcano Hekla is in the study area but had erupted in 1970; from its eruption history, it was not expected to erupt again for at least many decades. Thus, the network was not designed to provide strong constraints on volcanic activity at Hekla; the nearest site (BUR) is 15 km from the volcano and the others are

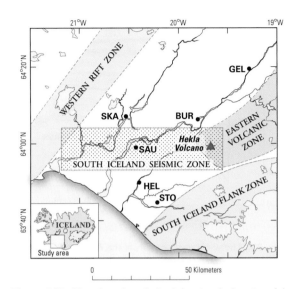

Figure 9.18. Map of southern Iceland showing the location of the borehole strainmeter network, Hekla Volcano, and south Iceland seismic zone.

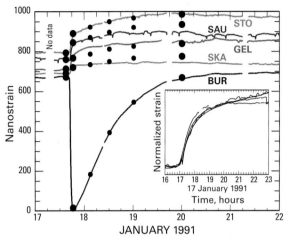

Figure 9.19. Borehole strain data for a 5-day period beginning 1 day prior to the 17 January 1991 eruption of Hekla Volcano, Iceland. Expansion is positive. Earth tides and atmospheric-pressure-induced strain changes have been removed. *Inset* (covering 7 hours, with amplitudes normalized) shows that even the 4 more-distant sites undergo expansion in a remarkably coherent manner. Large solid circles show strain changes calculated from modeling the eruption in two stages. Stage 1 (deflation of a deep reservoir together with dike formation) takes place during the interval from initiation of magma movement until the output from the strainmeter at BUR reaches a minimum. During the following two days (stage 2), when the eruption was essentially complete, the data can be modeled solely by continued deflation of the deep source. Smaller circles show results of calculations for intermediate times during stage 2. These three points correspond to pressure decreases of 30%, 60%, and 80% of the total change of 14 MPa during stage 2 (Linde *et al.*, 1993). All times are local.

at distances of 35–45 km, although they do cover a wide range of azimuths. But the eruption in 1970 apparently marked a change in eruptive nature from infrequent (∼100 years) large eruptions to small eruptions every 10 years or so. At the time of the 1980 eruption, only analog chart records of strain were available and these had significant gaps. But the eruptions in 1991 and 2000 were both well recorded on telemetered digital records.

Analysis of the January 1991 eruption was reported by Linde *et al.* (1993). All the more distant strainmeters showed extensional strain changes during the approximately two days during which the eruption was most active (Figure 9.19). However, the nearest station (BUR) drew a more complicated record, with large rapid contraction during the first two hours (Figure 9.20) followed by expansion for the next two days. These changes were modeled in terms of a deflating point spherical pressure source (Mogi source) together with a dike that propagated from a depth of 4 km to the surface (Figure 9.20). The strike and length of the dike were determined from surface observations of the eruption. From an expanded timescale plot (Figure 9.21), it is clear that the strain changes at BUR start about 30 minutes before the eruption. These pre-eruptive changes start with the initiation of dike propagation at depth. Thus, in the case of Hekla, the vertical speed of dike propagation is about 8 km hr^{-1}, indicating that the process is driven by gas rather than high-viscosity magma.

The February 2000 eruption of Hekla (Agustsson

et al., 2000; Linde and Sacks, 2000) was somewhat smaller than that in 1991, but the strain records were remarkably similar (Figure 9.22), indicating that both eruptions took place with essentially the same geometry. After the 1991 eruption, seismographs were installed closer to Hekla with one station 2 km from the summit. Historically, Hekla has remained seismically quiet except during eruptions. In 2000, the seismograph 2 km from the summit recorded earthquakes at Hekla starting about 90 minutes before the eventual eruption. The alert caused R. Stefansson (chief of the geophysics group at the Iceland Meteorological Office) to come to the office on a Saturday afternoon where he closely monitored the seismic and strain activity. As the earthquake activity increased, he issued an unofficial alert and requested the meteorologists to determine the wind velocity at altitudes of 10 km and above. The strain data are telemetered to the office at 50 samples per second with 20-bit resolution. When Stefansson saw the BUR strain decreasing as in 1991 (Figure 9.23), he issued a formal prediction and warning. This was broadcast to

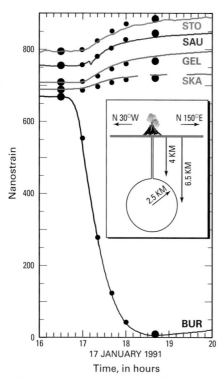

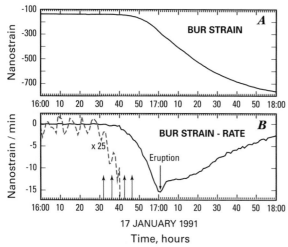

Figure 9.20. Borehole strain data for a 4-hour period that includes the onset of the 1991 eruption of Hekla Volcano. *Inset* shows the model used to calculate points overlaid on the curve. Symbols are the same as Figure 9.19. The exact onset time of the eruption is not known from direct observations owing to inclement weather. However, airport radar observations set the time between 17:00 hr and 17:10 hr (when the plume was already 11.5 km high), and farmers near the volcano reported seeing activity at ~17:05 hr (Linde et al., 1993). All times are local.

Figure 9.21. Strain data for 2 hours on 17 January 1991 from station BUR (A), and its time derivative (B), on an expanded timescale.

the public and to the Civil Aviation authorities, the latter because Hekla eruptions always send an ash plume to heights greater than 10 km within minutes of the eruption. This successful prediction was

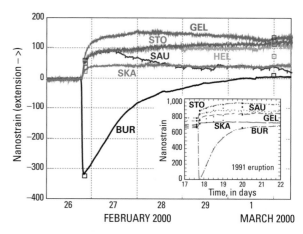

Figure 9.22. Borehole strain data from the February 2000 eruption of Hekla Volcano, together with comparable data from the same five stations for the 1991 eruption (*inset*).

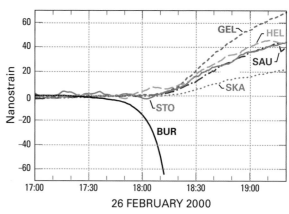

Figure 9.23. Borehole strain data for a 2.5-hour period including the onset of the 2000 eruption of Hekla. Times are local.

possible only because the strain changes are a direct indication of magma movement, and because the eruption repeated the pattern that had been recorded in the 1991 eruption.

9.5.4 Eruption of Usu Volcano, Japan, March 2000

Usu is a dacitic stratovolcano on the island of Hokkaido, Japan, which has been active episodically since the 17th century. During its penultimate eruption in 1977–1978, manual groundwater level measurements were being made in a 370 m deep well 2 km east of the summit (Watanabe, 1983). Following a pumice eruption in December 1977, the water level in this well was found to have risen 37 m relative to 1972 measurements. Following this finding, more frequent measurements of water level showed it to be falling steeply, with some small rises, especially accompanying felt earthquakes.

Usu volcano erupted again on 31 March 2000. Prior to this most recent eruption, a number of CGPS stations had been installed around the volcano. Groundwater levels in a number of wells were also being recorded, some specifically for the purpose of detecting crustal deformation, and others to monitor conditions in the area's geothermal aquifers.

Shibata and Akita (2001) report large changes in groundwater level preceding the eruption. They observed decreasing groundwater levels beginning as early as six months prior to the eruption, which they attributed to an increase of cracking near the surface induced by an influx of magma into the subvolcanic fracture system. Matsumoto et al. (2002) reported water level increases shortly before the eruption in slightly more distant wells. Three days after the eruption began, the water level in one well suddenly increased and began to spout, representing a fluid pressure increase of more than 1 MPa, possibly due to the addition of thermal fluid through new cracks created during the eruption.

9.5.5 Spreading of the western Pacific sea floor on the Juan de Fuca Ridge

Davis et al. (2001) describe 4 years of hourly fluid pressure data measured in subsea boreholes penetrating young (<3.6 Ma) oceanic crust near the axis and on the eastern flank of the Juan de Fuca Ridge, offshore the northwestern coast of North America (Figure 9.24). The boreholes are open to permeable basalts overlain by as much as several hundred meters of relatively impermeable sediments, so that confined conditions prevail at the depths where fluid pressure is being monitored. Fluid pressure records from these subsea boreholes contain fluctuations caused by tidal changes in the load of the overlying ocean (about 2.6 km deep), changes in atmospheric pressure, and other changes in ocean loading due to fluid circulation patterns. As for pressure records from continental boreholes, these effects can be removed using linear regression and harmonic analysis to reveal smaller signals and to constrain the ratio of fluid pressure change to crustal strain.

In 1999, a seismic swarm initiated on 8 June with a M 2.8 earthquake. The swarm included 8 events with magnitudes up to 4.1 detected by onshore seismic networks, and several hundred events detected by the US Navy SOSUS array. In three boreholes, 25 to 101 km from the ridge axis, fluid pressure rises of 0.1–1.6 KPa (1–16 cm of water) occurred abruptly

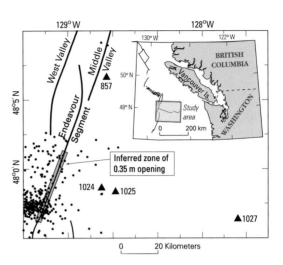

Figure 9.24. Map showing the location of CORK borehole observatory sites (triangles), the Endeavour segment of the Juan de Fuca Ridge, and the inferred surface projection of a spreading event (modified from Davis et al., 2001). Dots represent earthquakes recorded using the US Navy SOSUS array for the period 7 June–17 July 1999 (from NOAA/Pacific Marine Environmental Laboratory, Newport, Oregon).

with the first event and then continued more gradually over the succeeding hours to days, with similar time histories but time scales lengthening with increasing distance from the ridge (Figure 9.25). Davis et al. (2001) estimated strain sensitivities of 0.3–0.4 m per microstrain for these boreholes, based on physical properties of the formation and tidal response, and inferred initial strains of 0.5 microstrain contraction at the nearest site and 0.024 microstrain contraction at the most distant site. These strains are far larger than can be accounted for by any earthquakes in the swarm. Davis et al. (2001) modeled the observed strains with 0.35 m of opening perpendicular to the ridge, extending from the surface to a depth of 3 km and approximately 40 km along the strike of the ridge.

The longer rise times for the fluid pressure disturbances at increasing distances from the ridge axis would not be expected if the fluid pressure changes were simply proportional to poroelastic strain; in such case, all sensors would have the same time history as the strain imposed by the spreading event. Instead, Davis et al. (2001) showed that the time histories of the fluid pressure changes are well matched by assuming 1-D diffusion of the fluid pressure pulse induced by the spreading event. This observation illustrates an important difference between the observations made by a strainmeter and those made by a fluid pressure sensor. In this case, the fluid pressure records do not tightly constrain the

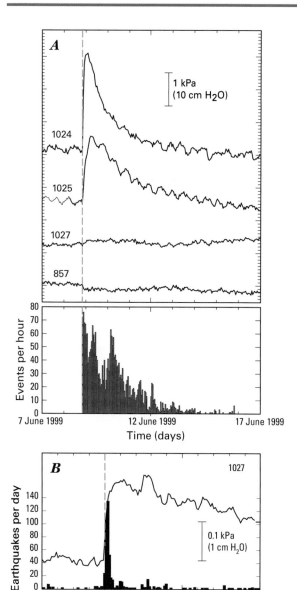

Figure 9.25. (A) Formation pressure for the 10-day period 7 June–17 June 1999 from sites 1024, 1025, 1027, and 857 with responses to tidal, barometric, and oceanographic loading removed, and number of earthquakes per hour. (B) Expanded-scale 140-day record of formation pressure from site 1027 (tidal, barometric, and oceanographic loading removed), with number of earthquakes per day. Pressure data from Davis et al. (2001); seismicity from NOAA/Pacific Marine Environmental Laboratory, Newport, Oregon.

detailed time history of the spreading event. Nevertheless, the results are remarkable in that a fluid pressure change of 2 cm of water, with a rise time of about 10 days, was clearly detected at site 1027,

100 km from the source (Figure 9.25). Moreover, the strain causing the fluid pressure change could be quantified, and the data clearly show crustal deformation beyond that represented by the earthquakes themselves, strongly implying the occurrence of a spreading event.

9.6   SUMMARY

Borehole strainmeters and groundwater level sensors in wells are increasingly being deployed near active volcanoes, and the examples described here demonstrate the promise of these techniques for enhancing predictive capabilities and learning about volcanic processes at depth. Noise is reduced by observing in a borehole environment. The resulting capacity to detect small signals promotes safety through automatic operation and installation at greater distances from volcanic vents. Moreover, signals originating from deep magma movements can be detected, and datasets that continue through the eruption can be obtained. The detailed time histories of deformation permit the ascent of magma to be tracked continuously in time.

Borehole strainmeter and fluid pressure monitoring installations can have somewhat greater initial costs than other geodetic techniques. A borehole must be available, and usually must be specially drilled if a strainmeter is to be installed. The first several months of strainmeter data may not be useful, due to grout curing, borehole creep, and instrument adjustments. For groundwater level data, it is also desirable to record data for a long enough period to determine the usual responses to precipitation and climate, prior to attempting interpretation in terms of crustal strain. For these reasons, these techniques are only useful in a volcanic crisis response if the instruments have been deployed at least several months ahead of time.

Borehole-strain measurements are most useful when made within a network of geodetic stations where absolute crustal movements also can be determined. When deployed well ahead of an eruption as part of an integrated volcano-monitoring network, crustal strain measurements in boreholes provide invaluable information about volcanic processes, and have already been used to issue a public warning in advance of an eruption.

CHAPTER 10

Hydrothermal systems and volcano geochemistry

Robert O. Fournier

The upward intrusion of magma from deeper to shallower levels beneath volcanoes obviously plays an important role in their surface deformation. This chapter will examine less obvious roles that hydrothermal processes might play in volcanic deformation. Emphasis will be placed on the effect that the transition from brittle to plastic behavior of rocks is likely to have on magma degassing and hydrothermal processes, and on the likely chemical variations in brine and gas compositions that occur as a result of movement of aqueous-rich fluids from plastic into brittle rock at different depths. To a great extent, the model of hydrothermal processes in sub-volcanic systems that is presented here is inferential, based in part on information obtained from deep drilling for geothermal resources, and in part on the study of ore deposits that are thought to have formed in volcanic and shallow plutonic environments.

The material presented here is adapted from an article that I had published in the journal *Economic Geology* (Fournier, 1999). That article emphasized ore-forming processes that are likely to occur as a result of emplacement and degassing of magmatic bodies at relatively shallow depths in volcanic systems. Here I emphasize the deformation resulting from these hydrothermal processes and the expected variations in compositions of discharged gases. I will begin by reviewing the factors that influence the brittle–plastic transition in the Earth's crust, emphasizing sub-volcanic or shallow plutonic conditions. I will then discuss the accumulation of exsolved magmatic fluids in plastic rock and tie together various coupled physical and chemical phenomena that result from a decrease in fluid pressure from near lithostatic to near hydrostatic in an environment of transition from plastic to brittle behavior.

10.1 THE HYDROLOGIC IMPORTANCE OF BRITTLE–PLASTIC PHENOMENA

The maximum depth of occurrence of earthquakes that result from shear failure in the crust marks the transition from brittle to plastic behavior in the lithosphere (e.g., MacElwane, 1936; Byerlee, 1968). Because the onset of plastic flow of rocks is highly dependent on temperature, it is not surprising that the bottoming of seismicity occurs at very shallow depths beneath large, hot, and presently active geothermal fields, such as The Geysers and Clearlake Highlands (California, USA) (Majer and McEvilly, 1979; Sibson, 1982), the Imperial Valley (California, USA) (Gilpin and Lee, 1978), and Yellowstone National Park (Wyoming, USA) (Smith and Braile, 1984, 1994; Miller and Smith, 1999). In addition to limiting seismic activity, the onset of plastic flow closes pre-existing interconnected pore spaces and fractures (Brace, 1972), thereby restricting the depth of circulation of meteoric water into the crust at hydrostatic pressure (pressure imparted by the weight of an overlying column of water). On the other hand, there is evidence from fluid inclusions that aqueous liquids are found deep in the crust at temperatures sufficient for rocks to behave plastically (Roedder, 1984). Petrologists and geochemists generally have assumed that, in these deep, hot environments, pore-fluid pressure (P_f) equals the lithostatic load or vertical stress (S_v) (e.g., Turner, 1981).

Many deep geothermal exploration wells in

continental crystalline rocks have encountered meteoric-derived fluids at hydrostatic pressure at temperatures up to about 350–360°C. To date, the few deep wells drilled to temperatures greater than 370–400°C either have produced gas-rich brines at greater than hydrostatic pressures or have encountered little permeability (e.g., wells discussed in Fournier, 1991). These observations indicate that the brittle–plastic transition commonly occurs at about 370–400°C within presently active continental hydrothermal systems. The well data also show that fluids at greater than hydrostatic pressure may accumulate in quasi-plastic rock, and that a narrow zone or shell of relatively impermeable material commonly separates two very different hydrologic domains.

10.2 THE BRITTLE–PLASTIC TRANSITION

10.2.1 General considerations

Figure 10.1 schematically shows the general relations for initiating failure of materials in the brittle and plastic regions of the Earth's crust. In the brittle region, the stress difference required to cause shear failure of a pre-existing open crack increases with increasing depth, and is relatively independent of temperature, rock type, and strain rate. It is, however, highly dependent on the coefficient of friction, on the orientation of the fracture with respect to the stress field, and on pore-fluid pressure P_f. Commonly, P_f in the crust is expressed relative to the vertical stress S_v by the relation $\lambda = P_f/S_v$. For an average fluid density of $1\,\mathrm{g\,cm}^{-3}$ and average rock density of $2.6\,\mathrm{g\,cm}^{-3}$, $\lambda = 0.38$ for hydrostatic P_f conditions. Most rocks have a coefficient of friction of about 0.6 to 0.8, according to measurements by Byerlee (1978).

In Figure 10.1, the line from the origin at zero depth through point A shows the stress difference $(\sigma_1 - \sigma_3)$ required to activate fault movement on an existing open crack (a crack having no cohesive strength) that is oriented at an optimum angle of about 26 degrees to the direction of application of the maximum principal stress, when P_f is assumed to equal the hydrostatic pressure. Here σ_1 is the maximum principal stress and σ_3 is the least principal stress. The stress difference required to cause shear failure increases dramatically when the direction of the maximum principal stress, with respect to the plane of the fracture, departs by more than 5 to 10 degrees from an optimum

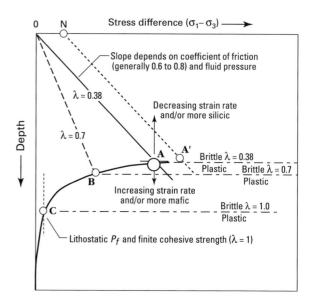

Figure 10.1. Stress difference $(\sigma_1 - \sigma_3)$ versus depth, showing schematically the conditions for brittle shear failure along a fracture in the upper crust at selected ratios of hydrostatic pressure to lithostatic load (λ values), and plastic deformation at deeper levels. See text for discussion.

angle of about 26 degrees (Sibson, 1985, 1990). When all the existing fractures are unfavorably oriented with respect to the direction of the maximum principal stress, increasing the stress difference may overcome the cohesive strength of the rock and cause a new crack to develop at an optimum angle before the stress difference becomes great enough to cause shear failure along any of the unfavorably oriented pre-existing open fractures (Sibson, 1985). Also, in the event that an optimally oriented fracture regains cohesive strength as a result of cementation by vein formation, a greater stress difference is required to cause shear failure than would have been required before cementation. For example, in Figure 10.1 the dashed line from point **N** through point **A'** shows the stress differences versus depth required to renew shear failure when $\lambda = 0.38$ along a fracture that has cohesive strength given by point **N**.

In contrast to the conditions for brittle failure, the stress difference required to initiate plastic deformation is highly dependent on temperature, strain rate, and rock type (material constants), and it is little affected by confining pressure. However, the presence of water allows plastic behavior at lower temperatures compared with the behavior of dry rock (Carter and Tsenn, 1986).

The general law governing steady-state plastic or ductile deformation can be expressed approximately by the equation:

$$\dot{\varepsilon} = A(\sigma_1 - \sigma_3)^n e^{-Q/RT} \qquad (10.1)$$

where $\dot{\varepsilon}$ is the strain rate, R is the gas constant, T is absolute temperature, $(\sigma_1 - \sigma_3)$ is the stress difference acting on the material, and A, Q, and n are material coefficients that change with rock type (Turcotte and Schubert, 2002). The stress difference required to cause plastic deformation at a given strain rate decreases exponentially with depth (curve $A'ABC$ in Figure 10.1) because temperature generally increases in direct proportion to depth. For a given externally imposed strain rate, the onset of plastic behavior for more mafic rocks, such as basalts and gabbros, occurs at higher temperature (deeper in the crust) than for shales and rocks rich in quartz. Even more important, according to (10.1), an increase in strain rate allows brittle behavior of all rock types to occur at higher temperatures (greater depths) in the crust.

For a given externally imposed strain rate, increasing the stress difference in the brittle region causes shear failure to occur before the stress difference becomes great enough to cause plastic deformation. In contrast, in the plastic region, at a given externally imposed strain rate, increasing the stress difference causes plastic deformation before the stress difference becomes great enough to cause shear failure. Therefore, the brittle-to-plastic transition occurs where the brittle and plastic deformation curves intersect. In Figure 10.1, point A marks the depth of the brittle-to-plastic transition at a specified strain rate when hydrostatic pressure prevails, or $\lambda = 0.38$. In reality, because of rock inhomogeneities the brittle-to-plastic transition may occur within a depth interval, rather than at a specific depth. Also, the brittle–plastic transition will occur at a greater depth along the plastic deformation curve (curve ABC in Figure 10.1) by an increase in P_f, or when $\lambda > 0.38$, such as at point B when $\lambda \approx 0.7$.

Theoretically, shear failure could occur on open fractures having no cohesive strength in response to vanishingly small stress differences when $P_f = S_v$, and the brittle–plastic boundary would occur very deep in the crust at a very high temperature. However, in nature open fractures quickly become chemically healed or cemented by vein deposition and by sintering of adjacent grains as temperature increases with depth, so that a finite stress difference is required to activate shear failure of faults even where $P_f = S_v$ (e.g., the dashed vertical line passing through point C in Figure 10.1).

Although brittle failure occurs only when the stress difference reaches some particular value, plastic flow is always occurring in rocks to some extent, even at low temperatures or in response to small stress differences (10.1). However, where temperatures are relatively low and/or the stress differences are very small, the rate at which this plastic flow occurs may be so slow that it is not evident over time spans of thousands to millions of years.

10.2.2 Brittle–plastic transition in an active volcanic environment

Because the stress difference required to initiate brittle shear failure depends mainly on depth, while that required to initiate plastic deformation depends mainly on temperature, it is necessary to know or assume a reasonable depth–temperature profile in order to portray where the brittle–plastic transition is likely to occur within a given natural system. In non-volcanic situations, temperatures generally increase fairly regularly with depth, with gradients of about $25–35°C\,km^{-1}$. In contrast, very hot rock (perhaps molten or partly molten intrusive magmatic bodies) is likely to be present at relatively shallow depths beneath presently active volcanoes, and convective hydrothermal systems tend to develop above such hot rock. Therefore, when portraying the transition from brittle to plastic conditions around shallow crystallizing magmatic bodies in a sub-volcanic environment, it is necessary to take these complexities into account.

Figure 10.2 is for a situation where there is a convecting hydrothermal system above a silicic magmatic body that is crystallizing at a depth of about 3.5 to 4 km. Note that for a basaltic or gabbroic system the brittle–plastic transition would likely occur at about $500–600°C$ when the strain rate is about $10^{-14}\,s^{-1}$ instead of at about $370–400°C$, which appears to be appropriate for shallow silicic systems. In Figure 10.2, an average temperature gradient of $125°C\,km^{-1}$ was used from the land surface to the depth where $375°C$ is attained (the maximum temperature likely for a hydrothermal system at hydrostatic pressure), and a temperature gradient of $500°C\,km^{-1}$ was used at depths where the temperature exceeded $375°C$. The $125°C\,km^{-1}$ gradient allows for significant transfer of heat by hydrothermal convection. In reality, within a vigorously convecting hydrothermal system in brittle,

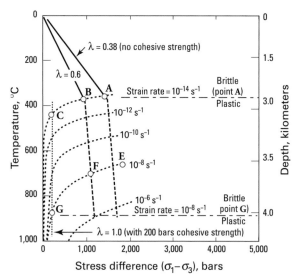

Figure 10.2. Stress difference $(\sigma_1 - \sigma_3)$ versus temperature, showing effects of strain rate, value of λ, and cohesive strength on the brittle-to-plastic transition. An average temperature gradient of 125°C km^{-1} was used from the land surface to the depth where 375°C is attained (the maximum temperature likely for a hydrothermal system at hydrostatic pressure), and a temperature gradient of 500°C km^{-1} was used at depths where the temperature exceeded 375°C. See text for discussion.

permeable rock, there would be little change in temperature from the bottom of hydrothermal circulation up to the point where boiling starts with declining hydrostatic pressure. Thereafter, temperatures would decline along the boiling point curve.

The assumed 500°C km^{-1} temperature gradient in Figure 10.2, where heat is transferred mainly by conduction from the magmatic heat source to the base of the hydrothermal system, is greater than a 320°C km^{-1} gradient measured in a deep well (WD-1) at the Kakkonda geothermal field in Japan (Ikeuchi et al., 1996), and less than the calculated 700–800°C km^{-1} gradient beneath the hydrothermal convection system at Yellowstone National Park (Fournier and Pitt, 1985; Fournier, 1989). The rheologic properties of wet quartz diorite (Carter and Tsenn, 1986) were used for the plastic region, and a coefficient of friction of 0.75 was used for shear failure in the brittle region.

10.2.3   Brittle behavior of normally plastic rock at high strain rates

Figure 10.2 shows calculated conditions for brittle and plastic behavior in a shallow, hot plutonic environment (temperature gradients specified above) when $\lambda = 0.38$, 0.6, and 1.0 at selected strain rates ranging from 10^{-14} s^{-1} to 10^{-6} s^{-1}. In regions of

active crustal deformation, strain rates generally range from about 10^{-14} s^{-1} to 10^{-15} s^{-1} (Pfiffner and Ramsay, 1982), and strain rates of 10^{-12} s^{-1} to 10^{-11} s^{-1} may be attained for very short times during periods of active faulting (Sibson, 1982). In volcanic environments, very high rates of deformation have been observed during dome emplacement and during collapse accompanying eruptions (Dzurisin et al., 1983). When the strain rate is 10^{-14} s^{-1} and hydrostatic fluid pressure prevails ($\lambda = 0.38$), the brittle-to-plastic transition in Figure 10.2 is at a depth of about 3 km and at a temperature of about 360°C (point **A**). Increasing fluid pressure to greater than hydrostatic moves the brittle-to-plastic transition point along curve **ABC** to higher temperatures (greater depths) and lower required stress differences, as discussed above.

The main point that Figure 10.2 illustrates is that a temporary increase in the strain rate to $>10^{-14}$ s^{-1} in initially plastic rock will allow brittle behavior at relatively low stress difference at very high temperatures. For example, temporarily increasing the strain rate to 10^{-8} s^{-1} allows brittle failure to occur at temperature and stress-difference conditions limited by curve **EFG**. For lithostatic P_f conditions, and rock material that has about 200 bars cohesive strength, fracturing will occur in response to a stress difference of 200 bars, and the brittle–plastic transition is moved to a depth of about 4 km where the temperature is about 870°C (point **G** in Figure 10.2). Mechanisms for temporarily increasing the strain rate in volcanic environments will be discussed subsequently.

The calculated conditions for shear failure at high strain rates shown in Figure 10.2 are very approximate because there would be changes in the material constants at very high temperatures where partial melting occurs. However, textural and structural features observed in many magmatic bodies support the general conclusion that shear failure of very hot material can occur. For example, in some plutonic bodies, fractures, veins, and dikes appear to have formed and to have been subsequently offset before the surrounding magmatic material was completely solidified, implying shear failure of crystal-rich magma that behaved in a brittle manner. Some specific examples are: (1) shearing of the partly molten Half Dome quartz monzonite of the Sierra Nevada batholith that occurred at the time of emplacement of alaskitic, aplitic, and pegmatitic dikes (Fournier, 1968); (2) the formation of very high temperature 'A' quartz veins in the El Salvador, Chile, porphyry copper deposit without parallel walls and generally before the rock was able to sustain continuous brittle fracture (Gustafson and Hunt, 1975); and (3) alkali feldspar seams cutting plagioclase phenocrysts, but with no trace of the seams cutting intervening fine-grained, aplitic-textured quartz and alkali feldspar at El Salvador (Gustafson and Hunt, 1975) and at Ely, Nevada (Fournier, 1967). I interpreted the above textures found in porphyry copper deposits as indicating that fracturing and K-feldspar replacement of plagioclase took place before the viscous, glassy groundmass had crystallized and solidified, at the time of a pressure quench of the porphyry that supercooled the residual magma by tens of degrees (Fournier, 1967). The most likely reason for these high-temperature shear failures (brittle behavior) in normally plastic rock is a very short-lived increase in strain rate, coupled with a situation in which $P_f \leq S_v$ so that shearing could occur in response to a very small stress difference. The system would be expected to return to a plastic condition almost immediately thereafter because the shear movement would diminish differential stress. However, even though such fractures may be quickly closed, while they last they may be important avenues for rapid movement of magmatic fluids from deeper to shallower levels through normally plastic rock and then into the normally brittle domain.

10.3 DEVELOPMENT OF PLASTIC ROCK AROUND SHALLOW INTRUSIVE BODIES

In the initial stages of development of a volcanic complex, magmas intruding upwards to depths less than 10 km encounter brittle country rock[1] having temperatures appropriate for normal thermal gradients in the crust (about 25–35°C). Hydrostatic P_f would probably prevail in these rocks to depths of at least 6–8 km, and possibly to >10–12 km. Under such conditions, the much cooler surrounding rock chills the margins of magmatic bodies that are intruded upward relatively quickly. A thin rind or shell of plastic rock develops immediately around the magmatic body. The thermal gradients

---

[1] Country rock refers to the rock intruded by and surrounding an igneous intrusion, or enclosing or traversed by a mineral deposit. The term is somewhat less specific than host rock, a body of rock that surrounds (hosts) other rocks or mineral deposits (e.g., a pluton containing xenoliths, or any rock in which ore deposits occur).

across the magma-country rock interface through this plastic rind will be very steep (Cathles, 1977; Norton, 1982), and heat will be transferred from the magmatic body in part by the escape of magmatic gases, but mostly by conduction. Efficient removal of heat from the system by hydrothermal convection at hydrostatic pressure occurs only in rock that has a temperature less than about 375°C. At higher temperatures, cavities or interconnected networks of open fractures with considerable vertical extent are mechanically unstable, as explained in Section 10.4.1. Section 10.4 as a whole addresses how hydrothermal fluids are stored in, and move through, rock at temperatures above 375°C.

As volcanic activity continues, a succession of dikes, sills, and small plutonic bodies come to rest below and within the enlarging volcanic edifice. Each of these magmatic bodies is intruded into country rock that has been heated by previous intrusives, so that there is a progressive increase in temperature at given depths beneath vigorously active volcanoes, and a corresponding increase in the volume of plastic rock at relatively shallow depths. Eventually, a considerable volume of rock can become hot enough to behave plastically at depths as shallow as 1.5 km. This has considerable importance in regard to the degassing of solidifying magmas, and the type(s) of associated hydrothermal activity that develops (discussed below).

10.4 STORAGE OF HYDROTHERMAL FLUID IN AND MOVEMENT THROUGH PLASTIC ROCK

10.4.1 Accumulation in horizontal lenses in plastic rock when and where $\sigma_3 = S_v$

Very high fluid pressures develop when water-bearing magmas crystallize in a closed system, because there is a net increase in volume (Burnham, 1967, 1979, 1985). Many other investigators (e.g., Morey, 1922; Norton and Cathles, 1973; Phillips, 1973; Whitney, 1975) also have noted this fact, and it is commonly assumed that the aqueous fluids that exsolve from crystallizing magmas become concentrated beneath a carapace at the top of newly crystallized subsolidus rock. Rupturing and brecciation of the overlying rock has generally been attributed to an increase in fluid pressure as crystallization progresses, as depicted in illustrations in Burnham (1979, 1985). This model is plausible. However, it seems unlikely that individual cavities or fractures beneath a carapace in plastic rock at the top of a crystallizing igneous intrusion will maintain much vertical extent over the long periods of time required for significant quantities of fluid to exsolve by fractional crystallization. This is because deformation in response to buoyancy in large bodies of plastic rock results in the lithostatic load becoming the least principal stress ($\sigma_3 = S_v$), and fluid-filled fractures or cavities are likely to be deformed by plastic flow into relatively thin, horizontal overlapping lenses in which $P_f = S_v$ (Bailey, 1990).

Consider a hypothetical situation in which a spherically shaped cavity is filled with fluid that is significantly less dense than the surrounding rock, and that cavity is enclosed within hot rock that behaves as a hydrostatic medium. The maximum P_f within the fluid-filled sphere is controlled by the maximum pressure exerted by the surrounding rock. Thus, the maximum P_f within the cavity is at its bottom, and this pressure is transmitted upward hydraulically through the fluid. Because the pressure gradient within the less dense fluid in the cavity is greater than that within the more dense surrounding rock, this results in the P_f at the top of the cavity exceeding S_v at that depth by the factor:

$$P_f = S_v + gh(\rho_r - \rho_f) \qquad (10.2)$$

where g is local gravitational acceleration, h is the vertical dimension of the cavity, ρ_r is the average density of the plastic rock surrounding the cavity, and ρ_f is the density of the fluid in the cavity. The sphere will respond to this pressure imbalance by flattening to minimize h.

In a natural system, the cavity may be of irregular shape, or even an interconnected network of fractures. Also, in a natural system at very high temperatures (>600–700°C), plastic flow in response to a very small stress difference (Figure 10.2) will force P_f at the bottom of the cavity (or fracture network) to approach the rock pressure S_v at that particular depth.

Thus, cavities or interconnected networks of open fractures with considerable vertical extent are mechanically unstable in rock that behaves plastically. Near their tops they will tend to deform by spreading out laterally, while the lower part of the cavity is squeezed shut. Horizontal fluid-filled lenses tend to evolve, and once formed they tend to remain stationary, or to expand horizontally with addition of more fluid, perhaps coming from still crystallizing magma. Thus, fluid-filled cavities with considerable vertical extent are likely to persist for only very short periods of time in silicic plastic rock immediately

above a crystallizing magma where temperatures exceed 600°C. In this situation, an exsolved fluid makes room for itself by lifting the overlying rock, and P_f is fixed at approximately lithostatic pressure independent of the degree of fractional crystallization of the magma. In silicic rocks at 400–500°C, hydraulically interconnected cavities with somewhat greater vertical extent might persist for slightly longer periods of time than at 600°C, because at lower temperatures greater stress differences are required to initiate plastic flow at moderate to high strain rates (10.1).

Fyfe *et al.* (1978) noted that horizontal lens-like veins or 'water sills' are common in epigenetic hydrothermal deposits beneath shale or other impervious materials, and concluded that they form where lithostatic P_f is attained. Good examples of flat-lying cavities that probably formed when exsolved magmatic fluid at lithostatic pressure began to accumulate in plastic rock are the relatively flat-lying 'B' veins in the El Salvador, Chile, porphyry Cu deposit, described by Gustafson and Hunt (1975). They are prominent structures exhibiting coarse-grained quartz crystals that grow inward from the walls. Fluid inclusions in these quartz crystals have filling temperatures as high as >600°C.

As mentioned previously, a few geothermal wells have been drilled to >400°C in crystalline rocks at the bottoms of presently active hydrothermal systems, and these wells either produce CO_2-rich hypersaline brine from plastic rock at pressures considerably above hydrostatic, or are nonproductive (Fournier, 1991). The hottest of the geothermal exploration wells to date (the Kakkonda, Japan WD-1 well) went through the brittle–plastic transition at a depth of 3.1 km at about 380°C. Above the transition, the temperature profile was typical of a convecting hydrothermal system. Below the transition, the temperature increased linearly to the bottom of the well at a depth of 3.7 km, where 500°C was attained (Ikeuchi *et al.*, 1996). Although this well did not encounter any 'productive' regions in plastic rock with sufficient permeability that brine could be produced at the wellhead, it did encounter hypersaline brine that entered the well near its bottom (Ikeuchi *et al.*, 1996). The estimated salinity of this brine is 52 weight percent NaCl equivalent. Unfortunately, the WD-1 well had to be abandoned soon after completion because of discharge of H_2S-rich gas.

The linear thermal gradient within the plastic region at Kakkonda is indicative of conductive transfer of heat. It also indicates a lack of sustained vertical permeability that otherwise would permit convective overturn of the hydrothermal fluids (brine) that were found to be present in the plastic rock. Thus, it appears that upward break-throughs of fluid by hydraulic fracturing from a lower lens to an overlying lens (if they occur) are rare. However, a change in the local stress field from one in which the least principal stress is the litho-static load to one in which $S_v > \sigma_3$, would immediately destabilize the situation and promote upward hydraulic fracturing.

In contrast to the silicic systems, in basaltic or gabbroic systems seismic activity can open fractures having considerable vertical extent in rock at temperatures up to 500–600°C, because the brittle–plastic transition occurs at much higher temperatures at comparable strain rates. However, where convective flow occurs in these fractures, chemical processes will tend to close or compartmentalize them relatively quickly at temperatures >400°C. This prevents sustained inflow of meteoric water and outflow of magmatic fluid (discussed later). It is noteworthy that a deep geothermal energy-production well (NJ-11) drilled into basaltic rock at Nejavellir, Iceland, showed that a transition from hydrostatic pressure to greater than hydrostatic pressure occurred within a relatively short distance at about 370–400°C (Steingrimsson *et al.*, 1990; Fournier, 1991).

10.4.2   Significance of accumulation of fluid in plastic rock at near lithostatic P_f

Where the rinds or shells of plastic rock that surround the still molten portions of plutonic bodies are thin (there is only a very narrow zone separating magma from brittle rock in which hydrostatic pressure prevails), there is little room for the accumulation of exsolved magmatic fluids at less than magmatic temperatures in plastic rock adjacent to the intrusive. In this situation the crystallizing and cooling magmatic body would have to serve as the receptacle for its own exsolved fluids, or lose them into the surrounding, hydrostatically pressured domain. A rapid loss of magmatic fluids across the relatively thin shell of plastic rock around the intrusive can be initiated relatively easily by various mechanisms discussed later. This would probably result in a decrease in P_f at the interface between that exsolved fluid and its water-saturated parent magma. A decrease in confining P_f would, in turn, result in an increase in the rate of volatile evolution from that water-saturated magma

(Burnham, 1967, 1979), possibly leading to a volcanic eruption.

The situation is very different when a large volume of plastic rock surrounds a crystallizing magmatic body. Because of gravitational settling within the plastic material, σ_3 generally becomes equal to S_v, and aqueous-rich magmatic fluids are able to exsolve and move away from a crystallizing magma into the large body of surrounding plastic rock only after attaining $P_f = S_v$. The exsolving fluids make room for themselves by an upward bulging of the overlying rock. In this situation, the P_f of the aqueous fluid that is exsolved is fixed at about lithostatic pressure throughout the crystallization process. Thus, the presence at a shallow depth of a large body of plastic rock, in which $\sigma_3 = S_v$, prevents exsolution of magmatic fluid until $P_f = S_v$, and allows an upward moving magma to retain its dissolved water and gas to a shallower depth than might otherwise be the case. A delay in the onset of rapid vesiculation and degassing of an upward intruding magma, in turn, would influence the explosive characteristics of any ensuing volcanic activity, and also would have a profound effect on the chemical nature of the exsolved fluids (discussed below).

It follows from the above discussion that when and where $\sigma_3 = S_v$ in hot plastic rock, that rock can act as a large receptacle for the temporary storage and cooling of fluids exsolved from newly intruded batches of crystallizing magmas. These fluids, which may accumulate over a long interval of time, become available to act as mineralizing agents in the formation of epithermal mineral deposits (Fournier, 1999). The long time that it takes to develop a sizable body of plastic rock at a relatively shallow depth beneath a volcanic complex, and to accumulate a large amount of magmatic fluid in that rock can explain why the initiation of epithermal ore deposition commonly takes place only after a long interval of preceding 'barren' volcanic and plutonic activity in a given region (e.g., deposits described in Silberman, 1983; Heald *et al.*, 1987).

10.4.3   Rapid upward movement of fluid through plastic rock when $\sigma_3 < S_v$

Where a large body of plastic rock is present in which σ_3 generally equals S_v, the accumulation of fluids in horizontal lenses is likely to be disrupted only during relatively short-lived episodes of high-angle shear failure that occur at high strain rates, when the rock behaves in a brittle manner, as discussed previously. In contrast, where the intruding magmatic body is a dike emplaced into cooler rock, tensile stresses may develop within and parallel to its walls as a result of rapid inward cooling. In this situation, σ_3 in the still plastic margins of the dike rock would act perpendicular to the dike wall. In addition, the magnitude of σ_3 would likely be significantly less than that of S_v while at $>400°C$.

If dike rock in tensile stress is cool enough to behave elastically and undergo hydraulically induced failure, a magmatic fluid exsolved from the still-molten interior of the dike (or from deeper in the system) would make room for itself by first creating and then enlarging hydraulic fractures approximately perpendicular to σ_3 (parallel to the dike wall) when $P_f \approx \sigma_3 < S_v$. As soon as the vertical dimension of such a fracture attains a particular length, depending on the fracture toughness of the surrounding rock, fluid in the hydraulic fracture will start to move upward by forcibly opening the top leading edge while simultaneously closing the lower edge (Secor and Pollard, 1975; Pollard, 1976). Thus, vertical lens-like bodies of fluid are transported upward relatively quickly and efficiently until they encounter a different stress field (perhaps where $\sigma_3 < S_v$ in the overlying plastic domain, or where open shear fractures exist in the brittle domain). An additional restriction for upward flow by this mechanism is that the least principal stress gradient with depth must be greater than the fluid pressure gradient. Where the reverse is true, hydraulic fracturing may move fluid downward rather than upward (Pollard, 1976).

10.5 SELF-SEALING AT THE BRITTLE–PLASTIC INTERFACE

Given that fluids at hydrostatic pressure circulate through brittle rock where permeability is maintained by recurrent seismic activity, and that magmatic fluids tend to accumulate in plastic rock at lithostatic pressure where $\sigma_3 < S_v$, the nature of the interface between these two different hydrologic regimes is of considerable importance in modeling hydrothermal activity in sub-volcanic systems. As discussed previously, deep-drilling results suggest that the transition from hydrostatic to much greater than hydrostatic P_f commonly occurs across a relatively narrow zone a few meters to a few tens of meters wide. This zone, which occurs where temperatures are about 380–400°C, can be characterized as a self-sealed zone where permeability is decreased both by mineral deposition from

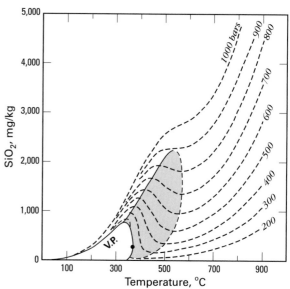

Figure 10.3. Calculated solubilities of quartz in water as a function of temperature at the vapor pressure of the solution (solid line labeled **V.P.**), and along isobars ranging from 200 to 1,000 bars (dashed lines). Stippled pattern shows a region of retrograde solubility in which the solubility of quartz decreases with increasing temperature at constant pressure (i.e., the slopes of the dashed solubility curves are negative in this region). From Fournier (1985).

circulating fluids, and by the onset of plastic flow in silicic rocks that decreases permeability and raises P_f.

In the brittle regime, the slow heating of relatively dilute solutions in contact with quartz at pressures ranging from about 34 to 900 bars results first in dissolution of quartz, and then precipitation of quartz with further heating above 340–400°C (depending on depth) (Kennedy, 1950). This is illustrated in Figure 10.3, which shows calculated solubilities of quartz at the vapor pressure of the solution and at selected isobars. It also shows the region of retrograde solubility where precipitation occurs upon heating at constant P_f (stippled area). Note in Figure 10.3 that at a P_f of 300 to 400 bars (a depth of about 3 to 4 km at hydrostatic P_f) the onset of quartz deposition with heating of a dilute solution would occur at about 370 to 390°C as it moved toward a heat source. Increasing salinity and increasing pressure (at greater depths of circulation) move the point of maximum quartz solubility to $>400°C$ (Figure 10.4(A) and (B)). However, at $>400°C$ silicic rocks appear to become sufficiently quasi-plastic for this also to be a significant factor in limiting the time that fractures are likely to remain open for fluid flow.

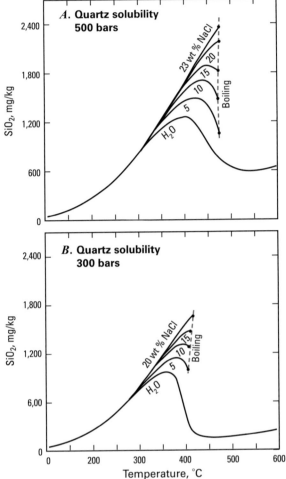

Figure 10.4. Comparison of calculated solubilities of quartz in water and in NaCl solutions at (A) 500 bars pressure, and (B) 300 bars pressure. From Fournier (1983).

Coming to the brittle–plastic transition zone from the other direction (fluid moving from the higher P_f region in plastic rock into the lower P_f region in brittle rock), there is a large potential for massive precipitation of silica, from both dilute and highly saline fluids, with rapidly decreasing pressure at >400°C (Figure 10.4(A) and (B)). Decompression is also likely to result in massive evaporative boiling of brine (discussed later), which causes additional supersaturation with respect to quartz. Silica supersaturation may become sufficiently great that amorphous silica precipitates. Evidence that precipitation of amorphous silica occurs in this environment is the common occurrence of veins filled mainly with equant, anhedral quartz grains, which are best explained as having formed by solid-state nucleation and growth within an amorphous-silica substrate

(Fournier, 1985).[2] In general, this texture seems to be indicative of a 'pressure quench' that occurs as a result of fluid moving quickly from a region in which lithostatic P_f prevails into a region in which hydrostatic P_f prevails.

In the above discussion, emphasis was placed on quartz deposition as a major factor in the formation of a self-sealed zone, because quartz veins commonly occur in hydrothermal systems. Other commonly observed vein minerals, including carbonates, sulfates, sulfides, oxides, and other silicates, also might play major roles in the development and/or re-establishment of a self-sealed zone, particularly in basaltic or gabbroic systems.

10.6 MECHANISMS FOR BREACHING THE SELF-SEALED ZONE AND DISCHARGE OF >400°C FLUID INTO COOLER ROCK

In a sub-volcanic hydrothermal–magmatic system, there is likely to be a precarious balance between processes that episodically create permeability and those that diminish permeability within the self-sealed zone separating the hydrostatic and greater than hydrostatic P_f regimes. One process that may result in a breach of the self-sealed zone is continued degassing of crystallizing magma and accumulation of the evolved fluid beneath a carapace until that carapace becomes stretched sufficiently to rupture by tensile failure, as postulated by many investigators (e.g., Norton and Cathles, 1973; Phillips, 1973; Burnham, 1979, 1985; Shinohara et al., 1995).

Upward injection of a new pulse of magma from depth is probably one of the more important mechanisms for initiating breaches of a self-sealed zone. As discussed above, this can increase the strain rate within the overlying rock to such a degree that the brittle-to-plastic transition is temporarily moved to a deeper and hotter condition. Because fluids in the initially plastic rock would be at or near lithostatic P_f ($\lambda = 1$), the change from plastic to brittle behavior with increasing strain rate would result in breaching of the self-sealed zone by shear failure in response to a small stress difference (just sufficient

---

[2] Amorphous silica is a naturally occurring oxide of silicon (SiO_2) characterized by the absence of pronounced crystalline structure. It deposits at low temperatures from silica-bearing water and is commonly present at thermal springs. Anhedral quartz refers to silica grains which, although crystalline, have a rounded or indeterminate form. Equant grains have nearly the same diameter in all directions.

to overcome the cohesive strength of the rock). A new pulse of upward magma injection also would be a source of additional release of volatiles upon crystallization. In addition, it would heat previously existing brine trapped in horizontal lenses (probably inducing boiling), and the resulting rapid expansion of that fluid could cause an additional increase in the strain rate, possibly leading to failure of the overlying rock.

Another process that might trigger rupture of the self-sealed zone is a sector collapse of a portion of an overlying volcanic edifice. This might decrease abruptly the lithostatic load experienced by a water-saturated magma sufficiently to induce rapid vesiculation. In this event, the steam available to participate in any resulting volcanic eruption would be the sum of that exsolved from the crystallizing magma and that flashed from any brine that had previously resided in plastic rock beneath the self-sealed zone. Most sector collapses are probably the result of steepening of a volcano flank when magma is injected into the volcanic edifice. However, some sector collapses might be induced wholly or in part by a steepening of the volcano flank as a result of local accumulation of brine or steam in plastic rock beneath a shallow, dome-shaped, self-sealed zone. From a practical point of view, the breaching of a self-sealed zone and discharge of high-pressure steam into the overlying brittle portion of the system is just as likely to induce a sector collapse as it is to be the result of such a collapse. This is because the pressurization of joints, fractures, and more permeable layers within the volcanic edifice allows shear failure to occur at lower stress differences than was possible previously. Conceivably, the sudden draining of a summit caldera lake could have the same effect as a sector collapse in lowering the load on a shallow magma.

A slower process that is likely to lead eventually to a breach in a self-sealed zone is extraction of heat from the system by fluid convection through overlying brittle rock at a faster rate than heat can be supplied by conduction from below through plastic rock. This would lead to a progressive downward decrease in temperature and simultaneous expansion of the permeable region as a result of volumetric contraction and tensile cracking (Lister, 1974; Norton and Knapp, 1977; Norton and Knight, 1977; Carrigan, 1986). Thus, as rock temperatures decrease, the depth at which brittle failure can occur is lowered into previously plastic rock where fluids had been trapped at lithostatic pressure. The high P_f

in the previously plastic, but now brittle, rock facilitates shear failure of the self-sealed zone in response to a relatively small stress difference (Figure 10.2).

10.7 CHEMICAL CHARACTERISTICS OF FLUIDS IN A SUB-VOLCANIC ENVIRONMENT

10.7.1 Salinity variations and phase relations of aqueous fluids at >400°C

The flashing (i.e., sudden boiling) of brine to a much larger volume of superheated steam beneath volcanoes is an important process that can have both geochemical and geodetic consequences. In addition to affecting the flux of water and gases from the system, it can cause surface deformation and possibly trigger phreatic explosions or even phreatomagmatic eruptions.

Information about phase relations in the systems NaCl–H$_2$O and NaCl–KCl–H$_2$O over widely ranging hydrothermal conditions has been used extensively to model the physical and chemical characteristics of chloride-rich brines and gases that are likely to be exsolved from crystallizing magmas (e.g., Fournier, 1987; Nash, 1976; Cunningham, 1978; Eastoe, 1978; Bloom, 1981; Cline and Bodnar, 1994). Figure 10.5 shows phase relations in the system NaCl–H$_2$O when P_f is controlled by lithostatic pressure (assumed average specific gravity of overlying rock $= 2.5\,\mathrm{g\,cm^{-3}}$). The brittle–plastic boundary is shown at about 390°C (Fournier, 1991), but could be at a higher or lower temperature for the various reasons described previously.

To assess possible consequences of upward intrusion of magma into water-rich rocks beneath a volcanic edifice, it is important to understand the significance of the various curves and phase relations shown in Figure 10.5. Consider a stovetop experiment in which a NaCl solution is placed in an open pot and heated. Eventually the temperature of the solution becomes high enough for it to boil. As boiling progresses and steam is lost, the residual solution becomes more saline, which results in boiling at slightly increasing temperatures. Eventually, solid NaCl starts to precipitate. Thereafter, the temperature and concentration of dissolved NaCl in the remaining, boiling solution do not change with continued input of heat. As soon as the last of the solution evaporates, the temperature of the salt in the 'dry' pot starts to increase and finally melting of that salt occurs at about 800°C.

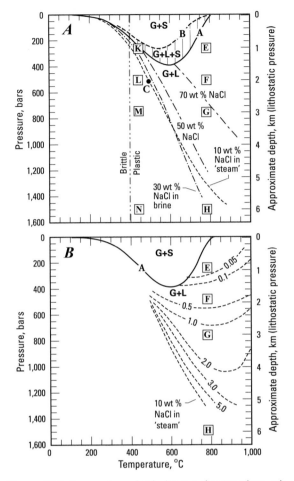

Figure 10.5. Temperature--depth diagram showing phase relations in the system NaCl--H_2O at lithostatic P_f as an analog for aqueous, chloride-rich fluids exsolved from crystallizing magma. **G** = gas, **L** = liquid, and **S** = solid salt: (A) Dot-dashed lines are contours of constant wt% NaCl dissolved in brine; short dashed line shows the boiling point curve for a 10 wt% NaCl solution at pressures and temperatures below its critical point (point **C**), and at pressures and temperatures above point **C** it shows the dew point curve (or condensation point curve) for 'steam' containing 10 wt% dissolved NaCl; curve **A** shows the three-phase boundary, **G** + **L** + **S**, for the system NaCl--H_2O; curve **B** shows the three-phase boundary, **G** + **L** + **S**, for the system NaCl--KCl--H_2O, with Na/K in solution fixed by equilibration with albite and K-feldspar at the indicated temperatures; the vertical, double dot-dashed line shows the approximate temperature of the brittle--plastic boundary when the strain rate is 10^{-14} s^{-1} (modified from figure 55.4 in Fournier, 1987)). (B) Dashed lines show dew point curves (or condensation point curves) for 'steam' containing indicated values of wt% dissolved NaCl; curve A shows the three-phase boundary, **G** + **L** + **S**, for the system NaCl--H_2O (modified from figure 55.6 in Fournier, 1987). See text for discussion.

Now, perform this experiment in a pressure cooker that allows steam to escape whenever the vapor pressure inside the cooker reaches a given value. The NaCl solution will start to boil at a higher temperature than in the open-pan experiment, and the salinity of the solution will be greater at the onset of deposition of solid salt (a higher solubility of NaCl at a higher temperature and vapor pressure). However, as soon as solid salt begins to precipitate, the temperature of the system is fixed as long as liquid (brine of fixed salinity) is present in the pressure cooker. When the last of the brine has evaporated as a result of continued heat input from the stove burner, the temperature of the precipitated salt increases while the vapor pressure of the steam in the pressure cooker remains constant (assuming steam can exit the system efficiently to compensate for the steam expansion resulting from heating). Finally, melting of solid NaCl occurs, but at a temperature less than 800°C because of the presence of steam at a moderate vapor pressure. Repeated experiments at successively higher fixed vapor pressures outline a stability field of steam (gas) plus solid NaCl. Thus, curve **A** in Figure 10.5(A) shows the position of the gas plus solid salt phase boundary in the simple system NaCl--H_2O. Note that, at vapor pressures greater than about 400 bars, there will be no precipitation of NaCl as a result of heating because the solubility of NaCl in the brine is essentially unlimited (a continuum from pure water to pure NaCl melt).

Curve **B** in Figure 10.5(A) shows the approximate position of the gas plus solid salt phase boundary when dissolved NaCl and KCl both are present and the Na/K ratio is fixed by the presence of coexisting albite and K-feldspar (Orville, 1963; Hemley, 1967). It is included to emphasize that the presence of salts other than halite (NaCl) will have a significant effect on the extent of the gas plus solid field in temperature–pressure space. Between curves **A** and **B** is a field of gas plus liquid plus solid salt (mainly halite). If only K-feldspar were present in contact with a mixed NaCl–KCl brine, the position of curve **B** would be moved to a shallower level at given temperatures. Also, in a natural system the positions of the boiling point and condensation point curves would be shifted upward or downward by addition of other dissolved constituents (e.g., downward by addition of $CaCl_2$ and non-condensable gases, such as CO_2).

In the remainder of this discussion, H_2O-rich vapor that is in equilibrium with brine at pressures >221 bars (the critical pressure of water) will be referred to as 'steam'. The quotation marks are used to emphasize that this 'steam' has a relatively high density compared with steam at 100°C and atmospheric pressure, and that it is capable of carrying significant concentrations of dissolved constituents. Quotation marks will not be used for steam at pressures less than 221 bars. Also, in the following discussion of brine and 'steam', bear in mind that non-condensable gases, such as CO_2, HCl, H_2S, and SO_2, preferentially partition into the 'steam' phase that evolves with brine at magmatic temperatures.

In Figure 10.5(A), the curves labeled 30 wt%, 50 wt%, and 70 wt% NaCl are boiling point curves for NaCl solutions having these salinities. Brine having any given salinity is not stable at P–T conditions to the right of its boiling point curve. The boiling process causes the brine to become more saline, which, in turn, results in creation of new boiling point curves, which are displaced successively to the right. Boiling may be induced by an input of heat from an external source, and/or by a lowering of P_f (perhaps during upward flow, or as a result of a breach in a self-sealed zone). Figure 10.5(A) illustrates that only highly saline brines may exist at high temperatures at relatively shallow depths in the Earth's crust. It shows that NaCl-rich magmatic fluids exsolving from a crystallizing magma will dissociate or unmix immediately to very saline brine and coexisting 'steam', and that the salinity of the resulting brine increases as depth decreases. For example, a fluid exsolving from magma at about 750 to 800°C at 1 km depth and confined by lithostatic pressure would dissociate to a brine containing so much salt dissolved in the water that it is more realistic to consider the liquid to be an NaCl melt that contains about 10–15 wt% dissolved water (Figure 10.5(A), square **E**). Figure 10.5(B) shows that the coexisting 'steam' or vapor would contain less than about 0.05 weight percent NaCl (Figure 10.5(B), square **E**).

In general, the solubility of NaCl in 'steam' at temperatures less than about 800°C decreases as the 'steam' expands with increasing temperature and with decreasing pressure. At a depth of 3 km and 750 to 800°C the brine at lithostatic P_f would contain about 65 wt% NaCl after dissociation (Figure 10.5(A), square **G**), and the coexisting 'steam' or vapor would contain about 1.25 wt% NaCl (Figure 10.5(B), square **G**). At a depth of 6 km and 750 to 800°C (square **H**) about 20 wt% NaCl can dissolve in the 'steam' at lithostatic P_f. This is likely to be greater than the salinity of aqueous fluids exsolving during the crystallization of most magmas (Hedenquist *et al.*, 1998). Therefore, in a simple NaCl–water model for the evolving magmatic fluid, no dissociation or splitting of that fluid into gas and highly saline brine would occur

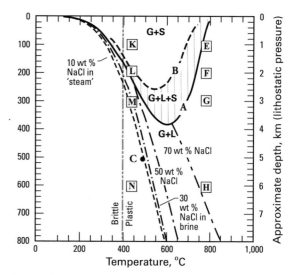

Figure 10.6. Temperature–depth diagram showing phase relations at hydrostatic P_f in the system NaCl–H_2O. Notation is the same as for Figure 10.5(A). See text for discussion.

from a silicic magma at that depth. However, in the real world, many different salts would be present in the evolving magmatic fluid, and it also would likely contain moderate to high concentrations of non-condensable gases, such as CO_2, N_2, S, and H_2S. The presence of these constituents could result in the splitting of the evolving magmatic fluid into gas and highly saline brine at a depth of about 6 km.

At depths greater than about 1.5 km, if the 'steam' and hypersaline brine remain in contact during cooling, and lithostatic P_f is maintained, some or all of the 'steam' will re-dissolve back into the brine, lowering the brine salinity while the bulk density of the brine plus 'steam' decreases as cooling progresses. Note that cooling of a fluid at lithostatic P_f at a depth of 1 km would cause halite to precipitate and persist in the temperature interval about 700 to 450°C, and then completely re-dissolve at lower temperatures (horizontal path from square **E** to square **K** in Figure 10.5(A)).

Figure 10.6 shows conditions when the saline magmatic fluids experience hydrostatic P_f during or shortly after their exsolution from the magma. As noted previously, this might be the situation when a very thin rind of chilled magmatic material separates degassing magma from the immediately surrounding cooler, brittle rock. Compared with Figure 10.5(A), in Figure 10.6 the field where solid salt is present extends about 2.5 times as deep, and brines are much more saline at comparable depths. Even at a depth of 6 km, the evolving saline magmatic fluid at 750 to 800°C will dissociate to a mixture of very saline brine containing about 70 wt% dissolved salt (circle **H** in Figure 10.6), and 'steam' containing about 0.5 to 1.0 wt% dissolved salt at 600 bars P_f (Figure 10.5(B)). Discharge of the saline magmatic fluid from the magma into cooler surrounding rock at depths less than about 3.5 to 4 km will result in precipitation of salt where hydrostatic P_f prevails.

10.7.2   Generation and behavior of HCl at high temperature and low P_f

Shinohara (1994) and Shinohara and Fujimoto (1994) showed experimentally that the concentration of HCl in a vapor in contact with brine at 600°C becomes greater as pressure is decreased from 2,000 to 400 bars, and it is well established that HCl tends to partition into a vapor phase in contact with boiling brine (Hemley *et al.*, 1992; Candela and Piccoli, 1995; Williams *et al.*, 1995). At 600°C in the system NaCl–H$_2$O, Fournier and Thompson (1993) measured an abrupt and a greater-than-predicted increase in the concentration of associated, non-reactive HCl in steam (denoted by HCl°) when NaCl began to precipitate at pressures below about 300 bars. This abrupt increase occurred because hydrolysis reactions that produced HCl° and NaOH by the reaction of NaCl with H$_2$O become important only at pressures sufficiently low for halite (and probably also NaOH) to precipitate. In addition, an order of magnitude more HCl° was obtained at comparable pressures and temperatures when quartz was present. Presumably this occurred because the quartz reacted with NaOH to form sodium silicate. Hydrolysis also occurs in the system CaCl$_2$–H$_2$O, yielding HCl° and Ca(OH)$_2$ at 380–500°C (Bischoff *et al.*, 1996). At 500°C, HCl° is generated and enters the vapor phase at pressures below about 520 bars, and at 380°C at pressures below about 230 bars. Also, in contrast to the system NaCl–H$_2$O, the hydrolysis in the system CaCl$_2$–H$_2$O proceeds without precipitation of CaCl$_2$. In natural systems, expected alteration products of hydrolysis reactions involving CaCl$_2$ are wairakite, zoisite, and prehnite (Bischoff *et al.*, 1996).

The above results indicate that significant concentrations of HCl° in the vapor or 'steam' phase are likely to be generated at >400°C whenever pressures are low enough for brine and/or 'steam' to enter the field where solid NaCl precipitates (see Figures 10.5 and 10.6).

10.7.3   Behavior of H$_2$S and SO$_2$ in sub-volcanic hydrothermal systems

With decreasing pressure, the SO$_2$/H$_2$S ratio increases in the vapor phase that exsolves from a crystallizing magma of given composition (Carroll and Webster, 1994). This is because H$_2$S reacts with water at high temperatures and low pressures, producing SO$_2$ and H$_2$ (Gerlach, 1993; Rye, 1993). However, the temperature and pressure at which H$_2$S is converted to SO$_2$ is dependent on the oxidation state of the system (Giggenbach, 1997). Qualitatively, magmas with identical initial oxidation states and dissolved sulfur concentrations will yield relatively H$_2$S-rich aqueous fluids when crystallization takes place deeper in the system (e.g., square **H** in Figure 10.5), and relatively SO$_2$-rich fluids when crystallization takes place at shallower depths (e.g., square **E** in Figure 10.5). During subsequent cooling of these magmatic fluids, the H$_2$S/SO$_2$ ratio is affected by the degree of fluid-rock interaction that occurs, staying about the same above 400°C if there is no reaction with wall rocks, or becoming H$_2$S dominant if there is reaction with the wall rocks that buffer oxygen fugacity (Ohmoto and Rye, 1979; Rye, 1993). Below about 400°C, SO$_2$ reacts with liquid water yielding H$_2$SO$_4$ and H$_2$S (Holland, 1965).

To summarize, the discharge of evolving magmatic fluid into plastic rock where it becomes trapped at lithostatic P_f favors the evolution of H$_2$S-rich and SO$_2$-poor fluids, because there is ample opportunity for SO$_2$ to be converted to H$_2$S as the fluid reacts with wall rock with declining temperature. Conversely, discharge of magmatic fluid across a narrow rind directly into brittle rock where hydrostatic P_f prevails at relatively shallow depths (especially less than about 2 km) strongly favors the formation and persistence of SO$_2$-rich 'steam'.

10.7.4   Decompression of the 'steam' phase

At temperatures >374°C and P_f >220 bars, 'steam' that forms by dissociation of a magmatic fluid to a mixture of hypersaline brine and gas would be called a supercritical fluid when decoupled or moved away from its parent brine. The types of decompression paths that may be taken by this supercritical fluid are shown in Figure 10.7, which is a graph of enthalpy versus pressure for pure water, contoured with selected isotherms. In Figure 10.7, point **B** is the critical point for pure water. Increasing

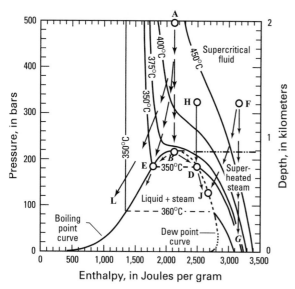

Figure 10.7. Pressure--enthalpy diagram for H_2O showing selected temperature contours. Arrows indicate adiabatic (vertical) and partly conductive (decreasing enthalpy with decreasing depth) cooling paths. See text for discussion.

concentrations of dissolved salt move the critical point to higher temperatures and higher pressures (Sourirajan and Kennedy, 1962). At pressures below the critical pressure (220 bars for pure water), there is a field of liquid water + steam. To the left of the critical point, the boiling point curve for water extends downward to lower pressures and enthalpies with decreasing temperatures. The condensation point curve (or dew point curve) extends downward to the right of the critical point. A field of liquid water is to the left of the boiling point curve, and a field of superheated steam is to the right of the dew point curve. For the probable composition of a natural fluid, the 2-phase field would be somewhat larger than the field shown in Figure 10.7.

Adiabatic decompression of 'steam' that has an initial enthalpy greater than about 2,800 Joules g^{-1} will bring that fluid into the field of superheated steam or gas (e.g., the path from point **F** toward **G** in Figure 10.7). Fluid following this path would discharge at the Earth's surface as a superheated steam vent. Along this path there is very great expansion of the fluid (resulting in underground deposition of the less volatile dissolved constituents), but no formation of a true liquid. If the 'steam' source region has near-magmatic temperatures at relatively shallow depths, the superheated steam discharged in fumaroles at the Earth's surface would likely have a relatively large SO_2/H_2S ratio, because there would be no liquid water present that otherwise would cause SO_2 to convert to H_2S and

H_2SO_4 at temperatures below about 400°C (Holland, 1965). Also, in the absence of liquid water, any HCl° or SO_2 initially present in the 'steam' remain unreactive. Therefore, there would be little acid alteration of the wallrock associated with the gaseous discharge.

Adiabatic decompression of fluids having initial enthalpies between about 2,060 and 2,800 Joules g^{-1} will result in the fluid moving into the field of superheated steam at a pressure below the critical pressure, and it will then intersect the dew point curve where liquid water is produced by condensation of previously superheated steam. An example would be point **H** in Figure 10.7 moving toward point **D**. At point **D**, the less-volatile dissolved constituents would partition into the newly forming liquid phase (point **E** in Figure 10.7) while the steam would retain most of the non-condensable gases. The newly condensed liquid water would be very acidic because HCl° carried in the superheated steam would immediately dissolve in that water and dissociate to reactive H^+ and Cl^-. In addition, SO_2 would react with the liquid water, producing H_2S and H_2SO_4 (Holland, 1965). The net result of this decompressional condensation would be a very corrosive liquid, initially unsaturated with respect to many of the minerals in the wallrock. Continued decompression increases the mass of liquid water relative to steam, but the volume occupied by the remaining steam steadily increases. This expansion of steam keeps fluid moving rapidly upward through the system and may result in brecciation at shallow levels. The steam discharged at the Earth's surface would still appear to be superheated, but SO_2 would appear to have been scrubbed from it relative to concentrations of other gases. A similar scenario would apply to fluids that initially have enthalpies greater than about 2,800 Joules g^{-1}, but cool partly by conduction during decompression so that the superheated steam intersects the dew point curve (e.g., path **F** to **J** in Figure 10.7). This would be a likely situation at the outer margins of a shallow, high-temperature system, away from the main discharge zone.

Deep, acid alteration that results from the above process can be important within active (or dormant) volcanoes, because the resulting clay minerals decrease the coefficient of rock friction and make it easier for faulting and landslides to occur.

The arrows from point **A** to point **B** in Figure 10.7 show the adiabatic decompression path of 'steam' initially at lithostatic P_f and about 425°C in plastic rock at a depth of 2 km. The 2-phase field of liquid + steam is intersected near the critical point for pure water (point **B**) at about 220 bars and 374°C. Additional decompression to about 165 bars would produce nearly equal masses of liquid water (point **E**) and steam (point **D**). Slight conductive cooling in addition to the decompressional cooling will cause supercritical fluid at point **A** to move first into the liquid only field and then intersect the 2-phase field on the boiling point curve, such as point **E** in Figure 10.7. Further decompression will follow along the well-known boiling point curve. With significant conductive cooling along the decompression path (a slower rate of flow) the initially supercritical fluid moves into and stays within the liquid only field with no boiling during ascent to the Earth's surface (e.g., the path from point **A** toward **L** in Figure 10.7). This cooling path would favor water–rock chemical equilibration along the flow path and likely result in the conversion of all SO_2 to H_2S (Ohmoto and Rye, 1979; Rye, 1993).

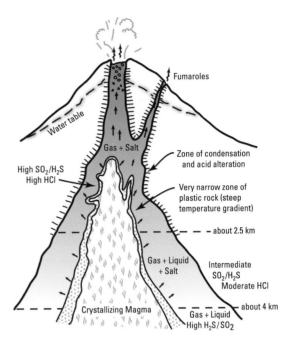

Figure 10.8. Schematic model of subsurface conditions during the early stage of development of a volcanic system. See text for discussion.

10.8 A GENERAL MODEL OF HYDROTHERMAL ACTIVITY IN A SUB-VOLCANIC ENVIRONMENT

The following model attempts to tie together the diverse physical and chemical processes that I have discussed previously. It is highly speculative and essentially the same model that I presented in Fournier (1999) where mechanisms of ore formation in silicic systems were stressed.

Figure 10.8 shows, very schematically, conditions in the early stages of development of a volcanic system. It was drawn with a silicic system in mind. Note that horizontal and vertical scales are not the same. Only a few magmatic bodies have come to rest at relatively shallow depths beneath the young and growing volcano, and they reside in a large volume of brittle rock having temperatures less than 400°C. The rind of plastic rock separating molten material from brittle rock is relatively narrow, so that gases evolved from the crystallizing magma either must reside within the parent intrusive body or escape into the hydrostatically pressured surrounding rock. It is likely that they would escape almost as soon as they evolve from the crystallizing parent magma, particularly if the intrusives are dominantly dike-like in character, allowing upward hydraulic fracturing in a tensile environment, as discussed above. There is a high potential for vesiculation within still molten portions of the

magma as a result of the effective confining P_f being less than lithostatic. A convecting hydrothermal system may or may not have developed in the surrounding rock. Any high-temperature, magmatic gases that happen to be discharged in and near the throat of the volcano are likely to be rich in SO_2 and HCl, and relatively poor in H_2S.

Figure 10.9(A) shows schematically the conditions after there have been sufficient injections of molten material to heat a significant volume of surrounding rock at relatively shallow depths to a temperature high enough (generally about 400°C in the silicic crust) for that rock to behave in a plastic manner at normal strain rates in a tectonically active region. The lake that occupies the small summit crater, and the vapor-dominated region within the volcanic pile, are common features of natural systems and affect subsurface P_f. They are not necessary for the model. Where a vapor-dominated system is present, its vertical dimension may be a few tens of meters or it may exceed 1 to 2 kilometers, extending down to the brittle–plastic transition, as may be the present situation at The Geysers geothermal field in California (Walters et al., 1992).

A depth scale is not shown in Figure 10.9(A) because the general model is applicable for tops of intrusions less than 1 km deep and deeper than 3 km, depending on many factors, including the topographic relief, the elevation of the water

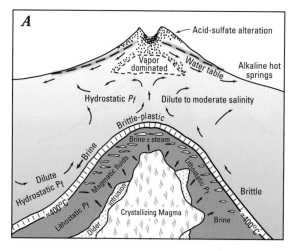

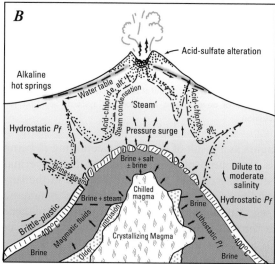

Figure 10.9. Schematic model of the transition from magmatic to epithermal conditions in a sub-volcanic environment where the tops of intruded plutons are at depths in the range 1 to 3 km. (A) The brittle-to-plastic transition occurs at about 370° to 400°C and dilute, dominantly meteoric water circulates at hydrostatic pressure in brittle rock, while highly saline, dominantly magmatic fluid at lithostatic pressure accumulates in plastic rock. (B) Episodic and temporary breaching of a normally self-sealed zone allows magmatic fluid to escape into the overlying hydrothermal system. See text for discussion.

table, the vertical extent of any vapor-dominated system, and the partial pressures of non-condensable gases that might be present. The narrow self-sealed zone develops by a combination of mineral deposition and plastic flow at the brittle–plastic boundary, as discussed previously. In Figure 10.9(A), this self-sealed zone separates a hydrostatically pressured hydrothermal system circulating through the cooler brittle rock and lithostatically pressured fluid residing in the plastic rock.

The large body of plastic rock serves as a reservoir for the accumulation of evolved magmatic fluids (hypersaline brine and 'steam') that accumulate in isolated, relatively thin, mechanically stable horizontal lenses or networks of fractures that have limited vertical interconnectivity. This accumulation may occur over long time intervals.

The water that circulates through the brittle rock at hydrostatic pressure where temperatures are less than about 360–370°C generally is relatively dilute (dominantly meteoric in origin) for continental volcanic systems. Details of the chemical variations found within that convecting hot spring system are not shown in Figure 10.9(A) because they have been discussed extensively by others (e.g., Oki and Hirano, 1970; Henley and McNabb, 1978; Henley and Ellis, 1983; Hedenquist and Lowenstern, 1994). While the self-sealed zone remains intact (Figure 10.9(A)), ongoing acid alteration is chiefly confined to near the Earth's surface where H_2S is oxidized to H_2SO_4.

In my *Economic Geology* article (Fournier, 1999, pp. 1206–1208), I described the general sequence of events that are likely to occur when there is a major rupture of the self-sealed zone with rapid escape of fluid into the normally brittle regime as follows: '*Episodically, major breaches of the self-sealed brittle–plastic transition zone occur ... [Figure 10.9(B)] ... as a result of one or more of the mechanisms discussed above. The pressure surge into the brittle region initiates faulting there, which increases permeability and facilitates the movement of the magmatic fluid upward and outward. The rapid escape of fluid from >400°C rock results in a temporary drop in P_f, vaporization of brine in confined spaces, and expansion of "steam". This expansion results in brecciation, increases the rate of expulsion of fluid across the initial brittle–plastic interface, and increases the stress difference and the strain rate within the >400°C material. This, in turn, allows a downward propagation of brittle fractures and brecciation into previously plastic rock. Rapid deposition of quartz, amorphous silica, and other minerals along channels of flow counteract the processes that increase permeability ...*'.

'*The rapid expulsion of "steam" and brine into the overlying, previously hydrostatically pressured part of the system results in displacement of the relatively dilute, meteoric-derived water upward and outward ahead of the advancing "magmatic" fluid ... [see Figure 10.9(B)]. This is an ideal environment for initiating many types of brecciation (Sillitoe, 1985). In particular, local heating and expansion of ground waters in small fractures adjacent to the main channels of fluid flow result in hydraulic fracturing and brecciation. Phreatic explosions are likely to occur above the most permeable channels of upward flow . . .'.*

Clearly, not all leakages across the brittle–plastic boundary are likely to be 'major', as described above. There probably is a wide spectrum of types of leakage, ranging from the slow diffusion of non-condensable gases (particularly H_2 and He) and minor discharges of steam through small fractures that are quickly resealed, up to large explosive events that vent to the surface. A volcanic eruption might even be initiated if there is sufficient 'unloading' of water-bearing magma.

The above model was formulated with long-lived andesitic volcanoes and large rhyolite caldera systems in mind. Rhyolitic caldera-forming systems are especially prone to producing large bodies of plastic rock at a shallow level in the crust. Geodetic movements possibly related to hydrothermal activity within large rhyolitic calderas are discussed in the next section.

In basaltic systems, the brittle–plastic transition is likely to occur at a significantly higher temperature than in the more silicic systems. However, a self-sealed zone separating fluids at hydrostatic pressure from fluids at greater than hydrostatic pressure is still likely to form at about 370–400°C because of mineral deposition, as discussed above. In this circumstance, self-sealing continues to limit the downward flow of meteoric water (or seawater) toward the heat source. It also allows P_f to become greater than hydrostatic on the high temperature side of the self-sealed zone, as was the situation encountered by the NJ-11 well in Iceland, mentioned previously. However, the maximum P_f that can be attained in the region between the self-sealed zone at about 400°C and the onset of plastic conditions at about 500–600°C may be considerably less than the lithostatic load. This is because stresses are likely to be tensile most of the time within the brittle upper parts of an oceanic shield volcano, or within a ridge-type spreading environment, such as is present in parts of Iceland.

Seismicity may repeatedly breach the self-sealed zone and open or re-open fractures that have relatively steep dips. There is uncertainty, however, about how long such fractures would remain open to hydrothermal circulation before becoming clogged by mineral deposition. Note in Figures 10.3 and 10.4 that the solubility of quartz decreases dramatically as pressure decreases at temperatures

above about 350°C, but is little affected by decreasing pressure at temperatures below about 350°C. Presumably many other vein-filling minerals will behave in a similar fashion. Fluids are likely either to be expelled upward and out of the 400–550°C part of the system by hydraulic fracturing, or to remain immobile in openings in the rock that may have considerable horizontal extent, but little vertical extent.

Tensile stresses are not likely to persist for long periods of time at >500–600°C deeper in basaltic systems where plastic conditions generally prevail. There it is likely that $\sigma_3 = S_v$ and exsolved magmatic fluids (brines and/or gases) may become trapped in horizontal lenses, as discussed previously. If such gas or brine-filled horizontal lenses were present within the plastic region, they could serve as guides for the diversion of upward moving basaltic magma into horizontal sill-like bodies. Repeated sill injections by this mechanism could contribute to the establishment and maintenance of the shallow magma chambers noted by others (Decker, 1987) at a depth of about 3 km beneath the respective summits of the Kīlauea and Mauna Loa Volcanoes.

Gases coming from regions where temperatures are less than 400°C will contain no SO_2. Gases escaping from the 400–550°C region through a breach in a self-sealed zone might begin to show traces of SO_2. However, the H_2S/SO_2 ratio likely would remain high. In contrast, gases emanating from the >600°C plastic region surrounding a shallow, upper level magma chamber would likely be rich in SO_2 relative to H_2S. If the conditions that bring about surface discharge of this gas also lead to eruption of basaltic magma from the shallow magma chamber, there would likely be a temporary excess of CO_2 and SO_2 relative to the mass of erupted basalt early in the eruption cycle.

10.9 UPLIFT AND SUBSIDENCE OF LARGE SILICIC CALDERAS

Relatively rapid rates of uplift and subsidence have been observed without accompanying volcanic eruptions within many large silicic calderas in various parts of the world (Newhall and Dzurisin, 1988; Chapter 7). It is possible that some or all of these movements could be the result of upward injection of magma and subsequent crystallization and cooling without involvement of hydrothermal fluids (all evolved magmatic gases or brines escape into the hydrostatically pressured domain without contributing to geodetic movements). It is also possible that some or all of these movements might be the result of ongoing crystallization and degassing of large batches of magma a few kilometers deep without any input of new magma from deeper in the system. This is because large silicic caldera-forming systems are ideal environments for the accumulation of evolved magmatic brines at lithostatic P_f in surrounding plastic rock, as discussed above. Thus, in a large caldera system a general uplift of the surface is expected as crystallization progresses, interspersed with deflation events whenever the evolved magmatic fluids leak from the plastic region into the hydrostatically pressured brittle region as discussed previously.

In the above scenario involving the storage within, and episodic leakage of fluids from plastic rock enveloping a crystallizing magma, the source regions for crustal inflation–deflation are approximately the same. Note, however, that the likelihood of fluid escaping by upward movement across the brittle–plastic boundary decreases as the distance between the source region for the fluid (the crystallizing magma) and the brittle–plastic boundary increases. In this situation, fluid escape becomes more likely by lateral expansion of horizontal brine-filled lenses until they breach the brittle–plastic boundary at the side of the magmatic intrusive (discussed above). A difficulty with this model is that deflation would be expected to start near the point of discharge across the brittle–plastic boundary at the side of the system and slowly work toward the center, while actual subsidence commonly seems to be symmetrically centered above the point of maximum inflation well within calderas, such as at Yellowstone (Chapter 7).

Another explanation offered here that may account for uplift and subsidence events that have been observed within some large silicic calderas involves episodic injection of magma into brine-filled horizontal structures in plastic rock. Consider a situation in which brine (or brine plus 'steam' at shallower depths) has accumulated in horizontal lenses deep within plastic silicic rock. Magma moving upward that encounters such a lens is likely to be diverted into the lens, forming a sill-like body. Heating of the fluid initially in the lens by the intruding magma can result in vigorous boiling, depending on the depth (lithostatic pressure), the salinity of the fluid in the lens, and the temperature of the intruding magma (Figure 10.5). If the action takes place deep within the plastic region, the least principal stress may remain equal to the lithostatic

load so that little brine or 'steam' escapes upward during the sill emplacement. Thus, inflation measured at the Earth's surface would be the result of a combination of (a) intrusion of sill-like bodies of magma (and evolution of its volatile components as it crystallizes), (b) thermal expansion of brine initially in the lens in plastic rock, and (c) boiling of that brine resulting in the formation of a much less dense 'steam' phase that also remains trapped in plastic rock. Subsequent cooling would result in solidification and contraction of the magma, and, possibly more importantly, condensation of all or most of the 'steam' back into brine with a corresponding volume decrease. The advantages of this mechanism are (a) the sources of surface inflation and deflation occupy essentially the same subsurface volumes, (b) there is a broad area of surface deformation, and (c) no leakage of hydrothermal fluid from the plastic into the brittle domain is required.

The above mechanism, which accounts for episodes of inflation–deflation by diverting upward intruding magmas into horizontal open fractures filled with brine, works best if the intruding magma is basalt, particularly at depths greater than about 5 to 6 km. This is because an intruding silicic magma, with a temperature in the range 700–800°C, would not be likely to cause much vaporization or dissociation of fluids that already are in the range 700–800°C, with the salinities expected at >5 km (Figure 10.5). An intruding basaltic magma at a depth of 6–8 km, having a temperature of 1,200°C or higher, would likely cause considerable dissociation of aqueous fluid into highly saline brine and considerably less dense 'steam'. As the thermal front moves upward and downward away from the sill, brine in other adjacent lenses also may boil or dissociate, lifting the overlying rock still farther.

Viscosity is another factor that is likely to be of importance; basaltic magma, which is much less viscous than rhyolitic magma, is more likely than rhyolite to move farther and more quickly into an open horizontal crack. The pressure surge resulting from vaporization of fluid in the lens also may result in a significant extension of the fluid-filled lens at the time of the intrusion. There is the possibility, however, that horizontal extension of such a lens might progress to and through the brittle–plastic transition at the side of the system, allowing depressurization by leakage into the brittle domain. Recent deflation followed by uplift of Yellowstone Caldera, measured by satellite radar interferometry, and confirmed by leveling surveys (Chapter 7; Wicks *et al.*,

1998; Dzurisin *et al.*, 1999) can be modeled nicely by injection of a basaltic sill at a depth of 8.5 ± 4 km, and may be an example of the process described in the preceding paragraphs.

Upward geodetic movements that are triggered by injection of magma sills with accompanying boiling of brine deep in silicic calderas require that there be a net uplift at the end of each cycle of inflation–deflation. Thus, over hundreds or thousands of years a general trend of uplift would be expected with episodic downturns between magmatic injections with relatively little leakage (or only slow and non-catastrophic leakage) of magmatic fluid into the hydrostatic domain. This is because the produced 'steam' re-condenses in the residual brine. If this model were operative, there would appear to be relatively little chance of triggering a phreatic volcanic eruption as a result of the inflation–deflation processes. On the other hand, if the inflation–deflation is mainly the result of hydrothermal processes in which deflation is the result of relatively rapid leakage of magmatic fluid from plastic rock into the brittle regime, there is an increased risk of the process 'running away' and triggering a major phreatic event, or even a volcanic eruption. Of course, within deforming silicic caldera systems, it is possible (perhaps even probable) that different processes may be occurring simultaneously at different levels, or that one process may be dominant over one period of time and another over a different span of time.

10.10 CONCLUSIONS

Hydrothermal processes can contribute to the deformation of volcanoes in a variety of ways, including uplift if fluids released from crystallizing magma are trapped underground at lithostatic pressure, or subsidence if these fluids escape into the hydrostatically pressured part of the system. Both the development of chemically self-sealed zones where subsurface temperatures are about 370–400°C, and the distribution of rock that behaves plastically beneath a volcano apparently play key roles in these processes.

In tectonically active regions, plasticity appears to become important in silicic volcanic systems at about 400°C, and in basaltic systems at about 500–600°C. The thickness of a plastic region around a crystallizing magma is of particular importance, because a large volume of plastic rock can serve as a receptacle for the storage at lithostatic pressure of brine and gas that separate from that magma over a long period of time. Where only a thin rind of plastic rock surrounds a crystallizing magma, evolving magmatic fluids are likely to escape into the hydrostatically pressured part of the system as crystallization progresses. There would likely be simultaneous rapid vesiculation in the magma because of the drop in confining fluid pressure.

Within plastic rock, brine and gas reside in horizontal lenses that may have considerable lateral, but little vertical interconnectivity. Upward-flowing magma that encounters such a lens might be diverted along the plane of weakness to form a sill. In such an event, there would be simultaneous flashing of brine in the lens to steam. Uplift of the surface results both from injection of the sill and flashing of brine to steam in a confined space. This is followed by partial deflation as the steam re-condenses back into the brine as the thermal anomaly dissipates by conduction. Where crystallizing magma that serves as a persistent source of brine and gas is present underground, episodic breaching and resealing of the self-sealed zone at the interface between the lithostatically and hydrostatically pressured parts of the hydrothermal system also can account for some episodic uplift and subsidence of volcanic systems.

Movement of hydrothermal fluids from a lithostatically pressured environment to a hydrostatically pressured environment will result in the flashing of brine to a much larger volume of superheated steam, or even a boiling dry of the brine with simultaneous precipitation of salts. The superheated steam that is evolved from an environment where solid Cl-rich salts have precipitated will be rich in HCl. Acid alteration will occur wherever this steam condenses. The development of clay minerals along permeable structures underground decreases coefficients of rock friction, thereby facilitating landslides and movements along faults.

CHAPTER 11

Challenges and opportunities for the 21st century

Daniel Dzurisin

Preceding chapters have described various geodetic and modeling tools that can be used to monitor volcano deformation, discussed examples of how those tools have been used to infer what might be happening beneath a few well studied volcanoes, and explored some emerging links between geodesy and other disciplines in volcanology. In this final chapter, I take a step back to consider some basic questions about the current state of volcano geodesy and try to glimpse its future – a future bright with the promise of real time, global surveillance but also clouded by increasing risk as populations continue to encroach on many of the world's dangerous volcanoes.

11.1 THE INTRUSION PROCESS: A COMPLICATED BUSINESS

The fact that volcanoes exist at all is really quite remarkable. Before magma pours onto the surface or explodes into the atmosphere, it must make its way through at least tens of kilometers of solid lithosphere, in some cases only after rising buoyantly through hundreds of kilometers of mantle. During its long ascent, magma cools, crystallizes, and interacts in various ways with host rock and groundwater. The fact that portions of the mantle melt and rise of their own accord to erupt at the surface strikes me as truly extraordinary.

What is *not* surprising is that the intrusion process profoundly affects both the rising magma and surrounding rock. For example, the confining pressure acting on a body of magma that rises from the base of Earth's crust (~30 km depth) to its surface decreases by about 4 orders of magnitude, from approximately 10 kbar to 1 bar. Volatiles such as water, carbon dioxide, and sulfur dioxide, which were contained in the magma at depth, exsolve as the magma decompresses and partially crystallizes. The net effect is to increase the viscosity of the magma by several orders of magnitude (Lejeune *et al.*, 1999). Until gas bubbles escape, they lower the average density of the magma and hasten its rise. When gases *do* escape, assuming they are relatively insoluble in groundwater and, therefore, can reach the surface (not true for sulfur dioxide, which can be effectively 'scrubbed' by solution in water), they provide an effective means to monitor the intrusion process. By measuring the types and quantities of magmatic gases released as a function of time, scientists can infer the presence of magma recently intruded into the upper crust, its intrusion rate, and approximate depth (e.g., Gerlach *et al.*, 1996).

The effect of the intrusion process on surrounding rock is equally profound. To accommodate intruding magma, the host rock *must* deform. To appreciate the inevitability of this requirement, consider that 1 km^3 of magma – a volume erupted somewhere on Earth's land surface about once every 10 years (Simkin and Siebert, 1994, p. 29) – would occupy a sphere of radius 620 m or a cylinder of radius 100 m extending from the surface to a depth of about 32 km. Creating space for either body or moving it through the crust is a prodigious task that cannot be accomplished without deforming the host rock.[1] If the intrusion

---

[1] The magnitude of surface deformation could be mitigated in several ways. For example, a large volume of magma might be stored at considerable depth beneath a volcano and then delivered to the surface through a relatively narrow conduit. Or a magma body might develop partly *in situ* by partial melting of host rock, so the net volume change in the vicinity of the body is less than the total volume of magma produced. Regardless of the details, at least *some* deformation of the host rock is inevitable, and models generally agree that the effect should be measurable at the surface in many cases.

rate is low enough or the temperature of the host rock is high enough, elastic or ductile deformation occurs and the intrusion process is virtually aseismic. At higher intrusion rates or lower temperatures, brittle failure of the host rock produces numerous earthquakes. It is not uncommon for small earthquakes (most of Richter magnitude $M \leq 3$, but a few as large as $M \sim 5$) to occur at rates exceeding 1 event per minute for hours or days during a rapid, shallow intrusion.

II.2 STRENGTHS AND WEAKNESSES OF GEODETIC MONITORING

The brittle component of volcano deformation (i.e., the part that produces earthquakes), has traditionally been the foundation for most volcano-monitoring strategies. As a monitoring target, earthquakes have several advantages over ground deformation. Earthquakes transmit seismic waves that can travel through the solid Earth, so even small events are detectable many kilometers from their source with relatively inexpensive seismometers. Deformation, in contrast, is a relatively local phenomenon that, until recently, has been measurable only close to its source or, farther away, only with sophisticated, expensive instrumentation. Also, the inversion of earthquake arrival-time data is a relatively straightforward, 3-D geometry problem that can be solved one event at a time for essentially two parameters: earthquake magnitude and location (ignoring path effects). Geodetic inversions generally require data spanning a longer time interval and with greater spatial density to determine the source location and also to distinguish among possible source geometries (e.g., sphere, ellipsoid, dike, fault dislocation). In the absence of spatially dense geodetic data, for example, an *inflating* dike might not be distinguishable from a *deflating* point source, because both can produce relative subsidence in a localized area. Measurements of absolute displacements, such as Global Positioning System (GPS) data, can distinguish between inflation and deflation sources but not among various source geometries unless the spatial density of the data is adequate. All of these factors combine to make effective geodetic monitoring of volcanoes more difficult than seismic monitoring.

On the other hand, earthquake monitoring alone has some important limitations at volcanoes. Because earthquakes are local phenomena, no single event provides information about the entire source region. Even the spatial distribution of earthquakes beneath a volcano is inherently ambiguous. Do earthquakes indicate where an intrusion *is* (i.e., where deformation of host rock is greatest), or where it *is not* (because magma itself does not fracture and therefore is aseismic)? Does an aseismic region beneath a volcano represent magma or hot, 'solid' rock (i.e., high-temperature, ductile rock that deforms aseismically under low strain rates)? Might magma be moving aseismically by invading pre-existing fractures or intruding hot, ductile host rock?

These ambiguities can sometimes be resolved with adequate geodetic monitoring. The surface-displacement field integrates the effect of the entire intrusion, even if the deformation is aseismic. Although steady-state magma movement through established conduits can occur without substantial deformation (e.g., at open-vent systems and during long-lived eruptions), fresh intrusions that precede most eruptions almost inevitably cause surface deformation that fingerprints the intrusion if the surface displacement field can be adequately characterized. This is especially true for intrusions within a few kilometers of the surface, which produce distinctive deformation patterns that reflect the location, shape, and strength[2] of the source (Dvorak and Dzurisin, 1997). Deeper sources produce less distinctive deformation patterns that contain less information about the shape of the source, but nonetheless reveal its location and strength (Dieterich and Decker, 1975; McTigue, 1987).

No single approach to volcano monitoring (e.g., seismology, geodesy, gas geochemistry) can answer all the questions posed by scientists and residents near a restless volcano. However, if the surface-deformation field can be measured with sufficient precision and spatial coverage, it can provide powerful constraints on the characteristics of a deforming magma body. Volcano geodesy, by virtue of the high information content of a well-characterized deformation field and the comparatively nascent state of deformation monitoring and modeling techniques, holds a large untapped potential for studying volcanic processes.

---

[2] Source strength is directly proportional to volume change, which depends on the pressure change of the source and on the elastic properties of the surrounding medium (typically expressed in terms of the shear modulus and Poisson's ratio) (Chapter 8).

II.3 WHY IS VOLCANO DEFORMATION SUCH AN ELUSIVE TARGET?

Deformation of host rock is virtually certain during magmatic intrusions, and models indicate that most intrusions are large enough to deform the ground surface by measurable amounts. Yet, measurements of deformation at volcanoes of intermediate composition (e.g., andesite–dacite stratovolcanoes or domes) are relatively rare, except during intense unrest when magma is already close to the surface or after an eruption has begun.[3] This paradox is partly explained by a combination of difficult logistics and complexity in the displacement field, both in space and time. By difficult logistics I mean that long repose intervals, difficult access, steep terrain, and limited human and financial resources encourage monitoring lapses during which any precursory deformation might escape detection. To appreciate the magnitude of the second factor, consider the complexity inherent in even the simplest model for volcano deformation: a spherical source in an elastic half-space that changes in volume, but not location, as a function of time (Mogi, 1958).

II.3.I This should be easy!

Predicted surface displacements for a magma body at depth $d = 5$ km that inflates by 10^7 m^3 are shown in Figure 11.1.[4] Explosive eruptions that produce a comparable volume of tephra, which are considered to be moderate to large in size and are capable of injecting ash into the stratosphere, are assigned a Volcanic Explosivity Index (VEI) of 3.[5] On average,

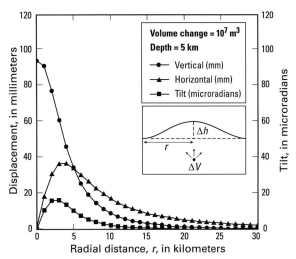

Figure 11.1. Predicted surface displacements and tilt as a function of radial distance r caused by a volume increase $\Delta V = 10^7$ m^3 in a spherical source buried 5 km deep in an elastic half-space (Mogi, 1958). According to the model, displacements and tilt scale linearly with volume change. So, for example, a volume change of 10^8 m^3 would produce displacements and tilts a factor of 10 greater than shown here.

eruptions of this magnitude occur on Earth several times per year (Simkin and Siebert, 1994, pp. 23–30). The maximum predicted surface displacements (vertical, horizontal, and tilt = ΔZ_{max}, ΔX_{max}, and $\Delta \tau_{max}$, respectively) are: $\Delta Z_{max} = 95$ mm at distance $r = 0$ km (directly above the source), $\Delta X_{max} = 37$ mm at $r = 3.5$ km, and $\Delta \tau_{max} = 16 \,\mu$rad at $r = 2.5$ km. According to the model, these results scale linearly with volume change. So a volume change of 10^9 m^3 (corresponding to a VEI 5 eruption) would produce maximum displacements and tilts 100 times larger than those shown in Figure 11.1, and a volume change of 10^6 m^3 (VEI 2) would produce displacements and tilts 10 times smaller.[6]

Taken at face value, these results seem encouraging. Even a small volume change of 10^6 m^3 at depth $d = 5$ km might be detectable with state-of-the-art geodetic techniques ($\Delta Z_{max} = 9.5$ mm, $\Delta X_{max} = 3.7$ mm, and $\Delta \tau_{max} = 1.6 \,\mu$rad). For volume changes of 10^7 m^3 or larger, which correspond to volumes erupted during most eruptions that pose serious hazards, surface deformation should be measurable even with classical geodetic instruments that are widely used for volcano monitoring (e.g.,

---

[3] Although relatively rare, there have been several recent discoveries using interferometric synthetic-aperture radar (InSAR) of deformation at otherwise quiescent andesite–dacite volcanoes, including Mount Peulik, Alaska (Lu *et al.*, 2002a) and Three Sisters, Oregon (Wicks *et al.*, 2002a,b) (Chapter 5). Deformation at active basaltic shields such as Kīlauea, Hawai'i, and at restless silicic calderas such as Yellowstone, Wyoming, and Long Valley, California, is relatively common.

[4] The magnitude of surface displacements depends on the change in source volume (or pressure) and on the elastic properties of the surrounding crust (e.g., shear modulus and Poisson's ratio; Chapter 8). In general, these parameters are poorly constrained near volcanoes. The results given here are based on typical values and, therefore, might differ from otherwise comparable results from other point-source models.

[5] VEI is a measure of the size of volcanic eruptions akin to the Richter magnitude scale for earthquakes. The VEI is a 0-to-8 index of increasing explosivity, each interval representing an increase of about a factor of ten. It combines total volume of explosive products, eruptive cloud height, descriptive terms, and other measures (Newhall and Self, 1982).

[6] All calculations assume Poisson's ratio $\nu = 0.25$, which means the uplift volume is 1.5 times the injected volume. For a discussion of the dependence of surface displacements on ν, see Section 8.4.1 or Delaney and McTigue (1994).

electronic distance meter (EDM), theodolite, bubble tiltmeters).

However, at least three factors combine to make such measurements more difficult in practice than implied by the model results. First, the spatial resolution of most geodetic datasets at volcanoes is inadequate for the task at hand. This shortcoming can be caused by difficult access (e.g., remote location, rugged terrain, high elevation, dense vegetation), unacceptable hazards to personnel in proximal areas, or limited resources (e.g., time, equipment, money, personnel). Second, the temporal resolution of most geodetic datasets is also inadequate, for some of the same reasons listed above and also because monitoring programs are difficult to fund and maintain during typically long repose intervals when little seems to be happening at the volcano. Politicians and funding agencies tend to be short-sighted when balancing future hazards mitigation against present-day 'necessities'. Finally, some geodetic techniques used to monitor volcanoes are not sufficiently precise to detect small surface changes that might precede many eruptions. In short, complete characterization of volcano deformation requires that measurements be made in the right place, at the right time, and with adequate precision – a difficult challenge using classical geodetic techniques.

11.3.2 Lessons from Mount St. Helens I: 1980

All three limiting factors played a role at Mount St. Helens prior to its catastrophic eruption on 18 May 1980. A brief review of that experience illustrates the difficulties faced by scientists trying to measure volcano deformation during a crisis, and also the potential of new space-based geodetic techniques (GPS and InSAR) for characterizing deformation in space and time.

In 1972, USGS geologist Donald A. (Don) Swanson first measured a 7.6-km EDM line from Smith Creek Butte to East Dome; it was the 115th year of quiescence at Mount St. Helens since the end of its previous eruption in 1857. The line was next measured on 10 April 1980, three weeks after the onset of an intense swarm of earthquakes beneath the volcano. In hindsight, the earthquakes were caused by the intrusion of a cryptodome, which caused the volcano's north flank to bulge outward at rates of 1.5–2.5 m day^{-1} before the edifice failed catastrophically on May 18 to produce the largest landslide in recorded history (Lipman *et al.*, 1981a) (Figure 11.2).

Figure 11.2. Mount St. Helens as it appeared 16 years after the catastrophic events of 18 May 1980. Hummocky deposit in the foreground is from a debris avalanche that truncated the volcano's north flank and summit area, choking the upper reach of the North Fork Toutle River and spawning a lahar that modified more than 120 km of channel in the Toutle, Cowlitz, and Columbia Rivers (Janda *et al.*, 1981). View is from the northwest. USGS photograph by David E. Wieprecht, June 1996.

In the preceding 8 years, the line between Smith Creek Butte and East Dome had contracted just 16 mm (i.e., less than the uncertainty in the measurement). The line was re-measured on 25 April 1980 (+15 mm), and again after the 18 May 1980 eruption (−18 mm). At the same time, repeated surveys of a local geodetic network (theodolite and EDM) established at the volcano in April 1980, revealed intense deformation of the volcano's north flank but only small changes elsewhere on the edifice (as much as 2.5 m day^{-1} versus a few mm day^{-1}). The authors concluded: '... *except for the bulge (on the volcano's north flank), Mount St. Helens was geodetically stable during this period*' (Lipman *et al.*, 1981a, p. 151).

The virtual absence of measured ground deformation beyond a 1.5 km × 2 km bulge on the north flank of Mount St. Helens during the 2 months preceding the 18 May 1980 eruption might be partly attributable to each of the three factors listed above. First, consider the EDM results for the line from Smith Creek Butte to East Dome. The essentially null result from 1972 to 10 April 1980, is the only geodetic measurement that began before the onset of intense, shallow seismicity on 20 March 1980. With the benefit of hindsight, let us first consider whether this measurement was made in the 'right place'.

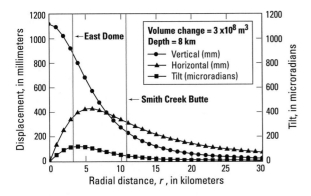

Figure 11.3. Surface displacements and tilt predicted by an elastic, point-source model (Mogi, 1958) for a source-volume increase of $3 \times 10^8 \, m^3$ at 8 km depth. These values correspond to the volume of magma erupted on 18 May 1980, and to the pre-March 1980 magma-storage depth, respectively, at Mount St. Helens. Vertical lines represent EDM stations East Dome and Smith Creek Butte, at radial distances of 3 km and 10.5 km from the volcano's summit. The line length between those stations was measured in 1972 and several times in 1980, with null results. The model predicts relatively large but nearly equal horizontal displacements at both stations, and, therefore, very little relative movement between them.

To a large degree, the right place is determined by the depth of the deformation source. Scandone and Malone (1985) concluded that the explosive eruptions at Mount St. Helens in 1980 were fed from a magma reservoir centered 9 km beneath the volcano (based on a model of surface deformation data for June–November 1980, which show cumulative subsidence), extending from 7 to 14 km (based on the distribution of earthquake hypocenters associated with the eruptions). Rutherford et al. (1985) based their independent estimate of the source depth, 7.2 ± 1 km, on the mineral phase assemblage found in May 18 pumice and on experimentally determined phase equilibria. Here, I assume a source depth of 8 km (Figure 11.3) and an erupted volume of $0.3 \, km^3$ (Pallister et al., 1992, p. 129). This source depth is appropriate for some unknown period before 20 March 1980, when magma stored at ~8 km depth began intruding toward the surface. Uncertainties in the erupted volume and the ratio of erupted to intruded volume correspond to a model uncertainty of roughly a factor of two.

Surface displacements predicted by the same elastic, point-source model discussed above for the case of $\Delta V = 0.3 \, km^3$ and $d = 8$ km are shown in Figure 11.3, together with the relative locations of East Dome (radial distance $r = 3$ km)

and Smith Creek Butte ($r = 10.5$ km). At those locations, the model predicts large, but essentially equal, horizontal displacements (344 mm and 327 mm, respectively). As a result, the distance between East Dome and Smith Creek Butte would contract by only 17 mm, which is very close to the observed value (16 mm) and less than the uncertainty in the EDM measurement. So, in hindsight, the EDM measurements made in 1972 and April 1980 might have been relatively insensitive to volume changes in the 8 km deep magma body that fed the 18 May 1980, eruption. As a result, substantial ground displacements could have gone undetected.

Of course, it is also possible that such large but similar displacements did not occur (i.e., the null EDM results taken at face value), and the source of the discrepancy between EDM measurements and modeling results is an inappropriate model (Section 11.3.3). This might be the case if most of the magma that erupted on 18 May 1980 accumulated in the crust before the first EDM measurements in 1972 and, therefore, the volume change used in the model is too large (see below). Alternatively, the deformation field might not have been radially symmetric, perhaps because the material properties of the crust beneath the volcano are heterogeneous. On the other hand, the discrepancy is not easily explained by inappropriate source geometry or Poisson's ratio, because these factors have relatively small effects on the surface-displacement field except for very shallow sources (Chapter 8; McTigue, 1987; Delaney and McTigue, 1994). I am not trying to defend the use of the Mogi model in this case, or to challenge the interpretation of Lipman et al. (1981a) that Mount St. Helens was stable from 1972 through April 1980, except for the growth of the north-flank bulge starting in March 1980. My point is just that there is another plausible explanation for the absence of relative motion between Smith Creek Butte and East Dome (i.e., both stations might have experienced large, but nearly equal, absolute displacements, which is consistent with a source at 8 km depth as indicated by the seismic and petrographic results).

Before moving on to other factors that might have influenced the EDM results at Mount St. Helens, consider what other guidance can be taken from the model results in Figure 11.3. Vertical displacements exceed horizontal displacements by a factor of three or more in proximal areas ($r < 3$ km). This suggests that height-change measurements near the summits of volcanoes should receive high priority, at least during periods of quiescence. Unfortunately, many summit areas are inhospitable to both human observers and *in situ* monitoring instruments, which makes such measurements difficult. For measurements of horizontal displacements, an implication of the model is that both proximal ($r < 3$ km) and distal ($r > 20$ km) stations are essential to adequately measure deep-seated deformation. Investing in such an expansive network might seem ill advised with limited resources, but it offers the potential to detect deep-seated magmatic deformation and thus to provide longer term warning of impending volcanic activity than is otherwise possible. For this application, GPS in both survey and continuous mode is especially promising and is being used at a growing number of volcanoes (Chapter 4).

A final observation from Figure 11.3 is that ground tilt and horizontal displacement measurements are inherently ambiguous, because the same changes occur at different distances from the source. Consider the observational puzzle posed by this double-valued relationship. Imagine that we measure 15 ± 1 cm of contraction on a radial EDM line with endpoints 1 km and 5 km from the summit. In the absence of other information, what can we conclude about the state of the volcano? Using a point-source model, we find that our result is consistent with 0.1 km^3 of inflation at 3 km depth. On the other hand, the model predicts the same result for 0.1 km^3 of *deflation* at 5 km depth. How can this be? In the first case, both stations move away from the summit but the closer one moves more, whereas in the second case both stations move toward the summit but the farther one moves more. Knowing that the volcano is *either* inflating *or* deflating is not very helpful, especially if lives or the local economy depend on the answer!

This example illustrates why EDM networks must be both spatially dense and extensive to adequately characterize volcano deformation, even in this simple case and neglecting such complications as measurement errors and unstable bench marks. Two important differences between EDM and GPS measurements are worth noting here. With just two GPS stations, it is possible to make an unambiguous determination of whether a volcano is inflating or deflating. Because GPS is not limited to lines-of-sight, we can directly measure the distance between two points on opposite sides of the volcano. If this distance increases, the volcano is inflating. If it decreases, the volcano is deflating. This assumes that the top of the source is not too

shallow, which would complicate the shape of the displacement field in proximal areas. GPS also has the advantage over EDM of measuring all three components of the displacement vector. If vertical movements are large enough to measure with GPS, they likewise provide unambiguous information (i.e., up = inflation, down = deflation). Though far from ideal, two GPS stations on opposite sides of a volcano are a relatively efficient and cost-effective means to monitor volcano deformation. A third station near the summit, where the largest vertical movements are likely to occur, is a worthwhile addition if logistical conditions permit. At the start of the 21st century, what was unthinkable in the 1970s and 1980s (i.e., monitoring 3-D ground displacements near the summit of a restless volcano continuously in near-real time) is now merely difficult. Within the next decade, GPS monitoring of many of the world's volcanoes seems destined to become routine.

Now let us return to Mount St. Helens and the null EDM results between East Dome and Smith Creek Butte from 1972 to April 1980. As we have seen, the locations of those stations made them insensitive to changes that might have occurred in an 8 km deep magma body beneath the volcano. There are at least two other explanations for the null results, which correspond to the other two complicating factors mentioned above. First, consider the possibility that the EDM measurements were made at the wrong time. If, for example, the 0.3 km^3 of magma erupted on 18 May 1980, accumulated at a constant rate from 1857 to 1980, then 93% of it had already arrived at 8 km depth before the first EDM measurement was made in 1972. The arrival of the final 7% between 1972 and 1980 could easily have gone undetected, especially if the Smith Creek Butte–East Dome line was insensitive to volume changes at 8 km depth.

A third factor that might have played a role in the null results at Mount St. Helens is the detection threshold for sparse (in both time and space) EDM measurements relative to the small amount of ground deformation that might have occurred from 1972 to April 1980. Lipman *et al.* (1981a, p. 148 and p. 151) estimated the uncertainty of their EDM measurements at Mount St. Helens to be ± 10 mm over distances of 2–4 km, and concluded that line-length changes on the order of $+16$ mm to -18 mm measured on the 7.5-km line from Smith Creek Butte to East Dome were statistically insignificant. As shown above, a Mogi-type model predicts only -17 mm of relative horizontal displacement between Smith Creek Butte and East Dome for the case of $0.3\,\text{km}^3$ of magma intruded at $8\,\text{km}$ depth. That means that, even if the model were exact and $17\,\text{mm}$ of contraction actually occurred between 1972 and April 1980, it might have gone undetected owing to uncertainty in the EDM measurements. Of course, the relative motion would have been even smaller if some of the magma that erupted on 18 May 1980, arrived at $8\,\text{km}$ depth prior to 1972.

A more effective approach might have been to focus on vertical deformation, even though vertical-angle measurements by theodolite are inherently less precise than EDM measurements over the distances involved at Mount St. Helens. For example, Lipman et al. (1981a, p. 148) estimated the uncertainty of their theodolite measurements to be ± 5–10 seconds of arc, which corresponds to ± 18–$36\,\text{cm}$ over the $7.5\,\text{km}$ distance from Smith Creek Butte to East Dome. The same Mogi model that predicts only $17\,\text{mm}$ of relative horizontal movement between those points predicts about $670\,\text{mm}$ of relative vertical movement (Figure 11.3). A change of that magnitude might have been resolvable by theodolite in 1980, and now it would easily be measurable with GPS or InSAR.

11.3.3   Lessons from Yellowstone

The importance of making frequent geodetic measurements, even during periods of apparent quiescence, is further illustrated by results of repeated leveling surveys across Yellowstone Caldera from 1976 to 1998 (Figure 11.4). Successive surveys revealed uplift from 1976 to 1984, subsidence from 1985 to 1995, and renewed uplift from 1995 to 1998, all with respect to bench mark 36 MDC 1976, a datum on the southeast caldera rim (Dzurisin et al., 1994; Dzurisin et al., 1999). In each case, the largest displacements occurred near DA3 1934, in the eastern part of the caldera near LeHardys Rapids. The net height change at DA3 1934 from 1976 to 1998 was only $10 \pm 5\,\text{mm}$, which corresponds to an average uplift rate of $0.5 \pm 1\,\text{mm}\,\text{yr}^{-1}$. On the other hand, intervening surveys reveal that DA3 1934: (1) rose $176 \pm 5\,\text{mm}$ from 1976 to 1984 ($22 \pm 1\,\text{mm}\,\text{yr}^{-1}$); (2) moved very little from 1984 to 1985 ($-2 \pm 1\,\text{mm}\,\text{yr}^{-1}$); (3) subsided $-188 \pm 5\,\text{mm}$ from 1985 to 1995 ($-19 \pm 1\,\text{mm}\,\text{yr}^{-1}$) at rates that varied from year to year between $-9 \pm 1\,\text{mm}\,\text{yr}^{-1}$ and $-35 \pm 1\,\text{mm}\,\text{yr}^{-1}$; and (4) rose $24 \pm 1\,\text{mm}$ from 1995 to 1998 ($8 \pm 1\,\text{mm}\,\text{yr}^{-1}$).

Clearly, infrequent surveys can produce a distorted view of actual ground motion if the displace-

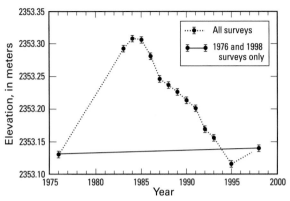

Figure 11.4. Elevation as a function of time at bench mark DA3 1934, near the center of Yellowstone Caldera, measured by repeated leveling surveys (Dzurisin et al., 1999). Changes are relative to bench mark 36 MDC 1976, located about 23 km southeast of DA3 1934 at Lake Butte, along the southeast caldera rim. Maximum uplift and subsidence along a level line across the caldera consistently occurred near LeHardys Rapids, within ~I km of DA3 1934. Solid line represents the displacement history that would have been inferred from just the two surveys in 1976 and 1998. Dotted line represents a more accurate, but still incomplete, history based on results of 14 surveys conducted during that interval.

ment rate fluctuates appreciably with time. The EDM measurements at Mount St. Helens in 1972 and April 1980, for example, could be analogous to the Yellowstone leveling results for 1984–1985 or 1976–1998 (i.e., not optimally timed to capture deformation that might have occurred between 1857 and 1980). A lesson to be learned from both experiences is that frequent, preferably continuous geodetic measurements are essential to ensure that important information is not missed, especially during periods of apparent quiescence.

Another lesson, this one from the Yellowstone InSAR results described in Chapter 5, is that information also can be lost as a result of inadequate spatial coverage. No matter how often the leveling traverse across the eastern part of the caldera had been measured, we could not have learned that the western and northern parts were behaving differently without also leveling there. Annual surveys of such a large area were not feasible, so it was not until we examined the caldera with InSAR that we realized its two resurgent domes were behaving at times like a giant see-saw, and the Norris area sometimes likes to play, too.[7] This is really the same lesson that Mount St. Helens taught us in 1980 (i.e., telltale signs easily can be missed if geodetic

---

[7] A see-saw is a plank balanced at the middle on which two children alternately ride up and down. The apparatus is also called a teeter-totter or teeterboard.

measurements are not made at the right time and place). Fortunately, getting it right has become much easier with the development of InSAR, continuous GPS (CGPS), and affordable, high-precision strainmeters and tiltmeters.

11.3.4 Lessons from Mount St. Helens II: 2004–2006 (continuing education)

A few years after dome-building activity at Mount St. Helens stopped in October 1986, it seemed unlikely that there would be another opportunity in my lifetime to study an erupting Cascade volcano. After all, it had been 123 years between the end of the previous eruptive episode at Mount St. Helens in 1857 and the start of the 1980s activity.[8] In the interim, there had been just one other eruption in the Cascades, at Lassen Peak from 1914 to 1917 (Loomis, 1926; Clynne et al., 1999). The odds were not in favor of me being around for the next one, whether at Mount St. Helens or elsewhere.

Nonetheless, on 23 September 2004, a swarm of small earthquakes began under the 1980–1986 lava dome at Mount St. Helens, signaling that the slumbering volcano was still restless. Similar swarms had come and gone several times during the 18-year hiatus since the October 1986 dome-building episode, so this latest swarm was interesting but not alarming. Seemingly true to past form, the seismicity intensified into the following day and then began to wane. At this point, the swarm was remarkably similar to one that occurred on 3–4 November 2001, which ended quietly.

This time, however, seismicity began to increase again on September 25. Event sizes increased to $M_{max} \sim 2.0$ and event rates also increased. We did a double take and wondered what the volcano had in store for us. The answer came quickly. The rate of seismic energy release increased in spurts through October 1, event sizes increased to $M_{max} \sim 3.5$, and earthquakes of $M > 2.5$ began occurring approximately one per minute. The

USGS David A. Johnston Cascades Volcano Observatory (CVO) issued a Notice of Volcanic Unrest on September 26, followed by a Volcano Advisory on September 29.[9] By this time, it was clear that something extraordinary was happening. Part of a glacier that had developed in the crater since the 1980s eruptive activity was bulging upward and breaking apart at a prodigious rate, producing a visible welt on the south crater floor (Figure 11.5). Seemingly against all odds, magma was again rising beneath the crater at Mount St. Helens (Dzurisin et al., 2005).

At midday on 1 October 2004, while a helicopter hovered nearby making thermal observations with a forward looking infrared radiometer (FLIR), a phreatic explosion from the growing welt sent steam and ash above the crater rim and pelted the western half of the 1980–1986 lava dome with ballistic ejecta (Figure 11.6). Mount St. Helens' first eruption of the 21st century was underway! The next day, a 50-minute episode of strong volcanic tremor prompted CVO to issue a Volcano Alert (Level 3), warning of the possibility of an imminent magmatic eruption. Another burst of tremor occurred on October 3, followed in the ensuing two days by more phreatic explosions. The three largest events on October 1, 4, and 5 lofted ash from 100s of m to $\sim$1 km above the vent. Several smaller events produced condensed steam plumes containing little or no discernible ash, or were unobserved but left thin ash deposits on the crater floor. Only the ashfall of October 5 affected populated areas: a light dusting of ash extended downwind (NNE) as much as 100 km to the northeast part of Mount Rainier National Park.

A lava spine was first seen emerging from the welt on October 11. During the next two weeks, the initial spine grew upward, several smaller spines appeared to the south, and surface temperatures in excess of 700°C were measured in fresh cracks by a helicopter-mounted FLIR. In late October, a whaleback-shaped extrusion emerged immediately southeast of the initial spine. This remarkable feature grew steadily through March 2005, at which time it had attained a volume of $\sim$45 × 10^6 m^3. The growing dome severely disrupted the east arm of the crater glacier by squeezing it against the east crater wall, forcing the glacier to

---

[8] The Goats Rocks eruptive period began about 1800 CE with the explosive eruption of dacitic pumice layer T (Mullineaux and Crandell, 1981), which produced a recognizable ash layer as far downwind as northern Idaho. From the 1830s to the mid-1850s, many minor explosive eruptions were observed by explorers, traders, and settlers in the area. Extrusion of the andesitic Floating Island lava flow was followed by growth of the dacitic Goat Rocks dome on the volcano's north flank, which ended in 1857. When the volcano reawakened in 1980, the Goat Rocks dome was deformed by the famous north-flank bulge and destroyed by the catastrophic landslide and eruption on May 18.

[9] The Mount St. Helens Emergency Response Plan defines three alert levels that differ from normal background activity: Level I, Notice of Volcanic Unrest (unusual activity detected); Level 2, Volcano Advisory (eruption likely but not imminent); and Level 3, Volcano Alert (eruption imminent or in progress).

Figure 11.5. The 1980 crater at Mount St. Helens as it appeared on 29 September 2004. View is to the southeast; Mount Adams, another Cascade volcano, is visible in the distance at upper left. Since the catastrophic landslide and eruption on 18 May 1980, which formed the horseshoe-shaped crater that is partly visible here, a dacite lava dome had grown episodically from December 1980 to October 1986, and as much as 240 m of glacier ice and rockfall debris had accumulated on the south crater floor. Large crevasses in the west arm of the glacier, near the bottom of the photo, had been developing for several years as the glacier moved northward. However, within a few days of the onset of a shallow earthquake swarm on 23 September 2004, new cracks appeared and a pronounced welt was visible on the south crater floor (ellipse).
USGS photo by Steve P. Schilling.

Figure 11.6. On 1 October 2004, eight days after the onset of a shallow earthquake swarm, this steam-and-ash eruption from the west margin of the uplifting welt (ellipse) heralded Mount St. Helens' awakening after 18 years of eruptive quiescence. View is to the southwest; part of the east crater rim is visible in the foreground.
USGS photo by John S. Pallister.

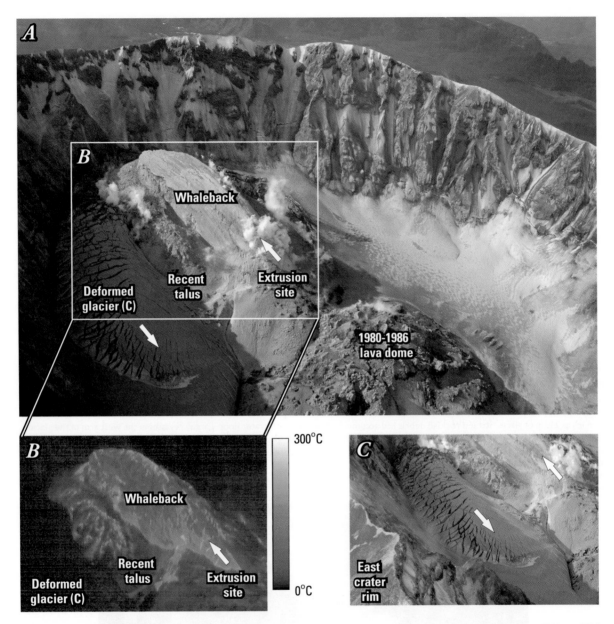

Figure 11.7. Visible (A, C) and FLIR (B) images of the 1980 crater and active lava dome ('whaleback') at Mount St. Helens on 15 March 2005. Views are to southwest; north is to lower right. Highest temperatures in the FLIR image correspond to the extrusion site at the north end of the whaleback and to a tongue of fresh talus. Lower, but elevated, temperatures delineate older portions of the whaleback south of the extrusion site, and dome material that was extruded earlier, bulldozed aside, and partly overridden by the active whaleback. (C) shows the east arm of the crater glacier, with numerous large crevasses and an arrow indicating the direction of accelerated movement. USGS photos and image by James W. Vallance.

surge northward (Figures 11.7 and 11.8). The growth rate of the dome slowed from $\sim 7\,\mathrm{m}^3\,\mathrm{s}^{-1}$ in October 2004 to $\sim 2\,\mathrm{m}^3\,\mathrm{s}^{-1}$ in February 2005, but the syncopated drumbeat of small (M 0–3) earthquakes continued at a rate of ~ 1 per minute. By December 2005, the beat had slowed to 1 earthquake every few minutes, but extrusion was continuing   at a rate of $\sim 0.7\,\mathrm{m}^3\,\mathrm{s}^{-1}$ and there

was no indication that the eruption was likely to end anytime soon.

Surely this latest eruption is yielding a treasure trove of geodetic data, right? You might reasonably expect that any ambiguity concerning premonitory far-field deformation, which lingered after the 1980s experience (Section 11.3.2), has been resolved with the benefit of hindsight and improved geodetic

Figure 11.8. The 1980 crater and lava domes as they appeared on 15 March 2005. Compare with the photo in Figure 11.5, which was taken on 29 September 2004. Mount Adams is visible in the distance in both photos. Dark ash on the east arm of the glacier and east crater wall, behind the domes in this view, is from a small explosion on 8 March 2005.
USGS photo by Steve P. Schilling, 11 March 2005.

monitoring techniques (GPS and InSAR). At least that is what I expected. In the Preface to this book, however, I wrote '... *volcanoes are humbling but very exciting places to study and learn*'. Suffice it to say that this eruption has been very exciting.

More than a year into the 2004–2006 eruption, we are scratching our heads and wondering how more than $50 \times 10^6 \, \mathrm{m}^3$ of dacite could have made its way to the surface with such scant geodetic consequences. To be sure, the crumbling glacier and rising welt on the south crater floor in late September and early October 2004 were unmistakable indications that magma was on the move, but those effects were confined to the immediate vicinity of the eventual outbreak. Earlier in this chapter (Section 11.3.1), I argued that a volume change of $10^7 \, \mathrm{m}^3$ at a depth of 5 km should produce measurable surface displacements out to radial distances of 10 km or more. If true, this means that CGPS station JRO1 at the Johnston Ridge Observatory,[10] ~9 km north of the vent, is well situated to record surface displacements associated with a crustal magma reservoir beneath Mount St. Helens. However, even in hindsight, there seems to be no evidence in the JRO1 record that the volcano inflated in the years, months,

or days prior to the swarm of shallow earthquakes that began on 23 September 2004. In that sense, the autumn of 2004 seems like a replay of the spring of 1980: sudden onset of intense, shallow seismicity followed within days by the appearance of a bulge at the surface, seemingly unheralded by premonitory far-field deformation. With apologies to Pete Lipman, Jim Moore, and Don Swanson (Lipman *et al.*, 1981a), maybe their null results for line-length changes from Smith Creek Butte to East Dome *should* be taken at face value (i.e., neither station moved much before the 18 May 1980 eruption, as they concluded at the time), instead of calling upon both stations having moved nearly the same large amount (as I suggested in Section 11.3.2).

By the way, that is Jim Moore on the front cover in a photo taken by Pete Lipman. Those guys did volcano geodesy the old-fashioned way, with boots on the volcano. In the process, they moved eruption prediction from the realm of seers and soothsayers onto a firm scientific foundation. During the quarter century since their work at Mount St. Helens, our tools have become more sophisticated and we have benefited from the ongoing revolutions in remote sensing and telecommunications. But these advances have come at the cost of sharpened intuition and serendipitous discoveries that only hands-and-eyes-on field work can offer. There is a limit to what can be learned from afar, even with 21st century tools and technology. One thing that scientists working

---

[10] The USGS David A. Johnston Cascades Volcano Observatory (CVO) and the USFS Johnston Ridge Observatory (JRO) commemorate USGS geologist Dave Johnston, who died in the 18 May 1980, eruption of Mount St. Helens near the current site of JRO.

at Mount St. Helens in the 1980s understood is that there is no substitute for being there. It is a lesson I have taken to heart, with a grateful nod to those who led the way.

Back to modern-day Mount St. Helens: the news from JRO1 in 2004–2006 was not entirely discouraging. Concurrently with the start of the earthquake swarm on 23 September 2004, JRO1 started moving southward, toward the volcano, at a rate of ~0.5 mm day^{-1}. The inward motion continued at a generally declining rate through June 2005, when the total southward displacement was ~20 mm.[11] At that time, JRO1's southward velocity was ~10 mm yr^{-1} and extrusion of gas-poor dacite lava was continuing. The situation had not changed appreciably by early 2006, when the volume of the new dome surpassed 75×10^6 m^3.

Unfortunately, JRO1 was the only dual-frequency CGPS station operating near the volcano in September 2004, when the premonitory earthquake swarm began. We started making campaign-style GPS observations on the edifice in late September, and worked with PBO and UNAVCO Inc. to install 9 new CGPS stations there in October.[12] In hindsight, based on the JRO1 record, about half of the total surface displacement that occurred from late September 2004 to June 2005 had already occurred when these stations came on-line. The campaign GPS measurements, by virtue of having started 1–2 weeks earlier, should have captured a larger fraction of the total signal. But the uncertainty in campaign measurements is inherently greater than for continuous measurements, and in this case any signal related to the eruption has been too small to emerge from the noise in the campaign data.

On the other hand, a clear deflation signal – consistent with the JRO1 observations – can be seen in data from several CGPS stations that were installed soon after the eruption began. The maximum horizontal and vertical-displacement rates for 17 October 2004 to 21 May 2005 are

small by volcano standards – about 20 mm yr^{-1} and 30 mm yr^{-1}, respectively – but the pattern of inward and downward displacements is unmistakable (Figure 11.9). A preliminary Mogi-type model places the best-fit pressure source ~13 km beneath the volcano. For comparison, recall from Section 11.3.2: (1) Scandone and Malone (1985) estimated the source depth for magma erupted at Mount St. Helens in 1980 to be 9 km based on a model of surface-deformation data for June–November 1980, and 7–14 km based on the distribution of earthquake hypocenters; and (2) Rutherford et al. (1985) proposed a source depth of 7.2 ± 1 km based on the mineral phase assemblage found in 18 May 1980, pumice. Preliminary results suggest that the 2004–2006 magma could have evolved from 1980s magma during the 18-year eruptive hiatus from 1986 to 2004, but many details (and probably some surprises) remain to be uncovered.

Circumstances at Mount St. Helens in 2004–2006 have contrived to limit the effectiveness of InSAR, the 'amazing geodetic camera' that was discussed with considerable fanfare in Chapter 5. First, the volcano became restless in late September, at the start of the normal wet season in the Cascades. Rain and snow are natural enemies of spatial coherence in interferograms, so the number of useful interferograms is small. Second, the Mount St. Helens crater is an extremely dynamic place: since the 1980s eruptive activity, persistent rockfalls and ice accumulation combined to form a glacier more than 240 m thick in the southern part of the crater (Section 6.7). Continuing accumulation and motion of the glacier, the latter under the combined influences of gravity and encroachment by the growing lava dome, are complicating factors that make it difficult to distinguish any deeper seated deformation source. Third, although deformation of the crater floor and glacier has been spectacular (Figures 11.5–11.7), it has been mostly confined to an area of only ~1 km^2, where movements have been very rapid (m d^{-1}). That is a difficult puzzle for InSAR to solve, even under the best of conditions. The CGPS results for October 2004 to May 2005 (Figure 11.9) suggest that widespread syn-eruptive deflation might be detectable with InSAR, but not until the spring snowpack melts and summer-to-summer interferograms are produced. This book was never intended to be a final word on anything, and in this case it surely is not. That is good news for present and future volcanologists – there is plenty of work still to be done!

The 2004–2006 eruption at Mount St. Helens

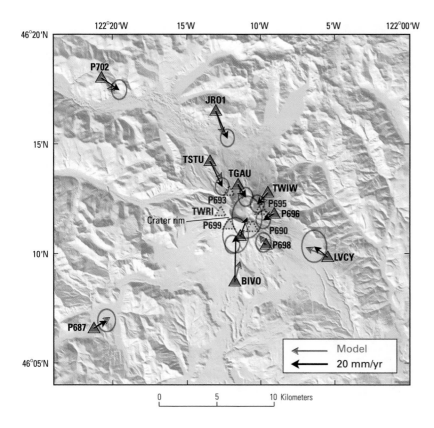

Figure 11.9. Horizontal surface-displacement rates at selected CGPS stations at Mount St. Helens, Washington, for the period 17 October 2004 to 21 May 2005. Observed station velocities are shown in black, model results in red. Error ellipses show 95% confidence limits. Data from five stations (dashed blue triangles) on the volcano's upper flanks were excluded, because they are biased by heavy rime-ice accumulation on the GPS antennas. The velocities are well-fit by a Mogi-type pressure source ~ 13 km beneath the volcano. The equivalent source-volume change is 16×10^6 m^3 in 7.25 months, corresponding to 26×10^6 m^3 yr^{-1}.

seems to be far from over, and we have yet to squeeze the pre-eruption GPS data as hard as possible in hopes of extracting meaningful information. Nonetheless, one thing is already clear. Any geodetic precursor to the eruption that might have extended much beyond the crater was small to nonexistent. Annual campaign-style GPS measurements at several stations on the 1980–1986 lava dome, which sits atop the 1980s vent, showed subsidence at rates of several cm yr^{-1} – presumably the result of cooling and compaction. Elsewhere in the crater, except on the glacier, any surface movements were too small to measure. The JRO1 CGPS station, ~ 9 km away, saw no premonitory signal that can be attributed to magmatic inflation. It started moving toward the volcano when seismicity commenced on September 23, presumably in response to a pressure drop as magma left a crustal reservoir and began grinding its way up the conduit.[13]

Faced with the slowly emerging signals from

CGPS stations on the volcano's flanks, combined with spectacular growth of the welt and whaleback extrusion in the crater, we set about to deploy sensors in the shadow of the growing dome. The area within a kilometer of the vent was occasionally in harm's way from small explosions or hot rock-falls, so we needed to keep our exposure time short. Enter the 'spider', a three-legged, multipurpose volcano-monitoring tool that can be helicopter-slung into place in a matter of minutes (Figure 11.10). A spider is a footlocker on legs, packed with batteries and outfitted with a radio, telemetry antenna, and one or more sensors. Equipped with a single-frequency GPS receiver, tiltmeter, geophone, accelerometer, volcanic gas sensor, digital camera, or temperature sensors (or some combination thereof), plus enough battery power to operate the package for weeks to months, a spider weighs in at ~ 100 kg – light enough to be slung into place with a small helicopter.

During the first nine months of the 2004–2006 eruption, GPS and seismic spiders were set on the growing whaleback to measure its velocity (up to 10 m day^{-1}) and to record earthquakes within a few hundred meters of their source. GPS spiders riding on the glacier, which is being shoved aside by the growing dome, tracked the passage of kinematic

---

[13] Among the distinctive features of the 2004–2006 eruption are persistent seismicity (~ 1 event per minute, many with similar waveforms, continuing for several months), and a ubiquitous layer of fault gouge, ~ 1 m thick, at the surface of a piston-like extrusion.

Figure 11.10. Helicopter-slingable monitoring instruments, dubbed 'spiders' when a three-legged base was added for stability on rough terrain, were developed and deployed for the first time during the 2004–2006 dome-building eruption at Mount St. Helens. (A) The basic design included a metal footlocker containing electronic equipment (e.g., single-frequency GPS receiver) and batteries, one or more sensors (e.g., GPS antenna, accelerometer), a digital telemetry system, and a means to attach the package to a remote-release sling line. Later versions included a rigid superstructure that could be snagged with a grappling hook for easy retrieval. (B) Variations on the basic design were developed for specific applications; in this case, two extended legs helped to level the package on the sloping surface of the whaleback (D). (C) Spiders were slung beneath a helicopter from a staging area near the volcano to the 1980 crater, where they were deployed on the 1980–1986 dome, the 2004–2006 dome, and the crater glacier. (D) The spider shown in (B) was deployed near the north end of the whaleback; the package (circle) included a single-frequency GPS receiver and an accelerometer.
USGS photos by Michael P. Poland (A, 12 October 2004), Steve P. Schilling (B, 14 January 2005; and C, 24 October 2004), and Daniel Dzurisin (D, 14 January 2005).

waves across the ice. Spiders on the 1980–1986 dome recorded innumerable earthquakes and tracked surface movement at rates of several $mm\,day^{-1}$ to a few $cm\,day^{-1}$. A few spiders have been snagged with a grappling hook, refurbished, and redeployed. One spider equipped with a digital camera and volcanic gas sensor was poised less than 100 m from the vent to record any changes in extrusion rate or gas-emission rate. Unfortunately, it was destroyed by a small explosion less than two days after it was deployed. Nonetheless, slingable sensor packages have proven to be a versatile and valuable addition to our monitoring tool kit at Mount St. Helens, and they likely will find their way onto other volcanoes as well.

11.4 CAPTURING VOLCANO DEFORMATION IN SPACE AND TIME

To track the intrusion process in space and time – and thus to anticipate eruptions before they are imminent – we need accurate geodetic measurements with high spatial and temporal resolution. At most volcanoes we have no a priori knowledge of the location or timing of future intrusions, so we must observe a very large area (to account for uncertainties in source depth and geometry) essentially all of the time. That might seem like an insurmountable problem, but it becomes tractable if we leverage the strengths of InSAR, CGPS, and other continuous deformation sensors. By combining repeated InSAR observations with continuous data from GPS stations, strainmeters, and tiltmeters, we should be able to determine both where and when volcanoes are deforming long before magma begins its final ascent to the surface.

II.4.1 Real-time, global surveillance: an achievable goal

Satellite-based InSAR is an essential component of any global volcano surveillance system. *In situ* geodetic sensors are critically important, too, but installing and operating them on all of Earth's 500+ active and potentially active volcanoes is a prodigious task that is not likely to be accomplished anytime soon. On the other hand, most of the world's volcanoes could be monitored effectively with a small constellation of InSAR satellites dedicated to natural-hazards mitigation. L-band would seem to be a better choice than C-band for such a mission, because many volcanoes are located in high-precipitation regions with lush vegetation. L-band's longer wavelength ($\sim$25 cm versus $\sim$5 cm for C-band) offers greater penetration and less sensitivity to atmospheric water vapor, which translates into more persistent radar coherence and better deformation interferograms. This advantage more than offsets the factor of $\sim$5 lower resolution relative to C-band, given the magnitude of typical eruption precursors (at least several centimeters of surface movement).

Another important requirement for a volcano InSAR mission is a relatively short orbital repeat cycle. The 35-day repeat cycle for ERS-1 and ERS-2 satellites is adequate for measuring long-term deformation at persistently restless systems like Yellowstone, but too long to be effective during a volcanic crisis. An L-band mission with a repeat cycle of a few days over key areas, including an efficient data retrieval and analysis system to support the production of timely deformation interferograms, would be a major step forward. The Phased Array L-band Synthetic-aperture Radar (PALSAR) aboard the Japanese Advanced Land Observing Satellite (ALOS), which was launched successfully on 24 January 2006, will operate at L-band but with a repeat cycle of 46 days.

As we have seen repeatedly in previous chapters, no single geodetic tool is adequate to measure the full spectrum of deformation patterns and histories presented to us by restless volcanoes. Unfortunately, there is no silver bullet in the volcanologist's tool kit. So an effective global volcano-monitoring system

must also include a variety of *in situ* sensors to provide highly precise, continuous, real-time geodetic information. Well-designed networks of well-installed and well-maintained GPS stations, strainmeters, and tiltmeters can maintain a constant vigil, even at long-dormant volcanoes. A comprehensive real-time monitoring program would be expensive and technically challenging, but recently there has been good news on several fronts in this regard. The cost of full-feature GPS receivers suitable for repeat surveys has fallen considerably, and low-power models designed specifically for continuous operation are now available at even lower cost. In cases where deformation is likely to be extreme or conditions are especially harsh, single-frequency ('L1-only') GPS receivers provide a useful alternative for as little as a few hundred dollars (Chapter 4). At the same time, extremely sensitive borehole strainmeters have repeatedly proven their value for volcano monitoring (Chapters 3 and 9), and their more widespread use promises to reduce the unit cost considerably. Finally, technical improvements have led to the widespread availability of inexpensive borehole tiltmeters capable of accurately measuring tilt changes of 10^{-7} radians or less over periods of up to several weeks (Chapter 3). Combined with growing recognition of the advantages of installing such instruments in boreholes, which can dramatically increase the signal-to-noise ratio by attenuating near-surface noise, these improvements have served to revitalize the role that tiltmeters can play in a state-of-the-art volcano-monitoring program.

Currently, a few dozen volcanoes are being monitored every 24–35 days with the C-band radars aboard ERS-2, Radarsat, and Envisat. This approach is not practical at most of the world's potentially active volcanoes, because interferometric coherence is destroyed by such factors as dense vegetation, frequent rainfall, or changing winter snow pack. Impressive results from the now-defunct JERS-1 satellite suggest that L-band radar is a superior choice in this regard and should receive strong consideration in the future. A coordinated, international program of frequent L-band and C-band observations of all suitable volcanoes worldwide, including the capability to observe a small number of high-priority targets daily for short periods of time, would almost surely lead to fundamental advances in our understanding of how volcanoes work. This, in turn, would improve volcanologists' ability to anticipate volcanic activity, mitigate volcano hazards, and save lives. Fortunately, new InSAR missions are under consideration by the space agencies of Canada, Europe, Japan, and the USA. Within a decade, routine InSAR monitoring of most of the world's volcanoes seems achievable.

Continuous volcano monitoring with *in situ* sensors requires a large investment of time and resources, both for initial installation and continuing operations. This is especially true if the monitoring network is adequate in terms of station density and coverage to fully characterize the deformation field for all plausible sources. For example, consider the implications of Figure 11.1, which shows the theoretical surface deformation produced when a Mogi-type source 5 km deep inflates by 10^7 m^3. The maximum vertical displacement occurs directly above the source, so at least one sensor, presumably GPS, should be installed as close to the volcano's summit as possible. For present purposes, we will ignore the logistical challenge presented by operating instruments near the summit of an active volcano, and also assume that the source is located directly beneath the summit. To characterize the falloff in uplift with increasing distance from the summit requires at least two additional well-placed GPS stations – say at distances of 5 km and 15 km from the summit. The patterns of horizontal displacement and tilt are more complicated than vertical-displacement patterns – both horizontal displacement and tilt increase from zero at the summit area outward to a fraction of the source depth, then decrease steadily to beyond 15 km (for a 5 km deep source). Let us assume that the 3 GPS stations installed to monitor vertical displacements are also suitably placed to characterize the horizontal displacement field. That leaves a minimum of three tiltmeters to characterize the tilt pattern (i.e., one near the summit where the tilt is modest, one in the area of maximum tilt 3–4 km from the summit, and one farther away in the range of 5–15 km from the summit).[14]

So far, we have included at least 3 GPS stations and 3 tiltmeters (or strainmeters, or both), which are manageable numbers for most volcano observatories. However, if we relax some of the unrealistic constraints we have imposed and consider a more likely situation, the requirements become much more daunting. For example, imagine that the

volcano of interest is located along an active fault zone, which might produce deformation that is neither centered at the volcano nor azimuthally symmetric about the summit area. Unless we understand the geometry of the fault zone extremely well and (or) are confident that we can separate tectonic effects from those related directly to the volcano (unlikely), we would be well advised to multiply the number of monitoring stations by a factor of 3–4 to account for possible asymmetry in the deformation field.

At this point, we have included 9–12 GPS stations and 9–12 tiltmeters or strainmeters – still reasonable numbers for a high-priority volcano. But now let us consider what other simplifying assumptions are implicit in our scenario, and their implications for a real-world volcano crisis. What if the source is not Mogi-like, but instead an inclined dike or sill? The resulting deformation patterns are more complex and, therefore, more stations are needed to distinguish among plausible source geometries. If the source is initially deeper than 5 km, more stations will be needed at distances greater than 15 km from the volcano, if for no other reason than to verify that measurable deformation is not occurring in distal areas. Conversely, if the source is shallower than 5 km, deformation will be concentrated closer to the summit and more stations are necessary in the near field. To discern any departure from elastic behavior by the host rock beneath the volcano, which is undoubtedly hot and fractured to varying degrees as a result of repeated intrusions, data from even more stations are required.

To honestly deal with a realistic scenario in which the depth, location, geometry, and volume (or pressure) change of the intruding magma body are treated as unknowns, along with the rheologic properties of the host rock and the time history of the intrusion, we need dozens of deformation sensors deployed in a network that extends at least 20 km in all directions from the volcano. Overlapping arrays of 10–20 CGPS stations, 5–10 borehole tiltmeters, and at least 5 borehole strainmeters – in most cases co-located with seismometers or other sensors – would probably be adequate to capture the evolving deformation field caused by a real-world intrusion. More than 5 strainmeters per volcano would be better, but their relatively high cost (~US$100,000 per station, including the cost of drilling a ~100 m deep borehole) is a limiting factor for the time being. The incredible sensitivity of these instruments to small strains plus some spectacular successes during two recent eruptions of Hekla

---

[14] This discussion could be phrased in terms of strain rather than displacement and tilt, to emphasize the importance of deploying borehole strainmeters in addition to GPS stations and tiltmeters. For brevity, I have omitted strainmeters here, even though I regard them as an essential component of any comprehensive volcano-monitoring program.

(Section 9.5.3; Linde *et al.*, 1993; Linde and Sacks, 2000; Agustsson *et al.*, 2000) and during several earthquake swarms associated with intrusions and a submarine eruption near the Izu Peninsula, Japan (Section 9.5.1; Koizumi *et al.*, 1999) make a compelling case for their increased use near volcanoes. Hopefully, innovative means to lower fabrication and installation costs will be forthcoming soon, perhaps through economies of scale, new instrument designs, or worthwhile tradeoffs between instrumental sensitivity and size (smaller, less-sensitive sensors could be installed in smaller diameter, less expensive boreholes) or between signal-to-noise and installation depth (shallower, less expensive installations might still provide acceptable signal-to-noise levels for some applications). As part of the Plate Boundary Observatory (PBO), clusters of 5–30 CGPS stations and 3–15 borehole strainmeters will be deployed at Long Valley, Yellowstone, and selected volcanoes in the Aleutian arc and Cascade Range starting in 2004.

Currently, only a few volcanoes are monitored nearly this well. These include, in order of increasing volcano number: Etna (Italy), Piton de la Fournaise (Réunion Island, Indian Ocean), Taal (Philippines), Sakurajima and Izu Oshima (Japan), Augustine (Alaska), Long Valley Caldera (California), Mount St. Helens (Washington), Kīlauea and Mauna Loa (Hawai'i), Popocatépetl (Mexico), Soufrière Hills Volcano (Montserrat, West Indies), and Hekla (Iceland).[15] Within the next few decades, though, it seems reasonable to expect that most volcanoes that threaten large populations will be adequately monitored. Compromises will continue to be necessary at many others, and a remote few will be monitored only from space for a long time to come. Nonetheless, for the first time in history, revolutionary tools such as GPS, borehole strainmeters and tiltmeters, InSAR and other remote-sensing techniques place the goal of global, real-time volcano surveillance clearly within reach.

II.4.2   On-the-fly volcano modeling

Even an ambitious geodetic monitoring program would be incomplete without an effective means to combine, analyse, and model various datasets in near-real time. There have been important developments in this regard recently (e.g., Owen *et al.*, 2000b; Segall *et al.*, 2001) and more seem to be on the horizon. Conceptually, diverse geodetic datasets (e.g., results of GPS, leveling, and microgravity surveys, interferograms, and continuous data streams from borehole tiltmeters, strainmeters, and other sensors) need to be combined and weighted according to their relative uncertainties. The appropriate weighting for such diverse model inputs as leveling, GPS, InSAR, and strainmeter observations has yet to be completely worked out, but in theory the issue should be manageable. The weighted data would then be compared with an a priori model that included all known or suspected deformation sources. The model parameters would be optimized by minimizing the misfit between model and data, using some appropriate statistic. Once the model had stabilized, each new piece of geodetic information would be compared with the model's expectations, as defined by predictive filtering or some similar technique, and the model would be updated in near-real time. Eventually, the model would converge to a robust 'virtual volcano' that could be constantly steered toward reality by automated, real-time data streams. Conversely, the model would identify weaknesses in the monitoring program that could be addressed by new sensors or techniques.

Most deformation models assume that the crust near volcanoes is flat, elastic, and homogeneous – simplifying assumptions that often yield satisfactory results when models are constrained only by relatively limited geodetic data. However, as geodetic measurements become more precise and denser in space and time, the limitations of simple inversion models will likely become apparent. Therefore, in anticipation of much denser, more precise geodetic datasets from PBO and other sources in the foreseeable future, more realistic models that include such features as non-idealized source geometry, non-elastic rheology (e.g., viscous or plastic, including temperature and strain-rate dependence), and inhomogeneous material properties (e.g., shear modulus, yield strength) need to be developed. New

---

[15] The Catalog of Active Volcanoes of the World (CAVW), a regional series of publications by the International Association of Volcanology and Chemistry of the Earth's Interior (IAVCEI), developed a volcano numbering system in the late 1930s that has been retained, where possible, by the Smithsonian Institution's Global Volcanism Program (Simkin and Siebert, 1994). The numbering scheme is geographic and hierarchical. The first two numerals identify the region, the next two identify the subregion, and the last two or three (after the hyphen) identify individual volcanoes in that subregion. For example, Mount St. Helens is the fifth volcano, numbered from North to South, in the first subregion (USA, west coast states) of the twelfth region (Canada and western USA) of the world. Its volcano number is therefore 1201–05.

geomechanical models based on numerical methods such as finite elements can accommodate such complexities and, if constrained by abundant, high-quality geodetic data in near-real time, might produce new insights into active processes near volcanoes.

The network-inversion filter (Segall and Matthews, 1997; Aoki et al., 1999) is an especially promising development in the field of automated, near-real time deformation modeling. The technique combines elements of linear inverse theory and discrete-time Kalman filtering to estimate the distribution of fault slip in space and time using data from dense, frequently sampled geodetic networks. Refinements under development include the addition of other sources of interest in volcanology (i.e., beyond the Mogi source and fault dislocation), such as an expanding sphere, ellipsoid, or pipe. In the near future, predictive modeling techniques like the network-inversion filter will make it possible not only to track volcano deformation in near-real time, but also to automatically detect anomalous departures from steady state unrest.

II.4.3 Implications for eruption forecasting and hazards mitigation

What might be gained if many of the world's volcanoes were intensively monitored for ground deformation? Longer term warnings of impending eruptions would be an important advance that seems achievable with currently available techniques. Key to such warnings is the ability to detect changes in a magma body while it is still deep enough to be aseismic (i.e., before an intrusion starts its final ascent toward the surface).

Past volcano-monitoring efforts have focused almost exclusively on this final phase of the intrusion process, when seismic and other forms of unrest are most apparent. This is a reasonable approach, given limited resources and the large number of potentially active volcanoes that need attention. In the past two decades, there have been some noteworthy successes in eruption prediction and hazards mitigation based on this approach, and more are likely as the experience gained at Mount St. Helens, Pinatubo, Rabaul, Montserrat, and elsewhere is applied to future volcanic crises around the world. We know that virtually all eruptions are preceded by at least a few hours of easily detectable seismic unrest, and that most volcanoes show clear signs of reawakening weeks to months in advance. Even a basic monitor-

ing program can ensure that future eruptions never come as a complete surprise, as happened at El Chichón Volcano, Mexico, in 1982 (Varekamp et al., 1984; Sigurdsson et al., 1984; Carey and Sigurdsson, 1986).

The foregoing applies to eruptions that quickly follow a long period of quiescence, but ongoing eruptions and prolonged periods of volcanic unrest pose a much bigger challenge. For example, the most damaging phases of the eruptions at Mount St. Helens in 1980, Rabaul in 1994, and Soufrière Hills Volcano, Montserrat, starting in 1995 were preceded by months to years of unrest that provided ample warning of the potential hazards. However, in all three cases, as the activity dragged on there was uncertainty over the eventual outcome that resulted in pressure on scientists and public officials to make forecasts accurate enough both to allow public access during 'safe' periods and also to ensure the public safety when the situations turned dangerous. The desire to minimize economic disruption and personal inconvenience during long periods of volcanic unrest is understandable, but unrealistic expectations are dangerous. Even if accurate probabilistic eruption forecasting were possible (e.g., 10% chance of a hazardous eruption within the next 24 hours), the uncertainty inherent in such a forecast and the high stakes involved argue for placing public safety above convenience in most cases.

Short-term eruption prediction and hazards mitigation will continue to pose difficult challenges for the foreseeable future, but continued progress is likely. An example of what is possible in this regard is the 26 February 2000, eruption of Hekla, which was predicted with amazing accuracy (Section 9.5.3; Agustsson et al., 2000). At 17:00 local time that afternoon, a seismometer within 2 km of the summit began recording a sequence of small earthquakes ($M < 1$), the likes of which are unknown at Hekla except as a prelude to eruptions. As the swarm intensified, the Civil Defense of Iceland was warned at about 17:15 of a probable, imminent eruption. At 17:47, a Sacks–Evertson borehole strainmeter 15 km from Hekla started recording increasing contraction. On that basis, a prediction was issued to the Civil Defense at 17:53 stating that an eruption was to be expected within 15–20 minutes and recommending that a warning be issued and broadcast to the public. The warning was broadcast and at 18:17 Hekla erupted explosively (4 minutes outside the prediction window, but who's counting?). Admittedly, this remarkable feat was only possible because a very similar sequence of events had been observed prior to Hekla's previous eruption in 1991, and the two eruptions turned out to be very similar. Nonetheless, the Hekla experience emphatically demonstrates that short-term eruption prediction is technically feasible and could become routine if sufficient resources are allocated for comprehensive, real-time monitoring of high-risk volcanoes.

Much less certain, but potentially important, is the possibility of a breakthrough in intermediate-term eruption forecasts based on earlier, deeper detection of magmatic intrusions. Geodesy and geochemistry hold the greatest promise in this regard, because experience has shown that intrusions are mostly aseismic except in the upper few kilometers of the crust. Long before they light up seismically, intrusions deform host rock and release volatiles. The effects at the surface are likely to be subtle, but if they can be recognized it might be possible to identify reawakening volcanoes years or even decades before magma approaches the surface and announces its presence with earthquakes or other forms of shallow-seated unrest. Such a long lead time would allow scientists to upgrade their monitoring and public education efforts at reawakening volcanoes, provide a strong impetus for government officials and the general public to develop emergency-response plans, and enable more effective hazards mitigation through land-use planning.

II.5 PIE-IN-THE-SKY VOLCANOLOGY

A constellation of InSAR-capable satellites imaging most of the world's volcanoes every few days, dense networks of GPS stations, strainmeters, and tilt-meters surrounding all worrisome volcanoes, and sophisticated computer models utilizing artificial intelligence to track changes beneath volcanoes in real time – what more could the future possibly hold?

Perhaps the greatest challenge facing volcanologists today is to move beyond recognition of the symptoms and possible outcomes of volcanic unrest to a quantitative understanding of the physical and chemical processes involved as a function of time and space. In the words of my friend and colleague Chris Newhall: '*Parameters that can be measured, such as microseismicity, ground deformation, and gas emission, are frustratingly indirect windows into the internal processes of volcanoes*' (Newhall, 2000, p. 353). Even the most comprehensive geodetic-monitoring program imaginable,

capable of imaging the entire deformation field continuously in space and time, would not directly address such fundamental questions as: Will the volcano erupt as a result of this intrusion? When will the current eruption end? Will the eruption be explosive or effusive? These and other questions might not be answerable even with correspondingly thorough seismic and gas-monitoring programs, because all three types of monitoring are sensitive to the symptoms of unrest rather than its causes. No practical means exist today to directly measure the parameters that primarily control whether and how a volcano will erupt (e.g., volatile content, viscosity, and supply rate of magma, strength of host rock, geometry of the conduit system). Most of what we measure is a second-order effect of our primary interest (i.e., the changing conditions in magmatic, tectonic, and hydrothermal regimes that are the root cause of eruptions).

How might we shift our focus from symptoms to causes? Can we hope to someday measure or infer parameters that directly affect the timing and course of eruptions? A good start has already been made in that direction and continued progress seems likely. For example, Denlinger and Hoblitt (1999) constructed a dynamic model for oscillatory behavior at Soufrière Hills Volcano that is based on sound physical principles (i.e., Newtonian flow of compressible magma through a conduit combined with a stick–slip condition along the conduit wall). They showed that, even if magma is forced into the conduit at a constant rate, stick–slip behavior results in oscillations in such parameters as pressure and magma-flow rate, which can drive cyclic eruptive behavior. A similar phenomenon is sometimes observed during high-pressure extrusion of industrial polymer melts. Models of this type can be used to constrain the time history of pressure, viscosity, and compressibility (hence bubble content) in a magma body that exhibits oscillatory behavior (as determined by conventional monitoring techniques).

This approach is inherently passive, because the model only applies when a volcano exhibits a specific type of behavior. At other times or at other volcanoes, the approach is blind to a wide variety of processes that generally are not cyclic. A more widely applicable approach would be to actively probe magma systems beneath volcanoes, even when they seem to be quiescent, and to continue probing throughout an entire eruption cycle. For example, active-source seismic tomography has been used extensively to explore for oil, but rarely to study the sizes, shapes, or conditions within magma bodies. The technique is expensive and magma bodies are difficult targets because they are generally small, highly attenuating and, in some cases, geometrically complex. Nonetheless, routine active-source seismic monitoring conceivably could be implemented in the foreseeable future, especially at selected high-risk volcanoes.

Newhall (2000) also recommends that more attention be paid to volcanoes' responses to natural disturbances such as distant earthquakes and Earth tides. For example, gas-saturated magma might be more prone to bubble formation and triggered seismicity in response to distant, large earthquakes than gas-poor magma (Linde et al., 1994; Johnston et al., 1995). Likewise, gas-rich magma might show greater volumetric and gravimetric responses to tidal compression and decompression cycles. Identifying such effects during any one event or cycle might be difficult, but continuous monitoring over an extended time period while conditions in a magma body change could conceivably yield important information about gas content, which strongly influences eruptive processes.

Better integration of seismic, geodetic, gas-emission, and other types of monitoring data would likely produce additional insights into magmatic processes and conditions. For example, geodetic data alone cannot distinguish between changes in volume and pressure (e.g., between enlargement of a magma body at constant pressure as a result of mass influx and pressurization at constant volume as a result of vesiculation). This ambiguity might be resolved by continuous monitoring of seismic activity (long-period (LP) earthquakes signal pressurization or fluid flow), ground deformation, gas emission (increased CO_2 indicates the arrival of fresh magma or the rupture of a sealed reservoir), and microgravity (Bouguer changes track subsurface density changes).

Still, for the near future, volcanology will remain a mostly inferential science until volcanologists are able to probe the roots of volcanoes directly to measure the essential parameters that control eruptive behavior and provide the basis for deterministic models (temperature, pressure, viscosity, magma supply rate, etc.). Scientific drilling into molten lava lakes (Kīlauea Iki, Hawai'i), young feeder dikes (Inyo Craters, California; Mt. Fugen, Unzen Volcano, Japan), and the peripheries of magma chambers (Long Valley Caldera, California) has been successfully accomplished at a few sites,

and has been earnestly proposed at several more (Hardee and Luth, 1980; Eichelberger et al., 1991). Drilling to typical magma-chamber depths of a few kilometers in cold rock is becoming routine, and high-temperature, high-pressure industrial processes point the way toward operating machinery and in situ sensors under magmatic conditions. Although the engineering challenges are daunting, the idea of drilling into an active magma body in the foreseeable future to emplace sensors and conduct experiments might not be too far-fetched. The payoff would almost surely be tremendous, akin to the advances that occurred when humans first flew into a hurricane, dove to the seafloor, or landed on the Moon. After all, those ideas once seemed outlandish, too!

II.6 A BRIGHT AND CHALLENGING FUTURE

The study of ground deformation near volcanoes holds tremendous potential, because magmatic intrusions almost always deform the ground surface long before they erupt. A complete characterization of the deformation field in space and time would include information about the location, volume, and shape of a subsurface magma body – information that could be used in conjunction with other datasets to better assess and mitigate volcano hazards. Until recently, this potential was not realized at most volcanoes because of difficult logistics, limited resources, the complexity of displacement fields produced by magma moving through a heterogeneous crust, and inadequate monitoring techniques. Point-to-point geodetic measurements, especially if the resulting data are sparse in space or time, are not enough. Modeling clearly demonstrates that, to distinguish among the full range of possible source locations and geometries, it is necessary to make precise measurements virtually 'everywhere, all the time'.

Fortunately, InSAR and continuous surveillance with GPS, strainmeters, and tiltmeters are important advances in that direction. Recent results hint at the possibility of achieving fundamental new insights into intrusive and eruptive processes through comprehensive geodetic monitoring and modeling. At the dawn of a new millennium, ground- and space-based geodesy offers a bright and challenging future to those who explore the unstable ground of restless volcanoes.

Glossary

For many of the following definitions, the author made use of the excellent material found in the American Geological Institute *Glossary of Geology* (1987) and the National Geodetic Survey *Geodetic Glossary* (1986).

ʻaʻā A Hawaiʻian term for lava flows that have a rough rubbly surface composed of broken lava blocks called clinkers. The incredibly spiny surface of a solidified ʻaʻā flow makes walking very difficult and slow. The clinkery surface actually covers a massive dense core, which is the most active part of the flow. As pasty lava in the core travels downslope, the clinkers are carried along at the surface. At the leading edge of an ʻaʻā flow, however, these cooled fragments tumble down the steep front and are buried by the advancing flow. This produces a layer of lava fragments both at the bottom and top of an ʻaʻā flow. *See also* **pahoehoe**.

accessory Term used to describe pyroclasts formed from fragments of an existing volcanic cone or earlier lavas during a volcanic eruption. *See also* **accidental** and **essential**.

accidental Term used to describe pyroclasts formed from fragments of non-volcanic rocks or from solid volcanic rocks not related to the erupting volcano. *See also* **accessory** and **essential**.

adjustment The process of finding, from a set of redundant observations, a set of best-fit values, in some prescribed sense, for the observed set of quantities or for quantities functionally related to them. For example, if each angle of a plane triangle is measured, it is unlikely that the sum of the three values will be exactly $180°$ as required by geometry. The mathematical process of calculating three angles similar to the measured values, but which do satisfy the requirement, is called adjustment. The most common interpretation of 'best-fit' is that the sum of the squares of differences between results obtained by measurement and results obtained by calculation shall be a minimum. *Geodetic adjustment* is the process of calculating best-fit values of geodetic quantities such as lengths, angles, directions, coordinates, etc., that characterize a geodetic network. Geodetic adjustment requires redundant observations, which are inherent in certain network configurations (e.g., groups of triangles, braced quadrilaterals, centered figures, or double-centered figures with common sides) (Bomford, 1980, pp. 3–6).

aerotriangulation A general term for the method used to determine x, y, and z ground coordinates based on photo-coordinate measurements, typically involving bridging control between photos comprising a strip of *stereo models* or a block of multi-model strips (Chapter 6). The addition of horizontal and/or vertical *control points* by photogrammetric methods, whereby the measurements of angles and/or distances on overlapping photographs are related into a spatial solution (stereo model) using the known viewing geometry and camera lens characteristics.

AKDA Acronym for Alaska Deformation Array, a network of continuously operating Global Positioning System (GPS) stations operated by the University of Alaska Geophysical Institute to study crustal deformation in Alaska, by NASA for orbit tracking, by the US Coast Guard for marine navigation, and by a private surveying company for general GPS base station use. Component of the Plate Boundary Observatory (PBO).

AKST Acronym for Alaskan Standard Time. AST = GMT – 9 hours. The AKST time zone includes all of the US state of Alaska except for the Aleutian Islands west of 169°30′W. The western Aleutians observe Hawaiian–Aleutian Standard Time (HST), one hour behind the remainder of the state.

albite A colorless or milky-white mineral of the *feldspar* group: $NaAlSi_3O_8$. It is a variety of *plagioclase* and an alkalai feldspar, representing the triclinic modification of sodium feldspar. Albite commonly occurs as phenocrysts in igneous rocks and is regularly deposited from hydrothermal solutions in cavities and veins. *See* **feldspar, phenocryst, plagioclase**.

aliasing Aliasing occurs when a signal is sampled discretely at a rate insufficient to capture the changes in the signal. The Nyquist sampling theorem states that, to avoid aliasing, the sampling frequency should be at least twice the highest frequency contained in the signal. To avoid aliasing a deformation field while sampling it with a geodetic survey, the station spacing should be half the dimension of the smallest feature of interest, or less. For example, to adequately resolve all deformation features larger than 1 km across, the station spacing should be 500 m or less. This requirement is often difficult to meet on volcanoes with rugged topography, except by using interferometric synthetic-aperture radar (InSAR).

almanac Approximate information about the orbital parameters and condition of each of the NAVSTAR or GLONASS satellites that comprise the Global Navigation Satellite System (GNSS). The almanac is kept up to date by ground controllers and broadcast by each of the satellites every 15 minutes. GPS receivers use the almanac information to determine which satellites are visible in what parts of the sky as a function of time.

altitude of ambiguity (InSAR) The amount of uncorrected topography that would produce exactly one fringe in a radar interferogram. For example, a 100-m altitude of ambiguity means that, for a given pair of radar images, a 100-m relative error within the digital elevation model (DEM) would produce one spurious fringe in the interferogram. The altitude of ambiguity is inversely proportional to the distance between the two vantage points from which the pairs of images were taken (the baseline distance). The shorter the baseline, the larger the altitude of

ambiguity. To produce a DEM from radar images, a small altitude of ambiguity is desirable because the topography will produce a large number of fringes. For ground deformation studies, a large altitude of ambiguity is preferable so the deformation interferogram will be insensitive to any topographic errors in the DEM.

ambiguity (GPS) *See* **phase ambiguity**.

amorphous Said of a mineral or other substance that lacks crystalline structure, or whose internal arrangement is so irregular that there is no characteristic external form (e.g., amorphous silica).

andesite Volcanic rock (or lava) that is characteristically medium dark in color, and contains 54 to 62 wt% silica and moderate amounts of iron and magnesium. *See also* **silica, basalt, dacite**, and **rhyolite**.

anhedral A morphological term referring to a crystal that has failed to develop its own planar crystal faces, or that has a rounded, irregular, or indeterminate form produced by the crowding of adjacent mineral grains during crystallization or recrystallization.

antenna swap A static initialization procedure for determining the integer ambiguities during a kinematic GPS survey. Two receivers track at least four satellites for several epochs, then trade places and track for a similar period before one of the receivers begins to move.

anti-spoofing (AS) A cryptographic technique employed by the US Department of Defense to deny full service to unauthorized users of the GPS. When AS is activated, the P-code is encrypted with the W-key, creating the Y-code that prevents an appropriately designed military GPS receiver from being fooled by an intentionally deceptive fake GPS signal. Anti-spoofing has been continuously activated since 31 January 1994.

aquifer A body of rock that is sufficiently permeable to enable favorable movement and storage of groundwater, yielding economically significant quantities of water to wells and springs. A *confined aquifer* is bounded by beds that are impermeable or distinctly less permeable than the aquifer itself. An *unconfined aquifer* is surrounded in part by permeable beds. Wells that are open to confined aquifers at depth can be used as strainmeters by monitoring water-level fluctuations caused by volumetric strain changes in the surrounding rock (Chapter 9).

array An ordered arrangement of monitoring instruments or measurement stations. *See also* **network**.

ascending node   The point in an orbit where a satellite crosses the Earth's equatorial plane from south to north. A satellite crosses the equator twice per orbit, once going north and again going south. These crossings are called the *orbital nodes*. The northbound crossing is the ascending node and the southbound crossing is the descending node.

aseismic   Not associated with an earthquake, as in aseismic slip along a fault or fault zone. The term is also used to describe an area not subject to earthquakes (e.g., an aseismic zone or volume). Aseismic volumes beneath active volcanoes are not uncommon, because high temperatures in the crust result in ductile behavior rather than brittle failure. Few earthquakes occur under such conditions unless the strain rate is high (e.g., during a rapid intrusion). The brittle–ductile transition commonly is shallower beneath active volcanoes than in adjacent crust, so the earthquake distribution likewise is shallower.

ash (volcanic)   Fine pyroclastic material, under 2.0 mm diameter and under 0.063 mm diameter for fine ash. The term usually refers to the unconsolidated material (e.g., airborne ash, airfall ash), but is sometimes also used for its consolidated counterpart, *tuff*.

ash flow   A ground-hugging density current, generally a hot mixture of volcanic gases, ash, pumice, scoria, and blocks traveling down the flanks of a volcano or along the surface of adjacent ground. Ash flows are produced by the explosive disintegration of viscous magma, by the explosive emission of gas-charged ash from a fissure or caldera ring-fault system, or by disintegration of hot blocks in a rockfall. The terms ash flow and *pyroclastic flow* are used interchangeably, often with descriptors to specifiy the flow's dominant constituent (e.g., pumiceous pyroclastic flow, lithic pyroclastic flow) or origin (e.g., rockfall block-and-ash flow, column-collapse pyroclastic flow).

ash-flow tuff   A *tuff* deposited by an ash flow, usually during a large, caldera-forming eruption. Well-known examples are the Bishop Tuff associated with the Long Valley Caldera and the Huckleberry Ridge Tuff, Mesa Falls Tuff, and Lava Creek Tuff associated with three successive calderas at the Yellowstone Plateau. *See* **tuff** and **ignimbrite**.

A-type earthquake   Seismic event with a clear P-wave and S-wave occurring beneath a volcano, generally at depths of 1–10 km. A-type earthquakes are indistinguishable from normal shallow tectonic earthquakes. In the original definition by Minakami (1961), they were contrasted to B-type earthquakes attributed to an extremely shallow focal depth. In recent years the contrast has been attributed to the source process, and A-type events are more often called volcano–tectonic (VT) or high-frequency (HF). *See also* **B-type earthquake**.

AVO   Acronym for the Alaska Volcano Observatory, a facility operated jointly by the United States Geological Survey (USGS), the Geophysical Institute of the University of Alaska Fairbanks (UAFGI), and the State of Alaska Division of Geological and Geophysical Surveys (ADGGS). AVO was formed in 1988 with offices in Anchorage and Fairbanks. The Observatory uses federal, state, and university resources to monitor and study Alaska's hazardous volcanoes, to predict and record eruptive activity, and to mitigate volcanic hazards to life and property. *See also* **CVO**, **HVO**, **LVO**, and **YVO**.

azimuth   Distance measured from the early azimuth line to a particular point in the footprint of an imaging radar system (i.e., the area illuminated by the radar beam) in the direction parallel to the flight path of the radar. Resolution in the azimuth direction improves as the antenna length decreases, but the antenna must be large enough to provide adequate signal-to-noise. Most imaging radars in Earth orbit use an antenna about 10-m long. *See* **ground range** and cf., **slant range**.

backsight   In leveling, a reading on a leveling rod held in its unchanged position on a survey point of previously determined elevation, known as a *turning point*, after the leveling instrument has been moved to a new position. The reading on a leveling rod on the next turning point is called the *foresight*. The elevation difference between bench marks is determined by taking an initial backsight on a starting bench mark and then leapfrogging a pair of leveling rods through a succession of backsights and foresights to the next mark.

bar   A unit of pressure in the cgs system equal to 10^6 dynes per square centimeter. Standard atmospheric pressure is taken to be 760 torr, or millimeters of mercury, which is equivalent to 1.01325 bar.

BARD   Acronym for Bay Area Regional Deformation Network. BARD is a network of about 67 continuously operating GPS receivers in northern California. The primary goal of the network is to monitor crustal deformation across the Pacific–North America plate boundary and in the San Francisco Bay Area for earthquake hazard-reduction studies and rapid earthquake emergency response assessment. Component of the Plate Boundary Observatory (PBO).

BARGEN Acronym for Basin and Range Geodetic Network, a network of continuously operating GPS stations in the western U.S.A. BARGEN is a collaborative project between the California Institute of Technology Division of Geological and Planetary Sciences and the Harvard–Smithsonian Center for Astrophysics Radio and Geoastronomy Division. The objective of BARGEN is to study and understand deformation in the ~1,000 km wide Basin and Range province, a major constituent of the Pacific–North American plate boundary zone. Historically, BARGEN was installed and operated as two sub-networks, the Northern Basin and Range (NBAR) network and the Yucca Mountain network. Component of the Plate Boundary Observatory (PBO).

barometer An instrument used to measure atmospheric pressure. Barometers are used as a means to correct some types of geodetic measurements, such as line lengths measured with an EDM, that are sensitive to atmospheric pressure.

basalt Volcanic rock (or lava) that is characteristically dark in color, contains 45 to 54 wt% silica, and is rich in iron and magnesium. *See also* **silica**, **andesite**, **dacite**, and **rhyolite**.

baseline In GPS and other types of surveying, a vector of coordinate differences between two points or an expression of the coordinates of one point with respect to the other. In the latter case, the coordinates of one of the points are assumed known and that point is referred to as a base or reference station. In common usage, the term often refers to the scalar distance between two points rather than to the vector. In SAR interferometry, the horizontal separation between vantage points for two overlapping SAR images, measured perpendicular to the SAR trajectories. To preserve the interferometric effect for orbiting SAR images, the baseline distance must be less than about 1 kilometer. Image pairs with long baselines are sensitive to topography and therefore a good choice if the goal is to produce a digital elevation model (DEM). Shorter baselines are preferable if the goal is to measure ground deformation, because the sensitivity to topographic errors is correspondingly lower. The National Geodetic Survey (1986, p. 21) prefers the spelling 'base line' for references to classical geodetic surveys and reserves 'baseline' as an adjective in Very Long Baseline Interferometry (VLBI). The latter spelling has been adopted by the GPS and InSAR communities, including NGS.

bench mark A relatively permanent, natural or artificial geodetic marker fixed to the Earth and bearing a marked point whose elevation above or below an adopted surface (datum) is known. Sometimes written 'benchmark'. A common type is an embossed and stamped disk of bronze or aluminum alloy, about 3.75 inches in diameter, with an attached shank about 3 inches in length, which can be cemented in bedrock, in a massive concrete post, or in the masonry of a substantial building or bridge abutment. A long metal rod driven to refusal is sometimes used to avoid near-surface soil creep and freeze–thaw effects. A well-defined point on the bench mark, typically a '.', '×', or '+', serves as the reference. The term is applied formally only to markers whose elevations are known, although it is used informally to refer to any point that serves as a reference for surveying purposes.

biotite A widely distributed and impotant rock-forming mineral of the mica group: $K(Mg,Fe^{+2})_3(Al,Fe^{+3})Si_3O_{10}(OH)_2$. It is generally black, dark brown, or dark green, and forms a constituent of crystalline rocks (either as an original crystal in igneous rocks of all kinds or a product of metamorphic origin in gneisses and schists) or a detreital constituent of sandstones and other sedimentary rocks. Biotite is useful in the potassium–argon method of age determination.

body wave (seismology) A seismic wave that travels through the interior of the Earth, with a propagation mode that does not depend on any boundary surface. A body wave may be either longitudinal (a P-wave) or transverse (an S-wave). *See also* **surface wave**.

bomb, volcanic A *pyroclast* that was ejected while viscous; many acquire rounded aerodynamic shapes during their travel through the air. Larger than 64 mm in size, and may be vesicular or hollow inside. Actual shape or form varies greatly, and is used in descriptive classification (e.g., breadcrust bombs, ribbon bombs, spindle bombs (with twisted ends), spheroidal bombs, and 'cow-dung' bombs). *See* **scoria**.

Bouguer correction A correction made to gravity data for the attraction of rock between the station and datum elevation (commonly sea level); or, if the station is below the datum elevation, for the rock missing between the station and datum. The Bouguer correction is $2\pi G\rho h$, where G is the universal gravitation constant, $6.67 \times 10^{-11}\,\mathrm{m^3\,kg^{-1}\,s^{-2}}$, ρ is the specific gravity of the intervening rock, and h is the difference in elevation between the station and datum. *See also* **free-air correction**.

bradyseism An episode of gradual surface uplift or subsidence, usually several kilometers to a few tens of kilometers across, caused by inflation or deflation of a magma reservoir or by pressurization or depressurization of a hydrothermal system. Bradyseisms can persist for several years, decades, or longer. The Phlegraean Fields Caldera near Naples, Italy, has exhibited bradyseisms, mostly steady subsidence punctuated by intense episodes of uplift, since the time of the Roman Empire. Bradyseisms are usually accompanied by swarms of small earthquakes and sometimes by larger earthquakes or a volcanic eruption.

breakline An x, y, z coordinate string representing the tops of ridges or bottoms of valleys and individual points (*break points*) representing peaks or objects of interest in a *stereo model*. Breaklines are collected manually and provided as input to software for producing a *digital elevation model* (DEM). During DEM creation, the manually collected points force the grid surface to closely follow these important inflections in the topography.

brine Water containing a high concentration of one or more dissolved salts is called a brine. In natural systems brines generally contain high concentrations of a few dissolved salts, particularly NaCl, KCl, and $CaCl_2$, and smaller concentrations of many different salts and metals, including Fe, Cu, Pb, Zn, Ag, Au, and Pt. Pore fluids in deep sedimentary basins, oil-field waters, and geothermal mineralizing fluids generally are brines. When a brine is evaporated at the surface of the Earth its salinity slowly increases until a point is reached at which precipitation of salts commences. With additional evaporation the continued precipitation of salts prevents the total concentration of the remaining dissolved salts in the brine from going above about 24 to 25 wt%. However, the solubilities of chloride-bearing salts generally increase as temperature and pressure increase going deeper into the Earth, so brines containing much more than 25 wt% dissolved salts may be encountered there. Such brines are referred to here as *hypersaline*. Hypersaline brines are likely to form where water-bearing fluids are released at very high temperatures and high pressures from crystallizing magmas a few to several kilometers deep beneath volcanoes. The composition of a particular hypersaline brine in a volcanic system is determined by the initial composition of the evolved magmatic fluid, the temperature and pressure at which the magma crystallizes, and by subsequent reactions of that brine with surrounding rock as temperature and pressure decrease (Chapter 10).

brittle Describes a rock that fractures at less than 3–5% deformation or strain. Most shallow earthquakes are associated with brittle-rock fracture. At depths greater than 5–10 km in active magmatic systems, temperature and pressure are sometimes high enough that aseismic, ductile deformation occurs instead. *See* **strain**.

broadcast orbit *See* **ephemerides**.

B-type earthquake Seismic event occurring under a volcano with emergent P-wave onset, no distinct S-wave, and a low-frequency content as compared with tectonic earthquakes of the same magnitude. Minakami (1961) used these characteristics, which differ from those of A-type earthquakes, to distinguish between the two types. He attributed the character of B-type earthquakes to an extremely shallow focal depth (<1 km). More recently the differences have been attributed to the source process and now B-type events are called more often long-period (LP) or low-frequency (LF). *See* **A-type earthquake**, **P-wave**, **S-wave**, and **emergent**.

bulk modulus A modulus of elasticity that relates a change in volume (volumetric strain) to the hydrostatic state of stress. It is the reciprocal of compressibility (also called compliance). The bulk modulus K of an elastic material is defined as the ratio of the pressure change ΔP to the resulting volumetric strain or dilatation, $K = \Delta P/\Delta$, where $\Delta = \Delta V/V$. *See also* **bulk modulus**, **modulus of elasticity**, **Poisson's ratio**, **shear modulus**, **compliance**, and **Young's modulus**.

C/A-code A pseudorandom-noise code used to modulate the carrier frequency of signals broadcast by NAVSTAR satellites, which form the heart of the US Global Positioning System (GPS). The coarse/acquisition or C/A-code repeats itself every millisecond, and the algorithm used to generate it is not classified. *See also* **P-code**, **W-key**, and **Y-code**.

cadastre An official register of the location, quantity, value, and ownership of real estate, compiled to serve as a basis for taxation. The cadastral concept is commonly extended to include geodetic control, topographic base maps, land information keyed to the base maps, and real estate files linked to the overlays, all of which serve to supply information for juridical, fiscal, environmental, and statistical purposes. Such a multipurpose cadastre is designed to support continuous, readily available, and comprehensive land-related information at the parcel level, and it provides a common framework for a geographic information system (GIS).

caldera A large, roughly circular or oval-shaped volcanic depression with a diameter many times larger than that of included volcanic vents; may range from 2 to 50-km across. Calderas form by collapse, erosion, or explosion. The largest calderas form when the roof of a magma chamber collapses during eruption of a large volume of silicic magma to form an extensive *ash-flow tuff*. Classic examples of collapse calderas and associated *tuffs* in the USA include Crater Lake (Wineglass Tuff), Long Valley Caldera (Bishop Tuff), Valles Caldera (Bandelier Tuff), and Yellowstone Caldera (Lava Creek Tuff). *See* **ash-flow tuff**, **tuff**.

carrier-beat phase The phase of the signal that remains when the incoming Doppler-shifted carrier signal from a GPS satellite is beat (i.e., the difference frequency signal is generated) with the reference frequency generated in a GPS receiver. The L1 and L2 carrier-beat phases are two GPS observables.

Cascadia The region between the Queen Charlotte Sound in British Columbia and Cape Mendocino in northern California including the Explorer plate, Juan de Fuca plate, Gorda plate, and inland about 300 km from the Cascadia Subduction Zone to include the Cascade Range.

cauldron subsidence A structure resulting from the lowering along a steep ring fracture of a more or less cylindrical block into a magma chamber; usually associated with *ring dikes*.

CCD *See* **charge-coupled device**.

CDMA Acronym for Code Division Multiple Access, a scheme employed by GPS in which all NAVSTAR satellites broadcast on the same two carrier frequencies but use different pseudo-random noise (PRN) codes to modulate the carrier signals. *See also* **FDMA**.

chain A measuring device used in land surveying, usually consisting of 100 links joined together by rings. The term is commonly used interchangeably with *tape*, although strictly a chain is a series of links and a tape is a continuous strip. In the USA, the chain is also a unit of length prescribed by law for the survey of public lands and equal to 66 feet or 4 rods. Ten square chains equals one acre. In areas where ground deformation is intense but hazards are acceptably low, a tape or chain can be a useful volcano-monitoring tool. Several of the dome-building eruptions at Mount St. Helens during 1980–1986 were predicted partly on the basis of tape measurements of the changing widths of cracks or the growth of thrust faults on the adjacent crater floor.

charge-coupled device (CCD) A semiconductor image-sensing device, usually in the form of a plane-rectangular matrix of microscopic, individual sensing elements, each of which corresponds to an image pixel. A CCD is typically placed at the focal plane of an optical imaging system, where it converts light intensity into electrical signals that represent the information contained within each pixel. CCDs are commonly used in digital cameras, telescopes, scanners, and bar code readers.

chip (GPS) The carrier signals broadcast by NAVSTAR and GLONASS satellites are modulated by binary pulse codes called pseudorandom noise (PRN) codes (i.e., noise-like sequences of binary values 0 or 1, corresponding to $+1$ or -1 states of the code, respectively, which repeat at fixed intervals). Each 0 or 1, or in some contexts the length of time required to transmit each 0 or 1, is called a chip. The number of chips transmitted per second is called the chip rate. For the GPS coarse/acquisition (C/A-) code, these values are approximately 977 ns and 1.023 MHz, respectively. The C/A-code is 1,023 chips long, so it repeats every millisecond and each chip corresponds to a travel distance at the speed of light of 293 m. *See* **pseudorandom-noise code**.

chip rate The number of chips in a binary pulse code transmitted per second. *See* **chip**.

chirp A brief radar pulse whose frequency increases linearly with time. This specific signal structure has useful properties for SAR signal processing (Chapter 5). If such a pulse were in the auditory frequency range, it would sound like a chirp.

choke ring (GPS) Concentric rings of metal around and below the ground plane of a GPS antenna, in some designs separated by microwave-absorbing foam, for the purpose of further reducing multipath effects produced by the geometry of the antenna relative to its local surroundings. *See also* **ground plane**.

circuit (surveying) A *level line* or series of lines that form a loop back to the starting point.

clinoscope A precursor to modern tiltmeters, the clinoscope was used to measure ground tilt at the Hawaiian Volcano Observatory (HVO) during the 1930s. The instrument consisted of a heavy, ring-shaped weight hung by a piano wire from a tripod. The weight dipped into a bath of automobile oil for damping, and a boom that extended through the hole in the weight magnified small deflections by a factor of ~ 50.

closure *See* **misclosure**.

code pseudorange Commonly referred to as pseudorange. A measure of the apparent propagation time from a GPS satellite to a receiver antenna, expressed as a distance. The apparent propagation time is determined from the time shift required to align a replica of the pseudo-random-noise code (PRN) that is generated in the receiver with the code received from a GPS satellite. The time shift is the difference between the time of signal reception (measured in the receiver time frame) and the time of transmission (measured in the satellite time frame). Code pseudorange is obtained by multiplying the apparent signal propagation time by the speed of light. Code pseudorange differs from the actual range by the amount that the satellite and receiver clocks are offset and as a result of propagation delays that occur in the ionosphere and troposphere. The absolute accuracy of positions determined from code pseudoranges is generally ± 5–50 m.

coda The concluding train of seismic waves that follows the principal waves from an earthquake. Coda waves are due to scattering and superposition of multipath arrivals. The coda from great earthquakes may persist for hours.

coherence Property of a radar *interferogram* related to the amount of change in the positions or radar-reflective properties of scatterers within corresponding pixels in the two images used to produce the interferogram. To measure surface displacements by radar interferometry, the deforming area must be mostly coherent; otherwise, phase-difference information is lost and the area is said to be 'incoherent'. Factors that destroy coherence include surface disruption by natural or cultural processes (e.g., erosion or agriculture); changes in soil moisture, which affect the dielectric constant and thus the radar-reflective properties of near-surface material; and changes in vegetation, snow cover, or surface properties that affect radar reflectivity. *See* **interferometry**.

compliance In elasticity, the ratio of strain to stress. Compliance is the reciprocal of modulus. For example, the reciprocal of shear modulus is shear compliance. The reciprocal of bulk modulus is simply called compliance or compressibility. *See also* **bulk modulus, compressibility, shear modulus, stress**, and **strain**.

compressibility The reciprocal of *bulk modulus*, also called *compliance*. For elastic materials, compressibility is the ratio of volumetric strain to stress, $\beta = \Delta/\Delta P$, where $\Delta = \Delta V/V$ and ΔP is the pressure change.

cone sheet A dike that is arcuate in plan and dips at 30–$45°$ toward the center of the arc. Cone sheets generally occur in concentric sets, which presumably converge at a magmatic center. They commonly are associated with *ring dikes* to form a ring complex. *See* **ring dike**.

constellation In GPS context, the group of NAVSTAR satellites that form the heart of the US Global Positioning System. Normally, this includes 21 active satellites and 3 orbiting spares arranged in 6 orbit planes. This configuration assures that at least 4 satellites are above the horizon at all times from any point on Earth. The fully deployed Russian GLONASS constellation comprises 24 satellites in 3 orbit planes. The latter system was not fully operational in 2006, but additional launches are planned.

constitutive equation A relation between two physical quantities that is specific to a material or substance, and does not follow directly from physical law. It is combined with other equations that do represent physical laws to solve some physical problem, like the flow of a fluid in a pipe. In solid mechanics, an equation that connects stress, elastic and/or plastic strain and possibly strain rate, experimental and material parameters. The simplest example of a constitutive equation is *Hooke's law*, which states that, for relatively small deformations of an object, the displacement or size of the deformation is directly proportional to the deforming force or load.

control point In photogrammetry, a station on the ground with a known position that is identified in a photograph and used to establish the positions of other points in a fixed coordinate system. A common type of control point is a *bench mark* surrounded by a photo target. For a photogrammetric survey, the locations of an array of bench marks are established geodetically, usually by *GPS*, and the photo targets are identified in a photograph of the area using a *stereoplotter*.

convergence angle (stereoscopy) *See* **parallactic angle**.

corner reflector An optical reflector designed in such a way that light incident on the reflector is redirected back to its source on a path parallel to the incidence direction. In one design, a corner is cut from a highly polished cube of glass in which the three intersecting planes are precisely perpendicular. Light enters the cube through the cut surface and is reflected back to the source by the polished cube faces. One use of a corner reflector is to return a light beam from an *EDM* (electro-optical distance meter), which measures the EDM-to-reflector distance by comparing the phases of outgoing and reflected beams. Also called a corner-cube reflector, corner cube, corner prism, or retrodirective prism.

country rock Rock enclosing or traversed by a mineral deposit, or intruded by and surrounding an igneous intrusion. The term is somewhat less specific than *host rock*.

cryptodome A very shallow intrusion of magma, usually into a volcanic cone, that causes severe deformation of the ground surface. Two well-known examples are the Showa–Shinzan cryptodome that formed at Usu Volcano, Japan during 1944–1945 and the Mount St. Helens cryptodome responsible for the north flank bulge and giant landslide of 18 May 1980. In the latter case, a dacite cryptodome that was growing within the Mount St. Helens edifice was torn apart by the May 18 landslide, which triggered a massive lateral explosion.

cumulo-volcano A dome-shaped volcano constructed of multiple domes and flows. Mammoth Mountain, located on the southwest rim of Long Valley Caldera in east-central California, is an excellent example.

Curie point The temperature above which thermal agitation prevents spontaneous magnetic ordering. At temperatures above the Curie point, ferromagnetism disappears and a substance becomes simply paramagnetic.

CVO Acronym for the USGS David A. Johnston Cascades Volcano Observatory, which was established in Vancouver, Washington, in response to renewed activity at Mount St. Helens in 1980. The Observatory was formally dedicated on 18 May 1982, and is named in memory of USGS geologist David A. Johnston, who died in the cataclysmic eruption of 18 May 1980. CVO monitors the Cascade volcanic arc in the US Pacific Northwest (Washington, Oregon, and northern California), assesses the associated volcanic and hydrologic hazards, and provides timely warnings of impending volcanic activity. It is also the home base of the Volcano Disaster Assistance Program (VDAP). *See also* **AVO**, **HVO**, **LVO**, and **YVO**.

cycle slip The result of an interruption in the integer carrier-phase count maintained by a GPS receiver while tracking a satellite, which introduces a cycle ambiguity. Cycle slips can be caused by sky-view obstructions that block the incoming satellite signal, by low signal-to-noise ratio due to such factors as noisy ionospheric conditions, multipath errors, or low satellite-elevation angle, and by incorrect signal processing by the receiver. Most cycle slips can be removed during data processing.

dacite Volcanic rock (or lava) that characteristically is light in color and contains 62 to 69 wt% *silica* and moderate amounts of sodium and potassium. *See also* **silica**, **andesite**, **basalt**, and **rhyolite**.

dashpot A mechanical damper; the vibrating part is attached to a piston that moves in a chamber filled with liquid. The L4C geophone uses a dashpot to damp the motion of its *proof mass* (Section 3.1.3 and Figure 3.2).

datum For mapping applications, a set of constants used to specify the coordinate system used for geodetic control. At least eight constants are needed to form a complete datum: three to specify the location of the origin of the coordinate system, three to specify the orientation of the coordinate system, and two to specify the dimensions of the *reference ellipsoid*. Before geocentric datums became possible, it was customary to define a geodetic datum by five quantities: the latitude and longitude of an initial point, the azimuth of a line from this point, and two parameters specifying the dimensions of a reference ellipsoid. In addition, specification of the *vertical deflection* at the initial point, or the condition that the minor axis of the ellipsoid be parallel to the Earth's axis of rotation, provided two more quantities. Such a datum, however, was still not complete because the origin of the coordinate system remained free to shift in one dimension. Different countries often used the same reference ellipsoid but with different offsets to account for this ambiguity. The adoption of geocentric reference ellipsoids, notably the World Geodetic System 1984 (WGS 84), greatly simplified the specification of a worldwide datum. For surveying purposes, a datum is any fixed or assumed position or element (such as a point, line, or surface) in relation to which others are determined. For example, the line between a theodolite station and distant point that serves as a reference for horizontal angle measurements. For leveling surveys, the height of a reference bench mark can be assumed to be fixed between surveys and thus serve as a datum for the purpose of determining relative height changes among other marks along a level line.

dB Abbreviation for decibel(s). In electronics and communications, the decibel is a logarithmic expression of the ratio between two signal power, voltage, or current levels. A decibel is one-tenth of a Bel, a seldom-used unit named for Alexander Graham Bell, inventor of the telephone. Decibels can be defined as $dB = 10 \times \log_{10}(P_1/P_2)$, where P_1 and P_2 represent signal powers, or equivalently, $dB = 20 \times \log_{10}(V_1/V_2)$, where V_1 and V_2 represent voltages. A value of 6 dB corresponds to the case where $V_1 = 2 \times V_2$ (i.e., a doubling of the input or output voltage). The dynamic range of many electronic systems, including sensors and telemetry systems used for volcano monitoring, are typically expressed in dB. For example, an analog-telemetry system with 40 dB of dynamic range is capable of handling input voltages that range over a factor of 100. Modern broadband seismometers have >100 dB of dynamic range and digital-telemetry systems used for seismic networks are typically 16-bit, which corresponds to 96 dB of dynamic range. Thus, both the sensor and telemetry are capable of handling input voltages that range over a factor of ~64,000 (i.e., $2^{16} = 65,536$ and $96 = 20 \times \log_{10}(63,096)$.

debris avalanche A rapid sliding and flowage of rock, snow, ice, and other debris mobilized by gravity. The term is used to refer both to the process and the resulting deposit. Volcanic activity, an earthquake, heavy rainfall, or rapid melting of snow can trigger debris avalanches originating at volcanoes. Some may occur without a recognizable trigger as a result of gradual destabilization of steep slopes by hydrothermal activity. The types of volcanic activity that cause debris avalanches include deformation of a volcanic edifice by intrusion, phreatic explosions, and slumping of a caldera wall (Ui, 1989). Debris avalanches range in size from small movements of loose debris on the surface of a volcano to massive failures of the entire summit or flank of a volcano. Volcanoes are especially susceptible to debris avalanches because: (1) they are composed of layers of weak, fragmented rocks often deposited on steep slopes; (2) hot acidic groundwater that forms as a result of hydrothermal activity can alter competent volcanic rocks to soft, slippery clay, which further destabilizes overlying material; and (3) intrusions and earthquakes, which are among the leading causes of debris avalanches, and relatively common at volcanoes. At least five large debris avalanches swept down the slopes of Mount Rainier during the past 6,000 years. The largest debris avalanche in historical time occurred at Mount St. Helens on 18 May 1980.

décollement A detachment structure above and below which the styles of deformation in the rocks are independent. The south flank of the Big Island of Hawai'i is underlain by a subhorizontal décollement at about 10-km depth that may correspond to the contact between the base of the volcano and underlying oceanic crust. Episodic slip along this décollement produces major earthquakes, such as the M 7.2 Kalapana earthquake of 29 November 1975.

decorrelation In SAR interferometry, term used to describe the situation in which two radar images of the same area cannot be registered on a pixel by pixel basis, owing to one of several factors. Spatial decorrelation occurs when the baseline between two images is too large or uncertain (i.e., the separation between vantage points is too great, or the trajectory (orbit) is poorly constrained). Temporal decorrelation occurs because the scattering properties of a target area change with time as a result of changes in vegetation, soil moisture, snow cover, erosion or deposition, etc. If the changes are so large and extensive that the images cannot be registered, they are said to be decorrelated.

deflation In volcanology, a term used to describe de-tumescence or shrinking of the ground surface, a volcanic edifice, or a subsurface magma body (the opposite of *inflation*). Deflation is a response to depressurization, which can have several causes including magma withdrawal (related to intrusion or eruption), volatile escape, thermal contraction, phase changes during crystallization, and tectonic extension. *See* **inflation**.

deflection of the vertical *See* **vertical deflection**.

deformation Any change in the shape or dimensions of a body resulting from stress. Deformation and *strain* are synonymous in most contexts, although deformation is a more general term. In geology, deformation refers broadly to the process of folding, faulting, shearing, compression, or extension of rocks as a result of various Earth forces.

DEM (digital elevation model) Any digital representation, usually in raster form, of the continuous variation of ground relief as a function of position on the Earth's surface. A typical DEM includes values for latitude, longitude, and elevation (or an equivalent set of parameters) at each raster point. Raster size might vary from 30 m or less for 1:24,000 or 1:62,500-scale topographic maps to 1 km for a global DEM. DTM (digital terrain model) is nearly synonymous with DEM. DTED (digital terrain elevation data) refers to a specific format for digital elevation data provided by the USGS. *See* **DTM**.

deviatoric stress Term used loosely for the stress difference that results in distortion of a body, synonymous with shear stress. However, Engelder (1994), in an article entitled 'Deviatoric stressitis: A virus infecting the earth science community', challenged common usage and provided a mathematical exposition of the stress matrix to which the term correctly refers. To avoid confusion, the term is used in this book only in a direct quote that is attributed to its original source.

diapositive A positive photograph on a transparent medium, usually polyester or glass.

dielectric constant Electrical property of matter that influences radar returns. Also referred to as complex dielectric constant. Materials with a higher dielectric constant are more reflective of electromagnetic energy. The dielectric constant consists of two parts (permittivity and conductivity) that are both highly dependent on the moisture content of the material considered. In the microwave region, most natural materials have a dielectric constant between 3 and 8 in dry conditions. Water has a high dielectric constant (80), at least 10 times higher than for dry soil. As a result, a change in moisture content generally causes a significant change in the dielectric properties of natural materials. Increasing moisture is associated with increased radar reflectivity.

diffusivity, hydraulic *See* **hydraulic diffusivity**.

diffusivity, thermal *See* **thermal diffusivity**.

dike A tabular intrusion of magma that cuts across the bedding or foliation of surrounding rock. Dikes are distinguished from *sills*, which are concordant (parallel) with bedding or foliation. Dikes typically dip steeply, whereas sills dip gently. However, the dip of an intrusive body is not diagnostic: a vertical intrusion along a vertical bedding or foliation plane is a sill, and a horizontal intrusive body that cuts across vertical bedding or foliation is a dike. *See* **sill**.

dilatometer Synonym for volumetric strainmeter. An extremely precise design (sensitivity ~1 part per trillion or 10^{-12}) that has been used successfully for volcano monitoring and other applications is the Sacks–Evertson volumetric strainmeter. The sensor consists of a stainless steel cylinder filled with silicon oil and rigidly attached to the walls of a borehole, preferably 100 m or more below the surface. Strain in the surrounding rock squeezes or stretches the walls of the container, causing fluid to flow out of or into it. The design is akin to a plastic water bottle with a narrow soda straw: squeezing the bottle even a small amount causes water to move up the straw a considerable distance, so a small amount of volumetric strain produces a large response that can be precisely measured. See Chapter 3 for a more complete description.

dilatation (or dilation) Deformation by a change in volume but not shape (i.e., volumetric expansion or contraction). The Sacks–Evertson volumetric strainmeter, or dilatometer, is sensitive to dilatational strain.

dithering The introduction of digital noise. This is the process the US Department of Defense used to add inaccuracy to satellite-clock signals from NAVSTAR satellites. Dithering of satellite-clock information and reduced accuracy of broadcast orbits were the means used to implement Selective Availability (SA). This practice, which was intended to withhold full military capability from civilian users, was discontinued in May 2000.

dome   *See* **cryptodome**, **resurgent dome**, **volcanic dome**, and **lava dome**.

dry tilt   Equivalent to *single-setup leveling*. The term dry tilt was coined by staff at the USGS Hawaiian Volcano Observatory in the late 1960s to distinguish this differential leveling technique from the *wet tilt* method, in which water hoses were used to connect small reservoirs attached to three concrete piers, thus forming a long-base watertube tiltmeter.

DTM (digital terrain model)   Similar to a DEM (digital elevation model), but may incorporate the elevation of significant topographic features and break lines that are irregularly spaced (e.g., ridge crests, valley bottoms) to better characterize the shape of bare Earth terrain.

ductile   Describes a rock able to sustain, under a given set of conditions, a 5–10% deformation before fracturing or faulting. Ductile behavior is favored at high temperature and pressure. In active magmatic systems, the transition from brittle to ductile behavior typically occurs at 5–10 km depth. This explains why most volcanic earthquakes occur within 10 km of the surface. For earthquakes to occur at greater depth, the strain rate must be exceptionally high.

earthquake magnitude   A measure of the strength of an earthquake, or the strain energy released by it, as determined by seismographic observations. Local magnitude M_L is based on the maximum amplitude of a seismogram recorded on a standard Wood–Anderson torsion seismograph. Although these instruments are no longer widely in use, M_L values are calculated using modern instrumentation with appropriate adjustments. Magnitudes determined at teleseismic distances are called body-wave magnitude M_b and surface-wave magnitude M_s. Moment magnitude M_w is a measure of strain-energy release, equal to the rigidity of the Earth times the average amount of slip on the fault times the area of fault that slipped. The local, body-wave, surface-wave, and moment magnitudes of an earthquake do not necessarily have the same numerical value, although any differences are usually small.

EBRY   Eastern Basin-Range (Wasatch Front) and Yellowstone Hotspot (Yellowstone–Snake River Plain) GPS Network (University of Utah). This network of continuously operating GPS stations focuses on the overall strain and deformation field of the Basin-Range province that includes Yellowstone and the track of the Yellowstone hotspot across the Snake River Plain as well as the 370 km long Wasatch fault zone. Component of the Plate Boundary Observatory (PBO).

EDM   Acronym for electro-optical distance meter, electronic distance meter, or electronic distance measurement. An EDM directs a narrow beam of amplitude-modulated, monochromatic light (modern EDMs use lasers) onto a distant reflector and measures the EDM-to-reflector distance by comparing the phases of outgoing and reflected beams. The result depends on the speed of light in air, which in turn depends on density (temperature and pressure) and water vapor concentration (humidity). If these parameters are measured from an aircraft flying along the line-of-sight during a measurement, or at the endpoints of the line being measured, corrections can be applied to compensate for differing atmospheric conditions between surveys. A better approach is to measure the line length with two or more light beams of different wavelengths. Atmospheric effects on the speed of light in air are known functions of wavelength, so the travel time difference between beams can be used to reference each measurement to a standard set of atmospheric conditions. The precision of a state-of-the-art 'two-color' laser EDM used by the USGS to monitor horizontal strain along the San Andreas fault system at Parkfield and at the Long Valley Caldera, both in California, is $[0.3^2 + (0.12 \times L)^2]^{1/2}$, or 1.3 mm for a 10 km long line.

EGNOS   Acronym for European Geostationary Navigation Overlay Service, a Satellite-Based Augmentation System (SBAS) for Europe. *See* **SBAS**.

elastic   Describes a rock in which strains are instantly and totally recoverable and in which deformation is independent of time. *See also* **plastic**.

electrical self-potential   *See* **self-potential**.

ellipsoid   A mathematical figure, also called the *reference ellipsoid* or ellipsoid of revolution, which is the best-fit surface to the mean shape of the solid Earth. The ellipsoid approximates the shape of the *geoid* but, unlike the geoid, is not affected by the inhomogeneous mass distribution within the Earth. The ellipsoid is widely used as the reference surface for horizontal coordinates (e.g., latitude and longitude) and for GPS observations. The difference between the ellipsoid and geoid, which is called the *geoid undulation* or *ellipsoid–geoid separation*, accounts for the difference between heights determined by leveling and GPS (Figure 2.1).

ellipsoid–geoid separation The vertical separation between the *geoid* and *ellipsoid*. For a global reference ellipsoid for the Earth, the geoid undulation can be as large as 100 m. Also called the *geoid undulation*. *See also* **ellipsoidal height**, **orthometric height**.

ellipsoidal height The perpendicular distance between a point and the *reference ellipsoid*. The relationship between ellipsoidal height h and *orthometric height H* is: $h = H + N$, where N is called the *geoid undulation* or *ellipsoid–geoid separation*. Heights determined by GPS are ellipsoidal, whereas heights determined by leveling are geoidal. The difference can be as large as 100 m.

emergent In seismology, term used to describe an earthquake signal with a gradual onset (i.e., a signal that 'emerges' from the background trace on a seismogram); also applied to an earthquake or seismic wave producing such a signal (e.g., *B-type earthquakes*, rockfalls). *See* **impulsive**.

endogenous Describes a volcanic dome that has grown primarily by expansion from within (i.e., by intrusion of magma into the dome edifice), as distinguished from an *exogenous* dome that was built primarily by extrusion of viscous lava onto the surface of the dome. The dacite dome that grew at Mount St. Helens from 1980 to 1986 was built both exogenously and endogenously. The proportion of endogenous growth increased with time, as the dome grew larger and thus better able to accommodate intruding magma internally. *See* **exogenous**.

ephemeris A set of orbital parameters for the NAVSTAR or GLONASS satellites. This information is determined by the control segments, updated periodically, and then broadcast by each of the satellites every 15 minutes. This version constitutes the so-called broadcast ephemeris (or broadcast orbits) and can be used to obtain timely positioning information. Better solutions are possible using improved ephemerides, which are available from several sources including the International GPS Service (IGS).

ephemerides Plural of *ephemeris*.

epicenter (seismology) The point on the Earth's surface vertically above the hypocenter (or focus) of an earthquake (i.e., above the point in the crust where a seismic rupture begins). *See also* **hypocenter**.

epigenetic Refers to: (1) ore bodies formed by

hydrothermal fluids and gases that were introduced into the host rocks from elsewhere, filling cavities in the host rock; or (2) mineralization that was deposited later than its immediate host rocks, typically forming a vein. The ore (mineralization) is younger than the host rocks.

epithermal Refers to: (1) a low-temperature, hydrothermal mineral deposit formed within about 1 km of the Earth's surface and in the temperature range of 50°C to 200°C, occurring mainly as veins; or (2) the environment or conditions in which such deposits form.

error, random An error produced by irregular causes whose effects on individual observations are governed by no known law that connects them with circumstances and so cannot be corrected by use of standardized adjustments. For geodetic leveling observations, the residual random error after all appropriate corrections have been applied to field data is proportional to the square root of the distance measured along the level line.

error, systematic An error whose algebraic sign and, to some extent, magnitude bears a fixed relation to some condition or set of conditions. In theory at least, a systematic error is predictable and, therefore, is not random; such errors are regular, and so can be determined a priori. Examples of systematic error in geodetic leveling observations include refraction error, rod-scale error, and pin settling. Refraction error can be partly removed using measured or estimated temperatures at two heights above the ground, typically 0.5 m and 1.5 m. Rod-scale error can be removed by having the leveling rods calibrated and applying the resulting corrections to field observations. The effect of pin settling (the tendency for a rod to settle slightly between the time it is read as a foresight in one setup and then as a backsight in the next) can be minimized by running adjacent sections in opposite directions.

essential Describes pyroclasts that are formed directly from magma, equivalent to the term *juvenile*. *See also* **accessory**, **accidental**, and **pyroclast**.

exogenous Describes a volcanic dome that has grown primarily by extrusion of viscous lava onto its surface, as distinguished from an endogenous dome that was built primarily by internal expansion (i.e., intrusion of magma into the dome edifice). The complex dacite dome that grew at Mount St. Helens from 1980 to 1986 was built both exogenously and endogenously. The proportion of endogenous growth increased with time, as the dome grew larger and thus better able to accommodate intruding magma internally. Likewise, the 2004–2006 dome at Mount St. Helens grew both exogenously and endogenously. *See* **endogenous**.

extensometer An instrument used for measuring small amounts of expansion or contraction. A simple design consists of a wire or rod stretched across a ground crack and attached at one end to a linear-displacement transducer, which produces a voltage proportional to the amount of opening or closing of the crack. Other designs use lasers or interferometers to measure very small changes in length.

extrusion The emission of lava onto Earth's surface; also, the rock so formed (e.g., a lava flow or a volcanic dome). Extrusive rocks, such as basalts, form when molten lava is expelled or flows onto the surface. The process is generally non-explosive.

Fast Fourier Transform (FFT) An algorithm for computing the Fourier transform of a set of discrete data values, thereby converting data from the time domain to the frequency domain. The FFT is often used in signal processing (e.g., an FFT algorithm can be used to co-register synthethic aperture radar (SAR) images accurately and efficiently).

FDMA Acronym for Frequency Division Multiple Access, a scheme employed by GLONASS, in which all satellites use the same pseudorandom noise (PRN) codes to modulate their carrier signals, but each satellite uses different carrier frequencies. *See also* **CDMA**.

feldspar A group of abundant rock-forming minerals of general chemical formula: $MAl(Al,Si)_3O_8$, where $M = K$, Na, Ca, Ba, Rb, Sr, or Fe. The most widespread of any mineral group, feldspars constitute $\sim 60\%$ of the Earth's crust.

ferromagnetism A type of magnetic order in which all magnetic atoms in a domain have their moments aligned in the same direction. Iron-rich basalts are commonly ferromagnetic. *See also* **paramagnetism**.

fiducial mark In photogrammetry, an index or point used as a basis of reference. One of typically eight index marks, four on each side and one in each of four corners, integral to a camera and lens (as on the metal frame that encloses the negative) that form an image on the negative or print such that lines drawn between opposing points inter-

sect at and thereby define the *principal point* of the photograph (Chapter 6).

firmware Software (programs or data) that has been written onto read-only memory (e.g., ROM, PROM, EPROM, EEPROM), where it is retained in the absence of electrical power. Firmware is used to control or add functionality to a computing device such as a printer, digital camera, or GPS receiver. Firmware can be thought of as 'hard software' that is retained until it is modified by the user.

FLIR Acronym for forward looking infrared radiometer, an airborne or vehicle mounted, electro-optical infrared imaging device that detects far-infrared energy, converts the energy into an electronic signal, and provides a visible image for day or night viewing. FLIR systems can be used to map variations in surface temperature and, with calibrated systems, to make quantitative temperature measurements remotely.

focal point (optics) A general term for the distance between the center, vertex, or rear *nodal point* of a lens (or the vertex of a mirror) and the point at which the image of an infinitely distant object comes into critical focus. The term must be preceded by an adjective such as 'equivalent' or 'calibrated' to have a precise meaning.

footprint For imaging radar systems, the area on the ground illuminated by the radar beam. The footprint of the ERS-1 and ERS-2 satellites is approximately 100-km long in the ground-range ('range') direction and 5-km wide in the azimuth direction. More generally, the instantaneous field-of-view of an airborne or satellite remote-sensing instrument. See *ground range, azimuth*.

foot plate *See* **turning plate**.

foreshortening A type of spatial distortion in radar images whereby terrain slopes facing a side-looking radar's illumination are mapped as having a compressed ground-range scale relative to flat-lying areas. The effect is more pronounced for steeper slopes and for radars using steeper incidence angles. Range-scale expansion, the complementary effect, occurs for slopes that face away from the radar. *See also* **layover** and **shadowing**.

foresight A reading taken on a level rod to determine the elevation of the point on which the rod rests. By taking a backsight at a bench mark of known elevation and leapfrogging a pair of rods ahead of the leveling instrument, the elevation difference between the starting bench mark and the next in a series can be determined with millimeter-scale accuracy. *See* **backsight**.

forward modeling *See* **inversion (numerical modeling)** and Chapter 8.

free-air correction A correction made to gravity data for the difference in elevation between the station and datum. The first term of the free-air correction is $0.3086 \, \text{mgal m}^{-1}$. *See* **Bouguer correction**.

fringe (radar interferometry) Color band corresponding to h_a meters of topographic relief in a radar interferogram that retains topographic information; or to half a wavelength of ground-range change in a topography removed interferogram, where h_a is the altitude of ambiguity for the interferogram. *See* **ground range**, **interferogram**, **altitude of ambiguity**.

FTP Acronym for File Transfer Protocol, an agreed-upon method for transferring files between computers on the Internet.

fumarole A volcanic vent from which gases and vapors are emitted. Fumaroles may occur along a fissure or in apparently chaotic clusters or fields. A sulfurous fumarole is sometimes called a *solfatara*, after the type locality at the Phlegraean Fields (Campi Flegrei) Caldera near Naples, Italy. A minor eruption occurred at Solfatara in 1198, and the vent still emits sulfurous gases. *See* **solfatara**.

gabbro The approximate intrusive equivalent of *basalt*.

Gal (gravity) A unit of acceleration used in gravity measurements. One $\text{Gal} = 1{,}000 \, \text{mGal} = 10^6 \, \mu\text{Gal} = 1 \, \text{cm s}^{-2}$. Earth's normal gravity is $980 \, \text{Gal}$.

Galileo A global-navigation satellite system being developed by the European Space Agency (ESA). The first Galileo satellite was launched from the Baikonur Cosmodrome in Kazakhstan on 28 December 2005. The constellation is expected to become operational in 2008. In addition to supporting navigation and geodesy, like its US and Russian counterparts (GPS and GLONASS, respectively), Galileo will provide a global Search and Rescue function. For navigation and geodetic applications, Galileo will be interoperable with GPS and GLONASS.

geodetic datum *See* **datum**.

geodetic moment A measure of the energy associated with rock deformation, such as slip on a fault or crustal uplift. The geodetic moment for a volume source producing a given deformation pattern is given by $M_0^{(g)} = a\kappa\Delta V$, where a is a

constant depending on source geometry, κ is an elastic modulus, and ΔV is the change in the source volume required to produce the observed deformation (Aki and Richards, 1980). For a Mogi source, for example, $M_0^{(g)} = (\lambda + 2\mu)\Delta V$, where ΔV is the volume change and λ is Lame's constant. For a shear dislocation (slip on a fault), the geodetic moment equals the *seismic moment*. In cases where the geodetic moment is much larger than the cumulative seismic moment, which is common at volcanoes, deformation is driven primarily by aseismic processes associated with mass transport (e.g., magma movement) (Hill *et al.*, 2003).

Geodimeter Acronym for geodetic distance meter and a trade name for a specific brand of electro-optical distance meter (EDM). In common usage, 'geodimeter' is sometimes used interchangeably with the more general term EDM.

Geodolite A trade name for a type of electro-optical distance meter (EDM).

geoid A theoretical surface that is perpendicular at every point to the direction of gravity (the plumb line), equivalent to the shape of a sea level surface extended through the continents. The geoid is widely used as the reference surface for vertical coordinates (heights). Its shape reflects the distribution of mass inside the Earth and therefore differs from that of the *ellipsoid*, which is used as the reference surface for horizontal coordinates and for GPS. The two surfaces can be related to one another if the local gravity field is well known.

geoid undulation *See* **ellipsoid–geoid separation**.

GEONET Acronym for GPS Earth Observation Network. A dense array of more than 900 continuous GPS stations distributed throughout Japan, operated by the Geographical Survey Institute of Japan. The primary goal of the array is to monitor crustal deformation associated with large earthquakes and other tectonic processes.

geophone *See* **seismometer**.

GIS Acronym for geographic information system, a computer system capable of assembling, storing, manipulating, and displaying geographically referenced information. In common usage, a GIS also comprises digital spatial datasets, including base maps and any other information that can be referenced to specific geographic coordinates (e.g., topographic maps, digital elevation models, geologic maps, sample localities and associated data, roads, streams, land ownership, etc.).

GLONASS   Acronym for GLObal NAvigation Satellite System. GLONASS is the Russian counterpart of the USA's GPS. Some receivers are capable of receiving and processing data from both constellations of satellites, although GPS is used much more widely.

GLORIA   Acronym for Geologic LOng-Range Inclined Asdic, a side-scanning sonar system used to image large areas of the sea floor.

GMT   Acronym for Greenwich Mean (or Meridian) Time, defined as the mean solar time at the Royal Greenwich Observatory in Greenwich near London, England, which by convention is at 0 degrees geographic longitude.

GNSS   Acronym for Global Navigation Satellite System, used to refer to any of, or the collective combination of, the operational spaceborne radio navigation systems (mainly GPS and GLONASS, soon to include Galileo).

GPS   Acronym for the US Global Positioning System, a worldwide positioning and navigation tool that makes use of a constellation of NAVSTAR satellites in Earth orbit and receivers on the ground, aboard aircraft, or other satellites. A receiver's 3-D position and velocity can be determined in an absolute reference frame by processing signals broadcast continuously by the satellites. A wide range of receiver types and data-processing schemes are available for various applications, including volcano monitoring. Positioning accuracy is typically 1–30 m for navigation and surveying applications, and a few millimeters for precise geodetic applications.

GPS receiver   An electronic device that receives and processes signals from NAVSTAR or other GNSS satellites to calculate positioning information (e.g., latitude, longitude, elevation) directly or, alternatively, to produce a dataset that can be further processed elsewhere to obtain such information. Receivers must be connected to specially designed antennas tuned to the GPS frequencies.

gravimeter (gravity meter)   An instrument for measuring variations in Earth's gravitational field, generally by registering differences in the weight of a constant mass as the gravimeter is moved from place to place, or in the time it takes a small mass to fall a known distance in a vacuum. See Chapter 3 for a discussion of various types and their applications to volcano monitoring.

groundmass   The material between the *phenocrysts* of a *porphyritic* igneous rock. It is relatively finer grained than the phenocrysts and may be crystalline, glassy, or both.

groundwater   Subsurface water within the zone of saturation or, loosely, all subsurface water as distinct from surface water. The interaction between groundwater and magma beneath volcanoes is important and sometimes violent, producing explosive phreatic or phreatomagmatic eruptions. Groundwater changes can also affect gravity measurements and produce localized subsidence of the ground surface.

ground plane (GPS)   A flat, electrically conductive surface at the base of a GPS antenna used to deflect errant signals (multipath) reflected from the ground and other nearby objects. *See also* **choke ring**.

ground range   Distance measured from the near-range line to a particular point in the footprint of an imaging radar system (i.e., the area illuminated by the radar beam) in the direction perpendicular to the flight path of the radar. The resolution of an imaging-radar system in the ground-range direction improves with higher radar frequency, greater signal bandwidth, and higher (more grazing) incidence angle. 'Range' is used synonymously in some cases, but ground range should not be confused with *slant range*. *See also* **azimuth**.

GST   Acronym for Galileo System Time, a time-scale generated by atomic clocks for the European Space Agency's Galileo global navigation system. GST is steered toward International Atomic Time (TAI), and is specified to be within 50 ns of TAI for 95% of the time over any yearly time interval.

Hertz (Hz)   Frequency unit in the International System of Units (SI), equal to one cycle per second, named after German physicist Heinrich Rudolph Hertz.

Holocene   In the geologic timescale, that part of the Quaternary period from the end of the Pleistocene, approximately 10,000 years ago, to the present time.

Hooke's law   Law of elasticity discovered by the English scientist Robert Hooke in 1660, which states that, for relatively small deformations of an object, the displacement or size of the deformation is directly proportional to the deforming force or load.

host rock   A body of rock that surrounds (hosts) other rocks or mineral deposits (e.g., a pluton containing xenoliths, or any rock in which ore deposits occur). *See also* **country rock**, **intrusion**, **pluton**, and **xenolith**.

hotspot A long-lived volcanic center, typically 100 to 200 km across and active for at least a few tens of millions of years, that is thought to be the surface expression of a rising plume of hot mantle material. Relative motion of a few centimeters per year between plumes and overlying tectonic plates give rise to hotspot tracks, which are alignments of large volcanic centers arranged in a regular age progression as successive centers drift over the plume. The Hawaii–Emperor volcanic chain and a series of large rhyolite calderas located along the eastern Snake River Plain and Yellowstone Plateau are both interpreted as hotspot tracks. In the first case, basalt generated by a plume rises buoyantly through relatively dense oceanic lithosphere to form a chain of volcanic islands and seamounts. In the second case, basalt encounters less-dense continental lithosphere, stalls, and assimilates crustal material to form large reservoirs of rhyolite magma in the mid-crust. Eruptions from these reservoirs eventually form a chain of large collapse calderas like the 640,000-year-old Yellowstone Caldera.

HST Acronym for Hawaiian Standard Time, also known as Hawaiian/Aleutian Standard Time. HST = GMT – 10 hours.

hummock A rounded or conical knoll, mound, or hillock. Hummocks are common and characteristic features of debris-avalanche deposits, such as the one emplaced at Mount St. Helens on 18 May 1980.

HVO Acronym for the USGS Hawaiian Volcano Observatory. HVO is the oldest of five volcano observatories operated solely or jointly by the USGS (*see also* **AVO**, **CVO**, **LVO**, and **YVO**). HVO was largely the creation of Dr. Thomas A. Jaggar (1871–1953), a Massachusetts Institute of Technology professor who established in February 1912 a center for volcanological research on the north rim of Kīlauea Caldera. Since then, HVO has been successively sponsored by the US Weather Bureau (1919–1924), the USGS (1924–1935), the National Park Service (1935–1947), and again the USGS (since 1947). HVO monitors the active volcanoes of Hawai'i, assesses the associated hazards, and provides timely information about volcanic activity to appropriate officials, agencies, and the general public.

hybrid earthquake A type of volcanic earthquake that shares attributes of both volcano–tectonic (VT) and long-period (LP) earthquakes. Hybrid events are thought to represent a combination of processes, such as a VT earthquake occurring near a fluid-filled cavity and setting it into oscillation. See also **long-period (LP) earthquake** and **volcano–tectonic (VT) earthquake**.

hydraulic diffusivity A parameter that relates the time rate of change in hydraulic head to the spatial variation of hydraulic head according to the diffusion equation, $(\partial h/\partial t) = D \times (\partial^2 h/\partial x^2)$, where h is hydraulic head and D is hydraulic diffusivity. In the Earth, fluid-pressure changes in response to strain occur instantaneously with the deformation. However, water's ability to flow within an aquifer causes spatial variations in pressure to equilibrate with each other over a timescale governed by the material's hydraulic diffusivity (Chapter 9). *See* **aquifer**.

hydrofracting Extension and propagation of fractures in rock by hydraulic pressure exerted by a fluid phase within the fractures, as during crystallization of a water-saturated igneous melt or by water injection into a rock mass.

hydrostatic pressure The pressure due to the weight of groundwater at higher levels in the zone of saturation. In highly fractured or porous rocks where groundwater flows freely, the total pressure approximates the hydrostatic pressure. *See also* **lithostatic pressure**.

hypersaline Refers to a *brine* containing much more than 25 wt% dissolved salts, which is the concentration at which precipitation of salts commences when a brine is evaporated at the Earth's surface. Hypersaline brines are likely to form where water-bearing fluids are released at very high temperatures and high pressures from crystallizing magmas a few to several kilometers deep beneath volcanoes (Chapter 10). *See* **brine**.

hypocenter (seismology) The point within the Earth where an earthquake rupture starts; also commonly termed the focus. *See* **epicenter**.

ignimbrite The rock formed by widespread deposition and consolidation of *ash flows*. The terms ignimbrite and ash-flow tuff are used interchangeably, especially in reference to large-volume deposits – either welded or non-welded – from explosive, caldera-forming eruptions. *See* **ash-flow tuff**, **welded tuff**.

image pyramid In softcopy photogrammetry, several linked image layers representing successively higher resolution versions of a given image area for each scanned photograph. Having multiple copies of the same image at different resolutions enables rapid identification and autocorrelation of *fiducial marks* for interior orientations and *pass points* for relative orientations. It also increases screen-refresh rates for pan and zoom operation of digital stereo images.

impulsive In seismology, term used to describe an earthquake signal with a sudden onset on a seismogram; also applied to an earthquake or seismic wave producing such a signal (e.g., *A-type earthquakes*). *See* **emergent**.

inflation In volcanology, a term used to describe tumescence (swelling) of the ground surface, a volcanic edifice, or a subsurface magma body. Inflation is a response to pressurization, which can have several causes including magma accumulation, exsolution of volatiles, geothermal processes, heating, and tectonic compression. *See* **deflation**.

InSAR (INSAR) Acronym for interferometric synthetic-aperture radar. This remote-sensing technique is capable, under favorable conditions, of imaging ground displacements over large areas (100 km × 100 km) to an accuracy of a few millimeters. The most common implementation of the technique makes use of two or more radar images of the same target area acquired during repeat passes of an Earth-orbiting, imaging-radar satellite. After corrections are made for viewing geometry, topography, and path-delay effects, any remaining differences in radar phase between the images are attributable to changes in satellite-to-ground *slant range* or, in other words, to displacements of the ground surface in the mean direction to the satellite at the times the images were acquired. Because radar waves are sinusoidal, slant-range changes that cause corresponding phase differences between images lead to characteristic interference patterns (fringes) when the images are combined. The number and pattern of fringes are essentially an image of the surface displacement field in the direction of the satellite.

interferometry (radar) A technique that uses the measured differences in the phase of the return radar signal between two satellite passes to detect slight changes on Earth's surface (Chapter 5). Two radar-phase images of the same target area, taken at different times and from slightly different vantage points, are flattened, co-registered, and differenced to produce an *inter-ferogram*. Resulting phase differences are represented by *fringes*.

interferogram (radar) An image produced by differencing two radar-phase images of the same target area. An interferogram can be used to produce a digital elevation model (DEM), to study subtle displacements of the ground surface, or to detect other forms of surface change (e.g., erosion, deposition, or a change in near-surface moisture content).

intrusion The process of emplacement of magma in pre-existing rock; magmatic activity; also, the igneous rock so formed within the surrounding rock. *See also* **extrusion**, **dike**, **sill**, **country rock**, and **host rock**.

Invar An alloy of nickel and iron, containing about 36% nickel, which has an extremely low coefficient of thermal expansion ($10^{-6}\,°C^{-1}$). It is used in the fabrication of surveying instruments such as level rods, first-order leveling instruments, and tapes.

inversion (numerical modeling) There are two general approaches to modeling geodetic and geophysical data, called forward and inverse methods. Forward models use physical principles and known (or assumed) material properties to calculate a theoretical system response. For example, the surface-displacement field caused by a given pressure change in a Mogi-type source embedded at a given location and depth in a semi-infinite half-space with given elastic properties can be calculated using a forward model (Chapter 8). An inverse model, on the other hand, starts with a generalized forward model and uses observations (data) with their associated uncertainties, plus known (or assumed) constraints called boundary conditions, to determine the best-fitting parameters for the model. This is accomplished by minimizing some model-to-data misfit criterion. For example, leveling or GPS data can be 'inverted' to determine the parameters of a Mogi source, including the optimal pressure change, location, and depth, that best-fits the data in a least-squares sense. Inversion problems can be either linear or nonlinear, depending on the physics involved. Nonlinear inversions are non-unique, and require stronger constraints (e.g., more and better data, or independent knowledge of the system) to achieve a satisfactory result.

IUSS US Navy's Integrated Undersea Surveillance Systems, which includes the Sound Surveillance System (*SOSUS*) for deep ocean surveillance. In October 1990, the Navy granted access to SOSUS data from the North Pacific to NOAA's Pacific Marine Environmental Laboratory (NOAA/PMEL) in Newport, Oregon, to assess the system's value in ocean environmental monitoring. NOAA/PMEL uses SOSUS to monitor seismicity around the northeast Pacific Ocean, including frequent earthquake swarms associated with volcanic activity along oceanic spreading centers.

juvenile *See* **essential**.

K-feldspar (K-spar) Potassium feldspar, an alkali *feldspar* containing the orthoclase molecule ($KAlSi_3O_8$). Forms a binary series with sodium feldspar ($NaAlSi_3O_8$). *See* **albite**.

kinematic GPS A form of GPS surveying in which the position of a moving receiver is tracked by means of uninterrupted carrier-phase measurements following successful solution of the integer ambiguities. This can be accomplished in a continuous mode where the receiver remains in motion for precise positioning of a vehicle, or in an intermittent mode where data are recorded only after a receiver is brought to a stationary point, and the observations while in motion are tracked as a way to maintain the integer ambiguities.

kinematic (on-the-fly) initialization In GPS, a form of ambiguity resolution that does not require that the receivers remain stationary for any length of time. Kinematic ambiguity resolution is suitable for carrier phase-based kinematic positioning applications such as aircraft navigation. Dual-frequency receivers capable of making both carrier phase and P-code pseudorange measurements are required. *See also* **static initialization**.

L1, L2 Primary carrier frequencies for NAVSTAR and GLONASS satellite signals. See Chapter 4 and Table 4.1 for details.

L3 This term has two distinct meanings in GPS terminology: (1) a military signal broadcast discontinuously at 1381.05 MHz by NAVSTAR satellites, and (2) a particular linear combination of the L1 and L2 carrier frequencies. See Chapter 4 for details.

L5 A third civilian carrier frequency to be broadcast in addition to L1 and L2 by Block IIF NAVSTAR satellites starting in 2006 and Block III satellites starting in 2012. See Chapter 4 for details.

lahar A watery slurry of volcanic rocks, mud, and other debris that surges downstream like rapidly flowing concrete (i.e., a volcanic mudflow). Lahars can be triggered by a variety of processes that commonly occur at volcanoes, including explosions and pyroclastic flows that melt snow or ice, landslides that mobilize water-saturated debris, and outbursts of lakes or glacial melt water that incorporate sediment from downstream valleys.

Lamé constants Two elastic parameters, λ and μ, which express the relationships between the components of stress and strain for linear elastic behavior of an isotropic solid; μ is identical with rigidity (also called modulus of rigidity, shear modulus, or torsional modulus), and λ is equivalent to the bulk modulus or stiffness K minus $2\mu/3$. Gabriel Lamé (1795–1870), considered by many to be the leading French mathematician of his time, made important contributions to the fields of mathematics, physics, thermodynamics, and applied mechanics. The Lamé constants, which are widely used for applications in solid mechanics, were not defined by Lamé. They were named after Lamé's death in recognition of his contributions to the field of mechanics.

lapilli A general term for pyroclastic ejecta in the size range from 2 mm to 64 mm. Fragments may be either solidified or still viscous when they land; thus there is no characteristic shape. They may be essential, accessory, or accidental in origin. *See* **pyroclast**.

lava dome *See* **volcanic dome**.

layover An extreme form of foreshortening in radar images in which the top of a reflecting object such as mountain is closer to the radar than are the lower parts of the object. The image of such a feature appears to lean toward the radar. The effect is more pronounced for radars that use steeper incidence angles. *See also* **foreshortening** and **shadowing**.

level, circular A spirit level in which the inner surface is circular and the graduations are concentric circles. This form of spirit level is used when precision is not critical, such as for plumbing a leveling rod or a compensator leveling instrument. Also called a universal level or bullseye level. *See* **level, spirit**.

level, digital A leveling instrument capable of automatically reading a barcode leveling rod and thus of determining elevation differences along a level line. Digital levels were introduced in the early 1990s and are now standard for geodetic leveling.

level, spirit A small, closed container of transparent material (usually glass), with the upper part of its inner surface curved, used to precisely establish a horizontal plane. The container is nearly filled with a fluid of low viscosity (alcohol or ether), with enough free space so that a bubble of air or gas always rises to the top of the container. The outer surface of its upper part carries an index mark or graduations. The term is also used to refer to a leveling instrument that contains a spirit level. *See* **leveling, spirit**.

leveling A surveying technique for determining relative height differences between fixed points on the Earth's surface (bench marks) by sighting through a leveling instrument (commonly called a 'level') to one or more graduated rods. Older 'spirit levels' use a bubble within a long vial of low-viscosity liquid such as ether or alcohol to indicate when the level is properly adjusted in a horizontal attitude. Most modern levels use an internal pendulum and self-leveling mechanism for this purpose. More generally, the process of finding elevations from a selected equipotential surface such as the geoid to points on the Earth's surface, or of finding elevation differences between points. Vertical motions (i.e., elevation changes) can be measured by repeated leveling surveys. Usually, leveling must be done either as the sum of incremental vertical displacements of a graduated rod (differential leveling) or by measuring vertical angles (trigonometric leveling). See Chapter 2 for a more thorough discussion of various forms of leveling and for definitions of related terms. *See also* **leveling, compensator**; **leveling, differential**; and **leveling, trigonometric**.

leveling, compensator Leveling carried out with a leveling instrument having a compensator, or self-leveling device, for making a line-of-sight horizontal. *See* **leveling instrument, compensator**.

leveling, differential Technique for determining the difference in elevation between two points by the sum of incremental vertical displacements of a graduated rod. The two usual methods of differential leveling are spirit leveling and compensator leveling, which differ in how the leveling instrument determines the horizontal line-of-sight. *See also* **leveling, spirit** and **leveling, compensator**.

leveling, double-run Leveling done by proceeding from starting point to final point and then returning to the starting point in the opposite direction. This and other precautions are intended to cancel or identify systematic errors as well as to reduce random errors. *See also* **leveling, single-run**.

leveling, geodetic Leveling to a high order of accuracy, usually extended over large areas, to furnish accurate vertical control for surveying and mapping. Repeated geodetic leveling surveys are used to measure vertical surface displacements in actively deforming areas.

leveling, single-run Leveling done by proceeding from starting point to final point without leveling back to the starting point. *See also* **leveling, double-run**.

leveling, single-setup A form of leveling in which the relative heights of a small number of bench marks forming an array are measured from a single instrument setup, usually for the purpose of determining ground tilt (Section 2.5.2). By repeating the measurements, relative vertical displacements among the marks can be determined and used to compute a local tilt vector. Typically, height differences among 3 marks spaced 30–50 m apart and forming an equilateral triangle are measured from the center of the array using a leveling instrument. This approach is sometimes referred to as tilt leveling or dry tilt (Yamashita, 1992) (Figure 2.28). Additional marks can be added for redundancy. The length of the leveling rods limits the maximum height difference within the array, so relatively flat ground is required. A variant on the technique called single-setup trigonometric leveling can be used where flat sites are unavailable. In this case, a total station is used to measure the height differences (Figures 2.29–2.31).

leveling, spirit Leveling with an instrument that depends on a spirit level for making its line-of-sight horizontal. A spirit level is attached to a telescope so that the axis of the level and the line of collimation of the telescope can be made parallel and the instrument can be adjusted so that its axis is horizontal. The difference of readings on leveling rods on two different points is the difference in elevation of the points. If the elevation of one point is known, the elevation of the other also becomes known. By repeating this procedure many times, the elevation of any point or series of points can be established from the elevation of a known point. *See* **level, spirit** and **leveling rod**.

leveling, trigonometric A surveying technique used to determine elevation differences by observing vertical angles and distances between points. A total-station instrument, which combines the capabilities of an EDM and theodolite, and specially designed reflector targets are very useful for this purpose. An advantage of trigonometric leveling is that the length of leveling rods does not limit the elevation difference that can be measured at each instrument setup, as is the case for differential leveling.

leveling instrument, compensator An automatic leveling instrument in which the line-of-sight is kept horizontal by a set of prisms or mirrors that swing freely in response to gravity to compensate for the non-horizontality remaining after the instrument has been leveled with a circular level.

leveling instrument, digital (digital level) *See* **level, digital**.

leveling instrument, spirit-level type A leveling instrument with a spirit level attached to the telescope for leveling the line-of-sight.

leveling rod A rod or bar designed for use in measuring a vertical distance between a point on the ground and the horizontal line-of-sight of a leveling instrument. A leveling rod, usually made of wood, has a flat face that is graduated in some linear unit and fractions thereof (meters or feet), or to which is attached a metallic strip that is so graduated. Modern leveling rods use an Invar strip for temperature stability. Some rods, which are designed for use with an automated, digital-leveling instrument (digital level), have a barcode pattern rather than evenly spaced graduations. *See* **Invar** and **leveling instrument, digital (digital level)**.

level line A set of measured elevation differences, presented in the order of their measurement, and the similarly ordered set of points to which the measurements refer. It is also customary to refer, loosely, to either the set of points or the set of measured differences as a level line. A level line or series of lines that form a loop back to its starting point is called a *circuit*.

lidar Acronym for light detection and ranging, a remote-sensing system similar to radar, in which laser light pulses take the place of microwaves. In addition to its many applications in atmospheric imaging, lidar can be used to measure the elevation of the ground surface and create digital elevation models with decimeter-scale accuracy, even in heavily vegetated terrain.

lithostatic pressure The vertical pressure at a point in the Earth's crust caused by the weight of the overlying column of rock or soil. *See also* **hydrostatic pressure**.

littoral Pertaining to the ocean environment between high water and low water (i.e., to the intertidal zone).

littoral cone An *ash* or *tuff* cone formed on a lava flow when it encounters a body of water, usually the sea. Such cones are the result of steam explo-sions that hurl into the air large amounts of ash, lapilli, and small bombs derived from the lava coming into contact with water.

long-period (LP) earthquake An earthquake with an emergent P-wave, no S-wave, and a dominant frequency in the range from 1 to 5 Hz. LP earthquakes are commonly associated with active volcanoes or hydrothermal systems. They are thought to be caused by fluid-pressurization processes, such as bubble formation and collapse, or by nonlinear fluid-flow through fractures or conduits. Similar waveforms can result from other seismic processes occurring at very shallow depth, for which attenuation and path effects play an important role. The term LP earthquake is usually synonymous with low-frequency earthquake.

Love wave A surface wave in which individual particles of material move back and forth in a horizontal plane perpendicular to the direction of wave travel; named after A.E.H. Love, the English mathematician who discovered it. *See also* **surface wave**, **Rayleigh wave**.

LVO Acronym for the USGS Long Valley Observatory. The USGS regional office in Menlo Park, California, serves as headquarters for LVO, which had its beginnings in 1980 when a prolonged episode of geophysical unrest began in the vicinity of the Long Valley Caldera, prompting the USGS to launch an intensive monitoring effort. In 1991, the Long Valley project was officially designated a USGS volcano observatory, joining HVO, CVO, and AVO (YVO was added in 2001). LVO monitors the Long Valley Caldera and nearby Mono-Inyo Craters volcanic chain for signs of volcanic and tectonic unrest, assesses the associated hazards, and provides timely information about volcanic activity to public officials, agencies, and the general public. *See also* **AVO**, **CVO**, **HVO**, and **YVO**.

M-code Military code-modulation structure, analogous to C/A-code and P-code, implemented on Block IIR-M and subsequent NAVSTAR satellite series starting in September 2005.

magnitude *See* **earthquake magnitude** and **Richter scale**.

manometer A device for measuring pressure differences, usually by the difference in height of two liquid columns. Manometers are used in fluid tiltmeters to sense vertical height changes between the ends of a fluid-filled tube (Chapter 3). A well that is open to a confined aquifer acts as a monometer when the water level in the well rises or falls in response to compressive or dilatational strain in the host rock (Chapter 9).

megaPascal (MPa) A unit of pressure. One MPa equals about 10 bar (atmospheres) or 150 pounds per square inch (psi).

microGal (μGal) *See* **Gal**.

microradian (μrad) A unit of tilt equal to 10^{-6} radians, which corresponds to a vertical rise or fall of 1 mm over a horizontal distance of 1 km. The magnitude of the solid Earth tide is of the order of 0.1 μrad. Most bubble tiltmeters used for volcano monitoring have sensitivities of 0.1 μrad or better. The best long-base tiltmeters have sensitivities of about 10^{-6} μrad, or one part per billion.

microseism A collective term for small motions in the Earth that are unrelated to an earthquake and that have a period of 1.0 to 9.0 s. They are caused by a variety of natural and artificial agents, including atmospheric events and oceanic-wave action.

milliGal (mGal) *See* **Gal**.

misclosure A cumulative measure of surveying error, equal to the amount by which a series of measurements fails to yield a theoretical or previously determined result. In leveling, the amount by which two values for the elevation of the same bench mark, derived by different surveys, by the same survey made along two different routes, or by independent observations, fail to exactly equal each other. For example, the misclosure is equal to the discrepancy between the net elevation difference measured around a circuit or double-run level line and the theoretical difference, which is zero. Allowable misclosures have been established for each order and class of leveling (Table 2.1). If the misclosure exceeds the allowable limit, the survey must be repeated.

modulus of elasticity The ratio of stress to its corresponding strain under given conditions of load, for materials that deform elastically, according to Hooke's law. *See also* **bulk modulus**, **Poisson's ratio**, **shear modulus**, and **Young's modulus**.

MPa *See* **megaPascal**.

MSAS Acronym for MTSAT Satellite-based Augmentation System (MTSAT being Multi-functional Transport Satellite), a Japanese Satellite-Based Augmentation System (SBAS) for Asia (see *SBAS*).

mudflow *See* **lahar**.

mud volcano An accumulation of mud and rock ejected by volcanic gases escaping from Earth's surface. Mud volcanoes range in height from a few centimeters to several meters. There are many mud volcanoes in Yellowstone National Park, both in the Mud Volcano locality and elsewhere. Mud volcanoes form in areas of active hydrothermal activity where shallow groundwater is present but not overly abundant. In wetter areas, hot springs and thermal pools are more common.

multi-look This term has two meanings in the field of synthetic aperture radar (SAR). The first applies to the processing of a SAR image from signals received at the SAR after scattering from the target area. At any instant, the return signal contains information from all of the resolution cells in the antenna footprint. The radar return from each cell is earmarked with range and azimuth (Doppler-shift) information, as discussed in Section 5.1.3. A *single-look complex* image is obtained by integrating the return signal across the entire spectrum of Doppler shifts that corresponds to the full synthetic aperture of the radar. The same data stream can be used to form several *multi-look* images by dividing the Doppler spectrum of the return signal into segments and assembling an image from each subset of the data, effectively using just a portion of the available synthetic aperture for each image. Stacking (averaging) multi-look images has the advantage of suppressing speckle, which for most applications is considered noise. Multi-look processing increases radiometric fidelity at the expense of spatial resolution, because the size of the synthetic aperture for each multi-look image is reduced relative to that for a full-resolution, single-look image. In SAR interferometry, 'multi-look' has a different meaning. It refers to spatial averaging of adjacent pixels in single-look complex images during the process of creating an interferogram. Multi-looking, in this sense, increases signal-to-noise in the images and can be used to form square pixels from rectangular ones. For example, the range and azimuth resolutions of ERS-1 and ERS-2 single-look complex images are about 25 m and 5 m, respectively. The resolution of ERS multi-look images, with 1×5 pixel averaging, is about 30 m. *See* **speckle**, **single-look complex**.

multipath errors These arise when incoming GPS signals are reflected from the ground surface or nearby objects into a receiving antenna, producing spurious carrier-phase data. They can be mitigated by choosing sites away from buildings or other reflectors and also by careful antenna design. The choke-ring design is especially effective in reducing multipath errors.

multiplet (earthquake) One of a group of seismic events, usually occurring in a swarm, with very similar signatures (codas) on seismograms. In some cases, multiplet codas are virtually identical – to the extent that the events appear to be clones of one another. Multiplets are thought to indicate repetitive, non-destructive excitation of the same source (e.g., opening and closing of a fluid-filled crack, bubble oscillation, or stick–slip motion across a resilient seismic patch on an otherwise freely slipping fault surface).

nadir The point where the direction of the plumb line at a given point on the Earth's surface, extended below the horizon, meets the celestial sphere. The nadir is directly opposite the zenith.

nadir, ground The point on the ground that is vertically underneath the perspective center of an airborne-camera lens system.

nadir, photograph The point at which a vertical line through the perspective center of an airborne-camera's lens system pierces the plane of the photograph. Also called nadir point.

nanometer (nm) A unit of length, equal to one billionth of a meter, or 10^{-9} meters. The wavelengths of ultraviolet radiation and visible light are short enough to be measured in nanometers. Most of the UV radiation that reaches the Earth from the Sun has wavelengths 100–400 nm, and visible light has wavelengths of 400–700 nm.

nanoTesla The SI unit for magnetic field strength equivalent to the gamma in the cgs system.

narrow lane A linear combination of the GPS carrier phases, Φ_{L1} and Φ_{L2}, such that $\Phi_{L1+L2} = \Phi_{L1} + \Phi_{L2}$. The corresponding wavelengths are: $\lambda_{L1} = 19.0\,cm$, $\lambda_{L2} = 24.4\,cm$, and $\lambda_{L1+L2} = 10.7\,cm$. Linear combinations of observables are useful for eliminating error sources that are inherent in GPS observations (Chapter 4).

NAVSTAR Acronym for Navigation Satellite Time and Ranging. A constellation of these satellites operated by the US Department of Defense is the heart of the US Global Positioning System (GPS).

network In geodesy, a group of surveying stations (recoverable points) that have been interconnected through measurements of their relative positions in such a way that the self-consistency of the measurements can be checked and adjusted. This is usually accomplished by incorporating closed loops or circuits in the design of the network or survey. A well-designed network should approximate the shape of the terrain it spans, and therefore can be used to measure deformation by repeated surveys. The term is sometimes used interchangeably with *array*, although the latter refers more correctly to an ordered arrangement of instruments, the positions of which may or may not have been established by surveying.

NEXRAD A network of advanced S-band Doppler radars operated by the National Weather Service, an agency of the National Oceanic and Atmospheric Administration (NOAA), throughout the USA. NEXRAD data, when processed, can be displayed in a mosaic map which shows precipitation patterns.

nodal point (optics) One of two points on the optical axis of a lens, or system of lenses, such that a ray emergent from the second point is parallel to the ray incident at the first. This first nodal point is also referred to as the front nodal point, incident nodal point, or nodal point of incidence. The second point is referred to as the rear nodal point, emergent nodal point, or nodal point of emergence.

normal fault A fault, typically with a dip of 45–90°, in which the hanging wall (overlying block) moves downward relative to the footwall and motion is primarily parallel to the dip.

nuées ardente A fast-moving, turbulent, ground-hugging cloud, sometimes incandescent, containing volcanic ash and other pyroclasts in its lower part. A nuées ardente commonly occurs when part of a growing lava dome becomes unstable and collapses, especially if the collapse occurs on or near a steep slope ('dome-collapse pyroclastic flow'). A type locality is Merapi Volcano on the island of Java, Indonesia, where persistent dome growth at the summit typically spawns nuées ardentes down the volcano's steep flanks. The lower part of a nuées ardente is comparable with an *ash flow* or *pyroclastic flow*, and the terms are sometimes used synonymously. In this usage, a nuées ardente can occur either as a primary emplacement process during an explosive eruption, or by secondary collapse of hot, previously extruded material. From the French term meaning 'glowing cloud'.

olivine An olive-green, grayish-green, or brown orthorhombic mineral: $(Mg,Fe)_2SiO_4$. Olivine is a common rock-forming mineral of relatively low-silica igneous rocks (e.g., gabbro, basalt, peridotite, dunite). It crystallizes early from a magma, weathers readily at Earth's surface, and metamorphoses to serpentine. *See* **gabbro** and **basalt**.

optical plummet A device used to center a surveying instrument precisely over a bench mark, by means of an eyepiece that allows viewing along the vertical collimation line. To center the instrument over a station, the instrument is first leveled, then centered and re-leveled. The process is repeated until the instrument is both level and centered over the desired point on the mark, as determined by looking through the eyepiece. An optical plummet permits centering to within $\pm 0.5\,$mm and is immune to wind effects, unlike a plumb bob that can be used for the same purpose.

orientation In photogrammetry, the process involving *fiducial marks* and *control points* by which overlapping aerial photographs or images are precisely arranged relative to one another to account for viewing geometry and produce a *stereo model* (Chapter 6). Interior orientation involves establishing the geometric configuration of the camera relative to the film image, which requires knowledge of the *projection center* and focal length of the camera–lens combination used to collect the photographs. Relative orientation re-establishes the photo positions and, consequently, the ray paths for adjacent photos in a stereo model. Exterior or absolute orientation allows a stereo model, cleared of all distortion, to be precisely oriented in space so that 3-D coordinate measurements in the model may be transformed to ground coordinates in a geodetic coordinate system.

orthometric height The distance of a point above the *geoid*, measured along the plumb line at the point. The relationship between *ellipsoidal height* h and *orthometric height* H is: $h = H + N$, where N is the *geoid undulation* (i.e., the difference between the ellipsoid and geoid at the point). Orthometric corrections are applied to precise-leveling measurements because level surfaces at different elevations are not parallel.

orthophotograph (orthophoto) A *rectified* photographic image prepared from an aerial photograph in which the displacements due to tilt and relief have been removed.

pahoehoe Hawai'ian term for basaltic lava that has a smooth, hummocky, or ropy surface. A pahoehoe flow typically advances as a series of small lobes and toes that continually break out from a cooled crust. *See* $'$a$'$ā.

PANGA Acronym for Pacific Northwest Geodetic Array, a network of continuously operating GPS stations operated by a consortium of universities and agencies in the USA and Canada. The cooperative project aims to improve understanding of the regional kinematics, tectonics, volcanism, and hazards associated with the Cascadia subduction zone. Component of the Plate Boundary Observatory (PBO).

pantograph An instrument for copying maps, drawings, or other graphics at a predetermined scale. Pantographs capable of adjustment for several scales are known as fixed-ratio pantographs.

parallactic angle The parallactic angle, also known as the convergence angle, is formed by the intersection of the left eye's line of sight with that of the right eye. The closer this point of intersection is to the eyes, the larger the convergence angle. The brain perceives the height of an object by associating depth at its top and its base with the convergence angles formed by viewing the top and base. X-parallax and parallactic angle are related. As x-parallax increases, so too does the parallactic angle. As the eyes scan overlapping areas between a stereo image pair, the brain receives a continuous 3-D impression of the ground. This is caused by the brain constantly perceiving the changing parallactic angles of an infinite number of image points making up the terrain. The perceived 3-D model is known as a stereoscopic model or stereo model (Section 6.3.6). *See* **parallax**, **stereoscopic model (stereo model)**, **x-parallax**.

parallax The apparent displacement of the position of an object with respect to a reference system, or to a set of points or objects, caused by an actual shift in the location of the observer. In conventional photogrammetry, an aircraft is used to acquire sequential overlapping vertical air photographs. The changing vantage point of the moving camera causes parallax shifts among control points and other features in adjacent photographs. The magnitude of the shift is proportional to the distance from the camera to the feature, and thus to the elevation of the feature. Parallax is responsible for the perceived stereo effect when properly aligned photos are viewed with a stereoscope. The magnitude of the shift can be contoured to produce a topographic map (Chapter 6). *See also* **x-parallax**, **y-parallax**.

paramagnetism A property of minerals or other substances that causes them to have a small positive magnetic susceptibility. Paramagnetic rock-forming minerals such as *olivine, pyroxene*, or *biotite* contain magnetic ions that tend to align along an applied magnetic field, but do not have a spontaneous magnetic order. *See also* **ferromagnetism**.

pass point A point whose horizontal and/or vertical position is determined from photographs by photogrammetric methods and which is intended for use as a supplemental *control point* in the *orientation* of other photographs.

P-code A pseudorandom-noise code used to modulate the L1 and L2 carrier signals broadcast by NAVSTAR satellites, which form the heart of the US Global Positioning System (GPS). The P-code repeats itself every 267 days, and the algorithm used to generate it is not classified. At the discretion of the Department of Defense, the classified Y-code can be substituted for the P-code. Whenever the Y-code is used, positioning capability is degraded for most civilian users. The Russian GLONASS satellites also use a P-code to modulate their carrier signals. *See also* **C/A-code**, **W-key**, and **Y-code**.

PDT Acronym for Pacific Daylight Time. PDT = GMT − 7 hours.

peg test (or **peg adjustment**) A procedure used to adjust a leveling instrument to correct any collimation error (i.e., departure of the line-of-sight from horizontal) by repeatedly measuring the elevation difference between two stable marks (pegs). The details of the test procedure differ for different instruments, but the following example is illustrative. First, the leveling instrument is set at the midpoint between two leveling rods spaced approximately 40 m apart (to simulate a typical sight distance during differential leveling) and the elevation difference between the rods is measured. Then the leveling instrument is moved to within a few meters of one of the rods and the elevation difference is measured again. The midpoint measurement is free of the effect of any collimation error, because such an error would affect both rod readings equally. Therefore, any discrepancy between the two measured elevation differences is attributable to collimation error, and the leveling instrument can be adjusted to make the two measurements match.

Pele's hair Thin strands of volcanic glass drawn out from molten lava have long been called Pele's hair, named after Pele, the Hawai'ian goddess of volcanoes. A single strand, with a diameter of less than 0.5 mm, may be as long as 2 m. The strands are formed by the stretching or blowing-out of molten basaltic glass from lava, usually from lava fountains, lava cascades, and vigorous lava flows (e.g., as *pahoehoe* lava plunges over a small cliff and at the front of an 'a'ā flow). Pele's hair is often carried high into the air during fountaining, and wind can blow the glass threads several tens of kilometers from a vent. Often found in association with *Pele's tears*.

Pele's tears Small bits of molten lava (*pyroclasts*) in fountains can cool quickly and solidify into glass particles shaped like spheres or tear drops called Pele's tears, named after Pele, the Hawai'ian goddess of volcanoes. They are jet black in color and are often found on one end of a strand of *Pele's hair*. Pele's tears may be tear-shaped, spherical, or nearly cylindrical; they are generally a few millimeters to ∼1 cm in size.

permeability The capacity of a porous rock, sediment, or soil for transmitting a fluid; it is a measure of the relative ease of fluid flow under unequal pressure.

phase ambiguity The integer number of phase cycles N in the GPS carrier signal between the time of transmission by a satellite and the time of arrival at a receiver. Also called the initial ambiguity at first observation, cycle ambiguity, or integer ambiguity. N can be included as an unknown parameter in the inversion of GPS data for receiver coordinates, a procedure called ambiguity resolution.

phase center A point in space near a GPS antenna that is analogous to the focal point of a parabolic mirror. It is the location of the phase center that is determined most directly by GPS. This location can be used to locate a nearby bench mark if the spatial relationship between the phase center and bench mark is known. This requires that we know: (1) the vector from a fixed point on the antenna to the phase center, and (2) the vector from the same fixed point to the bench mark. The first vector can be determined by careful testing and is well established for all antenna types used for geodetic purposes. The second is obtained by carefully centering the antenna over the bench mark and measuring the distance from the centering point to the fixed point on the antenna.

phase unwrapping A procedure for solving the 2π ambiguity inherent in radar interferograms, using one of several techniques to calculate the correct integer number of phase cycles to be added to each phase measurement. An unwrapped, flattened interferogram has a smooth range of pixel values corresponding to surface elevations or, in the case of a topography removed interferogram, to differential slant-range changes caused by surface deformation or path-delay anomalies.

phenocryst A conspicuous, usually large, crystal embedded in *porphyritic* igneous rock.

photogrammetry The science of surveying or mapping with the aid of photographs. Photogrammetry is the classical means for making topographic maps and precise measurements from overlapping aerial photographs. See Chapter 6 for a discussion of applications to volcano mapping and geodesy.

phreatic A type of volcanic eruption characterized by an explosion of steam, mud, or other accidental material. Phreatic eruptions are caused by the sudden heating and consequent expansion of groundwater by an igneous heat source, usually magma. *See also* **phreatomagmatic**.

phreatomagmatic A type of volcanic eruption driven both by magmatic gases and steam; typically, such eruptions are violently explosive and eject both essential and accidental material. They are caused by contact of magma with groundwater or shallow surface water.

piezomagnetic effect A change in the magnetization of rocks caused by a change in *deviatoric stress*. As a result, stress changes caused by tectonic or magmatic processes can be monitored by measuring the differential magnetic field among an array of magnetometers. *See* **deviatoric stress**.

pixel Short for 'picture element', the smallest part of a digital image. The number of pixels (width and height) in an image defines its size, and the number of pixels per inch (or centimeter) defines its resolution. Thus, an uncompressed color image with 1024×768 24-bit pixels displayed at a size of 3.4×2.6 inches (8.7×6.5 cm) has a resolution of about 300 pixels per inch (118 pixels per centimeter) and a file size of $(1024 \times 768$ pixels$) \times (24$ bits/pixel$)/(8$ bits/byte$) = 2,359,296$ bytes $\cong 2.3$ megabytes.

plagioclase A group of triclinic feldspars of general formula: $(Na,Ca)Al(Si,Al)Si_2O_8$. At high temperatures it forms a complete solid-solution series from Albite ($NaAlSi_3O_8$) to anorthite ($CaAl_2Si_2O_8$). Plagioclase minerals are among the most common rock-forming minerals, having characteristic twinning, and commonly display zoning.

plastic Describes a rock that undergoes permanent deformation of its shape or volume, without rupture. This type of deformation is characterized by a yield stress, which must be exceeded before deformation begins.

Pleistocene An epoch of the Quaternary period, after the Pliocene and before the Holocene. It began two to three million years ago and lasted until the start of the Holocene about 10,000 years ago.

plinian Refers to a type of explosive volcanic eruption in which a steady, turbulent stream of fragmented magma and magmatic gas is released at high velocity from a vent. Large volumes of tephra and tall eruption columns are characteristic. The *Volcanic Explosivity Index* (VEI) of plinian eruptions is typically in the range from 3 to 7, eruption column heights are 10 to 25 km above sea level or more, and the volume of tephra erupted is between 10^7 and 10^{12} m^3. Recent examples include the Mount St. Helens eruption of 18 May 1980, and the Mount Pinatubo eruptions of 12–15 June 1991. The term comes from an account of the 79 CE eruption of Vesuvius by *Pliny the Younger*, part of which is reproduced in the Preface of this book. *See* **tephra**, **Volcanic Explosivity Index (VEI)**, and **Pliny the Younger**.

Pliny the Elder *See* **Pliny the Younger**.

Pliny the Younger Gaius Plinius Caecilius Secundus (63–ca. 113 CE), better better known as Pliny the Younger, was a lawyer, author, and scientist of Ancient Rome. He was the nephew of *Pliny the Elder*, who is considered by many to be the greatest naturalist of antiquity. In a letter to the Roman historian Tacitus, Pliny the Younger described the 79 CE eruption of Mount Vesuvius in considerable detail; part of the letter is reproduced in the Preface of this book. Pliny the Elder had approached the volcano to investigate, and perished in the eruption. Volcanic eruptions of this type are referred to as *plinian*.

plumb bob A conical metal weight suspended by a cord used to project a point vertically in space for short distances. A plumb bob can be used to center a surveying instrument over a bench mark, but in most cases an optical plummet is preferable.

plume A large, buoyant mass of hot material rising through the mantle. Decompression causes a rising plume to partially melt, and additional melting occurs when the plume impinges on the base of the crust. Surface effects of the interaction between plume and crust include large-scale uplift and voluminous eruptions, typically over a period of several million years or longer. The resulting volcanic center is known as a *hotspot*, and its migration relative to the plume (caused by plate motion) produces a linear hotspot track. The Hawaiian Islands and Yellowstone Plateau are interpreted as hotspots, the tracks of which are apparent in the Hawaiian–Emperor chain and a string of silicic calderas along the eastern Snake River Plain. *See* **hotspot**.

pluton A large igneous intrusion formed within the crust, and later exposed at Earth's surface by erosion.

point positioning A GPS positioning mode, also called absolute positioning, in which the coordinates of a point are determined in an absolute reference frame with its origin at Earth's center of mass. *See also* **relative positioning**.

Poisson's ratio When a sample of material is stretched in one direction, it tends to get thinner in the other two directions. Poisson's ratio (ν) is a measure of this tendency. It is defined as the ratio of the lateral unit strain to the longitudinal unit strain in a body that has been stressed longitudinally within its elastic limit. For a perfectly incompressible material, Poisson's ratio would be exactly 0.5; for most crustal rocks, Poisson's ratio is approximately 0.25. Using this value, the Mogi (1958) elastic point-source model predicts that the surface-volume change is 1.5 times greater than the source-volume change. For the more general case of a spherical source, the surface-volume change exceeds the source-volume change by a factor of $2(1 - \nu)$, where ν is Poisson's ratio of the host rock (Delaney and McTigue, 1994). Poisson's ratio is named for Siméon-Denis Poisson (1781–1840), a French mathematician and physicist who made important contributions to pure mathematics, electricity and magnetism, and celestial mechanics.

porosity The percentage of the bulk volume of a rock or soil that is occupied by interstices, whether isolated or connected. Materials with large connected interstices have high *permeability*, while those with small isolated interstices have low permeability.

porphyritic Describes the texture of an igneous rock in which larger crystals (*phenocrysts*) are set in a finer grained *groundmass*, which may be crystalline, glassy, or both.

ppm An abbreviation for parts per million; 1 ppm is equivalent to 1 part in 10^6. For example, a linear strain of $1 \, \text{mm} \, \text{km}^{-1}$ can be expressed as 1 ppm.

predictive filter A mathematical algorithm that uses a stream of data as input to produce sequential estimates of future data values and of the errors made in predicting future values from previous ones. Predictive filtering algorithms are used in such diverse applications as digital-data compression and real-time GPS data processing.

principal point The geometric center of an aerial photograph, or the point where the optical axis of the lens meets the film plane in an aerial camera. The principal point is defined by the intersection of lines drawn between opposing *fiducial marks* (Chapter 6).

PRN code *See* **pseudorandom-noise code**.

proof mass The suspended weight in a *gravimeter* or *geophone*.

pseudorandom-noise (PRN) code A reproducible binary sequence with noise-like properties. PRN codes are used in spread-spectrum communications systems and in satellite ranging systems such as GPS. NAVSTAR satellites transmit two PRN codes: C/A-code and P-code. Each code is a fixed sequence of -1 or $+1$ states, corresponding to binary values 1 or 0. Each 0 or 1, or in some contexts the length of time required to transmit each 0 or 1, is called a chip. The sequence repeats at a constant interval and is used to modulate the carrier signals broadcast by global navigation satellites. Whenever the state of the C/A-code is -1 (binary value 1), the phase of the L1 carrier signal is shifted by 180 degrees. When the C/A-code state is $+1$ (binary value 0), the signal is unchanged (Figure 4.4).

pseudorange *See* **code pseudorange**.

PST Acronym for Pacific Standard Time. PST = GMT − 8 hours.

pumice A light, porous volcanic rock that forms during explosive eruptions. It resembles a sponge because it consists of a network of gas bubbles frozen amidst fragile volcanic glass and minerals. All types of magma (*basalt, andesite, dacite*, and *rhyolite*) will form pumice. It is often sufficiently buoyant to float on water and is economically useful as a lightweight aggregate and as an abrasive.

P-wave (seismology) A type of seismic body wave that involves particle motion (alternating compression and expansion) in the direction of propagation. It is the fastest of the seismic waves, traveling $5.5–7.2 \, \text{km} \, \text{s}^{-1}$ in the crust (much slower in highly fractured or unconsolidated near-surface materials at volcanoes) and 7.8–8.5 km s^{-1} in the upper mantle. P-waves can travel through solids, liquids, and gases; also known as a primary wave. *See also* **body wave, S-wave.**

pyroclast An individual particle ejected during a volcanic eruption, usually classified according to size. Examples are *lapilli* (2–64 mm) and blocks (>64 mm). *See* **lapilli, pyroclastics,** and **tephra.**

pyroclastic Pertaining to clastic rock material formed by volcanic explosion or aerial expulsion from a volcanic vent; also, pertaining to rock texture of volcanic origin. In the plural, the term is used as a noun. *See* **pyroclastics** and **pyroclast.**

pyroclastic flow A high-speed, ground-hugging density current composed of hot ash, rock fragments, and volcanic gas that moves down the flank of a volcano or along the surrounding ground surface during an explosive eruption, or when an active lava dome breaks apart and collapses. Pyroclastic flows typically travel in excess of 100 km hr^{-1}, have temperatures of several hundred degrees Celsius, and can travel up to 20 km from their source. They are capable of knocking down and incinerating virtually everything in their paths.

pyroclastics A general term for a deposit of *pyroclasts.*

pyroxene A group of dark rock-forming silicate minerals, closely related in crystal form and composition, and having the general formula: $ABSi_2O_6$, where A = Ca, Na, Mg, or Fe^{+2}, and B = Mg, Fe^{+2}, Fe^{+3}, Fe, Cr, Mn, or Al, with silicon sometimes replaced in part by aluminum. Colors range from white to dark green or black. Pyroxenes are a common constituent of igneous rocks.

quartz Crystalline silica, an important rock-forming mineral: SiO_2.

radar interferometry *See* **InSAR.**

radial lens distortion A common type of distortion that causes horizontal or vertical objects near the outer edges of a photograph to appear curved, either inward or outward. This distortion tends to be symmetric and radially distributed outward from the *principal point* of the lens. Photogrammetric lenses used with metric cameras are designed to produce very small distortions in aerial photographs, which can be largely removed using a *stereo plotter.* Modern, high-quality, wide-angle to normal focal-length photographic lenses with low-to-moderate radial dis-

tortion are easily calibrated and lend themselves well to terrestrial photogrammetry when paired with a non-metric camera (Chapter 6).

range (InSAR) *See* **ground range.**

Rayleigh wave (seismology) A surface wave in which individual particles of material move in an elliptical path within a vertical plane oriented in the direction of wave movement. It is named after Lord Rayleigh, the English physicist who predicted its existence. *See also* **surface wave, Love wave.**

real-aperture radar (RAR) A radar system where the antenna beamwidth is controlled by the physical length of the antenna; also known as brute force or non-coherent radar. The main advantages of real aperture radars are their simple design and modest data-processing requirements. However, it is impractical to design a RAR antenna long enough to produce high-resolution data. *See also* **synthetic-aperture radar (SAR).**

rectification In photogrammetry, the process of projecting a tilted or oblique aerial photograph onto a horizontal reference plane, the angular relation between photography and plane being determined by *control points.*

reference ellipsoid *See* **ellipsoid.**

reference frame In physics, the state of motion of an observer. An inertial reference frame is a member of a set of all reference frames moving uniformly with respect to one another, without relative acceleration. A reference frame is necessary but not sufficient to define a coordinate system, which also requires an origin point and an oriented set of orthogonal axes. In geodesy, the term is used to mean a coordinate system associated with a physical system, such as Earth as a whole. For example, the GPS reference frame is centered at Earth's center of mass and includes three orthogonal axes that are defined by convention. The first axis passes through the intersection of the Greenwich meridian and Earth's equatorial plane. The third axis is defined as the average position of Earth's rotation pole for the years 1900 to 1905. The second axis is orthogonal to the first and third axes. GPS heights are referenced to Earth's center of mass and commonly expressed relative to a reference ellipsoid.

reference system *See* **reference frame.**

relative positioning A GPS positioning mode, also called differential positioning, in which the coordinates of an unknown point are determined with respect to a known reference point, which is taken as the origin of a local coordinate system. *See also* **point positioning**.

resection A method in surveying by which the horizontal position of an occupied point is determined by drawing lines from the point to two or more points of known position. A common problem in resection is the three-point problem, when three known positions are observed to locate the occupied station.

relief displacement The geometric distortion on vertical aerial photographs caused by a combination of perspective and topography. Relief displacement is radial, and increases with distance, from the *principal point* of the photograph. For example, the tops of trees near the center of an aerial photograph appear directly over their trunks, whereas treetops near the edges of the photo are displaced radially outward from the trunks. The effect can be removed by photogrammetric methods to produce distortion-free *stereo models*.

resurgent dome The central highland in many large calderas formed by gradual upwarping of the caldera floor after caldera collapse as a result of renewed magma intrusion.

reticulite *See* **thread-lace scoria**.

Richter scale A numerical scale of earthquake magnitude devised in 1935 by seismologist Charles F. Richter at the California Institute of Technology. The magnitude of an earthquake is determined from the logarithm of the amplitude of waves recorded by seismographs, with adjustments for the variation in gain among various seismographs and in the distance between the seismographs and the epicenter of the earthquake. On the Richter scale, magnitude is expressed in whole numbers and decimal fractions. For example, a magnitude of 5.3 might be computed for a moderate earthquake, and a strong earthquake might be rated as magnitude 6.3. Because of the logarithmic basis of the scale, each whole number increase in magnitude represents a tenfold increase in measured amplitude; as an estimate of energy, each whole number step in the magnitude scale corresponds to the release of about 31 times more energy than the amount associated with the preceding whole number value. In theory, there is no upper limit to the magnitude of an earthquake, but the strength of Earth materials produces an actual upper limit of about 9. Very small earthquakes can have negative magnitude values. Earthquake magnitudes can be computed in different ways, which can

result in slightly different values for the same earthquake (e.g., local magnitude, body-wave magnitude, surface-wave magnitude). *See* **earthquake magnitude**.

rhyodacite Volcanic rock (or lava) that is intermediate in composition between *rhyolite* and *dacite*. *See* **dacite** and **rhyolite**.

rhyolite Volcanic rock (or lava) that characteristically is light in color, contains 69 or more wt% of *silica*, and is rich in potassium and sodium. Low-silica rhyolite contains 69 to 74 wt% silica. High-silica rhyolite contains 75–80 wt% silica. *See also* **silica**, **andesite**, **basalt**, and **dacite**.

rift zone In a volcanic setting such as Hawai′i or Iceland, a linear zone of fractures, vents, and collapse features associated with an underlying dike complex. Volcanic rift zones are an important aspect of the magmatic plumbing system for many basaltic volcanoes. Magma that rises from great depth to within a few kilometers of the surface can move laterally along a rift zone for tens of kilometers before erupting or stalling as an intrusion.

RINEX Acronym for receiver independent exchange format for GPS data. An ASCII based, receiver independent format designed to facilitate data exchange between different GPS receivers, software programs, and users. RINEX includes provisions for code pseudorange, carrier phase, and Doppler observations.

ring dike A dike that is arcuate or roughly circular in plan and is vertical or inclined away from the axis of the arc. Ring dikes are commonly associated with *cone sheets* to form a ring complex. *See* **cone sheet**.

SAR Acronym for synthetic-aperture radar. *See* **synthetic-aperture radar**.

SBAS Acronym for Satellite-Based Augmentation System, a set of geostationary satellites intended to augment GNSS use for civilian applications by broadcasting GPS look-alike signals that include real-time differential corrections and integrity information. These regional systems include: (1) WAAS (Wide-Area Augmentation System) for the USA; (2) CWAAS for Canada; (3) EGNOS (European Geostationary Navigation Overlay Service) for Europe; (4) MSAS (MTSAT Satellite-based Augmentation System, MTSAT being Multi-functional Transport Satellite) for Japan; (5) GAGAN (GPS and Geo Augmented Navigation system) for India (planned); and (6) SNAS (Satellite Navigation Augmentation System) for China (planned). WAAS and

EGNOS were operational in 2004; plans call for MSAS to become operational in 2006. SBAS is vital to providing the reliability and precision required by aviation and other real-time, precision-critical applications.

scale The ratio between linear distance on a map, chart, globe, model, or photograph and the corresponding distance on the surface being mapped. The scale of an aerial photograph is usually taken as the ratio of the focal length of the lens to the altitude of the camera above mean ground elevation. Common scales for topographic maps produced by the US Geological Survey include 1:24,000 (i.e., one inch on the map corresponds to 2,000 feet on the ground; 1:50,000, 1:62,500 (1 inch ≈ 1 mile), 1:100,000; and 1:250,000.

SCIGN Acronym for Southern California Integrated GPS Network, a dense array of continuous GPS stations in southern California operated by a consortium of agencies and universities for the purpose of monitoring regional tectonic strain. Component of the Plate Boundary Observatory (PBO).

scoria A vesicular, glassy lava rock of basaltic to andesitic composition, generally irregular in form, ejected from a vent during explosive eruption. The bubbly nature of scoria is due to the escape of volcanic gases during eruption. Scoria is typically dark gray to black in color, mostly due to its high iron content. The surface of some scoria may have a blue iridescent color; oxidation may lead to a deep reddish-brown color. Scoria is generally denser, darker, and more crystalline than *pumice*.

seiche A free or standing-wave oscillation of the surface of water in an enclosed or partly enclosed basin (such as a lake, bay, or harbor) that varies in period, depending on basin size, from a few minutes to several hours and in height from several centimeters to a few meters. Seiches are initiated mainly by local changes in atmospheric pressure, aided by winds, tidal currents, or earthquakes. They persist, pendulum fashion, for a time after the originating force ceases. Their effect can be removed from gauging records if the sampling frequency is high enough and the record is long enough to include at least a few complete cycles.

seismic moment A measure of the energy released by an earthquake or sequence of earthquakes. The seismic moment M_0 for an individual earthquake is $M_0 = \mu A d$, where μ is the shear modulus of the surrounding rock, d is the average slip across a fault, and A is the surface area of the fault. The seismic moment can be estimated from the waveforms produced by the earthquake as recorded by a network of calibrated seismometers, or from the results of a dislocation model. For a shear dislocation (slip on a fault), the seismic moment equals the *geodetic moment*.

seismogram The record made by a *seismograph* (i.e., a record of all seismic activity during a period of time, including background noise, body waves, and surface waves, from both natural and artificial events). *See also* **seismograph**, **seismometer**, and **geophone**.

seismograph An instrument that detects, magnifies, and records vibrations of the Earth, especially earthquakes. The resulting record is a *seismogram*. The term seismograph is sometimes used incorrectly as a synonym for *geophone*. A seismograph can include amplifiers, receivers and a recording device (such as a computer disk or magnetic tape) to record seismograms.

seismometer An instrument for monitoring earthquakes and other processes that cause ground motion, typically at frequencies greater than about 1 Hertz (cycles per second). The seismometer sensor, sometimes called a *geophone*, produces a voltage proportional to the displacement, velocity, or acceleration of ground motion. A seismometer is usually a damped oscillating mass connected to a fixed base and frame via a suspension (e.g., a spring). Such a damped mass-spring system is used to detect and measure ground motion relative to the suspended mass, which serves as an inertial reference. The motion of the base, which is fixed to the ground, with respect to the suspended mass is commonly transformed into an electrical voltage, which can be recorded on paper, magnetic tape, computer disk, or other recording medium. *Seismograph* is a term that refers to the seismometer as the sensor of the motion together with its recording device as a unit.

selective availability (SA) SA refers to the intentional corruption of GPS satellite clocks and the introduction of errors into the broadcast ephemerides by the US Department of Defense for the purpose of denying full system capability to unauthorized users. The fundamental frequency of the GPS clocks is 'dithered' (i.e., pseudorandom errors are introduced) and the quality of the broadcast ephemerides is reduced by truncating the orbital information in the satellite navigation message. The use of selective availability was discontinued permanently by order of US President Clinton on 1 May 2000. A related denial of service technique called *anti-spoofing* (AS) is still used.

self-potential The difference of potential (DC voltage) between two electrodes in good electrical contact with the ground at different locations. Self-potential results from currents generated by electrochemical or electrokinetic potential (streaming potential). A streaming potential commonly is generated when groundwater moves through permeable rock. Rock surfaces tend to capture excess negative charge from ions in solution, which is balanced by excess positive charge that develops in a layer of electrolytic groundwater near the rock–water interface. If a vertical pressure gradient forces groundwater to move relative to the rock, some of the excess positive charge is swept along with the water to produce a convection current. The resulting electrical charge at the ground surface is positive if the water rises and negative if it descends. Also called *spontaneous potential, electrical self-potential*, or *SP*.

setup In leveling, this refers to the placing, leveling, and reading of a leveling instrument between two leveling rods. By thus determining the elevation difference between the temporary points on which the rods rest (turning points), then leapfrogging the rods through a series of setups in the direction of a level line, the net elevation difference between bench marks can be determined.

shadowing The effect in side-looking radar images produced by steep topography or other obstructions that block down-range areas from being illuminated by the radar beam. Shadowed areas produce no radar backscatter and, therefore, appear dark in the radar image. The effect is more pronounced for radars that use shallow incidence angles. *See also* **foreshortening** and **layover**.

shear modulus A *modulus of elasticity* in shear, also called modulus of rigidity. The shear modulus G (or μ) is related to the bulk modulus K and *Poisson's ratio* ν by the equation $K = (2(1 + \nu)G)/3(1 - 2\nu)$.

shield volcano A volcano in the shape of a flattened dome, broad and low, built by flows of very fluid basaltic lava or, less commonly, by rhyolitic ash flows. Mauna Loa on the Big Island of Hawai'i is a classic basaltic shield volcano.

sidereal day The interval between two successive transits of a star over the meridian; the time required for Earth to rotate once on its axis, or approximately 86,166 s. The sidereal day is about 4 minutes shorter than the solar day (i.e., the interval between successive passages of the sun over the meridian, approximately 86,400 s) because of the orbital motion of Earth. NAVSTAR satellites repeat the same track and configuration over any point approximately every 24 hours, but four minutes earlier each day, as a result of this difference.

silica The chemically resistant dioxide of silicon: SiO_2. It is the most common substance on the surface of Earth and a common component of rocks and soils. Silica occurs in five crystalline polymorphs (the minerals quartz, tridymite, cristobalite, coesite, and stishovite); in cryptocrystalline form (chalcedony); in amorphous and hydrated forms (opal); in less pure forms (e.g., sand, diatomite, tripoli, chert, flint); and combined in silicates as an essential constituent of many minerals. Confusion can arise from the common practice of reporting the amount of SiO_2 in a rock sample as 'wt% SiO_2' or 'weight-percent silica'. For example, basalt contains 45–54 wt% SiO_2; andesite, 54–62 wt%; dacite, 62–69 wt%; and rhyolite, 69–80 wt%. These values refer to the *equivalent* amount of silicon dioxide (SiO_2) in the sample, and do not imply the presence of quartz or any other mineral form of silica. In fact, quartz is rare in most igneous rocks other than rhyolites and granites. On the other hand, the 'silica tetrahedron', a four-sided pyramid with an oxygen atom at each corner and a silicon atom (not silica) in the center, is the fundamental building block of the vast majority of minerals in Earth's mantle and crust. Because this structure, written as $(SiO_4)^{4-}$, has a negative four electronic charge, silica tetrahedra are linked together by positively charged atoms like sodium (Na), potassium (K), magnesium (Mg), calcium (Ca), and iron (Fe) to form minerals.

silica tetrahedron *See* **silica**.

sill A tabular igneous intrusion that is concordant (parallel) with the planar structure of the surrounding rock. Sills are distinguished from *dikes*, which cut across bedding or foliation. Sills typically dip gently, whereas dikes dip steeply. However, the dip of an intrusive body is not diagnostic: A vertical intrusion along a vertical bedding or foliation plane is a sill, and a horizontal intrusive body that cuts across vertical bedding or foliation is a dike. *See* **dike**.

single-look complex A type of radar image in which amplitude (intensity) information for each pixel is represented by the real part of a complex number and phase information is represented by the imaginary part. *See also* **multi-look**.

Skempton's coefficient A poroelastic constant B relating changes in average stress $\Delta\sigma_{kk}$ to changes in pore-fluid pressure Δp, $B = -\Delta p/\Delta\sigma_{kk}$. The minus sign is needed because pore pressure decreases with positive (extensional) stress. B is dimensionless with a value between 0 and 1.

slant range The straight-line distance from a sensor to a target (e.g., from an imaging radar antenna to a point in the antenna footprint (i.e., the area on the ground illuminated by the radar beam) or from a GPS satellite to a receiver). Not to be confused with *ground range*.

SLAR Acronym for side-looking airborne radar, a real-aperture radar system capable of imaging Earth's surface and acquiring precise topographic data from an aircraft or satellite through clouds and sparse vegetation. *See also* **SAR** and Chapter 5.

slowness (seismology) For seismic waves, the reciprocal of velocity, given in units of seconds per degree or seconds per kilometer; a large slowness corresponds to a low velocity. Most seismic tomography methods involve subdividing the medium into blocks and solving for slowness perturbations that cause predicted times to match observed arrival times better than an initial model.

SLR Acronym for side-looking radar, used interchangeably with *SLAR*.

softcopy stereoplotter (softplotter) A modern type of *stereo plotter* that utilizes digital images and software for translation of images instead of mechanical control of image movement, as is the case with analog stereo plotters. Softcopy systems evolved from analog systems as high-precision film scanners and high-speed computing platforms were developed, enabling creation and manipulation of high-resolution digital imagery in real time.

solfatara *See* **fumarole**.

solidus temperature (solidus) The temperature at which a rock first begins to melt. On a temperature-composition diagram, the locus of points in a system above which solid and liquid are in equilibrium and below which the system is completely solid.

SOSUS Acronym for SOund SUrveillance System, a component of the US Navy's Integrated Undersea Surveillance Systems (*IUSS*) network used for deep-ocean surveillance during the Cold War and more recently for scientific purposes including studies of low-frequency vocalizations from marine mammals and of undersea seismicity. SOSUS is useful for detecting and locating earthquakes associated with submarine volcanic activity.

speckle A scattering phenomenon produced by coherent systems, including synthetic-aperture radars (SARs), which arises because the resolution of the sensor is not sufficient to resolve individual scatterers. Physically speaking speckle is not noise, because the same imaging configuration leads to the identical speckle pattern. Speckle can be reduced by *multi-look* processing and spatial averaging. *See* **multi-look**.

spider Multipurpose volcano-monitoring station developed at the USGS David A. Johnston Cascades Volcano Observatory (CVO) for use during the 2004–2006 eruption at Mount St. Helens (Figure 11.9). A spider is essentially a footlocker on legs, outfitted with one or more sensors, a telemetry system, and batteries or solar panels. Equipped with a single-frequency GPS receiver, tiltmeter, geophone, accelerometer, volcanic gas sensor, digital camera, or temperature sensors, plus enough batteries to operate the station for weeks to months, a spider weighs in at ~100 kg – light enough to be slung into place with a small helicopter.

spontaneous potential *See* **self-potential**.

SPOT An acronym for the Systéme Pour l'observation de la Terre Earth observation satellite, developed by the French Centre National d'Etudes Spatiales (CNES). Its sensors operate in two modes, multispectral and panchromatic, at spatial resolutions of 2.5 m to 20 m. SPOT can observe the same area on the globe once every 26 days. The SPOT scanner normally produces *nadir* views, but it does have off-nadir viewing capability that allows one area on Earth to be viewed as often as every 3 days. This is quite useful for collecting data in the event of a natural or man-made disaster, where timeliness is crucial. It is also very useful in collecting stereo data from which elevation data can be extracted. *See* **nadir**.

SRTM Acronym for Shuttle Radar Topography Mission, which used dual C-band and X-band radars separated by a 60-m mast aboard the Space Shuttle to image 80% of Earth's land mass in February 2000. The separation afforded by the mast made interferometric observations possible with a single orbital pass. This contrasts with dual-pass satellite interferometry, in which a single radar is used to image a surface target on two different orbital passes. The primary objective of SRTM was to produce a worldwide digital elevation model (DEM) with $30 \, m \times 30 \, m$ spatial sampling, $16 \, m$ absolute vertical height accuracy, and $10 \, m$ relative vertical height accuracy.

static GPS A form of GPS surveying in which the position of a fixed point is determined by means of carrier-phase measurements. The integer ambiguities are resolved from an extended observation period through a change in satellite geometry.

static initialization The process of resolving the integer ambiguities at the start of a kinematic GPS survey. This can be accomplished by bringing the roving receiver to a known point momentarily, by keeping the roving receiver stationary long enough to obtain a static position, or by performing an antenna swap. *See also* **kinematic initialization**.

stereoscopic model (stereo model) The 3-D image formed when two overlapping photographs are viewed such that the left eye sees only the left photograph and the right eye sees only the right photograph. Stereo visualization is best achieved when the camera axes for both photos are parallel and image overlap is approximately 60%.

stereoscopic parallax *See* **parallax** and **x-parallax**.

stereo plotter (stereoplotter) A photogrammetric instrument used to reconstruct a distortion free *stereoscopic model* of the ground surface in which points, lines and surfaces can be constructed in 3-D space and projected in two dimensions as needed for production of topographic maps or *orthophotographs*.

strain Change in the shape or volume of a body as a result of stress, expressed as a dimensionless ratio of final to initial length, area, or volume. Strain and *deformation* are synonymous in many contexts, although the latter term is more general.

strainmeter An instrument designed to measure changes in the shape or volume of a body (deformation) as a result of stress. Most strainmeters are designed to measure either linear strain or volumetric strain. Linear strainmeters can be used to monitor changes in the width of a fracture or fracture system near a volcanic vent, for example. Borehole volumetric strainmeters are extremely precise instruments that can measure compression or dilation of Earth's crust, even

tens of kilometers away from a restless volcano or some other source of deformation.

streaming potential *See* **self-potential**.

stress The force per unit area acting on a surface within a solid body, commonly expressed in units of dynes per square centimeter. Mathematically, stress is represented by a tensor composed of nine values: three to specify the normal components and six to specify the shear components, relative to three mutually perpendicular reference axes.

sublittoral That part of the ocean environment between mean low tide and a water depth of approximately 100 meters.

surface wave (seismology) A seismic wave that travels along Earth's surface, or along a subsurface interface. Surface waves comprise the Rayleigh wave and the Love wave. Not to be confused with S-wave, a type of body wave. *See also* **Rayleigh wave**, **Love wave**, **body wave**.

SVN Acronym for Satellite Vehicle Number or Space Vehicle Number that is applied to the NAVSTAR satellite series, which constitute the space segment of the US Global Positioning System (GPS). For example, the operational series of 28 Block II NAVSTAR satellites are designated SVN 13 through SVN 40.

S-wave (seismology) A seismic body wave propagated by a shearing motion that involves oscillation perpendicular to the direction of propagation. Also known as a secondary wave, because it arrives later than the P-wave (primary wave). S-waves propagate at a speed of 3.0–4.0 km s^{-1} in the crust and 4.4–6.6 km s^{-1} in the upper mantle; they do not travel through liquids or through Earth's outer core. S-wave attenuation can indicate the presence of partial melt or geothermal fluid along the propagation path from the hypocenter of an earthquake to seismometers recording the resulting seismic waves. *See also* **P-wave**, **body wave**.

synthetic-aperture radar (SAR) An imaging radar system that takes advantage of the motion of a radar antenna to simulate the performance of a much larger antenna (Chapter 5). SAR systems take advantage of the long-range propagation characteristics of radar signals and the complex information processing capability of modern digital electronics to provide high-resolution imagery. Under favorable circumstances, successive SAR images of the same area can be combined to form interferograms that provide detailed spatial information about surface topography or deformation. *See also* **real-aperture radar (RAR)**.

TAI Acronym for Temps Atomique International (International Atomic Time), based on a continuous counting of the SI second by a large number of atomic clocks. TAI is currently (January 2006) ahead of UTC by 33 s (TAI − UTC = +33 s).

talus Rock fragments, usually coarse and angular, derived from and lying at the base of a cliff or very steep, rocky slope. Also, the outward sloping mass of such loose broken rock, considered as a unit. Hence, the terms talus cone, talus slope, and talus apron.

tare A discontinuity in data, indicating an error in measurement or computation rather than a sudden actual change in the quantity being measured. Tares in microgravity data can be identified and sometimes removed by repeatedly looping back to a reference station during the course of a survey. After the reference station readings are corrected for Earth tides and instrument drift, they should be repeatable within the expected measurement error. Any discrepancy among the readings might indicate a tare.

tellurometer An electronic instrument that measures distance by means of microwaves.

tephra A general term for fragments of volcanic rock and lava, regardless of size, that are blasted into the air by explosions or carried upward by hot gases in eruption columns or lava fountains; synomymous with *pyroclastics*. Tephra includes large dense blocks and bombs, and small light rock debris such as *scoria, pumice, reticulite*, and *ash*. See **pyroclastics**, **scoria**, **pumice**, **reticulite**, and **ash**.

theodolite A surveying instrument used for measuring angular distances between fixed points in both vertical and horizontal planes. The instrument can be rotated around two orthogonal axes so as to be sighted first on one point and then on another. By repeating vertical and horizontal angle measurements relative to the same fixed reference point at different times, any movement of the points between surveys can be determined in 3-D. The best theodolites are precise to 1 second of arc or better, which corresponds to 5 mm over a sight distance of 1 km. A transit theodolite, or transit, is designed so that the telescope can be reversed in its supports without being lifted from them, to facilitate making direct and reverse angle measurements. Any discrepancy between such measurements is attributable to circle graduation or collimation

error, which can be corrected by taking the mean of the two measurements.

thermal conductivity The time rate of transfer of heat by conduction, through unit thickness, across unit area for unit difference of temperature. A measure of the ability of a material to conduct heat. Typical values of thermal conductivity for rocks range from 3 to 15 mcal cm^{-1}s^{-1} $^{\circ}$C^{-1} or, equivalently, 1.3–6.3 W m^{-1} $^{\circ}$K^{-1}.

thermal diffusivity The thermal conductivity of a substance divided by the product of its density and specific heat capacity. For rocks, the common range of values is from 0.005 to 0.025 cm^2 s^{-1}.

thread-lace scoria A *scoria* in which the vesicle walls have burst and are represented only by an extremely delicate 3-D network of glass threads. Synonymous with *reticulite*. Thread-lace scoria is a product of vigorous fire fountaining of gas-rich mafic magma. *See also* **Pele's hair** and **Pele's tears**.

thrust fault A fault with a dip of 45° or less on which the hanging wall (overlying block) has moved upward with respect to the footwall. An otherwise similar fault with a steeper dip is called a high-angle reverse fault. Thrust faults form as a result of horizontal compression and are common near active volcanic vents and domes.

tiltmeter An instrument that measures slight changes in the tilt of the ground surface, usually in relation to a liquid-level surface or to the rest position of a pendulum. Two common designs are the watertube tiltmeter and bubble tiltmeter. A watertube tiltmeter, as the name implies, uses a long, U-shaped, water-filled tube to define a gravitational equipotential (i.e., level) surface. Changes in the tilt of the tube are measured by manually reading or otherwise sensing the position of the meniscus at each end of the tube. A bubble tiltmeter is based on the same principle, but in this case the tube or disk containing the bubble is much smaller (typically a few cm or less) and filled with an electrolyte. Bubble movements caused by tilt changes are sensed by electrodes projecting into the fluid and converted into an electrical signal that is proportional to the tilt change.

TIN *See* **triangulated irregular network**.

tomography (seismology) A technique to estimate the 3-D distribution of seismic wave velocities within a volume of the Earth by using numerous earthquake sources and an array of seismometers.

The basic scheme is first to locate and characterize a large set of earthquakes, which 'illuminate' the study region with seismic waves. Wave velocities along ray paths from each source to each seismometer are calculated from arrival times across the array. A computer synthesizes the resulting data to to construct a 3-D representation of the seismic velocity structure. Tomography is useful for delineating magma bodies, partial melt zones, and hydrothermal systems, which attenuate and slow seismic waves to a greater degree than the host rock.

TopSAR Acronym for topographic synthetic-aperture radar, a NASA/JPL airborne radar system designed for rapid production of digital terrain models (i.e., DEMs).

total station A surveying instrument that combines the functions of an EDM and theodolite and thus is capable of measuring both distance and azimuth between fixed points. Modern total stations offer a combination of high precision and electronic ease of use that makes them well suited to volcano-monitoring situations in which the hazard to field crews in proximal areas is considered low enough to be acceptable.

transit *See* **theodolite**.

traverse, first-order *Traverse* is a method of surveying in which lengths and directions of lines between points on Earth's surface are obtained by or from field measurements, and used in determining positions of the points. A *first-order traverse* is one that extends between adjusted positions of other first-order control surveys and conforms to the current specifications of first-order traverse. For surveys in the USA, see Federal Geodetic Control Committee (1984); for New Zealand, Bevin (2003); for Australia, ICSM (2000); for Canada, Surveys and Mapping Branch (1978).

tremor *See* **volcanic tremor**.

triangulation A surveying method for determining the directions and distances to, and the coordinates of, a network of points by means of bearings from two fixed points a known distance apart, using a theodolite or total station. Any relative movement among the points can be measured by repeating the survey at a later date. *See also* **trilateration**.

triangulated irregular network (TIN) A surface representation derived from irregularly spaced points and breakline features. The TIN dataset includes topological relationships between points and their proximal triangles. Each sample point has an x, y coordinate and a surface, or z-value. These points are connected by edges to form a set of non-overlapping triangles used to represent the surface. The term is used frequently in GIS applications.

tribrach A device used to level a surveying instrument on a tripod by means of a circular bubble and three height-adjustment screws. Many tribrachs have a built-in optical plummet for also centering the instrument over a bench mark. The bubble and optical plummet should be tested and adjusted periodically to ensure their accuracy.

trilateration A surveying method in which the lengths of the three sides of a series of triangles that share common points at their vertices are measured, usually with an EDM, and the angles between sides are computed from the measured lengths. Thus, the coordinates of each point in the network can be established relative to a known point, and any relative movement among the points can be measured by repeating the survey at a later date. *See also* **triangulation**.

tsunami A gravitational sea wave produced by any large-scale, short-duration disturbance of the ocean floor, principally by a shallow submarine earthquake, but also by a submarine debris flow or volcanic eruption. Tsunamis are characterized by great speed of propagation (up to $950 \, \text{km hr}^{-1}$), long wavelength (up to 200 km), long period (generally 10–60 min), and low amplitude on the open sea. They can pile up to heights of 30 m or more (hundreds of meters in the case of gigantic submarine landslides around the Hawai'ian Islands) and cause extensive damage on entering shallow water along an exposed coast, even thousands of kilometers from their source. In recorded history, together with lahars, tsunamis have caused more volcano-related deaths than all proximal volcanic hazards combined.

tuff A general term for all consolidated pyroclastic rocks. *See* **pyroclastic**, **ash**, and **ash-flow tuff**.

turning pin A metallic pin used to support a leveling rod at a turning point in leveling operations. In some cases, turning pins are preferred to turning plates except on hard surfaces such as roads. It is important that turning pins and turning plates be solidly placed to avoid *pin settling*, a type of systematic error in leveling surveys.

turning plate A metallic plate used to support a leveling rod at a turning point in leveling operations. Also called a *foot plate* or *turtle*.

turning point   A surveying point on which a leveling rod is placed, after a foresight has been made on it, and before the leveling instrument is moved so that a backsight can be made on it to determine the height of the instrument after resetting. A turning point is established to allow the instrument to be moved forward (alternately leapfrogging with the rod) along the level line without a break in the series of measured elevation differences. Usually, a turning plate or turning pin is used to establish a turning point.

turtle   *See* **turning plate**.

T-wave   Short-period (≤ 1 s) acoustic wave traveling in the ocean at the speed of sound in water. A 'T-phase' is sometimes identified in the records of earthquakes in which a large part of the path from epicenter to station is across the deep ocean. *See* **P-wave** and **S-wave**.

unit vector   A vector of unit length, sometimes called a directon vector. The unit vector $\hat{a}$ having the same direction as a given (nonzero) vector $\vec{a}$ is defined by $\hat{a} = \dfrac{\vec{a}}{|a|}$, where $|a|$ denotes the norm (i.e., magnitude or length) of $\vec{a}$. Thus, any vector $\vec{a}$ can be represented by the product of a unit vector $\hat{a}$ and a scalar magnitude $|a|$: $\vec{a} = \hat{a} \cdot |a|$. Unit vector notation is used for GPS and InSAR applications to denote the direction of surface displacement or the look angle of the radar, respectively.

UTC   Acronym for Coordinated Universal Time, the modern implementation of Greenwich Mean (or Meridian) Time (GMT).

USGS   Acronym for United States Geological Survey, an agency within the US Department of the Interior whose mission is to describe and understand the Earth; minimize loss of life and property from natural disasters; manage water, biological, energy, and mineral resources; and enhance and protect the quality of life. Among the agency's natural hazards programs is the Volcano Hazards Program (*VHP*), which strives to advance the scientific understanding of volcanic processes and to lessen the harmful impacts of volcanic activity. VHP is the locus of scientific research on volcanoes within the USGS.

VDAP   Acronym for the Volcano Disaster Assistance Program, which was established following the tragic 1985 eruption of Nevado del Ruiz Volcano, Colombia, in which over 23,000 people lost their lives. VDAP is a joint effort by the US Geological Survey and the Office of Foreign Disaster Assistance of the US Agency for International Development. The primary mission of this interagency program is to reduce eruption-caused fatalities and economic losses in developing countries. The principal components of VDAP are operational funding from OFDA, a small core group of scientists at the USGS David A. Johnston Cascades Volcano Observatory (CVO) in Vancouver, Washington, a large group of contributing scientists from CVO and other USGS offices, and a cache of portable volcano-monitoring equipment ready for rapid deployment.

VEI   Acronym for *Volcanic Explosivity Index*. *See* **Volcanic Explosivity Index**.

vertical deflection   (also called deflection of the vertical, deviation of the vertical, or deflection of the plumb line) The angle at a point on Earth's surface between the vertical, defined by gravity, and the direction of the normal to the reference *ellipsoid* through that point. The shapes of the *geoid* and reference ellipsoid differ as a result of mass inhomogeneities within the Earth. Leveling relies on gravity to determine the vertical, so heights measured in this way are relative to the geoid. Heights determined by GPS, on the other hand, are measured relative to the ellipsoid. The difference can be resolved if the local gravity field is well known.

VHP   Acronym for the USGS Volcano Hazards Program. The overall objectives of the Volcano Hazards Program are to advance the scientific understanding of volcanic processes and to lessen the harmful impacts of volcanic activity. The VHP monitors active and potentially active volcanoes, assesses their hazards, responds to volcanic crises, and conducts research on how volcanoes work to fulfill a Congressional mandate (P.L. 93–288) that the USGS issue 'timely warnings' of potential volcanic hazards to responsible emergency management authorities and to the populace affected. Thus, in addition to obtaining the best possible scientific information, the program works to effectively communicate its scientific findings to authorities and the public in an appropriate and understandable form. VHP operates five volcano observatories in the USA: the Alaska Volcano Observatory (AVO, in cooperation with the Geophysical Institute of the University of Alaska Fairbanks and the State of Alaska Division of Geological and Geophysical Surveys), the David A. Johnston Cascades Volcano Observatory (CVO),

Hawaiian Volcano Observatory (HVO), Long Valley Observatory (LVO), and Yellowstone Volcano Observatory (YVO, in cooperation with Yellowstone National Park and the University of Utah).

volcanic dome A steep-sided rounded extrusion of viscous lava forming a dome-shaped or bulbous mass above and around a volcanic vent. Many domes are built episodically through a combination of exogenous and endogenous growth. Synonymous with *lava dome*.

Volcanic Explosivity Index (VEI) A measure of the size of volcanic eruptions akin to the Richter magnitude scale for earthquakes. The VEI is a 0-to-8 index of increasing explosivity, each interval representing an increase of about a factor of ten. It combines total volume of explosive products, eruptive cloud height, descriptive terms, and other measures (Newhall and Self, 1982). Non-explosive eruptions are VEI 0. Small explosive eruptions with 10^4 to 10^6 m^3 of eruptive products and plume heights of 0.1 to 1 km are VEI 1. Moderate explosive eruptions with 10^6 to 10^7 m^3 of eruptive products and plume heights of 1 to 5 km are VEI 2. Large eruptions with 10^8 to 10^9 m^3 of eruptive products and plume heights of 10 to 25 km are VEI 4. Very large eruptions with 10^9 to 10^{12} m^3 of eruptive products and plume heights greater than 25 km are VEI 5–8. This last group includes mostly caldera-forming events such as the 1991 eruption of Pinatubo Volcano in the Philippines.

volcanic tremor A continuous seismic signal of long duration (minutes to days or longer) and dominant frequency in the 1–5 Hz range. Many investigators agree that tremor is a series of long-period earthquakes that originate from the same source and overlap in time. Harmonic tremor and spasmodic tremor are two common types of volcanic tremor. The former is a low-frequency, often-monotonic sinusoid with smoothly varying amplitude, and the latter is a higher frequency, pulsating, irregular signal. Harmonic tremor is commonly associated with near-surface magma flow or fountaining at basaltic volcanoes.

volcano–tectonic (VT) earthquake A high-frequency earthquake beneath a volcano caused by brittle failure or slip on a fault, indistinguishable in most ways from a tectonic earthquake in a non-volcanic environment. Volcano–tectonic earthquakes typically occur in swarms rather than as main shock–aftershock sequences. Most VT earthquakes have clear P- and S-waves, and dominant frequencies in the 5–15 Hz range. *See also* **long-period (LP) earthquake**.

WAAS Acronym for Wide-Area Augmentation System, a Satellite-Based Augmentation System (SBAS) for USA, Canada, and Puerto Rico. *See* **SBAS**.

welded tuff A glass-rich pyroclastic rock that has been indurated by the welding together of its glass shards under the combined action of the heat retained by particles, the weight of overlying material, and hot gases. It is generally composed of silicic pyroclasts and appears banded or streaky. *See* **tuff** and **pyroclast**.

wet tilt A procedure for measuring ground tilt using a portable long-base watertube tiltmeter. Water and air hoses were used to connect brass containers (pots) attached to three concrete piers separated by $\sim$30 m (Eaton, 1959). The water hoses were used to establish and maintain the same water level in all three pots. The air hoses assured equal pressure between the pots. Micrometers were used to measure the height of water in the pots simultaneously, so elevation differences between piers could be calculated precisely. The procedure was developed in the 1960s at the USGS Hawaiian Volcano Observatory.

W-key A secret Department of Defense encryption algorithm that is used in conjunction with the P-code to form the Y-code. The procedure, called anti-spoofing or AS, prevents an appropriately designed military GPS receiver from being fooled by an intentionally deceptive fake GPS signal. AS has been continuously activated since 31 January 1994.

wide lane A linear combination of the GPS carrier phases, Φ_{L1} and Φ_{L2}, such that $\Phi_{L1+L2} = \Phi_{L1} - \Phi_{L2}$. The corresponding wavelengths are: $\lambda_{L1} = 19.0$ cm, $\lambda_{L2} = 24.4$ cm, and $\lambda_{L1-L2} = 86.2$ cm. Linear combinations of observables are useful for eliminating error sources that are inherent in GPS observations (Chapter 4).

x-parallax (photogrammetry) From Wolf (1983, p. 159): An aerial camera exposing overlapping photographs at regular intervals of time obtains a record of positions of images at the instants of exposure. The change in position of an image from one photograph to the next caused by the aircraft's motion is termed *stereoscopic parallax, x-parallax*, or simply *parallax*. *See also* **y-parallax**.

xenolith A foreign inclusion in an igneous rock. Magma rising and accumulating beneath a volcano often incorporates fragments of the host rock that appear as xenoliths in the erupted lava.

y-parallax (photogrammetry) An essential condition that must exist for clear and comfortable stereoscopic viewing of overlapping aerial photographs is that the line joining corresponding points in the two photos be parallel with the direction of flight when the photos were acquired. When corresponding images fail to lie along a line parallel to the flight line, *y-parallax* is said to exist. Any slight amount of y-parallax causes eyestrain, and excessive amounts prevent stereoscopic viewing altogether (Wolf, 1983, p. 152). *See also* **x-parallax**.

Y-code Binary code modulation scheme for GPS signals formed by encrypting the P-code using a secret W-key. Y-code is the basis for the anti-spoofing (AS) feature of GPS. *See* **P-code** and **W-key**.

Young's modulus A modulus of elasticity in tension or compression, involving a change in length. *See also* **bulk modulus**, **modulus of elasticity**, **Poisson's ratio**, and **shear modulus**.

YVO Acronym for the Yellowstone Volcano Observatory, a cooperative scientific effort involving the US Geological Survey, National Park Service, and University of Utah. YVO is a long-term, instrument-based monitoring facility designed for observing volcanic and earthquake activity in the Yellowstone National Park region. The principal objectives of YVO are to: (1) provide seismic and GPS monitoring that enables reliable and timely warnings of possible renewed volcanism and related hazards in the Yellowstone region; (2) notify the NPS, other local officials, and the public of significant seismic or volcanic events; (3) improve scientific understanding of tectonic and magmatic processes that influence ongoing seismicity, surface deformation, and hydrothermal activity; (4) assess the long-term potential hazards of volcanism, seismicity, and explosive hydrothermal activity in the region; (5) communicate effectively the results of these efforts to responsible authorities and to the public, and (6) improve coordination and cooperation among the University of Utah, Yellowstone National Park, and the USGS. *See also* **AVO**, **CVO**, **HVO**, and **LVO**.

zenith The point at which a line opposite in direction from that of the plumb line at a given point on Earth's surface meets the celestial sphere. The zenith is directly opposite the *nadir*.

zenith, geocentric The point where a line from the center of Earth through a given point on its surface meets the celestial sphere. The term is sometimes used in astronomic work, but seldom appears in geodetic work. It should be used only in its entirely, because the single word, zenith, is reserved for designating the point determined by the direction of the plumb line.

zenith, geodetic The point where the normal to the reference ellipsoid, extended upward, meets the celestial sphere. The term should be used only in its entirety, because the single word, zenith, is reserved for designating the point determined by the direction of the plumb line.

References

Achilli, V., Baldi, P., Baratin, L., Bonini, C., Ercolani, E., Gandolfi, S., Anzidei, M., and Riguzzi, F. (1998). Digital photogrammetric survey on the island of Vulcano. *Acta Vulcanol.*, **10**, 1–5.

Agnew, D.C. (1986). Strainmeters and tiltmeters. *Rev. Geophys.*, **24**(3), 579–624.

Agnew, D.C. (2002). History of seismology. In: Lee, W.H.K., Kanamori, H., Jennings, P.C., and Kisslinger, C. (eds), *IASPEI International Handbook of Earthquake and Engineering Seismology*. Academic Press, San Diego, pp. 3–11.

Agustsson, K., Stefansson, R., Linde, A.T., Einarsson, P., Sacks, I.S., Gudmundsson, G.B., and Thorbjarndottir, B. (2000). Successful prediction and warning of the 2000 eruption of Hekla based on seismicity and strain changes. *EOS Trans. Amer. Geophys. Union*, **81**(48), Fall Meet. Suppl., Abstract F1337.

Aki, K. and Richards, P.G. (1980). *Quantitative Seismology: Theory and Methods*. W.H. Freeman and Co., San Francisco, 932pp.

Aki, K., Fehler, M., and Das, S. (1977). Source mechanism of volcanic tremor: Fluid-driven crack models and their application to the 1963 Kilauea eruption. *J. Volcanol. Geotherm. Res.*, **2**(3), 259–287.

Alber, C., Ware, R., Rocken, C., and Solheim, F. (1997). GPS surveying with 1 mm precision using corrections for atmospheric slant path delay. *Geophys. Res. Lett.*, **24**(15), 1859–1862.

Almendros, J., Chouet, B.A., and Dawson, P.B. (2001a). Spatial extent of a hydrothermal system at Kilauea Volcano, Hawaii, determined from array analyses of shallow long-period seismicity. Part 1: Method. *J. Geophys. Res.*, **106**(B7), 13565–13580.

Almendros, J., Chouet, B.A., and Dawson, P.B. (2001b). Spatial extent of a hydrothermal system at Kilauea Volcano, Hawaii, determined from array analyses of shallow long-period seismicity. Part 2: Results. *J. Geophys. Res.*, **106**(B7), 13581–13597.

Altamimi, Z., Sillard, P., and Boucher, C. (2002). ITRF2000: A new release of the International Terrestrial Reference Frame for earth science applications. *J. Geophys. Res.*, **107**(B10), 2214, doi: 10.1029/2001JB000561.

Altamimi Z., Angermann, D., Argus, D., Blewitt, G., Boucher, C., Chao, B., Drewes, H., Eanes, R., Feissel, M., Ferland, R., *et al.* (2001). The International Terrestrial Reference Frame and the dynamic Earth. *EOS Trans. Amer. Geophys. Union*, **82**(25), 273, 278–279.

Althausen, J., Baldwin, A., and Schwoppe, K. (2004). Current methods for developing digital terrain models. *Earth Observation Magazine*, **13**(7), November 2004, 5–7.

Amelung, F., Jonsson, S., Zebker, H., and Segall, P. (2000). Widespread uplift and 'trapdoor' faulting on Galapagos volcanoes observed with radar interferometry. *Nature*, **407**, 993–996.

American Geological Institute (1987). *Glossary of Geology* (3rd edition), Bates, R.L., and Jackson, J.A. (eds). American Geological Institute, Alexandria, Virginia, 788pp.

American Society for Photogrammetry and Remote Sensing (1996a). *Digital Photogrammetry – An addendum to the Manual of Photogrammetry*. ASPRS, Bethesda, MD.

American Society for Photogrammetry and Remote Sensing (1996b). *Close-range Photogrammetry and Machine Vision*. ASPRS, Bethesda, MD.

American Society for Photogrammetry and Remote Sensing (1997). *Manual of Photographic Interpretation* (2nd edition). ASPRS, Bethesda, MD.

American Society for Photogrammetry and Remote Sensing (1998). *Manual of Remote Sensing* (3rd edition). ASPRS, Bethesda, MD.

Anadóttir, T. and Segal, P. (1994). The 1989 Loma Prieta earthquake imaged from inversion of geodetic data. *J. Geophys. Res.*, **99**, 21835–21855.

Anderson, E.M. (1936). Dynamics of the formation of cone-sheets, ring-dykes, and cauldron-subsidences. *Proc. R. Soc. Edinburgh*, **56**, 128–157.

Anderson, J.G., Brune, J.N., Louie, J.N., Zeng, Y., Savage, M., Yu, G., Chen, Q., and dePolo, D.

(1994). Seismicity in the western Great Basin apparently triggered by the Landers, California, earthquake, 28 June 1992. *Bull. Seis. Soc. Am.*, **84**(3), 863–891.

Ando, M. (1979). The Hawaii earthquake of November 29, 1975: Low dip angle faulting due to forceful injection of magma. *J. Geophys. Res.*, **84**, 7616–7626.

Aoki, Y., Segall, P., Kato, T., Cervelli, P., and Shimada, S. (1999). Imaging magma transport during the 1997 seismic swarm off the Izu Peninsula, Japan. *Science*, **286**, 927–930.

Apple, R.A. (1987). Thomas A. Jaggar, Jr., and the Hawaiian Volcano Observatory. In: Decker, R.W., Wright, T.L., and Stauffer, P.H. (eds), *Volcanism in Hawaii* (USGS Prof. Paper 1350). US Geological Survey, Reston, VA, pp. 1619–1644.

Arnadottir, T., Segall, P., and Matthews, M. (1992). Resolving the discrepancy between geodetic and seismic fault models for the 1989 Loma Prieta, California, earthquake. *Bull. Seismol. Soc. Am.*, **82**, 2248–2255.

Arnet, F., Kahle, H.-G., Klingelé, E., Smith, R.B., Meertens, C.M., and Dzurisin, D. (1997). Temporal gravity and height changes of Yellowstone Caldera, 1977–1994. *Geophys. Res. Lett.*, **24**(22), 2741–2744.

Ashby, N. and Spilker, J.J. (1996). Introduction to relativistic effects on the Global Positioning System. In: Parkinson, B.W., Spilker, J.J., Axelrad, P, and Enge, P. (eds), *Global Positioning System: Theory and Applications* (Volume 1). American Institute of Aeronautics and Astronautics, Inc., Washington, D.C., pp. 623–695.

Ashjaee, J., Lorenz, R., Sutherland, R., Dutilloy, J., Minazio, J.-B., Abtahi, R., Eichner, J.-M., Kosmalska, J., and Helkey, R. (1989). New GPS Developments and Ashtech M-XII. *Proceedings of ION GPS-89, The Second International Technical Meeting of the Satellite Division of the Institute of Navigation, Colorado Springs, Colorado, September 27–29*, pp. 195–198.

Atwater, B.F. and Hemphill-Haley, E. (1997). *Recurrence Intervals for Great Earthquakes of the Past 3,500 years at Northeastern Willapa Bay, Washington* (USGS Prof. Paper 1576). US Geological Survey, Reston, VA, 108pp.

Bailey, R.A. (1989). Geologic map of Long Valley Caldera, Mono-Inyo Craters volcanic chain, and vicinity, eastern California. US Geol. Surv. Misc. Invest. Ser., Map I-1933, 1:62,500 scale, 2 sheets and pamphlet, 11pp.

Bailey, R.A. (2004). *Eruptive History and Chemical Evolution of the Precaldera and Postcaldera Basalt–Dacite Sequences, Long Valley, California: Implications for Magma Sources, Current Seismic Unrest, and Future Volcanism* (USGS Prof. Paper 1692). US Geological Survey, Reston, VA, 75pp.

Bailey, R.A., Dalrymple, G.B., and Lanphere, M.A. (1976). Volcanism, structure, and geochronology of Long Valley Caldera, Mono County, California. *J. Geophys. Res.*, **81**, 725–744.

Bailey, R.C. (1990). Trapping of aqueous fluids in the deep crust. *Geophys. Res. Lett.*, **17**, 1129–1132.

Bailey, A.L. and Hamilton, W.L. (1988). Holocene deformation deduced from shorelines at Yellowstone Lake, Yellowstone Caldera. Part II: The model. *EOS Trans. Amer. Geophys. Union*, **69**(44), 1472.

Balazs, E.I. and Young, G.M. (1982). Corrections applied by the National Geodetic Survey to precise leveling observations. *NOAA Technical Memorandum NOS NGS 34*, 12pp.

Baldi, P., Bonvalot, S., Briole, P., and Marsella, M., (2000). Digital photogrammetry and kinematic GPS for monitoring volcanic areas. Geophys. *J. Int.*, **142**(3), 801–811.

Balogh, A., Marsden, R.G., and Smith, E.J. (2001). *The Heliosphere near Solar Minimum: The Ulysses Perspective*. Praxis, Chichester, UK, 436pp.

Bamler, R., Eineder, M., and Breit H. (1996). The X-SAR single-pass interferometer on SRTM: Expected performance and processing concept. *Proc. EUSAR'96. Königswinter, Germany*, pp. 181–184.

Bamler, R. and Hartl, P. (1998). Synthetic Aperture Radar Interferometry. *Inverse Problems*, **14**, R1–54.

Bartel, B. (2002). Magma dynamics at Taal Volcano, Philippines, from continuous GPS measurements. MSc. thesis, Dept. of Geological Sciences, Indiana University, 168pp.

Battaglia, J., Got, J.-L., and Okubo, P. (2003). Location of long-period events below Kilauea Volcano using seismic amplitudes and accurate relative relocation. *J. Geophys. Res.*, **108**(B12), 2553, doi: 10.1029/ 2003JB002517.

Battaglia, M, Roberts, C., and Segall, P. (1999). Magma intrusion beneath Long Valley caldera confirmed by temporal changes in gravity. *Science*, **285**, 2119–2122.

Battaglia, M., Segall, P., Murray, J., Cervelli, P., and Langbein, J. (2003a). The mechanics of unrest at Long Valley Caldera, California. Part 1: Modeling the geometry of the source using GPS, leveling and 2-color EDM data. In: Sorey, M.L., McConnell, V.S., and Roeloffs, E. (eds), Crustal unrest in Long Valley Caldera, California: New interpretations from geophysical and hydrologic monitoring and deep drilling. *J. Volcanol. Geotherm. Res.*, **127**(3–4), 195–217.

Battaglia, M., Segall, P., and Roberts, C. (2003b). The mechanics of unrest at Long Valley Caldera, California. Part 2: Constraining the nature of the source using geodetic and micro-gravity data. In: Sorey, M.L., McConnell, V.S., and Roeloffs, E. (eds), Crustal unrest in Long Valley Caldera, California: New interpretations from geophysical and hydrologic monitoring and deep drilling. *J. Volcanol. Geotherm. Res.*, **127**(3–4), 219–245.

Beauducel, F., Briole, P., and Froger, J-L. (2000). Volcano-wide fringes in ERS synthetic aperture radar interferograms of Etna (1992–1998): Deformation or tropospheric effect? *J. Geophys. Res.*, **105**(7), 16391–16402.

Becker, J.M. and Bevis, M. (2004). Love's problem. *Geophys. J. Int.*, **156**, 171–178.

Behr, J., Bilham, R., and Beavan, J. (1992). Monitoring of magma chamber inflation using a biaxial Michelson tiltmeter in Long Valley Caldera, California. *EOS Trans. Amer. Geophys. Union*, **73**(43), 347–348.

Benioff, H. (1935). A linear strain seismograph. *Bull. Seis. Soc. Am.*, **25**, 283–309.

Bennett, R.A., Davis, J.L., and Wernicke, B.P. (1998). Continuous GPS measurements of contemporary deformation across the northern Basin and Range province. *Geophys. Res. Lett.*, **25**(4), 563–566.

Bennett, R.A., Davis, J.L., and Wernicke, B.P. (1999). Present-day pattern of Cordilleran deformation in the western United States. *Geology*, **27**(4), 371–374.

Benz, H. and Smith, R.B. (1984). Simultaneous inversion for lateral velocity variations and hypocenters in the Yellowstone region using earthquake and refraction data. *J. Geophys. Res.*, **89**, 1208–1220.

Berg, B. (1993). Locating global minima in optimizing problems by a randon-cost approach. *Nature*, **361**, 708–710.

Berrino, G., Corrado, G., Magliulo, R., and Ricardi, U. (1997). Continuous record of the gravity changes at Mt. Vesuvius (Boschi, E., ed.). *Annali di Geofisica*, **40**(5), 1019–1028.

Bevin, A.J. (2003). Accuracy Standards for Geodetic Surveys, SG Standard 1 version 1.1. Office of Surveyor-General, Land Information New Zealand, 27pp. Available online at: http://www.linz.govt.nz/ [accessed March 2006].

Bilham, R. (1989). Surface slip subsequent to the 24 November 1987 Superstition Hills, California, earthquake monitored by digital creepmeters. *Bull. Seis. Soc. Am.*, **79**(2), 424–450.

Bilham, R.G., Beavan, R.J., and Evans, K.F. (1981). Long baseline fluid-tube tiltmeter geometry and the detection of flexure and tilt. *Proceedings of the 9th International Symposium on Earth Tides, August 17–22, 1981, New York City, NY*, pp. 85–94.

Biot, M.A. (1941). General theory of three-dimensional consolidation. *Jour. Appl. Phys.*, **12**, 155–164.

Biot, M.A. (1956). General solutions of the equations of elasticity and consolidation for a porous material. *J. Appl. Mech.*, ASME, **23**, 91–96.

Bischoff, J.L., Rosenbauer, R.J., and Fournier, R.O. (1996). The generation of HCl in the system $CaCl_2$–H_2O: Vapor–liquid relations from 380–500°C. *Geochimica et Cosmochimica Acta*, **60**, 7–16.

Blong, R. and McKee, C. (1995). *The Rabaul Eruption 1994, Destruction of A Town*. Natural Hazards Research Centre, Macquarie University, Sydney, 52pp.

Bloom, M.S. (1981). Chemistry of inclusion fluids: Stockwork molybdenum deposits from Questa, New Mexico,

Hudson Bay Mountain, and Endako, British Columbia. *Economic Geology*, **76**, 1906–1922.

Bodin, P. and Gomberg, J. (1994). Triggered seismicity and deformation between the Landers, California, and Little Skull Mountain, Nevada, earthquakes. *Bull. Seis. Soc. Am.*, **84**(3), 835–843.

Bomford, G. (1980). *Geodesy* (4th edition). Oxford University Press, Oxford, UK, 855pp.

Bonaccorso, A. (1996). Dynamic inversion of ground deformation data for modeling volcanic sources (Etna 1991–1993). *Geophys. Res. Lett.*, **23**(5), 4261–4268.

Bonaccorso, A. and Davis, P.M. (1999). Models of ground deformation from vertical volcanic conduits with application to eruptions of Mount St. Helens and Mount Etna. *J. Geophys. Res.*, **104**(B5), 10531–10542.

Bonafede, M. (1990). Axi-symmetric deformation of a thermo-poro-elastic half space: Inflation of a magma chamber. *Geophys. J. Int.*, **103**, 289–299.

Bowden, G.B. (2003). Calibration of geophone microseismic sensors. Stanford Linear Accelerator Center, LCLS-TN-03-6, October 2003, 9pp.

Bower, D.R. (1983). Bedrock fracture parameters from the interpretation of well tides. *J. Geophys. Res.*, **88**(B6), 5025–5035.

Brace, W.F. (1972). Pore pressure in geophysics. In: Heard, H.C., Borg, I.Y., Carter, N.L., and Raleigh, C.B. (eds), *Flow and Fracture of Rocks* (Geophysical Monograph 16). American Geophysical Union, Washington D.C., pp. 265–273.

Brantley, S.R. (1990). *The eruption of Redoubt Volcano, Alaska, December 14, 1989–August 31, 1990* (USGS Circ. 1061). US Geological Survey, Reston, VA, 33pp.

Brown, J.M., Niebauer, T.M., Richter, B., Klopping, F.J., Valentine, J.G., and Buxton, W.K. (1999). A new miniaturized absolute gravimeter developed for dynamic applications. *EOS Trans. Am. Geophys. Union*, **80**(32), 355.

Bugge, T. (1983). Submarine slides on the Norwegian continental margin, with special emphasis on the Storegga area. *Publikas jon Institut for Kontinentalsokke Lundersookelser*, **110**, 152pp.

Bürgmann, R., Rosen, P.A., and Fielding, E.J. (2000). Synthetic aperture radar interferometry to measure Earth's surface topography and its deformation. *Ann. Rev. Earth Planet. Sci.*, **28**, 169–209.

Burnham, C.W. (1967). Hydrothermal fluids at the magmatic stage. In: Barnes, H.L. (ed.), *Geochemistry of Hydrothermal Ore Deposits*. Holt, Rinehart, and Winston, New York, pp. 34–76.

Burnham, C.W. (1979). Magmas and hydrothermal fluids. In: Barnes, H.L. (ed.), *Geochemistry of Hydrothermal Ore Deposits* (2nd edition). Wiley, New York, pp. 71–136.

Burnham, C.W. (1985). Energy released in subvolcanic environments: Implications for brecciation. *Economic Geology*, **80**, 1515–1522.

Byerlee, J.D. (1968). Brittle–ductile transition in rocks. *J. Geophys. Res.*, **73**, 4741–4750.

Byerlee, J.D. (1978). Friction of rocks. *Pure Applied Geophysics*, **116**, 615–626.

Cabral-Cano, E., Dixon, T., Mao, A., Finny, M., and Correa-Mora, F. (1996). Near real-time GPS deformation monitoring of Popocatépetl Volcano, central Mexico. *EOS Trans. Am. Geophys. Union*, **77**(46, suppl.), 809–810.

Candela, P.A. and Piccoli, P.M. (1995). Model ore-metal partitioning from melts into vapor and vapor/brine mixtures. In: Thompson, J.F.H. (ed.), *Magmas, Fluids, and Ore Deposits* (Mineralogical Association of Canada Short Course Series, **23**). Min. Assoc. of Canada, Ottawa, Ontario, pp. 101–122.

Caputo, M. (1979). Two-thousand years of geodetic and geophysical observations in the Phlegraean Fields near Naples. *Geophys. Jour. Royal Astron. Soc.*, **56**, 319–328.

Carey, S.N. and Sigurdsson, H. (1986). The 1982 eruptions of El Chichón Volcano, Mexico. Part 2: Observations and numerical modelling of tephra-fall distribution. *Bull. Volcanol.*, **48**, 127–141.

Carrigan, C.R. (1986). A two-phase hydrothermal cooling model for shallow intrusions. *J. Volanolc. Geotherm. Res.*, **28**, 175–192.

Carroll, M.R. and Webster, J.D. (1994). Solubilities of sulfur, noble gases, nitrogen, chlorine, and fluorine in magmas. In: Carroll, M.R. and Holloway, J.R. (eds.), *Volatiles in Magmas* (Reviews in Mineralogy, **30**). Mineral. Soc. Amer., Washington, pp. 231–279.

Carter, N.L. and Tsenn, M.C. (1986) Flow properties of continental lithosphere. *Tectonophysics*, **136**, 27–63.

Casadevall, T.J. (ed.) (1994). *Volcanic Ash and Aviation Safety: Proceedings of the First International Symposium on Volcanic Ash and Aviation Safety* (USGS Bull. 2047). US Geological Survey, Reston, VA, 450pp.

Casadevall, T.J., Johnston, D.A., Harris, D.M., Rose, W.I., Jr., Malinconico, L.L., Stoiber, R.E., Bornhorst, T.J., Williams, S.N., Woodruff, L., and Thompson, J.M. (1981). SO$_2$ emission rates at Mount St. Helens from March 29 through December, 1980. In: Lipman, P.W. and Mullineaux, D.R. (eds), *The 1980 Eruptions of Mount St. Helens, Washington* (USGS Prof. Paper 1250). US Geological Survey, Reston, VA, pp. 193–200.

Casadevall, T.J., Rose, W.I., Jr., Gerlach, T.M., Greenland, L.P., Ewert, J., Wunderman, R., and Symonds, R. (1983). Gas emissions and the eruptions of Mount St. Helens through 1982. *Science*, **221**, 1383–1385.

Casadevall, T.J., Stokes, J.B., Greenland, L.P., Malinconico, L.L., Casadevall, J.R., and Furukawa, B.T. (1987). SO$_2$ and CO$_2$ emission rates at Kilauea Volcano, 1979–1984. In: Decker, R.W.,

Wright, T.L., and Stauffer, P.H. (eds), *Volcanism in Hawaii* (USGS Prof. Paper 1350). US Geological Survey, Reston, VA, pp. 771–780.

Castle, R.O. (1987). Bulge bashing: Modern sophistry in action. *EOS Trans. Am. Geophys. Union*, **68**, 1506.

Castle, R.O., Church, J.P., Elliot, M.R., Gilmore, T.D., Mark, R.K., Newman, E.B., Tinsley, J.C., III, and Strange, W.E. (1983). The impact of refraction correction on leveling interpretations in Southern California. *J. Geophys. Res.*, **88**, 2508–2512.

Castle, R.O., Estrem, J.E., and Savage, J.C. (1984). Uplift across Long Valley caldera, California. *J. Geophys. Res.*, **89**, 11507–11516.

Cathles, L.M. (1977). An analysis of the cooling of intrusives by groundwater convection which includes boiling. *Economic Geology*, **72**, 804–826.

Cayol, V. and Cornet, F.H. (1997). 3-D mixed boundary elements for elastostatic deformation field analysis. *Int. J. Rock Mech. Min. Sci.*, **80**, 275–287.

Cayol, V. and Cornet, F.H. (1998). Effects of topography on the interpretation of the deformation field of prominent volcanoes: Application to Etna. *Geophys. Res. Lett.*, **25**(11), 1979–1982.

Cervelli, P. (2001) Estimating source parameters from deformation data, with application to the March 1997 earthquake swarm off the Izu Peninsula, Japan. *J. Geophys. Res.*, **106**, 11217–11238.

Cervelli, P. (2004). The threat of silent earthquakes. *Scientific American*, **290**(3), 86–91.

Cervelli, P.F. and Miklius, A. (2003). The shallow magmatic system of Kilauea Volcano. In: Heliker, C., Swanson, D.A., and Takahashi, T.J. (eds), *The Pu'u 'Ō'ō– Kupaianaha Eruption of Kilauea Volcano, Hawaii: The First 20 Years* (USGS Prof. Paper 1676). US Geological Survey, Reston, VA, pp. 149–163.

Cervelli, P., Segall, P., Johnson, K., Lisowski, M., and Miklius, A. (2002a). Sudden aseismic fault slip on the south flank of Klauea Volcano. *Nature*, **415**(6875), 1014–1018.

Cervelli, P., Segall, P., Amelung, F., Garbeil, H., Meertens, C., Owen, S., Miklius, A., and Lisowski, M. (2002b). The 12 September 1999 Upper East Rift Zone dike intrusion at Kilauea Volcano, Hawaii. *J. Geophys. Res.*, **107**(B7), ECV 1–13, doi: 10.1029/2001JB000602.

Chadwick, W.W., Jr., Swanson, D.A., Iwatsubo, E.Y., Heliker, C.C., and Leighley, T.A. (1983). Deformation monitoring at Mount St. Helens in 1981 and 1982. *Science*, **221**, 1378–1380.

Chadwick, W.W., Archuleta, R.J., and Swanson, D.A. (1988). The mechanics of ground deformation precursory to dome-building extrusions at Mount St. Helens 1981–1982. *J. Geophys. Res.*, **93**, 4351–4366.

Chouet, B.A. (1981). Ground motion in the near field of a fluid-driven crack and its interpretation in the study of shallow volcanic tremor. *J. Geophys. Res.*, **86**, 5985–6016.

Chouet, B.A. (1985). Excitation of a buried magmatic pipe: A seismic source model for volcanic tremor. *J. Geophys. Res.*, **90**, 1881–1893.

Chouet, B.A. (1986). Dynamics of a fluid-driven crack in three dimensions by the finite difference method. *J. Geophys. Res.*, **91**, 13967–13992.

Chouet, B.A. (1988). Resonance of a fluid-driven crack: Radiation properties and implications for the source of long-period events and harmonic tremor. *J. Geophys. Res.*, **93**, 4375–4400.

Chouet, B.A. (1992). A seismic model for the source of long-period events and harmonic tremor. In: Gasparini, P., Scarpa, R., and Aki, K. (eds), *Volcanic Seismology: International Association of Volcanology and Chemistry of the Earth's Interior (IAVCEI) Proceedings in Volcanology*. Berlin, Springer-Verlag, pp. 133–156.

Chouet, B.A. (1996a). Long-period volcano seismicity: Its source and use in eruption forecasting. *Nature*, **380**, 309–316.

Chouet, B.A. (1996b). New methods and future trends in seismological volcano monitoring. In: Scarpa, R. and Tilling, R.I. (eds), *Monitoring and Mitigation of Volcano Hazards*. Springer-Verlag, New York, pp. 23–97.

Chouet, B.A. (2003). Volcano seismology. In: Ben-Zion, Y. (ed.), *Seismic Motion, Lithospheric Structures, Earthquake and Volcanic Sources: The Keiiti Aki Volume*. (Pageoph Topical Volumes, **160**(3–4)). Birkhäuser Verlag, Basel, Switzerland, pp. 739–788.

Chouet, B.A., Page, R.A., Stephens, C.D., Lahr, J.C., and Power, J.A. (1994). Precursory swarms of long-period events at Redoubt Volcano (1989–1990), Alaska: Their origin and use as a forecasting tool. *J. Volcanol. Geotherm. Res.*, **62**, 95–135.

Chouet, B.A., Saccorotti, G., Martini, M., Dawson, P., De Luca, G., Milana, G., and Scarpa, R. (1997). Source and path effects in the wavefields of tremor and explosions at Stromboli Volcano, Italy. *J. Geophys. Res.*, **102**, 15129–15150.

Chouet, B.A., Dawson, P., and Kedar, S. (1998). Waveform inversion of very long period impulsive signals associated with magmatic injection beneath Kilauea Volcano, Hawaii. *J. Geophys. Res.*, **103**, 23839–23862.

Christiansen, R.L. (1984). Yellowstone magmatic evolution: Its bearing on understanding large-volume explosive volcanism. In: *Explosive Volcanism: Inception, Evolution, and Hazards. National Research Council Studies in Geophysics*. National Academy Press, Washington, DC, pp. 84–95.

Christiansen, R.L. (2001). *The Quaternary and Pliocene Yellowstone Plateau Volcanic Field of Wyoming, Idaho, and Montana* (USGS Prof. Paper 729-G). US Geological Survey, Reston, VA, 145pp.

Christiansen, R.L. and Peterson, D.W. (1981). Chronology of the 1980 eruptive activity. In: Lipman, P.W., and Mullineaux, D.R. (eds), *The 1980 Eruptions of Mount St. Helens, Washington* (USGS Prof. Paper 1250). US Geological Survey, Reston, VA, pp. 17–30.

Christiansen, R.L., Foulger, G.R., and Evans, J.R. (2002). Upper-mantle origin of the Yellowstone hotspot. *Bull. Geol. Soc. Am.*, **114**(10), 1245–1256.

Clarke, T.A. and Fryer, J.F. (1998). The development of camera calibration methods and models. *Photogrammetric Record*, **16**(91), 51–66.

Cline, J.S. and Bodnar, R.J. (1994). Direct evolution of brine from a crystallizing silicic melt at the Questa, New Mexico, molybdenum deposit. *Economic Geology*, **89**, 1780–1802.

Clynne, M.A., Christiansen, R.L., Felger, T.J., Stauffer, P.H., and Hendley II, J.W. (1999). *Eruptions of Lassen Peak, California, 1914 to 1917* (USGS Fact Sheet). US Geological Survey, Reston, VA, pp. 173–198.

Condor Earth Technologies (2004). Online at: http://www.3dtracker.com [Accessed April 2005].

Costantini, M. (1998). A novel phase unwrapping method based on network programming. *IEEE Transactions on Geoscience and Remote Sensing*, **36**, 813–821.

Couffer, J.C. (1956). The disappearance of Urvina Bay. *Natural History*, **65**, 378–383.

Crandell, D.R. and Mullineaux, D.R. (1978). *Potential Hazards from Future Eruptions of Mount St. Helens Volcano, Washington* (USGS Bull. 1383-C). US Geological Survey, Reston, VA, 26pp.

Crandell, D.R., Mullineaux, D.R., and Rubin, M. (1975). Mount St. Helens volcano: Recent and future behavior. *Science*, **187**, 438–441.

Crandell, D.R., Booth, B., Kazumadinata, K., Shimozuru, D., Walker, G.P.L., and Westercamp, D. (1984). *Source-Book for Volcanic-Hazards Zonation* (Natural hazards series 4). United Nations Educ. Sci. and Cult. Org., Paris, 97pp.

Crescentini, L., Amoruso, A., and Scarpa, R. (1999). Constraints on slow earthquake dynamics from a swarm in central Italy. *Science*, **286**, 2132–2134.

Crossley, D., Hinderer, J., Casula, O., Francis, O., Hsu, H. T., Imanishi, Y., Jentzsch, G., Kaarianen, J., Merriam, J., Meurers, B., Neumeyer, J., *et al.* (1999). Network of superconducting gravimeters benefits a number of disciplines. *EOS Trans. Amer. Geophys. Union*, **80**(11), 121, 125–126.

Cullen, A.B., McBirney, A.R., and Rogers, R.D. (1987). Structural controls on the morphology of Galapagos shields. *J. Volcanol. Geotherm. Res.*, **34**, 143–151.

Cunningham, C.G. (1978). Pressure gradients and boiling as mechanisms for localizing ore in porphyry systems. *U.S. Geological Survey Journal of Research*, **6**, 745–754.

Curlander, J.C. and McDonough, R.N. (1991). *Synthetic Aperture Radar: Systems and Signal Processing*. Wiley, New York, 647pp.

Damon, C. (2000). Translation of Pliny the Younger's Epistle 6.16 to the historian Tacitus, circa 79 A.D.

Available online at: http://www.amherst.edu/~classics/class36/ancsrc/01.html [Accessed April 2004].

Darby, D.J. and Williams, R.O. (1991). A new geodetic estimate of deformation in the central volcanic region of the North Island, New Zealand. *New Zealand J. Geol. Geophys.*, **34**, 127–136.

Davis, E.E., Wang, K., Thomson, R.E., Becker, K., and Cassidy, J.F. (2001). An episode of seafloor spreading and associated plate deformation inferred from crustal fluid pressure transients. *J. Geophys. Res.*, **106**, 21593–21964.

Davis, P.M. (1983). Surface deformation associated with a dipping hydrofracture. *J. Geophys. Res.*, **88**(B7), 5826–5834.

Davis, P.M. (1986). Surface deformation due to inflation of an arbitrarily oriented triaxial ellipsoidal cavity in an elastic half-space, with reference to Kilauea Volcano, Hawaii. *J. Geophys. Res.*, **91**, 7429–7438.

Davis, P.M., Hastie, L.M., and Stacey, F.D. (1974). Stresses within an active volcano: With particular reference to Kilauea. *Tectonophysics*, **22**, 363–375.

Davis, P.M., Pierce, D.R., McPherron, R.L., Dzurisin, D., Murray, T., Johnston, M.J.S., and Mueller, R. (1984). A volcanomagnetic observation on Mount St. Helens, Washington. *Geophys. Res. Lett.*, **11**(3), 233–236.

Davis, R.E., Foote, F.S., Anderson, J.M., and Mikhail, E.M. (1981). *Surveying Theory and Practice* (6th edition). McGraw-Hill Book Company, New York, 992pp.

Dawson, A.G., Long, D., and Smith, D.E. (1988). The Storegga slide: Evidence from eastern Scotland for a possible tsunami. *Marine Geology*, **82**, 271–276.

Dawson, P.B., Chouet, B.A., Okubo, P.G., Villasenor, A., and Benz, H.M. (1999). Three-dimensional velocity structure of the Kilauea Caldera, Hawaii. *Geophys. Res. Lett.*, **26**, 2805–2808.

Decker, R.W. (1987). Dynamics of Hawaiian volcanoes. In: Decker, R.W., Wright, T.L., and Stauffer, P.H. (eds), *Volcanism in Hawaii* (USGS Professional Paper 1350, Volume 2). US Geological Survey, Reston, VA, pp. 997–1018.

Decker, R. and Decker, B. (1998). *Volcanoes* (3rd edition). W.H. Freeman and Company, New York, ISBN 0716724405, 387pp.

Decker, R.W. and Wright, T.L. (1968). Deformation measurements on Mauna Loa Volcano, Hawaii. *Bull. Volcanol.*, **32**, 401.

Decker, R.W., Hill, D.P., and Wright, T.L. (1966). Deformation measurements on Kilauea Volcano, Hawaii. *Bull. Volcanol.*, **29**, 721–731.

del Nero, F. (ca. 1538). Lettera di Niccolo del Benino sul terremoto di Pozzuolo dal quale ebbe origine la Montagna Nuovo nel 1538. Manuscript translated and published in English in Horner, L. (1846). *Quart. Jour. Geol. Soc. London*, **3**(2), 19.

Delaney, P.T. (1982). Rapid intrusion of magma into wet rock: Groundwater flow due to pore pressure increases. *J. Geophys. Res.*, **87**(B9), 7739–7756.

Delaney, P.T. and McTigue, D.F. (1994). Volume of magma accumulation or withdrawal estimated from surface uplift or subsidence, with application to the 1960 collapse of Kilauea Volcano. *Bull. Volcanol.*, **56**, 417–424.

Delaney, P.T., Miklius, A., Arnadottir, T., Okamura, A.T., and Sako, M.K. (1993). Motion of Kilauea Volcano during sustained eruption from the Pu'u 'Ō'ō and Kupaianaha vents, 1983–1991. *J. Geophys. Res.*, **98**(B10), 17801–17820.

Delaney, P.T., Denlinger, R.P., Lisowski, M., Miklius, A., Okubo, P.G., Okamura, A.T., and Sako, M.K. (1998). Volcanic spreading at Kilauea, 1976–1996. *J. Geophys. Res.*, **103**(B8), 18003–18023.

De Meyer, F., Ducarme, B., and Elwahabi, A. (1995). Continuous gravity observations at Mount Etna (Sicily). *IUGG XXI General Assembly, Boulder, Colorado, July 2–14, 1995.*

De Netale, G. and Pingue, F. (1996). Ground deformation modeling in volcanic areas. In: Scarpa, R. and Tilling, R.I., (eds), *Monitoring and Mitigation of Volcano Hazards*. Sprinter-Verlag, Berlin, pp. 365–388.

Denlinger, R.P. and Hoblitt, R.P. (1999). Cyclic eruptive behavior of silicic volcanoes. *Geology*, **27**(5), 459–462.

Dewey, J. and Byerly, P. (1969). The early history of seismometry (to 1900). *Bull. Seis. Soc. Am.*, **59**(1), 183–227.

Dieterich, J.H. and Decker, R.W. (1975). Finite element modeling of surface deformation associated with volcanism. *J. Geophys. Res.*, **80**, 4095–4102.

Dixon, T.H., Mao, A., Bursik, M., Heflin, M., Langbein, J., Stein, R., and Webb, F. (1997). Continuous monitoring of surface deformation at Long Valley caldera, California, with GPS. *J. Geophys. Res.*, **102**, 12017–12034.

Donnelly-Nolan, J.M. (1988). A magmatic model of Medicine Lake Volcano, California. *J. Geophys. Res.*, **93**, 4412–4420.

Donnelly-Nolan, J.M., Champion, D.E., Miller, C.D., Grove, T.L., and Trimble, D.A. (1990). Post-11,000-year volcanism at Medicine Lake Volcano, Cascade Range, northern California. *J. Geophys. Res.*, **95**, 19693–19704.

Donnelly-Nolan, J.M., Champion, D.E., Miller, C.D., Grove, T.L., Baker, M.B., Taggart, J.E., Jr., and Bruggman, P.E. (1991). The Giant Crater lava field: Geology and geochemistry of a compositionally-zoned, high-alumina basalt to basaltic andesite eruption at Medicine Lake volcano, California. *J. Geophys. Res.*, **96**, 21843–21863.

Doukas, M.P. and Gerlach, T.M. (1995). Sulfur dioxide scrubbing during the 1992 eruptions of Crater Peak, Mount Spurr, Alaska. In: Keith, T. (ed.), *The 1992 Eruptions of Crater Peak Vent, Mount Spurr Volcano,*

Alaska (USGS Bull. 2139). US Geological Survey, Reston, VA, pp. 47–57.

Dracup, J.F. (1995). Geodetic surveys in the United States: The beginning and the next one hundred years, 1807–1940. *ACSM/ASPRS Annual Convention and Exposition Technical Papers, Bethesda MD, ACSM/ASPRS*, **1**, 24. Availalble online at: National Oceanic and Atmospheric Administration (NOAA), National Geodetic Survey, http://www.history.noaa.gov/stories_tales/geodetic1.html [Accessed April 2005].

Dragert, H., Wang, K., and James, T.S. (2001). A silent slip event on the deeper Cascadia subduction interface. *Science*, **292**(5521), 1525–1528.

Dragert, H. and Hyndman, R.D. (1995). Continuous GPS monitoring of elastic strain in the northern Cascadia subduction zone. *Geophys. Res. Lett.*, **22**, 755–758.

Dreger, D.S., Tkalcic, H. and Johnston, M.J.S. (2000). Dilational processes accompanying earthquakes in the Long Valley Caldera. *Science*, **288**, 122–125.

Dueholm, K.S. (1992). Geologic photogrammetry using standard small-frame cameras. In: Dueholm, K.S. and Pedersen, A.K (eds), Geological analysis and mapping using multi-model photogrammetry. *Grønlands Geologiske Undersøgelse Rapport*, **156**, 7–18.

Dueholm, K.S. and Pedersen, A.K. (1992). The application of multi-model photogrammetry in geology: Status and development trends. In: Dueholm, K.S. and Pedersen, A.K. (eds), Geological analysis and mapping using multi-model photogrammetry. *Grønlands Geologiske Undersøgelse Rapport*, **156**, 69–72.

Dueholm, K.S. and Pillmore, C.L. (1989). Computer-assisted geologic photogrammetry. *Photogrammetric Engineering and Remote Sensing*, **55**, 1191–1196.

Duffield, W.A., Stieltjes, L., and Varet, J. (1982). Huge landslide blocks in the growth of Piton de la Fournaise, La Reunion, and Kilauea Volcano, Hawaii. *J. Volcanol. Geotherm. Res.*, **12**, 147–160.

Dungan, M.A., Wulff, A., and Thompson, R. (2001). Eruptive stratigraphy of the Tatara–San Pedro Complex (36°S, southern volcanic zone, Chilean Andes): Reconstruction method and implications for magma evolution at long-lived arc volcanic centers. *J. Petrology.*, **42**(3), 555–626.

Dvorak, J.J. and Dzurisin, D. (1997). Volcano geodesy: The search for magma reservoirs and the formation of eruptive vents. *Rev. Geophys.*, **35**(3), 343–384.

Dvorak, J.J. and Okamura, A.T. (1987). A hydraulic model to explain variations in summit tilt rate at Kilauea and Mauna Loa Volcanoes. In: Decker, R.W., Wright, T.L., and Stauffer, P.H. (eds), *Volcanism in Hawaii* (USGS Prof. Paper 1350). US Geological Survey, Reston, VA, pp. 1281–1296.

Dvorak, J., Okamura, A., Mortensen, C., and Johnston, M.J.S. (1981). Summary of electronic tilt studies at Mount St. Helens. In: Lipman, P.W., and Mullineaux, D.R. (eds) *The 1980 Eruptions of Mount*

St. Helens, Washington (USGS Prof. Paper 1250). US Geological Survey, Reston, VA, pp. 169–174.

Dzurisin, D. (1992a). Electronic tiltmeters for volcano monitoring: Lessons from Mount St. Helens. In: Ewert, J., and Swanson, D.A. (eds), *Monitoring Volcanoes: Techniques and Strategies Used by the Staff of the Cascades Volcano Observatory, 1980–1990* (USGS Bull. 1966). US Geological Survey, Reston, VA, pp. 69–84.

Dzurisin, D. (1992b). Geodetic leveling as a tool for studying restless volcanoes. In: Ewert, J., and Swanson, D.A. (eds), *Monitoring Volcanoes: Techniques and Strategies Used by the Staff of the Cascades Volcano Observatory, 1980–1990* (USGS Bull. 1966). US Geological Survey, Reston, VA, pp. 125–134.

Dzurisin, D. (2003). A comprehensive approach to monitoring volcano deformation as a window on the eruption cycle. *Rev. Geophys.*, **41**(1), 1001, doi:10.1029/2001RG000107.

Dzurisin, D. and Yamashita, K.M. (1987). Vertical surface displacements at Yellowstone Caldera, Wyoming, 1976–1986. *J. Geophys. Res.*, **92**, 13753–13766.

Dzurisin, D., Anderson, L.A., Eaton, G.P., Koyanagi, R.Y., Okamura, R.T., Puniwai, G.S., Sako, M.K., and Yamashita, K.M. (1980). Geophysical observations of Klauea volcano, Hawaii. Part 2: Constraints on the magma supply during November 1975–September 1977. *J. Volcanol. Geotherm. Res.*, **7**, 241–270.

Dzurisin, D., Westphal, J.A., and Johnson, D.J. (1983). Eruption prediction aided by electronic tiltmeter data at Mount St. Helens. *Science*, **221**(4618), 1381–1383.

Dzurisin, D., Denlinger, R.P., and Rosenbaum, J.G. (1989). Cooling rate and thermal structure determined from progressive magnetization of the dacite dome at Mount St. Helens, Washington. *J. Geophys. Res.*, **95**, 2763–2780.

Dzurisin, D., Savage, J.C., and Fournier, R.O. (1990). Recent crustal subsidence at Yellowstone Caldera, Wyoming. *Bull. Volcanol.*, **52**, 247–270.

Dzurisin, D., Donnelly-Nolan, J.M, Evans, J.R., and Walter, S.R. (1991). Crustal subsidence, seismicity, and structure near Medicine Lake Volcano, California. *J. Geophys. Res.*, **96**, 16319–16333.

Dzurisin, D., Yamashita, K.M., and Kleinman, J.W. (1994). Mechanisms of crustal uplift and subsidence at the Yellowstone caldera, Wyoming. *Bull. Volcanol.*, **56**, 261–270.

Dzurisin, D., Wicks, C., Jr., and Thatcher, W. (1999). Renewed uplift at the Yellowstone caldera measured by leveling surveys and satellite radar interferometry. *Bull. Volcanol.*, **61**, 349–355.

Dzurisin, D., Poland, M.P., and Bürgmann, R. (2002). Steady subsidence of Medicine Lake volcano, northern California, revealed by repeated leveling surveys. *J. Geophys. Res.*, **107**(B12), 2372, doi:10.1029/2001JB000893.

Dzurisin, D., Vallance, J.W., Gerlach, T.M., Moran, S.C., and Malone, S.D. (2005). Mount St. Helens reawakens. *EOS Trans. Amer. Geophys. Union*, **86**(3), 25–29.

Eakins, B.W., Robinson, J.E., Kanamatsu, Toshiya, Naka, Jiro, Smith, J.R., Takahashi, Eiichi, and Clague, D.A. (2003). Hawaii's volcanoes revealed, U.S. Geological Survey Geologic Investigations Series I-2809 (poster) [URL http://geopubs.wr.usgs.gov/i-map/i2809].

Eastoe, C.J. (1978). A fluid inclusion study of the Panguna porphyry copper deposit, Bougainville, Papua New Guinea. *Economic Geology*, **73**, 721–748.

Eaton, J.P. (1959). A portable water-tube tiltmeter. *Bull. Seis. Soc. Am.*, **49**, 301–316.

Eaton, J.P. and Murata, K.J. (1960). How volcanoes grow. *Science*, **132**, 925–938.

Efron, B. and Tibshirani, R. J. (1993). *An Introduction to the Bootstrap*. Chapman and Hall, New York.

Eichelberger, J.C., Lysne, P.C., Miller, C.D., and Younker, L.W. (1985). Research drilling at Inyo domes, 1984 results. *EOS Trans. Amer. Geophys. Union*, **66**, 186–187.

Eichelberger, J.C., Hildreth, W., and Papike, J.J. (1991). The Katmai scientific drilling project, surface phase: Investigation of an exceptional igneous system. *Geophys. Res. Lett.*, **18**(8), 1513–1516.

Elachi, C. (1988). *Spaceborne Radar Remote Sensing: Applications and Techniques*. IEEE Press, New York.

Elsworth, D. and Voight, B. (1992). Theory of dike intrusion in a saturated porous solid. *J. Geophys. Res.*, **97**(B6), 9105–9117.

Emardson, T.R., Simons, M., and Webb, F.H. (2003). Neutral atmospheric delay in interferometric synthetic aperture radar applications: Statistical description and mitigation. *J. Geophys. Res.*, **108**(B5), 2231, doi:10.1029/2002JB001781.

Endo, E. (2005) *The Rabaul Volcano Observatory Real-Time GPS Upgrade* (USGS Open-File Rpt. 2005–1232). US Geological Survey, Reston, VA, 22pp.

Endo, E. and Iwatsubo, E.Y., (2000). Real-time GPS at the Long Valley caldera, California. *EOS Trans. Amer. Geophys. Union*, **81**(48), Fall Meet. Suppl., Abstract F320.

Engelder, T. (1994). Deviatoric stressitis: A virus infecting the Earth science community. *EOS Trans. Amer. Geophys. Union*, **75**(18), 209–212.

Ewert, J.W. (1992). A single-setup trigonometric leveling method for monitoring ground-tilt changes. In: Ewert, J., and Swanson, D.A. (eds), *Monitoring Volcanoes: Techniques and Strategies Used by the Staff of the Cascades Volcano Observatory, 1980–1990* (USGS Bull. 1966). US Geological Survey, Reston, VA, pp. 151–158.

Faller, J.E. (1967). Precision measurement of the acceleration of gravity. *Science*, **158**, 60–67.

Farr, T.G. and Kobrick, M. (2000). Shuttle Radar Topography Mission produces a wealth of data. *EOS Trans. Amer. Geophys. Union*, **81**, 583–585.

Farrar, C.D., Sorey, M.L., Evans, W.C., Howle, J.F., Kerr, B.D., Kennedy, B.M., King, C.-Y., and Southon, J.R. (1995). Forest-killing diffuse CO_2 emissions at Mammoth Mountain as a sign of magmatic unrest. *Nature*, **376**, 675–678.

Farrell, W.E. (1972). Deformation of the Earth by surface loads. *Rev. Geophys. Space Phys.*, **10**(3), 761–797.

Federal Geodetic Control Committee (1984). Bossler, J.D. (chairman), *Standards and specifications for geodetic control networks*, National Oceanic and Atmospheric Administration, Rockville, Maryland.

Ferrucci, F., Nannari, G., and Puglisi, G. (1994). Etna, ground deformation: Remote GPS operation for continuous measurements. *Acta Vulcanologica*, **6**, 22–24.

Fialko, Y., Khazan, Y., and Simons, M. (2001a). Deformation due to a pressurized horizontal circular crack in an elastic half-space, with applications to volcano geodesy. *Geophys. J. Int.*, **146**, 181–190.

Fialko, Y., Simons, M., and Khazan, Y. (2001b). Finite source modeling of magmatic unrest in Socorro, New Mexico, and Long Valley, California. *Geophys. J. Int.*, **146**, 191–200.

Fisher, R.V. and Heiken, G. (1982). Mt. Pelée, Martinique: May 8 and 20, 1902, pyroclastic flows and surges. *J. Volcanol. Geotherm. Res.*, **13**, 339–371.

Fisher, R.V., Smith, A.L., and Roobol, M.J. (1980)., Destruction of St. Pierre, Martinique by ash-cloud surges, May 8 and 20, 1902. *Geology*, **8**, 472–476.

Fisher, R.V., Heiken, G., and Hulen, J.B. (1997). *Volcanoes, Crucibles of Change*. Princeton University Press, Princeton, New Jersey, 317pp.

Flück, P., Hyndman, R.D., and Wang, K. (1997). Three-dimensional dislocation model for great earthquakes of the Cascadia subduction zone. *J. Geophys. Res.*, **102**(B9), 20539–20550.

Fournier, R.B. (1968). Mechanisms of formation of alaskite, aplite, and pegmatite in a dike swarm, Yosemite National Park, California. In: Coats, R.R., Hay, R.L., and Anderson, C.A. (eds.), Studies in volcanology. *Geological Society of America Memoir*, **116**, 249–273.

Fournier, R.O. (1967). The porphyry copper deposit exposed in the Liberty open-pit mine near Ely, Nevada. Part I: Syngenetic formation. *Economic Geology*, **62**, 57–81.

Fournier, R.O. (1983). Self-sealing and brecciation resulting from quartz deposition within hydrothermal systems. *Extended abstracts, 4th International Symposium on Water–Rock Interactions. Misasa, Japan*, pp. 137–140.

Fournier, R.O. (1985). The behavior of silica in hydrothermal solutions. In: Berger, B.R., and Bethke, P.M. (eds.), Geology and geochemistry of epithermal systems. Reviews in *Economic Geology*, **2**, 45–61.

Fournier, R.O. (1987). Conceptual models of brine evolution in magmatic-hydrothermal systems. In: Decker, R.W., Wright, T.L., and Stauffer, P.H. (eds), *Volcanism in Hawaii* (USGS Prof. Paper 1350, Volume 2). US Geological Survey, Reston, VA, pp. 1487–1506.

Fournier, R.O. (1989). Geochemistry and dynamics of the Yellowstone National Park hydrothermal system. *Ann. Rev. Earth Planet. Sci.* **17**, 13–53.

Fournier, R.O. (1991). The transition from hydrostatic to greater than hydrostatic fluid pressure in presently active continental hydrothermal systems in crystalline rock. *Geophys. Res. Lett.*, **18**(5), 955–958.

Fournier, R.O. (1999). Hydrothermal processes related to movement of fluid from plastic into brittle rock in the magmatic-epithermal environment. *Economic Geology*, **94**, 1193–1212.

Fournier, R.O. and Thompson, J.M. (1993). Composition of steam in the system NaCl-KCl-H_2O-quartz at 600degC. *Geochimica et Cosmochimica Acta*, **57**, 4365–4375.

Fournier, R.O. and Pitt, A.M. (1985). The Yellowstone magmatic-hydrothermal system, U.S.A. In: Stone, C. (ed.), *Geothermal Resources Council 1985 International Symposium on Geothermal Energy* (International Volume). GRC, Davis, CA, pp. 319–327.

Francis, O., Niebauer, T.M., Sasagawa, G., Klopping, F., and Gschwind, J. (1998). Calibration of a superconducting gravimeter by comparison with an absolute gravimeter FG5 in Boulder. *Geophys. Res. Lett.*, **25**(7), 1075–1078.

Francis, P. (1993). *Volcanoes: A Planetary Perspective.* Oxford University Press, New York, 443pp.

Francis, P.W., Wadge, G., and Mouginis-Mark, P.J. (1996). Satellite monitoring of volcanoes. In: Scarpa, R. and Tilling, R.I. (eds), *Monitoring and Mitigation of Volcano Hazards.* Springer-Verlag, New York, pp. 257–298.

Fung, Y.C. (1977). *A First Course in Continuum Mechanics* (2nd edition). Prentice Hall, Englewood Cliffs, NJ, 340pp.

Furuya, M., Okubo, S., Sun, W., Tanaka, Y., Oikawa, J., Watanabe, H., and Maekawa, T. (2003a). Spatiotemporal gravity changes at Miyakejima Volcano, Japan: Caldera collapse, explosive eruptions and magma movement. *J. Geophys. Res.*, **108**(B4), 2219, doi: 10.1029/2002JB0011989.

Furuya, M., Okubo, S., Kimata, F., Miyajima, R., Meilano, I., Sun, W., Tanaka, Y., and Miyazaki, T. (2003b). Mass budget of the magma flow in the 2000 volcano–seismic activity at Izu-islands, Japan. *Earth Planets Space*, **55**, 375–385.

Fyfe W.S., Price, N.J., and Thompson, A.B. (1978). *Fluids In the Earth's Crust.* Elsevier, New York, 383pp.

Gabriel, A.K., Goldstein, R.M., and Zebker, H.A. (1989). Mapping small elevation changes over large areas: Differential radar interferometry. *J. Geophys. Res.*, **94**, 9183–9191.

Gens, R. and Genderen, J.L. Van (1996). SAR interferometry: Issues, techniques, applications. *International Journal of Remote Sensing*, **17**(10), 1803–1835.

Gerlach, T.M. (1993). Oxygen buffering of Kilauea volcanic gases and oxygen fugacity of Kilauea basalt. *Geochimica et Cosmochimica Acta*, **57**, 795–814.

Gerlach, T.M. and Graeber, E.J. (1985). Volatile budget of Kilauea Volcano. *Nature*, **313**, 273–277.

Gerlach, T.M. and McGee, K.A. (1994). Total sulfur dioxide emissions and pre-eruption vapor-saturated magma at Mount St. Helens, 1980–1988. *Geophys. Res. Lett.*, **21**(25), 2833–2836.

Gerlach, T.M., Westrich, H.R., and Symonds, R.B. (1996). Pre-eruption vapor in magma of the climactic Mount Pinatubo eruption: Source of the giant stratospheric sulfur dioxide cloud. In: Newhall, C.G., and Punongbayan, R.S. (eds), *Fire and Mud: Eruptions and Lahars of Mt. Pinatubo, Philippines.* Philippine Institute of Volcanology and Seismology, Quezon City and University of Washington Press, Seattle, pp. 415–433.

Gerlach, T.M., Delgado, H., McGee, K.A., Doukas, M.P., Venegas, J.J., and Cardenas, L. (1997). Application of the LI-COR CO_2 analyzer to volcanic plumes: A case study, Volcan Popocatépetl, Mexico, June 7 and 10, 1995. *J. Geophys. Res.*, **102**, 8005–8019.

Gerlach, T.M., Doukas, M.P., Kessler, R., and McGee, K.A. (1999). Airborne detection of diffuse carbon dioxide emissions at Mammoth Mountain. *Geophys. Res. Lett.*, **26**, 3661–3664.

Gerlach, T.M., McGee, K.A., Elias, T., Sutton, A.J., and Doukas, M.P. (2002). Carbon dioxide emission rate of Kilauea Volcano: Implications for primary magma and the summit reservoir. *J. Geophys. Res.*, **107**(B9), 2189, doi: 10.1029/2001JB000407.

Giggenbach, W.F. (1997). The origin and evolution of fluids in magmatic-hydrothermal systems. In: Barnes, H.L., (ed.), *Geochemistry of Hydrothermal Ore Deposits* (3rd edition). Wiley Interscience, New York, pp.737–796.

Gill, P.E., Murray, W. and Wright, M.H. (1981). *Practical Optimization.* Academic, San Diego.

Gilpin, B. and Lee, T.-C. (1978). A microearthquake study of the Salton Sea geothermal area, California. *Seismological Society of America Bulletin*, **68**, 441–450.

Gladwin, M.T. (1984). High precision multi-component borehole deformation monitoring. *Rev. Sci. Instrum.*, **55**, 2011–2016.

Gladwin, M.T. and Hart, R. (1985). Design parameters for borehole strain instrumentation. *Pure and Applied Geophysics*, **123**(1), 59–80.

Goldstein, R.M., Zebker, H.A., and Werner, C.L. (1988). Satellite radar interferometry: Two-dimensional phase unwrapping. *Radio Science*, **23**(4), 713–720.

Got, J.-L. and Okubo, P. (2003). New insights into Kilauea's Volcano dynamics brought by large-scale relative relocation of microearthquakes. *J. Geophys. Res.*, **108**(B7), 2337, doi: 10.1029/2002JB002060.

Got, J.-L., Fréchet, J., and Klein, F.W. (1994). Deep fault plane inferred from multiplet relative relocation in the south flank of Kilauea Volcano, Hawaii. *J. Geophys. Res.*, **99**, 15375–15386.

Got, J.-L. and Coutant, O. (1997). Anisotropic scattering and travel-time delay analysis in Kilauea Volcano, Hawaii, earthquake coda waves. *J. Geophys. Res.*, **102**(B4), 8397–8410.

Got, J.-L., Okubo, P.G., Machenbaum, R., and Tanigawa, W. (2002). A real-time procedure for the progressive multiplet relative relocation at the Hawaiian Volcano Observatory. *Bull. Seismol. Soc. Am.*, **92**, 2019–2026.

Graham, L.N., Jr., Ellison, K.E., Jr., and Riddell, C.S. (1997). The architecture of a softcopy photogrammetry system. *Photogrammetric Engineering and Remote Sensing*, **63**(8), 1013–1020.

Gustafson, L.B. and Hunt, J.P. (1975). The porphyry copper deposit at El Salvador, Chile. *Economic Geology*, **70**, 857–912.

Gwinner, K., Hauber, E., Jaumann, R., Neukum, G. (2000). High-resolution, digital photogrammetric mapping: A tool for Earth science. *EOS Trans. Amer. Geophys. Union*, **81**(44), 513, 516, 520.

Gwyther, R.L., Gladwin, M.T., Mee, M.W., and Hard, R.H.G. (1996). Anomalous shear strain at Parkfield during 1993–1994. *Geophys. Res. Lett.*, **23**(18), 2425–2428.

Hagiwara, Y. (1977). The Mogi model as a possible cause of the crustal uplift in the eastern part of Izu Penninsula and the related gravity change. *Bull. Earthq. Res. Inst. U. Tokyo*, **52**, 301–309.

Hamilton, W.L. and Bailey, A.L. (1988). Holocene deformation deduced from shorelines at Yellowstone Lake, Yellowstone Caldera. Part I: Historical summary and implications. *EOS Trans. Amer. Geophys. Union*, **69**(44), 1472.

Han, S. and Rizos, C. (1997). Comparing GPS ambiguity resolution techniques. *GPS World*, **8**(10), 54–61.

Hannah, M.J. (1989). A system for digital stereo matching. *Photogrammetric Engineering and Remote Sensing*, **55**(12), 1765–1770.

Hanssen, R.A. (2002). *Radar Interferometry: Data Interpretation and Error Analysis* (Remote Sensing and Digital Image Processing, Volume 2). Kluwer Academic Publishers, Dordrecht/Boston/London, 328pp.

Hardee, H.C. and Luth, W.C. (1980). Continental scientific drilling: Comparative assessment of five potential sites for hydrothermal-magma systems. *EOS Trans. Amer. Geophys. Union*, **61**(46), 1148–1149.

Harding, D.J. and Berghoff, G.S. (2000). Fault scarp detection beneath dense vegetation cover: Airborne laser mapping of the Seattle fault zone, Bainbridge Island, Washington State. *Proceedings of the American Society of Photogrammetry and Remote Sensing Annual Conference, Washington, D.C.*, 9pp.

Harris, D.M., Rose, W.I., Roe, R., and Thompson, M.R. (1981). Radar observations of ash eruptions. In: Lipman, P.W., and Mullineaux, D.R. (eds), *The 1980 Eruptions of Mount St. Helens, Washington* (USGS Prof. Paper 1250). US Geological Survey, Reston, VA, pp. 323–333.

Harris, A., Johnson, J., Horton, K., Garbeil, H., Ramm, H., Pilger, E., Flynn, L., Mouginis-Mark, P.J., Pirie, D., Donegan, S., *et al.* (2003). Ground-based infrared monitoring provides new tool for remote tracking of volcanic activity. *EOS Trans. Amer. Geophys. Union*, **84**(40), 409, 418.

Harvey, B. (1996). *The New Russian Space Programme: From Competition to Collaboration*. Praxis Publishing, Chichester, UK, ISBN: 0-471-96014-4, 415pp.

Harvey, B. (2001). *Russia in Space: The Failed Frontier?* Praxis, Chichester, UK, ISBN: 1-85233-203-4, 352pp.

Harvey, B. (2003). *Europe's Space Programme: To Ariane and Beyond*. Praxis, Chichester, UK, ISBN: 1-85233-722-2, 373pp.

Hatch, R. (1990). Instantaneous ambiguity resolution. In: Schwarz, K.P., Lachapelle, G. (eds), *Kinematic Systems in Geodesy, Surveying and Remote Sensing*. Springer, New York, Berlin, Heidelberg, Tokyo, pp. 299–308 (Mueller II (editor): IAG Symposia Proceedings, **107**).

Hatch, R. and Euler H.J. (1994). Comparison of several AROF kinematic techniques. *Proc. ION GPS-94*, pp. 363–370.

Haugerud, R.A., Harding, D.J., Johnson, S.Y., Harless, J.L., Weaver, C.S., and Sherrod, B.L. (2003). High-resolution lidar topography of the Puget Lowland, Washington: A bonanza for Earth science. *GSA Today* (Geological Society of America), **13**(6) 4–10.

Hausback, B.P. and Swanson, D.A. (1990). Record of prehistoric debris avalanches on the north flank of Mount St. Helens Volcano, Washington. *Geoscience Canada*, **17**(3), 142–145.

Hayford, J.F. and Baldwin, A.L. (1908). Geodetic measurements of earth movements. In: Lawson, A.C. (ed.), *The California earthquake of April 18, 1906, Report of the State Earthquake Investigation Commission* (Volume 1). Carnegie Institute of Washington, Washington, D.C., pp. 114–145.

Heald, P., Foley, N.K., and Hayba, D.O. (1987). Comparative anatomy of volcanic-hosted epithermal deposits: Acid-sulfate and adularia-sericite types. *Economic Geology*, **82**, 1–26.

Hedenquist, J.W. and Lowenstern, J.B. (1994). The role of magmas in the formation of hydrothermal ore deposits. *Nature*, **370**, 519–527.

Hedenquist, J.W., Arribas, A., Jr., and Reynalds, T.J. (1998). Evolution of an intrusion-centered hydrothermal system: Far Southeast-Lepanto porphyry and epithermal Cu–Au deposit, Philippines. *Economic Geology*, **93**, 373–404.

Heliker, C. and Brantley, S.R. (2002). *The Pu'u 'Ō 'ō – Kupaianaha Eruption of Kilauea Volcano, Hawai'i, 1983 to 2003* (USGS Fact Sheet 144-02). US Geological Survey, Reston, VA, 2pp.

Heliker, C., Swanson, D.A., and Takahashi, T.J. (eds) (2003). *The Pu'u 'Ō 'ō – Kupaianaha Eruption of Klauea Volcano, Hawai'i: The First 20 Years* (USGS Prof. Paper 1676). US Geological Survey, Reston, VA, 206pp.

Hemley, J.J. (1967). Aqueous Na/K ratios in the system K_2O-Na_2O-Al_2O_3-H_2O [abstract]. Geological Society of America Abstracts with Program (1967 Annual Meeting, 94–95).

Hemley, J.J., Cygan, G.L., Fein, J.B., Robinson, G.R., and d'Angelo, W.M. (1992). Hydrothermal ore-forming processes in the light of studies in rock-buffered systems. Part I: Iron–copper–zinc–lead sulfide solubility relations. *Economic Geology*, **87**, 1–22.

Henderson, F.M. and Lewis, A.J. (eds) (1998). *Principles and Applications of Imaging Radar* (Volume 2): *Manual of Remote Sensing* (3rd edition). Wiley, New York.

Henley, R.W. and Ellis, A.J. (1983). Geothermal systems ancient and modern: A geochemical review. *Earth Science Reviews*, **19**, 1–50.

Henley, R.W. and McNabb, A. (1978). Magmatic vapor plumes and groundwater interaction in porphyry copper emplacement. *Economic Geology*, **73**, 1–20.

Hensley S., Munjy, R., and Rosen, P. (2001). Interferometric synthetic aperture radar (IFSAR). In: Maune, D.F. (ed.), *Digital Elevation Model Technologies and Applications: The DEM Users Manual*. American Society for Photogrammetry and Remote Sensing, Washington, D.C., pp. 143–206.

Herd, D.G., and Comité de Estudios Vulcanológicos (1986). The 1985 Ruiz Volcano disaster. *EOS Trans. Amer. Geophys. Union*, **67**(19), 457–460.

Herring, T.A. (1999). Geodetic Applications of GPS. *Proceedings of the IEEE*, **87**(1), 92–110.

Hildreth, W. (2004). Volcanological perspectives on Long Valley, Mammoth Mountain, and Mono Craters: Several contiguous but discrete systems. *J. Volcanol. Geotherm. Res.*, **136**, 169–198.

Hill, D.P. (1984). Monitoring unrest in a large silicic caldera, the Long Valley–Inyo Craters volcanic complex in east-central California. *Bull. Volcanol.*, **47**, 371–395.

Hill, D.P. (1992). Temperatures at the base of the seismogenic crust beneath Long Valley Caldera, California, and the Phlegraean Fields Caldera, Italy. In: Gasparini, P. and Scarpa, R. (eds), *International Association of Volcanology and Chemistry of the Earth's Interior (IAVCEI) Proceedings in Volcanology 3: Volcanic Seismology*, pp. 433–461.

Hill, D.P. (1996). Earthquakes and carbon dioxide beneath Mammoth Mountain, California. *Seismol. Res. Lett.*, **67**(1), 8–15.

Hill, D.P. and Prejean, S. (2005). Magmatic unrest beneath Mammoth Mountain, California. *J. Volcanol. Geotherm. Res.*, **146**(4), 257–283.

Hill, D.P., Bailey, R.A., and Ryall, A.S. (1985a). Active tectonic and magmatic processes beneath Long Valley Caldera, eastern California: An overview. *J. Geophys. Res.*, **90**, 11111–11120.

Hill, D.P., Wallace, R.E., and Cockerham, R.S. (1985b). Review of evidence on the potential for major earthquakes and volcanism in the Long Valley – Mono Craters – White Mountains regions of eastern California. *Earthquake Prediction Res.*, **3**, 571–594.

Hill, D.P., Ellsworth, W.I., Johnston, M.J.S., Langbein, J.O., Oppenheimer, D.H., Pitt, A.M., Reasenberg, P.A., Sorey, M.L., and McNutt, S.R. (1990). The 1989 earthquake swarm beneath Mammoth Mountain, California: An initial look at the 4 May through 30 September activity. *Bull. Seis. Soc. Am.*, **80**, 325–339.

Hill, D.P., Reasenberg, P.A., Michael, A.J., Arabasz, W.J., Beroza, G.C., Brumbaugh, D.S., Brune, J.N., Castro, R., Davis, S.D., dePolo, D.M., et al. (1993). Seismicity remotely triggered by the magnitude 7.3 Landers, California, earthquake. *Science*, **260**(5114), 1617–1623.

Hill, D.P., Johnston, M.J.S., Langbein, J.O., and Bilham, R. (1995). Response of Long Valley caldera to the $M_w = 7.3$ Landers, California, earthquake. *J. Geophys. Res.*, **100**, 12985–13005.

Hill, D.P., Dzurisin, D., Ellsworth, W.L., Endo, E.T., Galloway, D.L., Gerlach, T.M., Johnston, M.J.S., Langbein, J., McGee, K.A., Miller, C.D., et al. (2002a). *Response Plan for Volcano Hazards in the Long Valley Caldera and Mono Craters Region, California* (USGS Bull. 2185). US Geological Survey, Reston, VA, 57pp.

Hill, D.P., Dawson, P., Johnston, M.J.S., Pitt, A.M., Biasi, G., and Smith, K. (2002b). Very-long-period volcanic earthquakes beneath Mammoth Mountain, California. *Geophys. Res. Lett.*, **29**(10), 1370, doi:10.1029/2002GL014833.

Hill, D.P., Langbein, J.O., and Prejean, S. (2003). Relations between seismicity and deformation during unrest in Long Valley Caldera, California, from 1995 through 1999. In: Sorey, M.L., McConnell, V.S., and Roeloffs, E. (eds), Crustal Unrest in Long Valley Caldera, California: New interpretations from geophysical and hydrologic monitoring and deep drilling, *J. Volcanol. Geotherm. Res.*, **127**(3–4), 175–193.

Hofmann-Wellenhof, B., Lichtenegger, H., and Collins, J. (2001). *GPS: Theory and Practice* (5th edition). Springer-Verlag, New York, ISBN 3-211-83534-2, 382pp.

Hohle, J. (1996). Experiences with the production of digital orthophotos. *Photogrammetric Engineering and Remote Sensing*, **62**(10), 1189–1194.

Holcomb, R.T. and Colony, W.E. (1995). Maps showing growth of the lava dome at Mount St. Helens, Washington, 1980–1986. U.S. Geol. Surv. Misc. Invest. Ser., Map I-2359, 1:5000 scale, 1 sheet with text.

Holdahl, S.R. (1982). Recomputation of vertical crustal motions near Palmdale, California, 1959–1975. *J. Geophys. Res.*, **87**, 9374–9388.

Holland, H.D. (1965). Some applications of thermodynamical data to problems of ore deposits. Part II: Mineral assemblages and the composition of ore-forming fluids. *Economic Geology*, **60**, 1101–1199.

Husen, S., Smith, R.B., and Waite, G.P. (2004). Evidence for gas and magmatic sources beneath the Yellowstone volcanic field from seismic tomographic imaging. *J. Volcanol. Geotherm. Res.*, **131**, 397–410.

Hutchinson, R.A. (1992). Strong new thermal activity. *Bulletin of the Global Volcanism Network*, **17**(3), March 1992.

Hutchinson, R.A. (1993). Precursors to emergence of a major new mud volcano: Continuing unrest in the Yellowstone Plateau (abstract). *EOS Trans. Amer. Geophys. Union*, **74**, 592.

Hwang, P. (1991), Kinematic GPS for differential positioning: Resolving integer ambiguities on the fly. *Navigation*, **38**(1), 1–15.

ICSM (2000). Standards and Practices for Control Surveys (SP1). Inter-Governmental Committee on Surveying and Mapping (Australia), ICSM Publication No 1. Available online at: http://www.anzlic.org.au/icsm/publications/index.html [Accessed March 2005].

IGS (2004). International GPS Service: Monitoring Global Change by Satellite Tracking. Available online at: http://igscb.jpl.nasa.gov [Accessed February 2004].

Ikeuchi, K., Komatsu, R., Doi, N., Sakagawa, Y., Sasaki, M., Kamenosono, H., and Uchida, T. (1996). Bottom of hydrothermal convection found by temperature measurements above 500°C and fluid inclusion study of WD-1 in Kakkonda geothermal field, Japan. *Geothermal Resources Council Transactions*, **20**, 609–616.

Ishiguro, M. (1981). A Bayesian approach to the analysis of the data of crustal movements. *J. Geod. Soc. Japan*, **30**, 256–262.

Iverson, R.M., Reid, M.E., and LaHusen, R.G. (1997). Debris-flow mobilization from landslides. *Ann. Rev. Earth and Planet. Sci.*, **25**, 85–138.

Iverson, R.M., Schilling, S.P., and Vallance, J.W. (1998). Objective delineation of lahar-inundation hazard zones. *Bull. Geol. Soc. Am.*, **110**(8), 972–984.

Iwasaki, T. and Matsu'ura, M. (1982). Quasi-static crustal deformations due to a surface load: Rheological structure of the Earth's crust and upper mantle. *J. Phys. Earth*, **30**, 469–508.

Iwatsubo, E.Y. and Swanson, D.A. (1992). Methods used to monitor deformation of the crater floor and lava dome at Mount St. Helens, Washington. In: Ewert, J., and Swanson, D.A. (eds), *Monitoring Volcanoes: Techniques and Strategies Used by the Staff of the Cascades Volcano Observatory, 1980–1990* (USGS Bull. 1966). US Geological Survey, Reston, VA, pp. 53–68.

Iwatsubo, E.Y., Topinka, L., and Swanson, D.A. (1988). *Measurements of Slope Distances and Zenith Angles at Newberry and South Sister Volcanoes, Oregon, 1985–1986* (USGS Open-File Report 88-377). US Geological Survey, Reston, VA, 51pp.

Iwatsubo, E.Y., Topinka, L., and Swanson, D.A. (1992a). Slope–distance measurements to the flanks of Mount St. Helens, late 1980 through 1989. In: Ewert, J., and Swanson, D.A. (eds), *Monitoring Volcanoes: Techniques and Strategies Used by the Staff of the Cascades Volcano Observatory, 1980–1990* (USGS Bull. 1966). US Geological Survey, Reston, VA, pp. 85–94.

Iwatsubo, E.Y., Ewert, J.W., and Murray, T.L. (1992b). Monitoring radial crack deformation by displacement meters. In: Ewert, J., and Swanson, D.A. (eds), *Monitoring Volcanoes: Techniques and Strategies Used by the Staff of the Cascades Volcano Observatory, 1980–1990* (USGS Bull. 1966). US Geological Survey, Reston, VA, pp. 95–102.

Izett, G.A. (1982). *The Bishop Ash Bed and Some Older Compositionally Similar Ash Beds in California, Nevada, and Utah* (USGS Open-File Rpt. 82-582). US Geological Survey, Reston, VA, 47pp.

Izett, G.A., Wilcox, R.E., Powers, H.A., and Desborough, G.A. (1970). The Bishop ash bed, a Pleistocene marker bed in the western United States. *Quaternary Research*, **1**, 121–132.

Jachens, R.C. and Eaton, G.P. (1980). Geophysical observations of Kilauea volcano, Hawaii. Part 1: Temporal gravity variations related to the 29 November, 1975, $M = 7.2$ earthquake and associated summit collapse. *J. Volcanol. Geotherm. Res.*, **7**, 225–240.

Jachens, R.C., Spydell, D.R., Pitts, G.S., Dzurisin, D., and Roberts, C.W. (1981). Temporal gravity variations at Mount St. Helens, March–May 1980. In Lipman, P.W. and Mullineaux, D.R. (eds), *The 1980 Eruptions of Mount St. Helens, Washington* (USGS Prof. Paper 1250). US Geological Survey, Reston, VA, pp. 175–182.

Jaggar, T.A. and Finch, R.H. (1929). Tilt records for thirteen years at the Hawaiian Volcano Observatory. *Bull. Seismol. Soc. Am.*, **19**(1), 38–51.

Janda, R.J., Scott, K.M., Nolan, K.M., and Martinson, H.A. (1981). Lahar movement, effects, and deposits. In: Lipman, P.W. and Mullineaux, D.R. (eds), *The 1980 Eruptions of Mount St. Helens, Washington* (USGS Prof. Paper 1250). US Geological Survey, Reston, VA, pp. 461–478.

Jensen, R.A., and Chitwood, L.A. (1996). Evidence for recent uplift of caldera floor, Newberry Volcano, Oregon. *EOS Trans. Amer. Geophys. Union*, **77**(46), Fall Meet. Suppl., Abstract F792.

Johnson, D.J. (1992). Dynamics of magma storage in the summit reservoir of Kilauea Volcano, Hawaii. *J. Geophys. Res.*, **97**, 1807–1820.

Johnson, D.J. (1995). Gravity changes on Mauna Loa Volcano. In: Rhodes, J.M. and Lockwood, J.P. (eds), *Mauna Loa Revealed: Structure, Composition, History and Hazards* (Geophysical Monograph 92). American Geophysical Union, Washington, D.C., pp. 127–193.

Johnson, D.J., Sigmundsson, F., and Delaney, P.T. (2000). Comment on 'Volume of magma accumulation or withdrawal estimated from surface uplift or subsidence, with application to the 1960 collapse of Kilauea Volcano' by P.T. Delaney and D.F. McTigue. *Bull. Volcanol.*, **61**, 491–493.

Johnston, M.J.S. (1997). Review of electric and magnetic fields accompanying seismic and volcanic activity. *Surveys in Geophysics*, **18**, 441–475.

Johnston, M.J.S., Mueller, R.J., and Dvorak, J. (1981). Volcanomagnetic observations during eruptions, May–August 1980. In: Lipman, P.W. and Mullineaux, D.R. (eds), *The 1980 Eruptions of Mount St. Helens, Washington* (USGS Prof. Paper 1250). US Geological Survey, Reston, VA, pp. 183–189.

Johnston, M.J.S., Linde, A.T., and Agnew, D.C. (1994). Continuous borehole strain in the San Andreas fault zone before, during, and after the 28 June 1992, M_w 7.3 Landers, California earthquake. *Bull. Seismol. Soc. Am.*, **84**, 799–805.

Johnston, M.J.S., Hill, D.P., Linde, A.T., Langbein, J., and Bilham, R. (1995). Transient deformation during triggered seismicity from the 28 June 1992 $M_W = 7.3$ Landers earthquake at Long Valley volcanic caldera, California. *Bull. Seis. Soc. Am.*, **85**(3), 787–795.

Jónsson, S., Zebker, H., Cervelli, P., Segall, P., Garbeil, H., Mouginis-Mark, P., and Rowlands, S. (1999). A shallow-dipping dike fed the 1995 flank eruption at Fernandina Volcano, Galápagos, observed by satellite radar interferometry. *Geophys. Res. Lett.*, **26**(8), 1077–1080.

Jordan, R. and Kieffer, H.H. (1981). Topographic changes at Mount St. Helens: Large-scale photogrammetry and digital terrain models. In: Lipman, P.W. and Mullineaux, D.R. (eds), *The 1980 Eruptions of Mount St. Helens, Washington* (USGS Professional Paper 1250). US Geological Survey, Reston, VA, pp. 135–141.

Jordan, R.L., Caro, E.R., Kim, Y., Kobrick, M., Shen, Y., Stuhr, F.V., and Werner, M.U. (1996). Shuttle radar topography mapper (SRTM). In: Franceschetti, G., Oliver, C.J., Shiue, J.C., and Tajbakhsh, S. (eds), *Microwave Sensing and Synthetic Aperture Radar*. SPIE, Bellingham, pp. 412–422.

Jousset, P., Dwipa, S., Beauducel, F., Duquesnoy, T., and Diament, M. (2000). Temporal gravity at Merapi during the 1993–1995 crisis: An insight into the dynamical behavior of volcanoes. *J. Volcanol. Geotherm. Res.*, **100**(1–4), 289–320.

Kamo, K. and Ishihara, K. (1989). A preliminary experiment on automated judgment of the stages of eruptive activity using tiltmeter records at Sakurajima, Japan. In: Latter, J.H. (ed.), *Volcanic Hazards: Assessment and Monitoring, IAVCEI Proceedings in Volcanology 1*. Springer-Verlag, Heidelberg, pp. 585–598.

Kaplan, E. (ed.) (1996). *Understanding GPS: Principles and Applications*. Artech House, Boston, 570pp.

Kawakatsu, H., Ohminato, T., Ito, H., Kuwahara, Y., Kato, T., Tsuruga, K., Honda, S., and Yomogida, K. (1992). Broadband seismic observation at the Sakurajima Volcano, Japan. *Geophys. Res. Lett.*, **19**, 1959–1962.

Kawasaki, I., Asai, Y., Tamura, Y., Sagiya, T., Mikami, N., Okada, Y., Sakata, M., and Kasahara, M. (1995). The 1992 Sanriku-Oki, Japan, ultra-slow earthquake. *Journal of Physics of the Earth*, **43**, 105–116.

Kennedy, G.C. (1950). A portion of the system silica–water. *Economic Geology*, **45**, 629–653.

Kenner, S.J. and Segall, P. (2000). Postseismic deformation following the 1906 San Francisco earthquake. *J. Geophys. Res.*, **105**, 13195–13209.

Kersten, T. and Haering, S. (1997). Automatic interior orientation of digital aerial images. *Photogrammetric Engineering and Remote Sensing*, **63**(8), 1007–1011.

Klein, F. W., Koyanagi, R.Y., Nakata, J.S, and Tanigawa, W.R. (1987). The seismicity of Kilauea's magma system. In: Decker, R.W., Wright, T.L., and Stauffer, P.H. (eds), *Volcanism in Hawaii* (USGS Prof. Paper 1350). US Geological Survey, Reston, VA, pp. 1019–1185.

Kleinman, J.W. and Otway, P.M. (1992). Lake-level monitoring as a tool for studies of crustal deformation. In: Ewert, J. and Swanson, D.A. (eds), *Monitoring Volcanoes: Techniques and Strategies Used by the Staff of the Cascades Volcano Observatory, 1980–1990* (USGS Bulletin 1966). US Geological Survey, Reston, VA, pp. 159–179.

Koizumi, N., Tsukuda, E., Kamigaichi, O., Matsumoto, N., Takahashi, M., and Sato, T. (1999) Preseismic changes in groundwater level and volumetric strain associated with earthquake swarms off the east coast of the Izu Peninsula, Japan. *Geophys. Res. Lett.*, **26**(23), 3509–3512.

Kozlowski, J. (1998). Electronic total stations are levels too. Trimble Navigation Ltd., Available online at (North Carolina Department of Transportation, Division of Highways): http://www.doh.dot.state.nc.us/preconstruct/highway/location/support/support_files/library_doc/Trig_Leveling_PPT.pdf [Accessed April 2004], 14 pages plus 19 figures.

Krimmel, R.M. and Post, A. (1981). Oblique aerial photography, March–October 1980. In: Lipman, P.W. and Mullineaux, D.R. (eds), *The 1980 Eruptions of Mount St. Helens, Washington* (USGS Prof. Paper 1250). US Geological Survey, Reston, VA, pp. 31–51.

Krumm, F.W. (ed.), Schwarze, V.S. (ed.), Grafarend, E.W. (ed.), and Borghese, F.A. (2002). *Geodesy: The Challenge of The 3rd Millennium*. Springer-Verlag, Berlin, 473pp.

Kumagai, H. and Chouet, B.A. (2000). Acoustic properties of a crack containing magmatic or hydrothermal fluids. *J. Geophys. Res.*, **105**, 25493–25512.

Kumagai, H., Ohminato, T., Nakano, M., Ooi, M., Kubo, A., Inoue, H., and Oikawa, J. (2001). Very-long-period seismic signals and caldera formation at Miyake Island, Japan. *Science*, **293**, 687–690.

Lachenbruch, A.H., Sass, J.H., Munroe, R.J., and Moses, T.H., Jr. (1976). Geothermal setting and simple heat conduction models for the Long Valley Caldera. *J. Geophys. Res.*, **81**(5), 769–784.

LaCoste & Romberg (1998). LaCoste & Romberg G&D Meter Manual. Available online at: http://www.lacosteromberg.com/metermanuals.htm [Accessed April 2004].

LaHusen, R.G. and Reid, M.E. (2000). A versatile GPS system for monitoring deformation of active landslides and volcanoes. *EOS Trans. Amer. Geophys. Union*, **81**(48), Fall Meet. Suppl., Abstract F320.

Lanari, R., Lundgren, P., and Sansosti, E. (1998). Dynamic deformation of Etna volcano observed by satellite radar interferometry. *Geophys. Res. Lett.*, **25**(10), 1541–1544.

Landau, L.D. and Lifshitz, E.M. (1975). *Theory of Elasticity* (2nd edition). Pergamon Press, Tarrytown, NY.

Landau, L.D. and Lifshitz, E.M. (1986). *Theory of Elasticity. Course of Theoretical Physics, 7* (3rd edition). Pergamon Press, Oxford.

Lane, S.J., Chouet, B.A., Phillips, J.C., Dawson, Phillip, Ryan, G.A., and Hurst, E. (2001). Experimental observations of pressure oscillations and flow regimes in an analogue volcanic system. *J. Geophys. Res.*, **106**(4), 6461–6476.

Langbein, J. (1989). Deformation of the Long Valley caldera, eastern California, from mid-1983 to mid-1988: Measurements using a two-color geodimeter. *J. Geophys. Res.*, **94**, 3833–3850.

Langbein, J.O. (2003). Deformation of the Long Valley Caldera, California: Inferences from measurements from 1988 to 2001. In: Sorey, M.L., McConnell, V.S., and Roeloffs, E. (eds), Crustal Unrest in Long Valley Caldera, California: New interpretations from geophysical and hydrologic monitoring and deep drilling. *J. Volcanol. Geotherm. Res.*, **127**(3–4), 247–267.

Langbein, J., Linker, M., and Tupper, D. (1987a). Analysis of two-color geodimeter measurements of deformation within the Long Valley caldera: June 1983 to October 1985. *J. Geophys. Res.*, **92**, 9423–9442.

Langbein, J.O., Linker, M.F., McGarr, A.F., and Slater, L.E. (1987b). Precision of two-color geodimeter measurements: results from 15 months of observations. *J. Geophys. Res.*, **92**, 11644–11656.

Langbein, J., Hill, D.P., Parker, T.N., and Wilkinson, S.K. (1993). An episode of reinflation of the Long Valley caldera, eastern California: 1989–1991. *J. Geophys. Res.*, **98**, 15851–15870.

Langbein, J., Dzurisin, D., Marshall, G., Stein, R., and Rundle, J. (1995). Shallow and peripheral volcanic sources of inflation revealed by modeling of two-color geodimeter and leveling data from Long Valley caldera, California, 1988–1992. *J. Geophys. Res.*, **100**, 12487–12495.

Langbein, J., Gwyther, R.L., Hart, R.H.G., and Gladwin, M.T. (1999). Slip-rate increase at Parkfield in 1993 detected by high-precision EDM and borehole tensor strainmeters. *Geophys. Res. Lett.*, **26**(16), 2529–2532.

Larson, K.M., Freymueller, J.T., and Philipsen, S. (1997). Global plate velocities from the Global Positioning System. *J. Geophys. Res.*, **102**, 9961–9981.

Larson, K.M., Cervelli, P., Lisowski, M., Miklius, A., Segall, P., and Owen, S. (2001). Volcano monitoring using the Global Positioning System: Filtering strategies. *J. Geophys. Res.*, **106**(9), 19453–19464.

Latter, J.H. (ed.) (1989). *Volcanic Hazards, Assessment and Monitoring. IAVCEI Proceedings in Volcanology 1.* Springer Verlag, Heidelberg, 625pp.

Lauer, S. (1995). *Pumice and Ash: An Account of the 1994 Rabaul Volcanic Eruptions.* CPD Resources, Australia, 80pp.

Lees, J.M. and Crossen, R.S. (1989). Tomographic inversion for three-dimensional velocity structure at Mount St. Helens using earthquake date. *Journal of Geophysical Research*, **94**, 5716–5728.

Legat, K. and Hofmann-Wellenhof, B. (2000). Galileo or for whom the bell tolls. *Earth Planets Space*, **52**, 771–776.

LeGuern, G., Gerlach, T.M., and Nohl, A. (1982). Field gas chromatograph analyses of gases from a glowing dome at Merapi volcano, Java, Indonesia, 1977, 1978, 1979. *J. Volcanol. Geotherm. Res.*, **14**, 223–245.

Leick, A. (1990). *GPS Satellite Surveying.* Wiley, New York, 352pp.

Lejeune, A.M., Bottinga, Y., Trull, T.W., and Richet, P. (1999). Rheology of bubble-bearing magmas. *Earth and Planet. Sci. Lett.*, **166**(1–2), 71–84.

Linde, A.T. and Sacks, I.S. (2000). Real time predictions of imminent volcanic activity using borehole deformation data. *EOS Trans. Amer. Geophys. Union*, **81**(48), Fall Meet. Suppl., Abstract F1253.

Linde, A.T., Agustsson, K., Sacks, I.S., and Stefansson, R. (1993). Mechanism of the 1991 eruption of Hekla from continuous borehole strain monitoring. *Nature*, **365**, 737–740.

Linde, A.T, Sacks, I.S., Johnston, M.J.S., Hill, D.P., and Bilham, R.G. (1994). Increased pressure from rising bubbles as a mechanism for remotely triggered seismicity. *Nature*, **371**(6496), 408–410.

Linde, A.T., Gladwin, M.T., Johnston, M.J.S., Gwyther, R.L., and Bilham, R.G. (1996). A slow earthquake sequence on the San Andreas Fault. *Nature*, **383**, 65–68.

Lipman, P.W., Moore, J.G., and Swanson, D.A. (1981a). Bulging of the north flank before the May 18 eruption:

Geodetic data. In: Lipman, P.W. and Mullineaux, D.R. (eds), *The 1980 Eruptions of Mount St. Helens, Washington* (USGS Prof. Paper 1250). US Geological Survey, Reston, VA, pp. 143–155.

Lipman P.W., Nortin D.R., Taggart J.R., Brandt E.L., Engleman E.E. (1981b). Compositional variations in 1980 magmatic deposits. In: Lipman, P.W. and Mullineaux, D.R. (eds), *The 1980 Eruptions of Mount St. Helens, Washington* (USGS Prof. Paper 1250). US Geological Survey, Reston, VA, pp. 631–640.

Lipman, P.W., Lockwood, J.P., Okamura, R.T., Swanson, D.A., and Yamashita, K.M. (1985). *Ground Deformation Associated with the 1975 Magnitude-7.2 Earthquake and Resulting Changes in Activity of Kilauea Volcano, Hawaii.* (USGS Prof. Paper 1276). US Geological Survey, Reston, VA, 45pp.

Lipman, P.W., Normark, W.R., Moore, J.G., Wilson, J.B., and Gutmacher, C.E. (1988). The giant Alika debris slide, Mauna Loa, Hawaii. *J. Geophys. Res.,* **93**, 4279–4299.

Lipman, P.W., Rhodes, J.M., and Dalrymple, G.B. (1990). The Ninole basalt-implications for the structural evolution of Mauna Loa Volcano, Hawaii. *Bull. Volcanol.,* **53**, 1–19.

Lipman, P.W., Sisson, T.W., Ui, T., Naka, J., and Smith, J.R. (2002). Ancestral submarine growth of Kīlauea volcano and instability of its south flank. In: Takahashi, E., Lipman, P.W., Garcia, M.O., Naka, J., and Aramaki, S. (eds), *Hawaiian Volcanoes: Deep Underwater Perspectives* (Geophysical Monograph 128). American Geophysical Union, Washington, D.C., pp. 161–192.

Lister, C.R.B. (1974). On the penetration of water into hot rock. *Geophysical Journal of the Royal Astronomical Society,* **39**, 465–509.

Locke, W.W. and Meyer, G.A. (1994). A 12,000-year record of vertical deformation across the Yellowstone caldera margin: The shorelines of Yellowstone Lake. *J. Geophys. Res.,* **99**(B10), 20079–20094.

Lockhart, A.B., Marcial, S., Ambubuyog, G., Laguerta, E.P., and Power, J.A. (1996). Installation, operation, and technical specifications of the first Mount Pinatubo telemetered seismic network. In: Newhall, C.G. and Punongbayan, R.S. (eds), *Fire and Mud: Eruptions and Lahars of Mt. Pinatubo, Philippines.* Philippine Institute of Volcanology and Seismology, Quezon City and University of Washington Press, Seattle, pp. 215–223.

Logsdon, T. (1992). *GPS: Theory and Practice.* Van Nostrand Reinhold, New York.

Long, D., Smith, D.E., and Dawson, A.G. (1989). A Holocene tsunami deposit in eastern Scotland. *Journal of Quaternary Science,* **47**, 61–66.

Loomis, B.F. (1926). *Pictorial History of the Lassen Volcano.* Loomis Museum Assoc., 96pp.

Love, A.E. (1927). *A Treatise on the Mathematical Theory of Elasticity* (4th edition). Cambridge University Press, London.

Love, A.E. (1929). The stress produced in a semi-infinite solid by pressure on part of the boundary. *Phil. Trans. R. Soc. London A,* **667**, 377–420.

Lowry, A.R., Hamburger, M.W., Meertens, C.M., and Ramos, E.G. (2001). GPS monitoring of crustal deformation at Taal Volcano, Philippines. *J. Volcanol. Geotherm. Res.,* **105**(1–2), 35–47.

Lu, Z. and Freymueller, J. (1998). Synthetic aperture radar interferometry coherence analysis over Katmai Volcano group, Alaska. *J. Geophys. Res.,* **103**, 29887–29894.

Lu, Z., Mann, D., and Freymueller, J. (1998). Satellite radar interferometry measures deformation at Okmok Volcano. *EOS Trans. Amer. Geophys. Union,* **79**(39), 461, 467–468.

Lu, Z., Wicks, C., Power, J., and Dzurisin, D. (2000a). Ground deformation associated with the March 1996 earthquake swarm at Akutan Volcano, Alaska, revealed by satellite radar interferometry, *J. Geophys. Res.,* **105**, 21483–21495.

Lu, Z., Wicks, C., Dzurisin, D., Thatcher, W., Freymueller, J.T., McNutt, S.R., and Mann, D. (2000b). Aseismic inflation of Westdahl Volcano, Alaska, revealed by satellite radar interferometry. *Geophys. Res. Lett.,* **27**, 1567–1570.

Lu, Z., Mann, D., Freymueller, J., Meyer, D. (2000c). Synthetic aperture radar interferometry of Okmok volcano, Alaska 1: Radar observations. *J. Geophys. Res.,* **105**, 10791–10806.

Lu, Z., Wicks, C. Jr., Dzurisin, D., Power, J.A., Moran, S., and Thatcher, W. (2002a). Magmatic inflation at a dormant stratovolcano: 1996–1998 activity at Mount Peulik Volcano, Alaska, revealed by satellite radar interferometry. *J. Geophys. Res.,* **107**(B7), doi: 10.1029/2001/JB000471.

Lu, Z., Wicks, C., Power, J., Dzurisin, D., Thatcher, W., and Masterlark, T. (2002b). Interferometric synthetic aperture radar studies of Alaska volcanoes. *Proceedings of International Geoscience and Remote Sensing Symposium, Toronto,* pp. 191–194.

Lu, Z., Power, J.A., McConnell, V.S., Wicks, C., and Dzurisin, D. (2002c). Pre-eruptive inflation and surface interferometric coherence characteristics revealed by satellite radar interferometry at Makushin Volcano, Alaska: 1993–2000. *J. Geophys. Res.,* **107**, doi: 10.1029/2001JB000970.

Lu, Z., Masterlark, T., Power, J., Dzurisin, D., and Wicks, C. (2002d). Subsidence at Kiska Volcano, Western Aleutians, detected by satellite radar interferometry. *Geophys. Res. Lett.,* **29**(18), 1855, doi: 10.1029/2002GL014948.

Lu, Z., Masterlark, T., Dzurisin, D., Rykhus, R., and Wicks, C. Jr. (2003a). Magma supply dynamics at Westdahl Volcano, Alaska, modeled from satellite radar interferometry. *J. Geophys. Res.*, **108**(B7), 2354, doi: 10.1029/2002JB002311.

Lu, Z., Fielding, E., Patrick, M., and Trautwein, C. (2003b). Estimating lava volume by precision combination of multiple baseline spaceborne and airborne interferometric synthetic aperture radar: The 1997 Eruption of Okmok Volcano, Alaska. *IEEE Transactions on Geoscience and Remote Sensing*, **41**(6), 1428–1436.

Lu, Z., Rykhus, R., Masterlark, T., and Dean, K. (2004). Mapping recent lava flows at Westdahl Volcano, Alaska, using radar and optical satellite imagery. *Remote Sensing of Environment*, **91**, 345–353.

Lu, Z., Wicks, C., Jr., Kwoun, O., Power, J.A., and Dzurisin, D. (2005). Surface deformation associated with the March 1996 earthquake swarm at Akutan Island, Alaska, revealed by C-band ERS and L-band JERS radar interferometry. *Can. J. Remote Sensing*, **31**(1), 7–20.

Lynch, J.S. and Stephens, G. (1996). Mount Pinatubo: A satellite perspective of the June 1991 eruption. In: Newhall, C.G. and Punongbayan, R.S. (eds), *Fire and Mud: Eruptions and Lahars of Mt. Pinatubo, Philippines*. Philippine Institute of Volcanology and Seismology, Quezon City and University of Washington Press, Seattle, pp. 637–645.

Lyons, R.G. (2004). *Understanding Digital Signal Processing* (2nd edition). Prentice Hall, NJ, 688pp.

MacElwane, J. (1936). Problems and progress on the geologic–seismological frontier. *Science*, **83**, 193–198.

MacLeod, N.S., Sherrod, D.R., Chitwood, L.A., and Jensen, R.A. (1995). Geologic map of Newberry volcano, Deschutes, Klamath, and Lake counties, Oregon. USGS Misc. Invest. Ser., Map I-2455, 2 sheets plus 23 pages text.

Madsen, S. and Zebker, H.A. (1998). Imaging radar interferometry. In: Henderson, F. and Lewis, A.J. (eds), *Principles and Applications of Imaging Radar, Manual of Remote Sensing* (3rd edition). Wiley, New York, 866pp.

Majer, E.L. and McEvilly, T.V. (1979). Seismological investigation at The Geysers geothermal field. *Geophysics*, **44**, 246–269.

Malone, S.D., Boyko, C., and Weaver, C.S. (1983). Seismic precursors to the Mount St. Helens eruptions in 1981 and 1982. *Science*, **221**, 1376–1378.

Manley, C.R. (1993). Lava dome collapse causes pyroclastic flows. *EOS Trans. Amer. Geophys. Union*, **74**(27), 306–307.

Marquardt, D.W. (1963). An algorithm for least-squares estimation of non-linear parameters. *J. Soc. Ind. Appl. Math.*, **11**, 431–441.

Marshall, G.A., Langbein, J., Stein, R.S., Lisowski, M., and Svarc, J. (1997). Inflation of Long Valley caldera, California, Basin and Range strain, and possible Mono Craters dike opening from 1990–1994 GPS surveys. *Geophys. Res. Lett.*, **24**(9), 1003–1006.

Marson, I. and Faller, J.E. (1986). The acceleration of gravity: Its measurement and importance. *J. Phys. E. Sci. Instrum.*, **19**, 22–32.

Maruyama, T. (1969). Statistical elastic dislocations in an infinite and semi-infinite medium. *Bull. Earth Res. Inst. Tokyo Univ.*, **42**, 289–368.

Masson, D.G. (1996). Catastrophic collapse of the flank of El Hierro about 15,000 years ago and the history of large flank collapses in the Canary Islands. *Geology*, **24**, 231–234.

Masson, D.G., Watts, A.B., Gee, M.J.R., Urgeles, R., Mitchell, N.C., Le Bas, T.P., and Canals, M. (2002). Slope failures on the flanks of the western Canary Islands. *Earth-Sci. Rev.*, **57**, 1–35.

Massonnet, D. (1997). Satellite radar interferometry. *Scientific American*, **276**, 46–53.

Massonnet, D. and Feigl, K.L. (1998). Radar interferometry and its application to changes in the Earth's surface. *Rev. Geophys.*, **36**(4), 441–500.

Massonnet, D., Rossi, M., Carmona, C., Adragna, F., Peltzer, G., Feigl, K., and Rabaute, T. (1993). The displacement field of the Landers earthquake mapped by radar interferometry. *Nature*, **364**, 138–142.

Massonnet, D., Briole, P., and Arnaud, A. (1995). Deflation of Mount Etna monitored by spaceborne radar interferometry. *Nature*, **375**, 567–570.

Matsumoto, N., Sato, T., Matsushima, N., Akita, F., Shibara, T., and Suzuki, A. (2002). Hydrological anomalies associated with crustal deformation before the 2000 eruption of Usu Volcano, Japan. *Geophys. Res. Lett.*, **29**(5), 1057, doi: 10.1029/2001GL013968.

Matsumoto, K., Sato, T., Takanezawa, T., and Ooe, M. (2001). GOTIC2: A Program for computation of oceanic tidal loading effect. *J. Geod. Soc. Japan*, **47**, 243–248.

Maxwell, J.C. (1893). *A Treatise on Electricity and Magnetism*. Clarendon Press, Oxford.

McCann, G.D. and Wilts, C.H. (1951). *A Mathematical Analysis of the Subsidence in the Long Beach–San Pedro Area*. California Institute of Technology, CA, 117pp.

McGee, K.A. and Gerlach, T.M. (1998a). Airborne volcanic plume measurements using a FTIR spectrometer, Klauea Volcano, Hawaii. *Geophys. Res. Lett.*, **25**(5), 615–618.

McGee, K.A. and Gerlach, T.M. (1998b). Annual cycle of magmatic CO_2 in a tree-kill soil at Mammoth Mountain, California: Implications for soil acidification. *Geology*, **26**(5), 463–466.

McGee, K.A., Sutton, A.J., and Sato, M. (1987). Use of satellite telemetry for monitoring active volcanoes, with a case study of a gas-emission event at Kilauea Volcano, December 1982. In: Decker, R.W., Wright, T.L., and Stauffer, P.H. (eds), *Volcanism in Hawaii* (USGS Prof. Paper 1350). US Geological Survey, Reston, VA, pp. 821–825.

McGee, K.A., Gerlach, T.M., Kessler, R., and Doukas, M.P. (2000). Geochemical evidence for a magmatic

CO_2 degassing event at Mammoth Mountain, California, September–December 1997. *J. Geophys. Res.*, **105**, 8447–8456.

McGee, K.A., Doukas, M.P., and Gerlach, T.M. (2001). Quiescent hydrogen sulfide and carbon dioxide degassing from Mount Baker, Washington. *Geophys. Res. Lett.*, **28**(23), 4479–4482.

McKee, C., Talai, B., Lauer, N., Stewart, R., de Saint Ours, P., Itikarai, I., Patia, H., Lolok, D., Davies, H., and Johnson, R.W. (1995). The 1994 eruptions at Rabaul Volcano, Papua New Guinea. International Union of Geodesy and Geophysics, General Assembly, **21**, Week A, 448.

McNutt, S.R. (1996). Seismic monitoring and eruption forecasting of volcanoes: A review of the state-of-the-art and case histories. In: Scarpa, R. and Tilling, R.I. (eds), *Monitoring and Mitigation of Volcano Hazards*. Springer Verlag, New York, pp. 99–146.

McNutt, S.R. (2000a). Volcanic seismicity. In: Sigurdsson, H., Houghton, B., McNutt, S.R., Rymer, H., and Stix, J. (eds), *Encyclopedia of Volcanoes*. Academic Press, San Diego, CA, pp. 1015–1033.

McNutt, S.R. (2000b). Seismic monitoring. In: Sigurdsson, H., Houghton, B., McNutt, S.R., Rymer, H., and Stix, J. (eds), *Encyclopedia of Volcanoes*. Academic Press, San Diego, CA, pp. 1095–1119.

McTigue, D.F. (1987). Elastic stress and deformation near a finite spherical magma body: Resolution of the point source paradox. *J. Geophys. Res.*, **92**, 12931–12940.

McTigue, D.F. and Segall, P. (1988). Displacements and tilts from dip–slip faults and magma chambers beneath irregular surface topography. *Geophys. Res. Lett.*, **16**, 601–604.

Meertens, C.M. and Smith, R.B. (1991). Crustal deformation of the Yellowstone caldera from first GPS measurements: 1987–1989. *Geophys. Res. Lett.*, **18**, 1763–1766.

Meertens, C.M., Smith, R.B., Vasco, D.W. (1992). Subsidence and extension of the Yellowstone Plateau from GPS surveys, 1987–1991 (abstract). *EOS Trans. Amer. Geophys. Union*, **73**(43), 343.

Meertens, C.M., Smith, R.B., and Vasco, D.W. (1993). Kinematics of crustal deformation of the Yellowstone hotspot using GPS, *EOS Trans. Amer. Geophys. Union*, **74**(43), 63.

Meertens, C.M., Smith, R.B., Puskas, C.M. (2000). Crustal Deformation of the Yellowstone Caldera from campaign and continuous GPS surveys, 1987–2000. *EOS Trans. Amer. Geophys. Union*, **80**(46), Fall Meet. Suppl., Abstract F1388.

Merriam, R. and Bischoff, J.L. (1975). Bishop Ash: A widespread volcanic ash extended to southern California. *J. Sediment. Petrol.*, **45**, 207–211.

Metropolis, N.A. (1953). Equations of state calculations by fast computing machines. *Journal Chem. Phys.*, **21**, 1087–1092.

Meyer, G.A. and Locke, W.W. (1986). Origin and deformation of Holocene shoreline terraces, Yellowstone Lake, Wyoming. *Geology*, **14**(8), 699–702.

Miller, C.D. (1985). Holocene eruptions at the Inyo volcanic chain, California: Implications for possible eruptions in the Long Valley Caldera. *Geology*, **13**, 14–17.

Miller, C.D., Mullineaux, D.R., Crandell, D.R., and Bailey, R.A. (1982). *Potential Hazards From Future Volcanic Eruptions in the Long Valley–Mono Lake Area, East-central California and Southwest Nevada – A Preliminary Assessment* (USGS Circular 887). US Geological Survey, Reston, VA, 10pp.

Miller, D.S. and Smith, R.B. (1999). *P and S velocity structure of the Yellowstone volcanic field from local earthquake and controlled-source tomography. J. Geophys. Res.*, **104**, 15105–15121.

Miller, M.M., Dragert, H., Endo, E.T., Freymueller, J.T., Goldfinger, C., Kelsey, H.M., Humphreys, E.D., Johnson, D.J., McCaffrey, R., Oldow, J.S., *et al.* (1998). Precise measurements help gauge Pacific Northwest's earthquake potential. *EOS Trans. Amer. Geophys. Union*, **79**(23), 269, 275.

Miller, M.M., Melbourne, T., Johnson, D.J., and Sumner, W.Q. (2002). Periodic slow earthquakes from the Cascadia subduction zone. *Science*, **295**(5564), 2423.

Miller, T.P. and Chouet, B.A. (eds) (1994). The 1989–1990 eruptions of Redoubt Volcano, Alaska. *J. Volcanol. Geotherm. Res.*, **62**(1–4), 530.

Miller, T.P., McGimsey, R.G., Richter, D.H., Riehle, J.R., Nye, C.J., Yount, M.E., and Dumoulin, J.A. (1998). *Catalog of the Historically Active Volcanoes of Alaska* (USGS Open-File Report 98-582). US Geological Survey, Reston, VA, 104pp.

Mills, H.H. (1992). Post-eruption erosion and deposition in the 1980 crater of Mount St. Helens, Washington, determined from digital maps. *Earth Surface Processes and Landforms*, **17**, 739–754.

Mills, H.H. and Keating, G.N. (1992). Maps showing posteruption erosion, deposition, and dome growth in Mount St. Helens crater, Washington, determined by a geographic information system. U.S. Geol. Surv. Misc. Invest. Ser., Map I-2297, 4 sheets with text.

Mimatsu, M. (1962). *A Diary on the Growth of Showa-Shinzan*. Sobetsu Town Office, Sobetsu, Hokkaido, 202pp.

Minakami, T. (1960). Fundamental research for predicting volcanic eruptions. Part I: Earthquakes and crustal deformation originating from volcanic activities. *Bulletin of the Earthquake Research Institute* (University of Tokyo), **38**, 497–544.

Minakami, T. (1961). Study of eruptions and earthquakes originating from volcanos, I. *Int. Geol. Rev.*, **3**, 712–719.

Mindlin, R.D. (1936). Force at a point in the interior of a semi-infinite solid. *Physics*, **7**, 195–202.

Mindlin, R.D. and Cheng, D.H. (1950). Nuclei of strain in the semi-infinite solid. *J. Applied Phys.*, **21**, 926–930.

Misra, P. and Enge, P. (2001). *Global Positioning System: Signals, Measurements, and Performance.* Ganga-Jamuna Press, Lincoln, MA, ISBN: 0-9709544-0-9, 390pp.

Moench, A.F. (1985). Transient flow to a large-diameter well in an aquifer with storative semiconfining layers. *Water Resources Research*, **21**(8), 1121–1131.

Mogi, K. (1958). Relations between the eruptions of various volcanoes and the deformation of the ground surfaces around them. *Bull. Earthq. Res. Inst. U. Tokyo*, **36**, 99–134.

Monastersky, R. (1995). Attack of the vog: natural air pollution has residents of Hawaii all choked up. *Science News*, **147**(18), 284–285.

Moore, J.G. (1964). *Giant submarine landslides on the Hawaiian Ridge. Geological Survey Research 1964* (USGS Prof. Paper 501-D, D95-D98). US Geological Survey, Reston, VA.

Moore, J.G. and Albee, W.C. (1981). Topographic and structural changes, March–July, 1980: Photogrammetric data. In: Lipman, P.W. and Mullineaux, D.R. (eds), *The 1980 Eruptions of Mount St. Helens, Washington* (USGS Prof. Paper 1250). US Geological Survey, Reston, VA, pp. 123–134.

Moore, J.G. and Moore, G.W. (1984). Deposit from a giant wave on the island of Lanai, Hawaii. *Science*, **226**, 1312–1315.

Moore J.G., Clague, D.A., Holcomb, R.T., Lipman, P.W., Normark, W.R., and Torresan, M.E. (1989). Prodigious submarine landslides on the Hawaiian Ridge. *J. Geophys. Res.*, **94**(B12), 17465–17484.

Moran, S.C. (1994). Seismicity at Mount St. Helens, 1987–1992: Evidence for repressurization of an active magmatic system. *J. Geophys. Res.*, **99**(B3), 4341–4354.

Morey, G.W. (1922). The development of pressure in magmas as a result of crystallization. *Washington Acad. Sci. Jour.*, **12**, 219–230.

Morgan, L.A. and McIntosh, W.C. (2005). Timing and development of the Heise Volcanic Field, Snake River Plain, Idaho, western USA. *Bull. Geol. Soc. Am.*, **117**(3/4), 288–306, doi: 10.1130/B25519.1.

Mori, J. (1995). Volcano seismology, hazards assessment. *Rev. Geophys.*, Supplement, 263–267, July 1995, (U.S. National Report to International Union of Geodesy and Geophysics, 1991–1994).

Mori, T., Notsu, K., Tohjima, Y., and Wakita, H. (1993). Remote detection of HCl and SO_2 in volcanic gas from Unzen Volcano, Japan. *Geophys. Res. Lett.*, **20**, 1355–1358.

Mori, T., Notsu, K., Tohjima, Y., Wakita, H., Nuccio, P.M., and Italiano, F. (1995). Remote detection of fumarolic gas chemistry at Vulcano, Italy, using a FT-IT spectral radiometer. *Earth and Planet. Sci. Lett.*, **134**, 219–224.

Mouginis-Mark, P.J. and Domergue-Schmidt, N., (2000). Acquisition of satellite data for volcano studies. In: Mouginis-Mark, P.J., Crisp, J.A., and Fink, J. (eds), *Remote Sensing of Active Volcanism* (Geophysical Monograph 116). American Geophysical Union, Washington, D.C., pp. 9–24.

Mueller, R. J. and Johnston, M. J. S. (1998). Review of magnetic field monitoring near active faults and volcanic calderas in California, 1974–1995. In: Johnston, M. and Parrot, M. (eds), Electromagnetic effects of earthquakes and volcanoes. *Phys. Earth Planet. Interiors.*, **105**(3–4), 131–144.

Muffler, L.J.P., White, D.E., and Truesdell, A.H. (1971). Hydrothermal explosion craters in Yellowstone National Park. *Bull. Geol. Soc. Am.*, **82**, 723–740.

Mullineaux, D.R. and Crandell, D.R. (1981). The eruptive history of Mount St. Helens. In: Lipman, P.W. and Mullineaux, D.R. (eds), *The 1980 Eruptions of Mount St. Helens, Washington* (USGS Prof. Paper 1250). US Geological Survey, Reston, VA, pp. 3–15.

Murai, T. (1979). Photographs of volcanic smoke at Sakurajima, Japan. *J. Jpn. Soc. Photogram. Remote Sensing*, **18**(4), 3 [in Japanese].

Murray, T.L. (1992). A low-data-rate digital telemetry system. In: Ewert, J. and Swanson, D.A. (eds), *Monitoring Volcanoes: Techniques and Strategies Used by the Staff of the Cascades Volcano Observatory, 1980–1990* (USGS Bull. 1966). US Geological Survey, Reston, VA, pp. 11–23.

Murray, T.L., Ewert, J.W., Lockhart, A.B., and LaHusen, R.G. (1996a). The integrated mobile volcano-monitoring system used by the Volcano Disaster Assistance Program (VDAP). In: Scarpa, R. and Tilling, R.I. (eds), *Monitoring and Mitigation of Volcano Hazards.* Springer Verlag, New York, pp. 315–362.

Murray, T.L., Power, J.A., Davidson, G., and Marso, J.N. (1996b). A PC-based real-time volcano-monitoring data-acquisition and analysis system. In: Newhall, C.G. and Punongbayan, R.S. (eds), *Fire and Mud: Eruptions and Lahars of Mt. Pinatubo, Philippines.* Philippine Institute of Volcanology and Seismology, Quezon City and University of Washington Press, Seattle, pp. 225–247.

Murray, M.H., Marshall, G.A., Lisowski, M., and Stein, R.S. (1996c) The 1992 M = 7 Cape Mendocino, California, earthquake: Coseismic deformation at the south end of the Cascadia megathrust. *J. Geophys. Res.*, **101**(B8), 17707–17725.

Nakada, S., Nagai, M., Yasuda, A., Shimano, T., Geshi, N., Ohno, M., Akimasa, T., Kaneko, T., and Fujii, T. (2001). Chronology of the Miyakejima 2000 eruption: Characteristics of summit collapsed crater and eruption products. *J. Geogr.*, **110**, 168–180 [in Japanese with English abstract].

Nakiboglu, S.M. and Lambeck, K. (1982). A study of the Earth's response to surface loading with the application to Lake Bonneville. *Geophys. Jour. Royal Astron. Soc.*, **70**, 577–620.

Nash, J.T. (1976). *Fluid Inclusion Petrology: Data from Porphyry Copper Deposits and Application to Exploration* (USGS Prof. Paper 907-D). US Geological Survey, Reston, VA, 16pp.

National Geodetic Survey (1986). *Geodetic Glossary*, 274 p., supercedes 'Definitions of Terms Used in Geodetic and other Surveys', Coast and Geodetic Survey *Special Publication* 242, by Hugh C. Mitchell, 1948.

National Research Council (2001a). *Basic Research Opportunities in Earth Science*. Committee on Basic Research Opportunities in the Earth Sciences, Board on Earth Sciences and Resources, National Research Council, National Academy Press, Washington, D.C., 168pp.

National Research Council (2001b). *Review of Earth Scope Integrated Science*. Committee on the Review of EarthScope Science Objectives and Implementation Planning, Board on Earth Sciences and Resources, National Research Council, National Academy Press, Washington, D.C., 76pp.

Needham, J. (1959). *Science and Civilization in China* (Volume 3). *Mathematics and the Sciences*. Cambridge University Press, New York, pp 624–635.

Newhall, C.G. (2000). Balancing research and practical needs for volcanic eruption forecasts. In: Esaki, L. (ed.), *New Frontiers of Science and Technology* (Frontiers Science Series No. 31). Universal Academy Press, Inc., Tokyo, pp. 353–362.

Newhall, C.G. and Dzurisin, D. (1988). *Historical Unrest at Large Calderas of the World* (USGS Bull. 1855). US Geological Survey, Reston, VA, 1108pp.

Newhall, C.G. and Punongbayan, R.S. (1996). The narrow margin of successful volcanic-risk mitigation. In: Scarpa, R. and Tilling, R.I. (eds), *Monitoring and Mitigation of Volcano Hazards*. Springer Verlag, New York, pp. 807–838.

Newhall, C.G. and Self, S. (1982). The volcanic explosivity index (VEI): An estimate of explosive magnitude for historical volcanism. *J. Geophys. Res. (Oceans and Atmospheres)*, **87**, 1231–1238.

Newhall, C.G., Daag, A.S., Delfin, F.G., Jr., Hoblitt, R.P., McGeehin, J., Pallister, J.S., Regalado, T.M., Rubin, M., Tubianosa, B.S., Tamayo, R.A., Jr., et al. (1996). Eruptive history of Mount Pinatubo. In: Newhall, C.G. and Punongbayan, R.S. (eds), *Fire and Mud: Eruptions and Lahars of Mt. Pinatubo, Philippines*. Philippine Institute of Volcanology and Seismology, Quezon City and University of Washington Press, Seattle, pp. 165–195.

Newhall, C., Albano, S., Matsumoto, N., and Sandoval, T. (2001). Groundwater in volcanic unrest. *J. Geol. Soc. Phil.*, **56**, 69–84.

Niebauer, T.M., Hoskins, J.K., and Faller, J.E. (1986). Measurement and modeling, absolute gravity: A reconnaissance tool for studying vertical crustal motions. In: Proceedings of the Chapman Conference on vertical crustal motions. *J. Geophys. Res.*, **91**, 9145–9149.

Niebauer, T. M., Sasagawa, G.S., Faller, J.E., Hilt, R., and Klopping, F. (1995). A new generation of absolute gravimeters. *Metrologia*, **32**, 159–180.

NOAA (2004). NGS orbits. Available online at: http://www.ngs.noaa.gov/GPS/GPS.html [Accessed February 2004].

Norton, D.L. (1982). Fluid and heat transport phenomena typical of copper-bearing pluton environments. In: Titley, S.R. (ed.), *Advances in Geology of the Porphyry Copper Deposits Southwestern North America*. University of Arizona Press, Tucson, Arizona, pp. 59–72.

Norton, D. and Knapp, R. (1977). Transport phenomena in hydrothermal systems: Nature of porosity. *American Journal of Science*, **277**, 913–936.

Norton, D. and Knight, J.E. (1977). Transport phenomena in hydrothermal systems: Cooling plutons. *American Journal of Science*, **277**, 937–981.

Norton, D.L. and Cathles, L.M. (1973). Breccia pipes, products of exsolved vapor from magmas. *Economic Geology*, **68**, 540–546.

Ohminato T., Chouet, B. A., Dawson, P., and Kedar, S. (1998). Waveform inversion of very long period impulsive signals associated with magmatic injection beneath Kilauea Volcano, Hawaii. *J. Geophys. Res.*, **103**, 23839–23862.

Ohmoto, H. and Rye, R.O. (1979). Isotopes of sulfur and carbon. In: Barnes, H. L. (ed.), *Geochemistry of Hydrothermal Ore Deposits*. New York, Wiley Interscience., pp. 509–567.

Ohno, M., Sato, T., Notsu, K., Wakita, H., and Ozawa, K. (1999). Groundwater-level changes in response to bursts of seismic activity off the Izu Peninsula, Japan. *Geophys. Res. Lett.*, **26**(16), 2501–2504.

Okada, Y. (1985). Surface deformation due to shear and tensile faults in a half-space. *Bull. Seismol. Soc. Am.*, **75**(4), 1135–1154.

Okada, Y. (1992). Internal deformation due to shear and tensile faults in a half-space. *Bull. Seismol. Soc. Am.*, **82**(2), 1018–1040.

Okada, Y., and Yamamoto, E. (1991). Dyke intrusion model for the 1989 seismovolcanic activity off Ito, central Japan. *J. Geophys. Res.*, **96**(B6), 10361–10376.

Oki, Y. and Hirano, T. (1970). The geothermal system at the Hakone Volcano. *Geothermics*, Special Issue 2, **2** (part 2), 1157–1166.

Okubo, S., Yoshida, S., Sato, T., Tamura, Y., and Imanishi, Y. (1997). Verifying the precision of a new generation absolute gravimeter FG5: Comparison with superconducting gravimeters and detection of oceanic loading tide. *Geophys. Res. Lett.*, **24**(4), 489–492.

Olmsted, C. (1993). *Alaska SAR Facility Scientific SAR User's Guide*. ASF-SD-003, Fairbanks, AK, 53pp.

Orville, P.M. (1963). Alkali ion exchange between vapor and feldspar phases. *American Journal of Science*, **261**, 201–237.

Oswalt, J.S., Niclols, W., and O'Hara, J.F. (1996). Meteorological observations of the 1991 Mount

Pinatubo eruption. In: Newhall, C.G. and Punongbayan, R.S. (eds), *Fire and Mud: Eruptions and Lahars of Mt. Pinatubo, Philippines*. Philippine Institute of Volcanology and Seismology, Quezon City and University of Washington Press, Seattle, pp. 625–636.

Otway, P.M. (1989). Vertical deformation monitoring by periodic water level observations. In: Latter, J.H. (ed.), *Volcanic Hazards, Assessment and Monitoring, IAVCEI Proceedings in Volcanology 1*. Springer Verlag, Heidelberg, pp. 561–574.

Otway, P.M. and Sherburn, S. (1994). Vertical deformation and shallow seismicity around Lake Taupo, New Zealand, 1985–1990. *New Zealand J. Geol. Geophys.*, **37**, 195–200.

Owen, S., Segall, P., Freymueller, J.T., Miklius, A., Denlinger, R.P., Arnadottir, T., Sako, M.K., and Bürgmann, R. (1995). Rapid deformation of the south flank of Kilauea Volcano, Hawaii. *Science*, **267**(5202), 1328–1332.

Owen, S., Segall, P., Lisowski, M., Miklius, A., Denlinger, R., and Sako, M. (2000a). Rapid deformation of Kilauea Volcano: Global Positioning System measurements between 1990 and 1996. *J. Geophys. Res.*, **105**(B8), 18983–18998.

Owen, S., Segall, P., Lisowski, M., Miklius, A., Murray, M., Bevis, M., and Foster, J. (2000b). January 30, 1997 eruptive event on Kilauea Volcano, Hawaii, as monitored by continuous GPS. *Geophys. Res. Lett.*, **27**(17), 2757–2760.

Pallister, J.S., Hoblitt, R.P., Crandell, D.R., and Mullineaux, D.R. (1992). Mount St. Helens a decade after the 1980 eruptions: Magmatic models, chemical cycles, and a revised hazards assessment. *Bull. Volcanol.*, **54**, 126–146.

Parascondola, A. (1947). *I fenomeni bradisismici del Serapeo di Pozzuoli*. Stabilimento Tipografico G. Genovese, Naples, 156pp [English translation by M. Capuano and J. Dvorak].

Parkinson, B. and Spilker, J. (eds) (1996). *Global Positioning System: Theory and Applications* (Volume I and II). American Institute of Aeronautics and Astronautics, Washington, D.C.

Pauk, B.A., Power, J.A., Lisowski, M., Dzurisin, D., Iwatsubo, E.Y., and Melbourne, T. (2001). *Global Positioning System Survey of Augustine Volcano, Alaska, August 3–8, 2000: Data processing, Geodetic Coordinates and Comparison with Prior Geodetic Surveys* (USGS Open-File Rpt., 01-099). US Geological Survey, Reston, VA, 20pp.

PBO Steering Committee (1999). The Plate Boundary Observatory: Creating a four-dimensional image of the deformation of western North America. White paper providing the scientific rationale and deployment strategy for a Plate Boundary Observatory based on a workshop held October 3–5, 1999. ·

Pelton, J.R. and Smith, R.B. (1979). Recent crustal uplift in Yellowstone National Park. *Science*, **206**, 1179–1182.

Pelton, J.R. and Smith, R.B. (1982). Contemporary vertical surface displacements in Yellowstone National Park. *J. Geophys. Res.*, **87**, 2745–2761.

Pendick, D. (1993). Volcano watchers draft in radar. *New Scientist*, **16**, October 1993, 20.

Peterson, D.W. (1996). Mitigation measures and preparedness plans for volcanic emergencies. In: Scarpa, R. and Tilling, R.I. (eds), *Monitoring and Mitigation of Volcano Hazards*. Springer Verlag, New York, pp. 701–718.

Peterson, D.W. and Tilling, R.I. (1993). Interactions between scientists, civil authorities, and the public at hazardous volcanoes. In: Kilburn, C.R.J. and Luongo (eds), *Active Lavas*. UCL Press, London, pp. 339–365.

Pfiffner, O.A. and Ramsay, J.G. (1982). Constraints on geological strain rates: Arguments from finite strain rates of naturally deformed rocks. *J. Geophys. Res.*, **87**, 311–321.

Phillips, W.J. (1973). Mechanical effects of retrograde boiling and its probable importance in the formation of some porphyry ore deposits. *Institute of Mining and Metallurgy Transactions*, Sec. B, **82**, B90–B98.

Pierce, K.L. and Morgan, L.A. (1992). The track of the Yellowstone hotspot: Volcanism, faulting, and uplift. *Geol. Soc. Am. Mem.*, **179**, 1–53.

Pierce, K.L., Cannon, K.P., Meyer, G.A., Trebesch, M.J., and Watts, R.D. (2002). *Post-glacial Inflation–Deflation Cycles, Tilting, and Faulting in the Yellowstone Caldera Based on Yellowstone Lake Shorelines* (USGS Open-File Rpt. 02-0142). US Geological Survey, Reston, VA, 30pp.

Pingue, F., Troise, C., De Luca, G., Grassi, and Scarpa, R. (1998). Geodetic monitoring of Mt. Vesuvius Volcano, Italy, based on EDM and GPS surveys. *J. Volcanol. Geotherm. Res.*, **82**, 151–160.

Pitt, A.M. and Hutchinson, R.A. (1982). Hydrothermal changes related to earthquake activity at Mud Volcano, Yellowstone National Park, Wyoming. *J. Geophys. Res.*, **87**, 2762–2766.

Pitt, A.M., Weaver, C.S., and Spence, W. (1979). The Yellowstone Park earthquake of June 30, 1975. *Bull. Seis. Soc. Am.*, **69**(1), 187–205.

Poland, M.P., Bürgman, R., Dzurisin, D. (1997). Crustal extension and subsidence at Medicine Lake Volcano, Northern California. Geological Society of America, Cordilleran Section, 93rd annual meeting, Abstracts with Programs, Geological Society of America, **29**(5), 57.

Pollard, D.D. (1976). On the form and stability of open cracks in the earth's crust. *Geophys. Res. Lett.*, **3**, 513–516.

Pollard, D.D. and Holzhausen, G. (1979). On the mechanical interaction between fluid-filled fracture and the Earth's surface. *Tectonophysics*, **53**, 27–57.

Poupinet, G., Ratdomopurbo, A., and Coutant, O. (1996). On the use of earthquake multiplets to study fractures and the temporal evolution of an active volcano. *Annali di Geofisica*, **39**(2), 253–264.

Powell, W.B. and Pheifer, D. (2000). The electrolytic tilt sensor. *Sensors*, **17**(5), 120.

Power, J.A., Lahr, J.C., Page, R.A., Chouet, B.A., Stephens, C.D., Harlow, D.H., Murray, T.L., and Davies, J.N. (1994). Seismic evolution of the 1989–1990 eruption sequence of Redoubt Volcano, Alaska. In: Miller, T.P. and Chouet, B.A. (eds), The 1989–1990 eruptions of Redoubt Volcano, Alaska. *J. Volcanol. Geotherm. Res.*, **62**(1–4), 69–94.

Power, J.A., Stihler, S.D., White, R.A., and Moran, S.C. (2004). Observations of deep long-period (DLP) seismic events beneath Aleutian arc volcanoes, 1989–2002. *J. Volcanol. Geotherm. Res.*, **138**(3–4), 243–266.

Prejean, S.G. (2002). The interaction of tectonic and magmatic processes in the Long Valley caldera, California. Ph.D. dissertation, Stanford Univ., Stanford, Calif., 131pp.

Prejean, S., Ellsworth, W., Zoback, M., and Waldhauser, F. (2002). Fault structure and kinematics of the Long Valley Caldera region, California, revealed by high-accuracy earthquake hypocenters and focal mechanism stress inversions, *J. Geophys. Res.*, **107**(B12), 2355, ESE9, doi: 10.1029/2001JB001168.

Press, F. (1965). Displacements, strains and tilts at teleseismic distances. *J. Geophys. Res.*, **70**, 2395–2412.

Pritchard, M.E. and Simons, M. (2002) A satellite geodetic survey of large-scale deformation of volcanic centres in the central Andes. *Nature*, **418**, 167–171.

Punongbayan, R.S., Newhall, C.G., Bautista, Ma. L.P., Garcia, D., Harlow, D.H., Hoblitt, R.P., Sabit, J.P., and Solidum, R.U. (1996). Eruption hazard assessment and warnings. In: Newhall, C.G. and Punongbayan, R.S. (eds), *Fire and Mud: Eruptions and Lahars of Mt. Pinatubo, Philippines*. Philippine Institute of Volcanology and Seismology, Quezon City and University of Washington Press, Seattle, pp. 67–85.

Ramos, E.G., Laguerta, E.P., and Hamburger, M.W. (1996). Seismicity and magmatic resurgence at Mount Pinatubo in 1992. In: Newhall, C.G. and Punongbayan, R.S. (eds), *Fire and Mud: Eruptions and Lahars of Mt. Pinatubo, Philippines*. Philippine Institute of Volcanology and Seismology, Quezon City and University of Washington Press, Seattle, pp. 387–406.

Ray, R.G. (1960). *Aerial Photographs in Geologic Mapping and Interpretation* (USGS Prof. Paper 373). US Geological Survey, Reston, VA, 230pp.

Reid, H.F. (1910a). Permanent displacements of the ground. In: Lawson, A.C. (ed.), *The California Earthquake of April 18, 1906, Report of the State Earthquake Investigation Commission*. Carnegie Insitutute of Washington Publication No. 87, Washington, D.C., pp. 16–28.

Reid, H.F. (1910b). The Mechanics of the Earthquake. In: Lawson, A.C. (ed.), *The California Earthquake of April 18, 1906. Report of the State Investigation Commission* (Volume II). Carnegie Institution of Washington Publication No. 87, Washington, D.C., 192pp.

Remondi, B.W. (1984). Using the Global Positioning System (GPS) phase observable for relative geodesy: Modeling, processing, and results. Ph.D. dissertation, University of Texas at Austin, Center for Space Research, 360pp.

Remondi, B.W. (1986). Performing centimeter-level surveys in seconds with GPS carrier phase: Initial results. *Proceeding of the Fourth International Geodetic Symposium on Satellite Positioning, Austin Texas, April 28 – May 2*, **2**, 1229–1249.

Remondi, B.W. (1991). Kinematic GPS results without static initialization. National Information Center, Rockville, Maryland, NOAA Technical Memorandum NOS NGS-55, 25pp.

Remondi, B. W. and Brown, G. (2000). Triple differencing with Kalman filtering: Making it work. *GPS Solutions*, **3**(3), 58–64.

Rice, J.R. and Cleary, M.P. (1976). Some basic stress-diffusion solutions for fluid-saturated elastic porous media with compressible constituents. *Rev. Geophys. Space Phys.*, **14**(2), 227–241.

Richter, C.F. (1935). An instrumental earthquake scale. *Bull. Seism. Soc. Am.*, **25**(1), 1–32.

Richter, D.H., Waythomas, C.F., McGimsey, R.G., and Stelling, P.L. (1998). *Geologic map of Akutan Island, Alaska* (USGS Open-File Rpt. 98–135). US Geological Survey, Reston, VA, 22p (plus 1 plate).

Robertson, R.E.A., Aspinall, W.P., Herd, R.A., Norton, G.E., Sparks, R. Stephen J., and Young, S.R. (2000). The 1995–1998 eruption of the Soufriere Hills Volcano, Montserrat, WI. *Phil. Trans. Royal Soc., Mathematical, Physical and Engineering Sciences*, **358**(1770), 1619–1637.

Rodriguez, E. and Martin, J. (1992). Theory and design of interferometric synthetic aperture radars. *Proc. IEEE*, **139**(2), 147–159.

Roedder, E. (1984). *Fluid Inclusions. Reviews in Mineralogy* (Volume 2) Mineral. Soc. Amer., 644pp.

Roeloffs, E.A. (1996). Poroelastic methods in the study of earthquake-related hydrologic phenomena. In: Dmowska, R. (ed.), *Advances in Geophysics*. Academic Press, San Diego, pp. 135–195.

Roeloffs, E.A., Danskin, W.R., Farrar, C.D., Galloway, D.L., Hamlin, S.N., Quilty, E.G., Quinn, H.M., Schaefer, D.H., Sorey, M.L., and Woodcock, D.E. (1995). *Hydrologic Effects Associated with the June 28, 1992 Landers, California, Earthquake Sequence* (USGS Open-File Report 95-42). US Geological Survey, Reston, VA, 68pp.

Roeloffs, E., Sneed, M., Galloway, D.L., Quilty, E.G., Sorey, M.L., Farrar, C.D., and Armstrong, D. (1998). Water level changes in the Lookout Mountain well,

Long Valley caldera, California, caused by local and distant earthquakes. *EOS Trans. Am. Geophys. Union*, **79**(17), Spring Meet. Suppl., Abstract S365.

Roeloffs, E., Sneed, M., Galloway, D.L., Sorey, M.L., Farrar, C.D., Howle, J.F., and Hughes, J. (2003). Water-level changes induced by local and distant earthquakes at Long Valley Caldera, California. In: Sorey, M.L., McConnell, V.S., and Roeloffs, E. (eds), Crustal unrest in Long Valley Caldera, California: New interpretations from geophysical and hydrologic monitoring and deep drilling. *J. Volcanol. Geotherm. Res.*, **127**(3–4), 269–303.

Rogers, G. and Dragert, H. (2003). Episodic tremor and slip on the Cascadia subduction zone: The chatter of silent slip. *Science*, **300**, 1942–1943.

Rojstaczer, S. (1988). Intermediate period response of water levels in wells to crustal strain: Sensitivity and noise level. *J. Geophys. Res.*, **93**(B11), 13619–13634.

Rojstaczer, S. and Agnew, D.C. (1989). The influence of formation material properties on the response of water levels in wells to Earth tides and atmospheric loading. *J. Geophys. Res.*, **94**(B9), 12403–12411.

Rose, W.I., Kostinski, A.B., and Kelly, L. (1995). Real time C band radar observations of 1992 eruption clouds from Crater Peak/Spurr Volcano, Alaska. In: Keith, T. (ed.), *The 1992 Eruptions of Crater Peak Vent, Mount Spurr Volcano, Alaska* (USGS Bull. 2139). US Geological Survey, Reston, VA, pp. 19–26.

Rosen P., Hensley, S., Joughin, I., Li, F., Madsen, S., Ridriguez, E., and Goldstein, R. (2000). Synthetic Aperture Radar Interferometry. *Proc. IEEE*, **88**(3), 333–380.

Rubin, A.M. and Pollard, D.D. (1988). Dike-induced faulting in rift zones of Iceland and Afar. *Geology*, **16**, 413–417.

Rubin, A.M., Gillard, D., and Got, J.L. (1998). A reinterpretation of seismicity associated with the January 1983 dike intrusion at Kilauea Volcano, Hawaii. *J. Geophys. Res.*, **103**, 10003–10015.

Rundle, J.B. (1978). Gravity changes and the Palmdale uplift. *Geophys. Res. Lett.*, **5**, 41–44.

Rundle, J.B. and Whitcomb, J.H. (1986). Modeling gravity and trilateration data in Long Valley, California. *J. Geophys. Res.*, **91**, 12675–12682.

Rutherford, M.J., Sigurdsson, H., Carey, S., and Davis, A. (1985). The May 18, 1980, eruption of Mount St. Helens. Part 1: Melt composition and experimental phase equilibria. *J. Geophys. Res.*, **90**, 2929–2947.

Rutledge, D., Gnipp, J., and Kramer, J. (2001). Advances in real-time GPS deformation monitoring for landslides, volcanoes, and structures. *Proceedings of the 10th FIG International Symposium on Deformation Measurements, Orange, California, March 19–22, 2001*, pp. 110–121.

Rye, R.O. (1993). The evolution of magmatic fluids in the epithermal environment: The stable isotope perspective (SEG Distinguished Lecture). *Economic Geology*, **88**, 733–753.

Rymer, H. (1996). Microgravity monitoring. In: Scarpa, R. and Tilling, R.I. (eds), *Monitoring and Mitigation of Volcano Hazards*. Springer Verlag, New York, pp. 169–197.

Rymer, H. and Williams-Jones, G. (2000). Volcanic eruption prediction: Magma chamber physics from gravity and deformation measurements. *Geophys. Res. Lett.*, **27**, 2389–2392.

Rymer H., Cassidy, J., Locke, C.A., and Murray, J.B. (1995). Magma movements in Etna Volcano associated with the major 1991–1993 lava eruption: Evidence from gravity and deformation. *Bull. Volcanol.*, **57**, 451–461.

Sabins, F.F., Jr. (1999). *Remote Sensing, Principles and Interpretation* (3rd edition). W.H. Freeman and Company, New York.

Sacks, I.S., Suyehiro, S., Evertson, D.W., and Yamagishi, Y. (1971). Sacks-Evertson strainmeter, its installation in Japan and some preliminary results concerning strain steps. *Pap. Meteor. Geophys.*, **22**, 195–207.

Sarna-Wojicicki, A.M., Morrison, S.D., Meyer, C.E., and Hillhouse, J.W. (1987). Correlation of upper Cenozoic tephra layers between sediments of Western United States and the East Pacific Ocean, and comparison with biostratigraphic and megnetostratigraphic age data. *Bull. Geol. Soc. Am.*, **98**, 207–233.

Sasagawa, G., Klopping, F., Niebauer, T.M., Faller, J.E., and Hilt, R.L. (1995). Intracomparison tests of the FG5 absolute gravity meters. *Geophys. Res. Lett.*, **22**, 461–464.

Savage, J.C. (1975). A possible bias in the California State geodimeter data. *J. Geophys. Res.*, **80**(29), 4078–4088.

Savage, J.C. (1984). Local gravity anomalies produced by dislocation sources. *J Geophys. Res.*, **89**, 1945–1952.

Savage, J.C. (1988). Principal component analysis of geodetically measured deformation in the Long Valley caldera, eastern California. *J. Geophys. Res.*, **93**, 13297–13306.

Savage, J.C. and Clark, M.M. (1982). Magmatic resurgence in Long Valley caldera, California: Possible cause of the 1980 Mammoth Lakes earthquakes. *Science*, **217**, 531–533.

Savage, J.C. and Cockerham, R.S. (1984). Earthquake swarm in Long Valley caldera, California, January 1983: Evidence for dike inflation. *J. Geophys. Res.*, **89**, 8315–8324.

Savage, J.C., Cockerham, R.S., Estrem, J.E., and Moore, L.R. (1987). Deformation near the Long Valley caldera, eastern California, 1982–1986. *J. Geophys. Res.*, **92**, 2721–2746.

Savage, J.C., Lisowski, M., Prescott, W.H., and Pitt, A.M. (1993). Deformation from 1973 to 1987 in the epicentral area of the 1959 Hebgen Lake, Montana, earthquake ($M_s = 7.5$). *J. Geophys. Res.*, **98**, 2145–2153.

Scandone, R. and Malone, S.D. (1985). Magma supply, magma discharge and readjustment of the feeding system of Mount St. Helens during 1980. *J. Volcanol. Geotherm. Res.*, **23**, 239–262.

Scarpa, R. and Tilling, R.I. (eds) (1996). *Monitoring and Mitigation of Volcano Hazards*. Springer Verlag, New York, 841pp.

Schilling, S.P., Carrara, P.E., Thompson, R.A., and Iwatsubo, E.Y. (2002). Post-eruption glacier development within the crater of Mount St. Helens, Washington, U.S.A. *Geological Society of America Cordilleran Section Meeting Abstracts with Programs*, **34**(5), 91.

Schilling, S.P., Carrara, P.E., Thompson, R.A., and Iwatsubo, E.Y. (2004). Posteruption glacier development within the crater of Mount St. Helens, Washington, USA. *Quaternary Research*, **61**, 325–329.

Scott, W.E. and Gardner, C.A. (1990). Field trip guide to the central Oregon High Cascades. Part 1: Mount Bachelor–South Sister area, Oregon. *Geology*, **52**(5), 99–114.

Scott, W.E., Iverson, R.M., Schilling, S.P., and Fischer, B.J. (2001). *Volcano Hazards in the Three Sisters Region, Oregon* (USGS Open-File Rpt. 99-437). US Geological Survey, Reston, VA, 14pp.

Secor, D.T. and Pollard, D.D. (1975). On the stability of open hydraulic fractures in the earth's crust. *Geophys. Res. Lett.*, **2**, 510–513.

Seeber, G. (1993). *Satellite Geodesy: Foundations, Methods, and Applications*. Walter de Gruyther, New York, 531pp.

Segall, P. and Lisowski, M. (1990). Surface displacements in the 1906 San Francisco and 1989 Loma Prieta earthquakes. *Science*, **250**, 1241–1244.

Segall, P. and Matthews, M. (1997). Time dependent inversion of geodetic data. *J. Geophys. Res.*, **102**, 22391–22409.

Segall, P., Cervelli, P., Owen, S., Lisowski, M., and Asta, M. (2001). Constraints on dike propagation from continuous GPS measurements. *J. Geophys. Res.*, **106**, 19301–19317.

Segall, P., Jonsson, S., and Agustsson, K. (2003). When is the strain in the meter the same as the strain in the rock? *Geophys. Res. Lett.*, **30**(19), 1990, doi: 10.1029/2003GL017995.

Sen, B. (1951). Note on the stresses produced by nuclei of thermo-elastic strain in a semi-infinite elastic solid. *Quarterly Applied Mathematics*, **8**, 365–369.

Shapiro, I.I., Zisk, S.H., Rogers, A.E.E., Slade, M.A., and Thompson, T.W. (1972). Lunar topography: Global determination by radar. *Science*, **178**, 939–948.

Shibata, T. and Akita, F., (2001). Precursory changes in well water level prior to the March 2000 eruption of Usu Volcano, Japan. *Geophys. Res. Lett.*, **28**, 1799–1802.

Shinohara, H. (1994). Exsolution of immiscible vapor and liquid phases from a crystallizing silicate melt: Implications for chlorine and metal transport. *Geochimica et Cosmochimica Acta*, **58**, 5215–5221.

Shinohara, H. and Fujimoto, K. (1994). Experimental study in the system albite–andalusite–quartz–NaCl–H_2O at 600°C and 400 to 2000 bars. *Geochimica et Cosmochimica Acta*, **58**, 4857–4866.

Shinohara, H., Kazahaya, K., and Lowenstern, J.B. (1995). Volatile transport in a convecting magma column: Implications for porphyry Mo mineralization. *Geology*, **23**, 1091–1094.

Sibson, R.H. (1982). Fault zone models, heat flow, and the depth distribution of seismicity in the continental crust of the United States. *Seismological Society of America Bulletin*, **72**, 151–163.

Sibson, R.H. (1985). A note on fault reactivation. *Journal of Structural Geology*, **7**, 751–754.

Sibson, R.H. (1990). Rupture nucleation on unfavorably oriented faults. *Seismological Society of America Bulletin*, **80**, 1580–1604.

Sieh, K. and Bursik, M. (1986). Most recent eruption of the Mono Craters, eastern central California. *J. Geophys. Res.*, **91**, 12539–12571.

Sigmundsson, F., Durand, P., and Massonnet, D. (1999). Opening of an eruptive fissure and seaward displacement at Piton de la Fournaise volcano measured by RADARSAT satellite radar interferometry. *Geophys. Res. Lett.*, **26**(5), 533–536.

Sigurdsson, H., Carey, S.N., and Espindola, J.M. (1984). The 1982 eruptions of El Chichón Volcano, Mexico: Stratigraphy of pyroclastic deposits. *J. Volcanol. Geotherm. Res.*, **23**, 11–37.

Silberman, M.L. (1983). *Geochronology of Hydrothermal Alteration and Mineralization: Tertiary Epithermal Precious Metal Deposits in the Great Basin, in The role of Heat in the Development of Energy and Mineral Resources in the Northern Basin and Range Province* (Geothermal Resources Council Special Report No. 13). GRC, Davis, CA, pp. 287–303.

Sillitoe, R.H. (1985). Ore-related breccias in volcanoplutonic arcs. *Economic Geology*, **80**, 1467–1514.

Simkin, T. and Siebert, L. (1984). Explosive eruptions in space and time: Durations, intervals and a comparison of the world's active volcanic belts. In: *Explosive Volcanism: Inception, Evolution, and Hazards*. Studies in Geophysics, National Academy Press, Washington, DC, pp. 110–121.

Simkin, T. and Siebert, L. (1994). *Volcanoes of the World* (2nd edition). Geoscience Press, Inc., Tucson, AZ, 349pp.

Simons, M., Fialko, Y., Chapin, E., Hensley, S., Rosen, P.A., Shaffer, S., and Webb, F. (1999). Analysis of geodetic measurements of crustal deformation at Long Valley Caldera. *EOS Trans. Amer. Geophys. Union*, **80**(46), Fall Meet. Suppl., Abstract F1194.

Simons, W.J.F., Ambrosius, B.A.C., Noomen, R., Angermann, D., Wilson, P., Becker, M., Reinhart, E., Walpersdorf, A., and Vigny, C. (1999). Observing plate motions in S.E. Asia: Geodetic results of the GEODYS-SEA Project. *Geophys. Res. Lett.*, **26**(14), 2081–2084.

Singer, B.S., Thompson, R.A., Dungan, M.A., Feeley, T.C., Nelson, S.T., Pickens, J.C., Brown, L.L., Wulff, A.W., Davidson, J.P., and Metzger, *J.* (1997). Volcanism and erosion during the past 930 k.y. at the Tatara-San Pedro complex, Chilean Andes. *Geological Society of America Bulletin*, **109**(2), 127–142.

Slater, L.E. and Huggett, G.R. (1976). A multi-wavelength distance-measuring instrument for geophysical experiments. *J. Geophys. Res.*, **81**, 6299–6306.

Smith, R.B. and Arabasz, W.J. (1991). Seismicity of the Intermountain seismic belt. In: Slemmons, D.B., Engdahl, E.R., Zoback, M.L., and Blackwell, D.D. (eds), *Neotectonics of North America* (Decade Map volume 1), Geological Society of America, 185–228.

Smith, R.B. and Braile, L.W. (1984). Crustal structure and evolution of an explosive silicic volcanic system at Yellowstone National Park. In: *Explosive Volcanism: Inception, Evolution, and Hazards*. National Academy Press, Washington, D.C., pp. 96–109.

Smith, R.B. and Braile, L.W. (1994). The Yellowstone hotspot. *J. Volcanol. Geotherm. Res.*, **61**, 121–187.

Smith, R.B. and Siegel, L.J. (2000). *Windows into the Earth: The Geologic Story of Yellowstone and Grand Teton National Parks*. Oxford University Press, Oxford, 242pp.

Smith, R.B., Reilinger, R.E., Meertens, C.M., Hollis, J.R., Holdahl, S.R., Dzurisin, D., Gross, W.K., and Klingele, E.E. (1989). What's moving at Yellowstone! The 1987 crustal deformation survey from GPS, leveling, precision gravity and trilateration. *EOS Trans. Amer. Geophys. Union*, **70**(8), 113–125.

Solheim, F., Vivekanandan, J., Ware, R., and Rocken, C. (1999). Propagation delays induced in GPS signals by dry air, water vapor, hydrometeors and other atmospheric particulates. *J. Geophys. Res.*, **104**(D8), 9663–9770.

SOPAC (2004). Scripps Orbit and Permanent Array Center. Available online at: http://sopac.ucsd.edu [Accessed February 2004].

Sorey, M.L., Kennedy, B.M., Evans, W.C., Farrar, C.D., and Suemnicht, G.A. (1993). Helium isotope and gas discharge variations associated with crustal unrest in Long Valley Caldera, California. *J. Geophys. Res.*, **98**, 15871–15889.

Sorey, M.L., McConnell, V.S., and Roeloffs, E. (2003). Summary of recent research in Long Valley Caldera, California. In: Sorey, M.L., McConnell, V.S., and Roeloffs, E. (eds), Crustal unrest in Long Valley Caldera, California: New interpretations from geophysical and hydrologic monitoring and deep drilling. *J. Volcanol. Geotherm. Res.*, **127**(3–4), 165–173.

Sourirajan, S. and Kennedy, G.C. (1962). The system H_2O–NaCl at elevated temperatures and pressures. *American Journal of Science*, **260**, 115–141.

Spada, G. (2003). The TABOO post-glacial rebound calculator. Retrieved 2005 from http://samizdat.mines.edu/taboo/

Stein, R.S. (1981). Discrimination of tectonic displacement from slope-dependent errors in geodetic leveling from southern California, 1953–1979. In: Simpson, D.W., and Richards, P.G. (eds), *Earthquake Prediction, An International Review* (Maurice Ewing Series 4). American Geophysical Union, Washington, D.C., pp. 441–456.

Stein, R.S., Whalen, C.T., Holdahl, S.R., Strange, W.E., and Thatcher, W. (1986). Saugus–Palmdale, California, field test for refraction error in historical leveling surveys. *J. Geophys. Res.*, **91**, 9031–9044.

Stein, S. and Wysession, M. (2003). *An Introduction to Seismology, Earthquakes, and Earth Structure.* Blackwell Publishing Ltd., Boston, MA, 512pp.

Steingrimsson, B., Gudmundsson, A., Franzson, H., and Gunnlaugsson, E. (1990). Evidence of supercritical fluid at depth in the Nesjavellir field. *Proceedings, 15th Workshop on Geothermal Reservoir Engineering, January 23–24, 1990, Stanford University*, pp. 81–88.

Steketee, J.A. (1965). On Volterra's dislocation in a semi-finite elastic medium. *Can. J. Phys.*, **36**, 192–205.

Stine, S. (1990). Late Holocene fluctuations of Mono Lake, eastern California. In: Meyers, P.A. and Benson, L.V. (eds), Paleoclimates: The record from lakes, ocean and land. *Paleogeography, Paleoclimatology, Paleoecology*, **78**(3–4), 333–381.

Stoiber, R.E. and Jepsen, A. (1973). Sulfur dioxide contributions to the atmosphere by volcanoes. *Science*, **182**, 577–579.

Stoiber, R.E., Malinconico, L.L., and Williams, S.N. (1983). Use of the correlation spectrometer at volcanoes. In: Tazieff, H. and Sabroux, J.C. (editors), *Forecasting Volcanic Events*. Elsevier, Amsterdam, pp. 425–444.

Strange, W.E. (1981). The impact of refraction correction on leveling interpretations in southern California. *J. Geophys. Res.*, **86**, 2809–2824.

Sun, R.J. (1969). Theoretical size of hydraulically induced horizontal fractures and corresponding surface uplift in an idealized medium. *J. Geophys. Res.*, **74**, 5995–6011.

Surveys and Mapping Branch (1978). Specifications and Recommendations for Control Surveys and Survey Markers. R.E. Moore, Director-General, Surveys and Mapping Branch, Geodetic Survey Division, Natural Resources Canada, Ottawa, Canada, 60 p. Available online at: http://www.geod.nrcan.gc.ca/index_e/products_e/stand_e/specs_e/specs.html [Accessed March 2005].

Sutton, A.J., McGee, K.A., Casadevall, T.J., and Stokes, J.B. (1992). Fundamental volcano-gas-study techniques: An integrated approach to monitoring. In: Ewert, J. and Swanson, D.A. (eds), *Monitoring Volcanoes: Techniques and Strategies Used by the Staff of the Cascades Volcano Observatory, 1980–1990* (USGS Bull. 1966). US Geological Survey, Reston, VA, pp. 181–188.

Swanson, D.A. (1992). The importance of field observations for monitoring volcanoes, and the approach of 'keeping monitoring as simple as practical.' In: Ewert, J. and Swanson, D.A. (eds), *Monitoring Volcanoes: Techniques and Strategies Used by the Staff of the Cascades Volcano Observatory, 1980–1990* (USGS Bull. 1966). US Geological Survey, Reston, VA, pp. 219–223.

Swanson, D.A. and Holcomb, R.T. (1990). Regularities in growth of the Mount St. Helens dacite dome, 1980–1986. In: Fink, J. (ed.), *The Mechanics of Lava Flow Emplacement and Dome Growth, International Association of Volcanology and Chemistry of the Earth's Interior Proceedings in Volcanology 2.* Springer Verlag, Berlin, pp. 3–24.

Swanson, D.A., Duffield, W.A., and Fiske, R.S. (1976). *Displacement of the South Flank of Kilauea Volcano: The Result of Forceful Intrusion of Magma into the Rift Zones* (USGS Prof. Paper 963). US Geological Survey, Reston, VA, 39pp.

Swanson, D.A., Lipman, P.W., Moore, J.G., Heliker, C.C., and Yamashita, K.M. (1981). Geodetic monitoring after the May 18 eruption. In: Lipman, P.W. and Mullineaux, D.R. (eds), *The 1980 Eruptions of Mount St. Helens, Washington* (USGS Prof. Paper 1250). US Geological Survey, Reston, VA, pp. 157–168.

Swanson, D.A., Casadevall, T.J., Dzurisin, D., Malone, S.D., Newhall, C.G., and Weaver, C.S. (1983). Predicting eruptions at Mount St. Helens, June 1980 through December 1982. *Science*, **221**(4618), 1369–1376.

Swanson, D.A., Dzurisin, D., Holcomb, R.T., Iwatsubo, E.Y., Chadwick, W.W., Jr., Casadevall, T.J., Ewert, J.W., and Heliker, C.C. (1987). Growth of the lava dome at Mount St. Helens, Washington (USA), 1981–1983. In: Fink, J.H. (ed), *Emplacement of Silicic Domes and Lava Flows* (Geological Society of America Special Paper 212). Geological Society of America, Boulder, CO, pp. 1–16.

Szeliga, W., Melbourne, T.I., Miller, M., and Santillan, V.M. (2004). Southern Cascadia episodic slow earthquakes. *Geophys. Res. Lett.*, **31**, L16602, doi:10.1029/2004GL020824.

Takahashi, E., Lipman, P.W., Garcia, M.O., Naka, J., and Aramaki, S. (2002). *Hawaiian Volcanoes: Deep Underwater Perspectives* (American Geophysical Union Monograph 128). American Geophysical Union, Washington D.C., 18pp.

Taylor, E.M., MacLeod, N.S., Sherrod, D.R., and Walker, G.W. (1987). Geologic map of the Three Sisters Wilderness, Deschutes, Lane, and Linn counties, Oregon, U.S. Geol. Surv. Misc. Field Studies Map, MF-1952.

Telford, W.M., Geldart, L.P., and Sheriff, R.E. (1990). *Applied Geophysics* (2nd edition). Cambridge University Press, Cambridge, ISBN 0-521-33938-3, 770pp.

Thatcher, W. (1974). Strain release mechanism of the 1906 San Francisco earthquake. *Science*, **184**, 1283–1285.

Thatcher, W. (1975). Strain accumulation and release mechanism of the 1906 San Francisco earthquake. *J. Geophys. Res.*, **80**, 4862–4872.

Thatcher, W. (2001). Silent slip on the Cascadia subduction interface. *Science*, **292**, 1495–1496.

Thatcher, W. and Massonnet, D. (1997). Crustal deformation at Long Valley Caldera, eastern California, 1992–1996 inferred from satellite radar interferometry. *Geophys. Res. Lett.*, **24**(20), 2519–2522.

Thatcher, W.R., Marshall, G.A., and Lisowski, M. (1997). Resolution of fault slip along the 470-km-long rupture of the great 1906 San Francisco earthquake and its implications. *J. Geophys. Res.*, **102**, 5353–5367.

Thatcher, W., Foulger, G.R., Julian, B.R., Svarc, J., Quilty, E., and Bawden, G.W. (1999). Present-day deformation across the Basin and Range province, western United States. *Science*, **283**, 1714–1718.

Theodossious, E.I. and Dowman, I.J. (1990). Heighting accuracy of SPOT. *Photogrammetric Engineering and Remote Sensing*, **56**(12), 1643–1649.

Thornber, C.R. (1997). *HVO/RVTS-1: A Prototype Remote Video Telemetry System for Monitoring the Kilauea East Rift Zone Eruption, 1997* (USGS Open-File Rpt. 97-537). US Geological Survey, Reston, VA, 18pp.

Tilling, R.I., Koyanagi, R.Y., Lipman, P.W., Lockwood, J.P., Moore, J.G., and Swanson, D.A. (1976). *Earthquake and Related Catastrophic Events, Island of Hawaii, November 29, 1975: A Preliminary Report* (USGS Circular 740). US Geological Survey, Reston, VA, 33pp.

Turcotte, D.L. and Schubert, G. (2002). *Geodynamics* (2nd edition). Cambridge University Press, ISBN 0521661862, 456pp.

Tsuji, H., Hatanaka, Y., Sagiya, T., and Hashimoto, M. (1995). Coseismic crustal deformation from the 1994 Hokkaido-Toho-Oki earthquake monitored by a nationwide continuous GPS array in Japan. *Geophys. Res. Lett.*, **22**(13), 1669–1672.

Turner, F. J. (1981). *Metamorphic Petrology: Mineralogical, Field, and Tectonic Aspects* (2nd edition). McGraw Hill, New York, 524pp.

Ui, T. (1989). Discrimination between debris avalanches and other volcaniclastic deposits. In: Latter, J.H. (ed.), *Volcanic Hazards: Assessment and Monitoring, IAVCEI Proceedings in Volcanology 1.* Springer Verlag, Heidelberg, pp. 201–209.

Ukawa, M., Fujita, E., Yamamoto, E., Okada, Y., and Kikuchi, M. (2000). The 2000 Miyakejima eruption: Crustal deformation and earthquakes observed by the NIED Miyakejima observation network. *Earth, Planets, Space*, **52**(8), xix–xxvi.

USGS Hawaiian Volcano Observatory (2002). Summary of the Pu'u 'Ō'ō–Kupaianaha eruption, 1983–present. Available online at: http://wwwhvo.wr.usgs.gov/kilauea/summary/main.html [Accessed January 2004].

Van der Laat, R. (1996). Ground-deformation methods and results. In: Scarpa, R. and Tilling, R.I. (eds), *Monitoring and Mitigation of Volcano Hazards.* Springer Verlag, New York, pp. 147–168.

Vanicek P., Castle, R.O., and Balazs, E.I. (1980). Geodetic leveling and its applications. *Rev. Geophys.*, **18**, 505–524.

Varekamp, J.C., Luhr, J.F., and Prestegaard, K.L. (1984). The 1982 eruptions of El Chichón Volcano (Chiapas, Mexico): Character of the eruptions, ash-fall deposits, and gas phase. *J. Volcanol. Geotherm. Res.*, **23**, 36–68.

Vasco, D.W., Smith, R.B., and Taylor, C.L. (1990). Inversion for sources of crustal deformation and gravity change at the Yellowstone Caldera. *J. Geophys. Res.*, **95**, 19839–19856.

Vine F.J. (1966). Spreading of the ocean floor: New evidence. *Science*, **154**, 1405–1415.

Vine F.J. and Matthews D.H. (1963). Magnetic anomalies over oceanic ridges. *Nature*, **199**, 947–949.

Voight, B. (1996). The management of volcano emergencies: Nevado del Ruiz. In: Scarpa, R. and Tilling, R.I. (eds), *Monitoring and Mitigation of Volcano Hazards.* Springer Verlag, New York, pp. 719–769.

Von Rebeur-Paschwitz, E. (1889). The earthquake of Tokio, April 18, 1889. *Nature*, **40**, 294–295.

Wadge, G., Francis, P.W., and Ramirez, C.F. (1995). The Socompa collapse and avalanche event. *J. Volcanol. Geotherm. Res.*, **66**, 309–336.

Waite, G.P. and Smith, R.B. (2002). Seismic evidence for fluid migration accompanying subsidence of the Yellowstone Caldera. *J. Geophys. Res.*, **107**(B9), 2177, doi: 10.1029/2001JB000586.

Waite, G.P. and Smith, R.B. (2004). Seismotectonics and stress field of the Yellowstone volcanic plateau from earthquake first-motions and other indicators. *J. Geophys. Res.*, **109**(B2), B02301, doi: 10.1029/2003JB002675.

Wallis, G.B. (1969). *One-Dimensional Two-Phase Flow.* McGraw-Hill, New York, 408pp.

Walsh, J.B. and Decker, R.W. (1971). Surface deformation associated with volcanism. *J. Geophys. Res.*, **76**, 3291–3302.

Walsh, J.B. and Rice, J.R. (1979). Local changes in gravity resulting from deformation. *J. Geophys. Res.*, **84**, 165–170.

Walters, M.A., Sternfeld, J.R., Haizlip, A.F., Drenick, A.F., and Combs, J. (1992). A vapor-dominated high-temperature reservoir at The Geysers, California. In: Stone, C. (ed.), *Monograph on The Geysers geothermal field* (Geothermal Resources Council Special Report No. 17). GRC, Davis, CA, pp. 77–87.

Wang, H.F. (2000). *Theory of Linear Poroelasticity with Applications to Geomechanics and Hydrogeology.* Princeton University Press, Princeton, NJ, 287pp.

Watanabe, H. (1983). Changes in water level and their implications to the 1977–1978 activity of Usu Volcano. In: Shimozuru, D. and Yokoyama, I. (eds), *Arc Volcanism: Physics and Tectonics, Proc. 1981 IAVCEI Symposium.* International Association of Volcanology and Chemistry of the Earth's Interior, Tokyo.

Watts, A.B. and Masson, D.G. (1995). A giant landslide on the north flank of Tenerife, Canary Islands. *J. Geophys. Res.*, **100**, 24487–24498.

Weaver, C.S. and Smith, S.W. (1983). Regional tectonic and earthquake hazard implications of a crustal fault zone in southwestern Washington. *J. Geophys. Res.*, **88**(B12), 10371–10383.

Webb, F.H., Bursik, M., Dixon, T., Farina, F., Marshall, G., and Stein, R.S. (1995). Inflation of Long Valley Caldera from one year of continuous GPS observations. *Geophys. Res. Lett.*, **22**, 195–198.

Weitkamp, C. (ed.) (2005). *Lidar: Range-Resolved Optical Remote Sensing of the Atmosphere.* Springer Verlag, NY, 460pp.

Wells, D.E., Beck, N., Delikaraoglou, D., Kleusberg, A., Krakiwsky, E.J., Lachapelle, G., Langley, R.B., Nakiboglu, M., Schwarz, K.P., Tranquilla, J.M., et al. (1987). *Guide to GPS Positioning.* Canadian GPS Associates, University of New Brunswick Graphic Services, Fredericton, New Brunswick, Canada. (The book is out of print but the material is available as Lecture Notes 58 from the Department of Geodesy and Geomatics Engineering, University of New Brunswick (http://gge.unb.ca/Pubs/Pubs.html, accessed May 2004)).

Wernicke, B.P., Friedrich, A.M, Niemi, N.A, Bennett, R.A., and Davis, J.L. (2000). Dynamics of plate boundary fault systems from Basin and Range Geodetic Network (BARGEN) and geologic data. *GSA Today*, **10**(11), 1–3.

Weston, T. (1990). Precision rectification of SPOT imagery. *Photogrammetric Engineering and Remote Sensing*, **56**(2), 247–253.

Westphal, J.A., Carr, M.A., Miller, W.F., and Dzurisin, D. (1983). An expendable bubble tiltmeter for geophysical monitoring. *Review of Scientific Instruments*, **54**, 415–418.

White, R.A. (1996). Precursory deep long-period earthquakes at Mount Pinatubo: Spatio-temporal link to a basalt trigger. In: Newhall, C.G. and Punongbayan, R.S. (eds), *Fire and Mud: Eruptions and Lahars of Mt. Pinatubo, Philippines.* Philippine Institute of Volcanology and Seismology, Quezon City and University of Washington Press, Seattle, pp. 307–327.

Whitmore, G.D. (1952). The development of photogrammetry. In: *Manual of Photogrammetry* (2nd Edition). American Society of Photogrammetry, Washington, D.C., pp. 1–16.

Whitney, J. A. (1975). Vapor generation in a quartz monzonite magma: A synthetic model with application to porphyry copper deposits. *Economic Geology*, **70**, 346–358.

Wicks, C. Jr., Thatcher, W., and Dzurisin, D. (1998). Migration of fluids beneath Yellowstone Caldera inferred from satellite radar interferometry. *Science*, **282**, 458–462.

Wicks, C.W., Dzurisin, D., Ingebritsen, S., Thatcher, W., Lu, Z., and Iverson, J. (2001). Magmatic activity beneath the quiescent Three Sisters volcanic center, central Oregon Cascade Range, USA, inferred from satellite InSAR. *EOS Trans. Amer. Geophys. Union*, **82**(47), Fall Meet. Suppl., Abstract F272.

Wicks, C.W. Jr., Dzurisin, D., Ingebritsen, S., Thatcher, W., Lu, Z., and Iverson, J. (2002a). Magmatic activity beneath the quiescent Three Sisters volcanic center, central Oregon Cascade Range, USA. *Geophys. Res. Lett.*, **29**(7), doi: 10.1029/2001GL014205.

Wicks, C.W., Dzurisin, D., Ingebritsen, S., Thatcher, W., Lu, Z., and Iverson, J. (2002b). Ongoing magma intrusion beneath the Three Sisters volcanic center, central Oregon Cascade Range, USA, inferred from satellite InSAR, GSA Cordilleran Section. *98th annual meeting, Abstracts with Programs*, **34**(5), 90–91.

Wicks, C. Jr., Thatcher, W. and Dzurisin, D. (2002c). Satellite InSAR reveals a new style of deformation at Yellowstone caldera. *EOS Trans. Amer. Geophys. Union*, **83**(47), Fall Meet. Suppl., Abstract F1348.

Wicks, C., Thatcher, W., and Dzurisin, D. (2003). Stress transfer, thermal unrest, and implications for seismic hazards associated with the Norris uplift anomaly in Yellowstone National Park. *EOS Trans. Amer. Geophys. Union*, **84**(46), Fall Meet. Suppl., Abstract F500.

Wicks, C.W., Thatcher, W., and Dzurisin, D. (2005). Uplift, thermal unrest, and an episode of magma intrusion at Yellowstone caldera. Resubmitted to *Nature*, June 2005.

Wiley, C.A. (1965). Pulsed Doppler radar methods and apparatus. United States Patent No. 3,196,436.

Williams, C.A. and Wadge, G. (1998). The effect of topography on magma chamber deformation models: Application to Mt. Etna and radar interferometry. *Geophys. Res. Lett.*, **25**(10), 1549–1552.

Williams, C.A. and Wadge, G. (2000). An accurate and efficient method for including the effects of topography in three-dimensional elastic models of ground deformation with applications to radar interferometry. *J. Geophys. Res.*, **105**(B4), 8103–8120.

Williams, H. and McBirney, A.R. (1979). *Volcanology*. Freeman, Cooper and Company, San Francisco, 397pp.

Williams, J. (2002). *GIS Processing of Gecoded Satellite Images*. Praxis, Chichester, UK, ISBN: 1-85233-368-5, 353pp.

Williams, T.J., Candela, P.A., and Piccoli, P.M. (1995). The partitioning of copper between silicate melts and two-phase aqueous fluids: An experimental investigation at 1 kbar, 800°C and 0.5 kbar, 859°C. *Contributions to Mineralogy and Petrology*, **121**, 388–399.

Williams-Jones, G. and Rymer, H. (2002). Detecting volcanic eruption precursors: A new method using gravity and deformation measurements. *J. Volcanol. Geotherm. Res.*, **113**, 379–389.

Wilson, L. (1980). Relationships between pressure, volatile content and ejecta velocity in three types of volcanic explosions. *J. Volcanol. Geotherm. Res.*, **8**, 297–313.

Wilson, C.J.N., Rogan, A.M., Smith, I.E.M., Northey, D.J., Nairn, I.A., and Houghton, B.F. (1984). Caldera volcanoes of the Taupo Volcanic Zone, New Zealand. *J. Geophys. Res.*, **89**, 8463–8484.

Wolf, P.R. (1983). *Elements of Photogrammetry (with air photo interpretation and remote sensing)* (2nd edition). McGraw Hill, Inc., New York, 628pp.

Wolf, P.R. and Dewitt, B.A. (2000). *Elements of Photogrammetry with Applications in GIS*. McGraw Hill, New York, ISBN 0-07-292454-3.

Wolfe, E.W. (ed.) (1988). *The Pu'u 'Ō'ō eruption of Kilauea Volcano, Hawaii: episodes 1 through 20, January 3, 1983, through June 8, 1984* (USGS Prof. Paper 1463). US Geological Survey, Reston, VA, 251pp.

Wolfe, E.W. and Hoblitt, R. (1996). Overview of the eruptions. In: Newhall, C.G. and Punongbayan, R.S. (eds), *Fire and Mud: Eruptions and Lahars of Mt. Pinatubo, Philippines*. Philippine Institute of Volcanology and Seismology, Quezon City and University of Washington Press, Seattle, pp. 3–20.

Wolfe, E.W., Garcia, M.O., Jackson, D.B., Koyanagi, R.Y., Neal, C.A., and Okamura, A.T. (1987). The Pu'u 'Ō'ō eruption of Kilauea Volcano, episodes 1–20, January 3, 1983, to June 8, 1984. In: Decker, R.W., Wright, T.L., and Stauffer, P.H. (eds), *Volcanism in Hawaii* (USGS Prof. Paper 1350). US Geological Survey, Reston, VA, pp. 471–508.

Woodcock, D. and Roeloffs, E. (1996). Seismically-induced water level oscillations in a fractured-rock aquifer well near Grants Pass, Oregon. *Oregon Geology*, **58**(2), 27–33.

Wright, J.W. (2000). Crustal deformation in Turkey from synthetic aperture radar interferometry. D. Phil. Thesis, Oxford University, 212pp.

Wright, T.L. (1984). Origin of Hawaiian tholeiite: A metasomatic model. *J. Geophys. Res.*, **89**, 3233–3252.

Yamakawa, N. (1955). On the strain produced on a semi-infinite elastic solid by an interior source of stress. *J. Seism. Soc. Japan*, **8**, 84–98.

Yamashina, K., Matsushima, T., and Ohmi, S. (1999). Volcanic deformation at Unzen, Japan, visualized by a time-differential stereoscopy. In: Nakada, S., Eichelberger, J.C., and Shimizu, H. (eds), Unzen eruption: magma ascent and dome growth. *J. Volcanol. Geotherm. Res.*, **89**(1–4), 73–80.

Yamashita, K.M. (1981). *Dry Tilt: A Ground Deformation Monitor As Applied to the Active Volcanoes of Hawaii* (USGS Open-File Rpt. 87-293). US Geological Survey, Reston, VA, 32pp.

Yamashita, K.M. (1992). Single-setup leveling used to monitor vertical displacement (tilt) on Cascades volcanoes. In: Ewert, J. and Swanson, D.A. (eds), *Monitoring Volcanoes: Techniques and Strategies Used by the Staff of the Cascades Volcano Observatory, 1980–1990* (USGS Bull. 1966). US Geological Survey, Reston, VA, pp. 143–149.

Yamashita, K.M. and Doukas, M.P. (1987). *Precise Level Lines at Crater Lake, Newberry Crater, and South Sister, Oregon* (USGS Open-File Rpt. 87–293). US Geological Survey, Reston, VA, 32pp.

Yang, X. and Davis, P.M. (1986). Deformation due to a rectangular tension crack in an elastic half-space. *Bull. Seis. Soc. Am.*, **76**(3), 865–881.

Yang, X., Davis, P.M., and Dieterich, J.H. (1988). Deformation from inflation of a dipping finite prolate spheroid in an elastic half-space as a model for volcanic stressing. *J. Geophys. Res.*, **93**(B5), 4289–4257.

Yang, X., Davis, P.M., Delaney, P.T., and Okamura, A.T. (1992). Geodetic analysis of dike intrusion and motion of the magma reservoir beneath the summit of Kilauea Volcano, Hawaii: 1970–1985. *J. Geophys. Res.*, **97**, 3305–3324.

Yu, E. and Segall, P. (1996). Slip in the 1868 Hayward earthquake from the analysis of historical triangulation data. *J. Geophys. Res.*, **101**, 16101–16118.

Zebker, H.A. and Lu, Y. (1998). Phase unwrapping algorithms for radar interferometry: Residue-cut, least-squares, and synthesis algorithms. *J. Opt. Soc. Am. A.*, **15**(3), 586–597.

Zebker, H.A., Amelung, F., and Jonsson, S. (2000). Remote sensing of volcano surface and internal processes using radar interferometry. In: Mouginis-Mark, P.J., Crisp, J.A., and Fink, J.H. (eds), *Remote Sensing of Active Volcanoes* (Geophysical Monograph 116). American Geophysical Union, Washington, DC, pp. 179–205.

Zebker, H.A., Rosen, P.A., Goldstein, R.M., Gabriel, A., and Werner, C.L. (1994). On the derivation of coseismic displacement fields using differential radar interferometry: The Landers earthquake. *J. Geophys. Res.*, **99**, 19617–19634.

Zilkoski, D.B., Richards, J.H., and Young, G.M. (1992). Special report: Results of the general adjustment of the North American Vertical Datum of 1988. *Surveying and Land Information Systems*, **52**, 133–149.

Zlotnicki, J., Ruegg, J.C., Bachelery, P., and Blum, P.A. (1990). Eruptive mechanism on Piton de la Fournaise volcano associated with the December 4, 1983, and January 18, 1984, eruptions from ground deformation monitoring and photogrammetric surveys. *J. Volcanol. Geotherm. Res.*, **40**, 197–217.

Zöllner, F. (1869). Ueber eine neue Methode zur Messung anziehender und abstossender Kräfte. *Ber. sächs. Akad. Wis. Math.-nat. Klasse*, **21**, 280–284.

Zumberge, J.F., Heflin, M.B., Jefferson, D.C., Watkins, M.M., and Webb, F.H. (1997). Precise point positioning for the efficient and robust analysis of GPS data from large networks. *J. Geophys. Res.*, **102**(B3), 5005–5017.

Index